ENVIRONMENTAL VIBRATIONS

ISEV 2005

BALKEMA – Proceedings and Monographs in
Engineering, Water and Earth Sciences

PROCEEDINGS OF THE 2ND INTERNATIONAL SYMPOSIUM ON
ENVIRONMENTAL VIBRATIONS – PREDICTION, MONITORING, MITIGATION
AND EVALUATION (ISEV 2005), OKAYAMA, JAPAN, 20–22 SEPTEMBER, 2005

Environmental Vibrations: Prediction, Monitoring, Mitigation and Evaluation (ISEV 2005)

Hirokazu Takemiya
*Department of Environmental and Civil Engineering,
Okayama University, Okayama-shi, Japan*

Taylor & Francis
Taylor & Francis Group

LONDON/LEIDEN/NEW YORK/PHILADELPHIA/SINGAPORE

Published by: Taylor & Francis/Balkema
P.O. Box 447, 2300 AK Leiden, The Netherlands
e-mail: Pub.NL@tandf.co.uk
www.balkema.nl, www.tandf.co.uk, www.crcpress.com

ISBN 0 415 39035 4

Printed in Great Britain

Environmental Vibrations – Takemiya (ed.)
© 2005 Taylor & Francis Group, London, ISBN 0 415 39035 4

Table of Contents

Building vibrations by traffic and other sources

Assessment of environmental vibration for living condition

Environmental Vibrations – Takemiya (ed.)
© *2005 Taylor & Francis Group, London, ISBN 0 415 39035 4*

Preface

The 21st century is stepping into a high-speed transportation system. The expansions of high-speed trains are going on in Euro trunks and Japanese Shinkansen; a new Shinkansen is going to open shortly in Taiwan; new lines are planned in China extensively. The road traffic also see increasing numbers on highways connecting cities, passing through urban areas, across different grounds.

Amid of such modern transportation revolution, there is growing a public concern and desire for keeping harmonious developments with the built-in environments, since the rising levels in vibration and noise caused significant detrimental effects to our community. The environmental vibrations, caused by train/traffic, construction activities and factory operation, and other man-made sources have been investigated, globally, for their prediction, control and mitigation, basically, to improve the quality of life in modern society, operation of sensitive machines and high-tech production.

The Japanese Geotechnical Society, Architectural Institute of Japan, Japan Society of Civil Engineering, Chinese Society for Vibration Engineering and Okayama University have organized an International Symposium on Environmental Vibrations: prediction, monitoring and evaluation (ISEV 2005) during Sept. 20–22, 2005. This symposium, sequel to the first one held at Hangzhou, China in 2003, provides the internationals forum to exchange knowledge, ideas and experience in this field.

In order to record this international exchange of experience, knowledge and research discussed during the conference and make it available to the research community and working professionals, the proceedings are presented in a book form which contains the invited and regular papers written and presented by widely respected academic professionals and engineering specialists. They count totally more than 80 papers from 13 countries. Thus, the book describes the current state-of-the art in addition to the expertises and practices in the related fields including Wave propagation in soils; Soil dynamics; Soil-structure dynamic interaction; Field measurement of environmental vibration caused by traffic, construction, machines, wind, etc.; Monitoring of environmental vibrations; Development of vibration mitigation measures; Evaluation of environmental vibrations; Vibration effects on human perception; Vibration effects on high-precision machines.

Hirokazu TAKEMIYA
Chairman of the ISEV 2005

Environmental Vibrations – Takemiya (ed.)
© 2005 Taylor & Francis Group, London, ISBN 0 415 39035 4

Organization

Chairman
Prof. Hirokazu Takemiya, Okayama University (JP)

Co-Chairman
Prof. Yunmin Chen, Zhejiang University (CN)
Honorary members
President, Prof. Iichiro Kohno, Okayama University (JP)
Prof. Richard Woods, University of Michigan (USA)

Organizing Members
Prof. Hirokazu Takemiya, Okayama University (JP)
Prof. Yunmin Chen, Zhejiang University (CN)
Prof. Y.B.Yang, National Taiwan University (TW)
Prof. Takeshi Goto, Hosei University (JP)
Prof. Takashige Ishikawa, Japan Women's University (JP)
Prof. Yasutoshi Kitamura, Kobe University (JP)
Prof. Kiyoshi Hayakawa, Ritsumeikan University (JP)
Dr. Kimitoshi Ashiya, Japan Railway Technical Research Institute (JP)
Prof. He Xia, Northern Jiaotong University (CN)
Prof. Jianqun Jiang, Zhejiang University (CN)
Dr. Masazumi Shioda, Tobishima Corporation, Ltd. (JP)
Dr. Sunao Kunimatsu, National Institute of Advanced Industrial Science and Technology (JP)
Prof. Toshiyuki Sugiyama, Yamanashi University (JP)
Prof. Akio Matsuura, Shibaura Institute of Technology (JP)

International Scientific committee members
Prof. Ladislav Friba (Cz)
Prof. Victor Krylov (UK)
Prof. G. Degrande (B)
Dr. Chris Jones (UK)
Prof. D.Le Houedec (F)
Dr. R. Massarsch (SN)
Assoc. Prof. Glenn Rix (USA)
Assoc. Prof. Xiangwe Zeng (USA)

Wave propagation

Environmental Vibrations – Takemiya (ed.)
© 2005 Taylor & Francis Group, London, ISBN 0 415 39035 4

Sub and super seismic train induced ground vibrations- theoretical considerations and test results

A. Bodare

Department of Civil and Architectural Engineering, Royal Institute of Technology (KTH), Sweden

K. Petek

Civil and Environmental Engineering, University of Washington, Seattle, USA

ABSTRACT: With the introduction of high speed trains in Sweden, high vibration levels were encountered in low embankments founded on soft, cohesive soils. In these cases, it appeared that the maximum amplification occurred when the train speed exceeded a critical value. A series of field tests were therefore carried out at Ledsgård, a site in south western Sweden near Gothenburg, in order to investigate the problem. In these tests, an X2000 high speed train was run at speeds ranging from 10 km/h to 200 km/hr, and a substantial increase of the peak values of the particle displacement and velocity was observed at train speeds of 140–160 km/hr. The shear-wave velocities varied between 40 m/s to 60 m/s in the upper layers of the soil, which means that the tests at higher train speeds were performed with super seismic speeds in relation to the shear wave velocity of the soil.

This paper presents theoretical considerations on ground motions in the track environment when a train runs at sub and super seismic speeds. Mathematical expressions are presented along with a conceptual model that illustrates the behaviour. The phenomenon of the "two sources" for super seismic train speeds is explained and displacement and velocity time histories from the model are discussed for various speeds. Recordings from the field measurements are also presented and compared with the results of the conceptual model.

1 INTRODUCTION

As high speed train traffic increases and train lines are increasingly built on less-desirable soil conditions, train induced ground vibrations under the track and surrounding environment are of growing interest and importance. From 1997–2000, a series of measurements were carried out at Ledsgård, a site on the West Coast Line in southwestern Sweden, in order to evaluate ground vibrations due to freight and high speed trains, and the effect of subsequent soil improvement using lime-cement columns. The project was a cooperation of the Swedish National Rail Administration, Banverket, with the Royal Institute of Technology in Stockholm and Chalmers University in Gothenburg, along with the Swedish Geotechnical Institute (SGI) and Norwegian Geotechnical Institute (NGI). These field tests and remediation work have provided the opportunity for extensive research and greater understanding of train induced ground vibrations.

While many others have developed advanced numerical models to investigate this type of problem, the idea behind this research is to assume a very simple but relevant excitation mechanism for a train moving on two tracks and to then study the consequential motion of the environment. The reason is to isolate and identify mechanical phenomena occurring in the environment of a traveling train. This method has been applied and presented earlier by Krylov (1994 and 1995), but here the initial assumptions are made in different way.

Therefore, mathematical expressions describing the effect of train induced ground vibrations in the track environment are developed in this paper and a conceptual model that utilizes the expressions is also discussed. Comparisons between the model and field results are made for validation, however the objective of this work is to generally describe the observed phenomena instead of reproducing it exactly. Plots of ground displacements and velocities at various positions in the track environment are considered. Differences in the sub and super seismic speeds are examined along with effects of position and attenuation. The observed behavior of the simplified model is then compared with field data. From the comparisons, the theory is validated but the model limitations due to the assumptions employed are also observed.

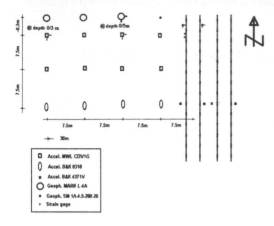

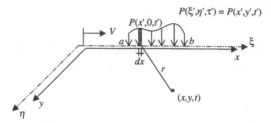

Figure 2. Diagram of load in moving system and observer in fixed system.

Figure 1. Instrumentation plan for the vibration measurements at Ledsgård, May 2000 (from Bahrekazemi (2001) with permission).

1.1 Ledsgård field tests

Instrumentation of the field tests consisted of accelerometers, seismometers, and strain gauges in a layout as shown in Figure 1. The railway embankment at this site consists of 1.3 m of crushed rock material. A dry crust of 1.5 m overlaying approximately 3 m of gyttja, soft organic clay, characterizes the soil at the Ledgård site. Below this lies a deposit of marine clay with shear strength increasing to bedrock at 70 m below the surface Bengtsson (1999).

2 THEORY

2.1 Kinematics

This analysis of train induced ground vibrations begins with considering a moving source and an observer at a fixed position in space. Suppose the source of the vibrations is moving in a straight line with a constant speed, relative to the ground. A fixed coordinate system (x, y) is introduced and the source moves along the x-axis. The coordinates of the source are denoted by a prime sign, i.e. $(x', 0)$. A second coordinate system (ξ, η), which moves with the velocity V, is introduced and time is τ. A diagram of the two systems is shown in Figure 2.

The two coordinate systems are related by the expressions:

$$\xi = x - Vt + x_p$$
$$\eta = y \tag{1}$$
$$\tau = t$$

The term x_p, which denotes the spacing between the coordinate systems at time $t = 0$, is set to zero without

losing generality. This means that the origins of the two co-ordinate systems pass each other at time $t = \tau = 0$. The relation between the time of excitation and the time of observation is then

$$t' = t - \frac{r}{c}; \quad r = \sqrt{(x - x')^2 + y^2} \tag{2}$$

where r is the distance traveled by a disturbance emitted at time t' at location $(x', 0)$ and received by an observer at time t at location (x, y). The velocity with which the disturbance propagates is called c. The location ξ' can be expressed as $x' - V(t - r/c)$. If the location of the source in the moving system ξ' and the location of the observer in the fix system are known, then the location of the emitted disturbance in the fixed system, x', can be solved by a second order equation. The two solutions will be:

$$x' = \frac{\xi - x + Vt}{1 - \beta^2} \pm \frac{\beta \cdot \sqrt{(\xi - x + Vt)^2 + (1 - \beta^2) \cdot y^2}}{1 - \beta^2} + x \tag{3}$$

and the expression for r is rewritten as

$$r = \left| \frac{1}{1 - \beta^2} \cdot \left(\sqrt{(\xi - x + Vt)^2 + (1 - \beta^2) \cdot y^2} - \beta \cdot (\xi - x + Vt) \right) \right| \tag{4}$$

Care must be taken when selected the sign of the location term, x' in Equation 3. For sub seismic load speeds, where the velocity ratio, β, defined as the source speed divided by the wave propagation velocity, is less than one ($\beta < 1$), the plus sign solution must be rejected. This is because the plus sign solution signifies a source that is located ahead of the actual source position, which cannot exist in reality. If the speed of the load is super seismic ($\beta > 1$), both solutions are valid. The conclusion must be that at super seismic load speeds the observation point is affected by two different sources at those times when there are waves passing the

4

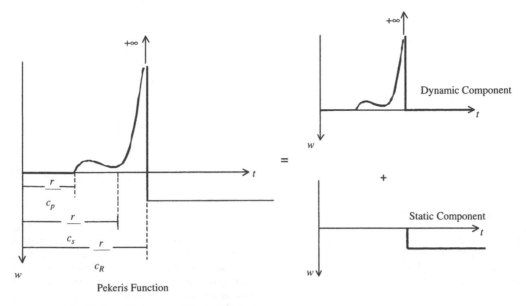

Figure 3. Pekeris function and decomposition into two components.

observer. In order to find these times the square root in Equation 4 is studied. The argument of the square root must be positive, which means

$$t \geq \frac{\sqrt{\beta^2 - 1} \cdot |y| - \xi + x}{V} = t_{app} \quad (5)$$

and t_{app} is called the time of appearance of the coordinate ξ. From this expression, a three zone effect is observed in the solution. If one considers a distributed load with a length ab and an observer located at some position r, the observer initially feels no effect until the first appearance time and this is called zone I. By setting ξ equal to b, the first appearance time, t_b, is found. At this point, the observer first receives waves from the source, beginning with signals from point b. Although it may seem counterintuitive that the observer experiences the end of the load first, it is seen from Equation 5 that because b is greater than a, t_b is less than t_a, and therefore point b appears at an earlier time that point a. The duration between t_b and t_a is zone II and in this time, the observer experiences signals from only a portion of the load. After the second appearance time t_a, the observer receives signals from the full load and this is called zone III. Again, because of the two solutions of Equation 3, two sources are felt by the observer in zones II and III.

2.2 Pekeris' solution

The Pekeris function, Pekeris (1955), approximated in Figure 3, is combined with the above expressions for

considering the train loading. This function describes the displacement of arbitrary points on the surface of a linear-elastic, homogenous half space under a step or Heavyside force, which is a force applied instantaneously and then held in place at a constant value. (Internal damping is therefore not considered.)

The Pekeris function, K(r, t), is thus the step response of a linear system. The displacement can then be expressed by a convolution integral as:

$$w(r,t) = \int_0^\infty K(r,\tau) \cdot \frac{\partial P(t-\tau)}{\partial \tau} \cdot d\tau \quad (6)$$

This Pekeris function is here considered to have two components as shown in Figure 3, a dynamic component due to the instantaneous loading and then a static component, such that K(r, t) = K_d(r, t) + K_s(r, t). It is further assumed that the static component is the major component and this portion is used here solely to describe the displacements from the train loading. Although it is an approximation to use only a portion of the load, it should be noted that this component is still a part of the exact solution. Additionally, the dynamic component contributes negatively to the total displacement and thus the simplification results in displacements that are too high.

The static part of the Pekeris function is in itself a step function, which can be written

$$K_s(r,t) = \frac{1-\upsilon}{2\pi} \cdot \frac{1}{Gr} \cdot H(t - \frac{r}{c_R}) \quad (7)$$

where the time independent part is the static displacement according to Boussinesq (1878) and first occurs at time t = r/c_R. The constant in Equation 7, $(1 - v)/2/\pi/G$, will hereafter be denoted A_0. Inserting Equation 7 into Equation 6 will give

$$w(r,t) = \frac{A_0}{r} \cdot P(t - \frac{r}{c_R})$$ (8)

In this simplified model the displacements at time t are proportional to the force but at a retarded time. The time difference is the travel time for the surface wave to reach the observer. The proportional constant, A_0/r, is inversely dependent on the distance. The propagation velocity, c, introduced earlier is considered here to be the Rayleigh wave velocity c_R, and hence the notation c implies Rayleigh wave velocity in the remainder of this paper.

Considering the displacement at a position (x, y) at time t, the expression becomes

$$w(r,t) = \frac{A_0}{r} \cdot p(x',0,t') \cdot dx$$ (9)

$$w(x,y,t) = \int_b \frac{A_0}{r} \cdot p(x',0,t') \cdot dx'$$ (10)

$$w(x,y,t) = \int_a \frac{A_0}{r} \cdot p\left(\xi',0,t - \frac{r}{c}\right) \frac{dx'}{d\xi'} d\xi'$$ (11)

where r is expressed in terms of ξ', as in Equation 4. From this, the expression for the sub seismic case becomes

$$w(x,y,t) = A_o \int_a^b \frac{p(\xi',t - \frac{r}{c})d\xi'}{\sqrt{(\xi'-x+Vt)^2 + (1-\beta^2)y^2}}$$ (12)

and in the super seismic case, three expressions for the three zones:
Zone I:

$$w(x,y,t) = 0$$ (13)

Zone II:

$$w(x,y,t) = A_o \left(\int_b^{\xi_i} \frac{p(\xi',t - \frac{r_-}{c}) \cdot d\xi'}{\sqrt{(\xi'-x+Vt)^2 + (1-\beta^2)y^2}} + \int_{\xi_i}^b \frac{p(\xi',t - \frac{r_-}{c}) \cdot d\xi'}{\sqrt{(\xi'-x+Vt)^2 + (1-\beta^2)y^2}} \right)$$ (14)

Zone III:

$$w(x,y,t) = A_o \left(\int_a^b \frac{p(\xi',t - \frac{r_-}{c}) \cdot d\xi'}{\sqrt{(\xi'-x+Vt)^2 + (1-\beta^2)y^2}} + \int_a^b \frac{p(\xi',t - \frac{r_-}{c}) \cdot d\xi'}{\sqrt{(\xi'-x+Vt)^2 + (1-\beta^2)y^2}} \right)$$ (15)

with the two roots for the distance, r, in the super seismic case defined as

$$r_{\pm} = \frac{1}{\beta^2 -1} \cdot \left(\beta \cdot (\xi'-x+Vt) \pm \sqrt{(\xi'-x+Vt)^2 + (1-\beta) \cdot y^2} \right)$$ (16)

Illustrations of the sub and super seismic behavior are shown in Figures 4 and 5b. In these figures, the observer stands at x = 0 and a time varying force, length ab, is represented by the series of parallel lines, whose position in time and space corresponds to the position on the respective axis. For the sub seismic case shown in Figure 4, the observer is constantly receiving signals. At time t_I, before the observer is overrun, the information path is the line ab called the Negative Branch. After the observer has been completely overrun, at time t_{III}, the signal received is the second line ab called the Positive Branch. While the source is passing the observer, at time t_{II} for example, both positive and negative signals are felt by the

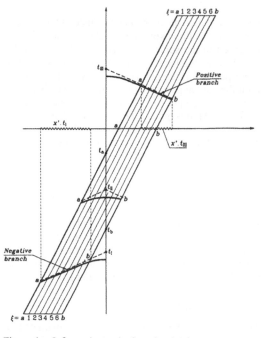

Figure 4.   Information paths for sub seismic case.

observer. A Doppler effect is very apparent for this case, where the length of the signal is shown on the x-axis by a squiggle line. Before being overrun, the source appears longer than it really is, while it appears shorter after being overrun.

In Figure 5a, a schematic diagram of a super seismic load in the moving system is shown in order to illustrate the three zone effect. In this figure, the lines extending from the ξ-axis mark the signal paths of the ends of the load and create the boundaries between the three zones. The corresponding information paths, in time-space coordinates, are shown in Figure 5b. Again the observer stands at position x = 0 and the three zones are labeled. Prior to time t_b (zone I), the observer does not receive any signal from the load source and the information path passes far ahead of the observer. After this time, t_b, the observer receives signals from a portion of the load, beginning with the end of the load (b) first and this is considered zone II. After time t_a, the observer receives two signals from the entire load (zone III), and again receives signals from the end portion of the load first. The two sources are clearly seen in the figure in all three zones and are denoted as the Positive and Negative branches. Again, the Doppler effect is observed here, with the lengths for one point in time marked by squiggle lines on the x-axis.

Two important observations are discovered in the equations presented above. The first is that the distance, y, normal to the track, is always accompanied by the factor $1 - \beta^2$ and it is found that when the velocity ratio β approaches 1, the influence of the distance from the track decreases. When $\beta = 1$, the y term is multiplied by 0 and drops out of the expressions. This means that when the velocity of the force is equal to the Rayleigh wave velocity ($\beta = 1$), there is no attenuation at all in the normal direction. This is possibly another factor that contributes to the strong ground vibrations obtained in soft soils.

The second observation is that when the super seismic force distribution in the moving system does not depend on the time then the contributions of the negative and the positive branch are equal. This is seen mathematically from Equations (14) and (15), where $p(\xi', t - r_{\pm}/c)$ becomes $p(\xi')$ and the equations for the displacement are rewritten

Zone II:

$$w(x,y,t) = 2A_o \left(\int_{\xi_t}^{b} \frac{p(\xi') \cdot d\xi'}{\sqrt{(\xi'-x+Vt)^2 + (1-\beta^2)y^2}} \right) \quad (17)$$

Zone III:

$$w(x,y,t) = 2A_o \left(\int_{a}^{b} \frac{p(\xi') \cdot d\xi'}{\sqrt{(\xi'-x+Vt)^2 + (1-\beta^2)y^2}} \right) \quad (18)$$

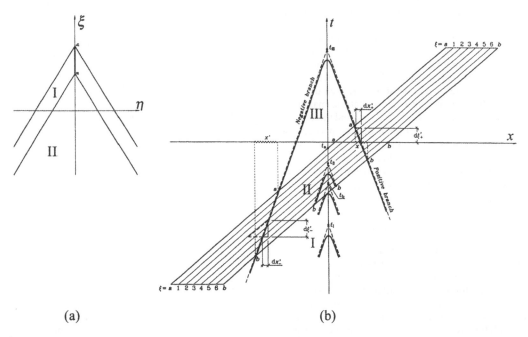

(a)   (b)

Figure 5.   (a) Schematic diagram of super seismic load in moving coordinate system (b) Information paths for super seismic case.

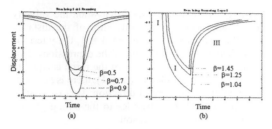

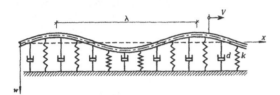

Figure 6. Time histories for a sub and super seismic constant force distribution.

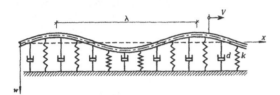

Figure 7. Bernoulli-Euler beam on a modified Winkler foundation.

Example

In order to see the behavior of the method presented, simple examples were initially examined. One is that of a constant force distribution moving in the positive x-direction with a velocity V and is presented here. Figures 6a and b show the displacement time histories at a position 15 m from the track for various β values. In the sub seismic case of Figure 6a, it is clear that as the load speed approaches the Rayleigh wave velocity (β approaches 1), the magnitude of the displacements increase significantly. It is also apparent that the width of the displacement trough decreases as β increases. It should be observed that the displacement time histories are symmetrical in the sub seismic case.

From these figures, it is clear that the maximum effect occurs at values closest to β = 1 and therefore in the super seismic case, the max displacements decrease as β increases and the load speed increases beyond the Rayleigh wave velocity. This is seen in Figure 6b. A very different shape is observed for this case: the time history is no longer symmetrical and the three zones are clearly seen.

2.3 Damped winkler foundation

With the train load considered as a portion of the Pekeris solution, the track and railway substructure are considered as a Bernoulli-Euler beam on a modified Winkler foundation, Winkler (1867), as shown in Figure 7. This system is an approximation of the field conditions, because in reality the track is only in contact with the ground surface through the ties or sleepers whereas this model assumes continuous contact.

By using the Bernoulli-Euler beam in the model, it is assumed that the sleeper spacing is very small in relation to the wavelength of the track.

For a load acting on the beam on a modified Winkler foundation, a differential equation is written to describe the behavior and includes a damping term, D. Kenney (1954) gave the solution for the displacement of all speeds of the load and from his solution, equations for the force distribution are derived. The force distribution, for a steady state in the moving system, is derived from:

$$p(\xi) = \kappa w + \delta \cdot \frac{\partial w}{\partial t} = \kappa w(\xi) - \delta V \cdot \frac{\partial w}{\partial \xi} \qquad (19)$$

such that

$$\xi < 0; \quad p(\xi) = \frac{P \cdot k_c}{2} \cdot \frac{\eta \cdot e^{\eta k_c \xi}}{b} \cdot \left(Q_1^- \cos\left(\alpha^- k_c \xi\right) + Q_2^- \frac{\gamma^-}{\alpha^-} \sin\left(\alpha^- k_c \xi\right) \right) \qquad (20)$$

$$\xi > 0; \quad p(\xi) = \frac{P \cdot k_c}{2} \cdot \frac{\eta \cdot e^{-\eta k_c \xi}}{b} \cdot \left(Q_1^+ \cos\left(\alpha^+ k_c \xi\right) + Q_2^+ \frac{\gamma^+}{\alpha^+} \sin\left(\alpha^+ k_c \xi\right) \right) \qquad (21)$$

where

$$Q_1^- = 1 - 2mD(\gamma^- + \eta); Q_2^- = 1 + 2mD(-\eta + \frac{\alpha^{-2}}{\gamma^-}) \quad (22)$$

$$Q_1^+ = 1 - 2mD(\gamma^+ - \eta); Q_2^+ = 1 + 2mD(\eta + \frac{\alpha^{+2}}{\gamma^+}) \quad (23)$$

The parameters are defined as:

$$m = \frac{V}{c_{cr}}; \qquad c_{cr} = \sqrt[4]{\frac{4EI \cdot \kappa}{\rho^2}}; \qquad D = \frac{\delta}{2\sqrt{\kappa\rho}}$$

$$(24\ a, b, c)$$

$$b = \eta^4 + (\eta m)^2 + \frac{1}{2}\left(\frac{mD}{\eta}\right)^2 \quad \gamma^\pm = -\frac{mD}{\eta^2} \pm \eta \quad \alpha^\pm = \sqrt{2m^2 + \eta^2 \pm 2 \cdot \frac{mD}{\eta}}$$

$$(25\ a, b, c)$$

where η is the positive real root to the characteristic equation

$$\eta^6 + 2m^2 \cdot \eta^4 + \left(m^4 - 1\right) \cdot \eta^2 - (mD)^2 = 0 \qquad (26)$$

The mechanical parameters are: EI is the bending stiffness of the Euler-Bernoulli beam, κ and δ are the stiffness and damping constants of the distributed

Table 1. Select embankment parameters used in model.

Thickness (m)	4
ρ_{beam} (kg/m)	10,300
EI (MN*m^2)	200×10^6
C_{cr} (m/s)	98

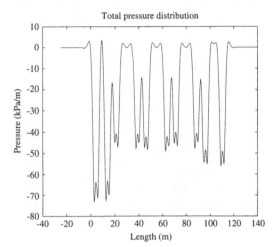

Figure 8. Sample Ledsgård X2000 pressure distribution used in conceptual model.

springs and dampers respectively. The linear density of the Euler-Bernoulli beam is ρ and D is the damping ratio of the beam (mass)-damper-spring system for a straight beam (i.e. when $\lambda \rightarrow \infty$).

The derived Kenney force equations are used in the simplified Matlab model to model the train loads acting on the railway substructures. Properties of the simplified system were selected based upon the embankment and foundation properties and are shown in Table 1. The soil is considered as a single layer in the model, whose properties are based upon the characteristics at the top of the marine clay layer. Internal damping is not considered here out of simplicity. Actual loading input from the X2000 trains used in the Ledsgård tests are implemented into the model as the derived Kenney forces, and a sample distribution is shown in Figure 8.

3 CONCEPTUAL MATLAB MODEL & TRAIN DATA

3.1 Model behavior

The conceptual model is used to explore various aspects of train induced ground vibrations and several interesting points are observed. The general model behavior is shown in Figure 9. In this figure, the displacements at a distance of 15 m from the track are plotted for various

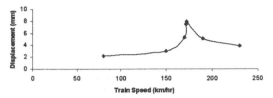

Figure 9. Displacements at position 15 m from track from various train speeds calculated in conceptual Matlab model.

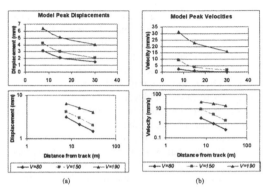

Figure 10. Peak displacements and velocities from conceptual Matlab model.

Table 2. Train speeds and corresponding β-values considered in model.

V	β	Type
80 km/hr	0.47	sub seismic
150 km/hr	0.87	sub seismic
190 km/hr	1.11	super seismic

train speeds. The maximum ground amplification occurs around 172 km/hr, which corresponds to the Rayleigh wave velocity, c. The train speed is hence sub and super seismic before and after this speed respectively. In the plot, it is clear that the zone of high amplification occurs over a relatively short range and that at speeds away from c, the increase is more gradual. It is also apparent that the super seismic displacements are greater than the sub seismic displacements.

Plots of peak ground displacements in linear and logarithmic scales are shown in Figure 10 for three selected train speeds at various distances from the track. The train speeds were selected based upon actual speeds of the Ledsgård field tests and attempt to capture a good range of beta values to observe the differences in sub- and super-seismic speeds. The speeds and their corresponding beta values are shown in Table 2.

In the linear scale plots, a characteristic decreasing polynomial shape of the peak displacements is observed, which reflects the rapid decrease in the displacement magnitudes with increasing distance from the track. This decreasing polynomial function translates to a linear shape in the log-log plots below. It is obvious that the peak displacements and the peak velocities increase substantially as the train speed increase. The slopes of the curves in the log-log plots decrease as the train speeds increase, according to the discussion above.

3.2 Comparison of recordings and conceptual model

Figures 11a, b show the linear and logarithmic plots of maximum displacements and velocities in the track environment at selected speeds for the field tests at Ledsgård, corresponding to the model output of Figures 10a, b. The peak ground displacements were calculated by double integration of filtered accelerometer signals from the field tests. It is important to note that this integration and the filtering of the signal introduce error into the analysis.

Comparing Figures 10 and 11, it is immediately apparent that the shapes of the peak displacement plots are very similar for the field and model conditions. This suggests that the theory developed, which assumes that the static component controls, is a reasonable description of the ground vibrations.

The measured peak displacements are however smaller than those calculated from the theory. At the position 7.5 m from the track the peak displacements are around half of the calculated and at longer distances the displacements are around 1/10 or smaller than those calculated. The smaller measured values can be explained by the fact that the theory exaggerates the positive values as previously noted. The difference could also be attributed to the simple foundation model used that does not consider internal damping or the stiffness reduction that occurs with increasing strains.

The measured peak velocities show the opposite behavior. They are in general higher than those calculated. At the position 7.5 m from the track they are more than 2 times higher for the train speed of 200 km/h.

They are around 6 times higher for the train speed of 150 km/h than the calculated values. This might be explained by omitting the dynamic part of the Pekeris' solution the dynamic part of the result is pressed down resulting in smaller peak velocities from the theory.

The influence of the train speed on the attenuation is not readily obvious. However there is a change, from 150 km/h to 190 km/h, of the slope in the log-log plot of the peak displacement that might be attributed to this phenomenon.

4 CONCLUSION

It is therefore found that simple expressions can be developed to describe train-induced ground vibrations in the environment around the track. Reasonable approximations of the behavior can be found by assuming that the static component of the force is dominant, however it is expected that considering the dynamic as well may be more accurate. It is seen clearly from the results that maximum amplification of the ground displacement occurs when the train velocity is equal to the Rayleigh wave velocity.

ACKNOWLEDGMENTS

We acknowledge the support from the Swedish National Rail Administration, particularly from A. Smekal, P. Zackrisson, and K. Adolfsson. Also the discussions with B. Andreasson, J&W, Gothenburg, P-E. Bengtsson, formerly at SGI, Linköping and teams from SNRA, SGI and NGI are appreciated.

REFERENCES

Krylov, V.V. 1994 Journal de Physique IV, Volume 4. On the theory of railway-induced ground vibrations.

Krylov, V.V. 1995 Applied Acoustics 44, pp.149–164. Generation of ground vibrations by superfast trains.

Bahrekazemi, M. 2001 Lime-cement columns as a countermeasure against train-induced ground vibration. Licentiate Thesis, Department of Civil Engineering and Building Sciences, Royal Institute of Technology, Stockholm, Sweden.

Bengtsson, P.E. et al. 1999 High speed lines on soft ground: Evaluation and analyses of measurements from the West Coast Line, Borlänge, Sweden: Banverket.

Pekeris, C.L. 1955 Proceedings of the National Academy of Science U.S.A., 41, pp.469–480. The seismic surface pulse.

Boussinesq, J. 1878 C. Rendus Acad. Sci. Paris, Vol. 86, pp. 1260–1263. Équilibre d'élasticité d'un solide isotrope sans pesanteur, supportant différents poids.

Winkler, E. 1867 Die Lehre von Elastizität und Festigkeit. Dominicus, Prague.

Kenney, J.T. 1954 Journal of Applied Mechanics, Vol 21, pp.359–364. Steady-state vibrations of beam on elastic foundation for moving load.

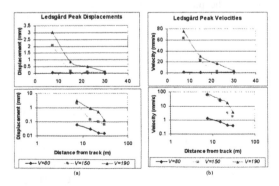

Figure 11. Peak displacements and velocities from Ledsgård field tests, May 2000.

Environmental Vibrations – Takemiya (ed.)
© *2005 Taylor & Francis Group, London, ISBN 0 415 39035 4*

Complex eigenvalue analyses of Love and Rayleigh waves

S. Morio
Professor, Maizuru National College of Technology, Department of Civil Engineering

Y. Kato
Research Associate, Maizuru National College of Technology, Department of Civil Engineering

S. Teachavorasinskun
Associate Professor, Chulalongkorn University, Department of Civil Engineering

ABSTRACT: A one dimensional FEM analysis technique is presented for layered soils. It aims to investigate the interaction and relationship between the body waves (P-wave and S-wave) and the surface waves (Rayleigh and Love waves). This technique is very simple, and a useful tool for the analysis of the surface waves due to traffic vibration. It is expected that this method can be applied to evaluate surface wave nature during the strong earthquakes as well.

1 INTRODUCTION

Due to recent expansion of the elevated highway networks, several countermeasures to reduce effects of traffic vibration have been proposed by related organizations. The main countermeasures can be divided into 3 categories, namely, measures that apply at sources, at wave passages and at destinations. Tools generally used in exploration of such aspect are model testing, numerical analysis, site monitoring and etc. Due to complication arisen from local site conditions; e.g., soil types and structures, there is no universal solution to reduce traffic vibration.

Tokunaga et al. (2000) conducted the field experiment at the T shape single pier girder highway. They found that (1) near the pier, the main component of vibration was P-wave, (2) with increasing distance from center of the pier, shear wave (S-wave) became dominant and (3) finally, when time passed by, the Rayleigh wave became the main cause of vibration. They proposed a new analytical technique and implemented into the two dimensional FEM analysis to identify the combination of body waves and surface wave. It represents the vertical distribution of Fourier displacement by the superposition of generalized Rayleigh wave mode shape inside the boundary of FEM model. Subsequently the participation factor for each mode can be computed. From the calculation result, it was found that Rayleigh wave was generated easier when the vibration source was at the ground surface than in the ground. Furthermore, low frequency vibration source generated more

Rayleigh wave. Using deeper cut-off wall may increase efficiency in vibration reduction especially for vibration having frequency of above 10 Hz, because Rayleigh wave with short wave length was diminished by deeper cut-off wall.

In the present study, a one dimensional analysis technique is presented for layered soils. It aims to investigate the interaction and relationship between the body waves (P-wave and S-wave) and the surface wave (Rayleigh and Love waves). This technique is very simple, and a useful tool for the analysis of the surface waves. By introducing the viscous damping, we can also investigate the dissipation characteristic of the waves. It is expected that this method can be applied to evaluate surface wave nature during the strong earthquakes as well.

2 ANALYTICAL TECHNIQUES

2.1 *Rayleigh wave*

Plane strain finite element analysis was conducted to study the characteristic of Rayleigh wave. By adopting the linear displacement shape in soil, the general equation for Rayleigh wave can be expressed as;

$$\{\delta\} = \alpha\{V\}e^{i(\omega t - \kappa x)} \tag{1}$$

where ω = angular frequency, κ = wave number in x direction, α = coefficient of mode shape.

Under the assumption given in Eqn.(1), the constitutive equation for Rayleigh wave can be written as;

$$\left([A]k^2 + i[B]k + [G] - \omega^2[M] \right)\{V\} = \{0\} \tag{2}$$

where matrices $[A]$, $[B]$ and $[G]$ are the stiffness matrices, $[M]$ is the mass density matrix.

In case when square meshes of n elements are used, each matrix shall have the dimension of $2n \times 2n$.

When ground properties along the depth can be determined, the wave number (k_s) and mode vector $(\{V_s\})$ can be obtained from Eqn.(2). The subscript s indicates sth mode shape, where s is in the range of $1\sim2n$. When wave number (k_s) is real, complex, imaginary and zero, the vectors $\{V_s\}$ is called real mode, complex mode, imaginary mode and zero mode, respectively. Among those, real and complex modes can propagate in the horizontal direction, while imaginary and zero modes do not propagate. Furthermore, in case of $k = 0$, Eqn.(2) expresses the characteristic of body wave (P and S waves). In the other words, the zero mode represents the natural mode of P and S waves. In this case, both horizontal and vertical components become real.

The real mode of Eqn.(2) represents so-called normal Rayleigh wave. When damping is zero, the horizontal component is real, while vertical component is imaginary. Namely, the phase different between the horizontal and vertical component is 90°. When both having the same sign, soil particle will rotate in clockwise direction. However, if they having opposite signs, then, soil particle will rotate in the counter clockwise direction. In real mode, the wave number (k_s) indicates the propagation characteristics. The phase velocity (C) and the group velocity (U) can then be written as;

$$C = 2\pi f / k \tag{3}$$

$$U = 2\pi \frac{df}{dk} \tag{4}$$

where f = frequency.

Eqn.(2) has been used by the soil-structural interaction FEM program called "FLUSH". Morichi et al.(1987) used the results of two dimensional finite element dynamic analysis to distinguish the combination between Rayleigh wave and body wave. However, Eqn.(2) can be adopted to simplify the computation of the above mention aspect in one-dimensional problem. With this method, the dissipation analysis for Rayleigh wave and the channel wave analysis can be easily done with good accuracy. Finally, the viscous damping can be easily introduced into the complex stiffness matrices, $[A]$, $[B]$ and $[G]$. Nevertheless, the difficulty arisen in the cause that

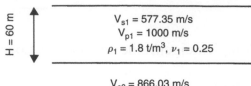

Figure 1. Analytical model.

FEM cannot be used to accurately simulate the problem of elastic half space.

2.2 Love wave

While Rayleigh wave has vertical and horizontal components, Love wave has only one horizontal component (normal to plane).

The constitutive equation for Love wave can be exhibited as shown in Eqn.(5)

$$\left([A]k^2 + [G] - \omega^2[M] \right)\{V\} = \{0\} \tag{5}$$

where matrices $[A]$, $[B]$ and $[G]$ are the stiffness matrices, $[M]$ is the mass density matrix.

In case when square meshes of n elements are used, each matrix shall have the dimension of $n \times n$. Similar to Rayleigh wave, Love wave has four modes; i.e., real mode, complex mode, imaginary mode and zero mode. Among those, real and complex modes can propagate in the horizontal direction. While, zero and imaginary modes do not propagate. If $k = 0$ is substituted into Eqn.(5), the equation for body wave (SH wave) is obtained. The real mode is generally defined so-called normal Love wave. In case of zero damping, the horizontal and vertical components are real.

3 CALIBRATION OF PROGRAM

The developed analytical program is calibrated against the exact solution obtained from the theory of wave propagation.

3.1 Calibration on propagation of Rayleigh wave

Figure 1 shows the model soil layers used for calibration. The top layer having the thickness of 60 m has smaller value of shear wave velocity than the lower layer. The lower layer (foundation layer) is assumed to the elastic half space. As depicted in the figure, V_s, V_p, ρ and ν stand for shear wave velocity, P-wave velocity, density and Poisson's ratio, respectively. The subscripts

12

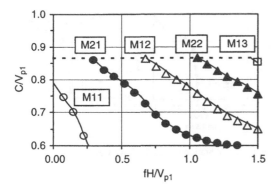

Figure 2. Dispersion curve of phase velocity.

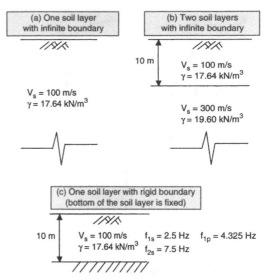

Figure 3. Analytical models.

0 and 1 refers to the foundation layer and soil layer, respectively. The coefficient of damping is set close to zero ($h = 0.001$). The FEM mesh is formed so that each element is smaller than one quarter of wave length of 30 Hz S-wave. As a consequence, there are 15 elements for the soil layer ($4\,\mathrm{m} \times 15\,\mathrm{m} = 60\,\mathrm{m}$) and another 15 elements for the foundation layer ($8\,\mathrm{m} \times 15\,\mathrm{m} = 120\,\mathrm{m}$). Furthermore, in order to simulate the effect of elastic half space, additional meshes with thickness $D = 1.5\lambda$, $\lambda =$ wave length) is added.

Figure 2 shows the dispersion of phase velocity (C). The horizontal axis is the dimensionless frequency ($= fH/V_{p1}$) while the vertical axis represents the dimensionless phase velocity ($=C/V_{p1}$). The open and close symbols are the calculation results of the present study, while the solid lines are the exact solutions adopted from Tajime et al. (1983). Rayleigh wave can be classified into several types. The analytical result of the present study well predicts the dispersion curves of Rayleigh wave types $M11$–$M13$ (Tajime et al. 1983).

The $M11$ mode is the fundamental mode. The area above the broken line shown in the figure refers to the area where P-wave from the top soil layer being transferred to S-wave in the lower foundation layer. This phenomenon is called P-wave Leaking mode. In this area, the line of $C = V_{p1}$ is the cut off frequency for $M1n$ and $C = V_{p2}$ is the upper bound for $M2n$. When the Poisson's ration of both layers approach 0.5, dispersion curve merges to that of the liquid wave. There are still unclear aspects of this Leaking mode which requires further investigation.

3.2 *Calibration on propagation of Love wave*

This shall be addressed in Section 5.2.

4 ANALYTICAL MODEL AND CONDITIONS

Figure 3 schematically shows the analytical model. Figure 3(a) shows the simple one single soil layer

extended to infinite with the shear velocity of 100 m/sec. Figure 3(b) shows the two layer systems, where the shear wave velocity of 100 and 300 m/sec are assumed for the top 10 m soil layer and the lower half space layer, respectively. While Fig. 3(c) stands for the one single soil layer on rigid foundation. The density of related soil layers are shown in the figure. In Rayleigh wave analysis, Poisson's ratio of 0.3 is adopted. In case of the one single soil layer on rigid foundation (Fig. 3(c)), the natural frequencies for S wave are 2.5 Hz (f_{1s}) and 7.5 Hz (f_{2s}) for the first and second mode, respectively. Similarly, the natural frequency for first mode of P-wave is 4.325 Hz (f_{1p}).

Damping is set close to zero ($h = 0.001$). The FEM mesh is prepared so that each element shall not greater than one quarter of wave length of the 30 Hz S wave. To simulate the influence of elastic half space, additional meshes with the thickness D ($=1.5\lambda$) is added at the bottom.

5 ANALYTICAL RESULTS

5.1 *One soil layer as elastic half space*

In case of the one single soil layer with infinite boundary shown in Fig.3(a), there is no existence of Love wave. There exists only fundamental mode of Rayleigh wave with phase velocity of 91.94 m/sec. There is no dispersion of Rayleigh wave. The result is in well accordance to theory of wave propagation.

Figure 4(a), (b) and (c) express the fundamental mode of Rayleigh waves with frequency of 2.5, 4.5 and 7.5 Hz, respectively. The solid symbol represents

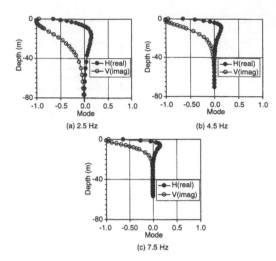

(a) 2.5 Hz   (b) 4.5 Hz

(c) 7.5 Hz

Figure 4.   Fundamental modes for Rayleigh wave.

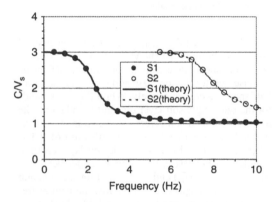

Figure 5.   Dispersion curves for Love wave.

the horizontal component while the open symbol represents vertical component of Rayleigh wave. As frequency increases, the displacement at shallow depth becomes more severe. Furthermore, at ground surface, both components have the same sign, therefore, the soil particle is rotated in the counter clockwise direction.

5.2   Result of Love wave analysis

The dispersion curve of Love wave from model of a soil on elastic half space (Fig.3(b)) is summarized in Figure 5. The vertical axis is the normalized phase velocity (C/V_s). The solid symbol represents result for fundamental mode ($S1$), while the open symbol represents other mode greater than 1 ($S2$). The exact solution from the theory of wave propagation is also drawn in the figure. It can be seen that the analytical result is well fit to the exact solution.

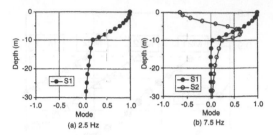

(a) 2.5 Hz   (b) 7.5 Hz

Figure 6.   Vibration modes for Love wave.

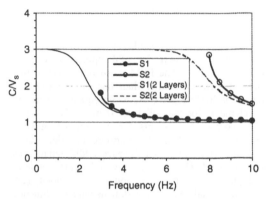

Figure 7.   Dispersion curves for Love wave.

$S1$ mode at frequency of 2.5 Hz is summarized in Figure 6(a). The real part of vector $\{V_s\}$ is plotted.

Since the frequency of wave is low, the displacement is penetrated down to the foundation layer. The displacement of the top soil layer is in the first mode. Figure 6(b) shows $S1$ and $S2$ mode at frequency of 7.5 Hz. Since the wave frequency is high, $S1$ mode of the top soil layer becomes typical first mode. For $S2$ mode, the displacement goes deep into the lower foundation layer and the top soil layer shakes in the second mode.

Figure 7 shows the dispersion of Love wave of soil on rigid base. The solid and open symbols stand for the $S1$ and $S2$ modes, respectively. The results obtained from case with soil on elastic half space are also plotted for comparison. It can be seen that close to frequency of 2.5 and 7.5 Hz, the phase velocity increases rapidly and starts to deviate from the results obtained from case of soil on elastic half space.

In the other word, at frequency of 2.5 and 7.5 Hz, $C = \infty$, $k = 0$ and soil expresses zero mode. And this is the characteristic of the body wave (SH wave).

Figure 8(a) and 8(b) shows $S1$ and $S2$ modes at frequency of 2.5 and 7.5 Hz. It is clear from the figure that the S1 mode is the first mode of SH wave and $S2$ model is the second mode of SH wave.

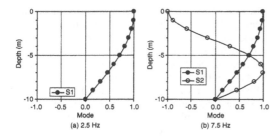

(a) 2.5 Hz

(b) 7.5 Hz

Figure 8.   Vibration modes for Love wave.

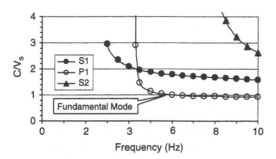

Figure 9.   Dispersion curves for Rayleigh wave.

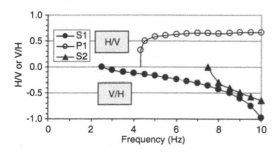

Figure 10.   Amplitude ratio for Rayleigh wave.

5.3   *Result of Rayleigh wave analysis*

The dispersion curve of Rayleigh wave for case of soil on rigid base is shown in Figure 9. *S1* and *S2* modes are identified as Rayleigh waves which express its zero modes at frequencies close to the natural frequencies for first and second modes of S wave. Similarly, *P1* mode is identified as wave which expresses its zero mode at frequency close to the natural frequency of P wave. *P1* mode is so-called fundamental mode of Rayleigh wave.

Figure 10 shows the amplitude ratio (horizontal amplitude by vertical amplitude (H/V) and vice versa) at the ground surface. For *S1* mode, the V/H = 0 at frequency of 2.5 Hz. Namely, at this frequency, there exists only horizontal component which is the natural

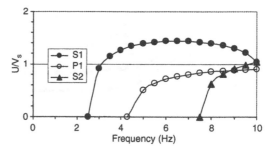

Figure 11.   Group velocity for Rayleigh wave.

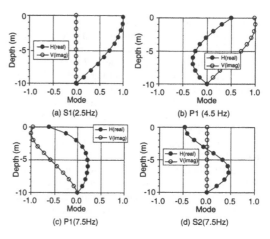

(a) S1(2.5Hz)

(b) P1 (4.5 Hz)

(c) P1(7.5Hz)

(d) S2(7.5Hz)

Figure 12.   Vibration modes for Rayleigh wave.

first mode of S wave. With increase in frequency, the vertical displacement becomes larger. Similarly, for *S2* mode at 7.5 Hz, there exists only horizontal component which is corresponding to the second mode of S wave. With increasing in frequency, the vertical displacement increases. For *P1* mode, H/V = 0 at frequency of 4.325 Hz, there exists only vertical component which is corresponding to the natural first mode of P wave. With increasing in frequency, the horizontal displacement increases. The elliptic particle motion slowly changes to be more circular-like with increasing in frequency. Since H/V > 0, both component having the same sign, consequently, soil particle is rotated in the counter clockwise direction.

Figure 11 shows the dispersion of group velocity (*U*) of *S1*, *S2* and *P1* modes. At frequency of 2.5, 4.325 and 7.5 Hz, group velocities of *S1*, *P1* and *S2* are zero. This implies that energy do not propagate in the horizontal direction.

Figure 12(a), (b), (c) and (d) show the *S1* mode at 2.5 Hz, *P1* mode at 4.5 Hz, *P1* mode at 7.5 Hz and S2 mode at 7.5 Hz, respectively. For *S1* mode at 2.5 Hz, there exists only horizontal component. This is

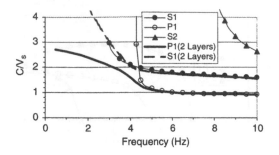

Figure 13.   Rayleigh wave dispersion curves.

corresponding to the first mode of S wave. For *P1* at 4.5 Hz, compared to Figure 4(b) of infinite soil layer, mode of displacement is greatly affected by the rigid base. Nevertheless, for *P1* mode at 7.5 Hz, due to it high frequency, the shape is similar to that obtained from single infinite layer (Fig.4(c)).

For *S2* mode at 7.5 Hz, there exists only horizontal component which is corresponding to the second mode of S wave.

Results obtained from case of soil on elastic half space (Fig.3(b)) are plotted in Figure 13 against *S1*, *S2* and P1 obtained from case of soil on rigid base. Similar to Love wave, for S wave at frequency close to 2.5 Hz and P wave at frequency close to 4.325 Hz, curves obtained from those two cases are deviated.

6   CONCLUSIONS

The present study adopted the FEM analysis with proper dispersion model to predict the surface wave characteristics. The main conclusions can be drawn as following;

1. With the presented analytical technique, the calculation results were in well accordance to the exact solution of Rayleigh and Love waves from theory of wave propagation.

2. In case of soil on rigid base, it was proved by the calculation result that there existed a fundamental mode of P wave at which $k = 0$ and $C = \infty$. With increasing in frequency, the horizontal component increased. Soil particle at ground surface was rotated in the counter clockwise direction.

3. Similarly, in case of soil on rigid base, there existed a fundamental of S wave at which $k = 0$ and $C = \infty$. With increasing in frequency, the vertical component increased.

4. Dispersion curves of Rayleigh wave and Love wave from case of soil of elastic half space started deviated from those obtained from case of soil on rigid base when the frequencies approached the fundamental values.

REFERENCES

Tokunaga, N., Morio, S., Iemura, H. and Nishimura, T. 2000. On the participation factor to the ground vibrations propagating from the urban viaduct, *Journal of Structural Engineering*, Vol.46A, 1703–1713. (in Japanese)

Lysmer, J. 1970. Lumped mass method for Rayleigh waves, *Bulletin of the Seismological Society of America*, Vol.60, No.1, 89–104.

Lysmer, J. and Drake, J. A. 1973. A Finite Element Method for Seismology Method in Seismological Physics, *Academic Press*, Chap.6, 180–216.

Morichi, S. Ohmachi, T. Toshinawa, T. and Miyai, A. 1987. A fundamental study on vibration models in an elastic layer characterized by Rayleigh Waves, *Journal of Structural Engineering*, Vol.33A, 631–644. (in Japanese)

Watanabe, K. and Okumura, H.1983. 2.6.6 Finite element analyses of surface waves propagating the layered material, Studies on composite material technology, The Institute of Industrial Science, Tokyo University, No.3, 184–188. (in Japanese)

Tajime, K. 1983. The story of surface waves, *Butsuri-Tanko* (Geophysical Exploration), Vol.36, No.1, 23–32. (in Japanese)

Environmental Vibrations – Takemiya (ed.)
© 2005 Taylor & Francis Group, London, ISBN 0 415 39035 4

A study on damping and mitigation performance of side surfaces of foundations on soft soil – Part 1: Forced vibration tests of foundation block with various embedment conditions

Y. Ikeda
Taisei Corporation, Tokyo, Japan

Y. Shimomura
Junior College of Nihon University, Chiba, Japan

M. Kawamura
Nihon University, Chiba, Japan

S. Ishimaru
Nihon University, Tokyo, Japan

ABSTRACT: To aim at progress of damping performance of foundations that will be built at soft ground, we have proposed an improved foundation work of backfilling a damping material into trenches dug along a foundation supported by improved soil medium. This damping material is a mixture of asphalt with crushed stones and rubber chips (MACSRC) and has itself high attenuation and mitigation performance. Not only to comprehend the attenuation ability of the improved foundation work quantitatively and qualitatively but also to verify the effectiveness of this work, we carried out forced vibration tests for two test blocks, which were constructed by a normal construction work and the above improved foundation work.

1 INTRODUCTION

According to investigation reports of The Great Hanshin Earthquake in 1995 and The Mid Niigata prefecture Earthquake in 2004, it has been recognized that structures that adopted the seismic isolation can improve performance of preventing not only loss of human life but also damage of structures and facilities. As base-isolated structures have an only effect on appropriate ground condition sites, soil improvement work has been necessary for weak ground. Improvement work of surrounding soil of structures is also important for enhancement of damping and mitigation performance of structures.

To aim at progress of attenuation and mitigation performance of side surface soil of foundations that will be constructed at soft ground, we have proposed an improved foundation work of backfilling a damping composite into trenches dug along a foundation supported by improved soil medium. This damping material is a mixture of asphalt with crushed stones and rubber chips (hereinafter call as MACSRC) and has itself high attenuation and mitigation performance.

To comprehend the attenuation ability of the improved foundation work and to verify the effectiveness of this work, we carried out forced vibration tests for two test blocks, which were constructed by a conventional construction work and the above improved foundation work. Here, the conventional construction work is a procedure to backfill the dug soil into trenches grubbed along a foundation. Hereinafter, the improved foundation work and the conventional construction work are called IF and CF, respectively.

In the experiments, IF and CF were constructed on the same axis line. First, we carried out forced vibration tests focusing on responses of both foundations and their cross interaction effects. In these tests, we changed conditions of trenches dug along IF. Basically, a shaking machine was mounted on CF. Next to confirm attenuation and mitigation abilities of MACSRC, we backfilled the trenches dug along IF with MACSRC and conducted an experiment of CF oscillated by the vibration generator. Finally, we remounted the vibration generator on IF whose trenches were paved by MACSRC and vibrated IF by the exciter to evaluate the effectiveness of the mitigation performance of

MACSRC. As an approach to global environmental issues, the rubber chips mixed with asphalt are recycling material.

2 GROUND CONDITION AT EXPERIMENTAL SITE AND FOUNDATION BLOCK

A layout of an experimental site is shown in Figure 1. The experimental site has been located at a vacant lot on the north of the experimental building (Joint Research Center for Environment Protection & Disaster Prevention City) in the Funabashi Campus of Nihon University in Chiba Prefecture, which is near Tokyo in Japan. Soil profile of the test site is illustrated in Figure 2. Ground surface 2.8 m shallower is loamy layer of the Kanto district and partially includes backfilling soil 0.1 m deep. Cohesive soil is distributed in the range of G.L.$-$2.8 m–G.L.$-$3.5 m. Tuff

fine sand can be seen in the range of G.L.$-$3.5 m–G.L.$-$5.3 m. G.L.$-$5.3 m deeper, silty fine sand and fine sand are distributed. Tips of the improved soil cement columns are located at G.L.-2.8 m, where loam and cohesive soil are distributed, and its value of the standard penetration test is five and under.

IF, which is 6 m away to east side from CF, has been built. Both foundations have been supported by four soil cement columns and the existing soil layers. The dimensions of the both blocks are 2.4 m $\times$ 2.4 m $\times$ 1.0 m. A diameter of a soil cement column is 600 mm and its length is 2.5 m. The test block and the soil cement columns are shown in Figure 3.

3 RESULTS OF MATERIAL TESTS

3.1 Soil cement columns

Two core rods, which had 90 mm in diameter and 2.5 m in length, were obtained from a soil cement column of the material age of 28 days. The core rod was then cut as specimens of 180 mm lengths for material tests. Stress–strain curves of the core samples of the soil cement column that were the highest and the lowest deformation moduli (E_{50}) among all of specimens (22 samples) are shown in Figure 4. The maximum values of stresses mean the unconfined compressive strength. Further, inclinations of the deformation modulus (E_{50max}, E_{50min}, the mean value and standard deviation σ) are also illustrated in Figure 4. The deformation moduli of all the specimens were 1250–1680 MPa. The unconfined compressive strengths of all the specimens became 2.0–3.8 times as large as the specified design strength, 1 MPa. Density and the unconfined compressive strength distributions against depth are shown in Figure 5. The densities distributed uniformly against variations of the depth and were in the range of 1.44 to 1.52 g/cm^3. In comparison of the results of the previous material tests of CF (Ishimaru et al. 2004), the unconfined compressive strengths of all the

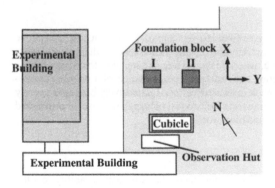

Figure 1. Schematic view of the experiment site.

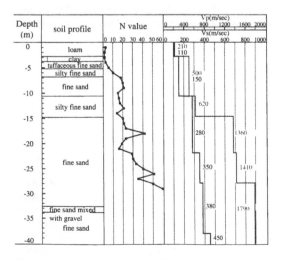

Figure 2. Soil profile of the test site.

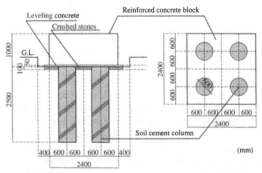

Figure 3. Foundation block on improved soil medium.

specimens were slightly small (1.97–3.81 MPa) and the deformation moduli (E_{50}) and densities were almost identical with them. Properties obtained by the unconfined compressive strength tests are shown in Table 1.

3.2 Rubber chips mixed with asphalt and crushed stones

In order to obtain basic material features of MAC-SRC paved into the trenches dug along IF, dynamic

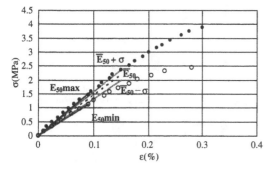

Figure 4. Stress–strain curves of core samples of soil cement column.

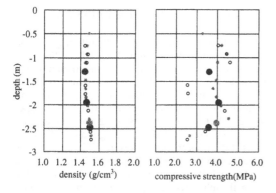

Figure 5. Density and unconfined compressive strength distributions against depth.

compressive triaxial tests have been conducted (Ishimaru et al. 2005b). Diameters and lengths of the specimens are about 100 mm and 190 mm, respectively. Taking into account a solution of global environmental problems, we mixed the rubber chips, which were recycling materials, with asphalt and crushed stones.

In the dynamic triaxial compressive tests, the specimens were consolidated by prescribed isotropic stresses and vertical pulsating strains were measured until approximately 0.1 percent strain levels. Examples of hysteresis loops of deviator stresses and axial strains for variations of mixed rates of the rubber chips to crushed stones, whose particle sizes are in the range of 5.0 mm to 2.5 mm, are shown in Figure 6. The mixed rate of the rubber chips has been defined by the ratio of the total weight (the rubber chips and the crushed stones) to the weight of the rubber chips.

According to the test results of the mix rates of the rubber chips, it was found that hysteresis loops of 12.5 percent mix rates of the rubber chips were relatively stable. It was revealed that the initial secant moduli of 12.5 percent mix rates of the rubber chips were undiminished and the equivalent damping constants increased, compared with no mixed rates of the rubber chips. Finally, we adopted MACSRC of 12.5 percent mixed rate of the rubber chips to backfill the trenches dug along IF. Blending of MACSRC is illustrated in Table 2. Strain dependence on the

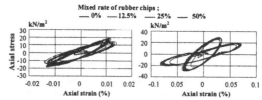

Figure 6. Examples of hysteresis of deviator stresses and axial strains for variations of mix rates of rubber chips and crushed stones.

Table 1. Results of unconfined compressive strength tests.

Test results	Deformation modulus E_{50} (MPa)	Density ρ(g/cm^3)	Unconfined compressive strength q_u (MPa)
Previous test			
Mean value	1456	1.46	2.86
Standard deviation	130	0.02	0.59
Present test			
Mean value	1462	1.48	3.77
Standard deviation	200	0.02	0.67

Table 2. Blending of mixture of asphalt with crushed stones and rubber chips.

Mixed rate: 12.5%	Mass ratio rate (%)	Density (g/cm^3)	Mass (gr)
Crushed stones	58.6	2.7	1983.6
Rubber tips	8.4	1.2	283.4
Slow curing	12	2.7	406.0
Fine sand	11	2.7	372.0
Filler	10	2.7	338.0
Total of aggregate	100		3383.0
Asphalt	7.5	1	274.3
Total (aggregate + asphalt)	–	–	3657.3

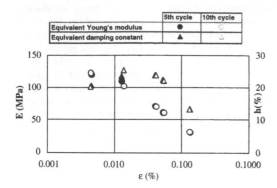

	5th cycle	10th cycle
Equivalent Young's modulus	●	○
Equivalent damping constant	▲	△

Figure 7. Strain dependence on equivalent Young's moduli and equivalent damping constants of specimens.

equivalent Young's moduli and the equivalent damping constants of the specimens whose mix rate is 12.5 percent is shown in Figure 7. Figure 7 represents a result of the experiments for various load levels with 50 kPa of confining pressure and 0.1 Hz of frequency. In the strain level of 0.01 percent and under, it was found that the equivalent Young's moduli and the equivalent damping constants became about 120 MPa and 20 percent, respectively.

4 SCHEDULE OF FORCED VIBRATION TESTS

We dealt with two test blocks in these experiments. To confirm progress of attenuation performance and cross interaction effects of the two test blocks, forced vibration tests were carried out under various conditions of the trenches dug along IF. Figure 8 shows a schedule of the forced vibration tests. The trenches excavated along CF, whose width and depth were about 0.4 m and 0.8 m, respectively, were backfilled by the dug soil until the ground surface level. The backfilled soil was tamped by the compaction work with a rammer. Conditions of the trenches excavated along IF from Stage 1 to Stage 3 in Figure 8 were as follows.

First, we backfilled the trenches along IF by the dug soil until the ground surface level (Stage 1). Then we excavated the trenches 0.3 m deep along IF (Stage 2). Next we made the trenches 0.5 m deep (Stage 3). At Stage 4, we filled in the trenches with MACSRC and carried out a forced vibration test to confirm mitigation ability of MACSRC. The vibration generator had been mounted on CF until Stage 4. We remounted the shaking machine on CF to IF and conducted an experiment focusing on attenuation performance of MACSRC (Stage 5).

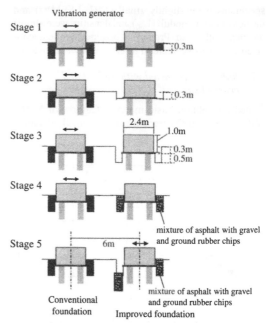

Figure 8. A schedule of forced vibration tests.

5 FORCED VIBRATION TEST RESULTS

While CF was oscillated by the vibration generator, forced vibration tests (from Stage 1 to Stage 4) related to various conditions of the trenches dug along IF were carried out. Amplitude and phase functions of the response displacement curves per unit exciting force for the exciting direction on the upper surface of IF on Stage 1, Stage 2 and Stage 4 are shown in Figure 9. Although magnitudes of amplitudes at the peak frequencies were different, the peak frequencies were almost identical on Stage 1, Stage 2 and Stage 4. It means that conditions of trenches dug along IF only affected magnitude of the amplitudes and did not have an effect on the peak frequencies. The shallower the depth of the trenches became, the smaller the magnitude of the amplitude at the peak frequencies were. Compared with Stage 1 and Stage 4, from fifteen to thirty percent of decrease in magnitude of amplitudes of IF on Stage 4 was found at the peak frequencies. It is expected that MACSRC would play an important role in progress of mitigation ability of the structures.

Before these experiments, we carried out other tests in which CF only existed (Ishimaru et al. 2005a). Acceleration resonance curves of CF in the previous tests are shown in Figure 10. Figure 10 also shows acceleration resonance curves of IF in Stage 5. Amplitude of Stage 4 at peak frequency, 11 Hz, reduced about forty percents compared with that of

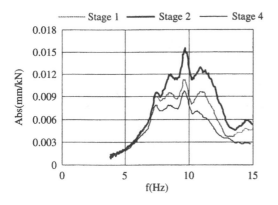

Figure 9. Resonance curves of IF in Stage 1, 2 and 4.

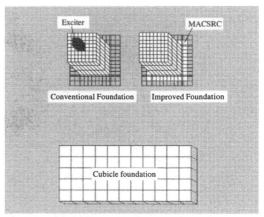

Figure 11. Models of hybrid approach.

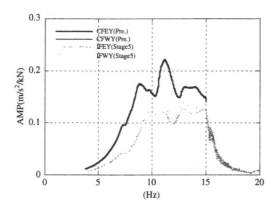

Figure 10. Responses of CF in previous test and IF in Stage 5.

Table 3. Properties of analysis model.

Depth (m)	ρ (t/m³)	Vs (m/s)	Poisson's ratio	Damping ratio
1.0	1.4	90	0.311	0.03
2.7	1.4	90	0.311	0.03
5.4	1.6	150	0.451	0.03
10.6	1.7	280	0.372	0.02
14.7	1.7	280	0.478	0.02
21.7	1.7	350	0.465	0.02
27.8	1.8	380	0.461	0.02
38.0	2.0	450	0.466	0.02
45.2	2.0	420	0.461	0.02

Material	S-wave Vs (m/s)	Density ρ (ton/m³)	Poisson's ratio	Damping ratio
Foundation block	–	2.4	0.20	0
Leveling concrete	–	2.4	0.20	0
Asphalt	163	1.68	0.35	0.20

the previous tests. As a result it was indicated that MACSRC backfilled into the trenches dug along the block had good attenuation performance. In the previous tests, the operative frequency was 15 Hz. Because the maximal exciting force in the present tests was 30 kN, amplitudes over 18 Hz were insufficiency.

6 SIMULATION ANALYSIS

6.1 Analysis model

After the end of the experiments, we carried out the three dimensional simulation analyses for Stage 1, 2 4 and 5 by a hybrid approach (Ikeda et al. 2003). In the hybrid approach, CF, IF and soil regions surrounding foundations including trenches are modelled by the three-dimensional finite element. The three-dimensional thin layer approach is applied to free field ground regions covered the above finite elements. Soil cement columns that support the foundations are modelled by hexagon elements. In Stage 2, we considered trenches along IF. In other Stages, both foundations were treated as partially embedment. As shown in Figure 1, there are some structures near the both test foundations. We modelled a cubicle (a power transformation vessel) foundation that locates at the south direction of CF. Bird's-eye view of the hybrid model in Stage 4 are illustrated in Figure 11. Properties of soil, foundations and MACSRC are shown in Table 3. We adopted the results obtained by the material test of MACSRC as the properties of MACSRC.

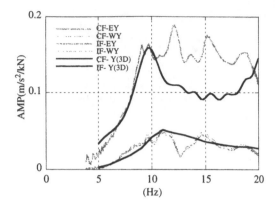

Figure 12. Comparison of response curves of CF and IF in Stage 1.

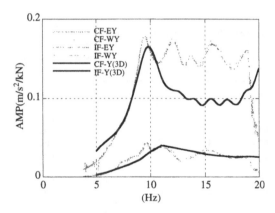

Figure 13. Comparison of response curves of CF and IF in Stage 4.

6.2 *Comparison of test and simulation results*

Resonance curves of acceleration per unit exciting force for the exciting direction on the upper surface of CF by experiments and analyses in Stage 1, and 4 are illustrated in Figures 12–13. At each Stage, the shaking machine was mounted on CF. The first peak frequency and its amplitude of resonance curves of by each analysis were good agreement with the experiment results. Resonance curves of CF by the analyses in frequency range over the first mode were estimated smaller than the experiments. On the other hand, amplitudes of the resonance curves of IF by the analyses were almost close to the average of the test results and both frequency dependencies were approximately the same.

Figure 14 shows a comparison with acceleration resonance curves of IF in Stage 1,2 and 4. Analyses are showing that amplitudes of resonance curves at the peak frequencies become smaller when depths of

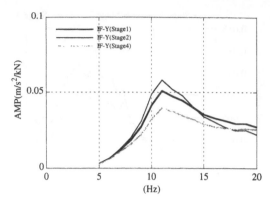

Figure 14. Comparison of response curves of IF in Stage 1, 2 and 4.

the trenches become shallower. Compared with the analysis results of Stage1 and Stage 4, magnitude of amplification of resonance curves at the peak frequency in Stage 4 became about 30% smaller than that of Stage 1. This tendency, that is, mitigation performance of MASCRC, corresponded with that of the experiments.

7 CONCLUSIONS

The material experiments of MACSRC, as a newly-adopted attenuation ingredient, were carried out to obtain their material properties. Forced vibration tests and their simulation analyses of the two test blocks supported by the improved soil medium were conducted to grasp fundamental dynamic feature of MACSRC.

The equivalent damping constant of MACSRC of 12.5% mix ratio adopted in the experiment in the low strain level were approximately 20%.

Compared with the experiment results, effectiveness in the attenuation performance of MACSRC backfilled into the trenches dug along the block was confirmed.

By the analysis and experiment results, it was also revealed that MACSRC provided mitigation performance for the foundation more than the grubbed soil backfilled into the trenches.

It is expected that MACSRC would exert not only attenuation but also mitigation ability against earthquakes or other external and internal forces.

ACKNOWLEDGEMENTS

This research was supported by Grant-in-Aid for Science Research (B), 15360303, 2003–2005, the

Ministry of Education, Culture, Sports, Science and Technology, Japan. Further, the research was conducted as a part of the Academically Promoted Frontier Research Program on "Sustainable City Based on Environment Preservation and Disaster Prevention" at Nihon University, College of Science and Technology (Head Investigator: Prof. Ishimaru, S.) under a grant from the Ministry of Education, Culture, Sports, Science and Technology, Japan. We would like to express our sincere appreciation to Prof. Iwai, S., College of Science & Technology, Nihon Univ., Mr. Hibino, S., Tenox Co., Mr. Miwa, S., Tokyu Construction and Mr. Kunisima, T., Takenaka Road Construction Co., LTD., for their useful suggestions and helpful advice in conducting the experiments.

REFERENCES

Ikeda, Y., Shimomura, Y., Adachi, H., Ogushi, Y., Nakamura, M. 2003. An experimental study of mock-up pile foundations Part 2 Dynamic cross interaction of foundations. Transactions of the 17th International Conference on Structural Mechanics in Reactor Technology (SMiRT 17).

Ishimaru, S., Hata, I., Shimomura, Y., Ikeda, Y., Ishigaki, H., Ogushi, Y. 2004. A feasibility study of new type seismic isolation – Composed system of piles covered by pipes and dampers with partial soil improvement. Proceedings of 13th World Conference on Earthquake Engineering. Paper No.2204.

Ishimaru, S., Shimomura, Y., Kawamura, M., Ikeda, Y., Hata, I., Ishigaki, H. 2005a. An experimental study on advancement of damping performance of foundations in soft ground, Part 1 Forced vibration tests of a foundation block constructed on improved soil medium. Transactions of the 18th International Conference on Structural Mechanics in Reactor Technology (SMiRT 18). K13-1 (to be published).

Ishimaru, S., Shimomura, Y., Kawamura, M., Ikeda, Y., Hata, I., Miwa, S. 2005b. An experimental study on advancement of damping performance of foundations in soft ground, Part 2 Experiment focusing on damping and antivibration performance of side surface of foundation blocks. Transactions of the 18th International Conference on Structural Mechanics in Reactor Technology SMiRT 18). K13-2 (to be published).

Environmental Vibrations – Takemiya (ed.)
© 2005 Taylor & Francis Group, London, ISBN 0 415 39035 4

A study on damping and mitigation performance of side surfaces of foundations on soft soil – Part2: Analysis of records of The Mid Niigata prefecture Earthquake in 2004 and its simulations

Y. Shimomura
Junior College of Nihon University, Chiba, Japan

Y. Ikeda
Taisei Corporation, Tokyo, Japan

I. Hata
Nihon University, Tokyo, Japan

S. Ishimaru
Nihon University, Tokyo, Japan

ABSTRACT: After the final forced vibration test, The Mid Niigata prefecture Earthquake in 2004 struck Niigata Prefecture on the evening of October 23rd, 2004. Fortunately, we obtained five records of the earthquake and its after quakes. In order to confirm the attenuation performance of MACSRC and the cross interaction effect of adjacent foundations for earthquakes, analyses of earthquake observation records and their simulation analysis were conducted. In comparison of the earthquake records and the simulation results, it was revealed that the damping performance of Improved foundation and the cross interaction effect of adjacent foundations worked effectively for not only forced vibrations but also earthquakes.

1 INTRODUCTION

After the final forced vibration test, we have continued seismic observations at the experimental site. In the final forced vibration test, we backfilled trenches dug along Improved foundation with a mixture of asphalt with crushed stones and rubber chips (here call as MACSRC) and shook Improved foundation by an exciter. In the seismography, the shaking machine that mounted on Improved foundation was removed.

The Mid Niigata prefecture Earthquake in 2004 struck Niigata Prefecture on the evening of October 23rd, 2004. It was M_{JMA} 6.8 earthquake. It was the most significant earthquake to affect Japan since the Great Hanshin Earthquake in 1995.

In order to confirm the attenuation performance of MACSRC and the cross interaction effect of adjacent foundations for earthquakes, analyses of earthquake observation records and their simulation analysis were conducted. In the simulation analyses, two-dimensional finite elements method and hybrid approach, which treated Conventional and Improved

Foundations and their adjacent soil region as three-dimensional finite elements and soil free field region as thin layer approach, were adopted. For modeling of the simulation analysis, two foundations, the soil region surrounding the foundations and ground points where seismographs were installed were taken into consideration. In the hybrid approach, we evaluated responses that were induced by not only x component of earthquake input motions but also y component to take into account the three-dimensional effect.

2 SEISMOGRAPH INSTALLATION AT EXPERIMENTAL SITE

A layout of an experimental site is illustrated in Figure 1. The experimental site is a back lot of the experimental building (Joint Research Center for Environment Protection & Disaster Prevention City) in a campus of Nihon University at Chiba Prefecture in Japan. As shown in Figure 2, one of seismographs was installed on the floor of basement of pit in an experimental Building. (Ishimaru et al. 2004, Ishimaru et al. 2005a) Two of them were located on the table of a base-isolation structure and another was installed on the basement of the base-isolation structure. Locations of seismographs beneath the experimental building and the base-isolation structure were at G.L.-7 m and G.L.-22 m. Because the shear wave velocity at G.L.-38 m under the base-isolation structure has exceeded 400 m/s, it is recognized that this point has been the engineering bedrock. Three components (x, y and z) are available for each seismograph. Here, call the engineering bedrock point as S4.

At the experimental site, seven seismographs have been installed near two test blocks, which are constructed by a conventional construction work and an improved foundation work. The conventional construction work is a procedure to backfill the dug soil into trenches along a foundation and the improved foundation work means backfilling a damping composite, that is MACSRC, into trenches dug along a foundation (Ishimaru et al. 2005b, Ikeda et al. 2005). Hereinafter, they are called as CF and IF. Two

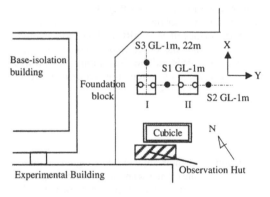

Figure 1. Schematic view of the experiment site.

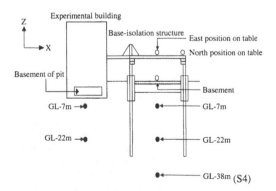

Figure 2. Seismograph installation.

seismographs were mounted at eastern and western sides on the surface of each foundation. Three seismographs near two foundations are located at G.L.-1.0 m. Each position is the midpoint of CF and IF (S1); 3 m eastern from the center of IF (S2); and 6 m northern from the center of CF (S3).

3 ANALYSIS OF OBSERVATION RECORDS

3.1 Time histories

After the final forced vibration test, we have continued seismic observations at the experimental site. In the final forced vibration test, we backfilled trenches dug along IF with MACSRC and shook IF by an exciter. In the seismography, the shaking machine mounted on IF was removed.

The Mid Niigata prefecture Earthquake in 2004 struck Niigata Prefecture on the evening of October 23rd, 2004. It was M_{JMA} 6.8 earthquake. It was the most significant earthquake to affect Japan since the Great Hanshin Earthquake in 1995. Location of the epicenter was in northwestern Honshu region, approximately eighty km south of Niigata city. The epicenter was beneath the Uonuma Hills. We obtained main shock of the earthquake and its after quakes. The epicentral distance from the epicenter to the experimental site at Funabashi city in Chiba prefecture is about 230 km.

Examples of acceleration time history records of two foundations and soil region near the foundations are illustrated in Figure 3. Acceleration time history records of the soil region near the foundations (x component) have analogous shapes of the principal motion close to those of two foundations. Except for S3, the principal motions of acceleration records of S1 and S2 are similar to those of the foundations.

Catalog of earthquake records is shown in Table 1. Magnitudes of the earthquakes observed at the site exceed M_{JMA} 6.0. Table 2 illustrates peak accelerations of time history records observed on both foundations and their surrounding soil. Here, y direction is equal to a parallel direction of CF and IF, and x direction is an orthogonal direction of the y direction. Symbol EX represents x component observed by seismograph on eastern side of foundations.

Peak accelerations of 29.4 cm/s² and 33.5 cm/s² were recorded at EX and WY of CF. IF had peak values of 27.8 cm/s² and 28.3 cm/s² at WX and WY. On both foundations, x components were smaller than y components. Compared with peak accelerations of both foundations, values of IF were smaller than these of Conventional one except for WX of IF.

In soil region near two foundations, peak acceleration of S1 and S3, for x component, are almost the same and S2 is greater than others. For y component, peak acceleration of S3 is greater than S1 and S2, and peak value of S1 is close to that of S2.

3.2 *Transfer functions*

Transfer functions of the ground points of S1, S2 and S3 to the ground point of S4, which is defined as the engineering bedrock, are shown in Figure 4. These transfer functions were calculated by the main shock of the earthquake. Peak frequencies of about 2.5 Hz, 6.0 Hz, 9.0 Hz, and 13.5 Hz can be seen in the transfer functions for x and y directions. These are the predominant frequencies of the soil region (from the first to the fourth modes), which is shallower or equal to G.L.-38 m. Especially, amplification of the fourth mode at S3 is greater than others.

Transfer functions of CF and IF to the ground point of S4 are displayed in Figure 5. Because eastern and western transfer functions of both foundations are almost identical, the former is only shown. In Figure 4, peaks can be seen at the predominant frequencies of the soil region. For x direction, the transfer functions of both foundations are approximately identical. On the other hand, for y direction, the transfer function of IF is smaller than that of Conventional one over 9.0 Hz. This tendency is found not only the main shock but after ones.

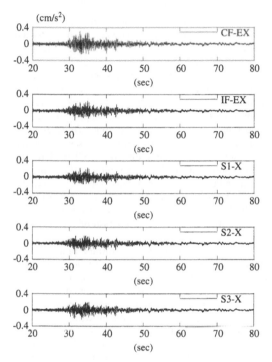

Figure 3.   Examples of acceleration time history records.

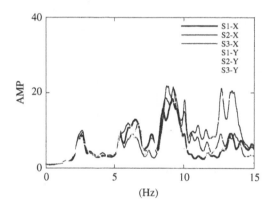

Figure 4.   Transfer functions of S1, S2 and S3 to S4.

Table 1.   Catalog of earthquake records.

No.	Occurrence time (M/D/H/m)	Hypocenter	Magnitude	Depth (km)	I_{JMA}
1	10/23/17/56	Chuestu of	6.8	13	II(III)
2	10/23/18/04	Niigata Pref.	6.3	9	II
3	10/23/18/12		6	12	II
4	10/23/18/35		6.5	14	II
5	10/27/10/41		6.1	12	II

Table 2.   Peak accelerations (cm/s²).

Obs. points	CF	IF	S1	S2	S3
x-comp.	29.4	27.8	22.1	20.1	22.2
x-comp.	33.5	28.3	23.7	23.7	26.5

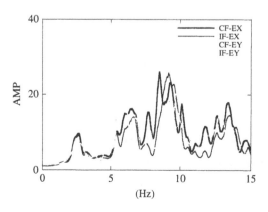

Figure 5.   Transfer functions of CF and IF to S4.

4 SIMULATION ANALYSIS

4.1 *Two-dimensional analysis model*

Simulation analyses by the 2-dimensional finite element approach were carried out, so that fundamental vibration characteristics of earthquake records of CF, IF and soil region surrounding both foundations can be grasped. In modeling of two-dimensional analysis, we took into account y direction that is equal to a parallel direction of CF and IF. The ground points of S1 and S2 in the soil region near two foundations and the ground point of S4, which locates G.L.-38 m beneath the base-isolation structure, were also modeled. Energy transmitting boundaries at both side edges of analysis model and dashpot mat beneath the model were taken into account. Two-dimensional model is shown in Figure 6. Properties of soil and foundations in analysis are illustrated in Table 3. The transfer functions of two foundations and soil points to the steady state

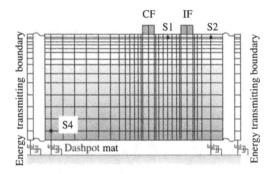

Figure 6.   Two-dimensional analysis model.

Table 3.   Properties of analysis model.

Depth (m)	ρ (t/m³)	Vs (m/s)	Poisson's ratio	Damping ratio
1.0	1.4	90	0.311	0.03
2.7	1.4	90	0.311	0.03
5.4	1.6	150	0.451	0.03
10.6	1.7	280	0.372	0.02
14.7	1.7	280	0.478	0.02
21.7	1.7	350	0.465	0.02
27.8	1.8	380	0.461	0.02
38.0	2.0	450	0.466	0.02
45.2	2.0	420	0.461	0.02

Material	S-wave Vs (m/s)	Density ρ (ton/m³)	Poisson's ratio	Damping ratio
Foundation block	–	2.4	0.20	0
Leveling concrete	–	2.4	0.20	0
Asphalt	163	1.68	0.35	0.20

harmonic incident wave defined at the bottom of the dashpot mat underneath the analysis model were evaluated. To compare with the transfer function estimated by the observed records, we calculated transfer functions of two foundations, the points of S1 and S2 to the point of S4 by the analysis model.

4.2 *Three-dimensional analysis model*

Three-dimensional analysis by hybrid model, which was carried out for simulation analysis of the forced vibration tests, was also applicable to the simulation of earthquake responses (Ikeda et al. 2005). Figure 7 shows three-dimensional analysis model. In analysis model, two foundations including their adjacent soil region were modeled by three-dimensional finite elements and thin layer approach was applied to the free field region that surrounds the above finite element domain. Focusing on horizontal responses, we carried out analyses for x and y directions. CF and IF, the ground points of S1, S2 and S3 in the soil region near the foundations, the basement of cubicle, and the ground point of S4, which locates G.L.-38 m underneath the base-isolation structure, were taken into consideration as same as simulation of forced vibration tests.

The transfer functions of arbitrary points to the steady state harmonic incident wave defined at the bottom of the dashpot mat that is supporting thin layer elements were obtained. We also calculated transfer functions of the foundations or the soil points to the point at G.L.-28 m so as to compare with observation records as well as the two-dimensional analyses. In three-dimensional analysis, we took into account responses that are induced by not only x component but also y component of earthquake input motions to obtain appropriate results (Ikeda et al. 2004).

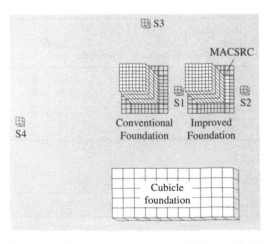

Figure 7.   Three-dimensional analysis model.

5 COMPARISON WITH OBSERVATION AND ANALYSIS

5.1 Transfer functions of both foundations

Transfer functions of CF and IF to the ground point of S4 calculated by the main shock of earthquake record and 3-dimensional analysis of x component are shown in Figure 8. Peak frequencies of 2.5 Hz, 6.0 Hz, 9.0 Hz and 13.0 Hz appeared on the transfer functions of observation record can be seen on results of three-dimensional analyses. The third mode's amplitude of CF of the earthquake record corresponds approximately with that of IF, which is the first coupled mode of foundations and ground. Results of analysis agree with the tendency of the observation.

Figure 9 illustrates transfer function of both foundations to the ground point of S4 calculated by the main shock of y component of earthquake record, 2-dimensional analysis and 3-dimensional one. Two-dimensional analysis and three-dimensional analyses

estimated peak frequency of the third mode lower than the observation. According to transfer functions of the observation, amplification of the first mode was small. Although two-dimensional analysis provided large amplification, amplification of the first mode of three-dimensional analysis agreed well with the observation. It is found that the third mode's amplification of IF of the observation is smaller than that of Conventional one. By results of three-dimensional analysis, the same trend can be seen at the first coupled mode of foundations and ground. It might be caused by attenuation performance of MASCRC or cross interaction of both foundations.

As a result, both analyses correspond approximately with the observation. Especially, it is confirmed that three-dimensional analysis provides detailed feature of the earthquake records.

5.2 Transfer functions of ground points

Transfer functions of the ground points S1, S2 and S3 to the point of S4 calculated by the main shock of earthquake record and 3-dimensional analysis of x component are displayed in Figure 10. Peak frequencies of the analysis are in agreement with these of the observation. At the third mode of the observation, magnitude of amplitude of S1 and S2 are larger than that of S3 and the three-dimensional analysis corresponds with this tendency. According to the observation, the fourth mode's amplitude of S3 of the earthquake record is greater than those of S1 and S2. On the other hand, no difference of amplitude of analysis at the fourth mode can be seen.

Transfer functions of the ground points S1, S2 and S3 to the point of S4 estimated by the main shock of y component of observation record, 2-dimesional analysis and 3-dimensional analysis are illustrated in Figure 11. Two-dimensional and three-dimensional analysis was able to represent all of predominant

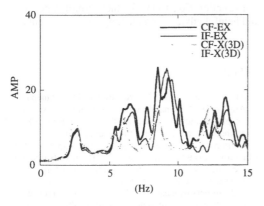

Figure 8.   Comparison of transfer functions of CF and IF of observations and analyses for x direction.

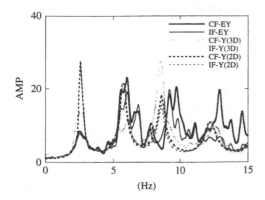

Figure 9.   Comparison of transfer functions of CF and IF of observations and analyses for y direction.

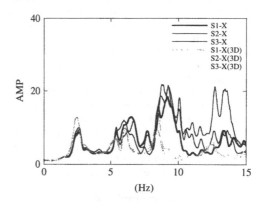

Figure 10.   Comparison of transfer functions of S1, S2 and S3 of observations and analyses for x direction.

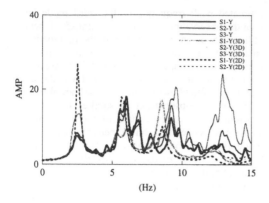

Figure 11.   Comparison of transfer functions of S1, S2 and S3 of observations and analyses for y direction.

frequencies of the earthquake. The first mode's amplitude of two-dimensional analysis is considerably higher than observation and three-dimensional analysis. transfer functions of S1 and S2 obtained by 2-dimensional analysis are identical until the third mode. Three-dimensional analysis can express the tendency that the fourth mode's amplitude of S3 is larger than that of S1 or S2.

6   CONCLUSIONS

After the final forced vibration test, we obtained five records of the Mid Niigata prefecture Earthquake in 2004. Simulation analyses by the two-dimensional finite element approach and the three-dimensional hybrid one were carried out, so that fundamental vibration characteristics of earthquake records of CF, IF and soil region surrounding both foundations can be obtained.

On both foundations, peak accelerations of time history records of x component were smaller than those of y component. Compared with peak accelerations observed at both foundations, values of IF were smaller than these of Conventional one. Acceleration time history records of the soil region near the foundations (x direction) have analogous shapes of the principal motion close to those of two foundations.

It is found that the amplification of transfer function of IF of the observation at the first coupled mode of foundations and ground points is smaller than that of Conventional one. By results of three-dimensional analysis, the same trend can be seen at the first coupled mode of foundations and ground points. It might be caused by attenuation performance of MASCRC and dynamic influence of the cross interaction of both foundations.

In comparison of the earthquake records and the simulation results, it was revealed that the damping performance of IF and the cross interaction effect of adjacent foundations worked effectively for not only forced vibrations but also earthquakes.

ACKNOWLEDGEMENTS

This research was supported by Grant-in-Aid for Science Research (B), 15360303, 2003–2005, the Ministry of Education, Culture, Sports, Science and Technology, Japan. Further, the research was conducted as a part of the Academically Promoted Frontier Research Program on "Sustainable City Based on Environment Preservation and Disaster Prevention" at Nihon University, College of Science and Technology (Head Investigator: Prof. Ishimaru, S.) under a grant from the Ministry of Education, Culture, Sports, Science and Technology, Japan. We would like to express our sincere appreciation to Prof. Iwai, S., College of Science & Technology, Nihon Univ., Mr. Hibino, S., Tenox Co., Mr. Miwa, S., Tokyu Construction and Mr. Kunisima, T., Takenaka Road Construction Co., LTD., for their useful suggestions and helpful advice in conducting the experiments.

REFERENCES

Ikeda, Y., Shimomura, Y., Nakamura, M., Haneda, O., Arai, T. 2004. Dynamic influence of adjacent structures on pile foundation based on forced vibration tests and earthquake observation. Proceedings of 13th World Conference on Earthquake Engineering. Paper No. 1869.
Ishimaru, S., Hata, I., Shimomura, Y., Ikeda, Y., Ishigaki, H., Ogushi, Y. 2004. A feasibility study of new type seismic isolation – Composed system of piles covered by pipes and dampers with partial soil improvement –. Proceedings of 13th World Conference on Earthquake Engineering. Paper No. 2204.
Ishimaru, S., Shimomura, Y., Kawamura, M., Ikeda, Y., Hata, I., Ishigaki, H. 2005a. An experimental study on advancement of damping performance of foundations in soft ground, Part 1 Forced vibration tests of a foundation block constructed on improved soil medium. Transactions of the 18th International Conference on Structural Mechanics in Reactor Technology (SMiRT 18). K13-1 (to be published).
Ishimaru, S., Shimomura, Y., Kawamura, M., Ikeda, Y., Hata, I., Miwa, S. 2005b. An experimental study on advancement of damping performance of foundations in soft ground, Part 2 Experiment focusing on damping and antivibration performance of side surface of foundation blocks. Transactions of the 18th International Conference on Structural Mechanics in Reactor Technology (SMiRT 18). K13-2 (to be published).
Ikeda, Y., Shimomura, Y., Kawamura, M., Ishimaru, S. 2005. A study on damping and mitigation performance of side surfaces of foundations on soft soil – Part1 Forced vibration tests of foundation block with various embedment conditions. ISEV2005, (to be published).

Soil dynamics

Environmental Vibrations – Takemiya (ed.)
© 2005 Taylor & Francis Group, London, ISBN 0 415 39035 4

Characteristics of stresses and settlement of ground induced by train

Y.M. Chen
Department of Civil Engineering, Zhejiang University, Hangzhou, Zhejiang, China

C.J. Wang
City College of Zhejiang University, Hangzhou, Zhejiang, China
Department of Civil Engineering, Zhejiang University, Hangzhou, Zhejiang, China

Y.P. Chen & B. Zhu
Department of Civil Engineering, Zhejiang University, Hangzhou, Zhejiang, China

ABSTRACT: By introducing the concept of equivalent stiffness, solutions of stresses in the ground induced by train are obtained based on the model of an Euler-Bernoulli beam on the elastic half-space subjected to the moving load. The characteristics of stresses in the ground including the spatial stress distribution and the stress path are studied. It is found that stresses induced by the train increase with the increasing of the moving speed of the train, and their values will be considerable if the moving speed of the train approaches the Rayleigh-wave speed of the soil. The stress path of the soil is very complex even for those elements beneath the central motion line of the load. Dynamic triaxial tests are carried out to study the dynamic properties of soft clay under train-induced stresses, in which those stresses are simplified as cyclic sinusoidal loadings. Dynamic strength of natural clay and remolded clay as function of the number of cycles is presented. Dynamic strength of the former decreases rapidly with the increasing of confining pressure and then goes to an asymptotic value, but the influence of confining pressure on the dynamic strength of the latter is limited. Based on the theoretical results and the limited minimum cyclic strength obtained by tests, the settlement of ground surface induced by train is estimated. Large settlement might happen due to destruction of the soil structure, accumulation of the deformation and enlargement of the disturbed area.

1 INTRODUCTION

In the last decays, the problem of the train-induced vibration became serious with the increasing of the train velocity. It is necessary to take into account the coupling of dynamic vibration between the track and the supporting soil to study this problem. Many in-site tests have shown that the vibration of the track-ground system will increase rapidly, that is to say, the famous Mach-effect arises, if the train velocity is higher than the Rayleigh-wave (*R*-wave) velocity (V_R) of the supporting soil. Hence, the train velocity should be limited below the *R*-wave velocity of the soil. In parts of TGV line, where the ground soil is soft, the train velocity is limited in the lowest critical velocity to avoid the coupling of dynamic vibration between the track and the soil (Dieterman & Metrikine, 1996).

The finite element method is versatile and powerful to analyze the vibration of ground induced by the moving train. But the analyses are usually very time-consuming for this three-dimensional problem. Even though the 2.5 D finite/infinite element method is an alternative choice to overcome this problem in some degree (Yang et al. 2001 and Yang et al. 2003), many degrees of freedom are needed to obtain satisfied solutions. It should be careful to use the above methods to simulate the wave propagation in the soil induced by the moving train. The Winkler foundation model used to model the interaction of the track and the ground greatly simplifies the problem (Kenney, 1954; Achenbach & Sun, 1965; Kerr, 1972; Xie W. P. et al., 2002; Chen Y. M. et al., 2003, 2005). Unfortunately, this simple model cannot represent the wave propagation between the foundation and the superstructure. Moreover, it is difficult to determine the spring stiffness in practice. The problem of a moving load over a beam resting on the elastic half-space has first been studied by Filippov (1961) as a model for the train-subsoil interaction. It was shown that the critical velocity of the train is almost same as the *R*-wave velocity, which is much lower than that of the Winkler model. After that, much work based on this research has been carried out (Labra, 1975; Dieterman & Metrikine, 1996, 1997).

Some in-site measurements have shown that considerable settlement would happen due to the long-term train running, even if the train speed is low. The main reason is that softening of saturated soft clay happens due to train-induced dynamic stresses. The softening of saturated soft clay subjected to cyclic loading may be caused by: (1) excess pore pressure increasing; (2) clay structure remolding due to the change of the stress path; (3) disturbance of the original structure of clay.

In order to study the train-induced vibration and the long-term settlement of ground, it is necessary to further study stresses in the ground including the stress distribution, the stress path and so on. In this paper, the train-embankment-ground system is simplified as an Euler-Bernoulli beam (E-B beam) resting on the elastic half-space under moving loads. Solutions of contact force on the ground surface and stresses in the ground are obtained based on the concept of equivalent stiffness of the half-space. Spatial stress distribution and stress paths due to the moving train are computed. To analyze the dynamic characteristics of the soil under the moving train, dynamic strength of soft clay is studied by dynamic triaxial tests. Long-term settlement of ground is estimated based on the theoretical results and tested results.

2 MODEL AND SOLUTIONS

As shown in Figure 1, let a concentrated load P_0 be applied on an E-B beam resting on the elastic half-space and be moving along the mid-line of the beam with a constant velocity V_0. A right-handed system of rectangular Cartesian coordinate axes x, y and z is introduced herein, with the positive x-axis coincident with the line of motion and oriented in the direction of motion, and with the positive z-axis pointing into the half-space. It is assumed that the contact between the beam and the half-space is smooth. Hence the shear stress is zero at their interface, and the normal stress between the beam and the half-space is uniformly distributed over the width of the beam. It is further assumed that the load has been moving for a

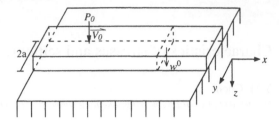

Figure 1. Analysis model of E-B beam and half-space system.

considerable long time and all impulse disturbances are neglected.

2.1 *Contact force on the beam-ground interface*

Contact force between the beam and the ground surface is

$$f(x,t) = -\frac{P_0}{4\pi a} \int_{-\infty}^{\infty} \frac{\left(2\pi\mu\chi(\xi V_0,\xi) - ic V_0\xi\right)e^{i\xi x_v}}{D(\xi V_0,\xi) + ic\xi V_0 - 2\pi\mu\chi(\xi V_0,\xi)} d\xi$$

(1)

where $D(\xi V_0,\xi) = -E_b I\xi^4 + \rho_b A\xi^2 V_0^2$, describes the free vibration of beam; E_b is the Young's modulus of beam; $I = ab^3/6$, is the moment of inertia, with a and b are the half width and the height of the beam, respectively; ρ_b and $A = 2ab$ are the mass density and the cross-sectional area of the beam; c is the viscosity coefficient; μ is the shear modulus of the half-space; $x_v = (x - V_0 t)$ is the moving coordinate; $2\pi\mu\chi(\xi V_0,\xi)$ is the equivalent stiffness which can be referred to Dieterman & Metrikine, 1996.

The real part of $\chi(\xi V_0,\xi)$ is symmetrical and the imaginary part is asymmetrical with respect to ξ, i.e.

$$Re(\chi(\xi V_0,\xi)) = Re(\chi(-\xi V_0,-\xi))$$
$$Im(\chi(\xi V_0,\xi)) = -Im(\chi(-\xi V_0,-\xi))$$

Using the symmetry of the equivalent stiffness, letting $\xi' = \xi'b$, which means twice of the characteristic length given by Suiker (1998), and substituting ξ' into Eq. (1), one gets

$$f = f^{sym} + f^{asym}$$

(2)

$$f^{sym} = -\frac{P}{2\pi a b}\int_0^{\infty}\frac{\left(\beta^2\left(Re(\chi)(D^{*}(\xi')-\beta^2 Re(\chi))+Im(\chi)(\alpha_s\lambda\xi'-\beta^2 Im(\chi))\right)-\alpha_s\lambda\xi'(\alpha_s\lambda\xi'-\beta^2 Im(\chi))\right)\cos(\xi' x_v/b)}{(D^{*}(\xi')-\beta^2 Re(\chi))^2+(\alpha_s\lambda\xi'-\beta^2 Im(\chi))^2}d\xi'$$

(3)

$$f^{asym} = -\frac{P}{2\pi a b}\int_0^{\infty}\frac{\left(\beta^2\left(Re(\chi)\alpha_s\lambda\xi'-Im(\chi)D^{*}(\xi')\right)+\alpha_s\lambda\xi'(D^{*}(\xi')-\beta^2 Re(\chi))\right)\sin(\xi' x_v/b)}{(D^{*}(\xi')-\beta^2 Re(\chi))^2+(\alpha_s\lambda\xi'-\beta^2 Im(\chi))^2}d\xi'$$

(4)

where $D^*(\xi') = (\alpha_1^2\,\xi'^2 - \xi'^4/12)/l$, $\alpha_1^2 = \rho_b V_0^2/E_b$, $\alpha_s^2 = V_0^2/V_s^2$, $V_s = \sqrt{\mu/\rho}$ is the shear wave velocity in the half-space, $l = b/2a$ is the aspect ratio of the beam, $\beta^2 = 2\pi\mu/E_b$ and $\lambda = cV_s/E_bb$.

The contact force induced by the train can be given by superposing contact forces induced by concentrated loads as

$$f_t = \sum_{i=1}^{n} f_i(x - x_i, t) \qquad (5)$$

where $f_i(x - x_i,t)$ is the contact force induced by the i th wheel load.

2.2 Stresses in the ground

As shown in Figure 2, the constant distributed force on the ground surface moves along the x-axis at the train speed (Wang et al. 2003). In view of this, the stress components, at any point in the elastic half-space, can be obtained by integrating the basic solutions of stresses induced by a moving concentrated load as

$$\sigma(x_0,y_0,z_0,t) = -\int\sigma^*(x_0-x,y_0-y,z_0,t)dP(x,y,t)$$
$$= -\int_{-L/2}^{L/2}\int_{-a}^{a}\sigma^*(x_0-x,y_0-y,z_0,t)q(x,y,t)dxdy \qquad (6)$$

where $\sigma^*(x_0 - x, y_0 - y, z_0, t)$ is the stress produced by a unit moving force on the surface of the half-space, and its expression was given by Eason (1965) and Lasing (1966); L is the effective length of force. Here compression stress is positive and tension stress is negative. What should be noticed is that Eq. (6) is valid only when the moving speed of the load is lower than the Rayleigh wave speed in the half-space, which represents the most normal and important case in practice.

3 NUMERICAL CALCULATIONS

In order to study the characteristics of stresses in the ground induced by the train, numerical analyses were carried out in this study. Figure 3 is the train-embankment-ground system. When $t = 0$, the train center is on the origin point. The train load is simplified as a series of point loads moving on a beam at the train speed. Some physical properties of the train, the embankment and the ground are: $\rho_b = 1900\,\text{kg/m}^3$, $E_b = 30000\,\text{MPa}$, $a = 2\,\text{m}$, $b = 0.3\,\text{m}$, $P_1 = P_2 = \ldots = P_{24} = 160\,\text{kN}$, $\mu = 10\,\text{Mpa}$, $\nu = 0.45$, $\rho = 1800\,\text{kg/m}^3$ and $V_S = 74.54\,\text{m/s}$.

3.1 Spatial distribution of stresses

Figure 3 also includes the corresponding contact force whose peaks are for wheels of the train. The contact force moves together with the train at the train speed.

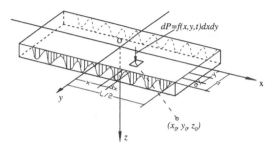

Figure 2. Stresses in ground induced by force between ground surface and embankment.

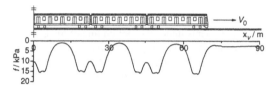

Figure 3. Train and its induced contact force between embankment and ground surface.

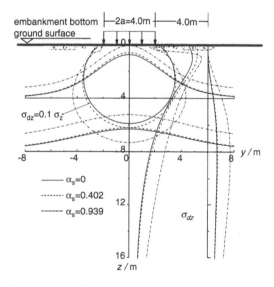

Figure 4. Distribution of σ_{dz} in the plane of $x_v = 0$ for different train speeds.

The vertical normal stress distribution under such force in the plane of $x_v = 0$ is calculated for train velocities of 0 m/s ($\alpha_s = 0$, means train is static on the embankment), 30 m/s ($\alpha_s = 0.402$) and 70 m/s ($\alpha_s = 0.939$) as shown in Figure 4. These stresses are also compared with the value of 0.1 time of the normal self-weight stress in the figure. It is found that stresses for $\alpha_s = 0.402$ are close to those induced by static loads. The train speed affects heavily on the

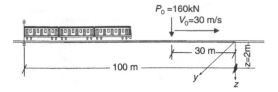

Figure 5. Sketch map of the soil element.

stresses in the ground, and the influenced depth is large especially for the high speed train.

3.2 Stress path

For soil elements under the motion line of load, shear stress τ_{xy} and τ_{xz} are always zero, that is to say σ_{dy} is always one of the principal stresses. Only stress path in the x-z plane is needed to analyze and it varies relatively simple. In this section the stress paths of the soil element with 2 m depth subjected to a moving concentrated load and a moving train as shown in Figure 5, are analyzed respectively. When $t = 0$, the soil element is 30 m away from the concentrated load and 100 m away from the train center (see Figure 5), respectively.

Figure 6 shows the change of the stress state with time and distance from the load. Dynamic stresses induced by the moving load and total principal stresses are shown in Figures 6a and 6b, respectively. And Figure 6c presents the change of the principal stress angle. The principal stress angle is defined as the angle from the x-axis to the principal stress axis. The clockwise direction is defined as positive but the anticlockwise direction is negative, and this angle is confined between $-\pi/2$ and $\pi/2$. The upper coordinate in Figure 6 is the moving coordinate, which means the distance between the soil element and the load. Because the initial stress state of the soil element is taken as hydro-static with three isotropic initial principal stresses equal to the self-weight of soil, the rotation of principal stress axes is only dependent on the dynamic stresses induced by the moving load. Here σ_1 (the maximum principal stress) is taken for example to show how the principal stress axes rotate. At the beginning σ_1 is coincident with x-axis, then it rotates clockwise with the positive angle. When the load is above the point exactly, τ_{zx} becomes zero and σ_1 is in coincidence with the z-axis. Keeping moving the load, σ_1 deviates from the z-axis clockwise with the negative angle until it is parallel to the x-axis and comes back to the initial stress state. So σ_1 rotates through 180 degree from the beginning to the end. It is just symmetric for the load approaching and departing from the analyzed soil element. In Figure 6c, the sudden change of the stress angle from $\pi/2$ to $-\pi/2$ means that the principal stress rotates across the z-axis clockwise.

Figure 7 shows the relationship between the partial dynamic stress and the shear stress. The present results shown in Figure 7b are compared with those of Zhou et al. 2001 shown in Figure 7a. Stresses of σ_{dz}, σ_{dy} and τ_{zx} correspond to σ_v, σ_h and τ_{vh}, respectively. Both curves in Figure 7 seem like an apple, which verifies the present method computing induced stresses by the moving wheel load. Several particular stress states in Figure 7b are taken for example to show how their stress state changes continually during the load moving. For point A, where x_v is infinite, the load is far away from the soil element, at which the soil element is at its initial stress state and all dynamic stresses are zero. With the load moving to the soil element, the horizontal shear stress increases faster than the partial dynamic stress does and becomes dominant. For point B, where x_v is about 3.75 m, the partial dynamic stress is zero and the soil element is at a simple shear state. Keeping moving the load to the soil element, the horizontal shear stress and the partial dynamic stress increase continually and the former reaches its maximum value at point C, where x_v is about 2.1 m. Then the horizontal shear stress decreases but the partial dynamic stress keeps on increasing and becomes dominant sooner or later. For point D, where x_v is zero and the load is above the soil element exactly, the shear stress is zero and the partial dynamic stress reaches its maximum value, at which the soil element is in a triaxial compressive state. During the departing of the load, change of the stress state of soil element is just reverse to the above one. In the whole process of load moving, the soil element turns from the simple shear state to the triaxial compressive state and finally back to the original simple shear state.

Based on the above results, the stress path induced by train for the same soil element can also be analyzed. Figure 8 gives the change of the principal stress angle against time and the coordinate. It is shown that the main part of the curve is periodical. At the beginning σ_1 is parallel to x-axis. When train moves near to the soil element σ_1 rotates clockwise with the positive angle until it is coincident with the z-axis. After switching across the z-axis twice, it is parallel to z-axis again and then the soil element is within the range of train length. σ_1 goes on rotating clockwise until it is parallel to the x-axis and then the main cycles begin. In each cycle σ_1 rotates clockwise from the x-axis to the z-axis, switches across the z-axis twice and goes on rotating clockwise back to the x-axis. The process is similar to the former one. Five main cycles later, the principal stresses rotating process is just reverse to that of the beginning.

The relationship between the partial stress and the shear stress induced by train is shown in Figure 9. It is much more complicated than that of the moving concentrated load shown in Figure 7 due to the superposition of loads. There are also five main periods in the

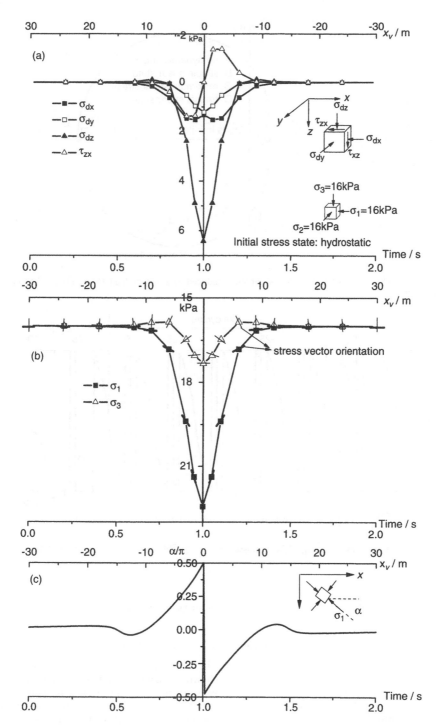

Figure 6. The change of stress state of a soil element with 2 m depth beneath the load motion line induced by a moving load (a) dynamic stresses, (b) total principal stresses, (c) principal stress angle.

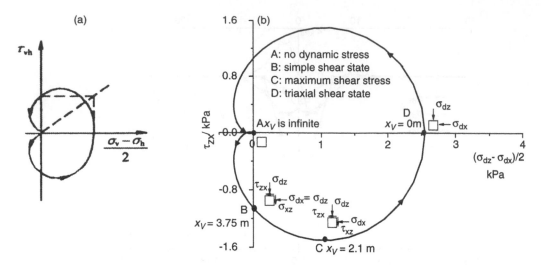

Figure 7. Relationship between partial stress and shear stress induced by a moving load, (a) results of Zhou et al. (2001), (b) the present results.

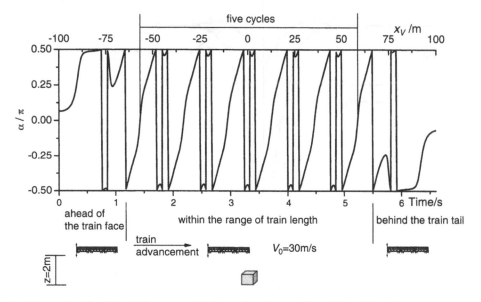

Figure 8. Change of angle of principal stress σ_1 with time and moving coordinate induced by train.

plot. The physical meaning of each point on the continuous curve can refer to Figure 7.

4 DYNAMIC TRIAXIAL TESTS ON PROPERTIES OF SOFT SOIL UNDER CYCLIC LOADINGS

Soft clay, such as silt and silty clay, is of generally characteristics of high water content, high void ratio and low strength. Further, its soil.

4.1 Design of tests

The basic physical parameters of soil samples are shown in Table 1. Compressive consolidation tests were carried out on Xiaoshan natural clay and remolded clay using KTG-98 automatic air pressure consolidation apparatus, and compressive curves are shown in Figure 10 from which it is found that Xiaoshan natural clay is well structured. The pre-consolidation pressure σ_{c0} and yielding stress σ_{cy} are 80 kPa and 160 kPa, respectively.

Table 1. Basic physical parameters of Xiaoshan clay.

γ kN/m^3	e_0	w %	G_s	w_p %	w_l %	I_p %	I_l %	c_{cu} kPa	ϕ_{cu}
16.14	1.738	62.3	2.734	26.5	53	26.5	1.069	15.5	15°

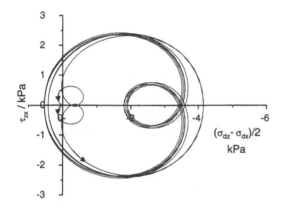

Figure 9. Relationship of partial stress and shear stress induced by train.

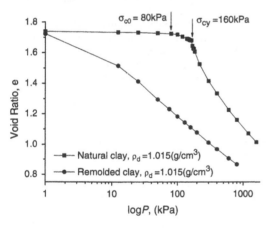

Figure 10. Compression curves of natural, remolded Xiaoshan clay.

Dynamic stresses in ground induced by the traffic load such as the train are different from those induced by the earthquake or the wave. The load in the former case is not sinusoidal but a kind of impulsive one. Yang & Zhou found that the excess pore water pressure grew slower under the impulsive load than that induced by the sinusoidal load through dynamic tests on railway embankment soil (Gong et al. 2001). Not only the magnitude of train-induced stresses varies but principal axes rotate. Yamada et al. (1983),

Ishihara et al. (1980) and Towhata et al. (1985) studied dynamic properties of sand under rotating shear stresses. It was shown that shear stress rotation induced extra plastic volume strain and caused excess pore pressure increasing faster in undrained tests. For the simplicity and easy operation, train-induced stresses were simulated by regular sinusoidal load and the frequency was taken as 1 Hz in the present tests. In order to consider the influence of the soil structural characteristics, confining pressure was chosen as $\sigma'_c < \sigma_{c0}$, $\sigma'_c = \sigma_{c0}$, $\sigma_{c0} < \sigma'_c < \sigma_{cy}$, $\sigma'_c = \sigma_{cy}$ and $\sigma'_c > \sigma_{cy}$ respectively. Dynamic strength and the limited minimum cyclic strength of natural and remolded Xiaoshan clay were studied. Based on the tested results, how post-settlement of soft clay ground under the traffic load develops is analyzed and the concept of critical depth is presented as below.

For comparison natural and remolded Xiaoshan clay specimens were used. Remolded specimens were control by the dry density. According to the water content and the density of natural specimens, their dry weight is 97.5 g and the dry density is 1.015 g/cm^3. The diameter and height of specimens are 3.91 cm and 8.0 cm, respectively. Having prepared specimens, saturated it by vacuum pumping, consolidated it on the triaxial test apparatus for 24 hours under different confining pressure, and then started undrained cyclic tests with the 1 Hz cyclic axial loading. Computer was used to record the axial dynamic loading, the axial deformation and the pore water pressure changing with number of cycles. The multi-functional triaxial test apparatus HX-100 is used to carry out such dynamic tests with the stress controlled loading. The 5% peak-peak strain is chosen as the specimen failure criterion.

4.2 Tested results and analysis

Figure 11 is the strength curves of natural Xiaoshan clay after 24-hour isotropic consolidation for different confining pressure. For each confining pressure, the dynamic strength decreases to a common asymptotic value, which was defined as the limited minimum cyclic strength $\sigma_d/2\sigma'_c)_{min}$ by Procter & khaffaf and its twice value was called limited dynamic stress ratio. This limited minimum cyclic strength means that no matter how many cycles are performed on the soil it will not fail if the dynamic stress ratio $\sigma_d/2\sigma'_c$ is

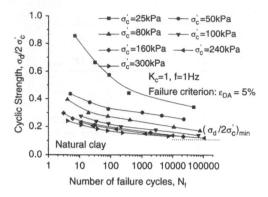

Figure 11. Dynamic strength curves of natural clay under different confining pressure.

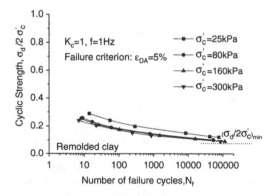

Figure 12. Dynamic strength curves of remolded clay under different confining pressure.

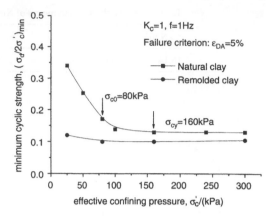

Figure 13. Curves of limited minimum cyclic strength of natural and remolded clay under different confining pressure.

smaller than this value. The dynamic strength of natural clay under confining pressure of 25 kPa is much higher than that of other confining pressure. With the increasing of confining pressure, the dynamic strength curves drops and approaches an asymptotic curve at last. Because soil is over-consolidated under confining pressure of 25 kPa, the soil structural characteristics is dominant and the external dynamic stress is small, the corresponding dynamic strength is relatively large. But with the increasing of confining pressure, the soil structure is weaker and weaker and dynamic stress is larger and larger, so that the dynamic strength of the soil goes down continuously.

Figure 12 is the dynamic strength curves of remolded clay under four confining pressure. It is shown that all the curves approach the limited minimum cyclic strength. Under the same confining pressure the limited minimum cyclic strength of remolded clay is smaller than that of natural clay. All dynamic strength curves of remolded clay are very close

except for that under confining pressure of 25 kPa. Because structure of remolded clay has been destroyed so that agglutination between soil particles is very weak and it takes effect only under small confining pressure. The large confining pressure breaks this kind of agglutination and the structural strength of the soil can be neglected.

Figure 13 is the limited minimum cyclic strength of natural and remolded specimens vs. confining pressure. The limited minimum cyclic strength of natural specimens is higher than that of the remolded. But with the increasing of the confining pressure, their limited minimum strength go to an asymptotic value respectively for both natural and remolded specimens.

5 SETTLEMENT OF GROUND SURFACE INDUCED BY TRAIN

As mentioned in the previous section, the limited minimum cyclic strength of natural clay decreasing with the increasing of confining pressure will approach to an asymptotic value at last. The limited minimum cyclic strength of natural clay is always higher than that of remolded clay. Due to relatively high confining pressure for natural clay, its structure will be destroyed but this phenomenon only occurs in local area where soil is very similar with the remolded. In other area soil is little disturbed and structural characteristics are still obvious. For the remolded clay, however, the specimen-preparing process has broken the connection of particles completely. As a result, its limited minimum cyclic strength is smaller than that of natural clay. If the cyclic loading acts on clay continuously, such as in the long-term train running, dynamic stresses on softening area will shift to the undisturbed area, softening

40

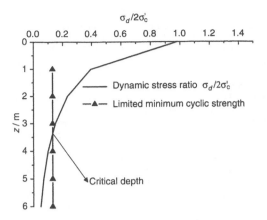

Figure 14. Sketch of determination of critical depth induce by traffic loading.

will happen in this area and then dynamic stresses will shift to other undestroyed area. Again and again, softening area is developing continuously until large area of natural clay is disturbed. Once the soil structure is destroyed considerable deformation will happen. So soil deformation grows with the enlarging of failure area in the progressive process of soil structural damage. That is the reason why the post-settlement of ground is considerable large and develops continuously for a long time.

In practical engineering ground soil is often disturbed such as over preloading or is even remolded such as dredger fill. For the safety, a certain range of soil under the embankment is always considered as disturbed or remolded soil, and its limited minimum cyclic strength is taken as that of destroyed natural clay or remolded clay. Figure 14 gives curves of theoretical dynamic stress ratio and the limited minimum cyclic strength (0.13 for natural clay after destruction and 0.1 for remolded clay) from tests vs. depth. In Figure 14 σ'_c is effective soil self-weight plus additional stress induced by embankment weight. It is shown that at a small depth dynamic stress ratio is bigger than the limited minimum cyclic strength. For the soil with the depth over a certain value termed the critical depth as shown in Figure 14, the dynamic stress ratio is larger than the limited minimum cyclic strength. Below the critical depth the dynamic effect of the loading can be neglected and cannot result in the soil failure. But over this depth it is necessary to take into account that the accumulation of soil deformation under long-term action of the loading. In Figure 14 the critical depth corresponding to the limit minimum strength of 0.13 and 0.1 is 3.35 m and 3.9 m, respectively. According to the present tests, the soil strain will be about 10% if the soil structure is

destructed. So within the critical depth deformation of ground may approach 33.5 cm or 39 cm. The post-settlement depends on soil properties, the soil disturbance range, the train load and the train velocity. Generally, the strain of the soil will be smaller than 1% if the soil structure is undisturbed, which means the settlement of ground will be very small. Therefore, the structure of soil plays an important role for the settlement of ground induced by the train. In a word, once the soil structure is destroyed, large deformation of soil will happen and structure-destroyed area will be enlarged under the cyclic loading, which results in the ground settlement larger and larger. In engineering applications, soil above the critical depth should be treated to avoid the unacceptable settlement.

6 CONCLUSIONS AND DISCUSSION

Based on the model of an Euler-Bernoulli beam resting on the elastic half-space subjected to the moving load, characteristics of stresses in the ground including the spatial stress distribution and the stress path are presented. Dynamic triaxial tests are carried out to study the dynamic properties of soft clay under train-induced stresses, and the dynamic strength of natural clay and remolded clay as function of the number of cycles is given. Based on the theoretical results and the obtained limited minimum dynamic strength of the clay, the settlement of ground surface induced by train is estimated. The following conclusions can be drawn in this study

(a) Stresses induced by a low speed train are close to those induced by static loadings. These stresses increase with the increasing of the moving speed of the train, and their values will be considerable if the moving speed of the train approaches the Rayleigh-wave speed of the soil. The stress path of the soil is very complex even for those soil elements beneath the central motion line of the train loads.

(b) For the structured Xiaoshan clay, dynamic strength of the natural soil decreases rapidly with the increasing of confining pressure and then goes to an asymptotic value, but the influence of confining pressure to the dynamic strength of the remolded soil is limited. There is a common limited minimum dynamic strength for both natural soil and remolded soil.

(c) Continuous destruction of soil structure and accumulation of the plastic deformation subjected to cyclic loadings within range of the critical depth will result in the considerable long-term post-settlement of ground.

In view of that large amount of computer storage and data preparation time are required in some numerical methods, the present analytical method is an alternative choice to study dynamic characteristics of the ground induced by the moving train though the analyzed model is simple. Stress states and the rotation of principal stress axes of the soil during the train moving are complex. Just the soil element on the plane of $y = 0$ is analyzed in this study for the simplicity and to save length of the paper, but the other soil elements can also be studied by the proposed model.

Some basic dynamic characteristics of the soil under sinusoidal loadings have been investigated. The following work will be carried out to study deformation and strength of the soil subjected the partial dynamic stresses on the two-dimensional dynamic triaxial test apparatus and the hollow-cylindrical torsional test apparatus, which can simulate the real state of the soil more closely.

ACKNOWLEDGEMENT

This research project is supported by the Research Fund for PhD Student of Chinese College through Grant No. 20040335083.

REFERENCES

Achenbach, J.D. & Sun, C.T. 1965. Moving load on a flexibly supported Timoshenko beam. *Int. J. Solids Structures*, Vol.1, 355–370.

Chen, Y.M., Wang, C.J., Ji, M.X. & Chen, R.P. 2003. Train-induced ground vibration and deformation. In: Yunmin Chen, Hirokazu Takemiya (ed.), *Environmental Vibration prediction, monitoring and evaluation*. Hangzhou, 2003. China Communications Press. 158–174.

Dieterman, H.A. & Metrikine, A.V. 1996. The equivalent stiffness of a half-space interacting with a beam. Critical velocities of a moving load along the beam. *Eur. J. Mech. A/Solids*. 15(1), 67–90.

Dieterman, H.A. & Metrikine, A.V. 1997. Steady-state displacements of a beam on an elastic half-space due to a uniformly moving constant load. *Eur. J. Mech. A/Solids*. 16(2), 295–306.

Eason, G. 1965. The stresses produced in a semi-infinite solid by a moving surface force. *Int. J. Engng. Sci.*, 2: 581–609.

Filippov, A.P. 1961. Steady-state vibrations of an infinite beam on elastic half-space subjected to a moving load. *Izvestija AN SSSR OTN Mehanika I Mashinostroenie*, 6, 97–105 (translated from Russian).

Gong, Q.M., Liao, C.F., Zhou, S.H. & Wang, B.L. 2001. Testing study of dynamic pore water pressure under train loading. *Chinese Journal of Rock Mechanics and Engineering (in Chinese)*. 20(A01), 1154–1157.

Ishihara, K. & Yamazaki, F. 1980. Cyclic simple shear tests on saturated sand in multi-directional loading. *Soils and Foundations*. 20(1), 45–59.

Kerr, A.D. 1972. Steady-state vibrations of beam on elastic foundation for moving load. *Int. J. Mech. Sci,*, 14, 71–78.

Kenney, J.T. 1954. Steady-state vibrations of beam on elastic foundation for moving load. *Journal of Appl. Mech.*, 76, 359–364.

Labra, J.J. 1975. An axially stressed railroad track on an elastic continuum subjected to a moving load. *Acta Mechanica*, 22, 113–129.

Lansing, D.L. 1966. The displacements in an elastic half-space due to a moving concentrated normal load. *NASA technical report*, TR R-238.

Procter, D.C. & Khaffaf, J.H. 1984. Cyclic Triaxial Tests on Remoulded Clays. *Journal of Geotechnical Engineering*, ASCE, Vol.110, No.10, 1431–1445.

Towhata, I. & Ishihara, K. 1985. Undrained strength of sand undergoing cyclic rotation os principal stress axes. *Soils and Foundations*. 25(2). 135–147.

Towhata, I. & Ishihara, K. 1985. Shear work and pore water pressure in undrained shear. *Soils and Foundations*. 25(3). 73–84.

Wang, C.J., Chen, Y.M. & Wen, S.Y. 2003. Train-induced cyclic loads in an elastic half-space. In: Li, X.B., Lok Tat Seng, Liu D.H.(ed.) *Proceedings of the 5th Asia-Pacific conference on shock and impact loads on structures*. Changsha, 2003, Singapore: CI-Premier Pte Ltd. 361–368.

Wang, C.J. & Chen, Y.M. 2005. Analysis of stresses in ground induced by train. *Chinese Journal of Rock Mechanics and Engineering (in Chinese)*. 24(7), 1178–1186.

Xie, W.P., Hu, J.W. & Xu, J. 2002. Dynamic response of track-ground systems under high velocity moving load. *Journal of Rock Mechanics and Engineering (in Chinese)*, 21(7), 1075–1078.

Yamada, Y. & Ishihara, K. 1983. Undrained deformation characteristics of sand in multi-directional shear. *Soils and Foundations*. 23(1), 61–79.

Yang, Y.B. & Hung, H.H. 2001. A 2.5 D finite/infinite element approach for modeling visco-elastic bodies subjected to moving load. *International Journal for Numerical Methods in Engineering*. 51, 1317–1336.

Yang, Y.B., Hung, H.H. & Chang, D.W. 2003. Train-induced wave propagation in layered soils using finite/infinite element simulation. *Soil Dynamics and Earthquake Engineering*. 23, 263–278.

Zhou, J., Bai, B. & Xu, J.P.(ed.) *Soil Dynamic Mechanics Theory and Calculation*. 2001. Beijing: China Architecture Industry Press.

Environmental Vibrations – Takemiya (ed.)
© *2005 Taylor & Francis Group, London, ISBN 0 415 39035 4*

Experimental study on dynamic strain of structural soft clay under cyclic loading

Y.P. Chen, B. Huang & Y.M. Chen
Researching Institute of Geotechnical Engineering, Zhejiang University, Zhejiang, China

ABSTRACT: This paper presents the results of an experimental investigation performed to study the failure criterion of strain for natural and remolded clay from Xiaoshan under cyclic loading. The strain curves with cycle number of natural and remolded clay have all turning points under different amplitude of stress. The strain of turning points would be different if the stress amplitude change but they could be fitted well by a linear formula.

The appearance of turning points means the failure of clay's structure. It'd better determine the failure criterion according to the failure of soil's structure.

1 INTRODUCTION

Soft clay, including mud, muddy clay, muddy silt clay, etc, generally contains characteristic of high water content, large void ratio, low strength, thixotrophy and structure. Thick soft clay layers are widely spread on the coastal areas and some inland cities in China. In recent years, a large number of high buildings, high ways, high-speed railways, airport tracks and docks has been built in these areas. These facilities and buildings are constantly affected by cyclic loading and thus their deformation and stability has become an issue of concern. For this, it is very important to study the dynamic strain and failure criterion of saturated soft clay under the effect of cyclic loading.

Many scholars home and abroad have done research about the deformation characteristics of saturated soft clay under the effect of cyclic loading. However, their research mainly focuses on the stress-strain or pore pressure-strain relationships and little research has been done on the failure criterion of strain.

In this article, a natural clay from Xiaoshan is sampled for dynamic test under different confining pressure. The determination of strain failure criterion of soft clay is discussed.

2 LABORATORY TEST

2.1 *Soil description*

The natural soft clay used in the test is taken from Xiaoshan, using a blade to cut out tens of cults with dimension of 25 × 25 × 22 cm from the same depth and take them back to the lab for test. The purpose is to reduce the disturbance on the clay so that the sample clay has better homogeneity and is closer to the properties of the clay in situ.

The clay is a kind of typical soft clay and its basic properties are listed in Table 1.

2.2 *Odometer test*

The results of conventional oedometer test of natural and remolded clay are shown in Figure 1. The influence of clay structure on the compression characteristics of the clay is obvious. The compression index of the clay with natural structure is a variable. With the increase of vertical pressure, the structure of the natural clay is gradually damaged and its compression curves get closer to those of the remolded clay.

On the compression curves of natural clay there are two obvious stress points: pressure σ_{v0} and yield stress σ_{vy}. These two stress values are 80 kPa and 160 kPa respectively. σ_{v0} and σ_{vy} are two key stress values of the structural clay. The former is the maximum overburden pressure ever received during the sedimentation process of the clay and the latter is the cut-off point at which the clay structure is severely damaged.

The compression curve of remolded clay becomes a straight line The compression index is a constant value. And it does not have σ_{v0} and σ_{vy}.

Table 1. Statistical geotechnical properties of Xiaoshan's soft clay.

Bulk density γ (kN/m^3)	Water content w (%)	Specific gravity d_s	Plasticity Index I_p	Liquid index I_l
16.14	62.3	2.734	26.5	1.069

2.3 Triaxial test

In this test, in order to study the influence of confining pressure and clay structure on the properties of dynamic strain of the soft clay, triaxial cyclic test is conducted on Xiaoshan natural clay under seven confining pressures (25 kPa, 50 kPa, 80 kPa, 100 kPa, 160 kPa, 240 kPa and 300 kPa. See Figure 2), with the pre-pressure σ_{v0} and structural yielding stress σ_{vy} of the natural clay tested in oedometer as the cut-off points. Meanwhile, remolded clay is also used to conduct comparison tests under confining pressures of 25 kPa, 80 kPa, 160 kPa and 300 kPa respectively.

3 TEST RESULTS

3.1 Dynamic strain curves of natural clay

Figure 3 shows the relationship of strain vs cycle number under 80 kPa and 160 kPa confining pressures.

When the magnitude of dynamic stress is of a smaller value, the axial dynamic strain within the initial cycle remains a very small value. When the cycle reach a certain number, the deformation of the clay is dramatically increased and clay is damaged quickly within very few cycles, that showing brittle failure features. With the increasing of dynamic stress, the number of cycles needed for clay structure failure is constantly decreasing and the brittle failure features become less obvious. The curves of the strain-cycle number under other confining pressures also share similar regularities and thus they are not all illustrated in this article.

During the process of axial deformation, there is a turning point on each curve (the turning point is even more obvious when the failure cycle number is of a bigger value). We define this strain as ε_{tp}. The greater the dynamic stress is, the earlier the turning point appears. After the turning point appears, the axial deformation of the clay becomes faster and failure of the clay structure occurs within very few cycle numbers. Therefore,

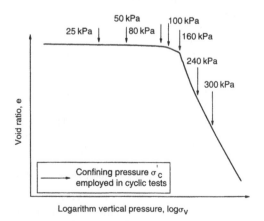

Figure 1. Compression curves of natural and remolded clay.

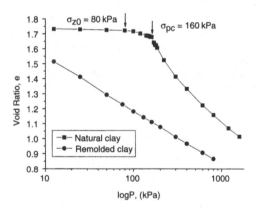

Figure 2. Schematic diagram of e–log(P) showing different effective confining pressure σ_c' employed in cyclic tests for natural soft clay.

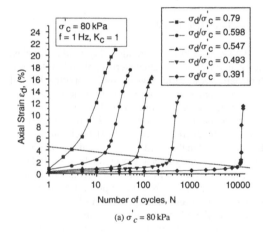

(a) $\sigma_c' = 80$ kPa

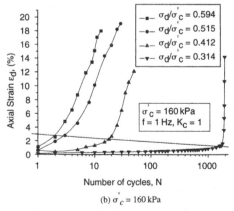

(b) $\sigma_c' = 160$ kPa

Figure 3. Curves of cyclic axial strain of Xiaoshan's natural clay with cyclic number under different dynamic stress. (a) $\sigma_c' = 80\,kPa$. (b) $\sigma_c' = 160\,kPa$.

the appearance of a turning point indicates that the clay structure would be severely damaged accompany with the loss of anti-shear strength of the clay. So it is appropriate to take ε_{tp} as the failure criterion of clay. From Figure 3, we can see that the strain corresponding to the severe damage of clay structure is not a constant but varies with the amplitude of cyclic stress. However, almost these turning points fall on the same straight line. If we define the cycle number corresponding to the strain turning point as failure cycle number N_f, then the straight line can be fitted with the following formula:

$$\varepsilon_{tp} = A * \lg N_f + B \qquad (1)$$

where A and B are fitted coefficients. The value of A and B under seven confining pressures are listed in Table 2.

The changes of fitted coefficients of the natural clay can be divided into 3 stages: When $\sigma'_c \leqslant 80$ kPa, A and B coefficients are of an approximate value and can be expressed by an average value as $\bar{A} = -0.859$, $\bar{B} = 4.463$. When $\sigma'_c \geqslant 160$ kPa, A and B coefficients are also of an approximate value and can likewise be expressed by an average value as $\bar{A} = -0.527$, $\bar{B} = 2.823$. While $80 < \sigma'_c < 160$ kPa, A and B can take the value by linear interpolation. Therefore, under different confining pressures, the $\varepsilon\varepsilon_{tp} - N_f$ relationship of the natural clay can be expressed as follows:

$$\begin{cases} \varepsilon_{tp} = -0.859 \lg N_f + 4.463, & \sigma'_c \leqslant \sigma_{v0} \\ \varepsilon_{tp} = R_1 \lg N_f + R_2, & \sigma_{v0} < \sigma'_c < \sigma_{vy} \\ \varepsilon_{tp} = -0.527 \lg N_f + 2.823, & \sigma'_c \geqslant \sigma_{vy} \end{cases} \qquad (2)$$

where

$$R_1 = -0.859 + 0.332 \frac{\sigma'_c - \sigma_{z0}}{\sigma_{vy} - \sigma_{v0}},$$

$$R_2 = 4.463 - 1.640 \frac{\sigma'_c - \sigma_{v0}}{\sigma_{vy} - \sigma_{v0}}$$

3.2 Dynamic strain curves of remolded clay

Figure 4 shows the dynamic strain-cycle number relationship of Xiaoshan remolded clay. Under different cyclic stress magnitude, the changes of the curves are similar to those of the natural clay. The relationship of

$\varepsilon_{tp} - N_f$ can also be fitted by formula(1). And Table 3 shows the value of A and B.

Table 3 shows that there are certain differences between A and B values when the confining pressure is of a very small value. Otherwise, A and B coefficients basically remain a constant value under the other three confining pressures which are of higher values. The constant values can be expressed by an average: $\bar{A} = -0.491, \bar{B} = 2.761$.

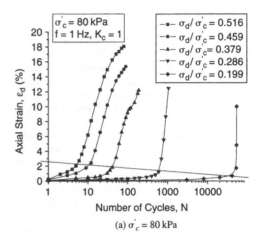

(a) $\sigma'_c = 80$ kPa

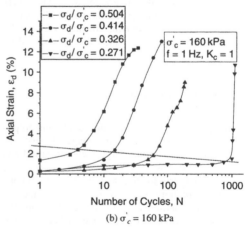

(b) $\sigma'_c = 160$ kPa

Figure 4. Curves of cyclic axial strain of Xiaoshan's remolded clay with cyclic number under different dynamic stress. (a) $\sigma'_c = 80$ kPa. (b) $\sigma'_c = 160$ kPa.

Table 2. A and B of natural clay under seven confining pressures.

Fitted coefficient	Confining pressure (kPa)						
	25	50	80	100	160	240	300
A	−0.865	−0.864	−0.847	−0.750	−0.523	−0.524	−0.533
B	4.418	4.447	4.525	3.891	2.807	2.826	2.837

45

Table 3. A and B of remolded clay under four confining pressures.

Fitted coefficient	Confining pressure (kPa)			
	25	80	160	300
A	−0.303	−0.480	−0.49	−0.504
B	1.885	2.755	2.764	2.764

3.3 Analysis on mechanics

From Figures 3 and 4, we can see that there is a turning point that indicates a sudden increase of the strain of natural and remolded clay under cyclic loading. This turning point is even more obvious when the failure cycle number is of a bigger value. This can be explained by the fact that even though there are obvious differences between natural clay and remolded clay in terms of structural characteristics and structural strength, they have both a kind of structure respectively. Before severely damaged in structure, the clay can resist the cyclic loading and the strain is of a smaller value. However, once severely damaged in structure, the clay cannot continue to bear the effect of cyclic loading and the clay deformation is dramatically increased. The greater the cyclic stress is, the quicker the damage is done to the clay structure. Therefore, the earlier the turning point of the strain appears, and the greater the corresponding strain value is.

3.4 Strain failure criterion for soft clay under cyclic loading

In actual project, the soil often endure different amplitude of cyclic stress before being damaged. In an earthquake, the clay goes through large cyclic stress and was damaged within few cycles, while the cyclic stress caused by long-term loading such as traffic vibrations or machine vibrations is usually of a smaller value and thus it should take more cycle to failure. Therefore, the strain failure criterion of clay also varies. For example, for an earthquake of 7 Richter scales, the equivalent cycle number is $N_f = 12$. According to Formula (2), the failure strain criterion of the clay under prepressure is $\varepsilon_{tp} = 3.54\%$. For long-term loading such as traffic vibrations, a greater cycle number is needed before the clay's structure failure. If we take $N_f = 10\,000$, then $\varepsilon_{tp} = 1.03\%$. So the strain failure criterion has obvious differences under different situation. In actual implementation, it is suggested that the failure criterion of soil be determined according to the equivalent cycle number of cyclic loading so that corresponding preemptive measures can be taken to prevent accidents caused by the failure of soil structure.

4 CONCLUSION

In this article, cyclic tests are conducted with Xiaoshan natural and remolded clay under different confining pressures. The test results indicate that:

(1) The strain curves of both natural clay and remolded clay under cyclic loading all exists a turning point which indicates the failure of soil structure and the sudden increase of the strain. Before the turning point appears, the strain development is at a low rate. After the turning point appears, the strain is dramatically increased. This turning point is even more obvious when the cyclic stress is of a little value.
(2) The turning points of strain under the same confining pressure but different cyclic stress magnitude will mostly fall on a same straight line, so the relationship between ε_{tp} and N_f could be fitted by a linear formula.
(3) The strain failure criterion is different under different cyclic stress. In actual project, it is suggested that the failure criterion of soil be determined according to the equivalent cycle number.

REFERENCES

Yasuhara, K. & Hirao, K. 1978. Strength and deformation of a saturated clay subjected to cyclic loading. *Proc. 5th Japan Earthquake Engineering Symp*: 729–736.
Yasuhara, K., Yamanouchi, T. & Hirao, K. 1982. Cyclic strength and deformation of normally consolidated clay. *Soils and Foudations*, 22(3): 77–91.
Robert, Y. L. & Fenggang Ma. 1992. Anisotropic plasticity model for undrained cyclic behavior of clays – I: Theory. *Journal of Geotechnical Engineering*, 118(2): 229–245.
Desai, C. S. & Wathugala, G. W. 1993. Constitutive model for cyclic behavior of clays – II: Applications. *Journal of Geotechnical Engineering*, 119(4): 730–748.
Yao Minglun & Nie shuanlin. 1994. A model for calculating deformation of saturated soft clay. *Journal of Hydraulic Engineering* No.7: 51–55.
Lo, K.Y. 1969. The pore pressure-strain relationship of normally consolidated undisturbed clays, Part I and Part II. *Canadian Geotechnical Journal* No. 6: 383–412.
Zhoujian. 2000. Elastoplastic study on dynamic strain of saturated soil. *Chinese Journal of Geotechnical Engineering* 22(4): 499–502.
Yasuhara, K., Hirao, K. & Hyde, A. FL. 1992. Effects of cyclic loading on undrained strength and compressibility of clay. *Soils and Foundations*, 32(1): 100–48.
Hyodo, M., Hyde, A. FL., Yamamoto Y. & Fujii T. 1999. Cyclic shear strength of undisturbed and remolded marine clays. *Soils and Foundations*, 39(2): 45–48.
Hyodo, M., Yasuhara, K. & Hirao, K. 1992. Prediction of clay behaviour in undrained and partially drained cyclic triaxial tests. *Soils and Foundations*, 32(4): 117–127.
Atilla, M. Ausai, & Ayfer E. 1989. Undrained behavior of clay under cyclic shear stress. *Journal of Geotechnical Engineering* 115(7): 968–983.

Environmental Vibrations – Takemiya (ed.)
© *2005 Taylor & Francis Group, London, ISBN 0 415 39035 4*

A developed dynamic elastoplastic constitutive model of soil based on the Davidenkov shear stress-strain skeleton curve

G.X. Chen & H.Y. Zhuang

Institute of Geotechnical Engineering, Nanjing University of Technology, Nanjing, P.R. China

ABSTRACT: Based on the Davidenkov one dimension dynamic shear strain-stress skeleton curve, a new dynamic shear stress-strain hysteresis curve of soils is constituted by using Mashing rules. The Davidenkov skeleton curve in the constitutive model of soil is corrected by using sectional functions, and upper limit shear strain amplitude is used as the sectional point. The modified skeleton curve can be approach to the upper limit shear stress when shear strain amplitude is approach to infinite. The equations used to calculate damping ratio of soil are deducted. How the fitting parameters A, B and γ_0 of the constitutive model of soil to influent the curves of dynamic shear modulus ratio and damping ratio versus shear strain amplitude($G/G_{max} \sim \gamma$ and $D \sim \gamma$) are studied in detail. Based on these studies, a new fitting method of the parameters A, B and γ_0 is advanced to fit the tested $G/G_{max} \sim \gamma$ and $D \sim \gamma$ curves. It is obvious that the new fitting method is better than the general methods to fit the tested $G/G_{max} \sim \gamma$ and $D \sim \gamma$ curves of soils in Nanjing city. The reference values of fitting parameters A, B and γ_0 are given to fit the $G/G_{max} \sim \gamma$ and $D \sim \gamma$ curves of soils tested by free vibration column apparatus.

1 INTRODUCTION

The dynamic strain-stress curves of soils under cyclic load have characteristics of nonlinear, hysteresis and accumulative. The study of one-dimension of dynamic shear strain-stress relation of soils in stress-strain hysteresis problem is the most essential content. The one-dimension dynamic strain-stress relation of soil under cyclic load with constant amplitude was put forward firstly by Mashing in 1926. The dynamic shear strain-stress hysteresis curves of soils are constituted by flowing rules: (1) the dynamic shear strain-stress skeleton curve of soils is hyperbola curve, (2) the shear strain-stress curve of soils under the first loading is according to skeleton curve, (3) the shear strain-stress curve of soils under reverse loading is two times of skeleton curve.

Constitutive relation of soils was formed in accordance with the Mashing rules. In 1964, Rosenblueth et al. modified the Mashing rules in order to restrict the reloading curves by 'external big circle' rule. In 1979, Pyke et al. modified the Mashing rules by 'n-times approach' to restrict the reloading curves.

The hyperbola skeleton curve used to fit the stress-strain relation of soils is discommodious and approximate. The hyperbola skeleton curve model has only one fitting parameters applied, it is sometimes difficult using it to fit the experimental curves of dynamic

shear modulus ratio versus shear strain ($G/G_{max} \sim \gamma$) of soils. On the other hand, it can not be used to fit experimental curves of damping ratio versus shear strain ($D \sim \gamma$) of soils. In order to fit the $G/G_{max} \sim \gamma$ curve better, in 1982, Martin et al. developed the Davidenkov skeleton curve of shear strain-stress relation. This model has three fitting parameters, A, B and γ_0, and can be used well to fit $G/G_{max} \sim \gamma$ curves of soils. But it has one disadvantage: when shear strain amplitude of soils is approach to infinite, the shear stress of soils is also approach to infinite. This point is conflicting to characteristics of soils.

Accordingly, The Davidenkov skeleton curve of soil shear strain-stress relation is corrected by using sectional functions, and a upper limit shear strain amplitude is used as the sectional point of the skeleton curve. The modified skeleton curve can be approach to a upper limit shear stress when shear strain amplitude of soils is approach to infinite. The equations used to calculate damping ratio of soil are derived. How the fitting parameters A, B and γ_0 to influent the curves of dynamic shear modulus ratio and damping ratio versus shear strain amplitude ($G/G_{max} \sim \gamma$, $D \sim \gamma$) of soils are studied in system. Based on these studies, a new fitting method is advanced to fit the tested $G/G_{max} \sim \gamma$ and $D \sim \gamma$ curves. It is obvious that the new fitting method is better than the general methods to fit the tested

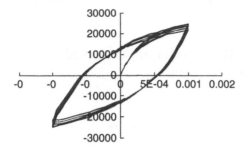

Figure 1. The dynamic shear stress-strain hysteresis curves given by Davidenkov model.

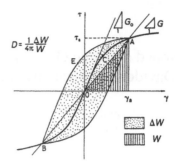

Figure 2. The damping ratio described by hysteresis curves.

$G/G_{max}\sim\gamma$ and $D\sim\gamma$ curves of representative soils in Nanjing city. The fitting parameter values A, B and γ_0 are given to fit the $G/G_{max}\sim\gamma$, $D\sim\gamma$ curves of soils tested by free vibration column apparatus.

2 IMPROVEMENT ON DAVIDENKOV SKELETON CURVE OF SHEAR STRAIN-STRESS RELATION

The relation of dynamic shear modular ratio versus the amplitude of shear strain of soils is expressed as follows by Hardin et al. (1972):

$$G/G_{max} = 1 - H(\gamma) \tag{1}$$

where

$$H(\gamma) = \frac{\gamma/\gamma_r}{1 + \gamma/\gamma_r} \tag{2}$$

Where γ_r is the referenced shear strain.
Martin et al. modified the expression (2) in 1982, the expression H(r) as follows:

$$H(\gamma) = \left\{ \frac{(\gamma/\gamma_0)^{2B}}{[1 + (\gamma/\gamma_0)]^{PB}} \right\}^A \tag{3}$$

where A, B and γ_0 are fitting parameters.
The relation of shear stress versus the amplitude of shear strain of soils can be expressed by following expression:

$$\tau(\gamma) = G \cdot \gamma = G_{max} \cdot \gamma \cdot [1 - H(\gamma)] \tag{4}$$

when $A = 1$, $B = 0.5$ and $\gamma_0 = \gamma_r$, the Davidenkov strain-stress skeleton curve is change to a hyperbola. According to Mashing rules, when $\gamma_0 = 0.1\%$, the hysteresis curves are shown in Fig. 1 for different fitting parameters A and B.
Damping ratio D is often used to describe the energy dissipation characteristics of soils under cyclic loading. According to the concept of damping ratio defined in the equivalent nonlinear model of soils, damping ratio D can be calculated as follows:

$$D = \Delta W / (4\pi W) \tag{5}$$

where, ΔW is the area of hysteresis loop, and the W is the area of the triangle as shown in Fig. 2. According to this definition, the damping ratio D of soils described by the hysteresis curves based on the Davidenkov shear stress-strain skeleton curve can be expressed as:

$$D = \frac{2}{\pi} \cdot \left\{ \frac{\gamma_c^2 - 2 \cdot \int_0^{\gamma_c} \gamma \cdot H(\gamma) d\gamma}{\gamma_c^2 \cdot [1 - H(\gamma_c)]} - 1 \right\} \tag{6}$$

In Eq. (6), the integral part can be solved by numerical method.
As far as soils are concerned, when $\gamma \to \infty$, $\tau(\gamma) \to \tau_{ult}$ where τ_{ult} is the upper limit shear stress of soils. However, in Eq.(4), when $\gamma \to \infty$, $\tau(\gamma) \to \infty$. This point disaccords to the shear stress-strain relation of soils. So, the Davidenkov stress-strain skeleton curve is modified as follows:
For different soils, each one has a upper limit shear strain γ_{ult}. When the shear strain γ exceeds to γ_{ult}, the soil will be damaged. After damaged, the soil will have very small increment of shear stress with shear strain increasing quickly. However, the damping ratio of soils will be increased largely. Sometimes, the shear stress of soils can be decreased obviously. These phenomena are named as the hardening and then softening characteristics of soils. It is difficulty to consider softening phenomena in the dynamic constitutive model of soils. So we don't consider the soil soft characteristics in the paper, and the modified Davidenkov shear stress-strain skeleton curve will be expressed in the form:

$$\tau(\gamma) = \begin{cases} G_{max} \cdot \gamma \cdot [1 - H(\gamma)] & \gamma_c \leq \gamma_{ult} \\ G_{max} \cdot \gamma_{ult} \cdot [1 - H(\gamma_{ult})] & \gamma_c > \gamma_{ult} \end{cases} \tag{7}$$

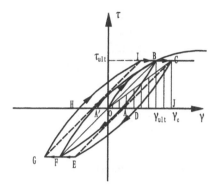

Figure 3. Dynamic shear stress-strain hysteresis curves given by corrected Davidenkov model.

$$\tau_{ult} = G_{max} \cdot \gamma_{ult} \cdot [1 - H(\gamma_{ult})] \tag{8}$$

where τ_c is the shear stress in the turning point of loading and unloading and the γ_c is the corresponding shear strain. The stress-strain hysteresis curves described by Eq. (7) are shown in Fig. 3.

When $\gamma_c \leq \gamma_{ult}$, the damping ratio of soils can be still calculated by formula (6). when $\gamma_c > \gamma_{ult}$, ΔW can be expressed as:

$$\Delta W = 8 \cdot \left(\int_{\gamma_c}^{\gamma_{ult}} \tau(\gamma) d\gamma - \frac{1}{2} \cdot \tau(\gamma_{ult}) \cdot \gamma_{ult} \right) + 8 \cdot \frac{1}{2} \cdot \tau(\gamma_{ult}) \cdot (\gamma_c - \gamma_{ult}) \tag{9}$$

W can be calculated by equation as:

$$W = \frac{1}{2} \cdot \tau(\gamma_{ult}) \cdot \gamma \tag{10}$$

Therefore, the equation to calculate damping ratio D of soils is given as:

$$D = \begin{cases} \frac{2}{\pi} \cdot \left\{ \dfrac{\gamma_c^2 - 2 \cdot \int_0^{\gamma_c} \gamma \cdot H(\gamma) d\gamma}{\gamma_c^2 \cdot [1 - H(\gamma_c)]} - 1 \right\} & \gamma_c \leq \gamma_{ult} \\[2ex] \frac{2}{\pi} \cdot \dfrac{2 \cdot \int_0^{\gamma_{ult}} \gamma \cdot [1 - H(\gamma)] d\gamma + \gamma_{ult} \cdot [1 - H(\gamma_{ult})] \cdot (\gamma_c - 2 \cdot \gamma_{ult})}{\gamma_{ult} \cdot [1 - H(\gamma_{ult})] \cdot \gamma_c} & \gamma_c > \gamma_{ult} \end{cases} \tag{11}$$

3 STUDY ON THE FITTING PARAMETERS A, B AND γ_0 OF SOILS IN THE NEW CONSTITUTIVE MODEL

There are three fitting parameters A, B and γ_0 in the new constitutive model of soils. In the modified Davidenkov model of shear stress-strain skeleton curve, there is an another fitting parameter γ_{ult}. The $G/G_{max} \sim \gamma$ and $D \sim \gamma$ curves of soils are popularly

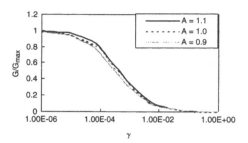

Figure 4. Influence of parameter A value on $G/G_{max} \sim \gamma$ curves of soils.

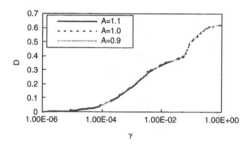

Figure 5. Influence of parameter A value on $D \sim \gamma$ curves.

tested by free vibration column apparatus or dynamic triaxial apparatus. In these two kind of tests, shear strain amplitude γ is often less than 5% which is less than upper limit shear strain γ_{ult}. Therefore, only three fitting parameters A, B and γ_0 can be ascertained by $G/G_{max} \sim \gamma$ and $D \sim \gamma$ curves of soils. Upper limit shear strain γ_{ult} can be also tested by dynamic strength test of soils. In this paper, the influence of fitting parameters A, B and γ_0 on the shapes of $G/G_{max} \sim \gamma$ and $D \sim \gamma$ curves of soils are studied only and some rules are given as follows:

(1) When the parameters B and γ_0 values are change-less and the parameter A value becomes more and more smaller, the whole $G/G_{max} \sim \gamma$ curves of soils moves downwards. At the same time, the whole $D \sim \gamma$ curves of soils moves upwards. These changes are shown in Fig. 4 and Fig. 5, respectively. It is obvious that the changes of parameter A value will have little influence on the shapes of $G/G_{max} \sim \gamma$ and $D \sim \gamma$ curves of soils.

(2) When the parameters A and γ_0 values are change-less and the parameter B value becomes more and more smaller, the $G/G_{max} \sim \gamma$ curves of soils moves downwards corresponding to smaller shear strain and moves upwards corresponding to bigger shear strain. However, the whole $D \sim \gamma$ curves of soils moves downwards obviously. These changes are shown in Fig. 6 and Fig. 7, respectively. It is obvious that the changes of

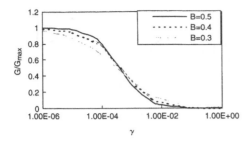

Figure 6. Influence of parameter B value on $G/G_{max}\sim\gamma$ curves.

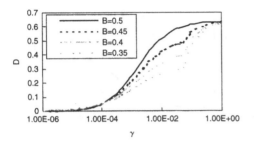

Figure 7. Influence of parameter B on $D\sim\gamma$ curves.

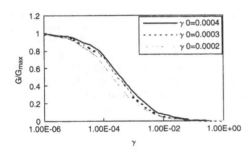

Figure 8. Influence of fitting parameter γ_0 on $G/Gmax\sim\gamma$ curves.

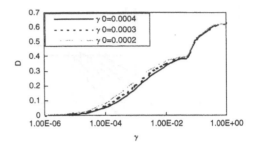

Figure 9. Influence of parameter γ_0 on $D\sim\gamma$ curves.

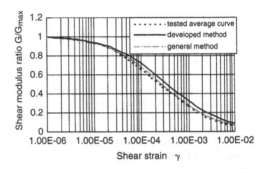

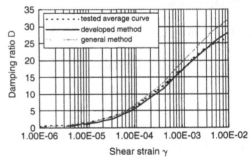

Figure 10. The $G/G_{max}\sim\gamma$ and $D\sim\gamma$ curves of clay fitted by different methods.

parameters B will have big effects on the shapes of $G/G_{max}\sim\gamma$ and $D\sim\gamma$ curves of soils.

(3) When fitting parameters A and B values are changeless, the parameter γ_0 value, as parameter A value, has similar influence on shapes of $G/G_{max}\sim\gamma$ and $D\sim\gamma$ curves of soils. These changes are shown in Fig. 8 and Fig. 9, respectively.

4 TEST AND FITTING RESULTS FOR $G/G_{MAX}\sim\gamma$ AND $D\sim\gamma$ CURVES OF SOILS IN NANJING CITY AND ITS NEIGHBOR

Large number of soil samples in Nanjing city and its neighbor are tested by free vibration column apparatus.

The average $G/G_{max}\sim\gamma$ and $D\sim\gamma$ curves of soils are obtained. In general fitting method, only the $G/G_{max}\sim\gamma$ curves of soils are fitted by constitutive model, and the $D\sim\gamma$ curves of soils are fitted by experiential equation. However, these soil samples are tested under the premise of isotropic consolidation. For large shear strain, the test values of G/G_{max} are smaller and the test values of D are larger compared to actual values. In this paper, the difference between test values and actual values are considered when the test average curves $G/G_{max}\sim\gamma$ and $D\sim\gamma$ of soils are fitted. In here, only the average curves and fitting curves of clay and sand are shown in Fig.10 and Fig.11, respectively. All values of fitting parameters of soils are shown in Table 1.

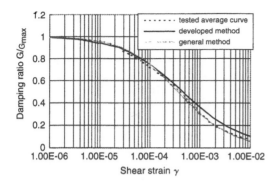

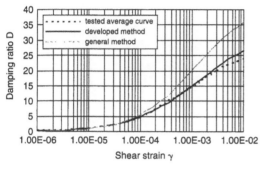

Figure 11. The $G/G_{max} \sim \gamma$ and $D \sim \gamma$ curves of sand fitted by different methods.

Table 1. Values of fitting parameters A, B and γ_0.

Fitting parameters	A	B	$\gamma_0 (\times 10^{-4})$
Mucky silty clay silty clay and silty Sand Interbedded	1.02	0.35	4.0
soil	1.05	0.345	3.5
Sand	1.1	0.35	3.8
clay	1.2	0.35	2.5
Silty clay	1.0	0.36	4.1
silt	1.0	0.375	4.1

5 CONCLUSIONS

(1) The Davidenkov skeleton curve of soil shear stress-strain relation is corrected by using sectional functions, and the upper limit shear strain amplitude is used as the sectional point. When soil shear strain amplitude $\gamma_c > \gamma_{ult}$, the Davidenkov skeleton curve is modified to be a horizontal line. A, B.

(2) How the fitting parameters A, B and γ_0 to influent the shapes of $G/G_{max} \sim \gamma$ and $D \sim \gamma$ curves of soils are studied. The results indicate that parameter B value is the key parameter which influents shapes of $D \sim \gamma$ curves of soils and parameters γ_0 and A values are the key parameters which influent shapes of $G/G_{max} \sim \gamma$ curve of soils.

(3) Values of fitting parameters A, B and γ_0 for several kind of soils in Nanjing city and its neighbor are given to fit the shapes of $G/G_{max} \sim \gamma$ and $D \sim \gamma$ curves tested by free vibration column apparatus.

REFERENCES

Xiaojun Li. 1993. Study on the method for analyzing the earthquake response of nonlinear site. Harbin: Institute of Engineering Mechanics, China Earthquake Administration.

Hardin B.O. & Drnevich V.P. Shear modulus and damping in soils design equations and curves[J]. Journal of Soil Mechanics and Foundation. ASCE, Vol.98, No.SM7, 1972.7: 603–642.

Martin P.P. & Seed H.B. One dimensional dynamic ground response analysis[J]. Journal of geotechnical engineering. ASCE, Vol.108, 1982.7: 935–954.

Guoxing Chen. & Xuezhu Liu. 2004. Testing study on ratio of dynamic shear modulus and ratio of damping for recently deposited soils in Nanjing and its neighboring areas [J]. Chinese Journal of Rock Mechanics and Engineering, 23(8): 1403–1410.

Jing Sun. 2004. Experiments and research on the application of dynamic shear strain modulus and damping ratio of rock and soils [D]. Harbin: Institute of Engineering Mechanics, China Earthquake Administration.

Environmental Vibrations – Takemiya (ed.)
© *2005 Taylor & Francis Group, London, ISBN 0 415 39035 4*

Research on lateral resistance of liquefying sand

S.L. Feng & J.H. Wang
Geotechnical Engineering Institute of Tianjin University, Tianjin, China

ABSTRACT: The bending moments along the model pile and the acceleration at the top of the model pile were measured during the shake table test of the model pile in saturated sand. And bending moments along the model pile are calculated by the method of the beam on nonlinear Winkler foundation and degradation p-y curves at the moments of 0.04s, 0.08s, 0.16s after the sand liquefied. Test and calculated bending moments are compared. Results show that the lateral resistance of the model pile degrades 90% if the relative density of sand strata ≤40% after the sand liquefied and 75% if the relative density of sand strata ≥50% after the sand liquefied.

1 INTRODUCTION

Pile foundation can adapt well to all kinds of soil conditions and load conditions. Big bearing capacity, good stabilization and little settlement and so on are its characteristics. So pile foundation is a kind of foundation widely used in offshore platforms. The lateral bearing behavior of pile foundation is commonly obtained through the nonlinear p-y curves method in designing the pile foundation of the offshore platform (Cheng 1998). Past engineering experience shows that the lateral bearing capacity of pile foundation in saturated sand decreases largely when the saturated sand liquefies under seismic loading and sometimes excessive lateral displacement of the superstructure can occur. And the superstructure can so much as be destroyed. Thus it is important to ascertain the p-y behavior of the pile in liquefying sand under seismic loading.

At present the pseudo-static method is mostly used to evaluate the lateral bearing behavior of pile foundation in liquefying sand. Architectural Institute of Japan (Architectural Institute of Japan 1988) and Japan Road Association (Japan Road Association 1980) codes propose that the lateral bearing capacity of pile foundation in liquefying sand should be scaled according to the position and liquefaction safety factor of liquefying. And the lateral bearing capacity of pile should be evaluated by the static method with the scaled parameters. But the rationality of the scaling of the parameters is not explained reasonably. Pile foundation aseismic codes in china also use the method of scaling the skin resistance of the pile and the elastic modulus of the soil (GB50011-2001 2001, China Railway Publishing House 1989, GB501-93 1993). Liu and Dobry's

research (Liu, L. & Dobry, R. 1995) has shown the lateral bearing capacity of pile foundation in saturated sand will decrease with the increase of the pore water pressure in soil: when the pore water pressure ratio in saturated sand is 1.0, the lateral bearing capacity of pile foundation is lowest. Wilson's further research (Wilson, D.W. 1998) has illustrated that the lateral bearing behavior of pile foundation in saturated sand relates to the density of the soil yet. Although researchers have some knowledge about the lateral bearing behavior of pile foundation in saturated sand through some tests in the laboratory, engineers and designers still feel that there is lack of sufficient test proof.

Until now, when engineers design the offshore platform pile foundation in saturated sand, first, it is judged whether the saturated sand could liquefy under seismic loading, if it is concluded that the saturated sand could liquefy, then it is followed that the lateral bearing capacity of pile foundation in saturated sand is zero. Because research shows that pile foundation in liquefying sand has certain lateral bearing capacity, it is obvious that the method of designing the offshore platform pile foundation in saturated sand might underestimate the lateral bearing capacity of pile foundation.

Based on the above analysis, the shake table model tests of pile foundation in saturated sand were done, and the bending moments of the model pile and the pore water pressure in saturated sand were measured. It is proposed that the static p-y curves of the pile should be scaled to determine the lateral bearing behavior of the pile in the liquefying sand. So the work in this paper could supply test proof and theory base to engineers to better use the lateral bearing capacity of the pile foundation in the liquefying sand.

Figure 1. Layout of the shake table model test.

2 SHAKE TABLE MODEL TESTS

The model tests used the large shake table in Institute of Engineering Mechanics, China Earthquake administration which is driven by hydraulic pressure and can simulate three dimension earthquake. The dead weight of the shake table is 20 tons and its maximum carrying capacity is 30 tons. Its length is 5 meters and its width is 5 meters. And its maximum displacement in the horizontal direction (x axis direction and y axis direction) is 80 millimeters. Its maximum displacement in the vertical direction (z axis direction) is 50 millimeters. And the maximum velocity in each direction is 15 centimeters per second. Under its maximum carrying capacity, its maximum acceleration in the horizontal direction (x axis direction and y axis direction) is 1 gravity acceleration and its maximum acceleration in the vertical direction (z axis direction) is 0.7 gravity acceleration. Its working frequency is 0.5 Hz to 40 Hz.

The model test box has inside dimensions of 1.6 m long, 0.9 m wide, by 1.3 m deep. Four same test boxes and model piles were made to measure the dynamic behavior of the pile in saturated sand under the same input shake and different relative density of sand. The relative density of sand in the test box is 20%, 30%, 40%, and 50%, and this paper would mainly list the results of the test in which the relative density of sand in the test box is 40%. The layout of the shake table model test is shown in Fig. 1. The model pile is an aluminum pipe 38.3 mm in diameter, 1500 mm long, with a 1.15 mm wall thickness. And the Young's modulus of the model pile is 64 MPa. Strain gauges were glued onto the exterior of the model pile along the shake direction to measure the bending moments of the pile. Fine sand is used to prepare the saturated sand layer. And the Grain size distribution curve of the fine sand is shown in Fig. 2. The structural system in the test box is illustrated in Fig. 3.

The test input shake is chosen as the sine wave shake with the shake frequency of 3 Hz and the shake

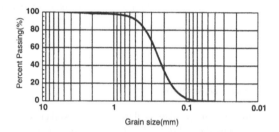

Figure 2. Grain size distribution curve of the sand.

amplitude of 0.3 g to simplify the question and be easy to study the lateral bearing behavior of pile foundation in saturated sand. Fig. 4 shows the acceleration time history of the shake table base. Fig. 5 illustrates the time history of pore pressure ratio at a position of 40 cm above the bottom of the test box in which the relative density of sand is 40%. And the time histories of bending moments at different positions of the pile are shown in Fig. 6. The acceleration time history of the superstructure mass at the top of the pile is shown in Fig. 7.

3 ANALYSIS OF THE LATERAL BEARING BEHAVIOR OF THE PILE

The degradation of the lateral bearing capacity of the pile is a dynamic process during the saturated sand is shaken to liquefy. Strictly speaking, this process should be simulated by dynamic calculation method. However, pseudo-static calculation method is used in the following for engineers' convenience to use the research results. The lateral bearing behavior of the pile is analyzed at the moment of 0.04s, 0.08s, 0.16s after the sand liquefied to find the way to describe the lateral bearing behavior of the pile in the liquefying sand.

The Beam on Nonlinear Winkler Foundation (BNWF) (Abghari, A., Chai, J. 1995) model is used

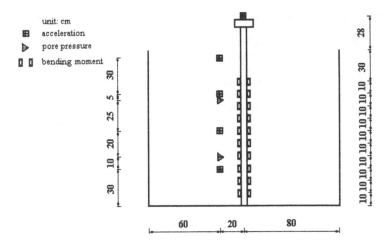

Figure 3. Model layout in the test box.

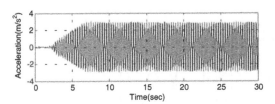

Figure 4. Acceleration time history of the shake table base.

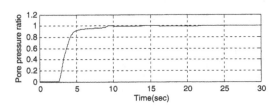

Figure 5. Time history of the pore pressure ratio in sand.

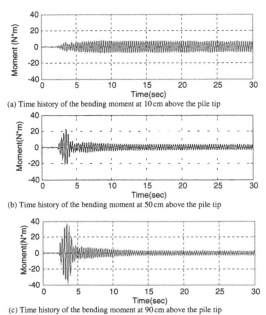

(a) Time history of the bending moment at 10 cm above the pile tip

(b) Time history of the bending moment at 50 cm above the pile tip

(c) Time history of the bending moment at 90 cm above the pile tip

Figure 6. Bending moments time histories at different positions of the pile.

to analyze the lateral bearing capacity of the pile in the liquefying sand, and the calculation model is shown in Fig. 8. In this model the soil-pile interaction is approximated using parallel nonlinear soil-pile (p-y) (Matlock et al. 1978) springs, and the relationship between the soil resistance force p and the pile displacement y under static force condition is obtained through the recommended practice of American Petroleum Institute (API) (American Petroleum Institute 2000), that is the static p-y curve which is simulated by the relationship between the force and the displacement of the nonlinear spring in the calculation model. Furthermore, in order to obtain the degradation p-y curve in the liquefying sand which describes the lateral bearing behavior of the pile, the soil resistance force p of

the static p-y curve is multiplied by a certain scalar factors. The new p-y curve (scalar factors applied to the p values) is used as the degradation p-y curve in the liquefying sand and is inputted into the calculation model. If the calculated bending moments are same to the test values at corresponding moment, it is thought that the scalar factors is right and the new p-y curve (scalar factors applied to the p values) can describe the lateral bearing behavior of the pile at corresponding moment; otherwise, the scalar factors need be adjusted to calculate again until a reasonable match between calculated and measured bending moments of the pile is obtained.

The moments of 0.04s, 0.08s, 0.16s after the sand liquefied are chosen respectively to calculate and analyze the lateral bearing capacity of the pile in

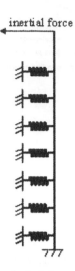

inertial force

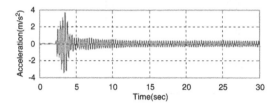

Figure 7. Acceleration time history of the superstructure mass at the top of the pile.

Figure 8. Schematic of calculation model.

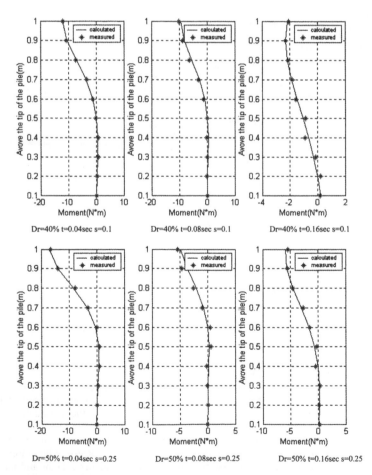

Figure 9. Test results and calculated results of bending moments of the model pile.

order to better understand the lateral bearing capacity of the pile in the liquefying sand.

Fig. 9 shows the measured and the calculated bending moments of the pile in order to compare the test results and the calculated results. We can see from the Fig. 9 that a reasonable match between calculated and measured bending moments of the pile is obtained when the scalar factors is right. And this shows that the degradation p-y curve which is obtained from the static p-y curve multiplied by a right scalar factors (scalar factors applied to the p values) can describe the lateral bearing behavior of the pile at corresponding time. Furthermore, we can know the following through analyzing the relationship between the relative density of the sand and the corresponding scalar factors: when the relative density of the sand ⩽40%, s = 0.1 i.e. the new p-y curve which is obtained from the static p-y curve multiplied by a scalar factors of 0.1 (scalar factors applied to the p values) can describe the lateral bearing behavior of the pile in the liquefying sand; when the relative density of the sand ⩾50%, s = 0.25 i.e. the new p-y curve which is obtained from the static p-y curve multiplied by a scalar factors of 0.25 (scalar factors applied to the p values) can describe the lateral bearing behavior of the pile in the liquefying sand.

4   CONCLUSIONS

In this paper the lateral bearing behavior of the pile in the liquefying sand is studied through the shake table test and the numerical simulation calculation of the pile-soil interaction. The time histories of the bending moments of the pile, the acceleration of the super-structure at the top of the pile and the pore pressure in the sand were obtained in the shake table model test. Based on the test results, it is proposed that the new p-y curve which is obtained from the static p-y curve multiplied by the right scalar factors (scalar factors applied to the p values) could describe the lateral bearing behavior of the pile in the liquefying sand. And the relationship between the relative density of the sand and the right scalar factor is illustrated by comparing and analyzing the calculated and measured bending moments of the pile at the moments of 0.04s, 0.08s, 0.16s after the sand liquefied. And results show that the lateral resistance of the model pile degrades 90%

if the relative density of sand strata ⩽40% after the sand liquefied and 75% if the relative density of sand strata ⩾50% after the sand liquefied. It is hoped that the research results in this paper could supply test proof and theory base to engineers to impersonally evaluate the lateral bearing capacity of the pile foundation in the liquefying sand.

ACKNOWLEDGMENT

This research was financially supported by Hi-tech Research and Development Program of China (Grant No. 2002AA615080).

REFERENCES

Abghari, A., Chai, J., 1995. Modeling of soil-pile-super-structure interaction in the design of bridge foundations. Geotechnical Special Publication, 1995, 51: 45–59.

American Petroleum Institute, 2000. Recommended Practice for Planning, Designing and Constructing Fixed offshore Platforms. America: American Petroleum Institute, 2000: 64–66.

Architectural Institute of Japan, 1988. Recommendations for design of building foundations. Architectural Institute of Japan, 1988.

Cheng Zekun, 1998. Analysis of High-pile Structures Based on P-Y Curve Method with Consideration of Interaction Between Pile and Soil. The Ocean Engineering, 1998, 16(2): 73–82.

China Railway Publishing House, 1989. Code for seismic design of railways. Beijing: China Railway Publishing House, 1989.

GB501-93, 1993. Code for seismic design of structures.

GB50011-2001, 2001. Code for seismic design of buildings.

Japan Road Association, 1980. Specifications for highway bridges. Japan Road Association, 1980.

Liu, L. and Dobry, R., 1995. Effect of liquefaction on lateral response of piles by centrifuge model tests. National Center for Earthquake Engineering Research (NCEER) Bulletin, 1995, 9(1): 7–11.

Matlock, H., Foo, S.H., and Bryant, L.L., 1978. Simulation of lateral pile behavior, Earthquake Engineering and Soil Dynamics, ASCE, pp. 600–619.

Wilson, D.W., 1998. Soil-pile-superstructure interaction in liquefying sand and soft clay. America: University of California, Davis, 1998: 88–135.

Environmental Vibrations – Takemiya (ed.)
© 2005 Taylor & Francis Group, London, ISBN 0 415 39035 4

Discussions on dynamic triaxal tests of intact loess

Y.S. Luo, J.Y. Liu & J. Li
Northwest A&F University, Yangling, Shannxi, China

D.Y. Xie
Xi'an University of Technology, Xi'an, Shannxi, China

ABSTRACT: Based on reviewing and analyzing the results of previous study on loess, surrounding the peculiarities of loess itself, the methods of dynamic triaxal tests and experiment data analyses for intact loess are developed. The results show that the ameliorated apparatus and methods can improve the efficiency of dynamic triaxal tests and precision of data analyses, and studies on intact loess dynamic characteristics at different areas can be carried through under different densities and different moistures. Furthermore, with the dynamic triaxal tests data of intact loess at Xi'an, Lanzhou and Taiyuan three typical loess regions of China, the current outcomes of intact loess dynamic characteristics are summarized. Finally, a viewpoint of considering the effect of structure characteristics in studying the loess dynamic characteristics in a large range is presented, in which the loess dynamic characteristics at different area are compared and studied systematically.

1 INTRODUCTION

As a zonal soil distributed more widely in the north regions of China, loesses are famous for their stratigraphical intactness and huge thickness. Along with the development of economy construction, the engineering problems due to dynamic loading action at loess are have been focused on. Using the methods of soils dynamics to study the dynamic characteristics of loess in-depth has very important significance not only for reducing and preventing disasters but also for the design and evaluation of projects related to resist vibration. Because the intact loess is the primary objects of project construction, a lot of tests on loess dynamic characteristics have been put up, while the lack of uniform standards for tests and analyse and the contact disciplinarian outcomes for different loess areas are the obvious shortage at present.

In this paper, based on reviewing and analyzing the previous study results of loess, surrounding the peculiarities of loess itself, the methods of dynamic triaxal tests and experiment data analyses for intact loess are improved on. Farther more, studies of intact loess dynamic characteristics of intact loess at three typical loess regions of China such as Xi'an, Lanzhou and Taiyuan are carried through under initial moistures, and trying to find out the variations regularity and contacts among different loess regions of loess dynamic parameter used for computing the dynamic responses and analyzing the characteristics of deformation and

Table 1. Physical character index of loess samples.

Samples source	Dry density/ g/cm^3	Void ratio	Grain composing/%		
			>0.05 (mm)	0.05~ 0.005 (mm)	<0.005 (mm)
Xi'an loess	1.26	1.151	9.8	61.8	28.4
Lanzhou loess	1.45	0.862	18.0	72.0	10.0
Taiyuan loess	1.34	1.015	27.8	58.2	14.0

strength. The physical character index of loess samples as Table 1 shows.

2 METHODS OF DYNAMIC TRIAXAL TESTS

For saturated loess or near saturated loess, besides the sample preparation technique and concretion stabilization standard, dynamic triaxal tests generally can be operated by referencing the item of vibration triaxal test for saturated sand at soils test regulations of PRC (SL237-1999). There are many different methods of dynamic triaxal test of unsaturated soil: According to the different shape of exerting dynamic load, the dynamic load can be divided into equivalent sine wave load and random earthquake wave load; Based on means of exerting load, load methods of dynamic triaxal test can be classified as gradual loading method under the same consolidation condition and the single step loading method under different consolidation condition. Ordinarily, in dynamic elastic modulus and dynamic damping ratio test, gradual loading method is adopted while the single step loading method is used in dynamic shear strength and deformation test. Loess, as a kind of loose deposit with great void and under compaction, because of different contributing factor and geologic period appears obvious district-distribution features, and has obvious divergence of properties. Even the samples that are taken from the same layer don't appear uniform clearly; further more, those samples have been interfered by alien factors such as sampling and transportation. Those will bring about larger scatter on experimental data. Especially, in the dynamic triaxal test of unsaturated soil, the exerting load isn't easy to determine when using the single step loading method as the scales of dynamic stress which can cause damage of different water content samples are pretty wide. To solve to this problem better, the gradual loading method can be adopted in studying the dynamic strength and deformation test just like the dynamic elastic modulus and dynamic damping ratio test. In this paper, the dynamic triaxal test of loess uses gradual loading method under the condition of equivalent sine wave load, which to exert dynamic loading that is firm number of vibration (number = 10), step by step and from low to high, on a sample that has been steady after consolidation deformation under the condition of undrained still to damage and write down the total experiment process. Through analyzing and processing those data the dynamic modulos, dynamic shear strength, dynamic damping ratio and dynamic deformation parameters can be obtained at the same time. A curve of dynamic stress and strain can be obtained by testing one sample (Wang Lanmin et al, 1991). That can degrade scatter of test result that is otherwise caused by using much more samples; further, amount of work is cut down greatly and the relationship of different dynamic parameters of loess comes to be closer; then the efficiency of data process gets enhanced.

3 THE MEETING PROBLEM IN DATA PROCESS AND IMPROVEMENT MEANS

In dynamic triaxal test, when axial dynamic stress of loess sample is compression tress or is tensile stress whose value is lower than deviator consolidation pressure, the axial dynamic stress is true. Then, dynamic deformation and residual deformation of compression orientation are both true. But, when axial dynamic stress is tensile stress whose value is higher than the deviator consolidation pressure, the sample can break away from vibrating axle, then it would be difficult to occur the compression-stretch deformation. At that time what the axial tensile-compress senor measured

isn't true axial stress that exerts on this sample already. And what the axial displacement senor measured is bigger than the true value far away. As matter of fact that value is the displacement of vibrating axle. The representative record of dynamic stress and dynamic deformation is shown as fig.1 (the orientation of compression as positive).

From fig.1 we can see that dynamic stress and dynamic strain of loess sample are not symmetric relatively with balanced location under the condition of deviatoric compression, the amplitude value of dynamic pressure is larger than that of tensile stress obviously, and, the amplitude value of deformation of compression orientation is far lower than that of tensile orientation. Obviously, deformation process is not reasonably as to samples. It is only to be the reflection of displacement of vibration axle. It is evident that using this untrue value to analyze the dynamic strength, it is impossible to draw a actual conclusion. Apparently, as a matter of fact, to the side of tensile, the tensile stress also affect the dynamic characteristic of loess, but this kind of affection is neglected by value of dynamic deformation, which is distorted seriously on the side of tensility and it is very difficult to considered about that in the actual process of analyzing the

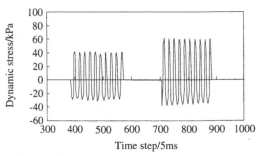

(a) Record of dynamic stress

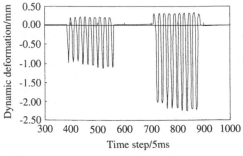

(b) Record of dynamic deformation

Figure 1. The representative record of dynamic stress and dynamic deformation (compression as positive; Luo yasheng, 2000).

data of dynamic test. As usually considered, the main reason of creating damage of ground earth layer or upper composition is shear wave that is propagated upward by the bedrock's movement. So we separate the process of pressure and tensile stress and caused deformation by both from the test writing and only use the data of pressure and deformation that can reflect on actual circumstances of test to analyze dynamic characteristic of loess, that should be more reasonable. In this paper, experimental data is arranged with this method.

4   ANALYZING EXPERIMENTAL DATA

4.1   Stress–strain curve of loess

The properties of stress–strain curve are decided by load form and constraint condition of border. Lots of experiential data of dynamic characteristic of undisturbed loess have shown that affected by dynamic loading the stress– strain relation (that is constitutive relation) of undisturbed loess presents hyperbolic curve shape and has pretty obvious nonlinearity property. That can be depicted well by Hardin-Drnevich's hyperbolic curve model. This paper carries out dynamic triaxal test about Xi'an, Lanzhou and Taiyuan this three representative loess under different moisture content state. Test result shows that the relation of ε_d/σ_d and ε_d presents linear relation well, that illustrates the relation of dynamic stress and dynamic strain of undisturbed loess under different moisture state can be depicted well by using hyperbolic curve yet. For instance, we can see that in fig. 2.

The results shows that the difference of water content has obvious affects on relation curve of dynamic stress and strain of undisturbed loess. The lower water content of sample is, the larger dynamic stress that is relative to same dynamic strain is; with the water

content from low to high or with consolidation stress ratio from little to large, relation curve of dynamic stress and strain has a tendency of degradation and becoming unhurried. To the undisturbed loess of different representative area, difference of loess's structure has a great affect on property of dynamic strengthen. Under the same condition of other situations, the effect of initial water content and consolidation pressure on curve of dynamic stress and strain is not necessarily same, so it is not enough to take dry density as the only one measurement criterion.

4.2   Dynamic elastic modulos of loess

The dynamic elastic modulos expresses the relation of resumptive part of dynamic stress and dynamic strain, which studies how much dynamic stress is required to create unit dynamic strain. The dynamic elastic modulos of loess can be obtained from the relation of dynamic stress and dynamic strain of undisturbed loess. On the base of definition of dynamic elastic modulos $E_d = \varepsilon_d/\sigma_d$, the dynamic elastic modulos can be got form the reciprocal of longitudinal intercept of $\varepsilon_d/\sigma_d \sim \varepsilon_d$ curve. Through working up the experimental data, the relation curve of maximal dynamic elastic modulos E_{max} and water content w of loess of Xi'an, Taiyuan and Lanzhou under different consolidation pressure can be obtained as what shows in fig. 3.

The results shows that under the same consolidation pressure, the maximal dynamic elastic modulus E_{max} of loess decrease with the increase of water content w. To loess of different area, the effect of water content w and consolidation pressure on the maximal dynamic elastic modulus E_{max} is not necessarily the same. The maximal dynamic elastic modulus E_{max} expresses complex effects of water content, consolidation pressure, stress history and structural index

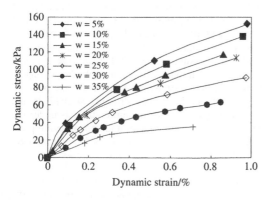

Figure 2.   The relation of dynamic stress and dynamic strain of Xi'an loess under different moisture state ($k_c = 1.69$, $\sigma_3 = 200$ kPa; Luo yasheng, 2000).

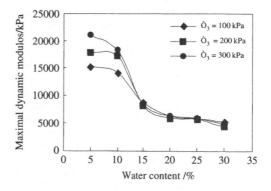

Figure 3.   The relation curve of maximal dynamic elastic modulos and water content of loess of Lanzhou (Luo yasheng, 2000).

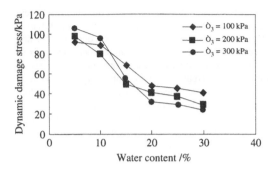

Figure 4. The relation curve of dynamic damage stress and water content of loess of Lanzhou (Luo yasheng et al, 2001a).

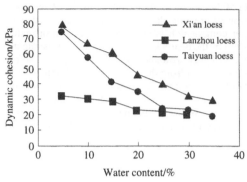

(a) The relation of c_d and w

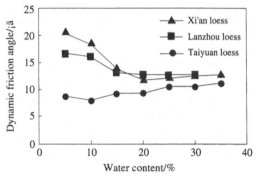

(b) The relation curve of ϕ_d and w

Figure 5. The relation curve of dynamic cohesion or dynamic friction angle and water content of loess (Luo yasheng et al, 2001a).

and so on. Among them, the effect of structural index is the most projecting. In analyses and evaluation of dynamic elastic modulus of loess, we shall combine the characteristic of loess of different area respectively to approach differently.

4.3 Dynamic strength of loess

Dynamic strength of soil is changeable following the rate effect of dynamic load action and circle effect. Usually, dynamic strength of soil is defined as the required dynamic stress that causes strain of sample to reach some one assigned damage strain ε_f which is created by one settled pressure under N number of times of action to and from. To loess, it is best to choose stress of the obvious transition point of stress-strain-curve as damage stress, which adopts yield criterion. Under the lower moisture state, dynamic damage stress of loess shall choose this stress that is holded by loess when it creates brittleness damage. Further more, under the higher moisture state, it is better to choose the stress of the obvious transition point of stress-strain-curve as dynamic damage stress of loess. According to this criterion, this paper obtains relation curve of water content and dynamic damage stress of loess of Xi'an, Taiyuan and Lanzhou under the different consolidation pressure, as presebted in fig. 4. Further, the value of dynamic strength parameters, c_d and ϕ_d, of Xi'an, Taiyuan and Lanzhou's loess can be ascertained. Finally, the relation curve of c_d and ϕ_d and water content w can be obtained as fig. 5 shows.

The results shows that the damage dynamic stress of loess σ_{df} under the same consolidation pressure, damage dynamic stress diminishes with the increase of water content: When water content is smaller the stress–strain curve becomes more precipitous; but when water content is higher the stress–strain curve becomes smoother and its limited water content is the plastic limit of loess. To loess of different area, the effect of consolidation pressure on damage dynamic

stress of loess is different due to the difference of loess's density of various areas, soil structure and stress history before water content of sample is less than limited water content. However later as water content of sample is larger than limited water content, effect of preconsolidation pressure and primary structure and so on essentially disappear. But effect of consolidation pressure shows in one voice that the larger consolidation pressure is, the smaller damage dynamic stress is. The amount of content embodies strong or weak the structure strength is to great extent: the more clay content is the bigger structure strength is and also the bigger damage dynamic stress is under the same initial moisture content and consolidation pressure conditions. In indexes of loess's dynamic strength of different areas, the dynamic cohesion becomes smaller with broadness of water content. When water content of sample is less than limited water content, the alteration law of the dynamic friction angle and water content is different because of differences of loess's density of various areas, soil structure and stress history and so on; But when water content of sample exceeds limited water content, there appears no more obvious variability with broadness of water content and not much differences of loess of different areas.

4.4 Dynamic damping of loess

Dynamic damping expresses the nature of energy loss because of internal resistance of loess and is a very important index of dynamic characteristics of soil. Under the action of repeated load, deformation of loess and other soils includes two parts: elastic deformation and plastic deformation. At low dynamic load, the deformation appears elastic. With the dynamic load increasing the plastic deformation broadens, and its relation of dynamic stress and dynamic strain has characteristics of nonlinearity and sluggishness. Usually, in the dynamic triaxal test, first, exert dynamic load gradually and measure the dynamic stress and strain at each step; then, choose some representative cycles and draw stress-strain hysteresis loops respectively; finally, calculate the damping ratio.

So far a good law concerning damping property of undisturbed loess has not been obtained. Writer holds the reason why not obtained is that there exist two faults in the means of measurement of loess strain besides the effect of loess composition: one is the lower moisture loess is different form ordinarily saturated clay and sand. Under the action of dynamic load, the affected scope of the action extends top down gradually – during the test, the phenomenon we observe is that the top of sample changes wide while the bottom changes none. Under this situation, it is difficult to affect actuality to calculate the strain of sample with original length; this kind of deviation affects the model calculation equally, but the above-mentioned effect appears more noticeable only because research of damping ratio is only on one cycle of loading and unload and thus increment of strain changes very little – this could be the most important reason of causing the law of measurement larger dispersity. Another is transformation of stress orientation of the sample will appear with loading in the area of tensile half-lap, when the sample is exerted any type of to and form load under the condition of even pressure consolidation or exerted lager dynamic stress than the deviator under the condition of deviator stress, which may not cause the sample of saturated sand and soft clay to break away from piston because the sample can extend at radial direction when extruded. But as unsaturated loess deformation is affected by stress insensitively, the sample is unlikely to have extruding-extending deformation but be bound to break away from vibration piston, thus make the value of deformation of tensile half-lap, which is measured by displacement sensor, larger than the actual. Because the actual deformation is too complex to depart, only using measured deformation to calculate strain directly is impossible to figure out an obvious law of property of damping ratio.

Owing to the importance of dynamic property parameters in analyses of dynamic affect experiments of damping ratio of soil have been researched by many scholars (Wang Lanmin et al, 1991; Liu Baojian, 1994). Considering the encountering hardship in using regular method – using the area of hysteresis loop to get the damping ratio, a new means of requesting damping ratio through phase difference method has been proposed (Liu Baojian, 1994). A expression by suing the knowledge of soil kinetic as follows:

$$\lambda = \frac{1}{2}tg\varphi$$

In the expression, φ is the phase difference between the dynamic stress and strain; it is the angle of lag of strain in correspondence with stress in phase (describing with angle) and effects property of damping ratio. So, if the phase difference is measured, the value of damping ratio can be reached.

This paper has analyzed the damping ratio of Xi'an loess. The test result shows that the damping ratio of Xi'an loess is less than 0.25. Relatively with the same consolidation pressure, the effect of change of water content on damping ratio has no definite law to follow basically. The larger consolidation pressure is, the stronger dispersion degree becomes. And damping ratio appears a tendency of decreasing. This dispersion should have some links with the consolidation pressure, the density of soil and the change of soil structure. To loess which is of strong structure, the normal methods of using hysteresis loop to calculate the damping ratio is very difficult to produce satisfactory results, so further research is necessary. The method according to angle of lag may suggest a way of estimating the damping ratio.

4.5 Vibration collapse of loess

The vibration collapse deformation of loess (only to earthquake load, also called seismic collapse deformation) is expressed by residual deformation (also called coefficent of seismic collapse deformation) in soil kinetics. It is defined as additional strain value that remains by soil after stopping exerting dynamic load. Usually, it is expressed in form of height difference and the height ratio before and after actions of dynamic stress. It is closely related to number of times N.

At the moment, most of dynamic experiments of loess are conducted on dynamic triaxal with equivalent sine cycle load. The test result is expressed by a vibration collapse curve of residual strain ε_p and amplitude value of dynamic stress σ_d. Using the method in this paper's to straighten out test data respectively, the vibration collapse curve of Xi'an, Lanzhou and Taiyuan undisturbed loess can be obtained, as fig. 6 shows.

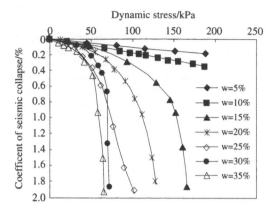

Figure 6. The vibration collapse curve of Xi'an loess under different moisture state ($k_c = 1.69$, $\sigma_3 = 100\,kPa$; (Luo yasheng et al, 2001b)).

The results shows that to same type of loess under same initial accepting force condition water content creates obvious effect on vibration collapse deformation of loess, which behaves as the larger water content is the larger vibration collapse deformation amount is in correspondence with identical dynamic stress. With little water content, the vibration collapse deformation nearly presents right line. With water content increasing continually curve characteristic of this relationship become more and more obvious while the dynamic stress, which is needed to attain dramatic evolution of vibration collapse deformation, becomes smaller and smaller. With same water content, residual strain becomes larger with dynamic stress increasing. When exerted little dynamic stress, residual strain of sample increases slightly and expresses nearly right line relationship. But when the dynamic stress reaches some limit point, residual strain presents nonlinearly and increases sharply till to damage. To loess of different areas, the dynamic stress to attain dramatic evolution of vibration deformation is obviously different even under same moisture and initial accepting force condition, and so is the effect of consolidation pressure on property of vibration collapse deformation. The stronger the structure is the harder vibration collapse damage takes place. The difference of the characteristics of structure of different loess can cause larger distinction of damage dynamic stress. Once loess takes place vibration collapse damage under the action of dynamic stress, follow-up vibration times of dynamic stress will cause dramatic evolution of vibration collapse deformation.

5 SOME OPINIONS AND PROPOSALS

1. Through the studies of loess dynamic characteristics at different areas such as Xi'an, Taiyuan and Lanzhou, it is undoubtedly right and necessary that the effect of water content be set at magnitude status in the study of loess dynamic characteristics in one loess area. But for the loess's from different areas, it is insufficient merely to consider the influences of dry density and water content when studying the variation regular and relationship of loess dynamic characteristics. The different effect owing to loess structure characteristics in different areas must be considered as well. So taking the effect of structure characteristics into consideration is a key to studying the dynamic characteristics of loess in a large range and exhuming the dynamic property presented as well as variations of parameter.

2. The improvement on method of intact loess dynamic triaxial test is a transition method in this paper, and the best settling approach is through reconstructing the dynamic triaxial apparatus in grain to satisfy the trait of dynamic triaxial test for loess itself. The writer consider that the dynamic triaxial apparatus should to be studied from two aspects: On one hand, at the precondition not to change the characteristics of dynamic, hold vibration spindle contact well with sample at the test process, thereby to measure the deformation of sample and dynamic loading accurately, so as to solve the precision testing of dynamic strength and dynamic module as well as dynamic damp radically; On the other hand, when studying the vibration collapse of intact loess, the fact of lateral pressure coefficient in soils body increase with the increase of deformation of vibration collapse should be reflected at the process of dynamic triaxial test, and vibration collapse deformation may be aggrandized by using methods of intact loess vibration test at present.

REFERENCES

Wang Lanmin, Zhang Zhenzhong, Wang Jun, et al, 1991. A test method of dynamic strength of loess under random seismic loading. Northwestern seismological journal. China. Vol. 13(3): pp: 50–55.

Liu Baojian, 1994. Experimental studies of the damping ratio of soil. Symposium on geomechanics and engineering. Shaanxi sci-tech press. China. pp: 121–124.

Luo Yasheng, 2000. Test analysis of dynamic characteristic and parameter of some typical loess in China. A thesis submitted to Xi'an University of Technology for the Degree of Master of engineering. China.

Luo yasheng, Xie Dingyi, Chen Cunli, 2001a. Test analysis of dynamic failure strength of loess under different moisture conditions. Journal of Xi'an University of Technology. China. Vol. 17(4): pp: 403–407.

Luo yasheng, Xie Dingyi, Dong Weimin, Chen Cunli, 2001b. Comparative analysis of the vibration deformation behavior of the loess from the different regions. Journal of Shaanxi water power. China. Vol. 17(1): pp: 4–7.

Environmental Vibrations – Takemiya (ed.)
© 2005 Taylor & Francis Group, London, ISBN 0 415 39035 4

Dynamic response of the underlying nearly saturated loess tunnel

X.M. Zhou, T.D. Xia & P. Xu
Institute of Geotechnical Engineering, Zhejiang University, Hangzhou, China

ABSTRACT: According to field investigation, a reasonable numeration model to loess tunnel is presented in this paper. The governing partial differential equations are presented based on the partially saturated soil theory of dynamics. The potential function theory and Laplace transform technique is used to solve the problem. The analytical solution of stresses, pore pressure and displacement induced by water pressure are derived in Laplace transform domain. Numerical results are obtained by inverse Laplace transformation and the influence of the saturation on dynamic response of the loess tunnel is discussed. It is shown that the influence of saturation on displacements is remarkable. The results presented in this paper are meaningful to the risk evaluation of loess tunnel.

1 INTRODUCTION

As a particular cacoethic geological phenomenon, the loess tunnel widely exists in western loess area of China, and has considerably hazarded the geological and natural environment of local area, such as soil and water loss etc. The classification, regularities of distribution and formation mechanics has been detailedly researched, and the loess tunnel is named "cavity erosion" (Xianmo Zhu, 1958, Binke Wang, 1988 & 1989). Moreover, due to the similarity between the loess tunnel and karst, the loess tunnel is also called "pseudo-karst" (Jingming Wang, 1996). Fig.1 shows the cross-section shape of loess tunnel. Especially, in recent years, with the development of loess tunnel, loess tunnel is seriously unfavorable to project safety passing the area, for example railway and highway engineering etc (Houtian Hu 1996, Xi'an Li 2004).

However, on the one hand, the loess tunnel is invisible; on the other hand, the geological strata often exhibit inhomogeneities of various types which are difficult to describe, either due to paucity of detailed physical measurements or due to their vast spatial complexity. So the former researches still stagnate in the observational research. To this date, a rather limited amount of work focuses on representing the quantitive results and the hydrodynamic action mechanics is also not very clear. Fortunately, the dynamic response theory is well studied. Biot presented the governing equations for acoustic propagation in porous media by using the relative displacement between the fluid and the solid matrix as an unknown (Biot, 1962).

Subsequently, many researchers have employed Biot's field equations for dynamic response problems in porous media. Thiruvenkatachar etc. discussed the dynamic response of a homogeneous isotropic and perfectly elastic half-space to time-dependent surface tractions over the boundary of a cavity in the form of a circular cylinder of infinite length at a finite depth below the plane boundary and parallel to it (Thiruvenkatachar et al, 1965). The radial displacements of liquid saturated porous medium with cylindrical cavity are obtained (Kumar et al, 1999).

This paper employs the former achievements and presents the mechanical numeration model of loess

Figure 1. Loess tunnel cross-section photo.

tunnel, which is generalized as the model of transient response of nearly saturated soil with cylindrical cavity of infinite length to time-dependent internal water pressure over the boundary of the cavity. The governing partial differential equations are derived based on the partially saturated soil theory of dynamics. The potential function theory and Laplace transform technique is used to solve the analytical solution of stresses, pore pressure and displacement induced by water pressure in Laplace transform domain. Numerical results are obtained by inverse Laplace transformation and are used to analyze the influence of the saturation on dynamic response of the loess tunnel.

2 MECHANICAL NUMERATION MODEL

2.1 Factors associated with loess tunnel formation and development

There are two type factors associated with loess tunnel formation. They are intrinsic factors and external factors respectively: the intrinsic factors are the essential characteristics of loess, include the granularity, structure, permeability, wetting-collapse, erosion-resisting, disintegrative and so on; and the thickness and structure of the stratum, land form, tectonic condition, hydrogeologic condition etc compose the external factors. During the process of loess tunnel formation, the intrinsic factors are essential factor. Meanwhile the external factors also play a key role to the formation and development of loess tunnel. Especially, among all the external factors, the action of atmospheric water is the most important. The loess tunnel is formed by hydrodynamic scouring action on the loess fracture.

2.2 Numeration model

A mechanical model for formation and evolution of the loess tunnel must be consistent with the hydraulic and soil mechanics properties of loess. It must also be consistent with field observations of the circumstances

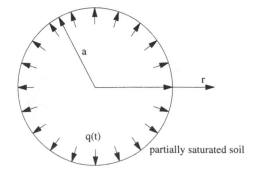

Figure 2.   Numeration geometry model.

under which new or induced cavity form. These field observations provide the most interesting and rigorous model constraints. According to the preceding analysis, during the formation and evolution of the loess tunnel, hydrodynamic action is the key. Combining with the field observation (see fig. 1), we can generalize the loess tunnel evolution as transient response problem of nearly saturated soil with cylindrical cavity of infinite length to internal water pressure over the boundary of the cavity. The geometry of the model see Figure 2. The rear section will solve the problem according the mechanical numeration model shown in Fig.2.

3 GOVERNING EQUATIONS

Biot's theory for dynamic poroelasticity is applied for the model shown in Fig.2, which can be considered as a two-dimensional plane strain axisymmetric problem. And the two independent variables are the radial solid displacement and the radial flow of the pore fluid relative to the solid matrix, which are denoted by $u_r(r,t)$ and $w_r(r,t)$, respectively.

When there are no body forces, the equations of motion can be written for the present case in terms of the displacements in the polar coordinates form as following.

For solid matrix:

$$\frac{\partial \sigma_r}{\partial r} + \frac{\sigma_r - \sigma_\theta}{r} = \frac{\partial^2}{\partial t^2}(\rho u_r + \rho_f w_r) \tag{1}$$

where σ_r and σ_θ = respectively, the total radial stress and the hoop stress; ρ = total density and $\rho = (1 - n)\rho_s + n\rho_f$; n = porosity; ρ_s = density of the solid grains and ρ_f = density of pore fluid.

For pore fluid:

$$-\frac{\partial p_f}{\partial r} = \frac{\partial^2}{\partial t^2}(\rho_f u_r + \frac{\rho_f}{n} w_r) + \frac{\rho_f g}{k_d} \frac{\partial w_r}{\partial t} \tag{2}$$

where p_f = pore pressure; g = gravity acceleration and k_d = dynamic permeability.

The pertinent stresses are:

$$\sigma_r = \lambda e + 2G \frac{\partial u_r}{\partial r} - \alpha p_f \tag{3a}$$

$$\sigma_\theta = \lambda e + 2G \frac{u_r}{r} - \alpha p_f \tag{3b}$$

$$\frac{\partial p_f}{\partial t} = \frac{\partial}{\partial t}(M\xi - \alpha Me) \tag{3c}$$

where λ and G = Lame constants of solid skeleton; $e = (\varpi u_r/\varpi r) + (u_r/r)$ and $\xi = -((\partial w_r/\partial r) + (w_r/r))$ are the dilatations of the solid and pore fluid; α and M = parameters accounting for the compressibility of grains and fluid, they can be given as

$$\alpha = 1 - \frac{K_b}{K_s} \quad M = \frac{K_s^2}{K_d - K_b} \quad K_d = K_s[1 + n(\frac{K_s}{K_f} - 1)]$$

where K_s and K_b = bulk moduli of grains and skeleton, respectively and K_f = bulk modulus of pore fluid. For the partially saturated soil, pore fluid is a mixture of water and air. One typical case of the partial saturation is when the degree of saturation is sufficiently high so that the air is embedded in pore water in the form of bubbles. For this special case, the bulk modulus of the homogeneous fluid K_f can be approximately expressed in terms of the degree of saturation as (Verruijt, 1969):

$$K_f = \frac{1}{\frac{1}{K_w} + \frac{1 - S_r}{P_0}} \tag{4}$$

where K_w = bulk modulus of pore water; S_r = the degree of saturation and P_0 = absolute pore pressure.

At this stage, introducing the Eqs. (3) and (4) into the Eqs.(1) and (2), the governing equations of the nearly water-saturated loess can be given in terms of displacements as

$$(\lambda + 2G + \alpha^2 M)\frac{\partial e}{\partial r} - \alpha M \frac{\partial \xi}{\partial r} = \frac{\partial^2}{\partial t^2}(\rho u_r + \rho_f w_r) \tag{5a}$$

$$\alpha M \frac{\partial e}{\partial r} - M \frac{\partial \xi}{\partial r} = \frac{\partial^2}{\partial t^2}(\rho_f u_r + \frac{\rho_f}{n} w_r) + \frac{\rho_f g}{k_d}\frac{\partial w_r}{\partial t} \tag{5b}$$

4 LAPLACE TRANSFORM SOLUTIONS

Here, we define two scalar potentials Φ_s (r,t) and $\Phi_f(r,t)$ (the subscripts s and f denote the correlation with solid skeleton and fluid, respectively) by:

$$u_r = \frac{\partial \Phi_s(r,t)}{\partial r}, \quad w_r = \frac{\partial \Phi_f(r,t)}{\partial r} \tag{6}$$

Substituting equation (6) into equation (5) and application of the Laplace transform to them with respect to t, we obtain the coupled equations:

$$[K][L]\{\bar{\Phi}\} = [M]\{\bar{\Phi}\} \tag{7}$$

where:

$$[K] = \begin{bmatrix} \lambda + 2G + \alpha^2 M & \alpha M \\ \alpha M & M \end{bmatrix} \quad [L] = \begin{bmatrix} \nabla^2 & 0 \\ 0 & \nabla^2 \end{bmatrix}$$

$$[M] = \begin{vmatrix} \rho s^2 & \rho_f s^2 \\ \rho_f s^2 & \frac{\rho_f}{n} s^2 + \frac{\rho_f g}{k_d} s \end{vmatrix} \quad [\bar{\Phi}] = \begin{Bmatrix} \bar{\Phi}_s \\ \bar{\Phi}_f \end{Bmatrix}$$

where s = the Laplace transform parameter; $\bar{\Phi}_s$ and $\bar{\Phi}_f$ are the Laplace transforms of Φ_s and, Φ_f, respectively.

If $\bar{\Phi}_s$ or $\bar{\Phi}_f$ is eliminated from these equations, both $\bar{\Phi}_s$ and $\bar{\Phi}_f$ satisfy the same equation

$$\{\nabla^4 - m\nabla^2 + n\}(\bar{\Phi}_s, \bar{\Phi}_f) = 0 \tag{8}$$

where

$$\nabla^4 = \nabla^2 \cdot \nabla^2 = \left(\frac{d^2}{dr^2} + \frac{1}{r}\frac{d}{dr}\right)^2;$$

$$m = \frac{A_1 s^2 + A_2 s}{(\lambda + 2G)M};$$

$$n = \frac{(\rho \frac{\rho_f}{n} - \rho_f^2)s^4 + \frac{\rho_f g}{k_d}s^3}{(\lambda + 2G)M};$$

$$A_1 = (\lambda + 2G + \alpha^2 M)\frac{\rho_f}{n} + M\rho - 2\alpha M \rho_f;$$

$$A_2 = (\lambda + 2G + \alpha^2 M)\frac{\rho_f g}{k_d};$$

By decomposing the equation (8), the solutions of the transformed equations are obtained as:

$$\begin{cases} \bar{\Phi}_s(r,s) = A_1 I_0(\beta_1 r) + B_1 K_0(\beta_1 r) + \\ \qquad\qquad A_2 I_0(\beta_2 r) + B_2 K_0(\beta_2 r) \\ \bar{\Phi}_f(r,s) = C_1 I_0(\beta_1 r) + D_1 K_0(\beta_1 r) + \\ \qquad\qquad C_2 I_0(\beta_2 r) + D_2 K_0(\beta_2 r) \end{cases} \tag{9}$$

where $A_i, B_i, C_i, D_i (i = 1,2)$ are arbitrary constants; I_n and K_n = the modified Bessel functions of the first and second kind of order n, respectively; β_1 and β_2 are the wave numbers associated with the two dilatational waves, and

$$\beta_1^2 = \frac{m - \sqrt{m^2 - 4n}}{2}, \quad \beta_2^2 = \frac{m + \sqrt{m^2 - 4n}}{2}$$

Since the domain in Figure 2 is unbounded, $I_0 (\beta_1 r)$ and $I_0 (\beta_2 r)$ are inadmissible in the expression for Φ_s and Φ_f. Therefore, the admissible solutions for Φ_s and Φ_f are

$$\begin{cases} \bar{\Phi}_s = B_1 K_0(\beta_1 r) + B_2 K_0(\beta_2 r) \\ \bar{\Phi}_f = D_1 K_0(\beta_1 r) + D_2 K_0(\beta_2 r) \end{cases} \qquad (10)$$

It can be shown that the constants B_i and D_i ($i = 1,2$) are linearly dependent. Substituting equations (10) into (5) yields the relation

$$D_i = \delta_i B_i \qquad (i = 1,2) \qquad (11a)$$

where

$$\delta_i = \frac{-\alpha M \beta_i^2 + \rho_f s^2}{M \beta_i^2 - \dfrac{\rho_f}{n} s^2 - \dfrac{\rho_f g}{k_d} s}, \quad (i = 1,2) \qquad (11b)$$

Thereafter, the general solutions in the Laplace domain for axisymmetric plane strain deformations of an infinite poroelastic medium can be expressed as

$$\bar{u}_r = -\beta_1 B_1 K_1(\beta_1 r) - \beta_2 B_2 K_1(\beta_2 r) \qquad (12a)$$

$$\bar{w}_r = -\delta_1 \beta_1 B_1 K_1(\beta_1 r) - \delta_2 \beta_2 B_2 K_1(\beta_2 r) \qquad (12b)$$

$$\bar{p}_f = -(\alpha + \delta_1) M \beta_1^2 K_0(\beta_1 r) B_1 - (\alpha + \delta_2) M \beta_2^2 K_0(\beta_2 r) B_2 \qquad (12c)$$

$$\begin{aligned} \bar{\sigma}_r = \{[\lambda + G + \alpha(\alpha + \delta_1)M]\beta_1^2 K_0(\beta_1 r) \\ + G\beta_1^2 K_2(\beta_1 r)\}B_1 + \{[\lambda + G + \alpha(\alpha + \\ \delta_2)M]\beta_2^2 K_0(\beta_2 r) + G\beta_2^2 K_2(\beta_2 r)\}B_2 \end{aligned} \qquad (12d)$$

$$\begin{aligned} \bar{\sigma}_\theta = \{[\lambda + G + \alpha(\alpha + \delta_1)M]\beta_1^2 K_0(\beta_1 r) - \\ G\beta_1^2 K_2(\beta_1 r)\}B_1 + \{[\lambda + G + \alpha(\alpha + \\ \delta_2)M]\beta_2^2 K_0(\beta_2 r) - G\beta_2^2 K_2(\beta_2 r)\}B_2 \end{aligned} \qquad (12e)$$

5 BOUNDARY CONDITIONS

In this study, a long cylindrical loess cavity of radius $r = a$ in infinite partially saturated loess subjected to axisymmetric water pressure applied at the cavity surface is considered as shown in Figure 2. Transient variation of loading intensity in the present study is shown in Figure 3. According to the field observation,

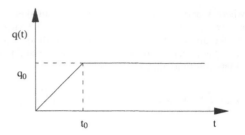

Figure 3.   Transient loadings considered in the analysis for a gradually applied step fluid pressure.

the cavity surface is a permeable surface and the loading is a gradually applied step fluid pressure. The boundary conditions corresponding to this loading case can be expressed as

$$\begin{cases} \bar{\sigma}_r(a,t) = 0 \\ \bar{p}_f(a,t) = \bar{q}(s) \end{cases} \qquad (13)$$

where $\bar{q}(s)$ is the Laplace transform of the applied fluid pressure as shown in Figure 3 and

$$\bar{q}(s) = \frac{q_0(1 - e^{-t_0 s})}{t_0 s^2} \qquad (14)$$

By substituting equations (13) into (14), a solution for the arbitrary coefficients B_1 and B_2 are obtained.

6 NUMERICAL RESULTS

To study the response in time domain, the Laplace domain solutions, presented in the preceding section, are converted into the time domain by using the numerical inversion method proposed by Durbin.

The properties of loess as given in Table 1 are employed in the numerical example.

Saturation effects of soils on time histories of solution at the cavity surface under applied water pressure are investigated first. The saturation effects on radial displacement and the hoop stress histories at $r = a$ are shown in Figures 4 and 5, respectively. The time history of the response is plotted with respect to non-dimensional time t^*, ($t^* = ta/\sqrt{\rho/G}$). It is noted that the maximum displacement and hoop stress at the cavity surface for each saturation case are attained at almost identical time instants. Four degrees of saturation are included to show the influence of saturation. Numerical results show that even a slight decrease from full saturation may have a significant effect on the radial displacement, but have a weak effect on hoop stress. The hoop stress remains compressive at

68

Table 1. The properties of the soil used in computation.

Quantity	Notation	Value
Density of grains (kg · m^{-3})	ρ_s	2700
Density of fluid (kg · m^{-3})	ρ_f	1000
Porosity	n	0.4
Dynamic permeability(m · s^{-1})	k_d	10^{-5}
Lame constant of solid skeleton (MP_a)	G	26
Bulk modulus of solid skeleton (MP_a)	K_b	38
Bulk modulus of solid grains (GP_a)	K_s	36
Bulk modulus of water (GP_a)	K_w	2.2
Transient water pressure (KP_a)	q_0	200

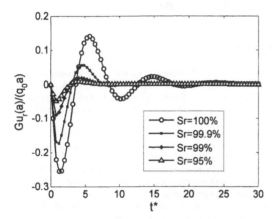

Figure 4. Effect of saturation on radial displacement histories at cavity surface.

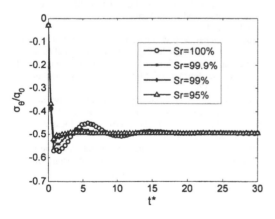

Figure 5. Effect of saturation on hoop stress histories at cavity surface.

all times at $r = a$, and the static conditions are observed at the cavity surface when $t^* > 20$.

Figures 6–9 show the variation of solution with radial distance for the pressurized cavity. Effect of time on the variation of solution with radial distance

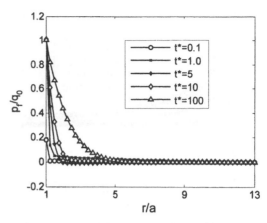

Figure 6. Effect of the time on radial variation of pore pressure.

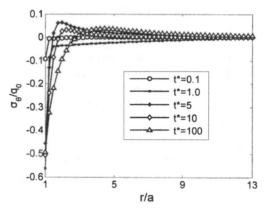

Figure 7. Effect of the time on radial variation of hoop stress.

is shown in Figures 6–8. Initially ($t^* < 1$), the response is localized to the region $a < r < 2a$ resulting in very high stress and pore pressure gradients and the displacement is observed only in the domain $a < r \leqslant 5a$. It shows that the hoop stress and pore pressure decreases rapidly with the radial distance at early time instants. As time progresses, wave fronts expand and the deformations are noted at distant points. The displacement at almost all points remains outward and a minor inward displacement may occur at some points at early time. Similarly, hoop stress and pore pressure are compressive inside the medium. Fig.9 investigates the effect of the degree of saturation on pore pressure at $t^* = 1$. It is shown that the degree of the saturation has a weak effect on pore pressure. The other variation of the pore pressure is same as above discussion and is not described repeatedly here.

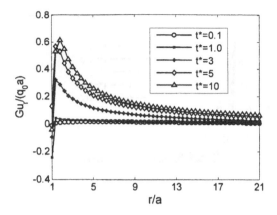

Figure 8. Effect of the time on radial variation of radial displacement.

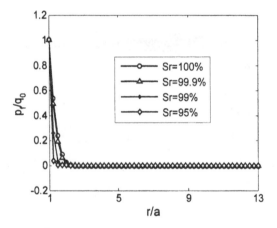

Figure 9. Effect of the saturation on radial variation of pore pressure ($t^* = 1$).

7 CONCLUSIONS

Based on field investigation, a reasonable numeration model to loess tunnel is presented. The analytical solution of stresses, pore pressure and displacement induced by water pressure are derived in Laplace transform domain. The potential function theory and Laplace transform technique is used to solve the problem. Numerical results are obtained by inverse Laplace transformation and the influence of the saturation on dynamic response of the loess tunnel is discussed. It is shown that the response is localized to the region $a < r < 2a$ resulting in very high stress and pore pressure gradients and the displacement is observed only in the domain $a < r \leq 5a$. As time progresses, wave fronts expand and the deformations are noted at distant points and the influence of saturation on displacements is remarkable. The results presented in this paper are meaningful to the risk evaluation of loess tunnel.

REFERENCES

Binke, Wang. 1988. Research on the cavity erosion of soil. Journal of soil and water conservation, 2:9–14

Binke, Wang. 1989. Research on main factors arousing cavity erosion. Journal of soil and water conservation, 3(3):84–90

Biot, M. A. 1962. Mechanics of deformation and acoustic propagation in porous medium, Journal of Applied Physics, 33(4):1482–1498

Durbin, F. 1974. Numerical inversion of Laplace transforms: an efficient improvement to Dubner and Abate's Method, The Computer Journal, 17(14):371–376

Houtian, Hu. 1996. Research on foundation tunnel of loess in HouXi railroad, Subgrade Engineering, 2:22–27

Jingming, Wang. 1996. Theory and application on the loess conformation joint, China Waterpower Press, China

Kumar, R. et al. 1999. Radial displacements of an infinite liquid saturated porous medium with cylindrical cavity, Computers and Mathematics with Applications, 37:117–123

Thiruvenkatachar, V.R. & Viswanathan, K. 1965. Dynamic response of an elastic half space with cylindrical cavity to time-dependent surface tractions over the boundary of the cavity, Journal of Mathematics and Mechanics, 14(4):541–572

Verruijt, A., 1969. Elastic storage of aquifers, Flow through porous media R.J.M de Wiest, ed., Academic, London

Xianmo, Zhu. 1958. The cavity erosion of loess area. J. Yellow River Construction, 22(3):118–123

Xi'an, Li. Research on the cause of formation of the loess tunnel and the hazard of highway engineering, Doctor Paper. Chang'an University, China

Environmental Vibrations – Takemiya (ed.)
© 2005 Taylor & Francis Group, London, ISBN 0 415 39035 4

Analysis on the 3-D elastic-plastic dynamic response of roadbed and natural soil layer under shock load

A.H. Zhou, H.R. Zhong & M.L. Lu
School of Civil and Architectural Engineering, Beijing Jiao tong University, Beijing, China

Y. Yuan
School of Civil & Hydraulic Engineering, Dalian University of Technology, Beijing, China

ABSTRACT: Based on the analysis of the influence of the reflected wave, the calculating area for studying the dynamic responses of roadbed and natural soil layer under shock load is determined. Then three-dimensional elastic-plastic dynamic responses of settlement and stress are simulated numerically. Some valuable conclusions have been drawn: the reflected wave's influence on settlement is very obvious and is relatively low on vertical stress; both the settlement and vertical stress for roadbed and the natural soil layer follow the same trend; the occurrence of maximum settlements lags behind the applied shock load, but the occurrence of the maximum vertical stress keeps synchronized with the applied shock load and both settlement and vertical stress occurred at the center of loading area.

1   INTRODUCTION

It is necessary to know the magnitude and time-space distribution of the settlement and vertical stress for certain roadbed and natural soil layer, under certain shock load for one project. To predict whether the task can be performed successfully or not, the three-dimensional elastic-plastic dynamic responses of settlement and vertical stress are simulated numerically with FEM method.

2   ANALYSIS OF THE INFLUENCE OF REFLECTED WAVE

In the dynamic calculation with FEM method, the calculating area of the model has effect on the results. Reflecting boundary can be used to eliminate the influence of reflected wave, but it wasn't set in most of software. Hence the cutting boundary was used commonly. Firstly it is necessary to analyze the influence of the reflected wave on the dynamic responses of the roadbed and natural soil layer subjected to the shock load and then determine the calculation area.

2.1   *Material parameters and boundary conditions*

Two-dimensional axis-symmetric model is adopted according to the load conditions and geometry of the

natural soil layer. Five cases are considered to investigate the influence of reflected wave. Case 1: Take 250 m in length along the horizontal direction of the soil layer; Case 2: Take 50 m in length along the horizontal direction of the soil layer; Case 3: Take 20 m in length along the horizontal direction of the soil layer; Case 4: Take 5 m in length along the horizontal direction of the soil layer; Case 5: Take 20 m in length along one horizontal direction of the soil layer but free in another horizontal direction. The symmetry axis of the model is restrained symmetrically. The bottom of the model is restrained vertically and the top is free. The models of the former four cases are restrained horizontally. The soil layer takes 20 m in depth for all the five cases. Drucker-Prager constitutive model is adopted to simulate soil layer's elastic-plastic characteristics under shock load. Material parameters of soil layer are as follows: elastic module = 48 MPa, Poisson ratio = 0.30, density = 1800 kg/m³, internal friction angle = 38°, cohesive strength = 48.24 kPa and damping ratio = 5%.

2.2   *Applied shock load*

The shock load, applied on the circular area of the soil layer whose diameter is 2 m, lasts 1.165 seconds. The maximum value of the load is 0.6 Mpa, occurring at 0.382 second. The time-history curve of the applied shock load is shown in Figure 1.

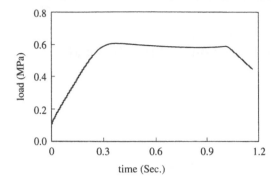

Figure 1. Time-history curve of shock load.

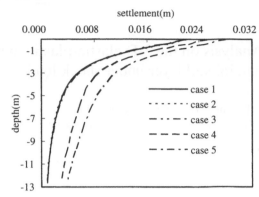

Figure 3. Settlements vs. depth along path two.

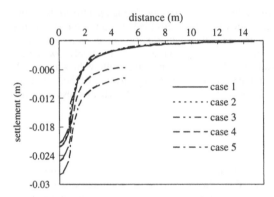

Figure 2. Settlements vs. distance from the loading centre along path one.

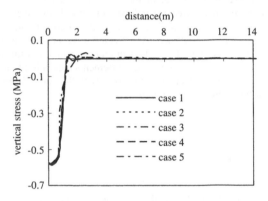

Figure 4. Vertical stresses vs. distance from the loading centre along path one.

2.3 Results

To analyze the influence of the reflected wave on the settlements and vertical stresses, two paths are set. Path one is horizontal, Path two is vertical. Both paths have passed the loading centre.

From Figure 2 to Figure 5, some conclusions can be obtained:

(1) The reflected wave's influence on settlements is very obvious. The settlement value for case 5 is much larger than that for case 4, which indicates settlements are affected by the lateral restraints. The former three cases have the same settlement, which can explain that the dimension defined above is larger enough to ignore the reflected wave's influence.

(2) Maximal settlement for all the five cases occurred at the loading centre. Maximum value for the former three cases are 2.2 cm, 2.5 cm and 2.8 cm for case 4 and case 5, respectively.

(3) The reflected wave's influence on vertical stress is very small. Maximum value for all the five cases is −0.59 MPa which occurring at the loading centre.

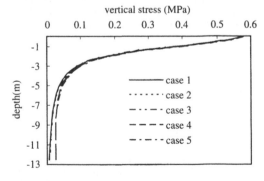

Figure 5. Vertical stresses vs. depth along path 2.

Result of settlements and vertical stresses for case 1, case 2 and case 4 were extracted.
From Figure 6 to Figure 11, it can be seen that:

(1) The reflected wave's influence on the dynamic responses of settlement after unloading is especially remarkable;

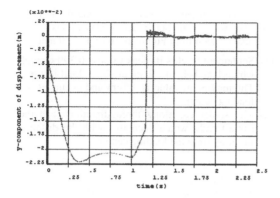

Figure 6.   The time-history curve of settlement at the loading centre for case 1.

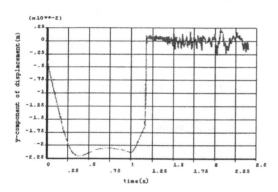

Figure 7.   The time-history curve of settlement at the loading centre for case 2.

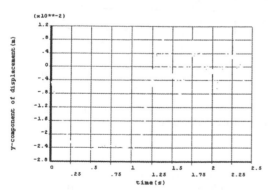

Figure 8.   The time-history curve of settlement at the loading centre for case 4.

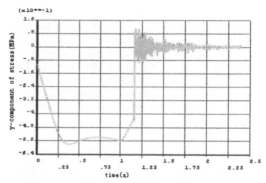

Figure 9.   The time-history curve of vertical stress at the loading centre for case 1.

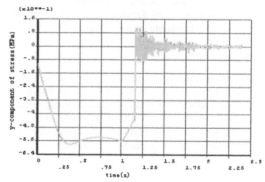

Figure 10.   The time-history curve of vertical stress at the loading centre for case 2.

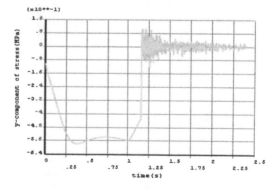

Figure 11.   The time-history curve of vertical stress at the loading centre for case 4.

(2) The occurrence of maximum settlements for all the five cases lags behind the applied shock load;
(3) The occurrence of maximum vertical stresses keep synchronized with the applied shock load.

3   CALCULATION MODEL

Based on the influence analysis of the reflected wave, the model dimension for calculating dynamic responses for roadbed is determined. The dimension is 61.23 m in the traverse direction of the road, 53.5 m in the longitudinal direction and 22.82 m in depth. Symmetry plane of the model is restrained symmetrically. The bottom of the model is restrained vertically and the top is free. The other two horizontal directions are restrained horizontally. Drucker-Prager Constitutive model is adopted as the above. The finite model is shown in Figure 12.

3.1 Material parameters

Road surface is made up of superficial layer, basal layer; and subgrade is composed of upper roadbed, lower roadbed, upper embankment and lower embankment. Under the subgrade is ground. The material used is considered as homogeneous, continuous, and isotropic and the stress and displacement on the interface are continuous. Material parameters of each layer are shown in Table 1 and damping ratio is 5%. Section of road and restraints are shown in Figure 13.

3.2 Results of settlement

To analyze the time distribution rule of settlement, the time-history results of settlements for partial nodes in the applied loading area are extracted. At the same time, three paths are set. Path one is transverse along the road, path two is longitudinal and path three is vertical. All the three paths have passed the loading centre. The settlement results for all the three paths are given when settlement reaches peak values.

It can be seen from Figure 14 to Figure 19 that:

(1) Settlement reaches its peak value at 0.390 second, which lags behind the input shock load in time. The maximum value is 3.9 cm and occurs at the loading center.

(2) The settlement along three directions has the same trends, which decreases with the distance far away from the loading center.

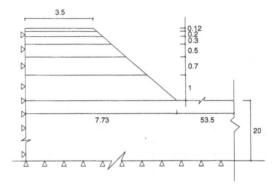

Figure 13.   Section of roadbed.

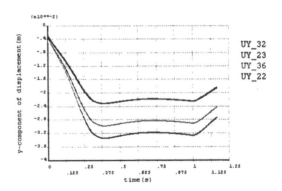

Figure 14.   The time-history curve of settlement for partial nodes in the loading area.

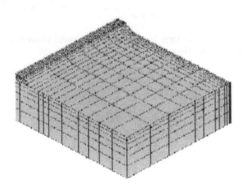

Figure 12.   The finite model.

Table 1.   Material parameter.

Soil layer	Elastic modulus (MPa)	Passion ratio	Density (kg/m³)	Angle of internal friction (degree)	Cohesive strength (kPa)
Superficial layer	200	0.24	1920	38	48.089
Basal layer	160	0.26	1910	36	43.124
Upper roadbed	48.0	0.30	1880	30	34.985
Lower roadbed	38.0	0.32	1860	26	38.441
Upper embankment	33.5	0.34	1840	24	36.806
Lower embankment	25.0	0.36	1820	22	36.360
Ground	18.0	0.41	1680	16	37.312

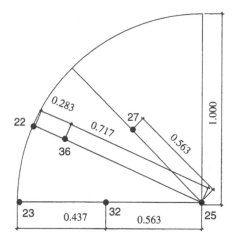

Figure 15.   The location of the partial nodes.

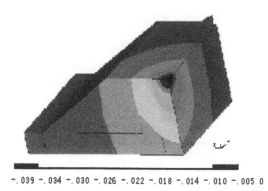

−.039 −.034 −.030 −.025 −.022 −.018 −.014 −.010 −.005 0

Figure 16.   Partial enlarged drawing of settlement at 0.390 second.

3.3 *Results of vertical stress*

Likely, to analyze the space distribution rule of vertical stress, the time-history results of the vertical stresses for partial nodes in the applied loading area are extracted and the results for all the three paths are given when vertical stresses reach their peak values.

It can be seen from Figure 20 to Figure 24:

(1) The occurrence of the maximum vertical stress keeps synchronized with the applied shock load, which is at 0.382 second.
(2) The maximum vertical stress is −0.686 MPa and occurs at the loading center. The distribution of the relatively larger vertical stresses is obviously concentrated on two areas. One is close to the loading center and the other is close to the verge of the loading.
(3) In the horizontal direction, the vertical stress reduces slowly first and then increases gradually, with the distance far away from the loading center. When getting close to the edge of loading, the vertical stress reaches another maximum, and then reduces rapidly. When leaving the loading

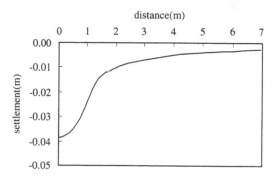

Figure 18.   Settlement vs. distance from the loading center along path two.

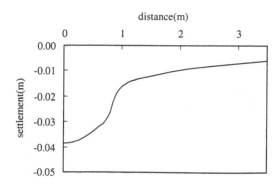

Figure 17.   Settlements vs. distance from the loading center along path one.

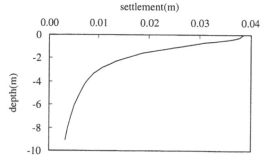

Figure 19.   Settlement vs. distance from the loading center along path three.

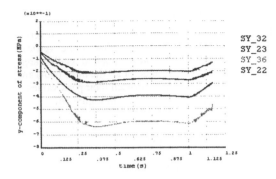

Figure 20. The time-history curve of vertical stress in the loading area of the road.

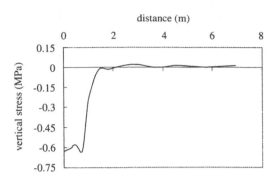

Figure 23. Vertical stress vs. distance from the loading center along path two.

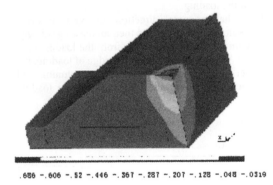

.686 −.606 −.52 −.446 −.367 −.287 −.207 −.128 −.048 −.0319

Figure 21. Partial enlarged drawing of vertical stress at 0.382 second.

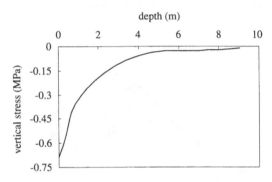

Figure 24. Vertical stress vs. distance from the loading center along path three.

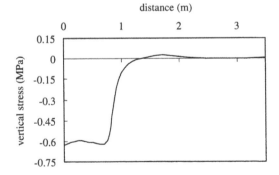

Figure 22. Vertical stress vs. distance from the loading center along path one.

area but not far from the loading area, it appears small tensile stress.

(4) In the vertical direction, the stress decreases gradually as the depth increases.

4 CONCLUSIONS

Through the above analysis, some main conclusions can be drawn:

1. The reflected wave's influence on settlement is very obvious, but on vertical stress is relatively low;
2. The reflected wave's influence on the dynamic response of settlement after unloading is especially remarkable;
3. Both the settlement and vertical stress for roadbed and the natural soil layer follow the same trend. The maximum value occurs at the loading center;
4. The trends of settlements along three directions are identical, which decrease as far away from the loading center;
5. In the loading area, relatively larger vertical stress is distributed obviously in two areas. One is close

to the loading center and the other is close to the verge of the loading area;

6. In the depth direction, the vertical stress reduces gradually as the depth increases;

7. It appears small tensile stress not too far away from the loading area;

8. The roadbed and natural soil layer can bear the load, but the settlement is quite large. So it is better not perform the task in view of safety of passing vehicles.

REFERENCES

Code for design of asphalt pavement of highway (JTJ-014-97):89. China Communications Press: China.

Code for design of highway roadbed (JTJ-013-95):11. Communications Press: China.

Code for design on subsoil and foundation of railway engineering (JTJ 10002.5-99):64. China Railway Publishing House: China.

Guo, D.Z. & Feng, D.C, 2001. Mechanics for lamellar elastic system:79. Harbin Institute of Technology Publication: Harbin, China.

Environmental Vibrations – Takemiya (ed.)
© 2005 Taylor & Francis Group, London, ISBN 0 415 39035 4

Measurement of reduction for ground response – Vibration test of soil-bag and 3D-FEM analysis

K. Yahata
Toyama Office of Building Technology, choufu-shi, Tokyo, Japan

T. Ishibashi, Y. Kiyota & K. Sakuraba
Kajima Technical Research Institute, choufu-shi, Tokyo, Japan

ABSTRACT: In order to clarify the availability of a soil-bag (breastwork) method for reducing response amplitude on the ground, two types of vibration tests were performed. In addition to the tests, simple shear tests to determine material properties were conducted. Through a parametric study on the vibration tests, the effective information for estimating dynamic properties such as elastic modulus, response amplification and complex features in the time history can be obtained. The damping factors are estimated based on the resonance curves and those are the range from 8% to 11%. The reason for the reduced occurrence is discussed. The grain size of the sands considerably affects the elastic modulus and the damping factor. While the amplification distribution along the heights and the amplitude spectrum ratios for the wide frequency range are simple, the dynamic behavior exhibits complexity. The 3D-FEM model is useful to reproduce the test results though through try and error.

1 INTRODUCTION

The purpose of this study is to clarify the reduction property of the soil-bag measure for the ground responses generated by a machine foundation and traffic vibration. As regard to the desirable measure, its rigidity for supporting the applied load is significant as well as the reduction characteristic. Matsuoka is a pioneering researcher in the soil-bag investigations and showed that a soil-bag was well suited to reduce the response amplitude. For instance, the investigations are shown in the papers (Matsuoka et al. 1999), (Matsuoka et al. 2004). In order to estimate the dynamic properties of the soil-bag within the frequency range from 10 Hz to 100 Hz, two types of vibration tests were performed. One was a vertically exciting test using a generator and the other was a shaking table test with vertical excitation. From the viewpoint of an actual ground, two vibration tests correspond to the models of a vibration source and an incident wave, respectively, as shown schematically in Fig. 1. Through the study with the tests, such material information as the damping factor, the elastic modulus and the density were obtained. The simulation analysis with 3D-FEM was carried out. The analytical model was available in reproducing the shaking table test results.

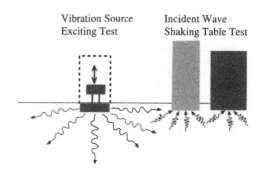

Figure 1. Schematic relation between vibration tests and ground-structure model.

2 OUTLINE OF VIBRATION TEST

There were two types of vibration tests as shown in Fig. 2. As for the parameters in the tests, the model heights and four types of the sand grain size were used.

2.1 Test model of soil-bag

The geometric dimensions of all soil-bag models were almost the same (rectangular prism with the base 43 cm × 39 cm and the height 8 cm). The mass

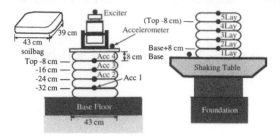

Figure 2. Models and vibration test system.

Table 1. Test models and tests.

| Models | | | | Test | |
Name		Height (cm)	Number of layers	Vibration test	Simple shear test
K2		40	5	Exciter 10–100 Hz	–
K3	K3-3	24	3	Exciter	done
	K3-5	40	5	10–100 Hz	
	K3-6	48	6	Shaking	
	K3-7	56	7	Table	
	K3-8	64	8	10–100 Hz	
K4		40	5	Exciter 10–100 Hz	–
Km		40	5	Exciter 10–100 Hz	done

Table 2. Properties of sands in test model.

Name	γ; Density (g/cm^3)	Ds; Mean grain size (mm)
K2	1.472	1.70
K3	1.447	1.18
K4	1.441	0.84
Km	1.741	–

of each soil-bag was made to be 20 kg. The variable-height models from 24 cm to 64 cm were tested by changing the numbers of the soil-bag as shown in Table 1. The vibration tests were conducted for four grain sizes separately as shown in Table 1. Each grain size and density of four cases is shown in Table 2. The cases of K2, K3 and K4 are ready-made, while Km is an order-made designed to let it have special natures.

A parametric study on the model height was carried out using K3 as shown in Table 1. After each soil-bag was set up on the base floor or on the shaking table as shown in Fig. 2, the compaction was applied uniformly to the soil-bag surface in order to make the soil-bag height approximately 8 cm.

Figure 3. Photo of test model using exciter.

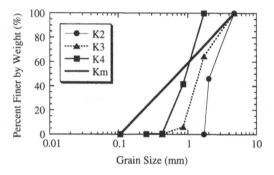

Figure 4. Distribution curve of grain size.

2.2 Vibration test and simple shear test

There were two types of the vibration tests. One was a vertically exciting test using a exciter in the frequency range of 10 Hz to 100 Hz. The force was produced by a exciter set up on the top of the model. A log-sweep wave was used as the forced vibration and the applied force amplitude was within the range of 30 N to 70 N. Fig. 3 is the photo of the model and the exciter. The other was a shaking table test with vertically excitation in the same frequency range. A log-sweep wave was also used and the input amplitude was 60 cm/sec^2. The responses were measured using accelerometers as shown in Fig. 2.

3 TEST RESULTS

3.1 Material properties and dynamic simple shear test (fig.4, fig.5)

The distribution curves of four grain-size sands are K2, K3, K4 and Km as shown in Fig. 4 where the abscissa is on a log scale. As pointed out before, Km was specially designed to linear distribution curve shown in Fig.4. Compared with other models, the density of Km is larger as shown in Table 2. Before

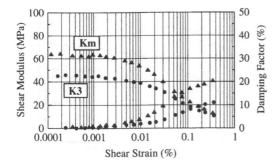

Figure 5. Dynamic characteristics of K3 and Km.

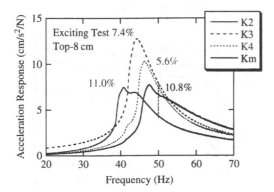

Figure 6. Response curves in exciting test.

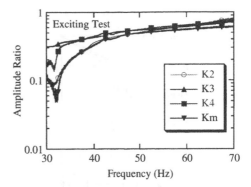

Figure 7. Spectrum ratios of Acc3 to Acc4 in responses.

the vibration tests, dynamic simple shear tests were performed and the overburden pressure was applied to estimate the dynamic characteristics of K3 and Km on the ground. The damping factor and shear modulus obtained from the tests are shown in Fig. 5. The shear modulus of Km is greater than that for K3.

3.2 Exciting tests for various sands (fig.6, fig.7)

Test models were four types of the soil-bags: K2, K3, K4 and Km. Five layer models were tested and the height was 40 cm. The accelerations response curves per unit force at the level of top-8 cm are shown in Fig. 6 and the damping factors were obtained from half power method applied to each peak frequency as shown in Fig. 6. The damping factors of K2 and Km are almost the same at 11% and greater than other two models. Apart from Km, the peak frequency strongly depends upon the grain size. As the grain size becomes fine, the peak frequency increases. However, the grain-size dependency of the damping factor is not clear. The dynamic property of Km might be available fairly for reducing the ground response because the stiffness and damping factor are greater than those of other models respectively. The difference between the

peak frequencies of K3 and Km agrees with the results in Fig. 5. The damping factors obtained from the tests are greatly different from the results for the small strain range below 0.001% in Fig. 5. Since the strain range obtained from acceleration in the tests is approximately 0.001%, the occurrence of the damping observed in the response is related to the dynamic characteristics of the soil-bag. In other words, the damping factor yielded by the vibration tests is different mechanism from that for the simple shear test.

Fig. 7 shows the four spectrum ratios of Acc3; top-16 cm; to Acc4; top-8 cm;. The general trends of K2 and Km are similar and that of K3 is close to the case of K4. The amplitude decreases as the frequency decreases and they are in the ranges from 0.5 to 0.8.

3.3 Excing test for various height of K3

There were four test models and all models are K3. The height parameter of the soil-bag was used in the excitation test as shown in Table 1. The models are denoted hereafter as K3-3 (3 layers), K3-5 (5 layers), K3-6 (6 layers), K3-7 (7 layers) and K3-8 (8 layers).

(a) Maximum acceleration distribution (Fig. 8)
The maximum ratio distributions of each acceleration to that at the height top-8 cm are shown in Fig. 8. The amplitudes decrease with the increasing distances from the top. The reduction trends from top-8 cm to top-16 cm are remarkable in all models, then the amplitudes decreases gradually toward the bottom. However, for the case of K8 the disturbance in the distribution appears in the ranges from top-16 cm to top-32 cm.

(b) Response curves (Fig. 9, Fig. 10)
Four cases of the response curves per unit force at the height of top-8 cm and those at four heights Acc4-Acc1; for K3-5 are shown in Figs. 9 and 10, respectively. Each damping factor also obtained and the peak frequencies are given in Fig. 9. The peak frequencies decrease with the increasing height. The

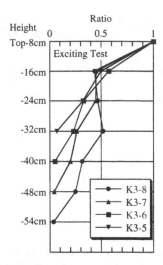

Figure 8. Acceleration ratio distributions (Ratio; Each height to Top-8 m).

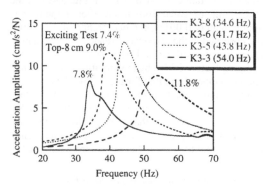

Figure 9. Response curves.

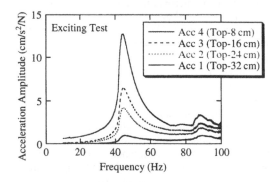

Figure 10. Response curves for K3-5.

damping factors are in the range from 7.4% to 11.8%. The response curves at the different heights for K3-5 are shown in Fig. 10. All these curves show the same trends in general. The amplitudes near the peak

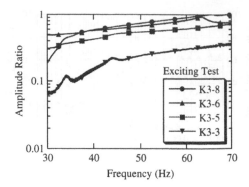

Figure 11. Spectrum ratios of Acc3 to Acc4 in responses.

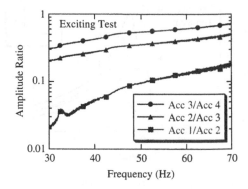

Figure 12. Spectrum Ratios for K3-5.

frequency correspond to those in Fig. 8, and therefore it may be explained that the amplitude distributes along the height as shown in Fig. 8 when the damping factor is 7.4%.

(c) Spectrum ratios (Fig. 11, Fig. 12)
Fig. 11 and Fig. 12 show the spectrum ratios and they are the same models in Fig. 9 and Fig. 10, respectively. The ratios of Fig.11 are the Acc3; top-16 cm; to Acc4; top-8 cm. The amplitude of K3-3 in Fig. 11 considerably differs from other three models because its height is low. The boundary condition of the bottom greatly influences the dynamic behavior for the case of a shallow model. As for Fig. 12, the amplitude of Acc1/Acc2 is small and greatly differs from other two cases. This may be due to the similar reason to the case of K3-3 in Fig. 11. The soil-bag is available method to reduce the response amplitude though its efficiency becomes little in the high frequency range for a high-rise model. Moreover, if the peak frequency occurs, the soil-bag method is available for decreasing the amplitudes because the spectrum ratio is not influenced by the existence of a peak frequency as shown in Figs. 11 and 12.

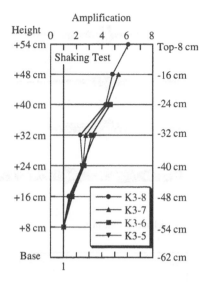

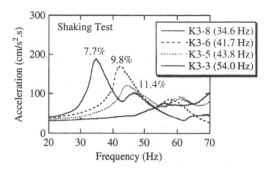

Figure 14. Response curves of variable heights for K3.

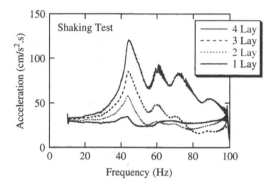

Figure 13. Acceleration ratio distributions (Ratio; Each height to Base + 8 cm).

Figure 15. Response curves of variable heights for K3-5.

3.4 Shaking table tests for various height of K3

The shaking table tests were carried for the model K3 as shown in Table 1 and the height parameter adopted in the tests was the same to the excitation tests described above.

(a) Maximum acceleration distribution (Fig.13)
The maximum ratio distributions of the acceleration are shown in Fig. 13. It is the ratio of each acceleration for the height base + 8 cm. All the distributions have similar trends and the maximum amplification is 6.0 for K3-8 which is the height model. In general, the amplification 6.0 is not a large number.

(b) Response curves (Fig. 14, Fig 15)
The response curves in Figs. 14 and 15 are the similar expressions to it in Figs. 9 and 10. However the amplitudes are derived from Fourier transformation of the acceleration. The peak frequencies are slightly different from those for Fig. 9 because the exciter mass was not there in the tests. For the cases of K3-5, K3-6 and K3-8, two peaks appear in the frequency range of 20 Hz-70 Hz, and therefore the response curves are more complex than those for the case shown in Fig. 9. It is not appropriate to apply half power method for the peak of K3-3 near 56 Hz because of the small amplitude. There is not a large discrepancy between the damping factors of Fig. 9 and Fig. 14 excepting K3. Through the discussion here, the damping factors in the tests may be estimated as in the range from 8% to 11%. Compared with the response curves in Fig. 10, Fig. 15 shows the complex features and the reason was described above.

The amplitudes near the first peak have close relation to those in Fig. 13.

4 ESTIMATION OF SHEAR WAVE VELOCITY (FIG. 16, FIG. 17)

It is significant to estimate the dynamic properties of the soil-bag, especially for elastic modulus, when estimating preliminary the reduction performance using an analytical procedure. PS logging is an effective method to obtain basic material information. The procedure how to perform PS logging for the soil-bag K3 with five layers is shown in Fig. 16. A small needlepoint flower holder was set up on the top of the soil-bag, and then hit its right and left sides independently. If the observation wave fronts due to right and left hitting are the same figures and the inverse signs, the test results could be regarded as being available data to estimate the shear wave velocity (Vs). The shear waves at each layer generated by hitting are shown in Fig. 17. Considering the actual thickness, the shear wave velocities for 4 Lay and 3 Lay were calculated as shown in Fig. 16. They are 79 m/s and 84 m/s, respectively. All the shear wave velocities for 5 layers are estimated based on the result of 4 Lay by

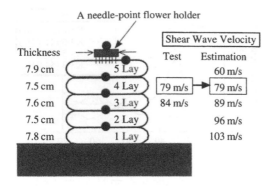

Figure 16. Method of PS logging and shear wave velocity from estimation.

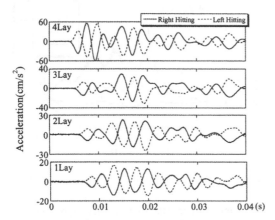

Figure 17. Shear waves in observation by hitting right and left sides.

taking into account the contribution due to the confined pressure. The velocities are noted in Fig. 16 and then each shear modulus is determined. For the case of 4 Lay; Vs = 79 m/s; it is 9.5 Mpa which is roughly 25% of the result in simple shear test as shown in Fig. 5.

5 ANALYTICAL RESULT (FIG. 18, FIG. 19, FIG. 20)

In order to simulate the results of the shaking table test, 3D-FEM was applied to the model K3. The model and material properties in the analysis were as follows: the model was five layers equal to the test model; 64 solid elements for one layer are the same shear modulus obtained from the shear wave velocity; density was 1.447 g/cm³ and Poisson ratio was 0.33. The presence of the bag and the material damping were neglected in the model. The input wave of the analysis shown in Fig. 18 was the wave in the shaking table test.

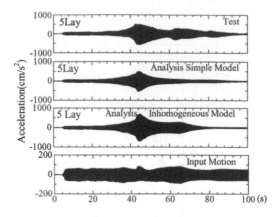

Figure 18. Acceleration time history.

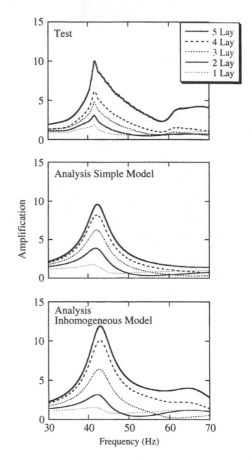

Figure 19. Transfer function.

Firstly, the peak frequency gained in the analysis was near 80 Hz using the material properties described above. Secondly, making the shear modulus small to simulate the peak frequency 43 Hz in the test,

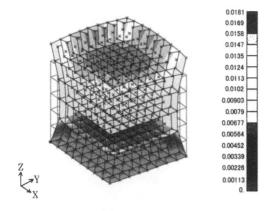

0.0181
0.0169
0.0158
0.0147
0.0135
0.0124
0.0113
0.0102
0.00903
0.0079
0.00677
0.00564
0.00452
0.00339
0.00226
0.00113
0.

Z
Y
X

Figure 20.   Vertical displacement contour Simple model.

the results denoted as simple model are shown in Figs. 18 and 19, which show the acceleration time histories at the 5 layer and the transfer functions of each layer to the input motion in simple model and the test, respectively. Though the peak frequency of simple model expresses the test result, the difference is seen between time history features of simple model and the test and the difference of the transfer functions between them also exits in the frequency range above 50 Hz. In order to understand the response behavior of the soil-bag, the vertical displacement contour near the peak frequency is shown in Fig. 20. The Fig. 20 gives information that the behavior of the outer elements along the sides is different from that for the inner elements. Therefore, the simple model might not represent such characteristic responses as a complex property appreciable in the time history.

For the purpose of reproducing the complexity in the test, the elastic modules of the outer elements and the inner elements were changed into greater for the outer and smaller for the inner than those of simple model. The results of inhomogeneous model are shown in Figs. 18 and 19. Its time history represents approximately the complex feature in the test and their agreement in the transfer function enhances for

the high frequency range. It is considered that the occurrence of the damping factor in the soil-bag during vibration has a close relation to the characteristic dynamic behavior different from it for the homogeneous solid. On the other hand, changing inversely the elastic modulus of the outer and inner, it was not to reproduce the test result.

6   CONCLUSION

From the quantitative viewpoint, the property of the amplitude reduction in the soil-bag responses for the wide frequency range was discussed based on the parametric test results. The damping factors of the soil-bag are regarded as being in the range from 8% to 11%. The grain size of the sands gives considerable influence on the elastic modulus and the damping factor. The occurrence of the reduction effect generated by the soil-bag during vibration is discussed. Through the vibration tests, the simple shear test and the PS logging test, the available information for estimating its dynamic properties such as the elastic modulus, the response amplification and the complex features in the time history, are clarified. The 3D-FEM model is useful to reproduce the test results approximately.

As the soil-bag response is very sensitive to the boundary condition, it would be necessary to study moreover about the availability for the reducing the response amplitude on an actual ground.

REFERENCES

Matsuoka, H. & Liu, S. 1999. Bearing capacity improvement by wrapping a part of foundation. Journal of Structural Mechanics and Earthquake Engineering. No.617/III-46, March, 235–250.
Matsuoka, H. & Muramatsu, D. 2004. Reduction of environmental ground vibration by soilbags. Journal of Geotechnical Engineering. No.764/III-67, June, 235–246.

Environmental Vibrations – Takemiya (ed.)
© 2005 Taylor & Francis Group, London, ISBN 0 415 39035 4

On the load transfer mechanism of Cement-flyash-gravel pile in soft clay by dynamic testing

S.Y. Liu & P. Ji
Institute of Geotechnical Engineering, Southeast University, Nanjing, China

Y.Q. Zhou, D.M. Ma & N.J. Lin
Department of Highway Construction, Lianyungang, China

M.J. Hu
Business School, Hohai University, Nanjing, China

ABSTRACT: The Cement-flyash-gravel (CFG) piles are widely used in the soft ground improvement in the highway construction of China. In most practical cases, the load transfer mechanism and design calculation theory are conducted empirically, which results in higher cost compared with other columns. This study is focused on the dynamic method in determining the bearing capacity and shaft resistance of CFG. The test site is selected in the Lianyungang city, China, where the soft marine clay is well known. Three CFG piles are specially prepared and complemented by high strain dynamic test, in which the bearing capacities are obtained and unit resistance along the pile are presented according to the stress wave theory. After the dynamic test, the static load tests are performed on the same location. Finally, the CFG pile design parameters are recommended.

1 INTRODUCTION

Lian-Yan Highway with length of 43 km is located in the sea plain of Lianyungang city, north of Jiangsu province. The soft ground soil is extensively distributed and belongs to marine deposit and alluvial marine deposit. These soft soils have high water content, low strength and high compressibility. The Cement-flyash-gravel(CFG) pile are used to improve this soft soil. It is important to quickly determine the capacity of the CFG pile in this large-scale engineering practice. In this paper, the dynamic load testing method is presented.

2 ENGINEERING GEOLOGY CONDITION

The segment K14 + 194 ~ K14 + 210 in Lian-Yan Highway was selected as testing section. The ground is composed of marine deposit.

The surface 2-1 soil layer is the yellow gray clay, mainly called hard shell layer, has the 2.3 m thickness. Below the surface layer are 2-2 soft gray silt layer and 2-3 gray-yellow mucky soil, with thickness of 7.7 m and 2.7 m, respectively. The bottom layer is brown yellow layer of leck. The main geotechnical properties of the layer 2-2 soil are summarized in Table 1.

Table 1. The main geotechnical properties of soft layer 2-2.

Layer	Index	Water content W%	Void ratio e	Liquid index I_L	Compressibility coefficient α_{1-2} MPa^{-1}	Shear modulus E_s MPa	Direct shear c kPa	Direct shear Φ	Consolidated quick direct shear c kPa	Consolidated quick direct shear Φ degree	Vane strength Cu kPa	Vane strength Cu' kPa	St
2-2	Min	44.7	1.14	1.19	0.92	1.15	3	1.0	3	0.0	4.33	1.44	1.6
~	~	~	~	~	~	~	~	~	~	~	~	~	~
	Max	78.3	2.13	2.47	2.73	2.48	10	5.4	27	16.9	12.2	5.39	3.4
	Average	62.3	1.66	1.98	1.70	1.66	7	3.1	11	7.4	8.34	3.32	2.6

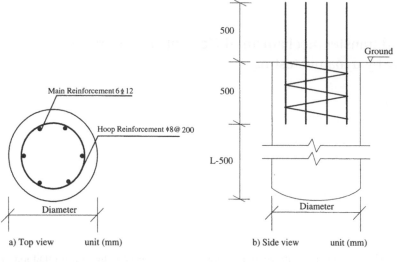

a) Top view   unit (mm)   b) Side view   unit (mm)

Figure 1.   The view of treatment on CFG pile.

Table 2.   CAPWAP result in CFG pile test 1.

Soil sgmnt no.	Dist. below gages m	Depth below grade m	Ru kN	Force in pile at Ru kN	Unit resist. respect to Depth kN	Unit resist. respect to Area kPa	Smith damping factor s/m	Quake mm
				355.0				
1	2.0	1.5	59.4	295.6	59.4	23.65	0.566	1.656
2	4.0	3.5	37.5	258.1	96.9	14.94	0.566	2.336
3	6.0	5.5	47.8	210.3	144.7	19.01	0.566	2.072
4	8.0	7.5	58.0	152.3	202.7	23.09	0.566	1.849
5	10.0	9.5	68.2	84.1	270.9	27.16	0.566	1.679
6	12.0	11.5	54.8	29.2	325.8	21.83	0.566	1.373
Average skin values			54.3			21.61	0.566	1.787
Toe			29.2			232.03	1.188	1.636

Table 3.   CAPWAP result in CFG pile test 2.

Soil sgmnt no.	Dist below gages m	Depth below grade m	Ru kN	Force in pile at Ru kN	Unit resist. respect to Depth kN	Unit resist. respect to Area kPa	Smith damping factor s/m	Quake mm
				259.9				
1	2.0	1.5	24.8	235.1	24.8	9.85	0.163	2.540
2	4.0	3.5	31.5	203.7	56.2	12.52	0.163	2.540
3	6.0	5.5	38.2	165.5	94.4	15.19	0.163	2.540
4	8.0	7.5	44.9	120.7	139.2	17.86	0.163	2.540
5	10.0	9.5	51.6	69.1	190.8	20.53	0.163	2.540
6	12.0	11.5	55.0	14.1	245.8	21.91	0.163	2.540
Average skin values			41.0			16.31	0.163	2.540
Toe			14.1			111.80	1.273	4.416

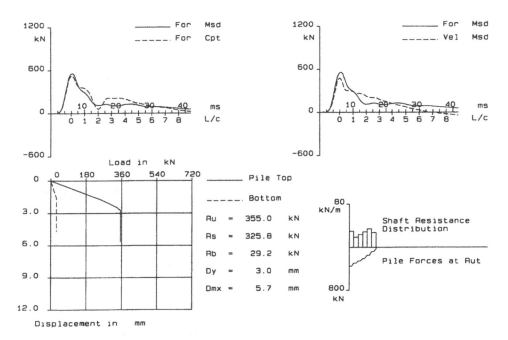

Figure 2.   The dynamic testing and analyzing curve of CFG pile test.

3   SINGLE PILE DYNAMIC LOAD TEST

The pile driving analyzer is adopted in this experiment. The system is composed of the hold computer (contain the signal collecting and enlarging system), strain force sensor, pezro-electric acceleration sensor, electric cable, etc. The hammer drop system is free falling with ram guide.

Because the CFG pile can not support the impaction caused by the falling hammer directly, reinforced concrete pile cap is considered, and it is thought that the cap will not affect the load transfer law of CFG pile. The particular project (see Fig.1) is as follows:

(1) The main reinforcement(6 Φ 12), 0.5 m long, should be installed at the top of the test CFG pile with the hoop reinforcement Φ8@200.
(2) The main reinforcement in the pile cap should be as same as in the top of the test pile, peeping out 0.5 m of the ground. Before manufacture, the broken layer and weak concretes at the top should be wiped off.
(3) The main reinforcement should be under concrete protection layer. Each has the same height.

(4) The pile cap should have the same cross-section with the test CFG pile.

The dynamic load test results are shown in Table 2 and Table 3. The testing and analyzing curve are shown in Fig.2 and Fig.3. The result collection of CFG pile dynamic testing is summarized in Table 4.

4   COMPARISON WITH STATIC LOAD TEST

After 10 days, when the dynamic load test was finished on the test CFG pile 1, the static load test was carried. After 14 days, the static load test was done on pile 2. The result is shown in Table 5 and Fig.4. A comparison of two test methods is illustrated in Fig.5 with the Q ~ S curve.

In the static load test, the CFG pile 1 ultimate capacity is 320 kN, the CFG pile 2 is 240 kN. According to the dynamic load test, analysed by the CAP-WAPC software, the ultimate capacity of the CFG pile 1 is determined as 355 kN, while for the CFG pile 2, it is 256 kN. It is shown that the result of CFG pile capacity in dynamic load test is consistent with that determined by the static load test.

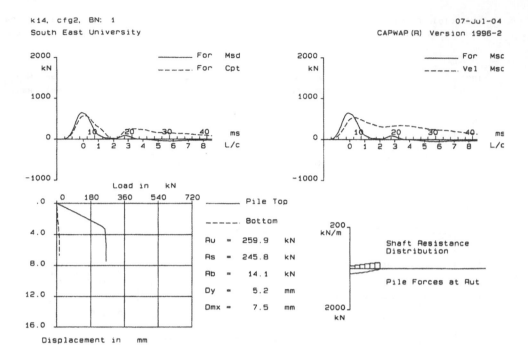

Figure 3. The dynamic testing and analyzing curve of CFG pile test.

Table 4. CFG pile dynamic test result collection.

Test no.	Length (m)	Diameter (mm)	Ultimate capacity (kN)		
			Total	Shaft friction	Tip resistance
CFG Pile 1	12	400	355	326	29
CFG Pile 2	12	400	260	246	14

Table 5. CFG pile static test result collection.

Test no.	Estimate ultimate capacity (kN)	Test ultimate load (kN)	Total settle-ment (mm)	Residue settle-ment (mm)	Ultimate capacity (kN)
CFG Pile 1	400	400	480.6	94.71	320
CFG Pile 2	400	280	112.01	104.03	240

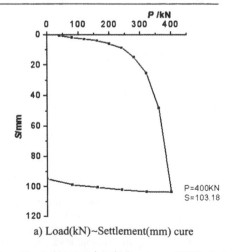

a) Load(kN)~Settlement(mm) cure   b) Settlement(mm)~Time(min.) cure

Figure 4a. The static test and analyzing curve of CFG pile 1.

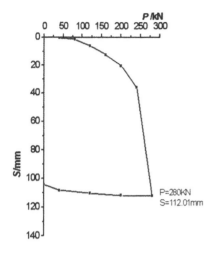 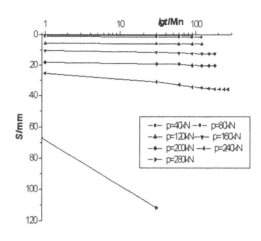

a) Load(kN)~Settlement(mm) cure   b) Settlement(mm)~Time(min.) cure

Figure 4b.   The static test and analyzing curve of CFG pile 2.

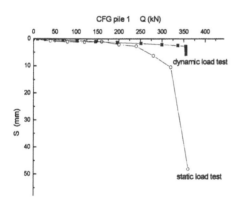 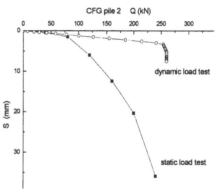

Figure 5.   A comparison between dynamic and static tests.

5   CONCLUSION

The testing result shows that with some treatment on the CFG pile cap, the dynamic load test can be applied to determining the bearing capacity of the CFG pile. The comparison of this method with the static load test indicates that the dynamic test provides a quick and reasonable method for determining the bearing capacity of the CFG pile.

REFERENCE

Xiu YouZai. The new dynamic test technique in pile experiment. Construct Industry Publisher, 2002.

Qian JiaHuan, Yin ZongZe. The soil principle and calculation. The Hydraulic Electric Publisher, 1999.

JTG/T F81-01-2004. Technical Specification of Dynamic Pile Tests for Highway Engineering. Chinese Transportation Ministry.

Environmental Vibrations – Takemiya (ed.)
© *2005 Taylor & Francis Group, London, ISBN 0 415 39035 4*

Nonlinear discrete shear-slice model for dynamic analysis of seismic response of traffic embankment under earthquake

M.T. Luan

State Key Laboratory of Coastal and Offshore Engineering, Dalian University of Technology, Dalian, China;
Institute of Geotechnical Engineering, School of Civil and Hydraulic Engineering, Dalian University of Technology,
Dalian, China; Institute of Rock and Soil Mechanics, The Chinese Academy of Sciences, Wuhan, China

Z. Li

State Key Laboratory of Coastal and Offshore Engineering, Dalian University of Technology, Dalian, China;
Institute of Geotechnical Engineering, School of Civil and Hydraulic Engineering, Dalian University of Technology,
Dalian, China

ABSTRACT: Presented in this paper is a nonlinear discrete shear-slice model based on equivalent lineariza-tion technique for nonlinear seismic response analysis of embankment which is divided into soil layers system with different shear moduli and damping ratios for each layer to simulate their non-inhomogeneous distribution along depth. The differential equations are solved mathematically to obtain the natural vibration characteristics and visco-linear elastic seismic response of the embankment. The resulting linear solutions are incorporated with equivalent linearization technique to consider the nonlinear dynamic behavior of soil. In order to make the mod-uli and damping ratios of all individual layers be compatible with their strain amplitudes, a series of linear iterative computations are performed to achieve a linear system which is overall equivalent to nonlinear system. Finally, numerical analyses are made for an example and the computed results are compared with the solutions obtained by finite element method to illustrate the rationality of the proposed method.

1   INTRODUCTION

The shear slice model was first developed by Mononobe (1936) for analysis of the linear-elastic seismic dynamic response of a homogeneous earth and rock-fill dam. Because of its simplicity, computational effectiveness, and certain reliability in evaluating both vibration char-acteristics and seismic dynamic responses along depth of the embankment, the shear slice model was exten-sively used in engineering practice including study of dynamic performances of dams and embankments or soil strata under earthquake shaking. Then this simpli-fied model has been extended to consider the non-homogeneity of soil modulus along depth of dam, two and three dimensional model with such boundary con-ditions as irregular river canyon and other complex conditions (Gazetas 1981; Dakoulas & Gazetas 1985; Gazetas 1987; Luan et al. 1989). In fact, the shear slice model based on linear-elastic constitutive relationship can not reproduce the nonlinear and non-elastic proper-ties of soils at moderate and large strain level. Therefore some shear-slice-based analysis models were developed to incorporate the nonlinearity and non-elasticity of

soils under dynamic loading. An effective procedure is developed by combining the conventional shear slice model (Gazetas 1981) with the equivalent lineariza-tion technique (Seed et al. 1969), in which a series of trial-and-error iterative computations are performed to make the modulus and damping ratio be compatible with dynamic shear strain amplitude and then an equiv-alent linear system to the nonlinear system is achieved. However, in this method, the average shear strain of whole dam was used as the iteration parameter, which is made to be compatible with the dynamic shear modulus and damping ratio, so that the compatibility of shear modulus and damping ratio with shear strain can not be satisfied at any depth of the dam. The analysis based on such a procedure may not yield an accurate estimation of dynamic response of dam to earthquake loading. Therefore some efforts have been made by incorporat-ing the shear slice model with the finite element method to reproduce nonlinear and non-elastic behavior of soils (Dakoulas & Gazetas 1985; Prevost et al. 1985). How-ever this type of combined models deserve more input data and are time-consuming, more sophisticated than conventional shear slice models, therefore this type of methods cannot play a predominate role in the nonlinear dynamic analysis of embankments.

In order to effectively consider the nonlinear behavior of soils in the shear-slice model, a discrete shear-slice model based on equivalent linearization technique for nonlinear analysis of seismic response of embankments is presented in this paper. In the proposed method, the discrete type of the shear-slice model is developed to consider the layered distribution of shear moduli and damping ratios of soils along depth. The differential equations for governing the vibration of embankment are solved mathematically under the assumption that the soils of embankment display visco-linear elastic behavior and the closed-form solutions for the natural characteristics and seismic response of the embankment are obtained by using mathematical method. The resulting linear solutions are incorporated with equivalent linearization technique which has been widely applied in seismic response of embankment or soil strata. A series of trial-and-error iterative computations are performed in order to make the moduli and damping ratios of all individual layers be compatible with their strain amplitudes. Then a linear system which is overall equivalent to nonlinear embankment system is achieved and the seismic response for this equivalent linearization system can be taken as a good approximation of nonlinear seismic response of embankment. Finally, numerical analysis are conducted for a given example and the computed results are compared with the solutions obtained by finite element method to illustrate the rationality of the proposed method.

2 DISCRETE SHEAR-SLICE MODEL

2.1 Vibration equation and its solutions

As shown in Fig. 1 is the typical profile of an embankment, and the embankment is divided into N soil layers with shear moduli of G_j, damping coefficients of c_j and mass densities of ρ_j ($j = 1, 2, ..., N$) respectively. For simplicity, the shear modulus is assumed to be constant in each soil layer. The mass density and damping coefficient can be different for each layer.

Under horizontal base excitation, only transverse displacement and shear deformation perpendicular to the axis of embankment will be initiated. For a differential soil slice with a thickness of dz in the j-th soil layer, the equation of motion with linear elastic stress-strain relationship can be formulated as

$$\frac{\partial^2 u_j}{\partial t^2} - \frac{G_j}{\rho_j}\left(\frac{\partial^2 u_j}{\partial z_j^2} + \frac{1}{z_j}\frac{\partial u_j}{\partial z_j}\right) = 0 \qquad (1)$$

in which $h_{j-1} \leq z_j \leq h_j$, $j \geq 2$; $h_0 \leq z_j \leq h_1$, $j = 1$, while u_j is the relative displacement of the j-th soil layer.

For a harmonic steady-state base excitation, the general solution of Equation 1 can be assumed to take the form of $u_j = U_j(z_j)e^{i\omega t}$, where ω is the natural vibration frequency of the embankment.

Substitution of Equation 2 into Equation 1 leads to

$$U_j'' + \frac{1}{z_j}U_j' + \left(\frac{\omega}{v_{sj}}\right)^2 U_j = 0 \qquad (2)$$

where $v_{sj} = (G_j/\rho_j)^{0.5}$ is shear wave velocity of j-th soil layer.

The solutions of Equations 2 can be given in the following closed form

$$U_j(z_j) = A_j J_0(k_j Z_j) + B_j Y_0(k_j Z_j) \qquad (3)$$

in which $k_j = \omega H_j/v_{sj}$, $Z_j = z_j/H_j$ and $H_j = h_j - h_{j-1}$ are non-dimensional frequency, non-dimensional depth and thickness of each soil layer respectively. $J_0(\ )$ and $Y_0(\ )$ are zero-order Bessel's functions of first and second kind respectively, while A_j and B_j are integration constants to be defined by the boundary conditions.

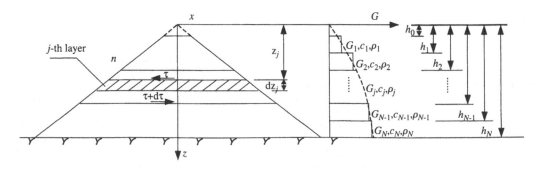

Figure 1. The discrete shear-slice model for dynamic response analysis of embankments.

2.2 Boundary conditions

For embankment profile shown in Fig. 1, the continuity conditions of displacement and shear stress between two adjacent soil layers yield the following conjunction conditions

$$U_j\big|_{z=h_j} = U_{j+1}\big|_{z=h_j} \qquad (4a)$$

$$G_j \frac{\partial U_j}{\partial z_j}\bigg|_{z=h_j} = G_{j+1} \frac{\partial U_{j+1}}{\partial z_{j+1}}\bigg|_{z=h_j} \qquad (4b)$$

The shear stress condition at the crest and displacement condition at the base of embankment can be formulated as

$$\frac{\partial U_1}{\partial z_1}\bigg|_{z=h_0} = 0 \qquad (5a)$$

$$U_N\big|_{z=h_N} = 0 \qquad (5b)$$

2.3 Natural vibration characteristics

A set of linear homogeneous algebraic equations with respect to $2N$ integration constants A_j and B_j can be obtained by substituting Equation 3 into Equation 4. Thereafter the successive relationship of A_j and B_j between adjacent soil layers can be given as following

$$\begin{Bmatrix} A_{j+1} \\ B_{j+1} \end{Bmatrix} = T^{j+1,j} \begin{Bmatrix} A_j \\ B_j \end{Bmatrix} = \begin{bmatrix} t_{11}^{j+1,j} & t_{12}^{j+1,j} \\ t_{21}^{j+1,j} & t_{22}^{j+1,j} \end{bmatrix} \begin{Bmatrix} A_j \\ B_j \end{Bmatrix} \qquad (6)$$

where $T^{j+1,j}$ is transfer matrix between the integration constants of $(j + 1)$-th and j-th soil layers which is given as below

$$T^{j+1,j} = \begin{bmatrix} t_{11}^{j+1,j} & t_{12}^{j+1,j} \\ t_{21}^{j+1,j} & t_{22}^{j+1,j} \end{bmatrix} \qquad (7)$$

Then the relation of integration constants between 1st soil layer and the N-th soil layer can be obtained

$$\begin{Bmatrix} A_N \\ B_N \end{Bmatrix} = T^{N,N-1} \cdots T^{2,1} \begin{Bmatrix} A_1 \\ B_1 \end{Bmatrix} = \begin{bmatrix} t_{11}^{N,1} & t_{12}^{N,1} \\ t_{21}^{N,1} & t_{22}^{N,1} \end{bmatrix} \begin{Bmatrix} A_1 \\ B_1 \end{Bmatrix} \qquad (8)$$

Together with boundary condition given by Equation 5, the resulting interrelations among the constants are written as

$$B_1 = -\frac{J_0'(k_1\lambda_1)}{Y_0'(k_1\lambda_1)} A_1 = \Delta_1 A_1 \qquad (9a)$$

$$B_N = -\frac{J_0(k_N\lambda_N)}{Y_0(k_N\lambda_N)} A_N = \Delta_N A_N \qquad (9b)$$

in which $\lambda_j = h_{j-1}/H_j$. The constants of all other layers can be represented by the constants of the first layer by substituting Equation 9 into Equation 6

$$\begin{Bmatrix} A_j \\ B_j \end{Bmatrix} = T^{j,1} \begin{Bmatrix} J_0'(k_1\lambda_1) \\ -Y_0'(k_1\lambda_1) \end{Bmatrix} \frac{A_1}{J_0'(k_1\lambda_1)} \qquad (10)$$

At the same time, the characteristic equation for defining natural frequencies of embankment can be obtained by considering Equation 5

$$f(\omega) = \left(t_{21}^{N,1} + \Delta_1 t_{22}^{N,1}\right) - \left(t_{11}^{N,1} + t_{12}^{N,1}\right)\Delta_N = 0 \qquad (11)$$

This equation has infinite roots through which the natural frequencies of embankment, ω_1, ω_2, ..., ω_n can be calculated. Then, the computed frequencies are substituted into Equation 10 to get all integration constants and further into Equation 3 to determine the mode shapes of embankment.

2.4 Seismic response analysis

Under the excitation of earthquake-induced ground motion at base of embankment denoted by $\ddot{u}_g(t)$, the differential equation of dynamic equilibrium of embankment can be written as

$$\frac{\partial^2 u_j}{\partial t^2} - \frac{G_j}{\rho_j}\left(\frac{\partial^2 u_j}{\partial z_j^2} + \frac{1}{z_j}\frac{\partial u_j}{\partial z_j}\right) = -\ddot{u}_g(t) \qquad (12)$$

The relative displacement $u_j(z_j; t)$ of embankment can be obtained by utilizing the mode superposition method and Duhamel's integration

$$u_j(z_j, t) = \sum_{n=1}^{\infty} P_n U_{jn}(z_j) D_n(t) \qquad (13)$$

in which P_n is defined as participation coefficient of the n-th order mode

$$P_n = \frac{\sum_{j=1}^{n} \rho_j \int_{\lambda_j}^{\lambda+\lambda_j} Z_j U_{jn}(Z_j) \mathrm{d} Z_j}{\sum_{j=1}^{n} \rho_j \int_{\lambda_j}^{\lambda+\lambda_j} Z_j U_{jn}^2(Z_j) \mathrm{d} Z_j} \quad (14)$$

$D_n(t)$ is the seismic response of a single degree of freedom system with frequency ω_n and damping ratio ξ_n

$$D_n(t) = -\frac{1}{\omega'_n} \int_0^t \ddot{u}_g(\tau) e^{-\xi_n \omega_n(t-\tau)} \sin \omega'_n(t-\tau) \mathrm{d}\tau \quad (15)$$

where $\omega'_n = \omega_n(1 - \xi_n^2)^{0.5}$ is the frequency of the system with damping ratio of ξ_n.

The absolute acceleration response at the deeoth z_j of embankment is given by

$$\ddot{u}_{aj}(z_j,t) = \ddot{u}_j(z_j,t) + \ddot{u}_g(t) = \sum_{n=1}^{\infty} P_n U_{jn}(z_j) D_{an}(t) \quad (16)$$

where

$$D_{an}(t) = \frac{1-2\xi_n^2}{\sqrt{1-\xi_n^2}} \omega_n \int_0^t \ddot{u}_g(\tau) e^{-\xi_n \omega_n(t-\tau)} \sin \omega'_n(t-\tau) \mathrm{d}\tau$$
$$+ 2\xi_n \omega_n \int_0^t \ddot{u}_g(\tau) e^{-\xi_n \omega_n(t-\tau)} \cos \omega'_n(t-\tau) \mathrm{d}\tau \quad (17)$$

The dynamic shear strain response at arbitrary depth of embankment is

$$\gamma_j(z_j,t) = \frac{\partial u_j(z_j,t)}{\partial z_j} = \sum_{n=1}^{\infty} \psi_{jn}(z_{jn}) D_n(t) \quad (18)$$

where $\psi_{jn}(z_j) = P_n \partial U_{jn}(z_j)/\partial z_j$ is the mode shape participation coefficient of the shear strain. Furthermore, other movement and mechanical parameters such as shear stress of embankment can be obtained by using their definition.

3 NONLINEAR SEISMIC RESPONSE

Under earthquake-induced transient and cyclic loading, soils behave nonlinearly and non-elastically. Therefore the nonlinear behavior of soils should be duly considered in the dynamic analysis of seismic response of embankment. However, full nonlinear analysis is sometime difficult to implement. In practice, the equivalent linearization technique proposed by Seed et al. (1969) are widely employed to consider soil nonlinear characteristics in which a series of trial-and-error iterative computations are performed to achieve a linear system which is overall equivalent to nonlinear system. In order to simulate the nonlinear seismic dynamic characteristics of embankments, the discrete shear-slice model (DSSM) proposed above is combined with the equivalent linearization technique to conduct the nonlinear seismic response analysis of embankment under a given earthquake shaking, then the seismic response for this equivalent linearization system can be taken as a good approximation of nonlinear seismic response of embankment.

In the conventional shear slice model, the average shear modulus on a horizontal plane from the crest to the base of the embankment is usually assumed to be exponential distribution of the depth as following

$$G(z) = G_0 \left(\frac{z}{H}\right)^m \quad (19)$$

in which G_0 is the average shear modulus at the base of the embankment and m is the empirical exponent, while the damping ratio is constant along depth of embankment. This type of non-homogeneous shear slice model was combined with the equivalent linearization technique to manifest nonlinear soil characteristics of embankments under earthquake shaking. In this method, in order to keep the simplicity of analytical solutions and the continuous distribution form of shear modulus, an overall-equivalent shear strain of the whole embankment is adopted to be representative of characteristic deformation parameter and used to define compatible dynamic shear modulus and damping ratio based on the given empirical curves of interrelating shear moduli and damping ratios and equivalent shear strain which are achieved experimentally. Then these equivalent visco-elastic dynamic parameters are used to conduct trial and error linear iteration. It is obvious that in the method, the full compatibility between the shear moduli and damping ratios with their shear strains at any depth of embankments cannot be accomplished. Therefore the solution may not be completely representative of nonlinear performance of seismic response of embankments. In order to overcome this shortcoming, the nonlinear shear-slice model (NSSM) is adopted to make the dynamic parameters used in the analysis be compatible with shear strain at any depth of embankment. In this method, the nonlinear behavior of embankment soils is fully simulated in the layer-wise while the simplicity of shear-slice model is reserved. The accuracy of the solution may be improved.

By using the nonlinear curves which relate shear moduli and damping ratios and equivalent shear strain directly obtained by experimental tests or the empirical functions given by curve-fitting of experiment data, the nonlinear seismic dynamic response analysis of embankments under earthquake shaking can be conducted with following steps:

(1) For each soil layer, the shear modulus and damping ratio which are compatible with the assumed initial shear strains can be chosen in the $G \sim \gamma$ and $\xi \sim \gamma$ curves.

(2) With use of these initial values of shear moduli and damping ratios, a linear dynamic response analysis is conducted by the proposed discrete shear-slice model to obtain time-history of seismic response of the embankment. From the time history of dynamic shear strains at all discrete layers, an equivalent cyclic shear strain can be achieved through the empirical equation proposed by Seed et al. (1969). The equivalent shear strain $\gamma_e = 0.65\,\gamma_{max}$ can be obtained empirically from the maximum value of shear strain, γ_{max}.

(3) Based on equivalent shear strain defined above, the corresponding shear modulus and damping ratios can be determined according to $G \sim \gamma$ and $\xi \sim \gamma$ curves. A new turn of linear iteration can be conducted by using the newly-defined dynamic parameters. The iterative computation will be fulfilled until the difference of equivalent shear strains between two adjacent iterations is smaller than a given allowable amount.

(4) The dynamic parameters in the final iteration are used to conduct a linear analysis to calculate the seismic response such as displacements, accelerations and shear strains as well as stresses. The linear solution from the final analysis can be taken as an approximation of nonlinear seismic response of embankment.

4 NUMERICAL ANALYSIS

As an example, the dynamic response of an embankment located on rigid base subjected to earthquake shaking is analyzed by using the proposed method. The height of the embankment is 40 m, the width of the crest is 8 m, the inclination of slope of both sides of the embankment is 1:2, and the mass density of soil is 2 t/m³. In order to illustrate the validity of the proposed method, the vibration characteristics of the embankment is analyzed and compared with the conventional shear-slice model in the range of elasticity. Then the nonlinear seismic dynamic responses of the embankment are computed and compared with the finite element method based on the equivalent linearization technique.

4.1 Vibration characteristics

As mentioned above, the distribution of shear modulus along depth of embankment usually takes the form of

Equation 19 in conventional shear-slice model. Here, it is chosen that the maximum shear velocity at the base of embankment to be 200 m/s, which corresponds to a maximum shear modulus of 80 MPa, and the exponent m to be 0.5. The embankment is divided into 5 soil layers along depth. Together with other data of embankment given above, the natural vibration characteristics of the embankment are computed using both the conventional shear model and the proposed method. Given in Table 1 are the natural frequencies of the first five modes, and the vibration shapes of the first three modes calculated by both methods are shown in Figure 2. It can be observed that for both natural frequencies and mode shapes, the proposed method gives very close solutions with conventional shear-slice model. For the dam with a continuous distribution of shear modulus and constant damping ratio along depth, based on discrete layered system with a certain number of layers, the discrete shear-slice model can give a precise prediction of natural vibration characteristics as the conventional shear slice model. However the proposed model can be adapted to more complex conditions compared with the conventional model.

4.2 Seismic response

In the seismic response analysis, the acceleration record of the Taft earthquake with the duration of 16 s is used

Table 1. The first five natural frequencies of the embankment (rad/s).

Mode	1	2	3	4	5
Conventional $G = G_o(z/H)^{0.5}$	10.9	22.6	34.4	46.1	57.9
DSSM with $N = 5$	10.4	21.9	33.8	45.8	58.0

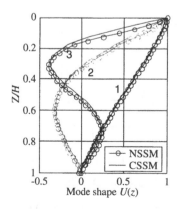

Figure 2. The first three mode shapes of the embankment.

as the input ground motion. The maximum acceleration is adjusted to be 0.2 g.

Before the dynamic analysis, the initial static stresses of the embankment are evaluated by the static finite element analysis. For simplicity, the linear-elastic stress-strain model is used, and the deformation modulus and Poison's ratio are 100 MPa and 0.4 respectively. The coefficient of lateral earth pressure at rest, K_0, equals to 0.45. The mean normal stress of the finite elements are calculated using equation $\sigma'_0 = [\sigma_x + (1 + K_0) \sigma_y]/3$. In the previous study (e.g. Gazetas 1981), the maximum shear modulus along depth of embankment usually has the form of Equation 19. While in the finite element analysis, the shear modulus $G_s = F(e, \text{OCR}) (\sigma'_0)^r$, where e and OCR are the void ratio and apparent over-consolidation ratio of soil element. It is given that $m = r + d$ by the study of Gazetas (1987) where d is a parameter which is dependent primarily on the size and stiffness of the core and the geometry of the embankment. In this example, d is about 0.1, r equals to 0.5 and therefore m is 0.6. Then the maximum shear modulus of the j-th soil layer can be evaluated with $G_{\text{max},j} = G_0(z_j/H)^{0.6}$.

In the dynamic response analysis, the nonlinear soil behavior under dynamic loads is described by the hyperbolic-type model proposed by Hardin and Drnevich (1972). The maximum initial shear modulus is evaluated by the empirical formulae of $G_{\text{max}} = 220(k_2)_{\text{max}}(\sigma'_0)^r$. In the computations, the maximum shear modulus coefficient $(k_2)_{\text{max}} = 100$, maximum damping ratio $\xi_{\text{max}} = 30\%$ and reference shear strain $\gamma_r = 0.0005$. The average shear modulus G_0 is calculated by averaging the maximum shear moduli of elements at the base of the embankment.

The maximum dynamic responses, relative displacements, absolute accelerations, dynamic shear strain and dynamic shear stresses along depth of the embankment computed by the proposed nonlinear shear-slice model and FEM are shown in Fig.3. It is indicated that except for shear stresses, all dynamic response obtained by the nonlinear shear-slice model is very close to the solution of FEM. The time histories and corresponding response spectra of absolute acceleration at the crest of dam obtained by both methods are respectively displayed in Fig.4 and Fig.5. It can be found that the absolute acceleration computed from NDSS agrees well in both peak value and phase with those computed from dynamic FEM.

Therefore, it can be concluded that the proposed nonlinear shear-slice model (NSSM) can rationally predict nonlinear seismic response of dam with a certain of accuracy. However the computational efforts will be considerably reduced compared with the nonlinear FEM.

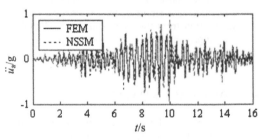

Figure 4. Time histories of absolute acceleration at the crest of the embankment from NSSM and FEM analysis.

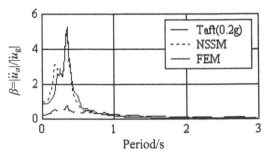

Figure 5. Amplitude coefficients of absolute acceleration at the crest of the embankment from NSSM and FEM.

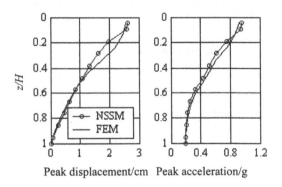

Peak displacement/cm   Peak acceleration/g

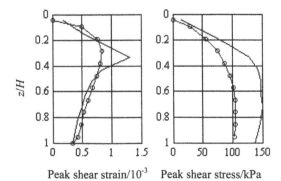

Peak shear strain/10^{-3} Peak shear stress/kPa

Figure 3. The peak response along depth of the embankment from NSSM and FEM analysis.

5 CONCLUDING REMARKS

An effective nonlinear procedure of seismic response of embankment is proposed in this paper by combining the discrete shear-slice model and equivalent linearization technique. The effectiveness and rationality are illustrated through an example analysis. The proposed method can not only well predict the natural vibration characteristics of the embankment with complex distribution profile of shear modulus and damping ratios along depth, but also give with a certain accuracy estimation of the acceleration and shear strains of embankment subjected to earthquake shaking. Compared with the conventional shear-slice model, the nonlinear seismic response of embankment can be assessed. Compared with the nonlinear dynamic FEM, the computational efforts can be reduced significantly. Therefore the proposed method can be used as an effective and yet reasonable tool for evaluating the dynamic response of embankment especially in the primary stage of design of the embankment.

REFERENCES

Dakoulas, P & Gazetas, G. 1985. A class of inhomogeneous shear models for seismic response of dams and embankments. *Soil Dynamics and Earthquake Engineering* 4(4): 166–182.

Dakoulas, P & Gazetas, G. 1985. Nonlinear response of embankment dams. *Proceedings of 2nd International Conference on Soil Dynamics and Earthquake Engineering, Springer-Verlag* 6/7, 5/29–44.

Gazetas, G. 1981. A new dynamic model for earth dams evaluated through case histories. *Soils and Foundations* 21(1): 67–78.

Gazetas, G. 1987. Seismic response of earth dams: some recent developments. *Soil Dynamics and Earthquake Engineering* 6(1): 2–47.

Seed H B et al. 1969. Analysis of shefield dam failure. *Journal of Soil Mechanics and Foundations Division, ASCE,* 95(SM6): 1435–1490.

Hardin, B O & Drnevich, V P. 1972. Shear modulus and damping in soils: measurement and parameter effects. *Journal of Soil Mechanics and Foundations Division, ASCE,* 98(SM6): 603–624.

Hardin, B O & Drnevich, V P. 1972. Shear modulus and damping in soils: design equations and curves. *Journal of Soil Mechanics and Foundations Division, ASCE,* 98(SM7): 667–692.

Maotian Luan et al. 1989. A simplified approach to evaluation of vibration of non-homogeneous embankments. *Journal of Dalian University of Technology* 29(4): 479–488.

Prevost J H et al. 1985. Nonlinear hysteretic dynamic response of soil systems. *Journal of Engineering Mechanics, ASCE,* 111(7): 882–897.

*Dynamic interaction of ground
and traffic structures*

Environmental Vibrations – Takemiya (ed.)
© *2005 Taylor & Francis Group, London, ISBN 0 415 39035 4*

The steady-state response of a periodically inhomogeneous model of a railway track to a moving load

A.V. Metrikine

Dept. of Civil Engineering and Geosciences, Delft University of Technology, Netherlands

ABSTRACT: This paper presents an efficient calculation method for prediction of the steady-state response of a three-dimensional model of a railway track to a uniformly moving axle load of a train. The track model includes rails, pads, sleepers and a layered ground. Due to the sleepers, the model is periodically inhomogeneous along the track. The method is based on the so-called periodicity condition, which requires any two points of a rail, which are separated by the sleeper distance, vibrate with a well-defined phase lag that depends on the sleeper distance, load velocity and load frequency. Employing the periodicity condition, a closed-form expression is derived for a dynamic stiffness of the ground under a sleeper. This dynamic stiffness relates the forces induced by all sleepers on the ground to the displacement of one sleeper. It is the same for all sleepers. To demonstrate the calculation method, a short analysis is carried out of the dynamic stiffness and the displacements of the rails for a sub-critical velocity of the load.

1 INTRODUCTION

The conventional railway lines can be classified as periodically inhomogeneous elastic structures. The periodicity takes place along the rails and is introduced by the sleepers. A number of methods have been proposed to study the dynamic response of such structures to moving loads. Mead (1971) and Jezequel (1980) based their approach on the Fourier series techniques. Bogach, Krzyzinski and Popp (1993) applied the Flouquet theorem. Vesnitskiy and Metrikine (1993, 1996) and Belotserkovskiy (1996) employed a so-called periodicity condition. Independently of the approach, in all these papers the railway line was considered as infinitely long and infinitely many sleepers were accounted for. Kalker (1996) proposed an approximate method, which accounts for a finite number of sleepers.

In the above mentioned papers, one dimensional models of the railway track were considered, which did not account for wave processes in the ground. In the last two decades a number of papers were published, in which periodically inhomogeneous models of the railway track were studied in a three-dimensional formulation. Aurby et al. (1994), Bode et al. (2000), Auersch (2005) and Takemiya (2005) have presented different approaches to the problem. Though different in many aspects, all these approaches do not make use of the fact that the steady-state response of a periodically inhomogeneous railway to a uniformly moving load must exhibit a certain periodicity property. This

property was employed by Metrikine and Popp (1999), Clouteau et al. (2000), Kruse and Popp (2001), Vostroukhov and Metrikine (2003) and Clouteau et al. (2005). In these papers, either Flouquet theorem or the periodicity condition was used, like in the above-mentioned studies of one-dimensional models.

Using the periodicity property seems to be more advantageous an approach, because it automatically results into the steady-state solution of the problem. However, this solution contains an infinite summation with respect to the sleepers. This summation cannot be performed numerically and, therefore, any direct numerical analysis requires an assumption on the number of sleepers to be accounted for. An attempt to circumvent this problem was undertaken by Metrikine and Popp (1999), who proposed a technique, which allows to compute the infinite summation quasi-analytically. This technique was slightly generalized by Vostroukhov and Metrikine (2003). In these papers, however, simplistic models of the ground were employed namely an elastic half-space and a viscoelastic layer fixed at the bottom. In both cases the ground response to a surface load can be found analytically in the frequency-wavenumber domain. This was used for summation of contributions of the sleepers.

This paper is the first step in generalization of the technique proposed by Metrikine and Popp (1999) to the case of a multilayered ground. The main result of this paper is that a general expression is obtained for the dynamic stiffness of multilayered ground in the steady state regime. This dynamic stiffness accounts for contributions of all sleepers and depends on the frequency of vibrations of the rails and on the load velocity (if the load magnitude is constant). Once the dynamic stiffness has been calculated, the displacement of the rails can be computed straightforwardly. Both the dynamic stiffness and the rails displacement are calculated in this paper and the effect of the interface conditions between the ground and the sleepers on these quantities is discussed.

2 MODEL AND GOVERNING EQUATIONS

We consider the steady-state response of a railway track to an axle load of a moving train. The model of the railway track is shown in Figure 1. It accounts for two rails, pads, sleepers and a layered ground. The rails are modeled by identical, infinitely long beams, which are assumed to move only vertically. The sleepers are modeled as identical rigid parallelepipeds, which are mounted to the surface of the ground assuming that the contact area is plane and horizontal. It is also assumed that the sleepers move only vertically. The beams and the sleepers interact through identical spring-dashpot systems, which model the pads. It is assumed that each pad provides a point-like support to one of the beams and that the pads are positioned along the beams periodically. The ground is modeled as a layered viscoelastic continuum, whose behavior is governed by the Kelvin-Voigt material model.

The axle load is represented by two point loads applied to the beams. These loads are assumed to be of the same constant magnitude and move along the beams with a constant speed.

The steady state vibrations of this model are governed by a system of equations, which consists of two equations for the beams, infinitely many identical equations for the sleepers, interface conditions at the surface of the layered ground, equations of motion for the ground and boundary conditions at the bottom of the

ground or at $z \to \infty$ (if the ground is represented by a layered half-space).

Since the model is symmetric with respect to the vertical plane $y = 0$ (this plane contains the centerline of the railway track), the vertical deflections of the beams must be identical. Employing the Euler-Bernoulli theory, the equation governing the motion of one of the beams can be written as

$$\rho^b A \partial_{tt} w^b + IE^b \partial_{xxxx} w^b = P\delta\left(x - Vt\right)/2$$
$$+ \sum_{k=-\infty}^{\infty} \left(c^s \partial_t + k^s\right)\left(w_k^s(t) - w^b(x,t)\right)\delta\left(x - kd\right) \quad (1)$$

where $w^b(x,t)$ is the vertical beam deflection, $w_k^s(t)$ is the vertical deflection of sleeper number k, $\rho^b A$ and E^b I are the mass per unit length and the bending stiffness of the beam, respectively, k^s and c^s are the stiffness and damping coefficient of one pad, respectively, P is the axle load, and $\delta(...)$ is the Dirac delta function.

The vertical motion of sleeper number k is governed by the following equation:

$$\int_{-a}^{a} \int_{-b+kd}^{b+kd} \sigma_{zz}\left(x, y, 0, t\right) dx\, dy = m^s\, \partial_{tt} w_k^s$$
$$+ 2\left(c^s \partial_t + k^s\right)\left(w_k^s(t) - w^b(kd, t)\right), \qquad k = 0, \pm 1, \pm 2 \ldots$$
$$(2)$$

where $2a$ and $2b$ are the dimensions of the contact area between the sleeper and the ground surface (see Figure 1), $\sigma_{zz}(x,y,0,t)$ is the normal traction at the ground surface, and m^s is the mass of one sleeper.

The interface and boundary conditions at the surface of the ground are given as

$$\{x,y\} \in \Omega_k, \ \Omega_k = \left\{|x - kd| \le b, |y| \le a\right\}:$$
$$w\left(x, y, 0, t\right) = w_s^k(t), \ u\left(x, y, 0, t\right) = v\left(x, y, 0, t\right) = 0 \quad (3)$$
$$k = 0, \pm 1, \pm 2, \ldots$$

$$\{x,y\} \notin \Omega_k:$$
$$\sigma_{zz}\left(x, y, 0, t\right) = \tau_{xz}\left(x, y, 0, t\right) = \tau_{yz}\left(x, y, 0, t\right) = 0 \quad (4)$$

where $\tau_{xz}(x,y,0,t)$ and $\tau_{yz}(x,y,0,t)$ are the shear tractions at the ground surface.

The equations of motion for layer j of the ground are

$$\hat{\mu}_j \Delta \underline{u}_j + \left(\hat{\lambda}_j + \hat{\mu}_j\right)\nabla\left(\nabla \underline{u}_j\right) = \rho_j\, \partial_{tt}\underline{u}_j, \ 1 \le j \le J \quad (5)$$

where $\underline{u}_j(x,y,z,t)$ is the vector-displacement, ρ_j is the mass density, $\hat{\lambda}_j = \lambda_j(1 + \gamma_j \partial_t)$ and $\hat{\mu}_j = \mu_j(1 + \gamma_j \partial_t)$ are differential operators that are used instead of the Lamé constants λ and μ to describe the layer material

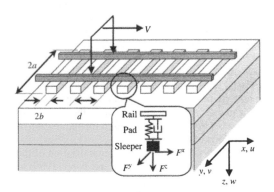

Figure 1. Model and reference system.

according to the Kelvin-Voigt phenomenological model. In these operators, γ_j characterizes material damping.

Equation (5) should be supplemented by interface conditions between the layers and a condition at the bottom of the ground. In what follows, continuity of the stresses and displacements is assumed at every interface between the layers. At the bottom of the viscoelastic ground the displacements are required to vanish independently of whether the last layer is considered to be of finite or infinite depth.

3 IDEA OF SOLUTION METHOD

3.1 Periodicity condition

To find the steady-state solution of the problem, the so-called periodicity condition will be used proposed by Vesnitskii & Metrikine (1993) and later used by a number of authors including Belotserkovskiy (1996), Metrikine & Popp (1999), Suiker et al. (2001), Vostroukhov & Metrikine (2003), Sheng et al. (2005).

The periodicity condition is a general tool, which significantly simplifies analysis of the steady-state response of infinitely long, periodically inhomogeneous systems to uniformly moving loads. For the problem at hand, the periodicity condition reads

$$w^b(x,t) = w^b(x+d,t+d/V)$$
$$w_k^s(t) = w_{k+1}^s(t+d/V) \qquad (6)$$
$$\underline{u}_j(x,y,z,t) = \underline{u}_j(x+d,y,z,t+d/V)$$

where d is the sleeper distance and V is the load velocity. The validity of the periodicity condition can be easily checked by direct substitution into Equations (1)–(5).

The periodicity condition, Equation (6), is physically transparent. It simply requires the steady-state response of a periodically inhomogeneous system to uniformly moving load of a constant magnitude be translated along the system to the distance d after each time period $T = d/V$. For example, the beam shape at $t = d/V$ can be obtained from that at $t = 0$ by shifting the former rightward by the sleeper distance.

The periodicity condition in the form given by Equation (6) is applicable if the moving load has a constant magnitude. Note, however, that it can be straightforwardly generalized to the case of harmonically varying loads of the magnitude $P \exp(i\Omega t)$, see Belotserkovskiy (1996), Suiker et al. (2001), Vostroukhov et al., Sheng et al. (2005). In this case, the periodicity condition for the beams, for example, would read

$$w^b(x,t) = w^b(x+d,t+d/V)\exp(-i\Omega d/V) \qquad (7)$$

While application of the periodicity condition to one- and two-dimensional systems is relatively straightforward, an intermediate step is to be made before this condition can be applied to the three-dimensional problem at hand. The aim of this step is to find the dynamic stiffness of the ground under each sleeper. In other words, the forces should be determined, with which the ground would react against a vertical motion of the sleepers. It is important to realize that in the steady-state regime these forces, as well as any other field variable (stresses, velocities), must satisfy the periodicity condition. This means that the force acting on the sleeper k from the ground at the time moment $t = 0$ must be the same as that acting on the sleeper $k + 1$ at $t = d/V$.

3.2 Decomposition of sleepers-ground interface

To find the dynamic stiffness at the sleepers-ground interface, one can use the Green's function (response to a surface point load) of the layered ground. Convoluting this function over the contact area, one could obtain a system of integral equations relating the stresses at the ground surface to the displacements of the sleepers. Solving such system of integral equations is not an easy task, especially in the case of a low material damping in the ground that corresponds to quasi-singular kernels of the integrals.

To avoid dealing with quasi singular integral equations, a different approximate approach is adopted in this paper. The contact area under each sleeper is decomposed into rectangular sub-domains, and the stresses beneath each sub-domain are assumed constant. The continuity between the displacements of the sleepers and those of the ground surface are required only in the midpoints of each sub-domain. Basically, this approach approximates the tractions beneath the sleepers as piecewise constant functions.

Thus, the contact area under each sleeper is decomposed into $M \times N$ rectangular sub-domains of the size $S = \Delta x \times \Delta y$, where

$$\Delta x = 2b/M, \ \Delta y = 2a/N \qquad (8)$$

The coordinates of the mid-points of these sub-domains read

$$x_{km} = kd - b + (2m-1)\Delta x/2, \ k = 0,\pm 1,\pm 2,...; \ 1 \le m \le M$$
$$y_n = -a + (2n-1)\Delta y/2, \ 1 \le n \le N \qquad (9)$$

Equation (9) shows that each sub-domain is specified by three indices: k,m,n. The first defines the sleeper, under which the sub-domain is located, whereas the second and the third indices define the position of the sub-domain under this sleeper.

It is assumed that within each sub-domain the stresses are uniform (coordinate-independent) and are given as

$$\{x, y\} \in \Omega_{kmn}, \quad \Omega_{kmn} = \{|x - x_{km}| \le \Delta x/2, |y - y_n| \le \Delta y/2\}:$$

$$\sigma_{zz} = F_{kmn}^z(t)/S, \quad \tau_{xz} = F_{kmn}^x(t)/S, \quad \tau_{yz} = F_{kmn}^y(t)/S \tag{10}$$

With this assumption, the stresses at the ground surface can be expressed as

$$\sigma_{zz}(x, y, 0, t) = \frac{1}{S} \sum_{k=-\infty}^{\infty} \sum_{m=1}^{M} \sum_{n=1}^{N} F_{kmn}^z(t) R_{kmn}(x, y)$$

$$\tau_{xz}(x, y, 0, t) = \frac{1}{S} \sum_{k=-\infty}^{\infty} \sum_{m=1}^{M} \sum_{n=1}^{N} F_{kmn}^x(t) R_{kmn}(x, y)$$

$$\tau_{yz}(x, y, 0, t) = \frac{1}{S} \sum_{k=-\infty}^{\infty} \sum_{m=1}^{M} \sum_{n=1}^{N} F_{kmn}^y(t) R_{kmn}(x, y)$$

$$\tag{11}$$

where

$$R_{kmn}(x, y) = H(\Delta x/2 - |x - x_{km}|) H(\Delta y/2 - |y - y_n|) \tag{12}$$

In Equation (12), $H(\ldots)$ is the Heaviside function.

Employing Equations (11) and (12), the double integral in Equation (2), which governs the vertical motion of the sleepers, can be replaced by a double sum to give

$$\sum_{m=1}^{M} \sum_{n=1}^{N} F_{kmn}^z(t) = m^s \partial_{tt} w_k^s$$

$$+ 2(c^s \partial_t + k^s)(w_k^s(t) - w^b(kd, t)), \quad k = 0, \pm 1, \pm 2 \ldots \tag{13}$$

Since the continuity between the displacements of the sleepers and those of the ground surface is required only at the midpoints of the sub-domains, the interface conditions, Equation (3), assume the following form:

$$w(x_{km}, y_n, 0, t) = w_s^k(t),$$

$$u(x_{km}, y_n, 0, t) = v(x_{km}, y_n, 0, t) = 0 \tag{14}$$

$$k = 0, \pm 1, \pm 2, \ldots; \ 1 \le m \le M, \ 1 \le n \le N$$

Now that the periodicity condition has been formulated and the sleepers-ground interface has been decomposed, the problem is to be transformed into the frequency domain.

4 TRANSFORMATION TO FREQUENCY DOMAIN

The integral Fourier transform is applied to transform the problem into the frequency domain. The Fourier image $\tilde{f}(\omega)$ of a function $f(t)$ is defined as

$$\tilde{f}(\omega) = \int_{-\infty}^{\infty} f(t) \exp(i\omega t) \, dt \tag{15}$$

The variables in the frequency domain are distinguished from those in the time domain by overhead tildes.

Transforming Equations (1), (13), (14), (4), (11) and (5), the statement of the problem is obtained in the frequency domain. This includes

• equation of motion of the beam:

$$-\rho^b A \omega^2 \tilde{w}^b + IE^b \partial_{xxxx} \tilde{w}^b = P \exp(i\omega x/V)/(2V)$$

$$+ \sum_{k=-\infty}^{\infty} (k^s - i\omega c^s)(\tilde{w}_k^s(\omega) - \tilde{w}^b(x, \omega)) \delta(x - kd) \tag{16}$$

• equations of motion of the sleepers:

$$\sum_{m=1}^{M} \sum_{n=1}^{N} \tilde{F}_{kmn}^z(\omega) = -m^s \omega^2 \tilde{w}_k^s$$

$$+ 2(k^s - i\omega c^s)(\tilde{w}_k^s(\omega) - \tilde{w}^b(kd, \omega)), \quad k = 0, \pm 1, \pm 2 \ldots \tag{17}$$

• continuity conditions at the sleepers-ground interface:

$$\{x, y\} \in \Omega_k, \quad \Omega_k = \{|x - kd| \le b, |y| \le a\}:$$

$$\tilde{w}(x_{km}, y_n, 0, \omega) = \tilde{w}_s^k(\omega),$$

$$\tilde{u}(x_{km}, y_n, 0, \omega) = \tilde{v}(x_{km}, y_n, 0, \omega) = 0 \tag{18}$$

$$k = 0, \pm 1, \pm 2, \ldots; \ 1 \le m \le M, \ 1 \le n \le N$$

• relationships between the stresses and the forces appalled to the sub-domains of the sleepers-ground interface:

$$[\tilde{\tau}_{xz}, \tilde{\tau}_{yz}, \tilde{\sigma}_{zz}]_{z=0}^T = \frac{1}{S} \sum_{k=-\infty}^{\infty} \sum_{m=1}^{M} \sum_{n=1}^{N} [\tilde{F}_{kmn}^x, \tilde{F}_{kmn}^y, \tilde{F}_{kmn}^z]^T R_{kmn}(x, y) \tag{19}$$

• boundary conditions at the free surface of the ground:

$$\{x, y\} \notin \Omega_k:$$

$$\tilde{\sigma}_{zz}(x, y, 0, \omega) = \tilde{\tau}_{xz}(x, y, 0, \omega) = \tilde{\tau}_{yz}(x, y, 0, \omega) = 0 \tag{20}$$

• equations of motion of the layered ground:

$$\mu_j \Delta \underline{u}_j + (\lambda_j + \mu_j) \nabla (\nabla \cdot \underline{u}_j) = \frac{-i\omega^2 \rho_j}{1 - i\omega \gamma_j} \underline{u}_j, \quad 1 \le j \le J \tag{21}$$

This set of equations should be completed by the periodicity condition, Equation (6). In the frequency domain this condition reads

$$\tilde{w}^b(x,\omega) = \tilde{w}^b(x+d,\omega)\exp(-i\omega\,d/V)$$
$$\tilde{w}^i_k(\omega) = \tilde{w}^i_{k+1}(\omega)\exp(-i\omega\,d/V) \tag{22}$$
$$\underline{u}_j(x,y,z,\omega) = \underline{u}_j(x+d,y,z,\omega)\exp(-i\omega\,d/V)$$

As mentioned before, the forces $\tilde{F}^x_{kmn}$, $\tilde{F}^y_{kmn}$ and $\tilde{F}^z_{kmn}$ have to satisfy the periodicity condition too (since they are proportional to the stresses at the ground surface). Therefore, the force-vector

$$\tilde{\underline{F}}_{kmn} = \left[\tilde{F}^x_{kmn}, \tilde{F}^y_{kmn}, \tilde{F}^z_{kmn}\right]^T \tag{23}$$

has to satisfy the following equation:

$$\tilde{\underline{F}}_{kmn}(\omega) = \tilde{\underline{F}}_{(k+1)mn}(\omega)\exp(-i\omega\,d/V) \tag{24}$$

Thus, as follows from Equations (22) and (24), the vertical displacement of the sleepers and the forces $\tilde{\underline{F}}_{kmn}$ (as well as all other field variables) can be sought for in the form

$$\tilde{w}^i_k(\omega) = C(\omega)\exp(i\omega k\,d/V)$$
$$\tilde{\underline{F}}_{kmn}(\omega) = \underline{C}_{mn}(\omega)\exp(i\omega k\,d/V) \tag{25}$$

where

$$\underline{C}_{mn}(\omega) = \left[C^x_{mn}(\omega), C^y_{mn}(\omega), C^z_{mn}(\omega)\right]^T \tag{26}$$

As shown in the next section, Equation (25) is of great help for deriving the dynamic stiffness of the ground under the sleepers.

5 DYNAMIC STIFFNESS OF THE GROUND

The dynamic stiffness is conventionally defined as the ratio of the amplitude of the harmonic force to the amplitude of the steady-state displacement caused by this force. This definition does not specify whether the displacement is considered at the same point where the load is applied. Nor the spatial distribution of the force is specified, which is necessary for the dynamic analysis of distributed systems.

Thus, a more specific definition is needed of the dynamic stiffness we are looking for. For the problem at hand it is customary to define the dynamic stiffness of the ground under one of the sleepers as the ratio of

the sum of *all forces* which the load exerts on the ground (through the sleepers) to the displacement of this particular sleeper. This definition, though quite specific, is very convenient since the so-defined dynamic stiffness is the same for all sleepers, as will be shown in this section.

To derive the dynamic stiffness of the ground the Helmholtz decomposition of the displacement vector $\underline{u}_j(x,y,z,t)$ into a scalar potential $\varphi_j(x,y,z,t)$ and a vector potential $\underline{\psi}_j(x,y,z,t)$ is used, accompanied by the conventional constraint for the vector potential (Achenbach 1973):

$$\underline{u}_j = \nabla\varphi_j + \nabla\times\underline{\psi}_j, \quad \nabla\cdot\underline{\psi}_j = 0, \quad 1\le j\le J \tag{27}$$

In terms of these potentials, and after transformation into the frequency domain, the equations of motion of the ground read

$$\tilde{c}^2_{jL}\Delta\tilde{\varphi}_j = -\omega^2\tilde{\varphi}_j, \quad \tilde{c}^2_{jT}\Delta\underline{\tilde{\psi}}_j = -\omega^2\underline{\tilde{\psi}}_j, \quad \nabla\cdot\underline{\tilde{\psi}}_j = 0 \tag{28}$$

where

$$\tilde{c}^2_{jL} = c^2_{jL}(1 - i\omega\gamma_j), \quad \tilde{c}^2_{jT} = c^2_{jT}(1 - i\omega\gamma_j) \tag{29}$$

and $c_{jL} = ((\lambda_j + 2\mu_j)/\rho_j)^{1/2}$ and $c_{jT} = (\mu_j/\rho_j)^{1/2}$ are the propagation velocities of the dilatational waves (*P*-waves) and shear waves (*S*-waves).

5.1 Ground response in the frequency domain

The next step of analysis is transformation of the equations governing the ground motion into the wavenumber domain. This is done employing the integral Fourier transform, which is defined as

$$\tilde{f}^{(k_x,k_y)}(k_x,k_y,z,\omega) = \int_{-\infty}^{\infty}\int_{-\infty}^{\infty}\tilde{f}(x,y,z,\omega)\exp(-ik_x x - ik_y y)\,dx\,dy \tag{30}$$

Application of this transform to Equations (28) results in

$$\left(\partial_{zz} + \omega^2/\tilde{c}^2_L - k^2_x - k^2_y\right)\tilde{\varphi}^{(k_x,k_y)}_j = 0$$
$$\left(\partial_{zz} + \omega^2/\tilde{c}^2_T - k^2_x - k^2_y\right)\underline{\tilde{\psi}}^{(k_x,k_y)}_j = 0 \tag{31}$$

$$ik_x\tilde{\psi}^{(k_x,k_y)}_{jx} + ik_y\tilde{\psi}^{(k_x,k_y)}_{jy} + \partial_z\tilde{\psi}^{(k_x,k_y)}_{jz} = 0 \tag{32}$$

Transformation of the surface tractions and displacements into the frequency-wavenumber domain can be accomplished using the expressions for these quantities in terms of the scalar and vector potentials, which

107

can be found, for example, in Achenbach (1973). The result of this transformation is given as

$$z_{j-1} \leq z \leq z_j, \quad z_j = \sum_{i=0}^{j} h_i :$$

$$\tilde{\sigma}_{zz}^{(k_x,k_y)} = \tilde{\lambda}\left(\partial_{zz} - k_x^2 - k_y^2\right)\tilde{\varphi}_j^{(k_x,k_y)}$$

$$+ 2\tilde{\mu}\left(\partial_{zz}\tilde{\varphi}_j^{(k_x,k_y)} + \partial_z\left(ik_x\tilde{\psi}_{jy}^{(k_x,k_y)} - ik_y\tilde{\psi}_{jx}^{(k_x,k_y)}\right)\right)$$

$$\tilde{\tau}_{xz}^{(k_x,k_y)} = \tilde{\mu}\left(2ik_x\partial_z\tilde{\varphi}_j^{(k_x,k_y)} + \partial_z\left(ik_y\tilde{\psi}_{jz}^{(k_x,k_y)} - \partial_z\tilde{\psi}_{jy}^{(k_x,k_y)}\right)\right) \quad (33)$$

$$+ ik_x\left(ik_x\tilde{\psi}_{jy}^{(k_x,k_y)} - ik_y\tilde{\psi}_{jx}^{(k_x,k_y)}\right)\right)$$

$$\tilde{\tau}_{yz}^{(k_x,k_y)} = \tilde{\mu}\left(2ik_y\partial_z\tilde{\varphi}_j^{(k_x,k_y)} - \partial_z\left(ik_x\tilde{\psi}_{jz}^{(k_x,k_y)} - \partial_z\tilde{\psi}_{jx}^{(k_x,k_y)}\right)\right)$$

$$+ ik_y\left(ik_x\tilde{\psi}_{jy}^{(k_x,k_y)} - ik_y\tilde{\psi}_{jx}^{(k_x,k_y)}\right)\right)$$

$$\tilde{u}_j^{(k_x,k_y)} = ik_x\tilde{\varphi}_j^{(k_x,k_y)} + ik_y\tilde{\psi}_{jz}^{(k_x,k_y)} - \partial_z\tilde{\psi}_{jy}^{(k_x,k_y)}$$

$$\tilde{v}_j^{(k_x,k_y)} = ik_y\tilde{\varphi}_j^{(k_x,k_y)} - ik_x\tilde{\psi}_{jz}^{(k_x,k_y)} + \partial_z\tilde{\psi}_{jx}^{(k_x,k_y)} \quad (34)$$

$$\tilde{w}_j^{(k_x,k_y)} = \partial_z\tilde{\varphi}_j^{(k_x,k_y)} + ik_x\tilde{\psi}_{jy}^{(k_x,k_y)} - ik_y\tilde{\psi}_{jx}^{(k_x,k_y)}$$

where, h_j is the thickness of layer j and $h_0 = 0$.

The general solution of Equations (31) can be written as

$$z_{j-1} \leq z \leq z_j, \quad 1 \leq j \leq J-1:$$

$$\tilde{\varphi}_j^{(k_x,k_y)} = A_{1+8(j-1)}e^{R_{jL}(z-z_{j-1})} + A_{2+8(j-1)}e^{R_{jL}(z_j-z)}$$

$$\tilde{\psi}_{jx}^{(k_x,k_y)} = A_{3+8(j-1)}e^{R_{jT}(z-z_{j-1})} + A_{4+8(j-1)}e^{R_{jT}(z_j-z)}$$ (35)

$$\tilde{\psi}_{jy}^{(k_x,k_y)} = A_{5+8(j-1)}e^{R_{jT}(z-z_{j-1})} + A_{6+8(j-1)}e^{R_{jT}(z_j-z)}$$

$$\tilde{\psi}_{jz}^{(k_x,k_y)} = A_{7+8(j-1)}e^{R_{jT}(z-z_{j-1})} + A_{8+8(j-1)}e^{R_{jT}(z_j-z)}$$

where

$$R_{jL} = \sqrt{k_x^2 + k_y^2 - \omega^2/\tilde{c}_{jL}^2}, \quad R_{jT} = \sqrt{k_x^2 + k_y^2 - \omega^2/\tilde{c}_{jT}^2} \quad (36)$$

The general solution for the bottom layer $j = J$ should be written in the form given by Equation (35) if this layer is of a finite depth or in the following form if this layer is approximated by a half-space:

$$\tilde{\varphi}_j^{(k_x,k_y)} = A_{1+8(J-1)}e^{R_{jL}(z_{j-1}-z)}, \quad \tilde{\psi}_{jx}^{(k_x,k_y)} = A_{2+8(J-1)}e^{R_{jT}(z_{j-1}-z)}$$

$$\tilde{\psi}_{jy}^{(k_x,k_y)} = A_{3+8(J-1)}e^{R_{jT}(z_{j-1}-z)}, \quad \tilde{\psi}_{jz}^{(k_x,k_y)} = A_{4+8(J-1)}e^{R_{jT}(z_{j-1}-z)} \quad (37)$$

In Equations (37) the real parts of R_{JL} and R_{JT} are assumed positive.

The general solution of Equations (31) contains $8 \times J$ unknown constants if the bottom layer has a finite depth and $8 \times J - 4$ constants if this layer is

modeled by a half space. To find these constants, the corresponding number of linear algebraic equations has to be formulated. The following conditions are to be employed to derive these equations: (a) the constraint for the vector potential, Equation (32), which provides $2 \times J$ equations (note that if the bottom layer is a half-space, the number of equations given by the constraint becomes $2 \times J - 1$), (b) continuity of the stresses and displacements at the interfaces between the layers, which provides $6 \times J - 6$ equations, (c) the balance of stresses at the surface of the ground, Equation (19) (first it is to be transformed into the wavenumber domain), which provides 3 equations, (d) in the case that the bottom layer has a finite lengths, the condition of vanishing displacements at the bottom, which provides 3 equations.

The only nonstandard equations among those mentioned above are the balances of forces at the ground surface, Equations (19). Transforming these equations into the wavenumber domain, we obtain

$$\tilde{\sigma}_{zz}^{(k_x,k_y)}\Big|_{z=0} = Q_z, \quad \tilde{\tau}_{xz}^{(k_x,k_y)}\Big|_{z=0} = Q_x, \quad \tilde{\tau}_{yz}^{(k_x,k_y)}\Big|_{z=0} = Q_y$$

(38)

where vector $\mathbf{Q} = [Q_x,Q_y,Q_z]^T$ is given as

$$\mathbf{Q} = \frac{4}{Sk_xk_y}\sin\left(\frac{k_x\Delta x}{2}\right)\sin\left(\frac{k_y\Delta y}{2}\right)$$

$$\times \sum_{k=-\infty}^{\infty}\sum_{m=1}^{M}\sum_{n=1}^{N}\tilde{\mathbf{F}}_{kmn}\exp\left(-ik_xx_{km} - ik_yy_n\right)$$

(39)

and vector $\tilde{\mathbf{F}}_{kmn}$ is defined by Equation (23).

Thus, if the forces $\tilde{\mathbf{F}}_{kmn}$ were known, the coefficients A in Equations (35) could be readily found using a standard solver of a system of linear algebraic equations with complex coefficients.

Note that the approach presented here does not make use of the transfer matrix as proposed by Thomson (1950) and Haskel (1953) and recently applied for studying the dynamic response of a layered ground to high-speed trains by Grundmann et al. (1999), Sheng et al. (1999), Lombaert et al. (2001), Takemiya (2003). In the opinion of the author both approaches are equally applicable as long as the number of layers is not too high.

Since the forces $\tilde{\mathbf{F}}_{kmn}$ are unknown, a 3 ×3 receptance matrix (one may also call it Green's matrix, as a generalization of the Green's function) has to be derived first. This matrix gives the displacements of the ground surface (in the frequency-wavenumber domain) caused by the unit force applied to the ground surface at $x = y = 0$. To compute the receptance matrix for each combination of ω, k_x and k_y, the linear system of algebraic equations with respect to the coefficients A has to be solved numerically three times. Each time only one component of the surface tractions in Equations (38) should be considered equal to unity while the remaining two should be set to zero. For example, to compute the first column of the transfer matrix, which gives the displacements caused by the unit force acting in the x-direction, the stresses at the surface are to be taken as

$$\left.\bar{\sigma}_{zz}^{(k_x,k_y)}\right|_{z=0} = 0, \quad \left.\bar{\tau}_{xz}^{(k_x,k_y)}\right|_{z=0} = 1, \quad \left.\bar{\tau}_{yz}^{(k_x,k_y)}\right|_{z=0} = 0 \qquad (40)$$

Each computation will result in an array of the coefficients A, which being substituted into Equations (35) will define the potentials and, consequently the displacements described by Equations (34). The latter should be computed for $z = 0$ and $j = 1$. In this manner, the receptance matrix $\underline{\underline{q}}$ can be computed for any combination of ω, k_x and k_y. The off-diagonal elements of this matrix satisfy the following relationships (the first and second indices correspond to the row number and the column number, respectively):

$$a_{12} = a_{21}, \quad a_{13} = -a_{31}, \quad a_{23} = -a_{32} \qquad (41)$$

Employing the receptance matrix, the displacements of the ground surface in the frequency-wavenumber domain can be found under the original stress field defined by Equations (38). These displacements read

$$\tilde{\underline{u}}^{(k_x,k_y)}\left(k_x,k_y,0,\omega\right) = \underline{\underline{q}}\,\tilde{\underline{Q}} \qquad (42)$$

Applying to Equation (42) the inverse Fourier transform with respect to k_x and k_y, the following expression is obtained for the vector-displacement of the ground surface in the frequency domain:

$$\bar{\underline{u}}\left(x,y,0,\omega\right) = \frac{1}{4\pi^2}\int_{-\infty}^{\infty}\int_{-\infty}^{\infty}\underline{\underline{q}}\,\underline{Q}\exp\left(ik_x x + ik_y y\right)dk_x\,dk_y \qquad (43)$$

This equation will be used in the next section to form a system of linear algebraic equations, the solution of which will allow to find the dynamic stiffness of the ground.

5.2 System of algebraic equations for deriving the dynamic stiffness of the ground

To derive the dynamic stiffness of the ground, the forces $\tilde{F}_{kmn}^z$ should be calculated, see Equation (17). To this end, Equation (43) should be considered at the mid-points of the sub-domains, into which the sleepers-ground interface is decomposed:

$$\bar{\underline{u}}\left(x_{pr},y_s,0,\omega\right) = \frac{1}{4\pi^2}\int_{-\infty}^{\infty}\int_{-\infty}^{\infty}\underline{\underline{q}}\,\underline{Q}\exp\left(ik_x x_{pr} + ik_y y_s\right)dk_x\,dk_y \qquad (44)$$

where

$$x_{pr} = pd - b + (2r-1)\Delta x/2, \quad p = 0,\pm1,\pm2,...; \; 1 \le r \le M$$
$$y_s = -a + (2s-1)\Delta y/2, \; 1 \le s \le N \qquad (45)$$

are the coordinates of the mid-points and the new integers p,r and s are introduced to avoid confusion with the dummy summation indices in Equation (39).

In accordance with Equations (18), the horizontal displacements of the mid-points of the sub-domains must be zero while the vertical deflections should be equal to the vertical displacement of the corresponding sleeper. Therefore, Equation (44) simplifies to

$$\left[0,0,\tilde{w}_s^p\right]^T = \frac{1}{4\pi^2}\int_{-\infty}^{\infty}\int_{-\infty}^{\infty}\underline{\underline{q}}\,\underline{Q}\exp\left(ik_x x_{pr} + ik_y y_s\right)dk_x\,dk_y \qquad (46)$$

Equation (46) can be simplified significantly by using Equation (25), which is, in fact, a form of the periodicity condition. Substituting the expression for the vector $\underline{Q}$, Equation (39), and using the expressions for $\tilde{w}_s^p$ and $\tilde{F}_{kmn}$, which are given by Equation (25), Equation (46) can be rewritten as

$$\left[0,0,C\right]^T = \sum_{m=1}^{M}\sum_{n=1}^{N}\underline{J}^{mnprs}\,\underline{C}_{mn} \qquad (47)$$
$$p = 0,\pm1,\pm2,...; 1 \le r \le M, 1 \le s \le N$$

where

$$\underline{J}^{mnprs} = \frac{1}{\pi^2 S}\sum_{k=-\infty}^{\infty}\int_{-\infty}^{\infty}\int_{-\infty}^{\infty}\underline{\underline{q}}\,\exp\left(ik_x\left(x_{pr}-x_{km}\right)+ik_y\left(y_s-y_n\right)\right)$$
$$\times\frac{1}{k_x k_y}\sin\left(\frac{k_x\Delta x}{2}\right)\sin\left(\frac{k_y\Delta y}{2}\right)\exp\left(i\frac{\omega}{V}(k-p)d\right)dk_x\,dk_y \qquad (48)$$

The latter expression can be further simplified by introducing new indices $\tilde{k} = k - p$, $\tilde{m} = m - r$ and $\tilde{n} = n - s$. Using these indices and omitting tildes in the final expressions, we can rewrite Equations (47) and (48) as

$$\left[0,0,C\right]^T = \sum_{m=1-r}^{M-r}\sum_{n=1-s}^{N-s}\underline{J}^{mn}\,\underline{C}_{(m+r)(n+s)} \qquad (49)$$
$$1 \le r \le M, 1 \le s \le N$$

109

where

$$\underline{\underline{J}}^{mn} = \frac{1}{\pi^2 S} \sum_{k=-\infty}^{\infty} \int_{-\infty}^{\infty} \int_{-\infty}^{\infty} \underline{\underline{a}} \, \exp\left(ik_x\left(-kd - m\Delta x\right) - ik_y n\Delta x\right)$$

$$\times \frac{1}{k_x k_y} \sin\left(\frac{k_x \Delta x}{2}\right) \sin\left(\frac{k_y \Delta y}{2}\right) \exp\left(i\frac{\omega}{V} kd\right) dk_x dk_y \quad (50)$$

It is important to realize what a great deal simplification has been achieved by reducing Equation (46) to Equation (49). This simplification is mainly concerned with the number of unknowns to be determined from these equations. It can be readily seen that Equation (46) contains $3 \times M \times N \times (2K_{max})$ unknowns ($2K_{max}$ is the number of sleepers to be taken into account instead of infinity), while Equation (49) contains only $3 \times M \times N$ unknowns. This is a very significant reduction since every coefficient in the equations in question is a sum of double integrals. The reduction of the number of equations is possible because the *steady-state* behavior of the *infinitely long* system is considered. In this case, the dynamic reaction of the ground against any sleeper is the same save the phase shift $\exp(i\omega k \, d/V)$, which is imposed by the periodicity condition.

The system of algebraic equations, Equation (49), can be solved to give ratios $3 \times M \times N$ ratio $\underline{C}_{mn}/C$. As shown in the next section, the dynamic stiffness of the ground can be easily calculated once these ratios are known.

5.3 Analysis of the dynamic stiffness

In this paper we pay attention only to the vertical dynamic stiffness of the ground, since the sleepers are considered to vibrate vertically. Note, however, that Equation (49) is general and its solution allows to determine all three components of the dynamic ground stiffness.

To derive an analytical expression for the vertical component of the dynamic stiffness, as it is defined in the beginning of Section 5, it is customary to consider Equation (17), which describes the vertical motion of sleeper k in the frequency domain. As follows from the periodicity condition, Equation (22), and from Equation (25), the beam displacement at the contact points with the pads, the vertical deflection of the sleepers and the forces $\tilde{F}^z_{kmn}(\omega)$ can be expressed as

$$\tilde{w}^b\left(kd, \omega\right) = C^b\left(\omega\right)\exp\left(i\omega k \, d/V\right)$$

$$\tilde{w}^s_k\left(\omega\right) = C\left(\omega\right)\exp\left(i\omega k \, d/V\right) \quad (51)$$

$$\tilde{F}^z_{kmn}\left(\omega\right) = C^z_{mn}\left(\omega\right)\exp\left(i\omega k \, d/V\right)$$

Substituting these expressions into Equation (17), we obtain the following equation, which is independent

of index k:

$$\sum_{m=1}^{M}\sum_{n=1}^{N} C^z_{mn} = -m^s\omega^2 C + 2\left(k^s - i\omega c^s\right)\left(C - C^b\right) \quad (52)$$

This equation clearly shows that the vertical dynamic stiffness of the ground is given as

$$\chi_z\left(\omega\right) = -\sum_{m=1}^{M}\sum_{n=1}^{N} C^z_{mn}/C \quad (53)$$

This stiffness is the same for all sleepers. It is important to note that this stiffness depends not only on the frequency but also on the load velocity.

Solving Equation (49) numerically, the dynamic stiffness $\chi_z(\omega)$ can be found for a given frequency. The most time-consuming part in doing so is calculation of the matrices $\underline{\underline{J}}^{mn}$ given by Equation (50). This can be done either by employing a standard integration routine (in this case transformation of the integrands into the polar system of coordinate is advisable) or by using the fast Fourier transform. Either approach requires that the infinite sum in Equation (50) is replaced by a finite sum from $-K_{max}$ to K_{max}. The physical meaning of $2K_{max}$ is transparent. It is the number of sleepers, which are taken into account (instead of infinitely many) to find the dynamic stiffness of the ground. It is not easy to determine how big should be K_{max} to approximate sufficiently well the dynamic stiffness, which would be calculated using Equation (50). Especially problematic it is when the load moves with a velocity comparable or larger than the Rayleigh wave velocity. Therefore, a preliminary analytical evaluation of Equation (50) proposed by Vostroukhov & Metrikine (2003) seems to be promising, since it allows to calculate the infinite sum in Equation (50) analytically. In this paper, however, no use is made of this technique, since it has not yet been generalized to the case of multilayered ground.

Evaluating Equation (50) numerically, it is customary to notice that truncating the sum and using the symmetry with respect to the summation index k, it can be rewritten to the following form

$$\underline{\underline{J}}^{mn} = \frac{1}{\pi^2 S} \sum_{k=0}^{K_{max}} \int_{-\infty}^{\infty} \int_{-\infty}^{\infty} \frac{1}{k_x k_y} \underline{\underline{a}} \, \exp\left(-ik_x m\Delta x - ik_y n\Delta x\right)$$

$$\times \sin\left(\frac{k_x \Delta x}{2}\right) \sin\left(\frac{k_y \Delta y}{2}\right) \left(\delta_{0k} + 2\cos\left(kd\left(\frac{\omega}{V} - k_x\right)\right)\right) dk_x dk_y$$

$$(54)$$

where δ_{0k} is the Kroneker delta, and matrices $\underline{\underline{J}}^{mn}$ possesses the following symmetry properties:

$$J_{11}^{m(-n)} = J_{11}^{mn}, \quad J_{12}^{m(-n)} = -J_{12}^{mn}, \quad J_{13}^{m(-n)} = J_{13}^{mn},$$
$$J_{22}^{m(-n)} = J_{22}^{mn}, \quad J_{23}^{m(-n)} = -J_{23}^{mn}, \quad J_{33}^{m(-n)} = J_{33}^{mn} \quad (55)$$

The symmetry of the remaining three terms can be retrieved using Equations (41). Note that there is no symmetry with respect to index m. This is because matrices $\underline{\underline{J}}^{mn}$ are inherently related to the moving load, which introduces a kind of anisotropy in the $x-$ direction.

To demonstrate the frequency dependence of the dynamic stiffness, the ground is modeled by a visco-elastic layer, which is fixed at the bottom. The parameters of the layer are

$$E_1 = 4.1 \times 10^7 \, \text{N} \, \text{m}^{-2}, \ v_1 = 0.3, \ \rho_1 = 1960 \, \text{kg} \, \text{m}^{-3},$$
$$\gamma_1 = 5 \times 10^{-4} \, \text{s}, \ h_1 = 10 \, \text{m} \tag{56}$$

where E_1, v_1 are the Young's modulus and the Poison's ratio of the layer and the other notations are explained before. The size of the sleepers and the load velocity are chosen as

$$a = 1.35 \, \text{m}, \ b = 0.135 \, \text{m}, \ V = 70 \, \text{m} \, \text{s}^{-1} \tag{57}$$

This load velocity is about 80% of the Rayleigh wave velocity.

Figure 1 shows the real and imaginary parts of the dynamic stiffness for 2 cases. In the first case the stresses under each sleeper are considered to be constant, e.g. $M = 1$ and $N = 1$ in Equation (53). The second case corresponds to $M = 1$, $N = 11$, which implies that the interface under each sleeper is broken into eleven subdomains, aligned along the y-axis (this makes each subdomain almost a square). In both cases 31 sleepers are accounted for, e.g. $K_{max} = 15$ in Equation (54).

Figure 2 shows that the dynamic stiffness tends to zero at $\omega \approx 25 \, \text{rad} \, \text{s}^{-1}$. This frequency is the cut-off frequency of the second mode of the layer. In the frequency band below this critical frequency the real part of the dynamic stiffness is positive while the imaginary part is close to zero. This corresponds to the elastic type of the ground reaction. In the frequency band higher than the critical frequency the imaginary part grows significantly showing that vibrations of the sleepers excite waves in the ground. The real part changes the sign in this band thereby showing that the type of the ground reaction changes from visco-elastic to visco-inertial. In the low frequency band the effect of the number of subdomains is perceptible but does not exceed 50%. With the growing frequency this effect becomes more pronounced. Therefore, the higher the frequency of ground vibrations, the finer should be the interface discretization.

The dynamic stiffness presented in Figure 2 was calculated taking into account contributions of 31 sleepers ($K_{max} = 15$). To check whether this choice is reasonable, the absolute value of the dynamic stiffness is plotted in Figure 3 as a function of K_{max} for three frequencies.

Figure 3 shows that at very low frequencies, represented by $\omega = 3 \, \text{rad} \, \text{s}^{-1}$, the dynamic stiffness converges monotonically and is calculated accurately by accounting for 25–35 sleepers. At, $\omega = 24 \, \text{rad} \, \text{s}^{-1}$ which is just below the critical frequency, the dynamic stiffness does tend to a certain value but the convergence is not monotonic. In the higher frequency range, represented by $\omega = 95 \, \text{rad} \, \text{s}^{-1}$, the convergence worsens tremendously. This is due to the wave field generated by each sleeper. This field extends to a large distance making a large number of sleepers contribute into the dynamic stiffness.

Thus, choosing the number of sleepers to be taken into account, one has to be very careful especially in

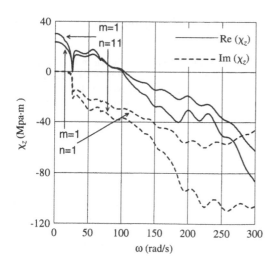

Figure 2. Dynamic stiffness versus frequency.

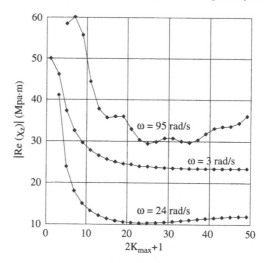

Figure 3. Effect of the number of accounted sleepers.

the relatively high frequency band that corresponds to the wave generation in the ground. To avoid this problem, it seems advisable to apply the technique proposed by Metrikine and Popp (1999) and enhanced by Vostroukhov and Metrikine (2003). This technique allows to accomplish summation over infinitely many sleepers using the contour integration method.

6   THE RAIL DISPLACEMENT

Once the dynamic stiffness of the ground has been calculated, the steady state displacement of the rails can be computed in the manner that is described in this section.

One can find an analytical expression for the beam displacement in the frequency domain. To this end, it is customary to rewrite Equation (16) in the following equivalent form:

$$-\rho^b A \omega^2 \tilde{w}^b + IE^b \partial_{xxxx}\tilde{w}^b = P \exp(i\omega x/V)/(2V) \tag{58}$$

$$\left[\tilde{w}^b\right]_{x=kd} = 0, \ \left[\partial_x\tilde{w}^b\right]_{x=kd} = 0, \ \left[\partial_{xx}\tilde{w}^b\right]_{x=kd} = 0 \tag{59}$$

$$IE^b\left[\partial_{xxx}\tilde{w}^b\right]_{x=kd} = \left(k^s - i\omega c^s\right)\left(\tilde{w}_k^s - \tilde{w}^b(kd)\right) \tag{60}$$

where the square brackets represent the difference $[f(x)]_{x=kd} = f(kd + 0) - f(kd - 0)$. In this form, the continuity of the beam displacement and slope and the balances of the moments and vertical forces at each pad are written in the explicit form.

The balance of vertical forces, Equation (60), can be rewritten making use of Equations (51) to give

$$IE^b\left[\partial_{xxx}\tilde{w}^b\right]_{x=kd} = \left(C - C^b\right)\exp(i\omega kd/V) \tag{61}$$

The general solution of Equation (58) in the interval $x \in [0,d]$ can be written as

$$\tilde{w}^b(x) = B_1 \exp(\alpha x) + B_2 \exp(-\alpha x) + B_3 \exp(i\alpha x) \\ + B_4 \exp(-i\alpha x) + B_0 \exp(i\omega x/V) \tag{62}$$

where

$$\alpha = \left(\rho^b A \omega^2 / IE^b\right)^{1/4}, \ B_0 = \frac{1}{2}PV^3/\left(IE^b \omega^4 - \rho^b A \omega^2\right) \tag{63}$$

The general solution in the interval $x \in [d,2d]$ can be obtained from Equation (62) by using the periodicity condition. The result reads

$$\tilde{w}^b(x) = \exp(i\omega d/V)\left(B_1 \exp(\alpha(x-d)) + B_2 \exp(\alpha(d-x))\right) \\ + B_3 \exp(i\alpha(x-d)) + B_4 \exp(i\alpha(d-x))\right) + B_0 \exp(i\omega x/V) \tag{64}$$

Substituting Equations (62) and (63) into the boundary conditions at $x = d$ (Equations (59) and (61) with $k = 1$), four linear algebraic equations can be obtained with respect to six unknowns: B_1, B_2, B_3, B_4, C and C^b. The remaining two equations are Equation (52), which, making use of the dynamic stiffness, can be written as

$$\chi_z(\omega) = -m^s \omega^2 + 2\left(k^s - i\omega c^s\right)\left(1 - C^b/C\right) \tag{65}$$

and the following relationship between C^b and the constants B_1, B_2, B_3, B_4:

$$C^b = B_1 + B_2 + B_3 + B_4 + B_0 \tag{66}$$

The latter equation can be readily obtained by comparing Equation (62) taken at $x = 0$ and the expression for $\tilde{w}^b(0)$ given by Equation (51).

Thus, provided that the dynamic stiffness $\chi_z(\omega)$ is known, the coefficients B_1, B_2, B_3, B_4 can be computed for a given frequency. Substituting these coefficients into Equation (62), the beam deflection $\tilde{w}^b(x)$ can be calculated in the interval $x \in [0,d]$. Outside this interval, the beam deflection can be readily calculated applying the periodicity condition, Equation (22). This implies that the beam deflection in the frequency domain has been found. The steady-state beam deflection in the time domain is to be found by applying the inverse Fourier transform over frequency:

$$w^b(x,t) = (2\pi)^{-1} \int_{-\infty}^{\infty} \tilde{w}^b(x,\omega)\exp(-i\omega t)d\omega \tag{67}$$

The beam displacement is shown in Figure 4 versus the spatial co-ordinate x at $t = 0$. To compute this deflection, parameters given by Equations (56) and (57) were used in calculations and, additionally,

$$IE^b = 1.22\times10^7 \ \mathrm{N\,m^2}, \ \rho^b A = 120.7 \ \mathrm{kg\,m^{-1}}, \\ m^s = 250\,\mathrm{kg}, \ k^s = 10^8 \ \mathrm{N\,m^{-1}}, \ c^s = 10^8 \ \mathrm{N\,s\,m^{-1}}, \ K_{max} = 15 \tag{68}$$

The beam displacement is plotted for two cases of the interface discretization namely, $M = 1, N = 1$ (dashed line) and $M = 1, N = 11$ (solid line).

Figure 4 shows that the effect of the interface discretization on the beam displacement is not very pronounced. This is because the beam displacement is

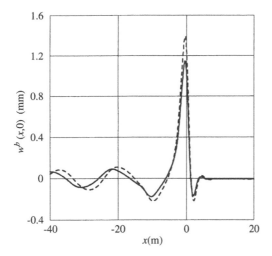

Figure 4. Rail displacement.

mainly governed by the low-frequency part of the dynamic stiffness. What is interesting to note is that the beam pattern is significantly asymmetric with respect to the load and that a wave can be observed in the beam behind the load. This wave is a consequence of the pronounced drop of the dynamic stiffness at the cut-off frequency of the second mode of the layer, see Figure 3. It the ground was modeled by a layered half-space with a modest variation of material properties, this wave would become less pronounced.

REFERENCES

Achenbach, J.D. 1973. *Wave propagation in elastic solids.* Amsterdam-London: NHPC.
Auersch, L. 2005. The excitation of ground vibration by rail traffic: theory of vehicle-track-soil interaction and measurements on high-speed lines. *Journal of Sound and Vibration* 284: 103–132.
Aurby, D., Clouteau, D. & Bonnet, G. 1994. Modelling of wave propagation due to fixed or mobile dynamic sources. In N. Chouw & G. Schmid (eds), *Wave 94*: 109–121. Berg-Verlag.
Belotserkovskiy, P.M. 1996. On the oscillations of infinite periodic beams subjected to a moving concentrated force. *Journal of Sound and Vibration* 193: 706–712.
Bode, C., Hirschauer, R. & Savidis, S.A. 2000. Three-dimensional time domain analysis of moving loads on railway tracks on layered soils. In N. Chouw & G. Schmid (eds), *Wave 2000*: 3–12. Rotterdam: Balkema.
Bogacz, R., Krzyzinski, T. & Popp, K. 1993. On dynamics of systems modeling continuous and periodic guideways. *Archives of Mechanics* 45: 575–593.
Clouteau, D., Degrande, G. & Lombaert G. 2000. Some theoretical and numerical tools to model traffic induced vibrations. In N. Chouw & G. Schmid (eds), *Wave 2000*: 13–28. Rotterdam: Balkema.
Clouteau, D., Arnst, M., Al-Hussaini, T.M. & Degrande, G. 2005. Freefield vibrations due to dynamic loading on a tunnel embedded in a stratified medium. *Journal of Sound and Vibration* 283 (1–2): 173–199.
Grundmann, H., Lieb, M. & Trommer, E. 1999. The response of a layered half-space to traffic loads moving along its surface. *Archive of Applied Mechanics* 69: 55–67.
Lombaert, G., Degrande, G. & Clouteau, D. 2001. The influence of the soil stratification on free field traffic-induced vibrations. *Archive of Applied Mechanics* 71(10): 661–678.
Jezequel, L. 1980. Analysis of critical speeds of a moving load on an infinite periodically supported beam. *Journal of Sound and Vibration* 73: 606–610.
Kalker, J.J. 1996. Discretely supported rails subjected to transient loads. *Vehicle System Dynamics* 25: 1–88.
Kruse, H. & Popp, K. 2001. A modular algorithm for linear, periodic train-track models. *Archive of Applied Mechanics* 71 (6–7): 473–486.
Mead, D.J. 1971. Vibration response and wave propagation in periodic structures. *Engineering for Industry – Transactions of ASME* 93: 783–792.
Metrikine, A.V. & Popp, K. 1999. Vibration of a periodically supported beam on an elastic half-space. *European Journal of Mechanics A/Solids* 18(4): 679–701.
Sheng, X., Jones, C.J.C. & Petyt, M. 1999. Ground vibration generated by a load moving along a railway track. *Journal of Sound and Vibration* 228(1): 129–156.
Sheng, X., Jones, C.J.C. & Thompson, D.J. 2005. Responses of infinite periodic structures to moving or stationary harmonic loads. *Journal of Sound and Vibration,* 282(1–2): 125–149.
Suiker, A.S.J., Metrikine, A.V. & de Borst R. 2001. Dynamic behavior of a layer of discrete particles. Part 2: Response to a uniformly moving, harmonically vibrating load. *Journal of Sound and Vibration* 240(1): 19–39.
Takemiya, H. 2003. Simulation of track-ground vibrations due to a high-speed train: the case of X-2000 at Ledsgard. *Journal of Sound and Vibration* 261(3): 503–526.
Takemiya, H. & Bian, X. 2005. Substructure simulation of inhomogeneous track and layered ground dynamic interaction under train passage. *Journal of Structural engineering –Transactions of ASME* (to appear).
Vesnitskii, A.I. & Metrikine, A.V. 1993. Transient radiation in a periodically non-uniform elastic guide. *Mechanics of Solids* 28(6): 158–162.
Vesnitskii, A.I. & Metrikine, A.V. 1996. Transition radiation in mechanics. *Physics – Uspekhi* 39(10): 983–1007.
Vostroukhov, A.V. & Metrikine, A.V. 2003. Periodically supported beam on a visco-elastic layer as a model for dynamic analysis of a high-speed railway track. *International Journal of Solids and Structures* 40(21): 5723–5752.
Vostroukhov, A.V., Metrikine, A.V., Vrouwenvelder, A.C.W.M., Merkulov, V.I., Misevich, V.N. & Utkin, G.A. 2003. Remote detection of derailment of a wagon of a fright train: theory and experiment. *Archive of Applied Mechanics* 73(1–2): 75–78.

Environmental Vibrations – Takemiya (ed.)
© *2005 Taylor & Francis Group, London, ISBN 0 415 39035 4*

Critical velocities of Timoshenko beam on an elastic half-space under moving load

C.J. Wang
City College of Zhejiang University, Hangzhou, Zhejiang, China
Department of Civil Engineering, Zhejiang University, Hangzhou, Zhejiang, China

Y.M. Chen & X.W. Tang
Department of Civil Engineering, Zhejiang University, Hangzhou, Zhejiang, China

ABSTRACT: The dispersion equation of Timoshenko beam resting on an elastic half-space subjected to a moving load is presented. According to the relative relation of wave velocities of the half-space and the beam, four cases with the combination of different parameters of the half-space and the beam, the system of soft beam and hard half-space, the system of sub-soft beam and hard half-space, the system of sub-hard beam and soft half-space, and the system of hard beam and soft half-space are considered. The critical velocities are studied using dispersion curves. It is found that critical velocities depend on the relative relation of wave velocities of the half-space and the beam. The Rayleigh wave velocity in the half-space is always a critical velocity. For the system of soft beam and hard half-space, wave velocities of the beam are also critical velocities. Besides the shear wave velocity of the beam, there is an additional minimum critical velocity for the system of sub-soft beam and hard half-space. While for systems of (sub-) hard beams and soft half-space, wave velocities of the beam are no longer the critical ones.

1 INTRODUCTION

Usually, the Winkler foundation model with independent springs is used to model the interaction of the track and the ground, which simplifies the problem greatly (Kenney, 1954; Achenbach & Sun, 1965; Kerr, 1972; Xie W. P. et al., 2002; Chen Y. M. et al., 2003). Unfortunately, this simple model cannot represent the wave coupling between the foundation and the superstructure, which is more and more obvious with the increasing of train speed. Furthermore, it is difficult to determine the spring stiffness in practice. The problem of a moving load over a beam on an elastic half-space has first been studied by Filippov (1961) as a model for the train-subsoil interaction. It was shown that the critical velocity of the train is almost same as the R-wave velocity, which is much lower than that of the Winkler model. Dieterman and Metrikine (1996, 1997) derived the equivalent stiffness of an elastic half-space interacting with an Euler-Bernoulli beam (E-B beam). The work found that the equivalent stiffness mainly depended on the frequency and the wave number of the beam. Furthermore, there are two critical velocities for E-B beam. One is the

R-wave velocity, the other is smaller than the R-wave velocity, termed the minimum critical velocity V_{cr}.

Though the analysis is simplified using the E-B beam, neglecting the effect of transverse shear deformation produces inaccurate bending solutions for deep beams or shear deformable beams such as the sandwich ones (Wang et al., 1997). The Timoshenko beam (T-beam) is a good choice to overcome this problem. The critical velocities of E-B beam are independent of wave velocities of the beam. However, using the finite element method (FEM), Suiker (1998) showed that the critical velocities are determined by the wave velocities of not only the half-space but the T beam. If the beam is soft compared to the half-space, wave velocities of the beam are critical velocities.

In this paper, dispersion equation of T beam on an elastic half-space under a moving load is derived by introducing the equivalent stiffness. According to the relative relation of wave velocities of the beam and the half-space, four cases are considered: the system of soft beam and hard half-space, the system of sub-soft beam and hard half-space, the system of sub-hard beam and soft half-space, and the system of hard beam and soft half-space. Using the dispersion curves, the critical velocities of the moving load are analyzed in details.

2 MODEL AND DISPERSION EQUATIONS OF T BEAM AND HALF-SPACE SYSTEM

As shown in Figure 1, let a concentrated load P_0 be applied to a T beam resting on an elastic half-space and be moving along the mid-line of the beam at a constant velocity V_0. A right-handed system of rectangular Cartesian coordinate axes x, y and z is introduced herein, with the positive x-axis coincident with the line of motion and oriented in the direction of motion, and with the positive z-axis pointing into the half-space. It is assumed that the contact between the beam and the half-space is smooth. Hence the shear stress is zero at the interface, and the normal stress between the beam and the half-space is uniformly distributed over the width of the beam. It is further assumed that the load has been moving for a considerable long time so that all transient disturbances can be neglected.

The governing equation of free vibration of the T beam resting on an elastic half-space is

$$E_b I \frac{\partial^4 w^0}{\partial x^4} - \rho_b I (\frac{E_b}{kAG_b}+1)\frac{\partial^4 w^0}{\partial x^2 \partial t^2} + \rho_b A \frac{\partial^2 w^0}{\partial t^2} + \frac{\rho_b^2 I}{kG_b}\frac{\partial^4 w^0}{\partial t^4}$$

$$+F(x,t) - \frac{E_b I}{kAG_b}\frac{\partial^2 F(x,t)}{\partial x^2} + \frac{\rho_b I}{kAG_b}\frac{\partial^2 F(x,t)}{\partial t^2} = 0 \qquad (1)$$

where w^0 is the vertical displacement of the beam, ρ_b is the mass density of the beam; E_b and G_b are its Young's modulus and shear modulus, respectively; $G_b = E_b/2$ $(1 + v_b)$ with v_b is the Poisson's ratio of the beam; $I = ab^3/6$, $A = 2ab$, are the moment of inertia and the area of the beam, respectively, with b is the beam height and k is the shear correction factor, $k = 10(1 + v_b)/$ $(12 + 11v_b)$ (Cowper, 1966); $F(x,t)$ is the force per length acting at the beam by the half-space.

According to the definition of the equivalent stiffness of half-space interacting with a beam, there is

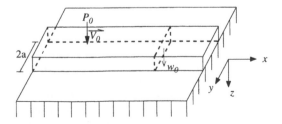

Figure 1. Analysis model of T beam and half-space system.

$$\int_{-\infty}^{\infty}\int_{-\infty}^{\infty} F(x,t)e^{i(\omega t-\xi x)}dtdx = 2\pi\mu h(\omega,\xi)\chi(\omega,\xi) \qquad (2)$$

where

$$h(\omega,\xi) = \int_{-\infty}^{\infty}\int_{-\infty}^{\infty} w^0(x,t)e^{i(\omega t-\xi x)}dtdx \qquad (3)$$

ω is the circle frequency, ξ is the wave number in x direction, μ is the shear modulus of the half-space, $2\pi\mu\chi(\omega, \xi)$ is the equivalent stiffness, for the expression of $\chi(\omega, \xi)$, please refer to Dieterman & Metrikine, 1996.

Applying 2D Fourier transform to Eq. (1) and using Eq. (2), one can give

$$h(\omega,\xi)\{D(\omega,\xi) - 2\pi\mu\chi(\omega,\xi)\} = 0 \qquad (4)$$

for T beam

$$D(\omega,\xi) = \frac{-E_b I\xi^4 + \rho_b I(1+\frac{E_b}{kG_b})\xi^2\omega^2 + \rho_b A\omega^2 - \frac{\rho_b^2 I}{kG_b}\omega^4}{1+\frac{E_b I}{kG_b A}\xi^2 - \frac{\rho_b I}{kG_b A}\omega^2}$$

Eq.(4) describes the free vibration of the T beam resting on an elastic half-space, in the Fourier domain. The expression of $D(\omega, \xi) - 2\pi\mu\chi(\omega, \xi)$ is the dispersion relation between the beam and the elastic half-space. The first member of this expression is the dispersion characteristics of the free beam and the second one describes the reaction of the half-space. And the dispersion equation of the T beam and the half-space system is

$$D(\omega,\xi) - 2\pi\mu\chi(\omega,\xi) = 0 \qquad (5)$$

There is $\omega = \xi V_{ph}$, V_{ph} is the phase velocity of waves in the beam-half space system, which is equal to the load velocity V_0 once there is a moving load on the system. Letting, $\xi' = \xi b$, which means twice of the characteristic length given by Suiker (1998), and substituting ξ' into Eq.(5), one gets

$$D^*(\xi') - \beta^2\chi(\xi V_{ph}/b,\xi'/b) = 0 \qquad (6)$$

where $D^*(\xi') = [-\xi'^4 + \alpha_1^2(1 + \gamma)\xi'^4 + 12\alpha_1^2\xi'^2 - \alpha_1^2\alpha_2^2\xi'^4]/(12 + \gamma\xi'^2 - \alpha_2^2\xi'^2)$, $\alpha_1^2 = V_{ph}^2/V_{bs}^2$, $\alpha_2^2 = V_{ph}^2/V_{bp}^2$, $V_{bp} = \sqrt{E_b/\rho_b}$ is the compression wave velocity in the beam, $V_{bs} = \sqrt{kG_b/\rho_b}$ is the shear wave velocity in the beam, $l = b/2a$ is the ratio of beam height to beam width, $\beta^2 = 2\pi\mu/E_b$, $\gamma = E_b/kG_b = (12 + 11v_b)/5$.

116

If V_{ph} is larger than V_R there exists the wave radiation. So it has to be satisfied that the load velocity $V_{ph} \in [0, V_R]$ for the dispersion equation. Then Eq. (6) can be rewritten as

$$D^*(\xi') + \beta^2 \xi' / 2l\alpha_s^2 I_1 = 0 \qquad (7)$$

where $-\xi'/2l\alpha_s^2 I_1$ is the equivalent stiffness for the load velocity smaller than V_R. $\alpha_S = V_{ph}/V_s$, $V_S = \sqrt{\mu/\rho}$ is the shear wave velocity in the half-space, ρ is the mass density of the half-space.

By solving Eq. (7) the relations between ξ' and wave velocities including the phase velocity, and the group velocity V_g can be derived. This relation is the dispersion relation. The group velocity is a physical parameter governing the velocity of the wave energy propagation. Its expression is

$$V_g = \frac{\partial \omega}{\partial \xi} = \frac{\partial(V_{ph} \cdot \xi)}{\partial \xi} = V_{ph} + \xi' \cdot \frac{\partial V_{ph}}{\partial \xi'} \qquad (8)$$

If $V_0 = V_{ph} = V_g$ the energy of radiated waves propagates with the same velocity as the moving load, for which the amount of radiated energy under the load increases towards infinity with the increasing of time. The corresponding velocity is the critical velocity.

3 DISPERSION CHARACTERISTICS OF T BEAM AND HALF-SPACE SYSTEM

According to the relative relation of wave velocities of the beam and the half-space, two sets of beam parameters and two sets of half-space parameters are considered as given in Table 1 and Table 2. Figure 2 shows the relative relation of wave velocities for the four cases.

Table 1. Material and geometry parameters.

T beam	$E_b(\mathrm{I})$	$30 \times 10^6 [\mathrm{N/m^2}]$ ($V_{bs} = 62.622$ m/s, $V_{bp} = 109.545$ m/s)
	$E_b(\mathrm{II})$	$1000 \times 10^6 [\mathrm{N/m^2}]$ ($V_{bs} = 361.551$ m/s, $V_{bp} = 632.456$ m/s)
	v_b	0.30 [–]
	ρ_b	2500 [kg/m³]
	$a \times b$	0.5×0.75
	γ	2.84 [–]
Half-space	$\mu(\mathrm{I})$	100×10^6 [N/m²] ($V_R = 207.376$ m/s, $V_S = 223.607$ m/s, $V_P = 418.33$ m/s)
	$\mu(\mathrm{II})$	10×10^6 [N/m²] ($V_R = 65.578$ m/s, $V_S = 70.711$ m/s, $V_P = 132.288$ m/s)
	v	0.3 [–]
	ρ	2000 [kg/m³]

3.1 System of soft beam and hard half-space

In this case the beam is very soft relative to the half space, and V_{bs} and V_{bp} of the beam are lower than the R-wave velocity of the half-space. Figure 3 shows the dispersion relations obtained from Eq. (7). Obviously, there are two modes in the system, one is coupled to the vertical displacement of the beam, and the other is coupled to the rotation of the beam. For each mode, Fig. 3 plots the relation between ξ' and the normalized phase velocity (the solid line) as well as the normalized group velocity (the dashed line) of the system.

Table 2. Parameters of four systems of beam and half-space.

System of beam and half-space	β^2[–]	l[–]
Soft beam and hard half-space: I-I	2.094	0.75
Sub soft beam and hard half-space: I-II	20.944	0.75
Sub hard beam and soft half-space: II-I	0.063	0.75
Hard beam and soft half-space: II-II	0.628	0.75

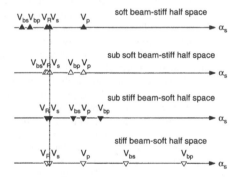

Figure 2. The relative relation of wave velocities in beam and those in half space in four cases.

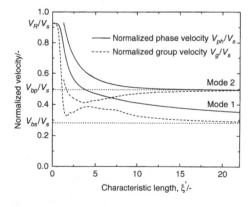

Figure 3. Dispersion curve of system of soft beam and hard half-space.

117

Both wave modes behave in a strongly dispersive manner ($V_{ph} \varkappa V_g$) over a large range of ξ'. For the long-wave limit ($\xi' \rightarrow 0$), the wave velocity of mode 1 approaches V_R. While for mode 2, the phase and group velocity curves are cut off at a wave limit with long wave length, but where $V_{ph} \varkappa V_g$, which is different to that given by Suiker (1998). Because the relatively long waves strongly penetrate the half-space, for a specific mode of the beam, the long-wave limit is fully determined by the half-space properties. Figure 3 also reveals that, for the short-wave limit ($\xi' \rightarrow \infty$), the wave velocity of the first mode approaches V_{bs}, while for the second mode it approaches V_{bp}. The reason is that infinite short waves cannot penetrate the half-space, so that the corresponding properties are then fully determined by the beam characteristics. So for the system of soft beam and hard half-space, there are three critical velocities, V_{bs}, V_{bp} and V_R.

3.2 System of sub-soft beam and hard half-space

The shear wave velocity of the beam V_{bs} is still lower than the R-wave velocity of the half-space, but V_{bp} is higher than the latter. Figure 4 shows their dispersion relations, where, in contrast to the case above, only one mode appears. This mode still behaves in a strongly dispersive manner ($V_{ph} \varkappa V_g$) over a large range of ξ'. Wave velocities approach V_R and V_{bs}, respectively, for the long-wave limit ($\xi' \rightarrow 0$) and the short-wave limit ($\xi' \rightarrow \infty$). Where $V_{ph} = V_g$, there is an additional minimum critical velocity, V_{cr}^m, which is smaller than V_R and V_{bs}. This minimum critical velocity has also been mentioned by Dieterman & Metrikine (1996), who discussed the case for a continuously moving load on an E-B beam resting on a half-space. But its critical velocities are independent of the wave velocities of the beam, because the study neglected the shear deformation of the beam. So for the system of sub-soft beam and hard half-space there are also three critical velocities: V_{cr}^m, V_{bs} and V_R.

3.3 System of sub-hard beam and soft half-space

Wave velocities of the beam are all higher than V_R and V_S for this case. Their dispersion curves are shown in Fig. 5 and only one mode exists. Wave modes will disappear if the characteristic length exceeds a certain value, which limits the domain ξ' considerably. Large stiffness parameters of the beam relative to the half-space results in the high wave velocities of the beam will never become critical even the short-wave limit ($\xi' \rightarrow \infty$) is approached. For the long-wave limit ($\xi' \rightarrow 0$) wave velocities all approach the R-wave velocity. For this case there is also a minimum critical velocity V_{cr}^m.

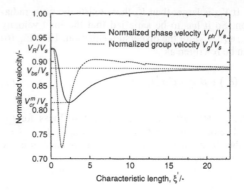

Figure 4. Dispersion curve of system of sub-soft beam and hard half-space.

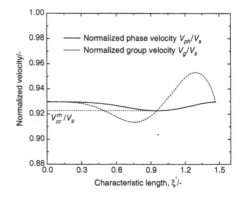

Figure 5. Dispersion curve of system of sub-hard beam and soft half-space.

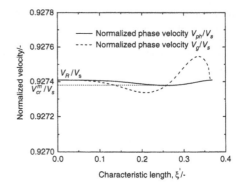

Figure 6. Dispersion curve of system of hard beam and soft half-space.

3.4 System of hard beam and soft half-space

It should be paid much attention to this case, especially for the high-speed train on soft soil in southeast of China. The wave velocities of the beam are much larger than those of the half-space. As shown in Fig. 6, the dispersion curves are very similar to those of the system of sub-hard beam and soft half-space. For the long-wave limit ($\xi' \to 0$) wave velocities all approach the R-wave velocity of the half-space. There is also a minimum critical velocity, and compared with the previous case, it is much closer to the R-wave velocity, $V_{cr}^m = 0.999967V_R$. The larger difference between wave velocities of the beam and the half-space is, the closer V_{cr}^m is to V_R. Suiker (1998) also showed that the two velocities will approach to a same value with the increasing of the beam stiffness continuously. For the latter two cases, the results are almost same, and we can take them both belong to the system of hard beam and soft half-space.

In one word, the critical velocities of system of the T beam and the half-space depend upon the relative relation of wave velocities of the beam and the half space. For systems of (sub-) hard beam and soft half-space, there are two critical velocities, V_{cr}^m and V_R. The harder the beam is relative to the half-space, the closer the two velocities are. For the system of soft beam and hard half-space, there are three critical velocities, V_{bs}, V_{bp} and V_R; but for the system of sub-soft beam and hard half-space, there are also three ones, V_{cr}^m, V_{bs} and V_R, which is not studied in the literature of Suiker (1998).

Because the influence of the Poisson's ratio, the mass densities of the beam and of the half space on the critical velocities is small, the critical velocities are mainly determined by the Young's modulus, the size of the beam and the shear modulus of the half-space. Eq. (7) shows that the influence of the beam size on the critical velocities depends on the ratio of height to width. With the fixed value of this ratio, the critical velocity will approach V_R with the continuously increasing of the Young's modulus of the beam. While it will approach V_{bs} with the continuously increasing of the shear modulus of the half space. So the up-limits of the critical velocities of the system of the T beam and the half-space system are V_R and V_{bs}.

4 CONCLUSIONS

The dispersion equation of T beam and half-space system is obtained by introducing the concept of equivalent stiffness of half-space interacting with a beam. The critical velocities are studied using dispersion curves for four beam-half space systems. The following conclusions can be drawn.

(a) The critical velocities of the T beam under the moving load depend on the relative relation of wave velocities of the beam and the half-space.
(b) The R-wave velocity of the half space is always a critical velocity.

(c) For the soft beam (relative to the half-space) the wave velocities of the beam are critical velocities. There is an additional minimum critical velocity except the system of soft beam and hard half-space.

ACKNOWLEDGEMENT

This research project is supported by the Research Fund for PhD Student of Chinese College through Grant No. 20040335083.

REFERENCES

Achenbach, J.D. & Sun, C.T. 1965. Moving load on a flexibly supported Timoshenko beam. *Int. J. Solids Structures*, Vol.1, 355–370.

Chen, Y.M., Wang, C.J., Ji, M.X. & Chen, R.P. 2003. Train-induced ground vibration and deformation. In: Yunmin Chen, Hirokazu Takemiya (ed.), *Environmental Vibration prediction, monitoring and evaluation*. Hangzhou, 2003. China Communications Press. 158–174.

Cowper, G.R. 1966. The shear coefficient in Timoshenko's beam theory. *J. Appl. Mech.*, Vol.33, 335–340.

Dieterman, H.A. & Metrikine, A.V. 1996. The equivalent stiffness of a half-space interacting with a beam. Critical velocities of a moving load along the beam. *Eur. J. Mech. A/Solids*, 15(1), 67–90.

Dieterman, H.A. & Metrikine, A.V. 1997. Steady-state displacements of a beam on an elastic half-space due to a uniformly moving constant load. *Eur. J. Mech. A/Solids.*, 16(2), 295–306.

Filippov, A.P. 1961. Steady-state vibrations of an infinite beam on elastic half-space subjected to a moving load. *Izvestija AN SSSR OTN Mehanika I Mashinostroenie*, 6, 97–105 (translated from Russian).

Kerr, A.D. 1972. Steady-state vibrations of beam on elastic foundation for moving load. *Int. J. Mech. Sci*, 14, 71–78.

Kenney, J.T. 1954. Steady-state vibrations of beam on elastic foundation for moving load. *Journal of Appl. Mech.*, 76, 359–364.

Labra, J.J. 1975. An axially stressed railroad track on an elastic continuum subjected to a moving load. *Acta Mechanica*, 22, 113–129.

Suiker, A.S.J., Borst, R. de & Esveld, C. 1998. Critical behavior of a Timoshenko beam-half plane system under a moving load. *Archive of Appl. Mech.*, 68, 158–168.

Wang, C.M., Yang, T.Q. & Lam, K.Y. 1997. Viscoelastic Timoshenko beam solutions from Euler-Bernoulli solutions. *J. Engng. Mech.*, 123(7), 746–748.

Xie. W.P., Hu. J.W. & Xu. J. 2002. Dynamic response of track-ground systems under high velocity moving load. *Journal of Rock Mechanics and Engineering (in Chinese)*, 21(7), 1075–1078.

Chen Y.M., Ji, M.X. & Huang, B. 2004. Effect of cyclic loading frequency on undrained behaviors of undisturbed marine clay. China Ocean Engineering, Vol. 18, No. 4.

Environmental Vibrations – Takemiya (ed.)
© *2005 Taylor & Francis Group, London, ISBN 0 415 39035 4*

Steady-state response of the plate on Kelvin foundation under moving loads

J.Q. Jiang & H.F. Zhou
Department of Civil Engineering, Zhejiang University, Hangzhou, Zhejiang, China

ABSTRACT: An infinite plate on Kelvin foundation subjected to moving loads was employed as the model to study the dynamic response of highway rigid pavement system under running vehicles. Four types of moving loads were considered, including constant and harmonic moving point load and rectangle load. In each case, the analytical solution for the steady-state displacement, a generalized double integral, was developed using Green's function method. Numerical evaluations of these integrals were followed using adaptive integration algorithm and the deflected shapes of the plate were obtained. The effect of load velocity and foundation damping on the deflection distribution, the maximum deflection, and its distance lag were investigated. It is found that there exists a critical velocity in the system which decreases as the foundation damping increases. The distribution and maximum of the deflection induced by moving rectangle load were compared with those induced by moving point load. No significant distinction was found.

1 INTRODUCTION

The plate on visco-elastic foundation is frequently used as a model to study the response of highway or runway pavement system under vehicle loads. In most previous studies (Westergaard 1925, Hung 1993), the vehicle loads have often been considered as static loads. However, the pavement systems in reality are subjected to moving loads. Dynamic response induced by a moving load may differ substantially from the static response induced by a static load as the velocity increases significantly. Furthermore, there are many pavement distresses those can only be explained through dynamic analysis of the pavement system. Hence, the problem of dynamic pavement response induced by moving vehicles has been one of increasing interest. Recently, works have been performed to study the dynamic response of pavement system subjected to moving vehicles, e.g., Kim and Roesset (1998), Liu et al. (2000), Hung & Thambiratnam (2002), Kim & McCullough (2003), Kim (2004a, b). However, few previous works were concerned with the effect of velocity and damping on the dynamic response. Therefore, the main efforts of this paper are concentrated on investigation of the key parameters to characterize the dynamic response of the pavement system caused by moving vehicles.

In this paper, an infinite plate on Kelvin foundation subjected to moving loads was employed as the model to study the dynamic response of rigid pavement system under moving vehicles. Four types of moving loads were considered, including constant and harmonic moving point load and rectangle load. In each case, the analytical solution for the steady-state displacement, a generalized double integral, was developed using Green's function method. Numerical evaluations of these integrals were followed using adaptive integration algorithm and the deflected shapes of the plate were obtained. The effect of load velocity and foundation damping on the deflection distribution, the maximum deflection, and its distance lag were investigated. It is found that there exists a critical velocity in the system which decreases as the foundation damping increases. The distribution and maximum of the deflection induced by moving rectangle load were compared with those induced by moving point load. No significant distinction was found.

2 GOVERNING EQUATIONS

The plate of infinite extent on Kelvin foundation was employed to model the highway or runway rigid pavement system. The concrete slab in the rigid pavement system was modeled using the Kirchhoff thin plate. The vertical stiffness and viscous damping of underlying layers in the rigid pavement system were modeled using Kelvin foundation. The governing equation for the plate resting on Kelvin foundation is given by

$$D\nabla^2\nabla^2 w + \mu w + \eta\frac{\partial w}{\partial t} + m\frac{\partial^2 w}{\partial t^2} = F \qquad (1)$$

where D is flexural rigidity of the plate defined by

$$D = \frac{Eh^3}{12(1-v^2)} \qquad (2)$$

where E, v and h are Young's modulus, Poisson's ratio and thickness of the plate, respectively; ∇^2 is the Laplacian operator; w is the displacement of the plate; μ and η are the vertical stiffness of underlying layers per unit of area, and viscous damping constant, respectively; m is the mass of the plate per unit of area.

Suppose that a force F moves at constant velocity c in the positive direction of x-axis from infinity to infinity. After a sufficient long time of force action, the dynamic response of the plate will have the steady-state form. The boundary conditions can be assumed that at infinite distance from the load the displacement as well as its derivatives is zero, i.e.

$$\begin{aligned} \lim_{x \to \pm\infty} \frac{\partial^n w}{\partial x^n} &= 0 \\ &\qquad\qquad n = 0,1,2,3\cdots \\ \lim_{y \to \pm\infty} \frac{\partial^n w}{\partial y^n} &= 0 \end{aligned} \qquad (3)$$

3 EXPRESSION FOR MOVING LOADS

3.1 A moving point source

Consider a concentrated load moving at constant velocity c in the positive direction of x-axis from infinity to infinity, the load can be expressed as

$$F(x,y,t) = f(t)\delta(x-ct)\delta(y) \qquad (4)$$

where $f(\cdot)$ is the amplitude variation function of the load; $\delta(\cdot)$ is the Dirac's function. For a moving point load with a constant magnitude of P, its expression is given as

$$F(x,y,t) = P\delta(x-ct)\delta(y) \qquad (5)$$

For a harmonic moving point load with amplitude of P and a circular frequency of ω_0, its expression is given as

$$F(x,y,t) = P e^{i\omega_0 t} \delta(x-ct)\delta(y) \qquad (6)$$

3.2 A moving area source

Because vehicle loads in practice will have a finite area over which they are distributed, the point load represents only an idealization, distributed loads were also studied in this paper. Consider a load arbitrary distributing over an area, which moves at constant velocity c in the positive direction of x-axis from infinity to infinity, the load can be expressed as

$$F(x,y,t) = f(t)g(x-ct,y) \qquad (7)$$

where $g(\cdot)$ is the distribution function of the load. For a load uniformly distributed over a rectangular area with constant load pressure and a resultant force of P, its expression is given as

$$F(x,y,t) = \frac{P}{4l_1 l_2} H\left(l_1^2 - (x-ct)^2\right) H\left(l_2^2 - y^2\right) \qquad (8)$$

where l_1 and l_2 are loaded length in the x and y directions; $H(\cdot)$ is the Heaviside's function. For a load uniformly distributed over a rectangular area with amplitude of the resultant force being P and a circular frequency of ω_0, its expression is given as

$$F(x,y,t) = \frac{P}{4l_1 l_2} e^{i\omega_0 t} H\left(l_1^2 - (x-ct)^2\right) H\left(l_2^2 - y^2\right) \qquad (9)$$

4 SOLUTIONS FOR THE DISPLACEMENTS

Green's function method is employed to find the analytical solution for the steady-state displacement of the plate under moving loads. The Green's function is the solution of Equation (1) by substituting a impulsive load acting at the origin of the coordinate system at time $t = 0$ with F in right hand of Equation (1). The displacement of the plate under specific moving load can be formulated as a convolution integral between specific moving load and Green's function, that is, the Duhamel's integral

$$\begin{aligned} w(x,y,t) = \int_S \int_{-\infty}^{\infty} & F(\zeta_1, \zeta_2, \tau) \\ & \times G(x-\zeta_1, y-\zeta_2, t-\tau; 0,0,0)\,d\tau dS \end{aligned} \qquad (10)$$

4.1 Green's function

Equation (1) is conveniently solved by integral transform. The double Fourier transform and its inverse adopted throughout are defined as

$$\begin{aligned} \bar{f}(k_1, k_2) &= \int_{-\infty}^{+\infty} \int_{-\infty}^{+\infty} f(x_1, x_2) e^{-i(k_1 x_1 + k_2 x_2)}\,dx_1 dx_2 \\ f(x_1, x_2) &= \frac{1}{4\pi^2} \int_{-\infty}^{+\infty} \int_{-\infty}^{+\infty} \bar{f}(k_1, k_2) e^{i(k_1 x_1 + k_2 x_2)}\,dk_1 dk_2 \end{aligned} \qquad (11)$$

The Laplace transform and its inverse are defined as

$$\tilde{f}(s) = \int_0^{+\infty} f(t) e^{-st} \, dt$$

$$f(t) = \frac{1}{2\pi i} \int_{a-i\infty}^{a+i\infty} \tilde{f}(s) e^{st} \, ds \tag{12}$$

By applying the double Fourier transform to Equation (1) with respect to variables x and y, one can obtain

$$D\left(k_1^2 + k_2^2\right)^2 \overline{w} + \mu \overline{w} + \eta \frac{\partial \overline{w}}{\partial t} + m \frac{\partial^2 \overline{w}}{\partial t^2} = \overline{F}_\delta \tag{13}$$

By applying the Laplace transform to the above equation with respective to variable t, one can obtain

$$D\left(k_1^2 + k_2^2\right)^2 \tilde{\overline{w}} + \mu \tilde{\overline{w}} + \eta s \tilde{\overline{w}} + ms^2 \tilde{\overline{w}} = \tilde{\overline{F}}_\delta \tag{14}$$

The transformed displacement can be obtained from the above equation

$$\tilde{\overline{w}} = \frac{1}{m} \frac{\tilde{\overline{F}}_\delta}{\left(s + p\right)^2 + \alpha^2} \tag{15}$$

where

$$p = \frac{\eta}{2m} \tag{16}$$

$$\alpha^2 = \alpha^2\left(k_1, k_2\right) = \frac{D}{m}\left(k_1^2 + k_2^2\right)^2 + \frac{\mu}{m} - p^2 \tag{17}$$

By applying inverse Laplace transform to the above equation, one can obtain

$$\overline{w}\left(k_1, k_2, t\right) = \frac{1}{m} \int_0^t \overline{F}_\delta\left(k_1, k_2, \tau\right)$$

$$\times e^{-p(t-\tau)} \frac{\sin\left(\alpha\left(k_1, k_2\right)\left(t - \tau\right)\right)}{\alpha\left(k_1, k_2\right)} d\tau \tag{18}$$

Finally, the displacement can be obtained using the double inverse Fourier transform

$$w\left(x, y, t\right) = \frac{1}{4\pi^2 m} e^{-pt} \times$$

$$\int_{-\infty}^{+\infty} \int_{-\infty}^{+\infty} \frac{\sin\left(\alpha\left(k_1, k_2\right)t\right)}{\alpha\left(k_1, k_2\right)} e^{i(k_1 x + k_2 y)} \, dk_1 dk_2$$

$$= G\left(x, y, t; 0, 0, 0\right) \tag{19}$$

So far, Green's function of the plate on Kelvin foundation has been obtained. It's the basis for deriving the displacement induced by moving loads.

4.2 Solution for the steady-state displacement

4.2.1 Steady-state displacement under constant moving point load

Substituting Equation (5) into Equation (10), one obtains

$$w = \frac{P}{4\pi^2 m} \int_{-\infty}^t \int_{-\infty}^{+\infty} \int_{-\infty}^{+\infty} e^{-p(t-\tau)}$$

$$\times \frac{\sin\left(\alpha\left(k_1, k_2\right)\left(t - \tau\right)\right)}{\alpha\left(k_1, k_2\right)} e^{i(k_1(x - c\tau) + k_2 y)} \, dk_1 dk_2 d\tau \tag{20}$$

By letting

$$I_1\left(k_1, k_2\right) = \int_{-\infty}^t e^{-p(t-\tau)} \frac{\sin\left(\alpha\left(k_1, k_2\right)\left(t - \tau\right)\right)}{\alpha\left(k_1, k_2\right)} e^{ick_1(t-\tau)} \, d\tau \tag{21}$$

Equation (20) can be rewritten as

$$w = \frac{P}{4\pi^2 m} \int_{-\infty}^{+\infty} \int_{-\infty}^{+\infty} I_1\left(k_1, k_2\right) e^{i(k_1(x - x_c) + k_2 y)} \, dk_1 dk_2 \tag{22}$$

where $x_c = ct$ is the x coordinate of the load location at time t. Equation (21) can be integrated by parts and the final result is as follow

$$I_1\left(k_1, k_2\right) = \frac{m}{D\left(k_1^2 + k_2^2\right)^2 - mck_1^2 + \mu - i\eta ck_1} \tag{23}$$

Substituting the above equation into Equation (22), one can finally obtain the steady-state displacement of the plate induced by constant moving point load

$$w = \frac{P}{4\pi^2} \times$$

$$\int_{-\infty}^{+\infty} \int_{-\infty}^{+\infty} \frac{e^{i(k_1(x - x_c) + k_2 y)}}{D\left(k_1^2 + k_2^2\right)^2 - mck_1^2 + \mu - i\eta ck_1} \, dk_1 dk_2 \tag{24}$$

4.2.2 Steady-state displacement under harmonic moving point load

Substituting Equation (6) into Equation (10), one obtains

$$w = \frac{P}{4\pi^2 m} \int_{-\infty}^t \int_{-\infty}^{+\infty} \int_{-\infty}^{+\infty} e^{i\omega_0 \tau} e^{-p(t-\tau)}$$

$$\times \frac{\sin\left(\alpha\left(k_1, k_2\right)\left(t - \tau\right)\right)}{\alpha\left(k_1, k_2\right)} e^{i(k_1(x - c\tau) + k_2 y)} \, dk_1 dk_2 d\tau \tag{25}$$

123

By letting

$$I_2(k_1,k_2) = \int_{-\infty}^{t} e^{-p(t-\tau)} \frac{\sin(\alpha(k_1,k_2)(t-\tau))}{\alpha(k_1,k_2)}$$

$$\times e^{-i(\omega_0 - ck_1)(t-\tau)} \, d\tau \tag{26}$$

Equation (25) can be rewritten as

$$w = \frac{P}{4\pi^2 m} e^{i\omega_0 t} \int_{-\infty}^{+\infty}\int_{-\infty}^{+\infty} I_2(k_1,k_2) e^{i(k_1(x-x_c)+k_2 y)} \, dk_1 dk_2 \tag{27}$$

where $x_c = ct$ is the x coordinate of the load location at time t. Similar to Equation (21), Equation (26) can be integrated by parts and the result is as follow

$$I_2(k_1,k_2)$$

$$= \frac{m}{D(k_1^2+k_2^2)^2 - m(\omega_0 - ck_1)^2 + \mu + i\eta(\omega_0 - ck_1)} \tag{28}$$

Substituting the above equation into Equation (27), one can finally obtain the steady-state displacement of the plate induced by harmonic moving point load. Due to the limited space, here it's omitted.

4.2.3 Steady-state displacement under constant moving rectangle load

Substituting Equation (8) into Equation (10), one can obtain the steady-state displacement of the plate induced by constant moving rectangle load

$$w = \frac{P}{4\pi^2 m} \int_{-\infty}^{+\infty}\int_{-\infty}^{+\infty} \frac{\sin(k_1 l_1)}{k_1 l_1} \frac{\sin(k_2 l_2)}{k_2 l_2}$$

$$\times I_1(k_1,k_2) e^{i(k_1(x-x_c)+k_2 y)} \, dk_1 dk_2 \tag{29}$$

where $I_1(k_1, k_2)$ is defined by Equation (23).

4.2.4 Steady-state displacement under harmonic moving rectangle load

Substituting Equation (9) into Equation (10), one can obtain the steady-state displacement of the plate induced by harmonic moving rectangle load

$$w = \frac{P}{4\pi^2 m} e^{i\omega_0 t} \int_{-\infty}^{+\infty}\int_{-\infty}^{+\infty} \frac{\sin(k_1 l_1)}{k_1 l_1} \frac{\sin(k_2 l_2)}{k_2 l_2}$$

$$\times I_2(k_1,k_2) e^{i(k_1(x-x_c)+k_2 y)} \, dk_1 dk_2 \tag{30}$$

where $I_2(k_1, k_2)$ is defined by Equation (28).

4.3 Steady-state displacement of the plate on Winkler foundation

Winkler foundation is the particular case of Kelvin foundation when there is no damping in the foundation. It's convenient to obtain the steady-state displacement of the plate on Winkler foundation from the counterpart of the plate on Kelvin foundation by letting $\eta = 0$.

4.3.1 Steady-state displacement under constant moving point load

By letting $\eta = 0$ in Equation (24), one can obtain the displacement of the plate on Winkler foundation induced by constant moving point load

$$w = \frac{P}{4\pi^2} \int_{-\infty}^{+\infty}\int_{-\infty}^{+\infty} \frac{e^{i(k_1(x-x_c)+k_2 y)}}{D(k_1^2+k_2^2)^2 - mck_1^2 + \mu} \, dk_1 dk_2 \tag{31}$$

4.3.2 Steady-state displacement under constant moving rectangle load

Similarly, one can obtain the displacement of the plate on Winkler foundation induced by constant moving rectangle load

$$w = \frac{P}{4\pi^2} \int_{-\infty}^{\infty}\int_{-\infty}^{\infty} \frac{\sin(k_1 l_1)}{k_1 l_1} \frac{\sin(k_2 l_2)}{k_2 l_2}$$

$$\times \frac{e^{i(k_1(x-x_c)+k_2 y)}}{D(k_1^2+k_2^2)^2 - mc^2 k_1^2 + \mu} \, dk_1 dk_2 \tag{32}$$

4.3.3 Critical velocity of the plate on Winker foundation

To examine the singularity of the integrals in Equations (31) and (32), one may let

$$D(k_1^2+k_2^2)^2 - mc^2 k_1^2 + \mu = 0 \tag{33}$$

From the above equation, the velocity can be obtained

$$c = \sqrt{\frac{D(k_1^2+k_2^2)^2}{m} \frac{1}{k_1^2} + \frac{\mu}{m}\frac{1}{k_1^2}} \geq \left(\frac{4\mu D}{m^2}\right)^{\frac{1}{4}} \tag{34}$$

Letting

$$c_{cr} = \left(\frac{4\mu D}{m^2}\right)^{\frac{1}{4}} \tag{35}$$

The integrals will have singularities when $c \geq c_{cr}$. Call c_{cr} the critical velocity of the plate on Winkler foundation.

124

5 NUMERICAL EVALUATIONS

At this stage, the analytical solutions for the steady-state displacements of the plate induced by moving loads have been made available. Numerical evaluations of these integrals are followed using adaptive integration algorithm. The material properties and geometry of the pavement models and dimensions of the axle loads used in this study are listed in Table 1. A wider range of damping value was used in this study to investigate the effect of damping.

The velocity and damping are divided by critical velocity and critical damping, respectively, to obtain relative velocity and relative damping.

$$\theta = \frac{c}{c_{cr}} \qquad (36)$$

$$\xi = \frac{\eta}{\eta_{cr}} = \frac{\eta}{2\sqrt{\mu m}} \qquad (37)$$

The corresponding critical velocity and critical damping are $439.5\,\mathrm{m \cdot s^{-1}}$ and $436.1\,\mathrm{kNs \cdot m^{-3}}$, respectively.

5.1 Deflection under constant moving point load

5.1.1 Deflected shape of the plate
Figure 1 shows the three-dimensional deflected shape of the plate in the vicinity of the load at the instant $t = 0$, the moment at which the load passes through the origin $(0, 0)$. To evaluate the effect of velocity on the distribution of deflection, deflected shapes of the plate subjected to loads moving at both sub-critical

Table 1. Properties of the rigid pavement models.

E $\mathrm{N \cdot m^{-2}}$	υ –	ρ $\mathrm{kg \cdot m^{-3}}$	h m	μ $\mathrm{N \cdot m^{-3}}$	P N	l_1 m	l_2 m
2.8×10^{10}	0.15	2.3×10^3	0.25	68.9×10^6	10.5×10^3	0.09	0.10

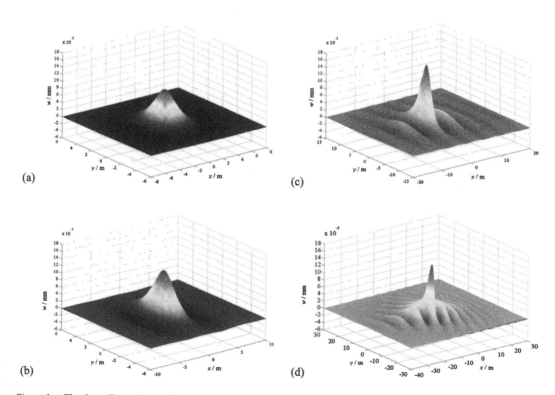

Figure 1.   The three-dimensional deflected shapes ($\xi = 0.1$): (a) $\theta = 0.05$; (b) $\theta = 0.80$; (c) $\theta = 1.00$; (d) $\theta = 1.20$.

and super-critical ranges were computed. As can be seen, the deflected shapes for loads moving in the sub-critical region differ substantially from those for loads moving in the super-critical region. To have a more comprehensive understanding on the differences of the deflected shape under various velocities, Figure 2 and Figure 3 plot the deflected shapes along x-axis and y-axis, respectively. The deflected shapes for no damping were also presented in these two figures so as to investigate the effect of damping.

From the deflected shape along the x-axis and y-axis, shown in Figure 2 and Figure 3, respectively, the following point can be concluded.

1. Without damping, the deflected shapes are symmetric with respect to the two axes that pass through the load. The maximum displacement occurs at the location of the load. With damping, the deflected shapes in the x direction are no longer symmetric. There is a lag between the position where the maximum deflection occurs and the location of the load. This lag is referred to as "distance lag". It depends on many factors, such as velocity, damping and so on. The effects of velocity and damping on the distance lag will de discussed in detail.

2. Without damping, the deflected region of the plate and the fluctuation frequency of the deflection in

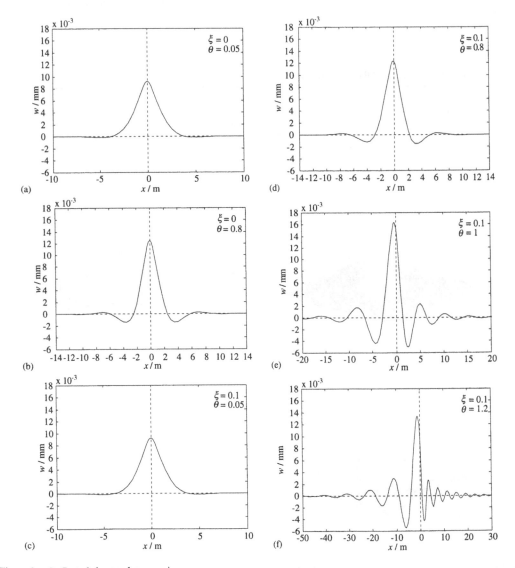

Figure 2. Deflected shapes along x-axis.

126

front of the load and those in rear of the load is the same. With damping, the deflected region in front of the load is less than that in rear of the load. The frequency of fluctuations in front of the load is larger than that in rear of the load.

3. As velocity increases, the deflected regions both in x and y directions are more widely spread, and more pronounced fluctuations occur both in x and y directions. Comparatively, the effect of velocity on fluctuations of the deflection is more pronounced in x direction than in y direction.

4. The maximum displacement doesn't increase monotonously with increasing velocity. The maximum

displacement value varies with velocity, damping and many other factors. The effects of velocity and damping on the maximum displacement will be discussed in detail.

5.1.2 *Effect of velocity on the maximum displacement*

The effect of velocity on the maximum displacement can be observed in Figure 4. When damping is small, shown in Figure 4 (a), the maximum displacement increases as velocity increases until it becomes close to the critical velocity, where the maximum displacement reaches a maximum. Beyond the critical velocity,

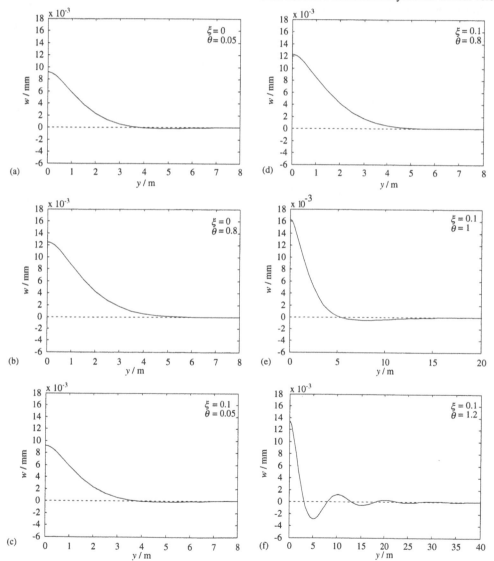

Figure 3.   Deflected shapes along y-axis.

the maximum displacement decreases with increasing velocity. As damping increases, shown in Figure 4 (b), the velocity where the maximum displacement reaches a maximum decreases, i.e., it becomes less than the critical velocity. With further increasing damping, shown in Figure 4(c), the maximum displacement decreases monotonously as velocity increases. Therefore, in this case, the maximum displacement induced by moving load is even less than that induce by static load. Define the velocity where the maximum displacement reaches a maximum as the critical velocity of the plate on Kelvin foundation. The variation of critical velocity with damping is shown in Figure 5. When the damping is small, the critical velocity of the

plate on Kelvin foundation is the same as that of the plate on Winkler foundation. As the damping increases, the critical velocity of the plate on Kelvin foundation decreases until it becomes zero.

5.1.3 Effect of damping on the maximum displacement

The effect of damping on the maximum displacement can be observed in Figure 6. As a general rule, the maximum displacement decreases as damping increases. However, the effect of damping on the maximum

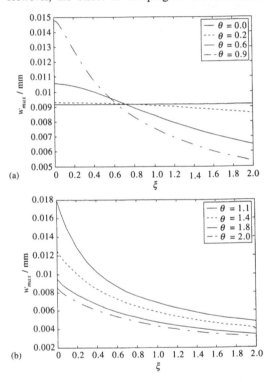

Figure 6. Effect of damping on the maximum displacement.

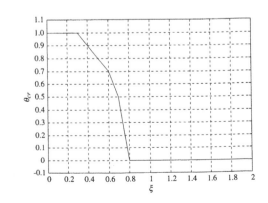

Figure 5. Relation between critical velocity and damping.

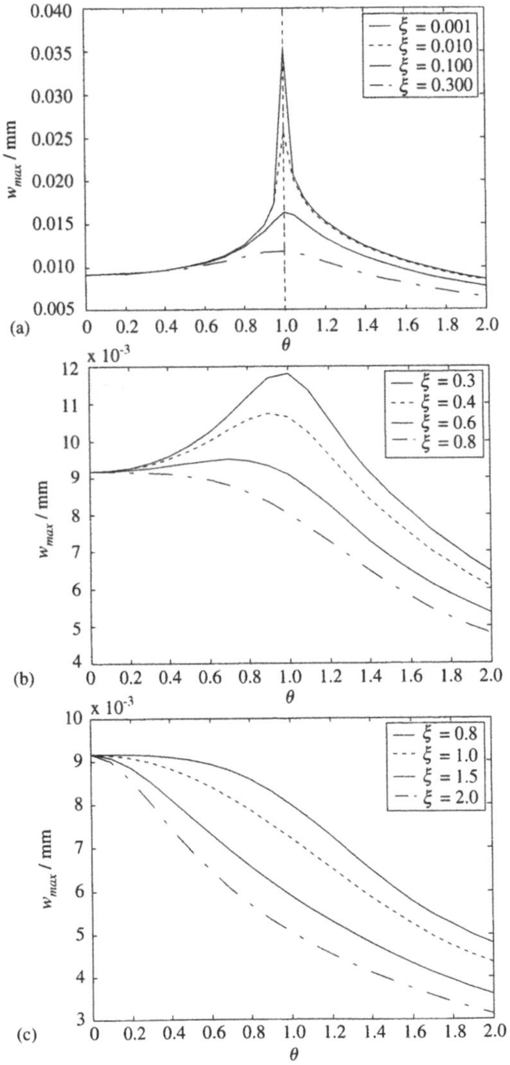

Figure 4. Effect of velocity on the maximum displacement.

displacement varies with velocity. For velocities in sub-critical range, the effect of damping on the maximum displacement becomes more pronounced as damping increases. For velocities in super-critical range, it gets slighter as damping increases.

5.1.4 Effect of velocity on the distance lag

Figure 7 shows the effect of velocity on the distance lag. As can be seen, the distance lag increases as velocity increases. However, the effect of velocity on the distance lag is related with damping. It becomes slighter as damping increases.

5.1.5 Effect of damping on the distance lag

Figure 8 shows the effect of damping on the distance lag. It is obvious that the effect of damping on the distance lag varies with velocity. For velocities in sub-critical range, the distance lag increases as damping increase. For velocities in super-critical range, the distance lag on the contrary decreases as damping increases.

5.2 Deflection under constant moving rectangle load

Because moving loads in practice will have a finite area over which they are distributed, and the point load

represents only an idealization, distributed loads were considered with the logical values of the tire print area. The main focus was concentrated on the comparison of the distribution and maximum of the displacement induced by moving point load and moving rectangle load. The deflected shapes of the plate subjected to a constant moving rectangle load were computed and compared with those for a constant moving point load. The resultant force of the moving rectangle load is equal to the magnitude of the moving point load. No significant differences were found in the deflected shapes of the plate. Therefore, the deflected shapes of the plate induced by moving rectangle load are not provided here to save the space. Table 2 lists

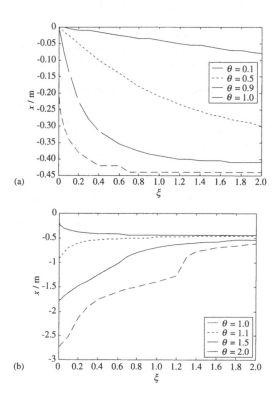

Figure 8. Effect of damping on the distance lag.

Table 2. Maximum displacements induced by moving rectangle load and moving point load.

θ	w_{max} (10^{-3} mm)		Δw_{max} (10^{-3} mm)	e (100%)
	MPL	MRL		
0.05	9.184	9.150	0.034	0.37
0.8	12.285	12.246	0.039	0.32
1.0	16.273	16.231	0.042	0.26
1.2	13.332	13.313	0.019	0.14

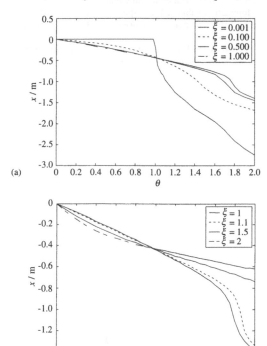

Figure 7. Effect of velocity on the distance lag.

the maximum displacements induced by moving rect-angle load and those induced by moving point load, where e is defined by

$$e = \frac{\Delta w_{max}}{w_{max}^2} \times 100\% = \frac{w_{max}^1 - w_{max}^2}{w_{max}^2} \times 100\% \qquad (38)$$

where w_{max}^1 and w_{max}^2 are the maximum displacement induced by point load and rectangle load, respectively. As can be seen, the difference in the maximum dis-placement induced by the two types of load is also negligible.

6 SUMMARY AND CONCLUSION

The dynamic response of rigid pavement system under moving vehicles was studied by employing the infinite plate on Kelvin foundation subjected to moving loads. Four types of moving loads were considered, including constant and harmonic moving point load and rectangle load. The analytical solutions for the displacements, generalized double integrals, were developed using Green's function method. Numerical evaluations of the integrals were followed using adap-tive integration algorithm and the deflected shapes of the plate were obtained.

The main efforts of the paper were concentrated on investigating the key parameters to characterize the dynamic response of the pavement system. The effect of velocity and damping on the deflection distribution, the maximum deflection, and its distance lag were investigated extensively. It is found that there exists a critical velocity in the system, which decreases as damping increases.

The distribution and maximum of the deflection induced by moving rectangle load were compared with those induced by moving point load. No significant dis-tinction was found except that the maximum deflec-tion induced by moving rectangle load is a bit less than that by moving point load.

REFERENCES

Hung, Y.H. 1993. Pavement analysis and design. Prentice Hall, New Jersey.

Hung, M.H. & Thambiratnam, D.P. 2002. Dynamic response of plates on elastic foundation to moving loads. ASCE Journal of Engineering Mechanics 128(9), 1016–1022.

Liu, C., McCullough, B.F. & Oey, H.S. 2000. Response of rigid pavements due to vehicle-road interaction. ASCE Journal of Transportation Engineering 126(3), 237–242.

Kim, S.M. & Roesset, J.M. 1998. Moving loads on a plate on elastic foundation. ASCE Journal of Engineering Mechanics 124(9), 1010–1016.

Kim, S.M. & McCullough, B.F. 2003. Dynamic response of plate on viscous Winkler foundation to moving loads of varying amplitude. Engineering Structures 25(9), 1179–1188.

Kim, S.M. 2004a. Influence of horizontal resistance at plate bottom on vibration of plates on elastic foundation under moving loads. Engineering Structures 26, 519–529.

Kim, S.M 2004b. Buckling and vibration of a plate on elastic foundation subjected to in-plane compression and mov-ing loads. International Journal of Solids and Structures 41, 5647–5661.

Westergaard, H.M. 1925. Stress in concrete pavements com-puted by theoretical analysis. Public Roads 7, 25–35.

Environmental Vibrations – Takemiya (ed.)
© 2005 Taylor & Francis Group, London, ISBN 0 415 39035 4

Dynamic simulation of vehicle-track coupling vibrations for linear metro system

Y.W. Feng, Q.C. Wei, L. Gao & J. Shi
School of Civil Engineering and Architecture, Beijing Jiaotong University, Beijing, China

ABSTRACT: Linear metro system is developed for urban transportation which is driven by linear induction motor (LIM). The vehicle mounted LIM interacts with the reaction plate (RP) fixed on the track system. A detailed model is built to analyze the vertical coupling vibrations characteristics of this system. The model consists of vehicle and track subsystem. The vehicle subsystem owns 10 DOFs. The track subsystem consists of three layers which are rail, track slab and rigid foundation. The track and the slab are both modeled by finite element method. The wheel/rail interaction is described by the Hertzian contact theory. The vertical attractive force between the LIM and its RP is modeled as special springs varied with the air gap. Dynamic simulation is carried out in Fortran program. Proper computing step has been determined through numerical test. The system dynamic characteristics and the air gap variations caused by the track irregularities have been studied.

1   INTRODUCTION

Linear metro system blends the design principles of conventional rapid transit system and automated people mover system. The car is driven by linear induction motor (see Figure 1), at the same time supported and oriented by wheels. It is a comparatively new urban transportation system which is economical, fast, quiet, comfortable and safe.

This system has been already used in 9 lines in 4 countries. The Beijing airport line and the Guangzhou metro line 4 have also decided to adopt this system, both of them have been designed and under construction.

In the dynamic model of linear metro system, the interaction between vehicle and track is coupling. The rails guide and sustain the wheels. The LIM suspended on the bogie interacts with the RP on the track and generates electromagnetic force between them. The longitudinal force is used for propulsion and braking, the vertical electromagnetic force varies with the air gap. All this made linear metro system unique of its dynamic characteristics.

There are mainly two countries develop the technology of the linear metro system, one is Canada and the other is Japan.

Fatemi (1993, 1996) from Canada carried out dynamic analysis of a new track for linear metro system in his doctor degree thesis, in which the vehicle and track system models were set up separately. He studied the influences of a number of parameters on

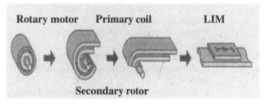

Figure1.   LIM formation theory.

the dynamic behavior of the track and the variation of the gap.

In China, the linear metro system has just been introduced, and the study on this system is only at its beginning phase. Pang (2004) did some qualitative dynamic study recently.

This paper establishes the vertical vehicle-track coupling dynamic model for linear metro system, carries out dynamic simulation computation and conducts some parametric studies. The vibration characteristics of this system are also analyzed.

2   SYSTEM MODEL

A detailed model is built to analyze the vertical coupling vibrations of the linear metro system. The model consists of vehicle subsystem and track subsystem. The vehicle subsystem is modeled as a combination of masses, springs and dampers with 10 DOF running

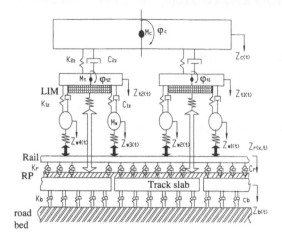

Figure 2. Vehicle-track coupling dynamic model.

Table 1. Basic parameters.

M_c, M_t	Car body mass, bogie mass
M_w, M_L	Wheel set mass, LIM mass
I_c, I_t	Car body inertia, bogie inertia
M_r, M_b	Rail unit length mass, track slab unit length mass
I_r, I_b	Rail inertia, track slab inertia
C_{1z}, C_{2z}	Primary suspension damping, secondary suspension damping
K_{1z}, K_{2z}	Primary suspension stiffness, secondary suspension stiffness
K_r, C_r	Rail pad stiffness, Rail pad damping
K_b, C_b	Stiffness under track slab, damping under track slab
l_c, l_t	Half of bogie distance, half of wheel set distance
l_r, l_b	Rail element length, track slab element length
h	Air gap between LIM and RP
z_c, z_{ti}, z_{wj}	Displacement of car body, bogie and wheel sets
ϕ_c, ϕ_{ti}	Pitching angular displacement of car body and bogie
z_{ri}, z_{bi}	Displacement of rail node, Displacement of track slab node
Φ_{ri}, ϕ_{bi}	Pitching angular displacement of rail node and track slab node
$F_i(t)$	Wheel/rail contact force
$F_i(h)$	Force between LIM and RP
$M_i(h)$	Force moment between LIM and RP
E_r, E_b	Rail elastic modulus, track slab elastic modulus

on the track with a constant velocity. The LIM is hung from the bogie, which is modeled as fixed tightly on the bogie. The track subsystem consists of three layers which are rail, slab and rigid foundation. Because the reaction plate is comparatively light and is fixed tightly on the slab, it is modeled as a part of the slab. The rail-pad under the rail and the emulsion asphalt under the concrete slab are modeled as dampers and springs. The rail is modeled as point supporting beam element, and the slab is modeled as continuous supporting beam element.

The interface between these two subsystems is the wheel/rail interaction described by the Hertzian non-linear elastic contact theory. The vertical attractive force between each LIM and its RP is modeled as special springs which are influenced by the air gap. When the air gap varies in a certain scope, the vertical attractive force between LIM and RP will increase while the air gap decrease, and vice versa.

Figure 2 shows system model and Table 1 shows the meaning of the symbols.

3 VERTICAL VEHICLE-TRACK COUPLING DYNAMIC EQUATIONS

3.1 Vehicle equations

(1) Bounce of the car body

$$M_c \ddot{z}_c + 2C_{2z}\dot{z}_c + 2K_{2z}z_c - C_{2z}\dot{z}_{t1} - C_{2z}\dot{z}_{t2}$$
$$-K_{2z}z_{t1} - K_{2z}z_{t2} = M_c g \qquad (1)$$

(2) Pitch of the car body

$$I_c \ddot{\varphi}_c + 2C_{2z}l_c^2 \dot{\varphi}_c + 2K_{2z}l_c^2 \varphi_c - C_{2z}\dot{z}_{t1} + C_{2z}\dot{z}_{t2}$$
$$-K_{2z}z_{t1} + K_{2z}z_{t2} = 0 \qquad (2)$$

(3) Bounce of the front bogie

$$M_t \ddot{z}_{t1} + (C_{2z} + 2C_{1z})\dot{z}_{t1} + (K_{2z} + 2K_{1z})\dot{z}_{t1}$$
$$-C_{2z}\dot{z}_c - K_{2z}z_c - C_{1z}\dot{z}_{w1} - K_{1z}z_{w1} - C_{1z}\dot{z}_{w2}$$
$$-K_{1z}z_{w2} - C_{2z}l_c\dot{\varphi}_c - K_{2z}l_c\varphi_c = M_t g + F_1(h) \qquad (3)$$

(4) Pitch of the front bogie

$$I_t \ddot{\varphi}_{t1} + 2C_{1z}l_t^2 \dot{\varphi}_{t1} + 2K_{1z}l_t^2 \varphi_{t1} - C_{1z}l_t\dot{z}_{w1} + C_{1z}l_t\dot{z}_{w2}$$
$$-K_{1z}l_t z_{w1} + K_{1z}l_t z_{w2} = M_1(h) \qquad (4)$$

(5) Bounce of the back bogie

$$M_t \ddot{z}_{t2} + (C_{2z} + 2C_{1z})\dot{z}_{t2} + (K_{2z} + 2K_{1z})\dot{z}_{t2}$$
$$-C_{2z}\dot{z}_c - K_{2z}z_c - C_{1z}\dot{z}_{w3} - K_{1z}z_{w3} - C_{1z}\dot{z}_{w4}$$
$$-K_{1z}z_{w4} + C_{2z}l_c\dot{\varphi}_c + K_{2z}l_c\varphi_c = M_t g + F_2(h) \qquad (5)$$

(6) Pitch of the back bogie

$$I_t \ddot{\varphi}_{t2} + 2C_{1z}l_t^2 \dot{\varphi}_{t2} + 2K_{1z}l_t^2 \varphi_{t2} - C_{1z}l_t\dot{z}_{w3} + C_{1z}l_t\dot{z}_{w4}$$
$$-K_{1z}l_t z_{w3} + K_{1z}l_t z_{w4} = M_2(h) \qquad (6)$$

(7) Bounce of the wheel sets

$$M_w \ddot{z}_{w1} + C_{1z}\dot{z}_{w1} + K_{1z}z_{w1} - C_{1z}\dot{z}_{t1} - K_{1z}z_{t1}$$
$$-C_{1z}l_t\dot{\varphi}_{t1} - K_{1z}l_t\varphi_{t1} = M_w g + F_1(t) \qquad (7)$$

$$M_w \ddot{z}_{w2} + C_{1z} \dot{z}_{w2} + K_{1z} z_{w2} - C_{1z} \dot{z}_{t1} - K_{1z} z_{t1}$$
$$+ C_{1z} l_t \dot{\varphi}_{t1} + K_{1z} l_t \varphi_{t1} = M_w g + F_2(t) \tag{8}$$

$$M_w \ddot{z}_{w3} + C_{1z} \dot{z}_{w3} + K_{1z} z_{w3} - C_{1z} \dot{z}_{t2} - K_{1z} z_{t2}$$
$$- C_{1z} l_t \dot{\varphi}_{t2} - K_{1z} l_t \varphi_{t2} = M_w g + F_3(t) \tag{9}$$

$$M_w \ddot{z}_{w4} + C_{1z} \dot{z}_{w4} + K_{1z} z_{w4} - C_{1z} \dot{z}_{t2} - K_{1z} z_{t2}$$
$$+ C_{1z} l_t \dot{\varphi}_{t2} + K_{1z} l_t \varphi_{t2} = M_w g + F_4(t) \tag{10}$$

3.2 Track equations

$$[M]_r \{\ddot{z}\}_r + [C]_r \{\dot{z}\}_r + [K]_r \{z\}_r = \{F\}_r$$
$$[M]_b \{\ddot{z}\}_b + [C]_b \{\dot{z}\}_b + [K]_b \{z\}_b = \{F\}_b \tag{11}$$

Where

$$\{z\}_r = \left\{ z_{r1} \quad \varphi_{r1} \quad z_{r2} \quad \varphi_{r2} \quad \cdots \quad z_{rn} \quad \varphi_{rn} \quad \cdots \right\}^T$$
$$\{z\}_b = \left\{ z_{b1} \quad \varphi_{b1} \quad z_{b2} \quad \varphi_{b2} \quad \cdots \quad z_{bn} \quad \varphi_{bn} \quad \cdots \right\}^T \tag{12}$$

In above equations, $[M]_r$, $[K]_r$, $[C]_r$ and $[M]_b$, $[K]_b$, $[C]_b$ are superposed by element mass matrix, element stiffness matrix and element damping matrix of rail element and slab element. F_r is the external load of the rail, includes wheel/rail interaction force and rail pad holding power. F_b is the external load of the slab, includes vertical interaction force between LIM and RP and the holding power under the slab.

3.3 Interaction between wheel and rail

The vertical interaction between wheel and rail followed the Hertzian nonlinear elastic contact theory. It is expressed as the following equation:

$$F = [\Delta z(t) / G]^{3/2} \tag{13}$$

Where G is the contact constant of wheel and rail. It is determined by the wheel type and the wheel radius; ΔZ_t is the elastic compression between the wheel and rail.

3.4 Interaction between the LIM and its RP

From the existing datum of the Japan metro association (Figure 3), we can see that the vertical electric magnetic

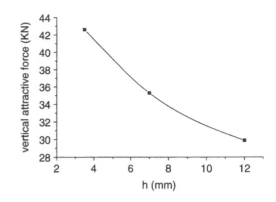

Figure 3. Relationship between vertical electric magnetic force and the gap.

force between the LIM and its RP decreased while the air gap increased. Figure 3 is obtained when the speed is 62 KM/h. In this article the relationship between them is modeled as special springs.

3.5 System equations

The whole system equations can be expressed as equation (14):

$$[M]\{\ddot{z}\} + [C]\{\dot{z}\} + [K]\{z\} = \{F\} \tag{14}$$

Where $[M]$, $[C]$, $[K]$, $[F]$ are the mass, damping, stiffness and external load matrices of the system.

$\{z\}$ is the displacement of this system. Newmark method has been used to solve this problem and computation is carried out in Fortran language.

4 SIMULATION RESULTS AND PARAMETRIC STUDIES

4.1 Vehicle and track parameters

Vehicle and track parameters used in this study are shown in Table 2.

4.2 Irregularity model

In this article, local cosine irregularity is used, which can be expressed as following:

$$Z_0(t) = \frac{1}{2} h (1 - \cos \frac{2\pi x}{\lambda}) \tag{15}$$

Where λ is the irregularity wavelength and h is the wave amplitude.

Table 2. Vehicle and track parameters.

Notation	Unit	Value	Notation	Unit	Value
M_c	kg	9171	E_r	N/m^2	2.1E11
M_t	kg	1971	I_r	m^4	3.09E-5
M_w	kg	323	M_r	kg/m	60.8
M_L	kg	650	l_r	m	0.625
I_c	kgm^2	12005	K_r	N/m	6 ×107
I_t	kgm^2	3200	C_r	N.s/m	3.625E4
K_{1z}	N/m	6.2E7	E_b	N/m^2	3.5E10
C_{1z}	N.s/m	5.4E5	I_b	m4	6.6875E-4
K_{2z}	N/m	2.13E4	M_b	Kg/m	2500
C_{2z}	N.s/m	1.67E5	l_b	m	3.6
l_t	m	0.85	K_b	N/m^3	1.25 E9
l_c	m	3.97	C_b	N.s/m^2	8.3 E4

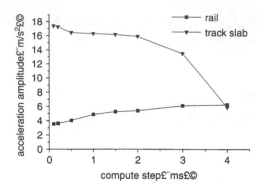

Figure 5. Computing step influences on rail acceleration and track slab acceleration.

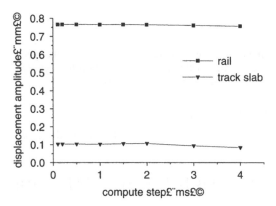

Figure 4. Computing step influences on rail displacement and track slab displacement.

4.3 Computing step test

For seeking the proper computing step of this program, numerical test has been done from 0.1 ms to 4 ms. The results are shown in Figure 4 and Figure 5. If the computing step is too small, it will waste computing time to get the results, and if it is too large, the results will not be correct.

From Figures 4 and 5, we can see that when the computing step is larger than 2 ms, the displacement and acceleration of the rail and the track slab are no longer stable. Acceleration changed more violently than displacement, especially the slab acceleration. It is concluded that the proper computing step of this system is 2 ms.

4.4 Cosine irregularity simulation results

Running speed is 90 km/h, computing step is 2 ms, and the cosine irregularity is put in the middle of the running length. Take $\lambda = 0.5$, and a = 1 mm as an example. The computing results are shown as following.

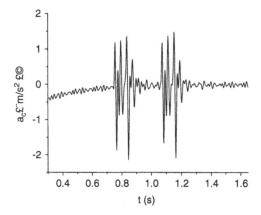

Figure 6 Influence of cosine irregularity on car body acceleration.

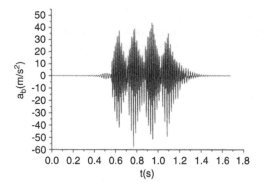

Figure 7. Influence of cosine irregularity on track slab acceleration.

Figures 6 and 7 are the vehicle and track dynamic characteristic curves. Figure 8 is the track and slab displacement amplitude variation with the irregularity amplitude.

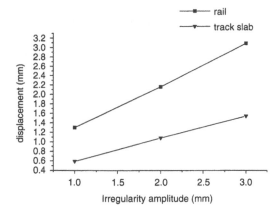

Figure 8. Influence of cosine irregularity amplitude on rail and slab displacement.

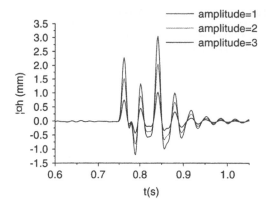

Figure 10. The influence of cosine irregularity amplitude on gap variation.

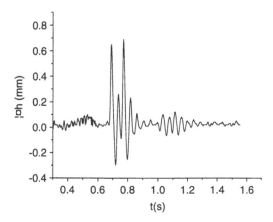

Figure 9. The influence of cosine irregularity on gap variation.

Compare figure 6 and 7 to the system response curve under no irregularity, we can see that both the car body and the slab acceleration have been increased obviously because of the local cosine irregularity.

Figure 8 shows that both the rail and slab displacement amplitude raise with the irregularity amplitude.

4.5 Air gap variation test

The air gap between LIM and its RP also changes under the irregularity. Figure 9 is the stable air gap variation curve in the middle point of the front bogie. Figure 10 shows the variation of the air gap with the irregularity amplitude.

In Figure 9, the air gap variation amplitude is 0.69 mm. It is within the normal variation scope of the system.

In Figure 10, the cosine irregularity amplitude are separately 1 mm, 2 mm and 3 mm, and the air gap variation amplitude are correspondingly 1.02 mm, 2.04 mm and 3.05 mm. It is concluded that the bigger the irregularity amplitude is, the bigger the air gap variation amplitude is, and they basically agreed with the direct ratio.

5 SUMMERY AND CONCLUSIONS

This article adopts dynamic finite element method and coupling vibration theory; models the interaction between the LIM and its RP as special springs, and builds the vertical vehicle-track coupling dynamic model.

The computing step test is carried out; the system dynamic characteristics and the air gap variation are discussed under cosine irregularity. The following conclusions are made on the basis of this study:

(1) The proper computing step for this program is 2 ms.
(2) The track slab acceleration and car body acceleration increased obviously under the action of local cosine irregularity.
(3) The rail displacement and slab displacement amplitude rise with the irregularity amplitude.
(4) The air gap variation amplitude rises with the irregularity amplitude, and they basically agree with the direct ratio.

REFERENCES

Fatemi, M.J. 1993. Resilient cross-tie track for a transit guideway.Canada:Queen's University.
Fatemi, M.J., Green, M.F. & Campbell, T.I.1996. Dynamic analysis of resilient crosstie track for transit system. *Journal of Transportation Engineering* 122(2):173–180.

Pang Shao-ming & Di Ming. 2004.Three-dimensional analyse of linear motor application on rail bound vehicle. *Electric Locomotives & Mass Transit Vehicles* 27(1):31–33.

Jin Xinmin. 1998.Application of linear induction motor in metro vehicles. *Electric Drive for Locomotive* 2:1–3.

Xia he. 2002.*Vehicle-Bridge Dynamic Interaction*. Beijing: Science Press.

Zhai wanming & Han weijun & Cai chengbiao. Dynamic properties of high-speed railway slab tracks.1999. *Journal of the China Railway Society*. 21(6):65–69.

Zhang geming. 2001.The track-bridge system dynamic analysis model and track irregularities control on quasi & high-speed railway: China Academy of Railway Sciences.

Environmental Vibrations – Takemiya (ed.)
© *2005 Taylor & Francis Group, London, ISBN 0 415 39035 4*

Dynamic analysis of vehicle-bridge-foundation interaction system

N. Zhang, H. Xia & J.W. Zhan

School of Civil Engineering & Architecture, Beijing Jiaotong University, Beijing, China

ABSTRACT: The effect of bridge foundation in the vehicle-bridge dynamic interaction system is studied in this paper. The contributions of the foundation are simplified into 6 translation and rotation springs by the m-method, then the additional stiffness of the substructure and foundation is added to the nodes at the bottom of the piers. As a case, a $54 + 90 + 90 + 54$ m continuous beam bridge is studied by computer simulation. The dynamic responses of the vehicle-bridge system with and without considering the effect of the foundation are calculated, and the contribution and the sensitivity of the foundation are analyzed.

1 INSTRUCTION

The substructure and foundation conditions play an important role in vehicle-bridge interaction system. Many experiments prove that the dynamic responses are quite discrepant between different substructures and foundations even for bridges with the same piers and beam structures under the same train passages. So it is necessary to consider the substructure and the foundation in studying the dynamic properties the train-bridge system.

Many cases proved the response of vehicle-bridge system could be seriously enlarged owing to the decrease of foundation stiffness. For example, the measurement of a 40 m-span steel plate girder supported by 28 m and 33 m high piers reported the 13 mm lateral displacement at mid-span under freight train at 68 km/h, which is much greater than the allowance 4 mm given in the Chinese Code. The site examination found the foundations are not fit to their design. Some measures are taken to strengthen the stiffness of the piers, and the displacements became only 1/3 of the value before its reinforcement (Gao 2001). The case illustrates that the vibration of the bridge can be controlled by strengthening the stiffness of the foundation.

In this paper, the dynamic interaction model of vehicle-bridge-foundation system is established, in which the detail pile foundation is modeled with the m-method in the Foundation & Substructure Code of China. The China-Star high-speed train passing the Liexiegou Bridge, a $54 + 90 + 90 + 54$ m continuous beam bridge on the Zhengzhou-Xi'an Special Passenger Railway is simulated. The dynamic responses of the train-bridge system are analyzed.

2 VEHICLE-BRIDGE-FOUNDATION INTERACTION SYSTEM

2.1 Vehicle element model

The vehicle model is established based on the following assumptions:

Each vehicle is considered as an independent element with 1 car body, 2 bogies and 4 wheel sets. The bodies, bogies and wheel-sets of the vehicle are all considered as rigid, which are linked by the suspensions composed of lateral and vertical springs and dampers. There are two suspension systems in vehicle model, as assumed by Diana (1989), Frýba (1999) and Yang (2002).

When the train runs at constant speed, the longitudinal movement of the vehicle is neglected in the element. A car body or a bogie has 5 DOFs in directions of Y, R_X, R_Z, Z and R_Y, a wheel set has 3 DOFs in directions of Y, R_X and Z, thus the vehicle model of 27 DOFs is established, in which the movements of wheel-sets are linked with the bridge and only the rest 15 DOFs are taken as independent ones in the vehicle equations (Xia 2002, 2005).

There are a lateral and a vertical springs and dampers at every side of wheel set and each side of bogie. So 8 lateral and vertical springs and dampers in the 1st suspension system, 4 lateral and vertical springs and dampers in the 2nd suspension system are considered in a vehicle element. The model of the vehicle element is shown in Figure 1.

By assuming the vibration amplitude of each component in is small and using the equilibrium conditions, the equations of motion can be derived:

$$[M_t]\{\ddot{v}_t\} + [C_t]\{\dot{v}_t\} + [K_t]\{v_t\} = \{P_{tb}\} \tag{1}$$

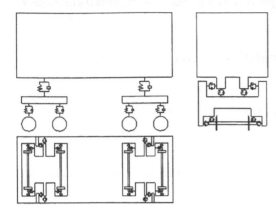

Figure 1.   Sketch of vehicle element.

where $[M_t]$, $[C_t]$ and $[K_t]$ are mass, damping and stiffness matrices, see Zhang (2003) and Xu (2004); $\{v_t\}$, $\{\dot{v}_t\}$ and $\{\ddot{v}_t\}$ are displacement, velocity and acceleration vectors of vehicle system; $\{P_{tb}\}$ is the force vector caused by wheel-rail force and can be expressed in terms of the mass of bogie and car body and the displacements and velocity of wheel set, defining $\{P_{tb}\} = \{P_1 P_2 \dots P_{15}\}^T$. Then each element of vector $\{P_{tb}\}$ can be expressed as follows.

For the DOFs of front ($i = 1$) and rear bogie ($i = 2$):

$$
\begin{cases}
P_{10i-9} = 2C_H(\dot{Y}_{2i+2} + \dot{Y}_{2i+3}) + 2K_H(Y_{2i+2} + Y_{2i+3}) \\
P_{10i-8} = 2C_H[b_1^2(\dot{R}_{2i+2} + \dot{R}_{X,2i+3}) - h_3(\dot{Y}_{2i+2} + \dot{Y}_{2i+3})] \\
\quad + 2K_H[b_1^2(R_{2i+2} + R_{X,2i+3}) - h_3(Y_{2i+2} + Y_{2i+3})] \\
P_{10i-7} = 2C_H q(\dot{Y}_{2i+2} - \dot{Y}_{2i+3}) + 2K_H q(Y_{2i+2} - Y_{2i+3}) \\
P_{10i-6} = M_1 g + 2C_V(\dot{Z}_{2i+2} + \dot{Z}_{2i+3}) + 2K_V(Z_{2i+2} + Z_{2i+3}) \\
P_{10i-5} = 2C_V q(-\dot{Z}_{2i+2} + \dot{Z}_{2i+3}) + 2K_V q(-Z_{2i+2} + Z_{2i+3})
\end{cases}
$$

For the DOFs of car body:

$$
\begin{cases}
P_9 = M_2 g \\
P_6 = P_7 = P_8 = P_{10} = 0
\end{cases}
$$

where K_V and K_H are vertical and lateral elastic coefficients at each side of bogie in the first suspension system; C_V and C_H are the corresponding damping coefficients; q = half of the distance between wheel sets; s is half of distance between bogies; b_1 = half spring span of the first suspension system; h_3 = vertical distance between bogie and wheel set.

2.2   Bridge system

The bridge model is analyzed by structural analysis program firstly and then the total mass, damping and stiffness matrices are created by the calculation results. The dynamic equation of bridge system is:

$$[M_b]\{\ddot{v}_b\} + [C_b]\{\dot{v}_b\} + [K_b]\{v_b\} = \{P_{bt}\} \tag{2}$$

where $[M_b]$, $[C_b]$ and $[K_b]$ are mass, damping and stiffness matrices; $\{v_b\}$, $\{\dot{v}_b\}$ and $\{\ddot{v}_b\}$ are displacement, velocity and acceleration vectors; $\{P_{bt}\}$ = the force vector caused by wheel-rail and determined by position, movement status of wheel sets.

The displacement of the bridge deck at any section in the cross-section in finite element analysis are usually identified in terms of lateral displacement Y_b, vertical displacement Z_b and torsional displacement R_{Xb} at the shear center of the structural member on which the vehicle runs.

The lateral, torsional and vertical forces given by wheel set j of bogie i corresponding to the deck displacement can be deduced in terms of the equilibrium conditions of the wheel and the relative position of the track to the shear center of the structural member on which the vehicle runs:

$$
\begin{aligned}
F_{Y,i,j} = {} & M_w \ddot{Y}_{2i+j+1} + 2C_H[\dot{Y}_{2i-1} - (-1)^j q\dot{R}_{Z,2i-1} \\
& - h_3 \dot{R}_{X,2i-1} - \dot{Y}_{2i+j+1}] + 2K_H[Y_{2i-1} - \\
& (-1)^j q R_{Z,2i-1} - h_3 R_{X,2i-1} - Y_{2i+j+1}]
\end{aligned} \tag{3a}
$$

$$
\begin{aligned}
F_{RX,i,j} = {} & I_w \ddot{R}_{X,2i+j+1} + 2C_V \cdot D \cdot b_1(-\dot{R}_{X,2i-1} \\
& + \dot{R}_{X,2i+j+1}) + 2K_V \cdot D \cdot b_1(-R_{X,2i-1} \\
& + R_{X,2i+j+1}) + h_4 \cdot F_{Y,i,j}
\end{aligned} \tag{3b}
$$

$$
\begin{aligned}
F_{Z,i,j} = {} & M_w(\ddot{Z}_{2i+j+1} + g) + 2C_V[\dot{Z}_{2i-1} \\
& + (-1)^j q \cdot \dot{R}_{Y,2i-1} - \dot{Z}_{2i+j+1}] \\
& + 2K_V[Z_{2i-1} + (-1)^j q \cdot R_{Y,2i-1} - Z_{2i+j+1}]
\end{aligned} \tag{3c}
$$

where M_w = mass of wheel set; I_w = iteria about the x-axis; D = rail gauge; g = gravity acceleration; h_4 = relative height difference between top of rail and the structural member on which the vehicle runs.

2.3   Foundation system

The foundation system is established by spatial beam elements and the surrounding soil is simplified into lateral and longitudinal springs.

The pilecap is simplified into a series of beam elements, which are much easier to deal with than brick elements. The soil surrounding the pilecap is simplified into springs perpendicular to the surface of the cap. In calculation, the surface of pilecap is divided

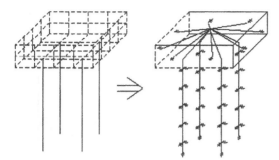

Figure 2. Model of foundation.

into some regions and the springs are put at center of each region. The elasticity coefficient of spring is defined as:

$$K = \int_A C_y dA \tag{4}$$

where C_y = foundation coefficient and A = area of certain region.

The beam elements consisting pilecap are considered as rigid bars because the deformation of the cap is neglected in calculation.

The rigid bars link from the upper surface center of pilecap to the center of each region of pilecap surface and the pile top.

The piles are simplified into a series of beam elements. The soil surrounding the pile and supporting the pile bottom are simplified into springs. The lateral and vertical elasticity coefficients of the springs are respectively defined as:

$$K_H = \int C_y \cdot W dL \tag{5}$$

$$K_V = C_0 \cdot A_P \tag{6}$$

where W = calculation width of the pile, as defined in the Code for Design on Subsoil & Foundation of Railway Bridges & Culverts; A_P = cross section area of pile body.

The model of foundation is shown in Figure 2.

The additional stiffness matrix is obtained by linear hypothesis. It is assumed that given forces at X, Y, Z, R_X, R_Y and R_Z directions act on the foundation at the center of pilecap upper surface, and the elements k_{ij} in the additional stiffness matrix is defined as the force in the jth direction caused by the unit displacement at the ith direction.

The spread foundation can be modeled as a pile foundation with only pilecap.

2.4 Model of track

Wheel hunting and track irregularity are two important self-excitations in the vehicle-bridge system in addition to the moving load of train. In this study, the wheel hunting displacement in the lateral direction is assumed as a sinusoid function with certain amplitude and a random phase:

$$Y_h(t) = A_h \sin\left(\frac{2\pi v t}{L_h} + \phi_h\right) \tag{7}$$

where A_h = hunting amplitude; L_h = hunting wavelength; ϕ_h = random phase of certain wheelset ranging between $0 \sim 2\pi$, v = speed of vehicle.

The track irregularities consist of the lateral irregularity $Y_s(x)$, vertical irregularity $Z_s(x)$ and irregularity $R_{Xs}(x)$. The measured track irregularities are used in this study.

In consideration of both the wheel hunting and track irregularities, the relation between the displacement of wheel set j under bogie i and that of same position on structural member on which the vehicle runs can be deduced as:

$$\begin{cases} Y_{w,i,j} = Y_b(x_{i,j}) + h_4 R_{Xb}(x_{i,j}) + Y_s(x_{i,j}) + Y_h(x_{i,j}) \\ R_{Xw,i,j} = R_{Xb}(x_{i,j}) + R_{Xs}(x_{i,j}) \\ Z_{w,i,j} = Z_b(x_{i,j}) + Z_s(x_{i,j}) \end{cases} \tag{8}$$

where $x_{i,j}$ = coordinate of wheel set j under bogie i along the bridge.

2.5 System combination

The vehicle model, bridge model and foundation model are coupled system. The interaction between vehicle and bridge system can be expressed by wheel-rail force defined in Equation (3) and their relation of movement defined in Equation (8).

The foundation is considered in the coupling system by considering its stiffness to the stiffness matrix of the bridge at the position of pier bottom. This can be realized by adding the additional stiffness matrix obtained in Part 2.3 to the corresponding DOF of the stiffness matrix of the bridge.

3 CASE STUDY

3.1 Vehicle information

The China Star train is adopted to simulate the train passing the bridge, which is formed by 1 tractor + 14 trailers + 1 tractor (see Figure 3). The train speeds in calculation are 160, 180, 200, 220, 240 and 270 km/h. The axle weight for tractor and trailer are 160 kN and

Figure 3. Composition of China Star Train.

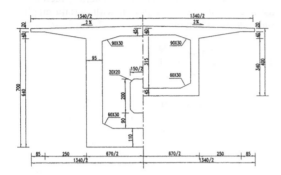

Figure 4. Cross section of beam.

Table 1. Natural frequencies of structure.

Items	Vertical/Hz	Lateral/Hz
Beam	1.43	2.30
Beam + pier	1.43	1.24
Beam + pier + foundation	1.45	1.22

146 kN respectively and the length of vehicle is 24.775 m for both tractor and trailer.

3.2 Bridge information

The China Star Train traveling on the Liexiegou Bridge on the Zhengzhou-Xi'an Special Passenger Railway is taken as an example in this paper.

The Liexiegou Bridge has a continuous beam with box section and the spans of 54 + 90 + 90 + 54 m. The cross section of the beam is shown in Figure 4. The deck is 13.4 m in width. The box is 6.7 m in width, and 7.0 m in height over pier and 4.0 m at mid-span. The piers have rectangular and hollow sections, with the heights of 34.5 m, 41.5 m and 32.5 m, respectively.

Thus the bridge system with beams and piers is modeled by the finite element method, based on which the natural frequencies are calculated and shown in Table 1.

3.3 Foundation information

Pile foundations are adopted for the bridge. There are 20 piles in each foundation, 57 m, 65 m and 65 m in length respectively, and 1.5 m in diameter. The pilecaps are cubic, with 4.0 m in height, 14.6 m in length and 18.6 m in width. The ichnography of the foundation is

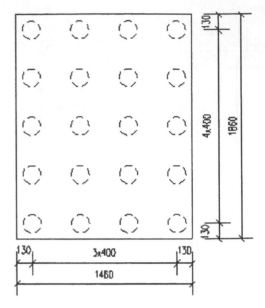

Figure 5. Ichnography of foundation.

Table 2. Elements in additional matrix.

Items*	Unit	1#	2#	3#
k_{11}	N/m	1.11E + 07	1.28E + 07	1.10E + 07
k_{22}	N/m	1.03E + 07	1.22E + 07	1.02E + 07
k_{33}	N/m	9.17E + 06	9.17E + 06	8.55E + 06
k_{24}	N/rad	4.53E + 07	5.26E + 07	4.44E + 07
k_{44}	N-m/rad	1.39E + 08	1.39E + 08	1.44E + 08
k_{15}	N/rad	6.72E + 07	7.01E + 07	6.54E + 07
k_{55}	N-m/rad	1.67E + 08	1.67E + 08	1.62E + 08
k_{66}	N-m/rad	1.52E + 08	1.52E + 08	1.47E + 08

*Elements not shown in this table are zero.

shown in Figure 5. Based the method above, the additional matrix of each foundation is obtained and given in Equation (9) and Table 2.

$$[K] = \begin{bmatrix} k_{11} & & & & & \\ k_{21} & k_{22} & & & sym. & \\ k_{31} & k_{32} & k_{33} & & & \\ k_{41} & k_{42} & k_{43} & k_{44} & & \\ k_{51} & k_{52} & k_{53} & k_{54} & k_{55} & \\ k_{61} & k_{62} & k_{63} & k_{64} & k_{65} & k_{66} \end{bmatrix} \quad (9)$$

3.4 Track information

Considering the transportation condition of the train and tracks, the random irregularity sample changed

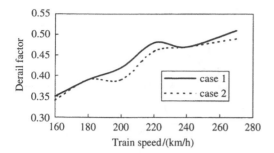

Figure 6.   Derail factor versus train speed.

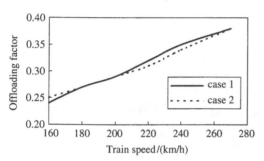

Figure 7.   Offloading factor versus train speed.

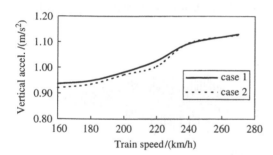

Figure 8.   Vertical acceleration of car body versus train speed.

from the Germany low disturbance spectrum and hunting of wheelset are used as system excitations.

The total length of the irregularity sample is 2600 m, with the maximum amplitude 4.51 mm in profile and 5.60 mm in vertical direction. The wavelength of hunting is $L_h = 15$ m and amplitude of hunting is $A_h = 3$ mm.

3.5   Vehicle response

The two cases with (case 1) and without (case 2) considering the effect of foundations are calculated, respectively. The distributions of maximum derail factors, offloading factors, vertical and lateral accelerations of car bodies versus train speed are shown in Figure 6 to Figure 9.

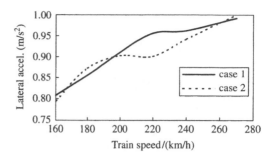

Figure 9.   Lateral acceleration of car body versus train speed.

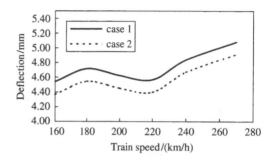

Figure 10.   Mid-span deflection versus train speed.

It can be concluded from the calculation results that all the 4 vehicle responses are increased with train speed, no matter with or without considering the effect of foundation. The difference between case 1 and case 2 is not obvious. In general, the responses of vehicles with the effect of foundation are a little larger than those without.

The maximum derail factor and offloading factor are 0.51 and 0.38; the maximum vertical and lateral acceleration are 1.13 m/s^2 and 0.99 m/s^2. The corresponding allowances in Chinese Code are 0.6 for derail factor, 0.8 for offloading factor, 1.3 m/s^2 and 1.0 m/s^2 for vertical and lateral car-body accelerations. So the response of vehicle system meets the requirement of Chinese Code.

3.6   Bridge response

The deflection, lateral displacement, vertical and lateral acceleration of mid-span of the bridge is shown in Figure 10 to Figure 13.

It can be seen from the figures that the bridge responses in the case with considering the foundation are also a little larger than those in the case without considering the foundation.

The deflection of the bridge at the beam mid-span is tend to increase and has a peak when train speed is about 180 km/h. It is due to the resonant vibration

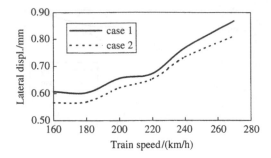

Figure 11. Mid-span lateral displacement versus train speed.

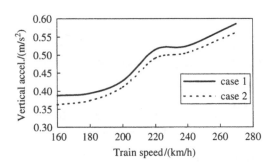

Figure 12. Mid-span vertical acceleration versus train speed.

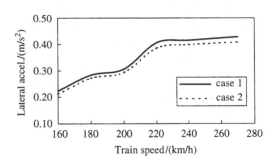

Figure 13. Mid-span lateral acceleration versus train speed.

between the bridge system and the vehicle system. The lateral displacement, vertical and lateral accelerations of the bridge at the beam mid-span are also increased with train speed.

The deflection and lateral displacement histories of the bridge at beam mid span when the train passing the bridge at speed of 270 km/h are shown in Figure 14 to Figure 17. It can be concluded that the values and the shapes of the histories with and without the effects of the foundation are very similar.

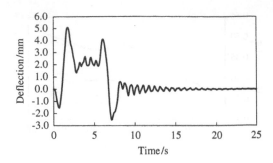

Figure 14. Deflection history with effect of foundation.

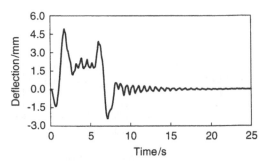

Figure 15. Deflection history without effect of foundation.

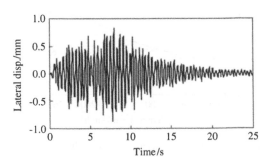

Figure 16. Lateral displacement with effect of foundation.

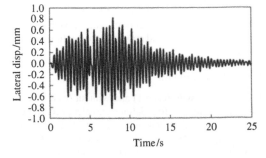

Figure 17. Lateral displacement without effect of foundation.

The maximum lateral displacement, vertical acceleration and lateral acceleration are 0.868 mm, 0.586 m/s^2 and 0.427 m/s^2, with the corresponding allowances 10 mm, 3.5 m/s^2 and 1.4 m/s^2 in Chinese Code, respectively. Also the responses of the bridge system meet the requirement of Chinese Code.

4 CONCLUSIONS

The following conclusions can be drawn up from the study and calculation results:

(1) The pilecap and pile are simplified into beam elements. The soil around is simplified into springs. Thus the additional stiffness matrix can be obtained by acting given forces in 6 directions and the elements k_{ij} in additional stiffness matrix is defined as the value of force in the jth direction, which causes unit displacement at the ith direction.
(2) The vehicle and bridge response with considering the foundation is a little larger than response without consideration of foundation.
(3) The vehicle and bridge responses meet the requirements of Chinese Code when the China Star Train passes the Liexiegou Bridge.

ACKNOWLEDGEMENT

This study is sponsored by the Natural Science Foundation of Beijing (No. 8042017) and the Key Research Foundation of Beijing Jiaotong University (No. 2004SZ005).

REFERENCES

Diana, G. & Cheli, F. 1989. Dynamic interaction of railway systems with large bridges. *Vehicle System Dynamics* 18: 71–106.
Frýba, L. 1999. *Vibration of Solids and Structures under Moving Loads*. London: Thomas Telford.
Gao, R. 2001. *Experimental report of No.35 bridge in Fengtai-Shacheng Railway*. Beijing: Beijing Jiaotong University.
Xia, H. & Zhang, N. 2005. Dynamic analysis of railway bridge under high-speed trains. *Computers & Structures*, in press.
Xia, H. 2002. *Dynamic Interaction of Vehicles and Structures*. Beijing: Science Press.
Xu, Y.L., Zhang, N. & Xia, H. 2004. Vibration of coupled train and cable-stayed bridge systems in cross winds. *Engineering Structures* 26: 1389–1406.
Yang, Y.B. 2002. *Theory of Vehicle-Bridge Interaction for High Speed Railway*. Taibei: DNE Publisher.
Zhang, N. & Xia, H. 2003. Theoretical analysis and experimental study on dynamic interaction of railway bridge and articulated trains. *China Railway Science* 24: 132–134.

Environmental Vibrations – Takemiya (ed.)
© *2005 Taylor & Francis Group, London, ISBN 0 415 39035 4*

Simulation of track vibration due to high speed train passing through track transition with irregularity and rigidity abrupt change

X. Lei

School of Civil Engineering, East China Jiaotong University, Nanchang, People's Republic of China

ABSTRACT: Analytical deformation formula for moving load passing through the track transition with irregularities and rigidity abrupt change has been derived by establishing the vibration differential equation of the track. The influences of the track irregularities, the track rigidity abrupt change and the train speeds on track vibration are investigated and the dynamic responses of the track under a single wheel set and a TGV high speed train passing through the track with different track irregularity amplitudes and the rigidity ratios are analyzed by using this formula and the superposition principle. The computations show that the track irregularities, the rigidity abrupt change and the train speeds have significant influences on track vibration in track transition. This influences increase with increases of the track irregularity amplitude, the rigidity ratio and the train speed.

1 INTRODUCTION

Transition regions are locations where a railway track exhibits abrupt changes in vertical stiffness (Kerr 1987). They usually occur at the abutments of open deck bridges, where a concrete tie track changes to a wooden tie track, at the ends of a tunnel, at highway grade crossing, at locations where rigid culverts are placed close to the bottom of the ties in a ballasted track, etc. Transition regions require frequent maintenance. When neglected, they will deteriorate at an accelerated rate. This may lead to pumping ballast, swinging or hanging cross ties, permanent rail deformations, worn track components, and loss of surface and gauge. These in turn may create a potential for a derailment. A large quantity of abutments of open deck bridges, road crossings and tunnel ends exist in Chinese railway lines. For instance, the total length of Beijing-Shanghai railway line is 1459 km and there are 1471 abutments of bridges and 490 road crossings on this line. In the past, the average speed of the trains in China usually was less than 80 km/h. The track transition problem is not serious in this situation. However, it is to be solved with the raising speed of trains in China. The maximum train speed possible is 160 km/h according to Chinese railway regulation (Min 1997).

A few literatures have focused on the vibration in track transition under moving vehicles (Kerr 1987; Kerr 1989; Moroney 1991; Kerr 1995; Lei 2004). Kerr (1987) presented a method for determining the track modulus using a locomotive or car on multi-axle trucks. Moroney (1991) discussed railroad track transition

points and problems. Kerr and Moroney (1995) investigated track transition problems and remedies. Lei and Mao (2004) studied dynamic response analyses of vehicle and track coupled system on track transition of conventional high speed railway by finite element method. Even though some achievements have been obtained studies on this field are still required to be undertaken.

To clarify the vibration generation mechanism of the track transition, an analytical formula of track deflection under a single moving load has been derived by establishing the differential equation of the track vibration for track transition with the track irregularities and the rigidity abrupt change in this paper. The influences of the track irregularities, the track rigidity abrupt change and the train speeds on track vibration are investigated and the dynamic responses of the track under a single wheel set and a TGV high speed train passing through the track with different track irregularity amplitudes and the rigidity ratios are analyzed by using this formula and the superposition principle.

2 MODEL OF TRACK VIBRATION UNDER A SINGLE LOAD PASSING THROUGH THE TRACK TRANSITION WITH THE TRACK IRREGULARITIES AND THE RIGIDITY ABRUPT CHANGE

To simplify the analyses, an analytical method is employed to study the influences of the track irregularities and the track rigidity abrupt change on track

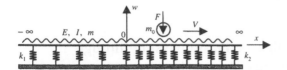

Figure 1. Model of track vibration under a single load passing through the track transition.

vibration under a single moving load with constant velocity V. The model of track vibration under a single load moving on track transition with the track irregularities and the rigidity abrupt change is shown in Fig. 1. The coordinate system is chosen such that the original of the coordinate system is in position of track rigidity abrupt change and the x axis is corresponding to the longitudinal direction of the track. The track foundation rigidity on the left side of the original of the coordinate system is k_1 and the track deflection is w_1 whereas the track foundation rigidity on the right side of the original of the coordinate system is k_2 and the track deflection is w_2. The track panel can be treated as a Eular-Bernoulli elastic beam of uniform mass m lying on a viscoelastic half-space. Let's take w_1 as an example, the following dynamic equation in case of ignoring damping can be used to describe its vertical deflections. (Belzer 1988)

$$EI\frac{\partial^4 w_1}{\partial x^4} + m\frac{\partial^2 w_1}{\partial t^2} + k_1 w_1 = -\left(F + m_0\frac{\partial^2 (w_1 + \eta)}{\partial t^2}\right)\delta(x - Vt)$$
(1)

where E and I are Young's modulus and the cross-sectional momentum of the track panel, x is the distance along the track, F is the wheel-axle load considered as a vertical point force, m_0 is the wheel-axle mass, V is the speed of the wheel, $\delta(x - Vt)$ is the Dirac delta function and η is the irregularity of the track which can be assumed as

$$\eta = a\left[1 - \cos\frac{2\pi(x - Vt)}{l}\right]$$
(2)

in which a is the amplitude of the track irregularity and l is the wave length of the track irregularity.

Equation (1) may be rewritten in a more convenient form

$$\frac{\partial^4 w_1}{\partial x^4} + 4\varepsilon_1^2\frac{\partial^2 w_1}{\partial t^2} + 4\gamma_1^4 w_1$$

$$= -\frac{1}{EI}\left(F + m_0\frac{\partial^2 (w_1 + \eta)}{\partial t^2}\right)\delta(x - Vt)$$
(3)

where

$$\varepsilon_1^2 = \frac{m}{4EI}, \quad \gamma_1^4 = \frac{k_1}{4EI}$$
(4)

The assumed stationary solution may be $w_1 = w_1(x - Vt)$ as the force term in (3) is the function of $(x - Vt)$. Introducing the variable $z = x - Vt$ we put equation (3) in the form

$$\frac{\partial^4 w_1}{\partial z^4} + 4\varepsilon_1^2 V^2\frac{\partial^2 w_1}{\partial z^2} + 4\gamma_1^4 w_1$$

$$= -\frac{1}{EI}\left(F + m_0\frac{\partial^2 (w_1 + \eta)}{\partial t^2}\right)\delta(z)$$
(5)

with the characteristic equation

$$p^4 + 4\varepsilon_1^2 V^2 p^2 + 4\gamma_1^4 = 0$$
(6)

The roots of this equation and, consequently, the solution depend on its coefficients. We therefore confined our further analysis to the case $V < \gamma_1/\varepsilon_1$.

Under the above condition the roots of (6) are given by

$$p = \pm\alpha_1 \pm j\beta_1$$
(7)

with

$$\alpha_1 = \left(\gamma_1^2 - V^2\varepsilon_1^2\right)^{\frac{1}{2}}, \quad \beta_1 = \left(\gamma_1^2 + V^2\varepsilon_1^2\right)^{\frac{1}{2}}$$
(8)

and the solution to (5) is

$$w_1(z) = e^{\alpha_1 z}\left(D_1\cos\beta_1 z + D_2\sin\beta_1 z\right)$$
$$+ e^{-\alpha_1 z}\left(D_3\cos\beta_1 z + D_4\sin\beta_1 z\right) + \varphi_1(z)$$
(9)

Here $\varphi_1(z)$ denotes a particular solution which depends on the force term and which can be, in fact, omitted, as the following consideration shows. The point $z = 0$ ($x = Vt$) indicates the position of the force. Then for $z \neq 0$ there is no external force and the solution is given by equation (9) with $\varphi_1(z) = 0$. Rejecting the terms, which show spatial growth of the waves, we get from equation (9)

$$w_1(z) = e^{\alpha_1 z}\left(D_1\cos\beta_1 z + D_2\sin\beta_1 z\right) \quad z < 0$$
(10)

where D_1, D_2 are the coefficients which are related with the boundary conditions.

146

In the same way, the dynamic equation for the track on the right side of the original of the coordinate system in case of ignoring damping can be expressed as

$$EI\frac{\partial^4 w_2}{\partial x^4} + m\frac{\partial^2 w_2}{\partial t^2} + k_2 w_2$$
$$= -\left(F + m_0\frac{\partial^2(w_2 + \eta)}{\partial t^2}\right)\delta(x - Vt) \quad (11)$$

Solution of the equation (11) is similar to (10), that is

$$w_2(z) = e^{-\alpha_2 z}(D_3 \cos\beta_2 z + D_4 \sin\beta_2 z) \quad z > 0 \quad (12)$$

with

$$\alpha_2 = \left(\gamma_2^2 - V^2\varepsilon_2^2\right)^{\frac{1}{2}}, \qquad \beta_2 = \left(\gamma_2^2 + V^2\varepsilon_2^2\right)^{\frac{1}{2}} \quad (13)$$

$$\varepsilon_2^2 = \frac{m}{4EI}, \quad \gamma_2^4 = \frac{k_2}{4EI} \quad (14)$$

where D_3, D_4 are the coefficients which are related with the boundary conditions too.

D_1, D_2, D_3 and D_4 may be found by matching above solutions and its derivatives up to the third order at $z = 0$, that is

$$w_1|_{z=0} = w_2|_{z=0} \quad (15)$$

$$\frac{\partial w_1}{\partial z}\bigg|_{z=0} = \frac{\partial w_2}{\partial z}\bigg|_{z=0} \quad (16)$$

$$\frac{\partial^2 w_1}{\partial z^2}\bigg|_{z=0} = \frac{\partial^2 w_2}{\partial z^2}\bigg|_{z=0} \quad (17)$$

$$-\frac{\partial^3 w_1}{\partial z^3}\bigg|_{z=0} + \frac{\partial^3 w_2}{\partial z^3}\bigg|_{z=0} = -\left(\frac{F}{EI} + m_0\frac{\partial^2(w_2 + \eta)}{\partial t^2}\right)\bigg|_{z=0} \quad (18)$$

This yields

$$w_1(z) = e^{\alpha_1 z}(C_1 \cos\beta_1 z + C_2 \sin\beta_1 z)C_3 \quad z < 0 \quad (19)$$

$$w_2(z) = e^{-\alpha_2 z}(C_1 \cos\beta_2 z + \sin\beta_2 z)C_3 \quad z \geq 0 \quad (20)$$

with

$$C_1 = \frac{2\beta_2(\alpha_1 + \alpha_2)}{(\alpha_1 + \alpha_2)^2 + \beta_1^2 - \beta_2^2}$$

$$C_2 = -\frac{\beta_2\left[(\alpha_1 + \alpha_2)^2 - \beta_1^2 + \beta_2^2\right]}{\beta_1\left[(\alpha_1 + \alpha_2)^2 + \beta_1^2 - \beta_2^2\right]}$$

$$C_3 = -\frac{F}{EI(G_1 C_1 + G_2 C_2 + G_3 + G)}$$

$$G = \frac{m_0 V^2}{EI}\left[(\alpha_2^2 - \beta_2^2)C_1 - 2\alpha_2\beta_2\right]$$

$$G_1 = -\alpha_1^3 - \alpha_2^3 + 3\alpha_1\beta_1^2 + 3\alpha_2\beta_2^2$$

$$G_2 = -3\alpha_1^2\beta_1 + \beta_1^3$$

$$G_3 = 3\alpha_2^2\beta_2 - \beta_2^3$$

The vertical vibration accelerations of the track can be derived by differentiating (19) and (20) with respect to t twice

$$\frac{\partial^2 w_1}{\partial t^2} = V^2 e^{\alpha_1 z}\left\{\left[(\alpha_1^2 - \beta_1^2)C_1 + 2\alpha_1\beta_1 C_2\right]\cos\beta_1 z + \left[(\alpha_1^2 - \beta_1^2)C_2 - 2\alpha_1\beta_1 C_1\right]\sin\beta_1 z\right\}C_3 \quad z < 0 \quad (21)$$

$$\frac{\partial^2 w_2}{\partial t^2} = V^2 e^{-\alpha_2 z}\left\{\left[(\alpha_2^2 - \beta_2^2)C_1 - 2\alpha_2\beta_2\right]\cos\beta_2 z + \left[(\alpha_2^2 - \beta_2^2) + 2\alpha_2\beta_2 C_1\right]\sin\beta_2 z\right\}C_3 \quad z \geq 0 \quad (22)$$

The dynamic equation of the track corresponding to equation (1) under multiple moving loads is

$$EI\frac{\partial^4 w_1}{\partial x^4} + m\frac{\partial^2 w_1}{\partial t^2} + k_1 w_1 =$$
$$-\sum_{i=1}^{N}\left(F_i + m_{0i}\frac{\partial^2(w_1 + \eta)}{\partial t^2}\right)\delta(x - a_i - Vt) \quad (23)$$

where F_i is the i th wheel-axle load, m_{0i} is the i th wheel-axle mass, a_i is the distance between the i th wheel set and the first wheel set of the train, and N is the total number of the wheel set.

Solutions of the equation (23) can be obtained by superposition of equation (19) to (22).

3 RELATIONSHIP BETWEEN THE TRACK FOUNDATION RIGIDITY AND YOUNG'S MODULUS OF THE FOUNDATION

In order to use the model of the track vibration under a single load moving on track transition with the track irregularities and the rigidity abrupt change it is necessary to find a relationship between the Young's modulus of the foundation and the track foundation rigidity. According to Vesic (1963) the Young's modulus of an elastic half-space can be related to the coefficient of subgrade reaction, k, by

$$k = \frac{0.65 E_s}{1 - v_s^2} \sqrt[12]{\frac{E_s B^4}{EI}} \qquad (24)$$

where k (MN/m^2) is the track foundation rigidity, E_s (MN/m^2) is the effective Young's modulus of the equivalent half space foundation, v_s is the Poisson's ratio and B is the width of the track, taken as the length of the sleepers in this case, normally $B = 2.5$–2.6 m.

4 INFLUENCES OF THE TRACK IRREGULARITIES AND THE TRACK RIGIDITY ABRUPT CHANGE ON TRACK VIBRATION

The dynamic responses of the track in transition due to the track irregularities and the track rigidity abrupt change depend on the track foundation rigidity k_1, k_2, the track bending rigidity EI, the amplitude of the irregularity a, the wave length l, the uniform mass of the track m, the wheel-axle load F, the wheel-axle mass m_0 and its speed V. In the computation, the uniform mass of the track m has to include the mass per unit length of the rail and sleepers and the mass per unit length of the ballast (Heelis et al. 1999).

In analyses of the track dynamic responses on the track transition, the conventional Chinese CWR track will be considered. The track is composed of 60 kg/m heavy continuous welded long rails with R.C. sleepers, 1760 pcs/km. Both of the thickness and the shoulder of the ballast are 30 cm and the density of the ballast is 2000 kg/m³. Let $E_s = 1000$ MN/m², it has $k = 118$ KN/m² according to equation (24). Parameters used in simulation are $EI = 13.25\,MN \cdot m^2$, $m = 2735$ kg/m, $m_0 = 2000\,kg$, $F = 170\,KN$.

In order to comprehensively understand the influences of the track irregularities and the track rigidity abrupt change on track vibration in transition, four cases of the rigidity ratios $k_2/k_1 = 1$, $k_2/k_1 = 2$, $k_2/k_1 =$

Table 1. Parameters used in simulation.

Parameter	$K_1(MN/m^2)$	$K_2(MN/m^2)$
Case 1	k_2	118
Case 2	$k_2/2$	118
Case 3	$k_2/5$	118
Case 4	$k_2/10$	118

5, $k_2/k_1 = 10$, shown as Table 1, and three kinds of the irregularity amplitude $a = 0.5$, 1, 2 mm with wave length $l = 2$ m will be studied in analyses.

4.1 Analyses of track vibration in transition under a single wheel set passing by

Now let's consider track vibration when a single wheel set passing by with the speed of $V = 60, 70, 80, 90$ m/s. Four cases of the rigidity ratios indicated in Table 1 and three kinds of irregularity amplitudes mentioned above will be investigated. Figure 2 are curves of the total track deflection and acceleration amplitudes versus moving load speeds for different track rigidity ratios and irregularity amplitudes. Here the total vibration amplitude is defined as the difference between the maximum vibration amplitudes and the minimum vibration amplitudes.

4.2 Analyses of track vibration in transition under a TGV high speed train passing by

Track vibration will be investigated again for a TGV high speed train passing by with the speed of $V = 90$ m/s. Figure 3 shows the model used in the simulation for track vibration under a TGV high speed train moving on track transition. The train is composed of one locomotive and four trailers. Parameters for TGV high speed trains can be found in reference (Lei 2002). The influences of the track irregularities and the track rigidity abrupt change on track vibration will be considered too. The time history of the vertical track deflection, velocity, acceleration and the wheel/rail interaction force for each case are demonstrated in Figure 4 to Figure 11.

From these computations, the following items can be obtained:

1 The track rigidity abrupt change has indeed influence on track vibration in transition. The larger the rigidity ratio is, the stronger the track vibrates, shown in Figure 2 and Figure 4 to Figure 11.
2 For specified rigidity ratio, the track dynamic response increases with the moving load speed.
3 The track irregularity has distinct influences on track vibration. The total track deflection amplitudes are

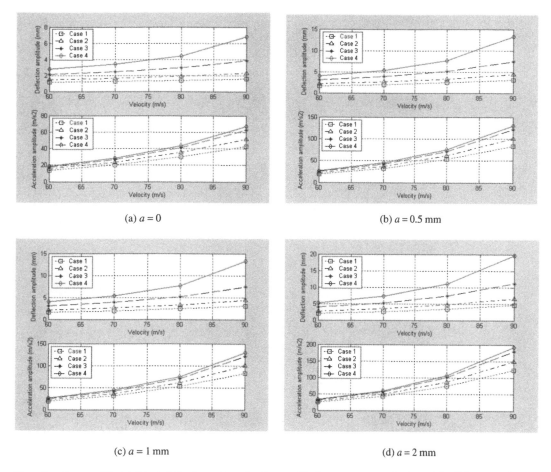

(a) $a = 0$ (b) $a = 0.5$ mm

(c) $a = 1$ mm   (d) $a = 2$ mm

Figure 2. Curves of the total track deflection and acceleration amplitudes versus moving load speed for different track irregularity amplitudes and rigidity ratios.

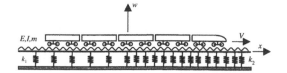

Figure 3. Model for track vibration under a TGV high speed train moving on track transition.

10.02, 13.23, 19.64 mm and the total acceleration amplitudes are 98.44, 129.93, 192.90 m/s² corresponding to track irregularity amplitude $a = 0.5$ mm, $a = 1$ mm, $a = 2$ mm, both of which are 1.47, 1.94 and 2.88 times as the total track deflection and acceleration amplitudes with $a = 0$ mm.

4 When the rigidity ratios are $k_2/k_1 = 2$, $k_2/k_1 = 5$ and $k_2/k_1 = 10$, the total deflection amplitudes are 1.44, 2.42 and 4.27 times, the total velocity amplitudes

are 1.26, 1.83 and 2.72 times, and the total acceleration or the wheel/rail interaction force amplitudes are 1.17, 1.38 and 1.43 times respectively as the total track deflection, velocity and acceleration or wheel/rail interaction force amplitudes with $k_2 = k_1$. The influences of the rigidity ratios on track vibration are significant when $k_2/k_1 \leqslant 5$ but increase slowly and tend to an asymptote after the rigidity ratio $k_2/k_1 > 5$.

5 The total vibration amplitude reflects the vibration scope of the track. The larger the total amplitude is, the stronger the track will vibrate. It is also worth noticing that the up and down oscillation amplitudes of the acceleration and the wheel/rail interaction force around the means of those amplitudes nearly tend to the same, shown as in Figure 7 and Figure 11. This may have an important implication on the degradation and liquefaction potential of the foundation material.

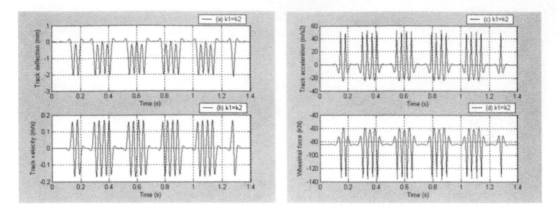

Figure 4.   Time history of the track deflection, velocity, acceleration and wheel/rail force ($k_1 = k_2$, $a = 0.5$ mm).

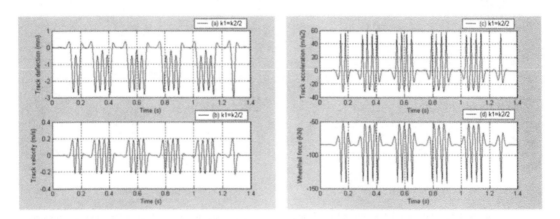

Figure 5.   Time history of the track deflection, velocity, acceleration and wheel/rail force ($k_1 = k_2/2$, $a = 0.5$ mm).

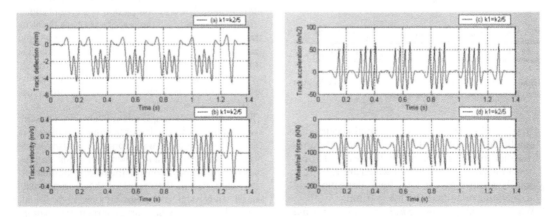

Figure 6.   Time history of the track deflection, velocity, acceleration and wheel/rail force ($k_1 = k_2/5$, $a = 0.5$ mm).

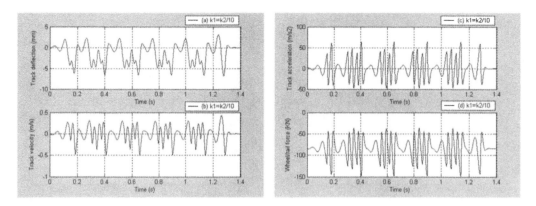

Figure 7. Time history of the track deflection, velocity, acceleration and wheel/rail force ($k_1 = k_2/10$, $a = 0.5$ mm).

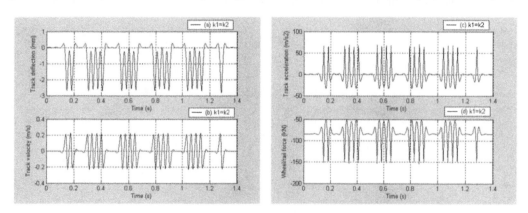

Figure 8. Time history of the track deflection, velocity, acceleration and wheel/rail force ($k_1 = k_2/5$, $a = 1$ mm).

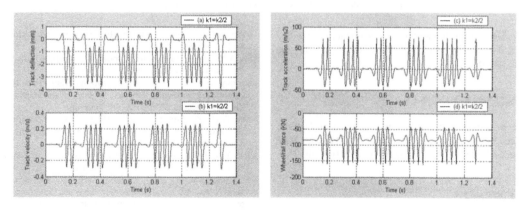

Figure 9. Time history of the track deflection, velocity, acceleration and wheel/rail force ($k_1 = k_2/2$, $a = 1$ mm).

5 CONCLUSIONS

The amplitudes of the track deflections, velocities, accelerations and wheel/rail interaction forces in track transition due to the track irregularities, the rigidity abrupt change and the train speeds, have been analyzed here by use of analytical solution and the superposition principle. The computations show that the track irregularities, the rigidity abrupt change and the train speeds in track transition have significant influences on track

151

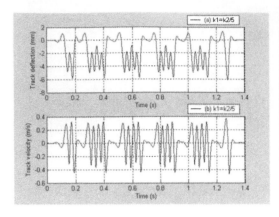
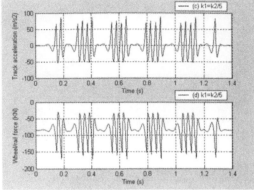

Figure 10. Time history of the track deflection, velocity, acceleration and wheel/rail force ($k_1 = k_2/5$, $a = 1$ mm).

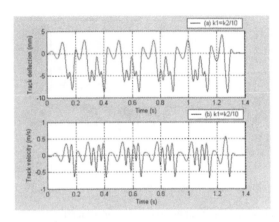
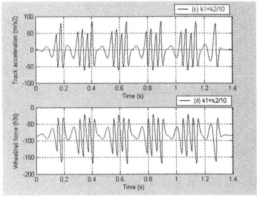

Figure 11. Time history of the track deflection, velocity, acceleration and wheel/rail force ($k_1 = k_2/10$, $a = 1$ mm).

vibration. This influences increase with increases of the track irregularity amplitude, the rigidity ratio and the train speed.

ACKNOWLEDGMENTS

The work reported herein was supported by Natural Science Foundation of China (50268001) and Natural Science Foundation of Jiangxi Province (0450012).

REFERENCES

Belzer, A. I. 1988. Acoustics of solids, Springer-Verlag, Berlin.
Heelis, M. E., Collop, A.C., Dawson, A.R. & Chapman, D.N. 1999. Transient effects of high speed trains crossing soft soil, Geotechnical Engineering for transportation Infrastructure, Barends et al. (eds), Balkema, Rotterdam, 1809–1814.
Kerr, A.D. 1987. A method for determining the track modulus using a locomotive or car on multi-axle trucks, Proceeding American Railway Engineering Association. 84(2):270–286.
Min, Y.X. 1997. Raising train speed for existing railway, Chinese Railway Publication.
Moroney, B.E. 1991. A study of railroad track transition points and problems, Master's Thesis, Department of Civil Engineering, University of Delaware.
Kerr, A.D. & Moroney, B.E. 1995. Track Transition Problems and Remedies, Bulletin 742-American Railway Engineering Association, Bulletin 742:267–297.
Kerr, A.D. 1989. On the vertical modulus in the standard railway track analyses, Rail International, 235(2): 37–45.
Lei X. and Mao L. 2004. Dynamic response analyses of vehicle and track coupled system on track transition of conventional high speed railway, Journal of Sound and Vibration, 271(3):1133–1146.
Lei X., 2002. New methods in railroad track mechanics and technology, Beijing, Chinese Railway Press, 38–42.
Vesic, A.S. 1963. Beams on elastic subgrade and the Winkler hypothesis, Proceedings 5th Int. Conf. soil Mech. Found. Engng, Paris, Vol.1 845–850.

Environmental Vibrations – Takemiya (ed.)
© *2005 Taylor & Francis Group, London, ISBN 0 415 39035 4*

Source characteristics and generation mechanism of road traffic vibrations and some remarks for mitigation

I. Masuda
Japan Transportation Consultants, Inc. Tokyo, Japan

R. Ishida
Chiba University, Chiba, Japan

ABSTRACT: Two different vibration transmissions from traffic are discussed with respect to the on-ground flat track and elevated viaduct track. The former vibration is generated by the forced motion by given surface displacement of the pavement. Consequently, the motion depends on vehicle types, cargo, running speed and the roughness. The latter vibration, besides the roughness of the road surface, the vertical girder motions of viaduct becomes important. The level differences at joints are also cause of vehicle vibrations. The viaduct vibrations emit waves from the fixed positioned foundations that propagate in ground. This paper interprets the vehicle load characteristics and vibrations generation mechanism, and then measures against such traffic induced vibrations.

1 INTRODUCTION

The traffic induced vibrations, being defined as the vibrations generated by the traffic on road, includes a variety of kinds of vehicles that run, depend on the loading condition and transmission in ground. The road system includes, on ground track, embankment type track, excavated type track, viaduct type track, tunnel type track. The assessment, prediction, mitigation measures for the environmental ground vibrations indicates a good correspondence to the similar issues in the earthquake engineering when traced back into the source fault mechanism and vibration propagation. It is essential to get through understanding of the traffic induced vibrations, the fundamental respective features involved from the source to perception.

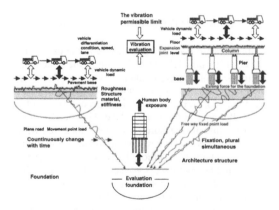

Figure 1. Vibration transmission from source for road traffic.

2 SOURCE CHARACTERISTICS TO VIBRATION ASSESSMENT

Figure 1 summaries some essential elements from the source characteristics of traffic to vibration assessment at the receiving houses, for both flat on ground track and elevated viaduct tracks.

A single track passage on the track is assumed for interpretation. The traffic induced vibration originates from the vehicle vibration on track. The vehicle vibration is amplified by the roughness of road surface. In the case on ground flat track, the surface condition of the pavement is essential element. In the case of the viaduct track the joint part becomes often more significant element for vibration generation in addition to the surface condition of the pavement.

The roughness of the road surface, more specifically the roughness of the pavement, and the level difference at the joints impose the forced motions according to the configuration. The vehicle motion occurs and reacts on to the road by the weigh and inertia. In the case of on ground flat track, the amplitudes and frequency characteristics are generated as the compound formula of the roughness of road surface and vehicle properties (cargo weight, running speed, truck type). In the case of the viaduct track, on the other hand, the interaction of vehicle and viaduct occurs so that this dynamic effects due to the structural dimensions like girder and piers of the viaducts becomes involved with the loading characteristics by vehicles.

The vehicles on ground flat track impose forces while moving along it, which is termed as moving loads. The relative position of moving loads and the observation points is essential factor for this kind of problem. In contrast, the vehicle loads on viaducts are transmitted only through their foundations on to the ground. The ground vibrations thus generated are the superposition of the wave field emitted from individual foundations.

The ground motion that are affected by the relative position between source and receiver points, the geological and topographical condition through which the wave propagates are input to nearby buildings through the kinematic soil-structure interaction due to the foundation size and stiffness contrast. Then, the receiver buildings response results accordingly. The building motion turns out an input in turn to the apparatuses on the floor. The vibration assessment of these vibration sensitive equipments is of importance against malfunctioning.

3 TRAFFIC LOAD CHARACTERISTICS

Figure 2 Traffic load, being evaluated at the pavement surface or bottom, defined as the sum of the static load due to the vehicle weight and the dynamic effect, the deviation from the static equilibrium position. The direct and most reliable method for prediction is set a load-cell device at the road surface. This needs a low of work while the measurement is very localized with instances of time. Therefore, conventionally, accelerometers are set on the vehicle body above the wheel springs and below them. Some calculation derives the estimates of dynamic load by using the acceleration measured and masses of vehicle elements. Assuming a mass-spring system for modeling a vehicle in that the weight is concentrated on the front and rear axles, we formulate the dynamic equilibrium by referring Figure 2, as

$$F = K_{TF}(U_{0F} - U_{1F}) + K_{TR}(U_{0R} - U_{1R}) + C_{TF}(\dot{U}_{0F} - \dot{U}_{1F}) + C_{TR}(\dot{U}_{0R} - \dot{U}_{1R})$$

$$= M_{TF}\ddot{U}_{1F} + M_{BF}\ddot{U}_{2F} + M_{TR}\ddot{U}_{1R} + M_{BR}\ddot{U}_{2R}$$

3.1 Natural frequency of vehicles

Figure 3 is the Fourier spectra of the accelerations at the above and below the positions of the front and rear wheel springs of a damp truck while it is running at the speed of 60 km/h. In view of this figure, it is noted that the natural frequency for the above wheel spring

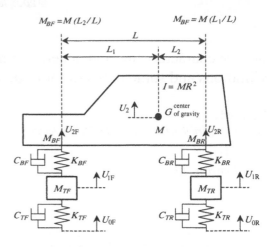

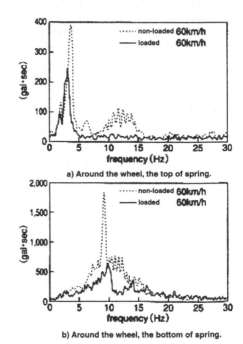

Figure 2. Vibration model of a automobile.

Figure 3. Fourier spectrum of a vehicle acceleration (Nishizaka & Fukuwa, 1997).

is located around 3 Hz, and that of the below wheel spring is around 10 Hz. Further, it is clearly noted that the increase of truck weight by the full cargo decreases the natural frequency of the above wheel spring body.

3.2 Running speed and loading characteristics

Vehicle vibration while running occurs according to the rough configuration of the road surface. The variation

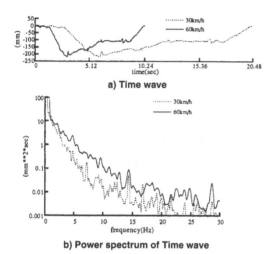

a) Time wave

b) Power spectrum of Time wave

Figure 4.   Road surface vibration vs. vehicle speed.

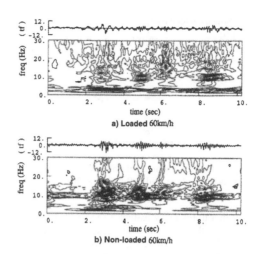

a) Loaded 60km/h

b) Non-loaded 60km/h

Figure 5.   Dynamic load wave and non-constant spectrum by the difference of loading situation (Nishizaka & Fukuwa, 1997).

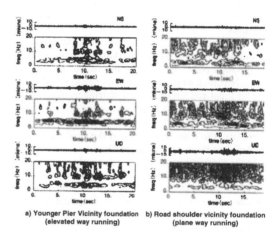

a) Younger Pier Vicinity foundation   **b) Road shoulder vicinity foundation**
(elevated way running)   **(plane way running)**

Figure 6.   Comparison of thee wave and non-constant spectrum of traffic road vicinity foundation that depends the road difference (non-loaded 60 km/h) (ISO 2000).

of the road roughness is a function of distance and converted into the time varying function by introducing the vehicle speed. The corresponding power spectral density results in Figure 4.b. Note here that the shape of the power spectral density of the input force is changed apparently by the vehicle speed. Thus, the increase of the vehicle speed leads the larger vibration amplitudes of vehicle response.

3.3 *Cargo weigh and loading characteristics*

An increase of cargo weight changes the vibration feature of vehicles even for the same types Figure 4 and Figure 5 show the time histories and their nonstationary spectra for the full cargo and no cargo loaded on the 10 tf type truck. It is interesting to note that the frequency features changes drastically by the cargo loaded on the truck. For the full cargo loaded situation, the natural frequency of the above wheel spring system is dominating the response, whilst for the no cargo loaded situation the natural frequency of the below wheel spring system becomes dominant.

3.4 *Road roughness and loading characteristics*

In view of Figure 5, it is confirmed that for the portion of smooth road surface, the pseudo-steady state dynamic loading can be observed where the natural frequency of vehicles are dominant; and at the difference existing portion at joins some impact loading appears.

4   VIBRATION FEATURES DUE TO TRAFFIC

Field experiments have been conducted by using a single dump truck at the on-ground flat road and at the viaduct as well. The running speed is set to 60 km/h in both tests. The ground responses are measured at both the roadside. Figure 6 is the time histories and their nonstationary power spectral densities.

In the test of on-ground flat road, the frequency contents are distributed evenly up to 20 Hz. In the test of viaduct, the low frequency response appears significantly below 5 Hz with high frequencies due to impact response at the fork joints. This is the consequence of the viaduct existence between traffic and ground. The viaduct works for the low frequency filtering due mainly to the girder response.

155

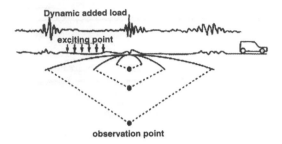

Figure 7. Position relation of the observation point and the exciting point in plane road traffic (AIJ, 2000).

Another noteworthy feature is the larger horizontal response in the perpendicular direction (EW component) to the viaduct axis. This shows a contrast to the feature of the on-ground flat road case where the vertical (UD component) is predominant. The comparable horizontal response, the commonly observed for the viaduct traffic needs some reconsideration to the current legislation for the traffic induced vibrations in which only the vertical components are considered.

Figure 7 illustrates schematically a loading and observation positions for the on-ground flat track. It should be noted that, as the observation point get away off the road edge, the attenuation rate between these points becomes small with the longer road length.

The local road roughness that triggers the maximum ground response is not always the origin on the measurement line, rather it is caused by other origins. Therefore, the smoothing work of the road surface is needed over the longer road length.

In the viaduct traffic case, the dynamic load amplitudes grow and the impulse-like response appears. The vibrations at the viaduct and on the ground include not only the direct effect from the nearest joints but also the effects by the fork joints at two span (100 m length) and 3-span (150 m length) girders. Therefore, the traffic induced vibration should be interpreted by the vibration emission from the foundations directly beneath the girder on which vehicles are passing, but also simultaneously from the foundations underneath the adjacent girders (See Figure 8).

5 VIBRATION REDUCTION MEASURES AT SOURCE AND ON FILED

The counter measures against traffic induced vibrations are taken in general by a road bureau in charge of it. They mostly carry out at the side of vibration generation or on the vibration propagation field. In this paper, some introduction is made but limited only for construction methods for the conventional measures. The environmental vibrations are measured in

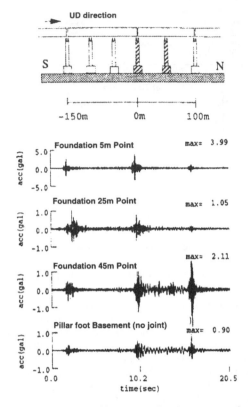

Figure 8. Pier top and bottom acceleration responses for a full loaded truck running at 60 km/h and dynamic load wave.

terms of the so-called vibrations level. The remarks in what follows use this indication.

5.1 Measures at source side

(1) Levelness of road

Since vibrations induced by traffic largely depend upon the smoothness of roads, minimizing road surface cracks and irregularities is one of the effective measures that can be taken at source against these vibrations. Figure 9 shows the result of the asphalt covering carried out for a large crack produced on the surface of a concrete paved road. The vibration reduction effect of this covering work was around 6 to 10 dB in terms of vibrations level, thus telling that leveling repair work is effective in reducing traffic induced vibrations. A significant vibration reducing effect can also be expected from a fitting work to repair drops or differences in road level. While asphalt pavement has an advantage over concrete pavement in road leveling, it yields to the latter in wear resistance, being prone to be rutted rather easily.

A level road (ground road) generally has differences in level due to manholes, which cause traffic

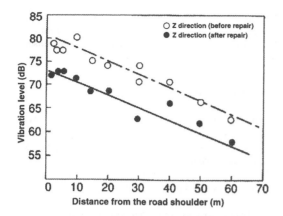

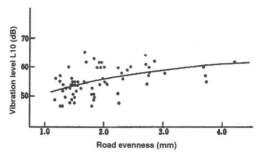

Figure 11. Road evenness and vibration level of road surface (Gyosei 1980).

Figure 9. Vibration comparison before and after repair of asphalt pavement over concrete pavement (Gyosei 1980).

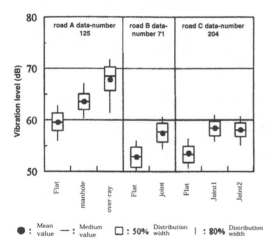

● : Mean value   — : Medium value   ▢ : 50% Distribution width | : 80% Distribution width

Figure 10. Maximum value of large car passage (outside lane) (Hirao & Yokota 1998).

induced vibrations. Shown in Figure 10 are the maximum values of vibrations level propagated by large-size vehicle traffic when they pass on manholes and other differences in pavement level.

Levelness of road σ is measured by a 3 m profilometer. The relationship between vibrations level and levelness of pavement is shown in Figure 11. The following expression is given when the relationship is calculated by logarithmic regression (Ishida et al. 2004):

$$L_{10} = 16.6 \log \sigma + 50.9 \quad \text{-------- Expression} \quad (1)$$

Expression (1) can be applied for practical use to reduce vibrations in repair works.

(2) Improvement in paving structure and soil (AIJ 2000)

The "vibration-reducing" pavement is best for improving the paving structure. Although still in the phase of research, there are various "vibration-reducing" paving methods, including (1) placing of a sheet between the surface and base courses; (2) adding to weight of the surface course using heavy density aggregates, etc.; (3) application of slab type surface course covered with rubber boards, and so on (Road 2004). The use of paving materials with high rigidity or high internal damping effect or the application of a thicker pavement is useful to reduce ground vibrations. On the other hand, the following methods have been proposed against a flimsy or weak ground: soil improving by chemico-piling construction method or similar methods that use a high-rigidity material in the circumference of pavement; sandwich construction method that puts a high-rigidity material course between concrete layers; and EPS block method that places a soft material below concrete slabs (about 4 to 10 dB can be reduced in terms of vibrations level). WIB (wave intercepting block) method, which makes use of the wave intercepting effect of these blocks against the stratified soil laying above a rigid firm ground, is also known widely.

(3) Fill-up structure (Gyosei 1980)

It is commonly known that less vibration is propagated by traffic on a fill-up road than on a ground level road. Shown in Figure 12 is the correlation between fill-up height and vibrations level. Seemingly, it is owing to distance damping because the rigidity of a well-compacted fill-up or embankment is usually higher than that of the ground, thus producing smaller vibrations and extending the wave propagation distance between the exciting and vibration-generating points.

(4) Traffic conditions

Road vibrations depend upon traffic conditions such as running car types, their weights and cruising speeds, and traffic volume. Vibration reduction can be expected from control of these traffic conditions, which will be

157

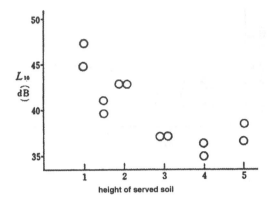

Figure 12. Correlation of height of served soil and vibration level (road edge) (Gyosei 1980).

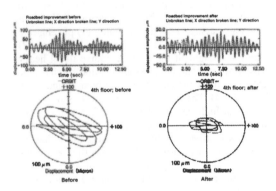

Figure 13. Displacement response of building (Kushida et al. 1992).

substantially effective in residential areas, and in precision industrial zones in which even micro vibrations could easily cause damage to their products. A difference of nearly 10 dB* is observed in traffic induced vibrations level between passenger cars and large-size vehicles. Differences in vibration response between loaded and empty large-size vehicles are becoming clear by the researches conducted by Ishida et al. (2004). Vibrations can be reduced also by changing the inflation pressure to change the spring constant of the tire. The 30% reduction from the standard inflation pressure reduces road vibrations by about 5 dB absolute.

These measures, however, are not practical because of the nature of roads that serve in the interest of the public.

Finally, shown in Figure 13 is an example of traffic induced vibrations at the 4th floor of a five story building that is located 50 meters away from an elevated bridge. The example shows vibration responses before and after road improvement work. What is noteworthy here is that the vibration response from the elevated bridge shows a characteristic horizontal vibration behavior, i.e., its horizontal amplitude is also decreased as a result of the road improvement work.

6 CONCLUSIONS

In this paper, we have discussed the exciting force characteristics and the vibration response propagated by road traffic with respect to two typical road types: ground level road and elevated road. We have also illustrated by example some vibration reduction measures at source and on propagation field.

The source of traffic induced vibration is usually found very close as compared with the hypocenter of an earthquake motion; even we can often identify it by visual check. In this sense, it is advantageous. Nevertheless, a source located close means that the characteristics of the source itself are predominant, because of which much more accuracy is required in the evaluation of traffic inducted vibration source characteristics than in the evaluation of the hypocenter of an earthquake motion.

If the external force at source (exciting force) is given in time series, consequently, ground vibrations at various points can be obtained in time series, they will allow you to create input motions for assessment of the impact of the vibrations under review on human body or health and for calculation of building response. As urban traffic networks advance in recent years, various types of roads, which are the source of vibrations, are being constructed out of necessity in the neighborhood of residential areas, laboratories, precision industrial zones, etc. in which vibrations are not welcome. In these circumstances (as the distance between the evaluation point and the source is becoming more and more closer), keen awareness of vibration sources will be increasingly important in the evaluation of vibration in the future. It is therefore of critical importance to create a reliable database using measured values of the exciting force of traffic induced vibrations, including their horizontal components, as well as illustrative examples of vibration reduction measures at source and on field.

REFERENCES

Architectural Institute of Japan The vibration by the present condition and criteria of the environment vibration evaluation regarding residence performance. 2000.6.25
Road environment research institute: The technology method of road environment assessment, volume 2nd, pp. 293–317, November, 2000
ISO: Mechanical vibration-Road surface profiles-Reporting of measured data, ISO8608, pp. 1–30, 1995
Kobayashi, A.: The automobile vibration, Toshoshupan, 1976

Nishizaka, R. and Fukuwa, N.: The research regarding the automobile traffic vibration problem in the researchplane road regarding the vehicle movement load characteristic in the traffic vibration problem(the 1), Structure engineering argument prose selection, Architectural Institute of Japan, No. 491, pp. 65–72, January, 1997.

Nishizaka, R., Shiozawa, N. and Fukuwa, N.: The foundation of the site that adjoin the plane roads and elevated road, analysis of the building vibration nature, structure engineering argument prose selection, Vol. 45B, pp. 93–101, March, 1999

The vibration regulation technology manual: Ministry of the Environment, Environment policy bureau, Gyousei 1980.11.1

The pavement committee: The pavement technology aimed for the environment improvements(1), Road, 2004.04

Hirao, Y. and Yokota, A.: The research regarding the pavement road surface condition and vibration by the road traffic vibration, The Acoustical Society of Japan, Noise and Vibration research paper N-99-44, 1998.06.25

The road traffic vibration prevention technology manual: Ministry of the Environment, Environment policy bureau, Gyousei 1980. 11.1

Ishida, R., Hanazato, T. and add 2: Symposium of the new technology of the prediction and measure by foundation vibration of the traffic vibration. The Japanese Geotechnical Society, 2004.05.13

Environmental Vibrations – Takemiya (ed.)
© 2005 Taylor & Francis Group, London, ISBN 0 415 39035 4

Ground vibrations induced by Shinkansen high-speed trains in view of viaduct-ground interaction

X.C. Bian & H. Takemiya
Okayama University, Okayama, Japan

ABSTRACT: The viaduct behaviors and the nearby ground motions under the high-speed train passage have been studied by using the computer simulation. Emphasized here is the soil-foundation-viaduct interaction analysis under moving axle loads. The solution method is to apply the dynamic substructure method in the frequency domain. The viaduct girders including track structure and pier supports are modeled by the three-dimensional beam-column elements. The supporting pile foundation and the near field of ground are discretized by the axisymmetric three-dimensional finite elements in a semi-analytical way. The nearby ground motions during train passage over viaduct have been calculated by superimposing the effects from the neighboring pile foundations. The main parameters affecting the viaduct vibrations are discussed. The results from this studies are validated from the field test data.

1 INTRODUCTION

Train viaducts supported by pile foundations are widely used for railway tracks at soft soil deposited area across urban areas. Increasing train speed in the last decades has benefited businesses and manufactures along railways; however, at the same time, some detrimental effects also have been brought by emitted vibrations to the alongside residential houses, sensitive equipments and high-tech production facilities. This problem has drawn serious concerns from people.

Dynamics of elevated tracks under the action of moving trains include two different issues: one is the interaction problem of vehicle and track in the high frequency range that has been analyzed by using a model of a sprung/unsprung mass on a beam. Roughness of rail surface due to the corrugation gives rise to the inertia motion of vehicles. The other issue is the interaction problem of train and supporting bridges in the low frequency range. In this regard, as an extension of the beam-on-Winkler-spring model, a double beam system representing rails on track and an extended viaduct on pillars respectively by the Winkler spring elements has been employed (Yoshioka & Ashiya, 1995). Apart from such modeling, there are many past investigations that have dealt with simply supported railway bridges. According to those recent works (Friba, 2001; Yang et al, 1999; Cheng et al, 2001; Song et al, 2003; Kwark et al, 2004), the train speed gives a critical frequency with respect to the span length and the sectional rigidity of bridges in view of the structural

resonance. The high-speed train on viaduct generates the structural borne vibrations. Therefore, the natural vibration modes of structures become important. The above double beam model has a shortcoming in this regard.

Train viaducts on soft ground are generally supported by buried foundations. The soil-structure interaction model should be considered since the global structural response is crucial to the structural response. Actually, it affects appreciably the dynamic responses of viaduct in low frequency range under the high-speed train loading. Thus structural vibrations are emitted from the fixed foundation positions that lead to significant wave propagation in ground, especially a dispersive waves in layered ground (Takemiya, 2003).

In this work, the author focused not only on the viaduct vibrations under high-speed train traffic but also on the resulting nearby ground vibrations. A case study is demonstrated by taking a high-speed Shinkansen train in Japan. The numerical computations interprets

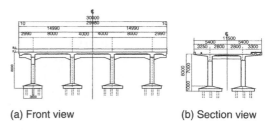

(a) Front view   (b) Section view

Figure 1. Geometry of Japan Shinkansen viaduct unit.

the involved dynamic features. The computation results are validated from the field measurement data.

2 FORMULATION FOR ANALYSIS

2.1 *Train loads*

Train loads with a series of wheel axles are formulated in the frequency domain (Takemiya, 2003) so that it is included straightforwardly into the structural equations. The geometry of wheel sets of a typical bogie-type train is illustrated in Fig. 2.

For a train comprising N numbers of cars, the vertical successive axle loads with a constant moving velocity c is described by

$$f_N = \sum_{n=1}^{N} f_n(x - ct) \qquad (1)$$

Each term at right-hand side indicates a single n-th car contribution whose detail expression is given by

$$f_n(x-ct) = P_{n1}\delta(x-ct+\sum_{s=1}^{n-1}L_s+L_0) + P_{n1}\delta(x-ct+a_n+\sum_{s=1}^{n-1}L_s+L_0)$$
$$+P_{n2}\delta(x-ct+a_n+b_n+\sum_{s=1}^{n-1}L_s+L_0) + P_{n2}\delta(x-ct+2a_n+b_n+\sum_{s=1}^{n-1}L_s+L_0) \qquad (2)$$

where $\delta(\cdot)$ is the Dirac's delta function to represent the moving axle loads, P_{n1} and P_{n2} at the front and rear bogies respectively, the notation L_n is the car length and L_0 is the distance to the reference position ahead of the first axle load position. The notations a_n, b_n are the distances between wheel axles as indicated in Fig. 2. The frequency domain representation of Eq. (2) becomes after the Fourier Transform, as

$$f_n^t(\omega,x) = \frac{1}{c}e^{i(\sum_{s=1}^{n-1}L_s+L_0+x)}$$
$$\times \left\{ P_{n1}(1+e^{ia_n\omega/c}) + P_{n2}(e^{i(a_n+b_n)\omega/c} + e^{i(2a_n+b_n)\omega/c}) \right\} \qquad (3)$$

Then, the whole train axle loads can be expressed as

$$f_N^t(\omega,x) = \sum_{n=1}^{N} f_n^t(\omega) = \sum_{n=1}^{N} \frac{1}{c}e^{i(\sum_{s=1}^{n-1}L_s+L_0+x)}$$
$$\times \left\{ P_{n1}(1+e^{ia_n\omega/c}) + P_{n2}(e^{i(a_n+b_n)\omega/c} + e^{i(2a_n+b_n)\omega/c}) \right\} \qquad (4)$$

It is postulated that the train axle loads are directly imposed on the girder surface. The continuous distribution $f^t_N(\omega, x)$ is discretized by the sleeper intervals, as indicated in Fig. 3. These lumped vertical loads are

$$F_{v1} = \int_{x_1}^{x_1+d_1/2} f_N^t(\omega,x)dx, \quad F_{v2} = \int_{x_2-d_1/2}^{x_2+d_2/2} f_N^t(\omega,x)dx,$$

$$F_{v3} = \int_{x_3}^{x_3-d_2/2} f_N^t(\omega,x)dx \qquad (5)$$

If the viaduct structure with a curvature of radius R, it imposes both the vertical and horizontal moving loads on the girder structure. The horizontal loads are assumed to be caused by the centrifugal action of train mass only so that it is derived as

$$F_{h,i}^t = F_{v,i}^t c^2 / gR \qquad (6)$$

in which, i is node index subject to train loads, $F_{v,i}^t$ is the vertical axle load defined in Eq. (5), c is the train speed and g is the gravity constant.

2.2 *Viaduct equations*

A generalized finite-element formulation is adopted to describe the motion of viaduct with multiple supports. The governing equation of the superstructure is formulated as

$$\mathbf{M\ddot{U}} + \mathbf{C\dot{U}} + \mathbf{KU} = \mathbf{F} \qquad (7)$$

where $\mathbf{M}$, $\mathbf{C}$ and $\mathbf{K}$ are the mass, damping and stiffness matrices, respectively. The Rayleigh matrix $\mathbf{C}$ is assumed

$$\mathbf{C} = \alpha\mathbf{M} + \beta\mathbf{K} \qquad (8)$$

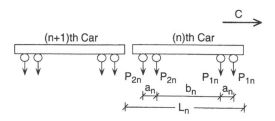

Figure 2.   Parametric description of train axle loads.

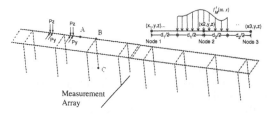

Figure 3.   Viaduct model under both vertical and horizontal distributed loads due to train passages (Points A, B and C are the nodes for observation).

where α and β are constants to give an internal damping coefficient $\lambda = 0.03$ for the fundamental frequency of the viaduct.

Under the harmonic load of frequency ω, the governing equation can be rewritten as

$$-\omega^2 \mathbf{M} \mathbf{U}^t + i\omega \mathbf{C} \mathbf{U}^t + \mathbf{K} \mathbf{U}^t = \mathbf{F}^t \tag{9}$$

For the inertial interaction between the viaduct-ground-pile foundation, a substructure method is implemented to couple them at their common interface. By separating the displacement and load vectors at superstructure nodes (denoted by the subscript 's') and at interface nodes (denoted by the subscript 'i'), Eq. (9) is rewritten as

$$-\omega^2 \begin{bmatrix} \mathbf{M}_{ss} & \mathbf{M}_{si} \\ \mathbf{M}_{is} & \mathbf{M}_{ii} \end{bmatrix} \begin{Bmatrix} \mathbf{U}_s^t \\ \mathbf{U}_i^t \end{Bmatrix} + i\omega \begin{bmatrix} \mathbf{C}_{ss} & \mathbf{C}_{si} \\ \mathbf{C}_{is} & \mathbf{C}_{ii} \end{bmatrix} \begin{Bmatrix} \mathbf{U}_s^t \\ \mathbf{U}_i^t \end{Bmatrix}$$
$$+ \begin{bmatrix} \mathbf{K}_{ss} & \mathbf{K}_{si} \\ \mathbf{K}_{is} & \mathbf{K}_{ii} \end{bmatrix} \begin{Bmatrix} \mathbf{U}_s^t \\ \mathbf{U}_i^t \end{Bmatrix} = \begin{Bmatrix} \mathbf{F}_s^t \\ \mathbf{F}_i^t \end{Bmatrix} \tag{10}$$

At the interface nodes of pier and foundation, the displacement compatibility with the foundation,

$$\mathbf{U}_i^t = \mathbf{U}_b^t \tag{11}$$

and the force equilibrium,

$$\mathbf{F}_i^t = -\mathbf{F}_b^t \tag{12}$$

are claimed. Now supposing the foundation impedance function $\mathbf{K}_f$ from the foundation-soil analysis in the next Section 2.4, the following relation is established.

$$\mathbf{K}_f \mathbf{U}_i^t = -\mathbf{F}_i^t \tag{13}$$

Assembling the Eq. (12) and Eq. (15) to derive the global dynamics equations of viaduct results in

$$-\omega^2 \begin{bmatrix} \mathbf{M}_{ss} & \mathbf{M}_{si} \\ \mathbf{M}_{is} & \mathbf{M}_{ii} \end{bmatrix} \begin{Bmatrix} \mathbf{U}_s^t \\ \mathbf{U}_i^t \end{Bmatrix} + i\omega \begin{bmatrix} \mathbf{C}_{ss} & \mathbf{C}_{si} \\ \mathbf{C}_{is} & \mathbf{C}_{ii} \end{bmatrix} \begin{Bmatrix} \mathbf{U}_s^t \\ \mathbf{U}_i^t \end{Bmatrix}$$
$$+ \begin{bmatrix} \mathbf{K}_{ss} & \mathbf{K}_{si} \\ \mathbf{K}_{is} & \mathbf{K}_{ii} + \mathbf{K}_f \end{bmatrix} \begin{Bmatrix} \mathbf{U}_s^t \\ \mathbf{U}_i^t \end{Bmatrix} = \begin{Bmatrix} \mathbf{F}_s^t \\ 0 \end{Bmatrix} \tag{14}$$

2.3 Pile foundation and near field ground

In order to formulate the dynamic interaction of the soil-pile foundation, the near field has been modeled together with pile foundation by the three-dimensional axisymmetric finite elements, while the far field that extends horizontally has been modeled by the thin

layer elements (Kausel & Roesset, 1976). The numerical model is depicted in Fig. 4.

In the modeling of pile group subject to the arbitrary foundation excitations, the semi-analytical solution of axisymmetric three-dimensional finite element with Fourier series expansion in azimuth is taken to advantage for the convenience of memory size and computation. The geometric and material properties of pile foundation and soil media are assumed prismatic. Hence, the displacements in cylindrical reference system can be expressed in terms of the Fourier amplitudes, as

$$-\omega^2 \mathbf{M}_n \mathbf{U}_n + i\omega \mathbf{C}_n \mathbf{U}_n + \mathbf{K}_n \mathbf{U}_n = \mathbf{P}_n^e + \mathbf{P}_n^i \tag{15}$$

for either a symmetric or an anti-symmetric motion. The notations $\mathbf{M}_n$, $\mathbf{C}_n$, $\mathbf{K}_n$ are the mass, damping and stiffness matrices corresponding to the Fourier amplitudes. $\mathbf{P}_n^e$ is the effective input force vector, and $\mathbf{P}_n^i$ is the internal force vector from the resistance of piles. The Eq. (15) is simply expressed as,

$$\mathbf{D}_s \mathbf{U}_n = \mathbf{P}_n^e + \mathbf{P}_n^i \tag{16}$$

where

$$\mathbf{D}_s = -\omega^2 \mathbf{M}_n + i\omega \mathbf{C}_n + \mathbf{K}_n \tag{17}$$

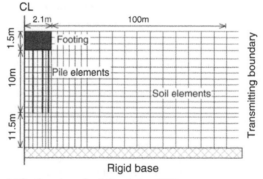

(a) Section view of soil-foundation FEM

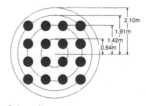

(b) Plan view of ring piles

Figure 4. Axisymmetric finite element model of pile foundation and surrounding soil.

163

Piles with circular cross sections are assumed to be placed in a concentric layout in the plan view. The behaviors of piles on the same radius from the center are approximated by the limited Fourier harmonics of $n = 0$ and 1 along the circumferential direction connecting the concerned piles axes. This assumption may be reasonable as far as the rings of piles keep almost the original axisymmetric configuration during the motion at any depth. In the analysis of the interaction problem between discrete ring-piles and the surrounding continuum soils, the pile elements and the neighboring soil element are assumed to have the common displacements. The piles are modeled by the 2-noded 3-dimensional beam elements, and the dynamic stiffness of nth Fourier harmonic expansion is superimposed onto the soil stiffness (Takemiya, 1985; 1986) to eliminate $\mathbf{P}_n^i$ in Eq. (16).

The pile-head impedance functions for different types of motions are condensed into the total impedance function for the motions of a rigid body footing portion, then

$$\mathbf{F}_b^t = \mathbf{K}_b \mathbf{U}_b^t \tag{18}$$

where, $\mathbf{F}_b^t$, $\mathbf{u}_b^t$ are the loads and displacements vectors as defined at the footing top.

3 SIMULATION RESULTS

3.1 Numerical model of Shinkansen viaduct

Herein, two neighboring units of the Japan Shinkansen viaduct in Fig. 1 are modeled for the numerical computation in view of a possible neighboring effect in response. The FEM model is depicted in Fig. 4. The pile foundation supports are replaced by the springs and dashpots with frequency dependant nature from the impedance functions for the pile-soil interaction. The weak connection between two neighboring units by rails is considered by the beam element of a proper elastic module. Three specific nodal points indicated in Fig. 3 are primarily focused for the response observation: first one is at the center of mid-span (Point A), second one is just on the pier top (Point B) and the third one is at the pier bottom (Point C).

3.2 Pile foundation impedance

The soil condition at the studied site is listed in Table 1. It is postulated that at the bottom of these soil layers a rigid bedrock exists. The corresponding axisymmetric finite element model of pile foundation and soil is depicted in Fig. 4(a). The ring arrangement of pile group is presented in Fig. 4(b). Three rings of piles of 10 meters long are constructed to support the footing and superstructure. The properties of the piles and footing are listed in Table 2.

Table 1. Soil properties.

Layer No.	Density [ton/m^3]	Vs [m/s]	Poison ratio	Damping [%]	Thickness [m]
1	1.5	141	0.45	5.0	1.0
2	1.6	70	0.3	5.0	5.5
3	1.7	275	0.4	5.0	0.5
4	2.0	173	0.3	5.0	0.5
5	1.7	310	0.4	5.0	7.5
6	1.8	256	0.3	5.0	2.0
7	1.9	320	0.3	5.0	6.0

Table 2. Foundation properties.

Density [ton/m^3]	2.4
Poisson ratio	0.167
Damping	0.05
Elastic modulus [kN/m^2]	3.0E7
Section area [m^2]	0.192
Effective section area[m^2]	0.144
Sectional inertia moment Ix [ton · m^2]	7.37E–4
Sectional inertia moment Iy [ton · m^2]) [ton · m^2]	7.37E–4
Sectional inertia moment Iz [ton · m^2] [ton · m^2]	1.47E–3

The far field in the ground model has been divided into 29 thin sublayers to be compatible with the finite element meshes in the near field zone. The computation results of pile foundation impedance functions are shown in Fig. 5 for the 6 degree of freedom components. These functions are found to be strongly frequency-dependent. Therefore, in order to evaluate the vibrations of viaduct structure accurately, the dynamics soil-structure interaction should be taken into account accordingly.

3.3 Eigenmodes analysis

Moving train with a given speed generates dynamic loading on the viaduct of several driving frequencies. These frequencies can shift to the higher range as the speed increases. Hence, there is a great possibility for one or more of these frequencies to become close to the natural frequencies of the viaduct structure. When the resonance occurs at either of these frequencies, the viaduct structure and the train body vibrate with big amplitudes that may not only lead the train operation to danger but also cause severe environmental disturbances alongside.

In order to check the structural borne vibrations, the eigen-modes are investigated for one unit viaduct. The first limited numbers of eigen-frequencies and the associated vibration modes have been computed by taking springs-dashpots supports (at $f = 0$) from the pile foundation analysis. The vibration shapes shown in Fig. 6 correspond to those for the first eight eigen-frequencies.

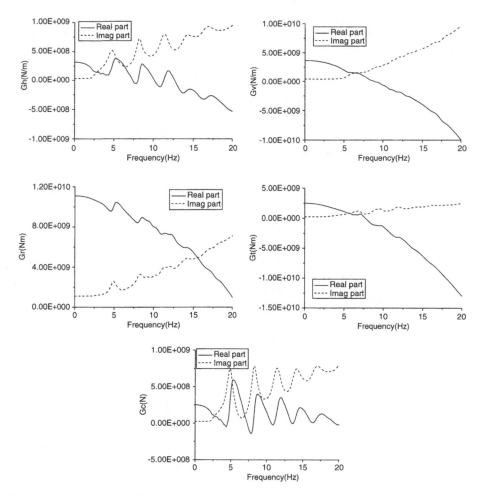

Figure 5. Impedance functions of pile foundation at footing top.

The fundamental vibration mode has a significant global twisting along the vertical axis with respect to the center of the mid span. The 2nd mode has a lateral sway in the perpendicular direction to the viaduct axis. It is therefore pointed out that the viaduct may show a substantial twisting response when loaded asymmetrically by trains. The 3rd mode is a global longitudinal motion in the viaduct direction, and the 4th mode has a significant girder sway motion at the center span along the viaduct axis whereas with a twisting behavior at the side spans. During the successive action of the axle loads by the train passage, the horizontal motions along the viaduct axis are expected to occur significantly.

3.4 *Viaduct response*

In the illustrative computation, the viaduct is excited by the Shinkansen train of 16 cars where the same axles load $P = 150$ kN is taken for each car. The moving speed of $c = 83.3$ m/sec (300 km/h) is assumed. The car length is 25 m, which gives the main driving frequency 3.33 Hz. In the same manner, some other driving frequencies can be determined from the combination of wheel set distances as 11.1 Hz, 33.3 Hz and so on. The comprehensive disposition of the natural frequencies of the viaduct and the train-loads geometry are primarily concerned to determine the critical speed for resonance for the train operation. Another important driving frequency may be due to the inverse of the viaduct span length 8 m against the train speed (c = 83.3 m/s), viz. 10.4 Hz.

The train is assumed to run either on the up-bound track or on the down-bound track, namely, in Fig. 1 these orientations correspond to the running respectively from left to right or vice versa on the track. The axle loads are then imposed on either side track beam in Fig. 3. This means that the loading position is 2.8 m off the longitudinal centerline of the viaduct, so that it is asymmetric about it. Because the viaduct motions are formulated in the frequency domain, an appropriate range of frequency should be chosen to include the motions of the major contributing structural modes for the response computation. Herein, the frequency up to 60 Hz is found to suffice the accuracy of analysis for the viaduct and ground dynamic interaction. In view of this requirement and the response time window for a train passes by a focused point completely, the Fourier transform points are chosen as 2048 with the time increment of 0.008 seconds, which leads the computation duration to be 16.4 seconds.

The Fourier spectra of viaduct deflections at the Point A in Fig. 3 due to the up-bound train running are plotted in Fig. 7. The response frequencies are noted mostly in the range below 15 Hz. At the frequency of 3.3 Hz, a high peak is generated that coincides with the main driving frequency from the individual car length of the Shinkansen against the speed. Besides the peak around this driving frequency, the second peak is found at the natural frequency, 6.54 Hz of the viaduct structure for the twisting mode. The 3rd significant peak appears at 10.4 Hz that is related to quotient of viaduct span length against train speed. The horizontal response in viaduct direction U_x is regarded as mostly induced by the dynamic behavior. However, the responses in the perpendicular direction to the viaduct U_y and in the vertical direction U_z include substantial quasi-static (zero frequency) component because of an asymmetric load distribution on the girder due to the train weights.

The global motions of the viaduct structure are affected not only by the structure itself, but also by the support conditions. From the comparison of responses at the pier bottom (at C) and at the span center of the girder (at A), the soil structure interaction effects can be recognized as significant and cannot be neglected.

3.5 Ground response

The ground motion caused by the viaduct vibrations is another important aspect of the present work. The

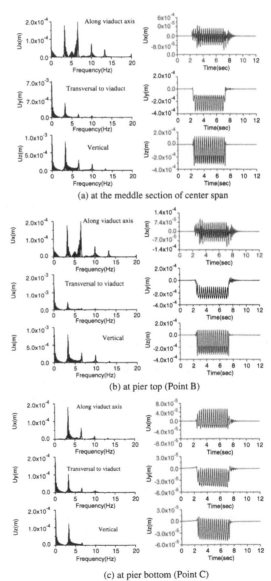

(a) at the meddle section of center span

(b) at pier top (Point B)

(c) at pier bottom (Point C)

Figure 7. Viaduct displacements for Shinkansen Nozomi, 300 km/h at middle section of the center span.

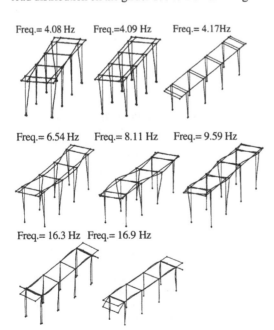

Freq.= 4.08 Hz   Freq.=4.09 Hz   Freq.= 4.17Hz

Freq.= 6.54 Hz   Freq.= 8.11 Hz   Freq.= 9.59 Hz

Freq.= 16.3 Hz   Freq.= 16.9 Hz

Figure 6. First eight natural frequencies and corresponding modes of viaduct structure.

dynamic substructure method in this paper can derive the interaction forces at the common interface nodes between the superstructure and the pile foundations. Thus obtained internal forces at the footing top are depicted in Fig. 8. It is noted that the vertical load is much larger than other two horizontal force components. From the moment loads, we also can find that the global twisting response of the viaduct is significant, although those locals at the individual pier bottom are quite small. The force in the viaduct direction (x-axis) and the bending moment around the lateral direction (y-axis) are relatively large, and both of these two components contribute to the horizontal motion of ground in x-direction as described below. Under these combined loads, the ground responses in the viaduct direction (x-axis) are expected to be significant.

The ground motions in the vicinity of the viaduct are of interest under the moving train loads from the

environmental preservation along it. Five positions on the ground surface in Fig. 3 are addressed for the response computation of vertical and two perpendicular horizontal components. The vibrations at these points are obtained by superposing the effects of the individual ground motions from the 16 adjacent pile foundations of the neighboring two viaduct units.

The time histories of velocity responses on the ground surface, including two horizontal components and the vertical one are depicted in Fig. 9. The distance between the foremost and the last rear wheel sets is 395 m, therefore the duration of complete passage of a whole train across a specific observation on the viaduct becomes 4.74 sec. The spindle forms at both ends of the wave motions are clearly noted at the far focused points, which are caused by the energy emission from multiple sources and the dissipative nature in ground. The simulation results give satisfactory accuracy for the major portion of the time history; however, the free motion of the viaduct is not as large as observed in the field measurement.

Next, in order to grasp the wave interference while the vibrations are propagating in field from the multiple foundations, a contour map is depicted for the velocity response of the ground surface. In the case of train tracks directly on ground in which the location of vibration source varies continuously along the axis, so that the wavenumber is continuous in the form of delta function of $\delta(k - \omega/c)$ and it is uniquely determined when integrated over the wavenumber by the ratio of frequency and train-speed ω/c. In contrast, viaduct tracks have the discrete wavenumbers as provided by additional effects by span-length S of adjacent foundations so that in the form of $\Sigma\delta(k - \omega/c - 2\pi n/S)$ ($n = 1,2,$). The total wave field is therefore summed up from the response contour maps with the certainly

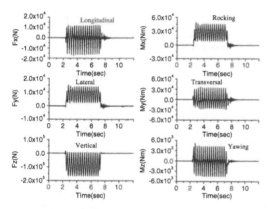

Figure 8.   Internal force actions on the footing top.

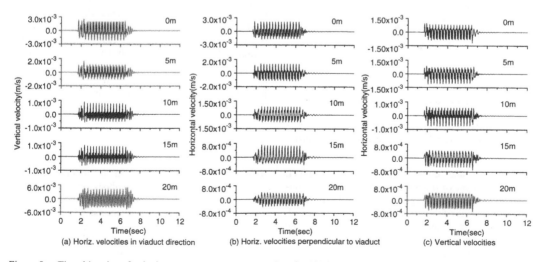

Figure 9.   Time histories of velocity responses on ground surface for Shinkansen Nozomi, 300 km/h.

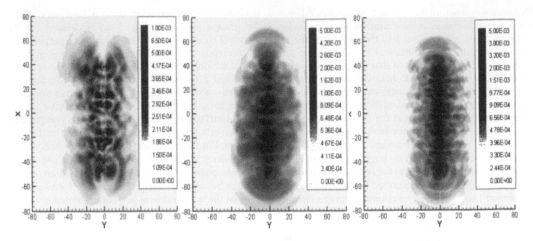

Figure 10. Contour maps for the maximum velocity responses on ground surface for two unit viaducts.

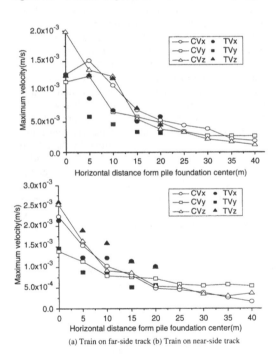

(a) Train on far-side track (b) Train on near-side track

Figure 11. Maximum velocity responses of ground surface (Prefix 'T' for field test data, 'C' for computation results).

determined space coordinate along the viaduct. However, in view of the relative intensities in all contour maps, the overlapping effect of wave fields of the consecutive units of the three-span viaducts are small. Hence, this proves the good matching between the present simulation and the measurement.

The computation results are depicted in Fig. 11 for the maximum velocity and compared with the available data from the field measurements. It is clearly noted that when the Shinkansen train runs on the near track to the focused points there are more intensified motions than those on the far track. The matching is good between computation and field measurement for the maximum vibration velocities, especially regarding the vertical component that shows a monotonically decreasing trend with the distance from the viaduct. In the vicinity of the viaduct the vertical component indicates the larger response than the horizontals. However, in the far distance the horizontal response along the viaduct grows larger than other components. Another important finding is the difference of responses when the train runs on the near or far track. In the case of the train passage on the near track to the observation, the horizontal responses, both along the viaduct and perpendicular to it, are predicted smaller than the measurement data. This difference should be attributed to the two folded reasons. One is the structural twisting vibration mode that leads different force action by the different track use. The other is the presence of another rows of foundation piles on the propagating wave field when the far track is used, which leads more wave scattering and then mitigation.

4 CONCLUSIONS

The vibrations induced by the high-speed Shinkansen train running on a viaduct on a soft ground are first investigated from the FEM analysis. Two units of 3-span continuous girders are employed in the present computation model. Both frequency spectrum and time history of structural response are investigated to evaluate the vibration characteristics and their intensities.

From the eigen-modes analysis of one unit of the viaduct, the most important eigen-modes affecting the structural borne responses are detected. Besides, the train load profile, the viaduct geometry and the ground properties at site are key parameters to characterize the total vibration features. Among them, of most importance is the train speed against the surface wave velocity in ground. The soil-structure interaction (SSI) effect is recognized to be crucial for accurately determining the viaduct vibrations and also assessing the alongside ground vibrations from the environmental viewpoint.

From the comparison of the simulated results and the field measurement data, the present solution method and developed computer program are validated. In the case of high-speed train passage, the horizontal ground motions along the viaduct become comparably large to the vertical responses. Furthermore, for the curved viaduct, the in-plane horizontal motion is also quite substantial on the ground close it because of the horizontal centrifugal force generation. These 3-dimensional responses should be taken into account when effective vibration-reduction measures are developed.

REFERENCES

Cheng, Y.S., Au, F.T.K. and Cheung, Y. K. 2001. Vibration of railway bridges under a moving train by using bridge-track-vehicle element, Engineering Structures, 23, 1597–1606.

Fryba, L. 2001. A rough assessment of railway bridges for high speed trains, Engineering Structures, 23(5), 548–556.

Kausel, E. and Roesset, J.M. 1981. Stiffness matrices for layered soils. Bull. Seism. Soc. Am., Vol. 71(6), pp. 1743–1761.

Kwark, J.W., Choi., E.S., Kim, Y. Kim, J.B.S. and Kim, S.I. 2004, Dynamic behavior of two-span continuous concrete bridges under moving high-speed train, Computers & Structures; 82(4–5), 463–474.

Song, M.K., Noh, H.C. and Choi, C.K. 2003. A new three-dimensional finite element analysis model of high-speed train–bridge interactions. Engineering Structures, 25, 1611–1626.

Takemiya, H. 1985. Three-dimensional seismic analysis for soil-foundation superstructure based on dynamic substructure method, Proc. JSCE, 1985; 3(1), 139–149.

Takemiya, H. 1986. Ring pile analysis for grouped piles subjected base motion. Structural Eng./Earthquake Eng, 3(1), 195–202.

Takemiya, H. 2003. Simulation of track-ground vibrations due to a high-speed train: the case of X-2000 at Ledsgard. Journal of Sound and Vibration, March., 261(3), 503–526.

Takemiya, H., Bian, X.C., Yamamoto, K. and Asayama, T. 2004. High-speed train induced ground vibrations: Transmission and mitigation for viaduct case, Proc. 8th Int. Workshop on Railway Noise, I/107–118.

Takemiya, H. and Bian, X.C. 2005. Substructure simulation for train track/layered ground interaction dynamics through discrete sleepers. Journal of Engineering Mechanics Division, EM7. ASCE

Yang, Y.B., Chang, C.H. and Yau, J.D. 1999. An element for analysing vehicle-bridge systems considering vehicle's pitching effect. International Journal for Numerical Methods in Engineering, 46(7), 1031–1047.

Yoshioka, O. and Ashiya, K. 1995. A dynamic model on excitation and propagation of Shinkansen-induced ground vibrations. Butsuri-Tansa, 48(5), 299–315.

Dynamics of traffic structures

Environmental Vibrations – Takemiya (ed.)
© 2005 Taylor & Francis Group, London, ISBN 0 415 39035 4

Measuring the bridge frequencies from the response of a passing vehicle

Y.B. Yang
Department of Civil Engineering, National Taiwan University, Taipei, Taiwan

C.W. Lin
Structural Department, China Engineering Consultants, Inc., Taipei, Taiwan

ABSTRACT: This paper intends to extract the fundamental frequency of a bridge from the dynamic response of a two-wheel cart towed by a light truck passing over the bridge. The truck serves as an exciter to the bridge, while the towed cart as a receiver of the bridge response. The response recorded using an accelerometer installed on the cart is processed by the Fast Fourier Transform (FFT), from which the bridge frequency is extracted as the source frequency of excitation to the cart. The feasibility of the idea in scanning the bridge frequency from the towed cart response is confirmed by the field tests. Such an approach is simple, efficient and highly mobilized, which can be applied to a wide range of bridge problems, if only the fundamental frequency is of interest.

1 INTRODUCTION

Among the dynamic properties of bridges, the frequencies of vibration, especially of the fundamental ones, represent a kind of information useful for detecting the integrity or health condition of the bridge (Ward 1984, Kato & Shimada 1986, Mazurek & DeWolf 1990). A general understanding is that a drop in the frequencies of vibration implies a deterioration in the overall stiffness of the bridge, whether it be caused by a damage or failure in any component, joint, or support of the bridge, or by a loss in tendon or cable forces, depending on the type of bridges considered. For the purpose of maintenance, it is desirable that the variation in frequencies of a bridge, as well as other dynamic properties, be frequently monitored, such that precautions can be undertaken to ensure the safe and normal functioning of the bridge.

Traditionally, various techniques have been employed to measure the frequencies of vibration of bridges, e.g., the ambient vibration method, forced vibration method, impact force method, etc. All these methods require installation of the vibration sensors, whether temporary or permanent, on the bridge for the purpose of recording the bridge response. As such, they have been referred to as the *direct approaches.*

In this paper, we shall introduce an *indirect approach* for measuring the bridge frequencies. Rather than working directly on the bridge response, we shall extract the frequencies of vibration from the vertical dynamic response of a vehicle passing over the bridge. A theoretical study of such an approach has been presented for

the first time by Yang et al. (2004), in which a vehicle modeled as a sprung mass moving over the bridge was used as the *exciter* to the bridge and simultaneously as a *receiver* of the bridge response. The bridge frequencies are then extracted from the vertical dynamic response of the moving vehicle by the Fast Fourier Transform (FFT) as the source frequencies of excitation.

This paper can be regarded as an experimental verification of the theory presented by Yang et al. (2004). Instead of using a single moving vehicle, a two-wheel cart towed by a light truck is used. The truck serves to *excite* the bridge through its movement over the bridge. The bridge that is set into motion will affect the dynamic response of the towed cart, which serves as a *receiver* or *scanner* of the bridge dynamic properties. The response recorded using an accelerator installed on the cart is then processed by the FFT to yield the bridge frequencies, which appear as the source frequencies of excitation to the cart.

If only the fundamental frequency is of concern, the present approach can be successfully used. It has the advantage of being simple, efficient, and highly mobilized, compared with the traditional direct approaches, as the test cart can be easily towed from one bridge site to the other with basically no preparatory works on the bridge to be measured. It is realized that the health condition of a bridge cannot be assessed by solely monitoring its fundamental frequency. Nevertheless, a routine monitoring of the variation of the fundamental frequencies of bridges can still reveal some essential useful information to bridge maintenance engineers. It is with this in mind that the technique is developed.

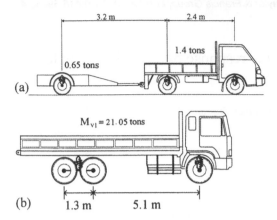

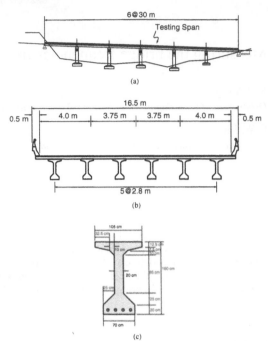

Figure 1. Test vehicles: (a) tractor-trailer, (b) heavy truck.

Figure 2. Bridge under test: (a) elevation, (b) cross section, (c) girder.

2   DESCRIPTION OF TEST VEHICLES

In this study, a small tractor-trailer system as shown in Figure 1(a) is adopted as test vehicle. The tractor is a four-wheel commercial light truck, which, when moving over the bridge, serves to excite the bridge into motion. The trailer is a small two-wheel cart, which will be excited by the bridge that is already in vibration, thereby serving as a receiver of the bridge motion. The vertical vibration of the cart can be regarded as measured from its own suspension system. The linkage connecting the truck and the cart is free to rotate in all directions, implying that no moments will be transferred in between.

Theoretically, the bridge frequencies will be contained in the dynamic response of the cart, as they represent the source frequencies of excitation. By recording the dynamic response of the cart during its passage over the bridge, the bridge frequencies can be extracted from the cart response, if the frequencies associated with the cart structure can be identified and excluded.

During the field test, a heavy truck with a net weight of 21.05 tons was also employed. It plays the role of "ongoing traffic" in one of the tests when the tractor-trailer was moving simultaneously over the bridge.

3   DESCRIPTION OF TEST BRIDGE

The test bridge, slightly curved, is composed of six spans of prestressed concrete girders resting on columns of various depths, as shown in Figure 2(a). However, the test span selected is straight and simply supported. This bridge is part of an important link used by trucks and tractor-trailers carrying heavy goods from the Keelung Harbor in northern Taiwan.

The bridge unit considered has a span length of 30 m and is composed of six prestressed I girders, placed at a center-to-center distance of 2.8 m (Figure 2). The cross section of the bridge has a total width of 16.5 m, overlaid with a concrete deck slab and AC pavement. The two ends of the bridge unit considered have an elevation difference equivalent to a longitudinal slope of 6.4%. The area and moment of inertia of each I girder section are $0.64\,m^2$ and $0.2422\,m^4$ respectively. The elastic modulus of the concrete is 29 GPa and the density is $2400\,kg/m^3$.

4   INSTRUMENTATION

Two kinds of vibration transducers were used in this study. One is the acceleration-type model AS-1GA made by Kyowa Inc. with a size of $1.4 \times 1.4 \times 1.8$ cm, as shown in Figure 3(a). The specifications of this sub-miniature low-capacity acceleration transducer include: rated capacity $\pm 9.807\,m/s^2$, response frequency range from DC to 40 Hz, and resonance frequency at 70 Hz. The other is the velocity-type transducer model VSE-15D made by Tokyo Shokushin Co., of which the specifications are as follows: response frequency range from 0.2 to 70 Hz and rated capacity range from 150 μ kine to 10 kine (1 kine = 1 cm/s), depending on the setting by software. The vibration transducer was used only in the ambient vibration test.

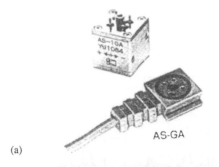

(a)

(b)

Figure 3. Instruments: (a) acceleration transducer, (b) central control system.

A portable ambient vibration monitoring system called the SPC51 system provided by the Tokyo Shokushin Co. was used in all the field tests (see Figure 3(b)). This system consists of a laptop computer, data processing unit, dual power supply switcher, and amplifiers and high frequency data scanner that are capable of dealing with 16 channels simultaneously.

In measuring the dynamic response of the cart during its movement, the vibration transducer was installed vertically at a position near the gravity center of the cart, while the data acquisition system was placed in the front seats of the light truck, for which the power supply was made available through a battery (see Figure 3(b)). For the purpose of comparison, an ambient vibration test was conducted for the bridge. In such a test, the vibration transducer was installed vertically at the center of the surface deck of the bridge of concern.

5 PLAN OF TESTING

Before conducting the cart response test, an ambient vibration test was conducted for the bridge free of any traffic for directly measuring the bridge frequencies, which will be used as the reference for comparison. In addition, a free vibration test was conducted for the test cart (trailer) at rest to determine its dynamic properties, including the vertical vibration frequency and damping

coefficient. Four key items are included in the present experimental study:

1. Record the dynamic response of the bridge to the moving action of the heavy truck. Such a response allows us to measure the bridge frequencies directly from the bridge response that serve as the basis of comparison.
2. During the passage of the heavy truck over the bridge, record the vertical response of the test cart placed at rest at the midspan of the bridge. The bridge frequencies extracted from the recorded cart response provides a means for evaluating the influence of the cart structure in transmitting the bridge frequencies.
3. Record the dynamic response of the cart towed by the light truck moving over the bridge. The bridge frequencies extracted from the cart response are exactly the ones desired.
4. Same as item 3, but in the meantime let the heavy truck move simultaneously over the bridge to play the role of on-going traffic. From the cart response recorded herein, the influence of ongoing traffic on the extracted bridge frequencies can be evaluated.

In each of the above tests, four different vehicle speeds are considered, so as to evaluate the effect of vehicle speed on the measured results.

6 EIGENVALUE ANALYSIS RESULTS

As shown in Figure 2, the bridge deck is supported by 6 girders. An eigenvalue analysis was conducted for a single girder with simple supports to obtain the first vertical frequency, which appears to be 3.761 Hz. The bridge was then model as a gird structure simply supported at the two longitudinal ends, in which the bridge deck was represented by three sets of horizontal bracings to account for its transverse stiffness contribution. The first two frequencies computed for the grid structure are 3.732 and 7.571 Hz associated with the first vertical and lateral modes, respectively.

7 EXPERIMENTAL RESULTS

7.1 Ambient vibration test

For the ambient vibration test, the total length of each record is 200 sec with a sampling rate of 100 Hz. Figure 4 shows the FFT of the recorded data, from which the peaks associated with the dominant frequencies of the bridge can be identified. In Figure 4(a), the lowest frequency can be read as 3.76 Hz, same as the one in Figure 4(b). However, the peak associated with the second frequency 7.60 Hz in the acceleration spectrum of part (a) was not well matched by the velocity spectrum of part (b) in terms of magnitude. This is due to the fact that the acceleration sensor was not mounted in a perfectly vertical position, because

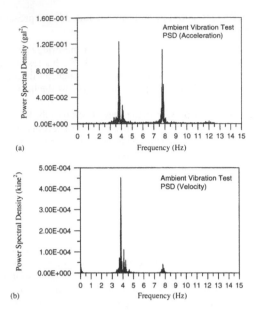

(a)

(b)

Figure 4. Power spectral density of ambient vibration records: (a) acceleration, (b) velocity.

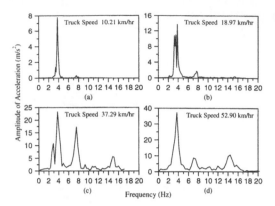

Figure 5. Acceleration spectra of bridge response to moving truck.

it was directly attached to the inclined pavement surface of the bridge. However, the velocity sensor was installed on a supporting base with compensation made for the inclination.

The fundamental vertical and lateral frequencies measured from the ambient vibration tests are identical to those of the eigenvalue analysis results, with an error of less than 1%. The purpose of conducting the ambient vibration tests is to obtain *directly* the frequencies of the bridge, which will be used as the base of comparison with those obtained from the test vehicle or by numerical simulations.

7.2 *Vehicle characteristic test*

The dynamic characteristics of the test cart (trailer) are investigated before it is actually put into use. The cart can be regarded in the extreme case as a single sprung mass system. By applying an initial vertical displacement on the cart and then releasing it suddenly, we can record the vertical acceleration response of the cart (not shown). By taking the average of four intervals between the adjacent peaks in the figure, the fundamental frequency of the trailer is found as 1.814 Hz, and the associated damping ratio ξ as 13% using the following formula (Berg 1988):

$$\xi = \frac{\delta/2\pi}{\left[1+\left(\delta/2\pi\right)^2\right]^{1/2}} \qquad (1)$$

where δ denotes the logarithmic value of the ratio of two adjacent peaks in the free-vibration response.

7.3 *Bridge response to moving heavy truck*

For comparison, the frequencies of vibration of the bridge will be measured *directly* from its response to the heavy truck moving at the four speeds 10.21, 18.97, 37.29 and 52.9 km/hr. The FFT of the bridge responses have been plotted in Figure 5, from which the first and second bridge frequencies are identified to be around 3.7 and 7.5 Hz, respectively, for all the speeds considered, which agree well with those computed from the eigenvalue analysis. The damping ratio of the bridge can be estimated using Equation (1) from the free decay response after the truck left the span in Figure 5, which is around 3%.

7.4 *Response of stationary trailer on bridge excited by a moving heavy truck*

It is essential to examine how the cart responds when it stays on the bridge under the action of other moving loads. To this end, the cart was placed at the midspan of the bridge, when the heavy truck moved at different speeds over the bridge. The FFT of the vertical acceleration responses recorded of the cart during the passage of the heavy truck at the four speeds: 10.21, 18.55, 37.8, and 52.32 km/hr, were plotted in Figures 6(a)–(d). As can be seen, the cart response was dominated primarily by the dynamic property of the bridge. Because of the damping effect of the cart, only the vibrations associated with the source frequency, i.e., the first frequency of the bridge, 3.7 Hz, remain on the cart.

From Figure 6, it can also be observed that the peak amplitude associated with the bridge frequency 3.7 Hz increases as the truck speed increases. By comparing Figure 6 with Figure 5, one observes that the bridge responses measured from the cart far exceed those

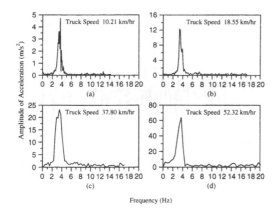

(a)

(b)

(c)

(d)

Frequency (Hz)

Figure 6. Acceleration spectra of stationary cart's response to moving truck.

measured directly from the bridge deck. The reason is that for a single DOF system, the response of a single DOF system will be amplified when the ratio of the excitation frequency to the structure frequency is greater than 0.7 for a damping ratio of 10% (Berg 1988). Clearly, the frequency ratio of the bridge (source) to the test cart (receiver) is a parameter that must be considered in the development of the measurement technique of concern.

7.5 *Response of a trailer hauled by a light truck with no on-going traffic*

Here, we shall proceed to examine whether the cart response still contains the bridge frequency, if the cart does not remain stationary, but is allowed to move over the bridge at certain speeds. In this regard, the cart will be towed by the light truck mentioned above, which serves to excite the bridge as well. To mark the beginning and ending time of the cart's movement over the bridge, two ropes were placed transversely across the bridge deck each at the entrance and exit of the span considered, each of which, when rolled over by the cart, will produce a *jump* in the recorded response.

For the cart moving at the speed of 13.0 km/hr, the number of peaks occurring within each time unit, as shown in Figure 7(a) is found to be 3 to 4. This implies that the cart response is dominated by the bridge frequency, which should be interpreted as the source frequency of excitation to the cart. In contrast, the response shown in Figure 7(b) for the speed 51.84 km/hr seems to be dominated by some high frequency components, from which the number of peaks occurring within each time unit cannot be visually counted.

Figures 7(a)–(b) reveal that as the vehicle speed increases, the period of the dominant vibration component becomes longer. This seems to be natural, since

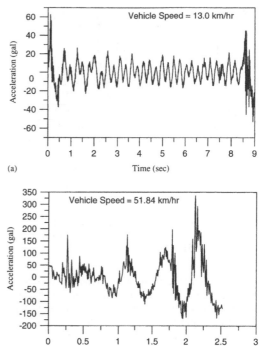

(a)

(b)

Figure 7. Dynamic response of moving cart at speeds: (a) 13.0 km/hr, (b) 51.84 km/hr.

the dynamic properties, especially the fundamental frequency, of the cart tend to be excited and reflected in the response when it travels at high speeds. On the other hand, one also observes from Figures 7(a)–(b) that as the vehicle speed increases, there is an increasing influence of the high frequency components in the response, mainly due to contributions associated with the mechanical parts of the cart, roughness in pavement, or minor collision in the connector between the tractor and trailer.

The FFT of the cart responses at the four speeds 13.0, 17.28, 35.13, and 51.84 km/hr were plotted in Figure 8. As was expected, the lowest peaks in this figure indicate the fundamental vibration frequency of the cart. Obviously, the peaks representing the fundamental frequency of the bridge can be identified from the results for the lowest three passing speeds. However, the same is not true for the highest passing speed, i.e., 51.84 km/hr, for which case the bridge frequency seems to be blurred by the involvement of high frequency components, related possibly to the pavement roughness and mechanical parts of the cart, which were excited by the relatively large amount of energy carried by the vehicle at higher speeds. For the vehicle moving at high speeds, we can also say that the bridge frequency becomes not so obvious in the

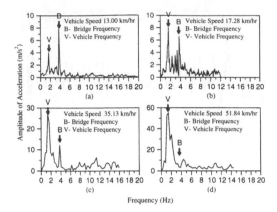

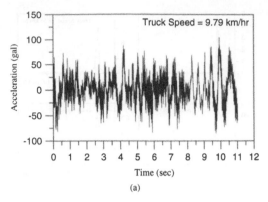

Figure 8. Acceleration spectra of cart response for tractor-trailer moving at: (a) 13.00 km/hr, (b) 17.28 km/hr, (c) 35.13 km/hr, (d) 51.84 km/hr.

cart response, as the input energy transmitted directly from the tractor is higher than that from the bridge. Therefore, to make the bridge frequency more visible in the cart response, the tractor-trailer should not be allowed to move at too high a speed.

As a side remark, the second frequency of the bridge that was identified by the ambient vibration test in Figure 4 is invisible from the cart response spectrums in Figure 8, as it was mixed with other high-frequency components.

7.6 *Response of a trailer hauled by a light truck with on-going traffic*

One essential question is whether the on-going traffic on a bridge can affect the dynamic response of the cart or trailer hauled by a tractor. To answer this question, we shall let the heavy truck proceed at a similar speed as the tractor-trailer to simulate the effect of on-going traffic, and then record the vertical response of the cart during such a passage. Figures 9(a)–(b) show the dynamic responses of the cart moving at the two speeds: 9.79 and 46.94 km/hr.

Compared with those in Figure 7, one observes that under similar vehicle speeds, the existence of an on-going moving truck has resulted in significant increase in the magnitude of the cart response. Such a phenomenon is considered *beneficial* and also reasonable, as the bridge (the source to the cart) was excited to a larger extent by the on-going truck.

From the FFT of the responses plotted in Figure 10 for the four different speeds: 9.79, 15.41, 38.56, and 46.94 km/hr, one observes that for the vehicles moving at speeds lower than 40 km/hr, the bridge frequency can be easily identified from the cart response. However, when the vehicle speed reaches 46.94 km/hr, the cart response was dominated primarily by its own

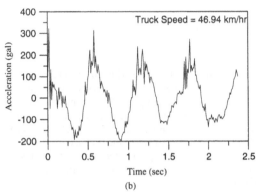

Figure 9. Dynamic response of moving cart (plus moving truck) at speeds: (a) 9.79 km/hr, (b) 15.41 km/hr, (c) 38.56 km/hr, (d) 46.94 km/hr.

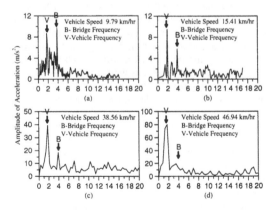

Figure 10. Acceleration spectra of cart response for tractor-trailer (plus heavy truck) moving at: (a) 9.79 km/hr, (b) 15.41 km/hr, (c) 38.56 km/hr, (d) 46.94 km/hr.

frequencies, while the bridge frequency cannot be easily identified. There exists another drawback for a vehicle to move at high speeds, as the acting time of the vehicle on the bridge is too short for the resolution of the fast Fourier transform to be kept at a proper level. Meanwhile, for the results shown in Figure 10, no visibility exists for the second frequency of the bridge, also due to involvement of high frequency components.

8 COMPARISON OF MEASURED RESULTS WITH NUMERICAL RESULTS

In this section, the experimental results will be compared with those by the finite element method. The finite element procedures developed in previous studies (Yang et al. 2004, Yang & Yau 1997) will be employed to simulate the two-wheel cart hauled by the light truck moving over the bridge, considering the effect of interaction between the bridge and moving vehicles. The bridge is modeled as an assembly of three-dimensional beam elements and the tractor-trailer as the model shown in Figure 11. With reference to the dynamic parameters for the tractor-trailer indicated in Figure 11, the following properties have been adopted in the present analysis:

(a) Tractor: $M_t = 1450\,\text{kg}$, $I_t = 1865\,\text{kg} \cdot \text{m}^2$, $k_f = 49\,\text{kN/m}$, $c_f = 0.26\,\text{kN} \cdot \text{s/m}$, $k_v = 63\,\text{kN/m}$, $c_v = 0.22\,\text{kN} \cdot \text{s/m}$, $m_{wf} = 110\,\text{kg}$, $m_{wr} = 125\,\text{kg}$, $d_1 = 0.8\,\text{m}$, $d_2 = 1.6\,\text{m}$, and $d_3 = 3.2\,\text{m}$.
(b) Trailer: $m_v = 650\,\text{kg}$, $k_v = 84\,\text{kN/m}$, $c_v = 2.85\,\text{kN} \cdot \text{s/m}$, and $m_{wv} = 50\,\text{kg}$.178

The step-by-step marching scheme adopted for solving the system equations in time domain is Newmark's β method, with the control parameters set as $\beta = 0.25$ and $\gamma = 0.5$ for unconditional numerical stability. No consideration is made for pavement roughness.

The maximum cart response computed by the finite element method have been listed in Table 1, along with those for the four vehicle speeds 13.0, 17.28, 35.13, and 51.84 km/hr. As can be seen, the maximum response of the vehicle increases as the vehicle speed increases. Except for the speed 13.0 km/hr, the maximum responses of the vehicle computed by the finite element simulation appear to be generally lower. The difference between the measured and numerical results in Table 1 can be attributed to the interaction of mechanical components and the existence of random pavement roughness in field tests.

On the other hand, significant difference exists in the phase angles between the experimental and numerical results, due to the fact that the initial conditions for both the bridge and vehicles were not taken into account in the numerical simulation. Figure 12 shows the FFT of the dynamic responses of the cart responses computed. Obviously, both the vehicle and bridge frequencies can be identified from the cart response, especially for the cases with higher vehicle speeds. From Table 2, we observe that the fundamental frequencies of the bridge

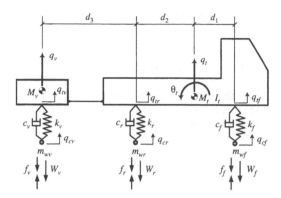

Figure 11. Mathematical model for tractor-trailer.

Table 1. Maximum cart response for various passing speeds (gal).

Methods	Vehicle speed (km/hr)			
	13.0	17.28	35.13	51.84
Numerical results	34.5	44.3	70.8	210.9
Experimental results	28.7	50.2	98.1	215.4

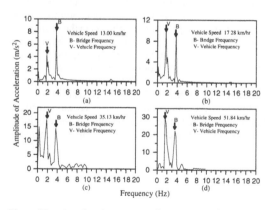

Figure 12. Acceleration spectra of cart response by numerical simulation.

Table 2. Fundamental bridge frequency extracted from cart response for various passing speeds (Hz)*.

Methods	Vehicle speed (km/hr)			
	13.0	17.28	35.13	51.84
Numerical results	3.741	3.748	3.750	3.752
Experimental results	3.728	3.727	3.734	N/A

obtained either numerically or by scanning agree with each other to the first decimal digit.

9 CONCLUDING REMARKS

The feasibility of using a passing vehicle to scan the bridge frequency was tested in this paper. The vehicle system considered is a tractor-trailer. A four-wheel light truck is selected as the tractor, which serves to excite the bridge. And a two-wheel cart is selected as the trailer, which in turn is excited by the bridge and therefore serves as the receiver or scanner of the bridge vibrations. The feasibility of employing the tractor-trailer system to scan the bridge frequencies has been verified through comparison of the measured results with those from other dynamic tests and numerical studies.

Based on the field tests, it is confirmed that for the tractor-trailer moving at speeds less than 40 km/hr, the bridge frequency can be easily identified from the response spectrum of the cart. However, as the cart speed increases, the bridge frequency will be blurred due to involvement of high-frequency components resulting from the cart structure and pavement roughness. The existence of ongoing traffic is considered beneficial since it tends to intensify the cart response.

For application of the present methodology to practical problems, the following are suggested: (1) The dynamic properties of the test cart be identified first. (2) The vehicle speed be kept lower for poorer road conditions in order to get better resolution in the frequency spectrum. (3) At least three runs of tests with different vehicle speeds be conducted before the bridge frequency is ascertained from the cart spectrum.

The methodology presented herein can be used for scanning the first bridge frequency. To scan the second and higher frequencies of the bridge requires further improvement in the level of resolution for the FFT and the level of accuracy of all the instrumentation leading to the FFT, such that the high frequency components from various sources can be filtered out. This may be considered as part of further research. It should be admitted that all the conclusions made above are strictly valid for the conditions, assumptions and instrumentation adopted in this study. Further studies will be carried out to identify the most suitable range of application of the approach proposed herein.

ACKNOWLEDGEMENTS

The authors like to express their cordial thank to Mr. Bun-Chi Kao and Dr. Yu-Chi Soong for their help in making it possible to conduct the experimental studies described in this paper on the test bridge. The assistance from Messrs. Sin-Hwa Su, Kou-Rei Liu, and Li-Ren Chen was greatly appreciated. This study is partly sponsored by the National Science Council of the Republic of China through grant No. NSC 93-2211-E-002-046.

REFERENCES

Berg, G.V. 1988. *Elements of Structural Dynamics*, Prentice-Hall International, Inc., Englewood Cliffs, N.Y.

Kato, M. & Shimada, S. 1986. Vibration of PC bridge during failure process. Journal of Structural Engineering, ASCE, 112(7):1692–1703.

Mazurek, D.F. & DeWolf, J.T. 1990. Experimental study of bridge monitoring technique. Journal of Structural Engineering, ASCE, 116(9):2532–2549.

Ward, H.S. 1984. Traffic generated vibrations and bridge integrity. Journal of Structural Engineering, ASCE, 110 (10):2487–2498.

Yang, Y.B., Lin, C.W. & Yau, J.D. 2004. Extracting the bridge frequencies from the dynamic response of a passing vehicle. Journal of Sound and Vibration; 272(3–5): 471–493.

Yang, Y.B. & Yau, J.D. 1997. Vehicle-bridge interaction element for dynamic analysis. Journal of Structural Engineering, ASCE; 123(11):1512–1518.

Environmental Vibrations – Takemiya (ed.)
© 2005 Taylor & Francis Group, London, ISBN 0 415 39035 4

Resonance analysis of train-bridge dynamic interaction system

W.W. Guo, H. Xia & N. Zhang
School of Civil Engineering & Architecture, Beijing Jiaotong University, Beijing, China

ABSTRACT: In this paper, the resonance mechanism and conditions of train-bridge system are discussed theoretically both in lateral and vertical directions. The railway bridge resonance can be excited by: the periodically loading of moving load series of the wheel-axle weights, the centrifugal forces and lateral winds of moving trains; the periodical actions of train vehicles excited by rail irregularities and hunting movements; the loading rates of moving load series of vehicles. The vehicle resonance can be induced by the periodical actions of regularly arranged bridge spans and their deflections. The resonant train speeds for the bridge system are deduced. The resonant vibration characteristics of some typical railway bridges subjected to moving trains are investigated. The numerical and experimental results show that the proposed method may help to well predict the resonance vibrations of the train-bridge system.

1 INTRODUCTION

With the rapid development of high-speed railways and heavy traffic flows, intensive vibrations of railway bridges induced by running trains have been investigated (Diana 1989, Fr"ba 1999, Xia 2002, Yang 1997). When a row of train vehicles travel through a railway bridge at some critical speeds, the resonant vibrations may occur and the dynamic responses of the bridge or the vehicles will be greatly amplified.

The resonance of train-bridge system is influenced by many factors, such as the periodically loading on the bridge of moving load series of the wheel-axle weights, the centrifugal forces and lateral winds of train vehicles; the harmonic forces on the bridge of the moving trains excited by rail irregularities and hunting movements; and the periodical actions on the moving vehicles of long bridges with identical spans and their deflections, and so on. The loading frequencies will change corresponding to different train speeds. The resonant vibrations occur when the loading frequencies coincide with the natural frequencies of the bridges or the train vehicles.

The strong vibrations induced by the resonance of train-bridge system not only directly influence the working state and serviceability of the bridge, but also result in the reduction of the stability and safety of the moving train vehicles, the deterioration of passenger riding comfort, and sometimes, may destabilize the ballast track on the bridge. Therefore, it is very necessary to analyze this problem and develop some methods that can be used to predict the resonant speeds of the running trains and to assess the dynamic behaviors of railway bridges in resonance conditions (Ju 2003, Xia 2005, Yang 2002).

2 FUNDAMENTAL THEORY

The resonance vibrations of the train-bridge system are very complicated. The discussions on the mechanisms and conditions of the train-bridge system resonance include the bridge resonance induced by moving train loads and the vehicle resonance excited by the deformation of the bridge.

2.1 *Vertical resonant vibration of bridge*

The mechanism of the resonant vibrations induced by moving train loads can be described as follows.

A simply supported beam with span L_b subjected to a series of concentrated constant loads P with uniform interval d_v and moving at speed V from left to the right-hand side is analyzed, to simulate the loading actions of a real train that consists of N cars of length l_v moving on the bridge (see Fig.1).

The equation of motion for the beam acted on by the moving load series can be written as

$$EI\frac{\partial^4 y(x,t)}{\partial x^4} + \overline{m}\frac{\partial^2 y(x,t)}{\partial t^2}$$

$$= \sum_{k=0}^{N-1}\delta\left[x - V(t - \frac{kd_v}{V})\right]P \qquad (1)$$

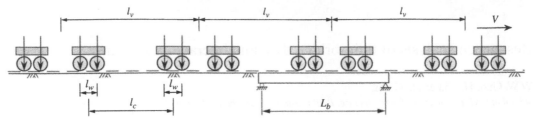

(a) Arrangement of wheel loads of a train.

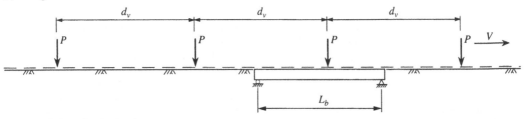

(b) Simplified moving load series.

Figure 1. Loading series of train vehicles on the bridge.

where E = elastic modulus; I = constant moment of inertia of the beam cross section; $\bar{m}$ = constant mass per unit length of the beam; $y(x, t)$ = vertical deflection of the beam at the point x and time t; N = total number of moving loads; and δ = Dirac function.

Equation (1) can be expressed in terms of the generalized coordinates as

$$\ddot{q}_n(t) + \omega^2 q_n(t) = \frac{2}{\bar{m}L_b} P \sum_{k=0}^{N-1} \sin \frac{n\pi V}{L_b} (t - \frac{kd_v}{V}) \quad (2)$$

The general solution of Equation (2) for the first vibration mode of the beam is

$$q(t) = \frac{2PL^3}{EI\pi^4} \frac{1}{1-\beta^2} \sum_{k=0}^{N-1} \left[\sin \bar{\omega} \left(t - \frac{k \cdot d_v}{V} \right) \right.$$
$$\left. - \beta \sin \omega \left(t - \frac{k \cdot d_v}{V} \right) \right] \quad (3)$$

where $\beta = \bar{\omega}/\omega$ is the ratio of loading frequency to the natural frequency of the beam; $1/(1 - \beta^2)$ is the dynamic magnification factor; $\bar{\omega} = \pi V/L_b$ is the exciting circular frequency of the moving loads; and ω is the natural circular frequency of the beam

$$\omega = \frac{\pi^2}{L_b^2} \sqrt{\frac{EI}{m}} \quad (4)$$

The vertical deflection of the beam can thus be expressed as

$$y(x,t) = \frac{2PL_b^3}{EI\pi^4} \frac{1}{1-\beta^2} \sin \frac{\pi x}{L_b} \cdot \left[\sum_{k=0}^{N-1} \sin \bar{\omega} \left(t - \frac{k \cdot d_v}{V} \right) \right.$$
$$\left. - \beta \sum_{k=0}^{N-1} \sin \omega \left(t - \frac{k \cdot d_v}{V} \right) \right] \quad (5)$$

The first term of the right side of Equation (5) represents the forced vibration of the beam due to the moving load while the second term the transient vibration. According to their different mechanisms, the vertical resonance of a simply supported beam subjected to moving load series can be divided into two types.

2.1.1 The 1st vertical resonance

The first vertical resonance vibration of the beam is induced by the transient response mechanism, which can be derived through transforming the triangle progression into the following form

$$\sum_{k=0}^{N-1} \sin \omega \left(t - \frac{k \cdot d_v}{V} \right) = \sin \omega t + \sum_{k=1}^{N-1} \sin \omega \left(t - \frac{k \cdot d_v}{V} \right)$$
$$= \sin \omega t + \frac{\sin \left[(N-1) \frac{\omega d_v}{2V} \right] \cdot \sin \left[\omega t - N \frac{\omega d_v}{2V} \right]}{\sin \frac{\omega d_v}{2V}} \quad (6)$$

When $\omega d_v/2V = \pm i\pi$, the second term of the equation becomes an indeterminate form 0/0, but by L'Hospital's rule, the limit solution is found to be

$$\lim_{\frac{\omega d_v}{2V}=\pm i\pi} \frac{\sin\left[(N-1)\cdot\frac{\omega d_v}{2V}\right]\cdot\sin\left[\omega t - N\cdot\frac{\omega d_v}{2V}\right]}{\sin\frac{\omega d_v}{2V}}$$

$$= (N-1)\sin\omega\left[t - N\cdot\frac{d_v}{2V}\right] \qquad (7)$$

Obviously, the extreme condition with physical significance for Equation (7) is

$$\frac{\omega d_v}{2V} = i\pi \qquad (i=1, 2, \cdots) \qquad (8)$$

It can be seen that each force in the moving load series may induce the transient response of the structure, and the successive forces form a series of periodical excitations. The responses of the structure will be successively amplified with the increase of the number of wheels traveling through the bridge.

The similar results can be obtained for higher modes of the bridge. Considering all of these modes and let $\omega = 2\pi f_{bn}$, the 1st resonant condition of the bridge under vertical loading series can be written as

$$V_{br} = \frac{3.6\cdot f_{bn}\cdot d_v}{i} \quad (n=1,2,\cdots, i=1,2,\cdots) \qquad (9)$$

where V_{br} = resonant train speed (km/h); f_{bn} = nth vertical natural frequency of the bridge (Hz); and d_v = intervals of the moving loads (m).

Equation (9) indicates that when a train moves on the bridge at speed V, the regularly arranged vehicle wheels may produce periodical dynamic actions on the bridge. The bridge resonance occurs if the loading period is close to the nth natural vibration period of the bridge. A series of resonances related to different natural frequencies may occur corresponding to different train speeds. This is called the 1st vertical resonant condition, which is determined by the time of the load traveling through the distance d_v.

2.1.2 The 2nd vertical resonance

From Equation (5) one can see that when the frequency ratio $\beta = 1$, i.e. $\omega_n = \bar{\omega}_n$, the dynamic magnification factor $D = 1/(1 - \beta^2)$ will become infinitive. At this time the resonant vibrations of the bridge is excited. For the simply supported beam under moving loads, $\bar{\omega}_n = n\pi V/L_b$, and the nth natural frequency of the beam $\omega_n = 2\pi f_{bn}$, the resonant train speed V_{br} can be described as

$$V_{br} = \frac{7.2\cdot f_{bn}\cdot L_b}{n} \quad (n=1, 2,\cdots) \qquad (10)$$

where L_b = length of the bridge span (m).

Equation (10) indicates that the bridge resonance occurs when the time of the train's traveling through the bridge equals to half or n times of the natural vibration period of the bridge. This is called the 2nd vertical resonant condition, which is determined by the loading rate of the moving loads.

2.2 Lateral resonant vibration of bridge

The lateral vibration of a bridge is analyzed as a system instead of a simply-supported beam, because of the coupled movement of the piers and beams.

In the analysis, the basic global lateral deformation of the bridge deck where the train runs on can be reasonably assumed as a sinusoidal wave with a half wavelength L_b and the frequency f_b. By the similar analysis method, but changing the vertical loads into the lateral forces transmitted through the wheels onto the bridge deck, the lateral resonant vibrations of the bridge can be divided into three types.

2.2.1 The 1st lateral resonance

The 1st lateral resonance of the bridge is induced by the periodically loading of the lateral moving load series formed by the centrifugal forces or the lateral wind pressures acting on the vehicle bodies. The resonant train speed can be deduced in the same way as in the first vertical resonance and written as

$$V_{br} = \frac{3.6\cdot f_{bn}\cdot d_v}{i} \qquad (n=1, 2, \cdots, i=1, 2,\cdots) \qquad (11)$$

which is basically the same as Equation (9) for vertical resonance, excepted that f_{bn} becomes the nth lateral natural frequency of the bridge.

Equation (11) indicates that the bridge resonance occurs if the loading period is close to the nth lateral natural vibration period of the bridge. Since the lateral frequency of the bridge system is usually lower than the vertical one, the resonant train speed for lateral vibration is also lower, which is of special significance for bridges with high piers.

2.2.2 The 2nd lateral resonance

The second lateral resonance of the bridge is induced by the loading rates of the lateral moving load series, such as the centrifugal forces or the lateral wind pressures, and so on, acting on the vehicle bodies on the bridge. The resonant train speed can be deduced in the same way as in the second vertical resonance and written as

$$V_{br} = \frac{7.2\cdot f_{bn}\cdot L_b}{n} \qquad (n=1, 2,\cdots) \qquad (12)$$

183

which is basically the same as Equation (10) for vertical resonance analysis, excepted that f_{bn} becomes the nth lateral natural frequency of the bridge.

Since the lateral frequency of the bridge system is usually much lower than the vertical frequency, the critical train speed for lateral resonance is also lower. This is called the 2nd lateral resonant condition, which is determined by the loading rate of the lateral moving loads. The 2nd lateral resonance analysis has special significance for bridges with high piers under moving load series induced by centrifugal forces or the lateral wind pressures.

2.2.3 The 3rd lateral resonance

The third lateral resonance is induced by the periodical actions on the bridge of the lateral moving load series excited by the track irregularities and wheel hunting movements. The resonant train speed can be determined by

$$V_{br} = \frac{3.6 \cdot f_{bn} \cdot L_s}{i} \quad (n=1, 2, \cdots, i=1, 2, \cdots) \quad (13)$$

which is basically the same as Equation (11) for the 1st lateral resonance, excepted that L_v is replaced by L_s which represents the dominate wavelength of the track irregularities or wheel hunting movements.

2.3 Application scopes of resonance conditions

Based on the analysis above, the resonant vibrations of bridges induced by moving trains can be further classified into three mechanisms. The first is related to the intervals of the moving load series, which form the periodically loading on the bridge. The second is induced by the loading rate, i.e. the relative moving speed of the train vehicles to the bridge. The third is excited by the track irregularities and wheel hunting movements.

In fact, the axle loads of a train are not always in uniform intervals as assumed in the resonance analysis for the fist mechanism. There exist several axle intervals in a real train: the total length of a car l_v, the center-to-center distance between two bogies l_c, the fixed distance between two wheel-sets l_w, and the different compositions of these distances. According to the relative lengths between the span or the length of a bridge and the above loading intervals, the application scopes of the resonant conditions can be further discussed as follows (see Fig. 2).

The figure shows that for a row of train vehicles, the arrangement of the axle loads is not in equidistance, and neither equal are the values of all the axle loads on the bridge. Therefore, a series of resonant vibrations may be excited when the train moving at various speeds on the bridge, and a series of corresponding resonant train speeds could be found. Therefore, the

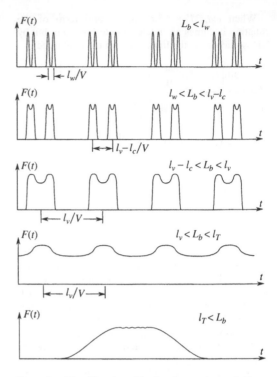

Figure 2. Time histories of load series moving on bridge.

precise resonance analysis usually depends on the simulation calculations of the train-bridge dynamic interaction system according to the real composition of the train and the wheel arrangement and loads of the vehicles.

2.4 Analysis of train resonance

As a row of train vehicles traveling over a bridge at speed V, the periodical actions on the vehicles can be excited by the deflections of the bridge that consists of a long series of identical spans. The loading frequency can be estimated as

$$f = V / L_b \quad (14)$$

The vehicle resonance occurs if this loading frequency coincides with the natural frequency of the train vehicles (ref Fig. 3), when the dynamic responses of the vehicle will be greatly amplified. The critical train speed can be written as

$$V_{vr} = 3.6 \cdot f_v \cdot L_b \quad (15)$$

where V_{vr} = critical train speed (km/h); f_v = natural vertical frequency of the vehicle (Hz); L_b = span length of the bridge (m).

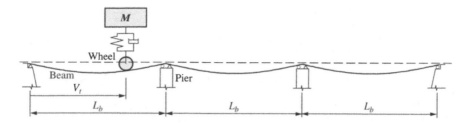

Figure 3.   Vehicle vibration induced by bridge deflection curves.

The excitation of bridge deflections on the vehicles is equivalent with the mass-spring system on the ground in harmonic vibrations. The transmissibility between the amplitudes of the mass and the deflection of the beam can be estimated as (Clough 2003)

$$TR = \sqrt{\frac{1+(2\xi\beta)^2}{(1-\beta^2)^2+(2\xi\beta)^2}} \qquad (16)$$

For a half vehicle model with the sprung mass M = 24 t, the equivalent spring stiffness k = 800 kN/m and the damping ratio ξ = 0.2, the natural frequency is calculated as 0.92 Hz. At the critical train speed, i.e. β = 1, the transmissibility can be calculated as TR = 2.69. It means that when the deflection of the beam is 2 mm, the amplitude of the vehicle will be 5.38 mm. Moreover, the resonance of vehicles will in turn enlarge the dynamic impact on the bridge.

The 1st vertical natural frequencies of the train vehicles are usually between 0.8~1.5 Hz. For the railway bridges with 20~40 m spans, the corresponding critical train speeds could be estimated as V_{vr} = 57~216 km/h. Therefore, it is better not to arrange long series of identical spans in the design of railway bridges, to prevent the vehicle resonance due to the bridge deflections. On the other hand, the stiffness of the bridge spans should also be controlled to minimize the influence of the bridge deflections on the running stabilities of the train vehicles.

3   CASE STUDY OF LATERAL RESONANCE

3.1   Lateral resonance induced by loading series

Several lateral resonant vibrations of the train-bridge system are analyzed. The moving load series are the lateral wheel loads induced by wind pressures acting on vehicle bodies. The resonant train speeds are evaluated by Equations (11) and (12). To verify the evaluated resonant conditions, the dynamic response curves of the bridges versus train speed are provided by the whole history simulations of train-bridge system (Guo 2004a, b). By the comparison of the resonant train speeds from the resonant conditions and the critical

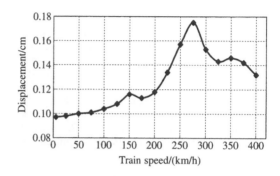

Figure 4.   Lateral displacements of steel beam *vs* train speed.

train speed from the simulated results, some useful conclusions have been obtained.

The simply supported bridges with medium spans and high piers are used in the analysis. The train concerned is composed of one locomotive followed by 18 passenger cars. The total length of each car is 26.57 m. The average wind velocity is 25 m/s.

3.1.1   Simple supported steel-plate beam

The lateral natural frequency of the 32 m steel-plate beam is 3.2 Hz. The resonant train speeds estimated by Equation (11) are

$$V_{br1} = \frac{3.6 \times f_{bn} \times d_V}{i} = \frac{3.6 \times 3.2 \times 26.57}{1} \approx 306 \, km/h$$

$$V_{br2} = \frac{3.6 \times f_{bn} \times d_V}{i} = \frac{3.6 \times 3.2 \times 26.57}{2} \approx 153 \, km/h$$

$$V_{br3} = \frac{3.6 \times f_{bn} \times d_V}{i} = \frac{3.6 \times 3.2 \times 26.57}{3} \approx 102 \, km/h$$

According to the estimated results above, the simulation analysis of the train-bridge system is carried out, with the calculation train speeds in the range of 5~400 km/h. Figure 4 shows the lateral displacements of the beam under various train speeds.

It can be seen in the figure that the lateral displacements of the bridge reach peaks when the train speeds are around 100 km/h, 150 km/h and 270 km/h, respectively. These critical train speeds are close to but slightly lower than the resonance train speeds estimated by Equation (11). Considering that the vibration frequencies of the bridge with loads are lower than their natural unloaded frequencies, the two results are in good coincidence.

Theoretically, the resonance train speeds can also be estimated by Equation (12), but the results are of little sense because the estimated train speeds are much higher than those in reality.

3.1.2 *Bridge with high piers*
Since bridges often contains piers to support the bridge girders, the dynamic effects cannot be ignored in studying the bridge resonance. A bridge system consists of two 32 m simply supported beams and a 56 m high pier is analyzed as an example. The modal analysis shows that the first three modes of the bridge are dominated by the lateral vibrations of the pier. With respect to the lateral frequencies 0.95 Hz, 2.52 Hz and 5.02 Hz, the resonant train speeds estimated by Equation (11) are

V_{br1}=91 km/h, V_{br2}=46 km/h, V_{br3}=31 km/h
V_{br1}=240 km/h, V_{br2}=120 km/h, V_{br3}=81 km/h
V_{br1}=481 km/h, V_{br2}=240 km/h, V_{br3}=161 km/h

The resonance train speeds also can be evaluated by Equation (13) with $L_b = 64$ m as

V_{br1}=7.2×0.95×64/1≈438 km/h
V_{br2}=7.2×2.52×64/2≈581 km/h
V_{br3}=7.2×5.02×64/3≈771 km/h

According to the results, the simulation analysis of the train-bridge system is carried out, with the calculation train speeds in the range of 10~900 km/h. Figure 5 shows the lateral displacements of the pier top under various train speeds. The numerical results show that the estimated resonant train speeds coincide with the critical train speeds obtained by the simulation analysis by the train-bridge system.

3.2 *Lateral resonance induced by vehicle movement*

Although the track irregularities and wheel hunting movements are of random properties, Equation (13) can be used to estimate the third lateral resonance of a bridge induced by their principal wavelengths.

Figure 6 shows an example: the distributions of the lateral displacements of the pier tops versus train speed, which were measured in the field experiments at two real bridges on the Chengdu-Kunming Railway (Xia 1984). The peak values appear at certain train

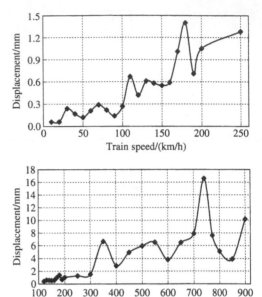

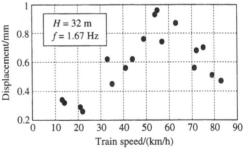

Figure 5. Lateral displacements of bridge pier *vs* train speed.

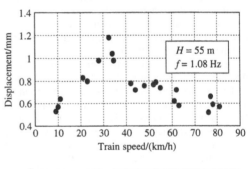

Figure 6. Lateral displacements of piers *vs* train speed.

speeds, which are in accordance with the estimated resonant train speeds of 33 km/h and 51.1 km/h, respectively, calculated by Equation (13) using the hunting wavelength $L_s = 8.5$ m of the wheels with worn tyres and the given pier heights $H = 55\,\text{m}$ and $32\,\text{m}$ and the corresponding frequencies $f = 1.08\,\text{Hz}$ and $1.67\,\text{Hz}$, respectively.

4 CONCLUSIONS

The resonant vibrations of train-bridge system can be divided into following types according to their generation mechanisms:

(1) Bridge resonance excited by periodically loading of moving load series of moving vehicles, due to the wheel intervals of the vehicles.
(2) Bridge resonance excited by the loading rate of moving load series of vehicles.
(3) Bridge resonance excited by the periodical actions of the running trains induced by rail irregularities, wheel flats and hunting movements.
(4) Vehicle resonance excited by the periodical actions of regularly arranged bridge spans and their deflections.

For bridge resonance analysis, the load series both consist of the vertical forces of the axle weights of the train, and the lateral forces transmitted from the wheels due to the centrifugal forces and wind pressures acting on the vehicles should be noticed.

ACKNOWLEDGMENTS

This study is sponsored by the National Natural Scientific Foundation of China (50478059) and the Key Research Foundation of BJTU (2004SZ005).

REFERENCES

Diana, G. & Cheli, F. 1989. Dynamic interaction of railway systems with large bridges. *Vehicle System Dynamics* 18: 71–106.

Clough, R.W. & Penzien, J. 2003. *Dynamics of Structures*. New York: McGraw Hill Inc.

Frýba, L. 1999. *Vibration of Solids and Structures under Moving Loads*. London: Thomas Telford.

Frýba, L. 2001. A rough assessment of railway bridges for high speed trains. *Engineering Structures* 23: 548–556.

Guo, W.W. & Xia, H. 2004. Dynamic responses of long suspension bridges and running safety of trains under wind action. *Proc. ISSST'2004*, Shang'hai: 1937–1944.

Guo, W.W. 2004. *Dynamic Responses of Long Span Bridges and Running Safety of Trains under Wind Action*. Beijing: Beijing Jiaotong University.

Ju, S.H. & Lin, H.T. 2003. Resonance characteristics of high speed trains passing simple supported bridges. *Sound & Vibration* 267: 1127–1141.

Xia, H. & Chen, Y.J. 1984. Dynamic response of high piers under train-loads and its influences on running vehicle stability. *Proc. ISGBS*, Lanzhou, China: 589–600.

Xia, H. 2002. *Dynamic Interaction of Vehicles and Structures*. Beijing: Science Press.

Xia, H. & Zhang, N. 2005. Dynamic analysis of railway bridge under high-speed trains. *Computers & Structures*, in press.

Yang, Y.B. 2002. *Theory of Vehicle-Bridge Interaction for High Speed Railway*. Taibei: DNE Publisher.

Yang, Y.B. & Yau, J.D. 1997. Vehicle-bridge interaction element for dynamic analysis, *Structural Engineering ASCE* 123(11): 1512–1518.

Environmental Vibrations – Takemiya (ed.)
© 2005 Taylor & Francis Group, London, ISBN 0 415 39035 4

Dynamic analysis of the light-rail-station viaduct under moving vehicle loads

D.S. Shan, Q. Li & X.W. Yang
Department of Bridge and Structural Engineering, Southwest Jiaotong University, Chengdu, China

ABSTRACT: In this paper, a dynamic analysis of light-rail-station viaduct under moving vehicles is presented. Moving vehicles are idealized as moving load-sequence. The overhead and multi-stories' light-rail-station viaduct is modeled by the three-dimensional finite element model. Then when applying the time-integration to the dynamic system, the equations of motion are efficiently solved based on these models. The proposed structure is then applied to a real overhead and multi-stories light-rail-station viaduct with a group of moving vehicles. The braking forces, attractive forces and lateral sway forces are also taken into accounts which correspond with the accelerated and decelerated processes of moving vehicles when they enter and depart the station. The dynamic responses of the light-rail-station viaduct and the riding comfort indices are computed. The results show that the formulation presented in this paper can efficiently and effectively predict the dynamic response of light-rail-stations viaducts under a normal operation condition.

1 INTRODUCTION

In Chongqing, there are three elevated express railways under programming. They are Line No.1, Line No.2 and Line No.3, which form a crossed express railway traffic network from the eastern to the western and from the southern to the northern. Line No.2 which is under construction now starts from Jiaochangkou going along Jialing River all the way to Xinshan Village with the overall length 17.6 km. And the monorail light rail is the first-of-all in China.

According to the geographical characteristics of the city, the elevated saddle-type monorail of transportation is used in Chongqing. Most of them are realized by using elevated rail-viaducts and station-viaducts. And because the hall, platform and the staircases between the hall and the platform of the station as well as the monorail girder where the monorail vehicles go through are all located on the same pier, the dynamic capability of the elevated light monorail station viaduct and the human comfort when passing the viaduct are both vital factors that have to be taken into consideration during the designing process.

The Yuan's Post elevated light monorail station viaduct is made up of platforms, halls and monorail girders. There are 7 spans of monorail and the composition of spans is 10 m + 5 × 20 m + 10 m, so the overall length of monorail girder is 120 m. The overall length of the platform is also 120 m. And the composition of

the hall is 6 × 10 m, and the overall length of the hall is 60 m. Using 3-D beam elements to present the elevated light monorail station viaduct, 1225 elements and 853 nodes can be obtained. When discrete structuring, the master-slave relationship between the monorail girder and the pier beam is also considered. Figure 1 is the discrete structure of the elevated light monorail station viaduct.

2 MODEL OF THE MOVING SADDLE-TYPE VEHICLE LOADS AND RIDE COMFORT

When a vehicle goes through the elevated light monorail station viaduct, the moving vehicles, the viaduct and the monorail are all involved into the vibration system. However, because the dynamic frequency of the track is much higher than that of the vehicle and the viaduct, the track vibration is usually overlooked (Shen 1998). Due to the couple between the moving vehicles and viaducts is weak, so the moving vehicle can be simplified to moving load-sequence model which at the same time overlooks the mass of the vehicle. That is, the vehicle is simplified into some moving concentrated forces and the vibration caused by these concentrated forces when going through the viaduct at certain speed becomes the concern. The model of the moving load-sequence takes all the possible factors that may affect the viaduct vibration into accounts including the axis weight, the distance and arrangement of the axes, the length of the vehicle etc.

2.1 Load-sequence model of moving saddle-type vehicle

When a saddle-type vehicle gets across the elevated light monorail station viaduct, even the static weight of the vehicle may arouse the variation of internal forces and displacements of the elevated light monorail station viaduct would be happened as time elapses. This kind of variation caused by static load is defined as static impulse. And when the variational period of the load-sequence coincides with the vibrating period of the elevated light monorail station viaduct system to certain degree, the maximum value of the dynamic system can be obtained (Fryba 1972 and 1996, Shan 1999a and b).

Figure 2 shows an elevated light monorail train runs on track in Chongqing Municipality. Using rubber tires, the saddle-type monorail vehicle travels on a remarkably smooth interface and the two moving wheels of the saddle-type vehicle are on the monorail girder (see Figure 3). Thus the vibration of elevated light

monorail station viaduct caused by moving saddle-type vehicles is different from the vibrations of the viaduct caused by other kinds moving vehicles. The major factors that cause vibration of elevated light monorail station viaduct is the moving load-sequence instead of the rail unevenness and the impact of the vehicles.

Referenced to the saddle-type monorail vehicles used above light-rails in Chongqing, a simplified model

Figure 2.   The saddle-type vehicle for Chongqing's elevated light monorail system.

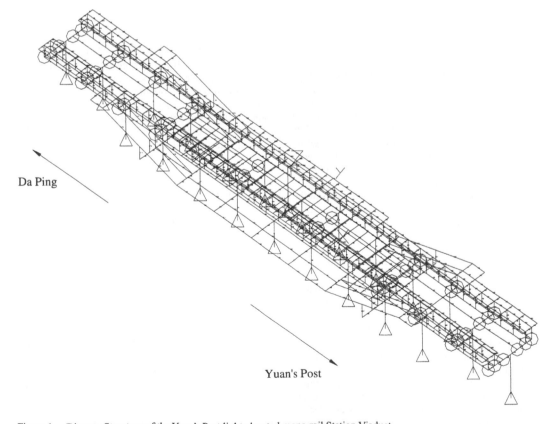

Da Ping

Yuan's Post

Figure 1.   Discrete Structure of the Yuan's Post light elevated mono rail Station Viaduct.

of moving load-sequence can be obtained as shown in Figure 4, in the figure the number 1, 2, 3 and 4 indicate the wheel-set of each saddle-type vehicle, number 1 denotes the front wheel-set, number 4 denotes the rear wheel-set, number 3 and 4 denotes the middle two wheel-sets. As shown in the figure 4, the distance from the first wheel-set to second wheel-set is 1.5 m, the distance from the third wheel-set to fourth wheel-set is also 1.5 m, and the distance from second wheel-set to third wheel-set is 8.1 m, so the length of the each vehicle is 12.85 m; the space between two adjacent vehicle is 1.75 m. And the weight of each wheel-set has an equivalent value, 110 kN.

A train is composed of 8 cars. What present in Figure 4 are just one of the cars and the axis connecting the fore car and the back car. The distance between the center of the car body and the top of track is 1.3 m. In order to simulate the real running situation of the saddle-type vehicles enter or departure from its elevated light monorail station viaduct, the braking forces and startup forces must be reckoned on. The swing forces and impact forces which caused by the moving saddle-type vehicles were also taken into consider. All of these forces were considered as those parameters multiple the weight of the wheel-set. And the parameters used in computing are list in Table 1.

2.2 Numerical method

Finite element method can be used to analyze the model of moving load-sequence because it is a general method which can help to make out the same calculating formula for different structures and is skilled at dealing with bridge structure with complicated configurations and complex sections. The elevated light monorail station viaduct dynamic equation is like this (Clough 2003, Cheng 2003):

$$[M]\{\ddot{y}\} + [C]\{\dot{y}\} + [K]\{y\} = \{f_v(t)\} \tag{1}$$

Where $[M]$, $[C]$ and $[K]$ are the mass, damping and stiffness matrices of the elevated light monorail station viaduct, the structural damping of the viaduct is assumed to be Raleigh damping. $\{\ddot{y}\}$, $\{\dot{y}\}$ and $\{y\}$ are the acceleration, velocity and displacement vectors of the viaduct, respectively; and $\{f_v(t)\}$ denotes the vector of equivalent nodal forces, consisting of forces from the wheels of the saddle-type vehicles to the monorail girder through the track. At each calculated step, the newly formed equivalent nodes forces vector should be changed along with the moving load-sequence.

The numerical method used here to deal with the vibration of the elevated light monorail station viaduct under the moving saddle-type vehicles is crucial. It should be low computational cost, stable and able to deal with the nonlinear behavior of the dynamic system. In this paper, Wilson-θ step-by-step integration method was adopted because it was unconditional stable integration procedure.

2.3 Ride comfort

With the increase of the living standard, people's demand for a sound riding atmosphere has been increasing, thus the issue of dynamic comfort appears to be more important. At present, there are various standards for evaluating the ride comfort (Shen 1998 and Zeng 1999). Unfortunately, a uniform standard has not been found yet home and abroad. The ride comfort investigated in this study refers to the comfort of human body vibration from head to foot transmitted

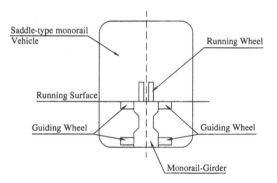

Figure 3. Saddle-type Vehicles and Cross-Section of monorail-Girder.

Table 1. Calculated parameters for these forces.

Category	Parameter
Brake force	0.15
Startup force	0.15
Swing force	0.25
Impact force	0.286

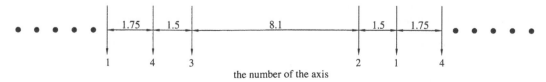

the number of the axis

Figure 4. The calculated model of load-sequence (unit: m).

by the light-rail-station viaduct vibration when the vehicle going through the viaduct. After a synthetical evaluation, the Comfort Standard for Japanese state railway is adopted (Zeng 1999). Table 2 shows the ride index when the ride quality equal to 1 and Table 3 show the classification of the grade for ride quality.

3 ANALYSES OF THE COMPUTED RESULTS

Based on the models presented above, the dynamic analysis was carried out under several calculated cases: 1) different speeds (40 km/h, 50 km/h, 60 km/h, 70 km/h, and 80 km/h), 2) the going upward enters and departs the station; 3) the going downward enters and departs the station; 4) the going upward and downward enter and depart the station simultaneously. Therefore, with 5 cases of speed and 3 cases of entering and departing the station, 15 (3 × 5) calculated cases have been taken into consideration. In addition, because the entering and departing is virtually a process of decelerating, stopping and accelerating, the speeds involved here all refer to even instant speeds (Chatterjee 1994).

3.1 Natural frequency of the light-rail-station viaduct

Based on the calculated model shown in Figure 1, the natural frequencies of the Yuan's Post elevated light

monorail station viaduct are worked out. Natural frequencies of the first five modes of the light-rail-station viaduct are listed in Table 4, and the corresponding dynamic modes can be seen in Figure 5.

According to Table 4 and Figure 5, it is obvious that the dynamic mode shape in the first mode are the transverse vibration of the right-side pier and the monorail girder; the dynamic mode shape in the second mode are the transverse vibration of the left-side pier and the monorail girder, most of which are the monorail girder transverse vibration; the natural frequencies from mode No.3 to mode No.5 are close to each other and most of the vibrations are transverse vibrations of the monorail girder. What's more, Figure 5 shows that in the first two modes, the transverse vibration at the platform layer is caused by the bending vibration of the piers on both sides, but then the vibration of the platform itself is not evident.

Actually, enough rigidity of the monorail girder has been ensured when designing it. Therefore, except for the transverse rigidity of the piers on both sides, which may be less rigid, rigidities of the other parts are more than that of the track girder, which is to say that the rigidities of the light-rail-station viaduct's platform, hall and staircases are enough.

3.2 Dynamic analyses of the saddle-type vehicle and the elevated light monorail station viaduct

The maximal accelerations of the transverse, longitudinal and vertical acceleration at the platform, the hall and the staircases of the elevated light monorail station viaduct can be seen in Table 5.

According to Table 5, no matter from which directions: longitudinal, transverse or vertical, the acceleration values at the platform layer are more than the values at the other two positions, among which the maximal vertical acceleration reaches the value: 0.057 g.

In addition, referenced to the Comfortable Standard for Japanese state railway (Zeng 1999), the comfortable indices of the platform, the hall and the staircases of the Yuan's Post station viaduct are all below 1.5 as has shown in Table 2, and belonging to the "good" grade.

Table 2. Ride index of the ride quality is 1.

Dynamic ways	Dynamic frequency (Hz)	Ride index
Left and right (transverse) vibration	0~0.1	0.08 g
	1~4	0.08 g
	4~12	0.02 g
	>12	0.0016 g
Backward and forth ward (longitudinal) vibration	4~5	0.025 g
	>15	0.0016 g
Upward and downward (vertical) vibration	1~6	0.02 g
	5~20	0.033 g
	>20	0.0016 g

Table 3. Classification of the ride quality.

Ride quality	Ride comfort	Classification
<1	very good	①
1~1.5	good	②
1.5~2.0	gentle	③
2.0~3.0	no good	④
>3.0	extremely bad	⑤

Table 4. Natural-vibration frequencies of the first 5 modes.

Mode	Frequency(Hz)
1	1.975
2	2.246
3	2.325
4	2.356
5	2.375

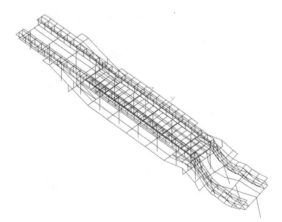

Figure 5a.   Mode No. 1.

In Figure 6, time-histories with the maximal longitudinal, transverse or vertical accelerations at the platform layer and their corresponding displacement's time-histories are made clear.

The time-histories of the longitudinal, vertical and transverse displacements responses of the elevated light monorail station viaduct at the corresponding maximal accelerations positions are shown in Figure 6b), Figure 6d) and Figure 6f) respectively. It is shown from the longitudinal displacement time-history that when the saddle-type vehicles run on the left-side of the viaduct, and the longitudinal displacements increase gradually, and when the saddle-type vehicles run on the right-side of the viaduct, the longitudinal displacements decrease gradually. The viaduct then has a free vibration when the train leaves from it. The maximal longitudinal displacement is 4.86 mm. As shown in Figure 6d), the transverse displacements are negative

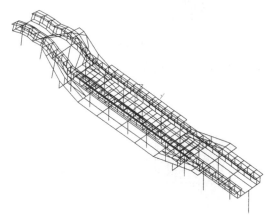

Figure 5b.   Mode No. 2.

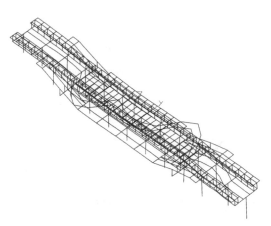

Figure 5d.   Mode No. 4.

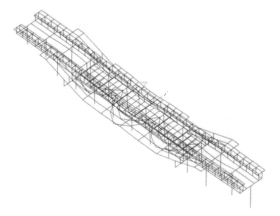

Figure 5c.   Mode No. 3.

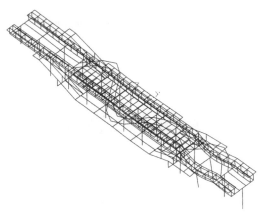

Figure 5e.   Mode No. 5.

Figure 5.   Natural dynamic modes for the first 5 modes.

when the vehicles run on the left-side and are positive when the vehicles run on the right-side. The maximal and the minimal transverse displacement are 1.76 mm and −1.26 mm, respectively.

It is shown from the vertical displacement time-history that when the saddle-type vehicles run on the left-side of the viaduct, the vertical displacements increase gradually, and when the vehicles run on the right-side of the viaduct, the vertical displacements

Table 5. Maximal accelerations at different positions (Unit: g).

Position	Longitudinal	Transverse	Vertical
Platform	0.033	0.032	0.057
Hall	0.021	0.005	0.012
Staircases	0.023	0.004	0.010

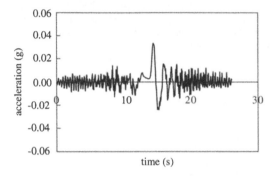

Figure 6a.　Longitudinal acceleration time-history.

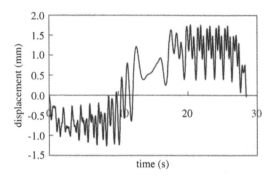

Figure 6d.　Transverse dynamic displacement time-history.

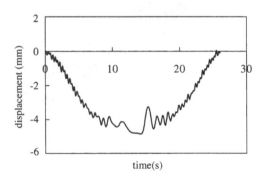

Figure 6b.　Longitudinal dynamic displacement time-history.

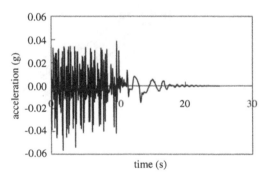

Figure 6e.　Vertical acceleration time-history.

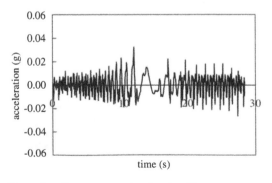

Figure 6c.　Transverse acceleration time-history.

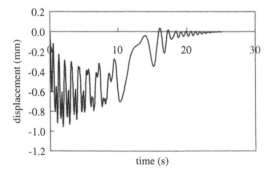

Figure 6f.　Vertical dynamic displacement time-history.

Figure 6.　Acceleration's time-histories with the maximum and the corresponding displacement's time-histories.

decrease gradually. The viaduct then has a free vibration when the train leaves from it. The maximal vertical displacement is little than 1 mm, and this maximum meet the requirement of the design.

According to Figure 6, the longitudinal displacement is the maximal displacement which is caused by the braking and starting of the vehicle when entering or departing the elevated light monorail station viaduct. But the maximal value of displacement is just 5 mm or so which indicates that the rigidities of every direction are adequate enough under the moving saddle-type vehicles. And the Figure also shows that the displacement's time-history will quickly decline after the vehicle departed the station with no amplified displacement.

4 CONCLUSION

Based on the simplified analysis method, the moving load-sequence method, for coupled vibration of moving saddle-type vehicle and elevated light monorail station viaduct, the vibration of the Yuan's Post elevated light monorail station viaduct has analyzed when saddle-type vehicles go through the station. And then the human comfortableness when riding on the elevated light monorail station viaduct has been evaluated. Finally, the self-vibration properties of it are obtained. The analyzed results show that the design of the Yuan's Post station viaduct is reasonable; the elevated light monorail station viaduct is of adequate rigidity and the whole dynamic capacity of which is comparatively good; the vibrations at every place of the viaduct meet the standard for human physical comfortableness.

REFERENCES

Ruili Shen, "Coupled Dynamic Study of bridges and vehicles for high-speed railway (Chinese)", A Dissertation submitted to Southwest Jiaotong University in Fulfillment of the Requirement for the Degree of Doctor of Philosophy, China,1998, 58~76

Haojiang Ding, Yiquan Xie, "Finite Element Method for Elastic and Plastic Mechanics (Chinese)", Second Edition, Publish of Mechanical Industry, China, 1989, 155~164

Qingyuan Zeng, Xiangrong Guo, "Dynamic Analytic Theory and Appliance of Time-varying System for Bridges and Vehicles (Chinese)", Publish of Chinese Railway, China, 1999, 119~127

Ray Clough, Joseph Penzien, "Dynamics of Structure", Second Edition (revised), Computer and Structures, Inc., U.S.A., 2003, 116~132

Deshan Shan, Qi Li and Yonglin Jiang, "Dynamic Response of Curved-girder under moving load", Journal of Southwest Jiaotong University, China, 1999a, vol.17, 163~172

Deshan Shan, Qiao Li, "Dynamic response of simply supported curved-girder under moving load series (Chinese)", Journal of the china railway society, China, 1999b, vol.23, No.3, 99~103

Ladislav Fryba, "Vibration of solids and structure under moving loads", Noordhoff International Publishing, Groningen, 1972, 309~324

Ladislav Fryba, "Dynamics of Railway Bridges", Institute of Theoretical and Applied Mechanics, Academy of Science of the Czech Republic, 1996, 32~45

Chen Yunmin, Takemiya Hirokazu, "Environmental Vibration: Prediction, Monitoring and Evaluation", China Communication Press, 2003, 242–266

P.K.Chatterjee, T.K.Datta and C.S.Surana, Vibration of Continuous Bridges under Moving Vehicles, Journal of Sound and Vibration (1994) 169(5), pp: 619~632

Environmental Vibrations – Takemiya (ed.)
© *2005 Taylor & Francis Group, London, ISBN 0 415 39035 4*

Study on the dynamics vibration character of the concrete slab track in track traffic system of linear induction motor

L. Liao, L. Gao & J.F. Hao

School of Civil Engineering and Architecture, Beijing Jiaotong University, Beijing, China

ABSTRACT: Based on the LIM(Linear Induction Motor) track traffic system of Guangzhou, vertical dynamic coupling model of LIM vehicle-concrete slab track and their dynamic differential equations are set up, including the model of vehicle, the model of vehicle-track interaction, the model of the rail and the concrete slab track. The control technology of gap between LIM and reaction-plate and the vibration characteristics of the concrete slab track are analyzed by the dynamics simulation testing. Reasonable suggestions are put forward for the design, construction and maintenance of the slab track and the track traffic system of LIM.

1 INTRODUCTION

LIM has been used in urban transit system, and the vehicle is driven by linear induction motor with short stator. It can be assumed that the rotating electrical machine is unfolded to a line. The stator (primary coil) is installed on the vehicle and the rotor (Reaction Plate) is installed on the track. The traveling field (traveling magnetic field) was generated when alternating current in stator (primary coil), and the same time, the induced voltage and current can be produced in rotor (RP). The propulsion and braking of vehicle can be realized by the magnetic interaction between the LIM and RP[1].

LIM is the development tendency of the URT (urban rail transit) in 21st. As a new sort of track transit system, LIM has many advantages such as slight vibration, low noise, the capacities of steep gradients and the adaptabilities to Sharp curves, and low fabrication, low energy consumption, good security, etc.

LIM track transit system is both the bearing platform of vehicle and the supporting point of non-adhesion driving. The rail-track relation is different from the wheel-rail interaction of the conventional adhesion driving and magnetic-rail interaction of maglev system, thus track transit system of LIM has characteristics in vehicle-track coupling dynamics.

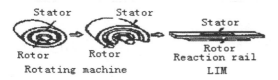

Figure 1. Theory of LIM.

At present, LIM system has been already used in 4 countries, including Canada, Japan, America and Thailand[2]. In China, this system will be utilized in Guangzhou, Beijing, Wuhan, Taibei, etc. Therefore it will be of great importance in theory and practice to carry out research in LIM systematically.

2 SYSTEM MODEL

2.1 Vehicle model of LIM

The vehicle is a vibrating system with many DOF (degrees of freedom). Vehicle, bogie and wheels are considered rigid, neglecting the elasticity and torsion deformation. Vehicle subsystem contains 10DOF (nodding and vertical floating of vehicle, bogie, and wheels)[3]. The connection between vehicle, bogie and wheels were simplified as spring-damping suspension.

2.2 Model of track structure

The whole track structure consists of three layers, which are rail, slab and rigid foundation. Rail can be considered as point supporting infinite beam, the vertical displacement and turn angle are taken into account only. The slab track can be simplified as infinite beam on continuous elastic foundation, because emulsion asphalt is laid between slab track and foundation, the slab track contacts the concrete continuously, and the thickness of slab track is much less than its width and length, and vibration of concrete foundation were neglected[4].

2.3 Coupling model of wheel-rail

The coupling interaction between vehicle and track system can be realized by contact rigidity of wheel-rail. In order to calculating, nonlinear spring was simplified as linear spring in this model[5,6]. It is expressed as following equation:

$$K_w = 1.5\left(\frac{1}{G}\right)P_0^{1/3} \tag{1}$$

Where G is the wheel/rail contact constant, which is determined by the wheel type and the wheel radius, P_0 is static loading of wheel.

2.4 Model of LIM-reaction

Each vehicle has two linear induction motors, and LIM is fixed on the bogie. The RP is fixed on the slab track vertically by bolt. The continuous magnetic interaction between LIM and RP (Fig.2)[7,8,9] is equivalent to the even distant spring and damping between LIM and the slab track.

2.5 Vertical coupling model of LIM vehicle-track

In this paper, vertical coupling model of vehicle and track is established (Fig.3). The parameters and the meanings are described as follows (Table 1).

2.6 Model of slab track step irregularity

According reference [10,11,12], slab track stepped irregularity is due to non-uniform of foundation. In

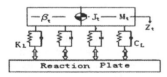

Figure 2. Model of magnetic interaction.

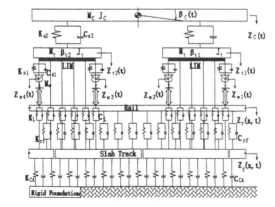

Figure 3. Vertical coupling model of LIM vehicle-track.

this article, slab track stepped irregularity can be expressed as following:

$$y(x) = \begin{cases} h(1-\cos(\frac{\pi x}{2\lambda})) \\ h \end{cases} \tag{2}$$

Where λ is the irregularity wavelength, h is the wave amplitude, and Nf is the number of the sinking slab track (Fig.4).

3 DYNAMIC EQUATION OF SYSTEMATIC VERTICAL VIBRATION

In this paper, the minimum potential energy theory is adopted to establishment vibration differential equation of LIM vehicle-track.

Table 1. Meanings of basic parameters.

Mc, Mt, Mw	Car body mass, bogie (Including LIM), Wheel set mass
Z_c, Z_{ti}, Z_{wj}	Displacement of car body, bogie and wheel sets
J_c, J_t	Car body inertia, bogie inertia
β_c, β_{ti}	Pitching angular displacement of car body and bogie
M_{0r}, M_{0b}	Rail unit length mass, track slab unit length mass
$E_r J_r$	Rail vertical flexural rigidity
$E_b J_b$	Track vertical flexural rigidity
I_r, I_b	Rail inertia, track slab inertia
K_{s1}, C_{s1}	Primary suspension stiffness, damping
K_{s2}, C_{s2}	Secondary suspension stiffness, damping
K_{rf}, C_{rf}	Rail pad stiffness, Rail pad damping
K_b, C_b	Stiffness under track slab, damping under track slab
l_c, l_t	Half of bogie distance, half of wheel set distance
l_s	Rail element length
l_f	Track slab element length
g	Air gap between LIM and RA
u_i, v_i	Displacement of rail node, Displacement of track slab node
β_i, θ_i	Pitching angular displacement of rail node and track slab node
K_{CA}, C_{CA}	Vertical rigidity and damping of emulsion asphalt under slab track

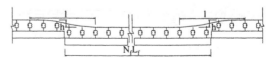

Figure 4. Model of slab track stepped irregularity.

3.1 Vehicle equations

The energy of vehicle system contains the kinetic energy of vehicle bogie and wheel by their nodding and floating motion, and the gravitational potential energy. The strain energy of the spring-damping between vehicle bogie and wheel.

First-order variation of vehicle kinetic energy T_c

$$\delta T_c = M_c \ddot{Z}_c \delta Z_c + J_c \ddot{\beta}_c \delta \beta_c + M_{t1} \ddot{Z}_{t1} \delta Z_{t1} +$$

$$J_t \ddot{\beta}_{t1} \delta \beta_{t1} + M_{t2} \ddot{Z}_{t2} \delta Z_{t2} + J_t \ddot{\beta}_{t2} \delta \beta_{t2} + \sum_{j=1}^{4} M_w \ddot{Z}_{wj} \delta Z_{wj}$$

(3)

First-order variation of vehicle strain energy U_c

$$\delta U_c = K_{s2} \left(Z_c - Z_{t1} + l_c \beta_c \right) \left(\delta Z_c - \delta Z_{t1} + l_c \delta \beta_c \right)$$
$$+ K_{s2} \left(Z_c - Z_{t2} - l_c \beta_c \right) \left(\delta Z_c - \delta Z_{t2} - l_c \delta \beta_c \right)$$
$$+ K_{s1} \left(Z_{t1} - Z_{w1} + l_t \beta_{t1} \right) \left(\delta Z_{t1} - \delta Z_{w1} + l_t \delta \beta_{t1} \right)$$
$$+ K_{s1} \left(Z_{t1} - Z_{w2} - l_t \beta_{t1} \right) \left(\delta Z_{t1} - \delta Z_{w2} - l_t \delta \beta_{t1} \right)$$
$$+ K_{s1} \left(Z_{t2} - Z_{w3} + l_t \beta_{t2} \right) \left(\delta Z_{t2} - \delta Z_{w3} + l_t \delta \beta_{t2} \right)$$
$$+ K_{s1} \left(Z_{t2} - Z_{w4} - l_t \beta_{t2} \right) \left(\delta Z_{t2} - \delta Z_{w4} - l_t \delta \beta_{t2} \right)$$

(4)

First-order variation of vehicle damping potential energy W_c

$$\delta W_c = C_{s2} \left(\dot{Z}_c - \dot{Z}_{t1} + l_c \dot{\beta}_c \right) \left(\delta Z_c - \delta Z_{t1} + l_c \delta \beta_c \right)$$
$$+ C_{s2} \left(\dot{Z}_c - \dot{Z}_{t2} - l_c \dot{\beta}_c \right) \left(\delta Z_c - \delta Z_{t2} - l_c \delta \beta_c \right)$$
$$+ C_{s1} \left(\dot{Z}_{t1} - \dot{Z}_{w1} + l_t \dot{\beta}_{t1} \right) \left(\delta Z_{t1} - \delta Z_{w1} + l_t \delta \beta_{t1} \right)$$
$$+ C_{s1} \left(\dot{Z}_{t1} - \dot{Z}_{w2} - l_t \dot{\beta}_{t1} \right) \left(\delta Z_{t1} - \delta Z_{w2} - l_t \delta \beta_{t1} \right)$$
$$+ C_{s1} \left(\dot{Z}_{t2} - \dot{Z}_{w3} + l_t \dot{\beta}_{t2} \right) \left(\delta Z_{t2} - \delta Z_{w3} + l_t \delta \beta_{t2} \right)$$
$$+ C_{s1} \left(\dot{Z}_{t2} - \dot{Z}_{w4} - l_t \dot{\beta}_{t2} \right) \left(\delta Z_{t2} - \delta Z_{w4} - l_t \delta \beta_{t2} \right)$$

(5)

First-order variation of gravitational potential energy W_{cg}

$$\delta W_{cg} = M_c g \delta Z_c + M_t g \delta Z_{t1} + M_t g \delta Z_{t2} + \sum_{j=1}^{4} M_w g \delta Z_{wj}$$

(6)

3.2 Rail

Rail can be considered as infinite Euler beam[13] (Fig.5). Element energy consists of kinetic energy,

Figure 5.   Point supporting beam.

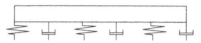

Figure 6.   Continuous supporting beam.

bending strain energy of rail element and damping potential energy of rail element.

First-order variation of rail kinetic energy T_r

$$\sum \delta T_r^e = \sum \{\delta \ddot{u}_r\}^{eT} \left[\int_0^{l_r} m_r [N]^T [N] dx \right] \{\ddot{u}_r\}^e$$

$$= \sum \{\delta \ddot{u}_r\}^{eT} [m]_r^e \{\ddot{u}_r\}^e$$

(7)

First-order variation of rail bending strain energy U_r

$$\sum \delta U_r^e = \sum \{\delta u_r\}^{eT} \left[\int_0^{l_r} E_r J_r [N'']^T [N''] dx \right] \{u_r\}^e$$

$$= \sum \{\delta u_r\}^{eT} [k]_r^e \{u_r\}^e$$

(8)

in which, $\{u_r\}$ is vertical displacement of rail node; $[m]_r^e$, $[k]_r^e$ is mass matrix and stiffness matrix of point supporting rail beam element.

According to reference [14], at present, linear scale damping is adopted in homogeneous structural damping–Rayleigh Damping.

3.3 Slab track

Slab track can be considered as continuous supporting beam (Fig.6), the kinetic energy of beam element is basically identical to that of the rail. However, the mass of beam element is different [11].

First-variation of slab track kinetic energy T_b:

$$\sum \delta T_b^e = \sum \{\delta \dot{v}_b\}^{eT} \left[\int_0^{l} m_b [N]^T [N] dx \right] \{\dot{v}_b\}^e$$

$$= \sum \{\delta \dot{v}_b\}^{eT} [m]_b^e \{\dot{v}_b\}^e$$

(9)

Strain energy of slab track element contains bending strain energy U_{21}^e of beam element and elastic potential energy of supporting spring.

First-variation of bending strain energy of slab track element U_{21}

$$\sum \delta U_{21}^e = \sum \{\delta v_b\}^{eT} \left[\int_0^{l} E_b J_b [N'']^T [N''] dx \right] \{v_b\}^e$$

$$= \sum \{\delta v_b\}^{eT} [k_{21}]_b^e \{v_b\}^e$$

(10)

First-variation of supporting spring elastic energy:

$$\sum \delta U_{22}{}^{e} = \sum \{\delta v_b\}^{eT} \left[\int_0^{l_t} K_{CA}[N]^T[N]dx \right]\{v_b\}^{e}$$

$$= \sum \{\delta v_b\}^{eT} [k_{22}]_b^{e} \{v_b\}^{e} \qquad (11)$$

Damping of slab track element contains structural damping (*Rayleigh damping*) and damping of slab element supporting. The following matrix of damping is as follows:

$$[c_2]_b^{e} = \frac{C_{CA}}{K_{CA}} [k_{22}]_b^{e} \qquad (12)$$

in which, $\{v_b\}$ is vertical displacement of slab element node, K_{CA}, C_{CA} are vertical rigidity and damping of Emulsion asphalt, $[m]_b^e$ $[k_{21}]_b^e + [k_{22}]_b^e$ are mass matrix and stiffness matrix of continuous supporting beam[13].

3.4 Wheel-rail coupling

First-variation of general strain energy in the contact spring between wheel and track:

$$\delta U_{wr} = \sum_{j=1}^{4} K_w \left(Z_{wj}(x,t) - Z_r(x_{pj},t) - Z_{r0}(t) \right)$$

$$\cdot \left(\delta Z_{wj}(x,t) - \delta Z_r(x_{pj},t) \right) \qquad (13)$$

in which K_w is instantaneous elastic rigidity; $Z_{wj}(t)$ is the displacement of j wheel at the time of t; $Z_r(x_{pj},t)$ is the displacement of rail under j wheel at the time of t; $Z_{r0}(t)$ is the rail irregularity[5]. In particular cases, when $\delta Z_t < 0$, wheel is separated from rail. In this article, $Z_{r0}(t)$ is caused by slab track stepped irregularity.

3.5 Vertical interaction between rail and slab track

Vertical interaction between rail and slab track is considered as the spring-damper, where K_{rb} and C_{rb} are rigidity and damping of magnetic interaction of rail and track are those of the fastenings.

First-variation of backing strap U_{rb}

$$\delta U_{rb} = K_{rb} \sum \left(\delta u_r - \delta v_b \right)\left(u_r - v_b \right) \qquad (14)$$

First-variation of damping force potential energy W_{rf}

$$\delta W_{rf} = C_{rf} \sum \left(\delta \dot{u}_r - \delta \dot{v}_b \right)\left(\dot{u}_r - \dot{v}_b \right) \qquad (15)$$

3.6 Magnetic interaction of LIM and track

The instantaneous gap between LIM and RP is equal to the displacement between rail and track. The gap decreasing between bogie and track can make the calculation simpley[7,8].

First-variation of potential energy:

$$\delta U_{zb} = K_L \sum_{i=1}^{2} \int (Z_{zi}(x,t) - Z_b(x,t) - Z_{b0}(t))dx$$

$$\cdot (\delta Z_{zi}(x,t) - \delta Z_b(x,t)) \qquad (16)$$

First-variation of damping potential energy:

$$\delta W_{zb} = C_L \sum_{i=1}^{2} \int (\dot{Z}_{zi}(x,t) - \dot{Z}_b(x,t) - \dot{Z}_{b0}(t))dx$$

$$\cdot (\delta \dot{Z}_{zi}(x,t) - \delta \dot{Z}_b(x,t)) \qquad (17)$$

in which, $Z_{zi}(x,t)$ is vertical displacement of bogie at the time of t; $Z_b(x,t)$ is track displacement under the bogie at the time of t; $Z_{b0}(t)$ is the value of slab track stepped irregularity.

3.7 System differential equation

The above equation (3~19) can be formed into coupling dynamic equation of LIM vehicle-slab track:

$$[M]\{\ddot{z}\} + [C]\{\dot{z}\} + [K]\{z\} = \{F\} \qquad (18)$$

in which, $[M],[C],[K]$ are mass matrix, stiffness matrix and damping matrix of the system. $\{F\}$ presents the force matrix corresponding to the slab track stepped irregularity.

$$\{z\} = \{Z_c, \beta_c, Z_{t1}, \beta_{t1}, Z_{t2}, \beta_{t2}, Z_{w1}, Z_{w2}, Z_{w3}, Z_{w4}$$

$$u_1, \beta_1, \cdots, u_i, \beta_i, \cdots, u_N, \beta_N$$

$$v_1, \theta_1, \cdots, v_i, \theta_i, \cdots, v_M, \theta_M\} \qquad (19)$$

$\{\ddot{z}\}$, $\{\dot{z}\}$, $\{z\}$ are the accelerate, velocity and displacement of the system.

Herz nonlinear elastic contact spring is replaced by linear spring in analysis of wheel-rail coupling interaction. Electromagnetic interaction between LIM and RP is deal with linear method. In this paper, *Newmark* numerical integration method is adopted.

4 EMULATION CALCULATION AND INTERPRETATION OF RESULTS

4.1 Parameter

In this paper, Japanese LIM vehicle, normal rail(*T60*), and normal slab track are used. Furthermore, space intervals between fastening and bolting are 0.6 m[3].

Table 2. System parameter.

Parameter (Unit)	Value	Parameter (Unit)	Value
M_c (kg)	31,994	M_r (kg/m)	60.64
M_t (kg)	3,333	E_rJ_r (kg $\cdot$ m^2)	6.5×10^6
M_w (kg)	1,650	K_{rb} (N/m)	6.0×10^7
J_c (kg $\cdot$ m^2)	2.1×10^6	C_{rb} (N $\cdot$ s/m)	3.625×10^4
J_t (kg $\cdot$ m^2)	3,200	M_b (kg/m)	833.3
K_{s2} (N/m)	4.0×10^5	E_bJ_b (kg $\cdot$ m^2)	2.34×10^7
C_{s2} (N $\cdot$ s/m)	4.0×10^4	K_{CA} (N/m)	1.25×10^9
K_{s1} (N/m)	2.5×10^6	C_{CA} (N $\cdot$ s/m)	8.0×10^4
C_{s1} (N $\cdot$ s/m)	2.5×10^5	K_L (N/m)	1×10^5
L_c (m)	5.57	C_L (N $\cdot$ s/m)	1.725×10^4
L_t (m)	1.0	Nf	2

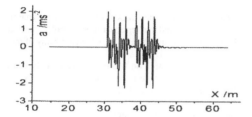

Figure 7. Vehicle acceleration.

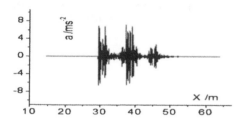

Figure 8. Vehicle acceleration.

In this paper, calculation can be carried out according to the vertical coupling dynamic model of LIM system, the parameters described in the above table and the *LVSTVDSS (LIM Vehicle–Slab Track Vertical Dynamic Simulation System)* programmed by *FORTRAN90*. The vehicle is running on the rail of 64.2 m at the speed of 100 km/h.

4.2 *Slab track stepped irregularity*

In the middle of slab track, take 1.8 m wavelength and 1 mm wave amplitude of slab track stepped irregularity as an example.

When the wave amplitude is 1 mm, the maximal vehicle acceleration is less than 1 ms^{-2}; when the wave amplitude is 2 mm, the maximal vehicle velocity is less than 1.9 ms^{-2} (Fig.7). The maximal acceleration value of slab track node is more than 6.5 ms^{-2} (Fig.8).

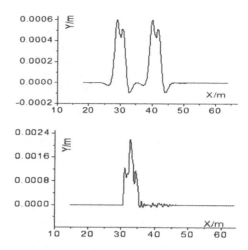

Figure 9. Displacement comparison of bogie and track.

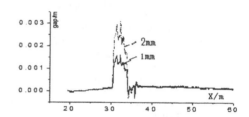

Figure 10. Influence of wave amplitude on gap variation.

In Fig. 9 vertical displacement of bogie is greater than that of slab node. The maximal displacement of slab track node is less than 0.6 mm, and the maximal displacement of bogie is more than 2.0 mm (Fig. 9). The vertical displacement variation of bogie has more effect on the gap variation between LIM and RP.

Slab track stepped irregularity has more effect on the slab acceleration, and the wave amplitude is greater, slab acceleration will be the bigger (Fig.10). The maximal gap variation value of 1 mm amplitude is 1.6 mm; when the wave amplitude is 2 mm, the maximal gap variation is 3.0 mm, which is more than 2 mm[15].

5 SUMMARY AND CONCLUSIONS

In this paper, differential equation and LIM-slab track dynamic coupling model have been established. With the use of software programmed in this research, calculation and analysis can be carried out and the following conclusions:

1) The vertical displacement of LIM is greater than that of slab track node; the effect caused by bogie vertical displacement on the gap variation is more.

2) The wave amplitude of slab track stepped irregularity is the greater. The displacement of bogie and corresponding gap variation will be greater.

3) Slab track stepped irregularity has more effect on the acceleration of slab, and the acceleration will get fold increase with the increase of amplitude.

4) Slab track stepped irregularity has more effect on the gap variation. The value of slab track stepped irregularity should be controlled within 2 mm[17], thus the safety in operation can be guaranteed.

REFERENCES

[1] Wu Junqing, Application and development of LIM in Skytrain of Canada, science and technology in subway 2003(2), p28–30

[2] Wei qingchao, Feng Yawei, Shi Hong, Transportation pattern and technical economy characteristics of LIM, urban rapid transit system, 2004.1(17), p48–53

[3] Zhai Wanming, Vehicle-track coupling dynamics (Version Two) [M], 2002

[4] Zhai Wanming, Zhao Chunfa, Cai Chengbiao, Comparison in the dynamic action on line and bridge of magnetic suspension and wheel-rail high speed, transaction of transportation and communication. 2001.1, p7–12

[5] Su Qian, A Spatial-timing–coupled Dynamic Model for Railway Subgrade of High Speed Line and its Application. Southwest Jiaotong University Doctor Degree Dissertation, 2001.9

[6] Li Chenhui, The Vibration Theory of Track Structure and its Application, Southwest Jiaotong University Doctor Degree Dissertation, 1996.5

[7] Cai Y, Chen S.S, 'VehiclePGuideway dynamic interaction in maglev system of dynamic system[J]', Measurement and Control, Trans, ASME. 1996, 118: 526–530

[8] Wu Fanyu, Gao Liang, Wei Qingchao, Dynamic Effect on Structure Characteristics of Elevated Guideway on Magnetic System, Engineering Mechanics, 2004.8

[9] Shi Jing, Wei Qingchao, Wu Fanyu, Study on Vibration of Beam of Magnetic Levitation Express Railway and its Control, China safety science journal, 2003.10

[10] Zhao Pingrui, Analysis of Slab Track's Dynamic Performance and Study of Parameter, Southwest Jiaotong University Master Degree Thesis, 2003.9

[11] Wang Qichang, High-speed railway in civil engineering, Southwest Jiaotong University Press House, 1999.9

[12] Zhang geming, The track-bridge system dynamic analysis model and track irregularities control on quasi & high-speed railway [D]. Beijing:China academy of railway sciences, 2001.2

[13] Wang Ping, Research on Wheel/Rail System Dynamic on Turnout, Southwest Jiaotong University Doctor Degree Dissertation, 1997.10

[14] Feng Wenxian, Chen Xin. Estimation method of damping matrix in structural vibration

[15] A.K. Wallace, J.H.Parker, C.E.Dawson. 'SLIP CONTROL FOR LIM PROPELLED TRANSI VEHICLES'. IEEE TRANSACTIONS ON MAGNETICS, VOL.MAG-16, NO.5, 1980.9, p710–712

Environmental Vibrations – Takemiya (ed.)
© 2005 Taylor & Francis Group, London, ISBN 0 415 39035 4

Coupled vibration analysis of X-style arch bridge and moving vehicles

Q. Li & D.S. Shan

Department of bridge and structural engineering, Southwest Jiaotong University, Chengdu, China

ABSTRACT: A study on the coupled vibration of X-style arch Railway Bridge and moving vehicles is presented in this paper. The bridge is represented with a 3-D finite element model. Each 4-axle vehicle is modeled by a 35 degrees-of-freedom dynamic system. The random irregularities of the tracks are generated from the power spectral density function under the given track condition. The couple between the bridge and train is realized through the contact forces between the wheels and track. Then applying time-integration and iteration techniques to the coupled system, the coupled equations of motion are efficiently solved. The proposed formulation and the associated computer program are then applied to a certain X-style arch railway bridge. The dynamic responses of the bridge-vehicle system are computed. The results show that the formulation presented in this paper can well predict dynamic behaviors of both bridge and train with reasonable computation efforts.

1 INTRODUCTION

The vibration caused by the moving vehicles is a principal problem in dynamic analysis of railway bridges (Cheng 2000, Shan 1999a, b, Chu 1980 and Garg 1984). In China, with the crucial demands for the construction of high-speed railways and the improvement of the speed for existing railways, a comprehensive research has been undertaken in recent years by which an extensive knowledge was obtained of the coupled vibration of moving vehicles and railway bridges. However, most of these studies are concentrated on the girder bridges with little attention paid to the interaction between X-style railway arch bridges and moving vehicles.

Based on the existing theories, the coupled vibration of the X-style railway arch bridge and moving vehicles is discussed in this paper. A three-dimensional finite element model is used to represent a certain X-style arch bridge. Each 4-axle vehicle in a train is modeled by a 35-D.O.F. dynamic system. The random irregularities of the track are generated from a power spectral density function under the given track condition. The dynamic interaction between the bridge and train is realized through the contact forces between the wheels and track. Then based on these models and applying the time-integration and iteration techniques to the coupled system, the coupled equations of motion are efficiently solved. To examine the proposed formulation together with the associated computer program, a certain X-style arch bridge carrying double-line railway within the bridge deck is taken as a numerical example.

2 BASIC DYNAMIC MODELS

The key steps to understand the coupled vibration between moving vehicles and X-style arch bridges are: to construct a suitable negotiation mathematical model of vehicle, to deduce proper dynamic equations of X-style arch bridge and to choose a right numerical integral method in time-domain of the vehicle–bridge system.

2.1 Dynamic model of train

A train usually consists of several locomotives, passenger cars, freight cars, or their combinations. Each vehicle is in turn composed of a car body, bogies, wheel-sets, and the connections between the three components. To simplify the analysis but with enough accuracy, the following assumptions are used in the modeling of the train in this study (Cheng 2000):

(1) The vehicle negotiates the bridge at an even speed; the structure of vehicle is symmetric.
(2) The car body, bogies and wheel-sets in each vehicle are regarded as rigid components, neglecting their elastic deformation during vibration (see Fig. 1).
(3) The connections between the car body and a bogie are represented by two linear springs and two viscous dashpots of the same properties in either the horizontal direction or the vertical direction (see Fig. 1). The stiffness and damping coefficients are denoted as k_{2ij}^h and c_{2ij}^h for the springs and dashpots in the jth bogie of the ith vehicle in

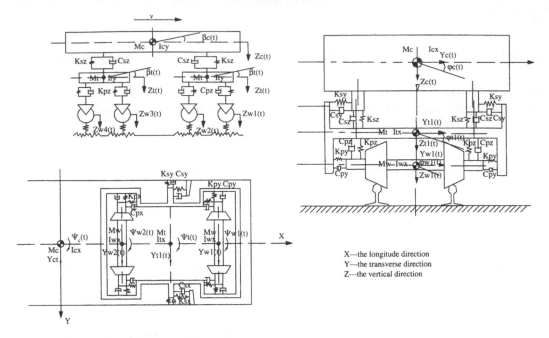

Figure 1. The typical model of the moving passenger car.

the horizontal direction and k_{2ij}^v and c_{2ij}^v for the springs and dashpots in the jth bogie of the ith vehicle in the vertical direction.

(4) The connections between a bogie and a wheelset are characterized as two linear springs and two viscous dashpots of the same properties in either the horizontal direction (k_{1ij}^h and c_{1ij}^h) or the vertical direction (k_{1ij}^v and c_{1ij}^v).

(5) The longitudinal, lateral and vertical suspension characteristics of each component are independent separately.

(6) The interactions among vehicles are ignored; in another word ignore the coupler forces and moments.

With the aforementioned assumptions and the dynamic characteristics of vehicles during negotiation, a typical vehicle model is developed as shown in figure 1.

For the sake of determining the vibration of the car body, the lateral displacement and the yaw angle must be taken into account; the rolling angle is reckoned in for studying the apparent unbalance motion of car body caused by the effect can't deficiency; On the other hand, due to the coupled vibration of bending and torsion for X-style arch, the model should also included the bouncing and pitching displacements. In a word, the car body has 5 degrees of freedom (D.O.F). They are defined as the later displacement y_c, the yaw angle β_c, the roll angle Φ_c, the bounce displacement z_c and the pitching displacement Ψ_c. The determination of

Table 1. Degrees of the curving model.

Component	Shift	Bounce	Roll	Pitch	Yaw
Car body	y_c	z_c	Φ_c	ψ_c	β_c
Front-bogie	y_{t1}	z_{t1}	Φ_{t1}	ψ_{t1}	β_{t1}
Rear-bogie	y_{t2}	z_{t2}	Φ_{t2}	ψ_{t2}	β_{t2}
1st wheel sets	y_{w1}	z_{w1}	Φ_{w1}	ψ_{w1}	β_{w1}
2nd wheel sets	y_{w2}	z_{w2}	Φ_{w2}	ψ_{w2}	β_{w2}
3rd wheel sets	y_{w3}	z_{w3}	Φ_{w3}	ψ_{w3}	β_{w3}
4th wheel sets	y_{w4}	z_{w4}	Φ_{w4}	ψ_{w4}	β_{w4}

the bogie's D.O.F is the same as the car body. The number of D.O.F of wheel-sets is determined by the model of creep-force. The nonlinear grand creep model is introduced in the paper, so all 5 degrees of freedom are considered except the longitudinal one. For a 4-axle 2-bogie vehicle studied in this paper, the total D.O.F is 35, as shown in table 1 (Cheng 2000 and Shan 1999a).

By assuming that vibration amplitude of each component in a vehicle is small and using the equilibrium conditions, the equations of motion for the ith vehicle can be derived as follows (Cheng 2000):

$$[M_v]\{\ddot{x}_v\} + [C_v]\{\dot{x}_v\} + [K_v]\{x_v\} = \{F\} \qquad (1)$$

where $[M_v]$, $[C_v]$ and $[K_v]$ denote the mass, damp and stiffness matrices of vehicle respectively; $\{x_v\}$ denotes

204

Table 2. The parameters of the PSD in Chinese high-speed railway.

Parameter	A	B	C	D
y_l	0.1270	−2.1531	1.5503	4.9835
y_r	0.3326	−1.3757	0.5497	2.4907
z_l	0.0627	−1.1840	0.6773	2.1237
z_r	0.01595	−1.3853	0.6671	2.3331

Parameter	E	F	G
y_l	1.3891	−0.0327	0.0018
y_r	0.4057	0.0858	−0.0014
z_l	−0.0847	0.0340	−0.0005
z_r	0.2561	0.0928	−0.0016

the column vector of general unknown displacement, $\{\dot{x}_v\} = \partial\{x\}/\partial t$ and $\{\ddot{x}_v\} = \partial^2\{x\}/\partial t^2$. $\{F\}$ denotes the column vector of exciting forces.

2.2 Modeling of track's irregularities

Many investigations have shown that the irregularity of the track is an important factor that affects the dynamic responses of both bridge and vehicle (Shan 1999a). The track geometry is defined in terms of four irregularities consisting of gage, cross level, alignment and vertical surface profile. These irregularities may be described as a realization of a random process that can be described by a power spectral density (PSD) function respectively. The following PSD functions were proposed by CARS (China Academy of Railway Science) for these track's irregularities in Chinese high-speed railway.

$$S(f) = \frac{A(f^2 + Bf + C)}{f^4 + Df^3 + Ef^2 + Ff + G} \qquad (2)$$

where $S(f)$ is the PSD function (m³/cycle) for the track-irregularity; f is the spatial frequency (cycle/m); A, B, C, D, F and G are the irregularity coefficient which are shown in the Table 2. In table 2, y_l and y_r are the elevation of left and right track respectively; z_l and z_r are the alignment of left and right track.

The track-irregularity is assumed to be a zero-mean stationary Gaussian random process. Therefore, it can be generated through an inverse Fourier transform.

$$y(x) = \sum_{k=1}^{N} \sqrt{2S(f = k\Delta f)\Delta f}\ \cos(2\pi fx + \theta_k) \qquad (3)$$

where θ_k is the random phase angle uniformly distributed from 0 to 2π.

2.3 Dynamic model of X-style arch bridge

When the X-style arch bridge carries a railway, the track will be laid on the bridge deck and the forces

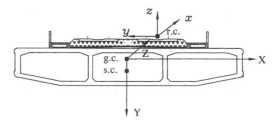

Figure 2. The coordinate transform.

from the wheels of a train will be transmitted to the bridge deck through the track. Since X-style arch bridges are considered here, this study assumes that there is no relative displacement between the track and bridge deck. The elastic effects of the track system are also neglected. The bridge is modeled as a three-dimensional system. The equation of motion for the bridge can be thus expressed as (Shan 1999a, Fryba 1972 and 1996 and Clough 2003)

$$[M_b]\{\ddot{x}_b\} + [C_b]\{\dot{x}_b\} + [K_b]\{x_b\} = \{F_w\} \qquad (4)$$

where $[M_b]$, $[C_b]$ and $[K_b]$ are the mass, damping and stiffness matrices of the bridge; the structural damping of the X-style arch bridge is assumed to be Raleigh damping. $\{\ddot{x}_b\}$, $\{\dot{x}_b\}$, and $\{x_b\}$ are the acceleration, velocity and displacement vectors of the bridge, respectively, and $\{F_w\}$ is the force vector, consisting of forces from the wheels of a train on the bridge deck through the track.

Since the contact points between the wheel-sets and track don't coincide with the gravity center or torsional center of sections, the general displacements of contact points should be transmitted from the X-style arch coordinate system to the rail coordinate system and the wheel-track contacted forces should be transformed form the rail coordinate system to deck coordinate system. The relations of displacements and coupled forces between the deck and the wheel-sets on the contact points are shown as following (see Fig. 2) (Shan 1999a).

$$y_w = -u_X - \theta_Z(d_h + d_y) + y_r \qquad (5a)$$

$$z_w = -(u_Y + \theta_Z e_x) + z_r \qquad (5b)$$

$$\Phi_w = \theta_Z + \Phi_r \qquad (5c)$$

$$\Psi_w = -\theta_X \qquad (5d)$$

$$\beta_w = -\theta_Y \qquad (5e)$$

$$M_X = -m_{wy} \tag{6a}$$

$$M_Y = -\left[m_{wz} + f_{wz}e_x + f_{wy}d_h\right] \tag{6b}$$

$$M_Z = m_{wx} - f_{wz}e_x - f_{wy}\left(d_h + d_y\right) \tag{6c}$$

$$F_X = -f_{wy} \tag{6d}$$

$$F_Y = -f_{wz} \tag{6e}$$

where y_w, z_w, Φ_w, Ψ_w and β_w are the displacement of the wheel-sets on the contact points (Tab.1.), u_X, u_Y, u_Z, θ_X, θ_Y, θ_Z are the general displacements of the X-style arch on the contact points. d_h is the vertical distance from the gravity center (g.c.) of the section to the contact point. d_y is the vertical distance between the gravity center and the torsional center (t.c.), e_x is the horizontal distance from the rail center (r.c.) to the gravity center (g.c.). y_r, z_r and Φ_r are the lateral, vertical and rolling irregularities of track. f_{wy}, f_{wz}, m_{wx}, m_{wy} and m_{wz}, are the wheel-track contact forces. Fx, Fy, Mx, My and Mz are the exciting forces of the X-style arch bridge.

2.4 Numerical method

The numerical method used here to deal with the coupled vibration of vehicle-bridge is crucial. It should be low computational cost, stable and able to deal with the nonlinear behavior of the dynamic system. In this paper, the vehicle-bridge system is divided into two separate subsystems, i.e. vehicle and bridge structure subsystems. Since the two separate subsystems are connected by the contact forces between wheels and rails, which affect the dynamic responses of the two subsystems, the solution process of the equations is iterative (Shan 1999a and b). Meanwhile, as the contact forces are nonlinear, the responses of the two subsystems are also nonlinear in spite of the linear material and geometrical properties of them. Based on the nonlinear excitation, a new algorithm is introduced to the new model of vehicles and X-style arch bridges in order to study the coupled vibration between moving vehicles and X-style arch bridges. And the software IDASVCB (Interaction Dynamic Analysis System of Vehicle and Curved-Bridge), corresponding to the new algorithm, is developed, which can deal with the coupled vibration of moving vehicles and X-style arch bridges.

3 NUMERICAL EXAMPLE

Take a new built X-style arch bridge as the example to discuss the coupled vibration between the X-style arch bridge and vehicles. With its calculated span 112 m,

the arch-axis shaped in catenary, rise span ratio 1/5, and the coefficient of arch-axis 1.347, the deck system is thus composed of a single box with three cells, 2.5 m high. The section of the arch-ring is in the form of a dumb-bell and its external diameter is 1.0 m. The angle of dip of the arch rib is 77 degree. And the bridge which formed by rigid tie-beam and rigid arch-rib belongs to the Nelson System Arch. The distance between suspenders which were made up of 127Φ7 parallel tendons is 8 m. The length of the overlapped segment between two arch-ribs is 8 m; and there are 4 transverse braces link the two arch-ribs, two of them near the vault used circular steel tubes, the other are K-style cross braces.

3.1 Computational model and study cases

The steel-tube concrete section of arch-rib is replaced with an equivalent cross-section in the study; and the equivalent methods are listed in *Specification for Design and Construction of Concreted-Filled Steel Tubular Structures* (CECS 28:90). They are:

$$EA = E_c A_c + E_s A_s \tag{7a}$$

$$EI_x = E_c I_{cx} + E_s I_{sx} \tag{7b}$$

$$EI_y = E_c I_{cy} + E_s I_{sy} \tag{7c}$$

$$EI_z = E_c I_{cz} + E_s I_{sz} \tag{7d}$$

$$EI_\omega = E_c I_{c\omega} + E_s I_{s\omega} \tag{7e}$$

where, EA, EI_x, EI_y, EI_z and EI_ω respectively denote the equivalent axial stiffness, the equivalent flexural stiffness in the direction of x, y and z and equivalent wrapping stiffness of the Concreted-Filled Steel Tubular arch-rib; $E_c A_c$, $E_c I_{cx}$, $E_c I_{cy}$, $E_c I_{cz}$ and $E_c I_{c\omega}$ respectively indicate the axial stiffness, the flexural stiffness in the direction of x, y and z and wrapping stiffness of the concrete filled in the steel tubular; $E_s A_s$, $E_s I_{sx}$, $E_s I_{sy}$, $E_s I_{sz}$ and $E_s I_{s\omega}$ respectively represent the axial stiffness, the flexural stiffness in the direction of x, y and z and wrapping stiffness of the steel-tube.

The main girder can be discreted by 3D beam elements, the concreted-filled steel tubular arc-rib can be discreted by 3D curved girder elements and the suspender can be discreted by 3D cable elements. In this way, 208 elements and 145 nods can be got. (Fig.3)

3.2 Moving vehicle and track-irregularities

A passenger train (one SS8 railway motor car drawing nine passenger cars) and a freight train (one SS8 railway motor car drawing nine C62 freight cars) are taken into

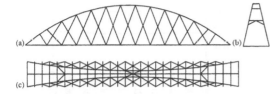

Figure 3. Computational model. (a) elevation view of model figure. (b) Side view of model. (c) Top view of model.

consideration in the study. The computing conditions are like this: the freight train passed the bridge single line and double line with each time the speeds of 60, 65, 70, 75 and 80 km/h respectively; the passenger train passed the bridge again single line and double line with each time the speeds of 80, 90, 100, 110, 120, 130, 140, 150, 160, and 170 km/h respectively. Based on the above mentioned computing conditions, a comparatively detailed analysis of the security and comfortableness of the while-driving train is made from the aspects of its lateral swing, derailment factor, lightening-factor of wheel-load, vertical and lateral acceleration of the car body, vertical and transverse, the performance criteria for cargo trains and the ride indices for passenger cars, etc.

The Track-unevenness-power spectral density recommended by China Academy of Railway Sciences is adopted during study. At the same time, the unevenness is worked out by artificial simulation by the standard for newly-built quasi high-railway. Due to the track-unevenness is the excited resource for the coupled vibration of vehicles and bridges, it is noteworthy that the difference of the track-unevenness for the going upward and going downward. Being a double line railway of this bridge the tropism of the track can directly affect the computing results, especially the result of transverse vibration of vehicles and bridges.

3.3 Dynamic responses of X-style arch bridge

3.3.1 Natural frequencies of the X-style arch bridge
Table 3 shows the first ten natural frequencies and mode shapes of the bridge. The natural frequencies are worked out by using general finite element software (GFEMS) and IDASVXAB (Interactive Dynamic Analysis System of Vehicle and X-Style Arch Bridge). According to the table, it is obvious that the frequencies which computed by the two software are similar and the mode shapes are identical.

3.3.2 Dynamic amplification factor
Dynamic amplification factor (DAF) can be defined as the ratio between the active internal force and the maximal static internal force of the same section under the same load. The relationship between the dynamic

Table 3. Natural frequencies (Hz) and mode shapes.

| Mode | Computed frequency | | Mode shapes |
	GFEMS	IDASVXAB	
1	1.1733	1.2185	Symmetrical lateral bending of arch-rib
2	2.0329	2.0444	Anti-symmetrical lateral bending of arch-rib
3	2.0639	2.0828	Anti-symmetrical vertical bending
4	2.5874	2.3368	Symmetrical vertical bending
5	2.8506	2.7787	Symmetrical lateral bending of arch-rib
6	3.1988	3.6842	Symmetrical lateral bending of main girder & arch-rib
7	3.7579	3.9504	Anti-symmetrical lateral bending of arch-rib
8	4.0566	4.0321	Symmetrical vertical bending
9	4.4776	4.7500	Symmetrical lateral bending of arch-rib
10	5.2261	5.4254	Torsion of main girder

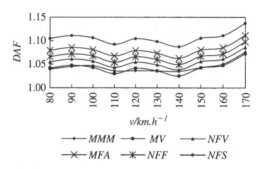

Figure 4. Variation of dynamic amplification factors as a function of velocities.

amplification factors of the main girder's mid-span, vault, foot of the arch, the suspender in middle-span and the traveling speed (v) of the vehicle can be seen in Figure 4. In the figure, the MMG indicates the moment of middle span of the main-girder, MV indicates the moment of the vault, MFA indicates the moment of the foot-section for arch, NFV indicates the axial force of vault, NFF indicates the axial force of the foot-section for arch and. NFS indicates the axial force of the suspender of the middle span. According to Figure 5, with the increase of the velocity the dynamic amplification factors doesn't always increase; in addition, the varying laws of dynamic amplification factors between the main girder and the arch, the suspender are not the same either with tiny change of the main girder's dynamic amplification factors, which is caused by the

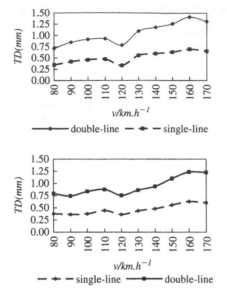

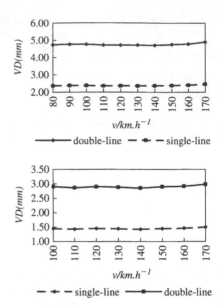

Figure 5. Variation of maximal transverse displacements as a function of velocities. (a) Mid-span of the main-girder. (b) Vault.

Figure 6. Variations of maximal vertical displacements as a function of velocities. (a) Mid-span of the main-girder. (b) The vault.

comparatively stronger rigidity of the main girder. In short, the variation of this bridge's dynamic amplification factors with the variation of velocity is quite remarkable with the maximal dynamic amplification factor is 1.1375 and the minimal dynamic amplification factor is 1.0347.

3.3.3 *Transverse displacements*

In Figure 5, the relationships between the transverse displacements (TD) of the middle-span of the main-girder and the vault and the traveling speed (v) of the vehicle are represented when passenger cars are grouped to run single-line or double-line. Generally speaking, the transverse displacements of the main girder's mid-span and the vault's don't always increase with the increase of the vehicle speed and the transverse displacements of the main girder are more remarkable than that of the vault. What's more, the transverse displacements of double-line trains are twice as much as that of single-line trains. The maximal transverse displacement (1.4 mm) of the main girder's mid-span occurs when a train runs double-line with the speed of 160 km/h and at the same time the maximal transverse displacement (1.2 mm) of the vault also occurs when the train runs double-line with the same speed. Thus, the transverse displacement of this bridge meets the standard in the *Chinese Criterion for Inspecting Railway Bridge*, that is, transverse displacement is required not to go beyond L/16.5 (6.79 mm). Therefore, the transverse rigidity of this bridge is qualified.

3.3.4 *Vertical displacement*

As has shown in Figure 6, with the increase of speed (v) no remarkable increases of the maximal vertical displacements of the middle-span of the main-girder and the vault occur when passenger cars are grouped to run single-line and double-line. But the absolute value of the main girder's maximal vertical displacement is a bit more than that of the arch's. In addition, the maximal vertical displacement of double-line trains is always about twice as much as that of single-line trains. The maximal vertical displacements (4.8 mm) of the main girder's mid-span occurs when a train runs double-line with the speed of 100 km/h and at the same time the maximal vertical displacement (1.5 mm) of the vault also occurs when the train runs double-line with the speed of 170 km/h.

3.4 *Responses of moving vehicles*

As to the vibration of moving vehicles, the security evaluation can be done by two parameters: derailment-factor (DF) and lightening-ration of wheel-loads (LR). Derailment-factor (DF) is the ratio of Q (lateral force of a single wheel) to P (vertical force of the same wheel). And lightening-ration of wheel-loads (LR) can be defined by the equation: $\Delta P = (P_s - P_d)/P_d$, where P_s is the static axis weight of a single wheel and P_d is the dynamic axis weight of the same wheel.

The extreme values of motor car's derailment-factor vary with the velocity variations of vehicles when

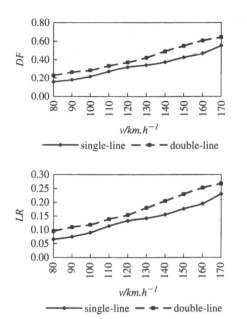

── single-line ── ● ── double-line

── single-line ── ● ── double-line

Figure 7. Variation of derailment-factors and lightening-rations of wheel-loads as a function of Velocities. (a) Derailment-factors versus velocities figure. (b) Lightening-rations of wheel-loads versus velocities.

passenger cars are grouped to run single-line and double-line as has shown in Figure 7a. According to the Figure, the derailment-factors increase with the increase of velocities; the derailment-factors may vary from 0.16 to 0.55 when the vehicles run single-line; and the derailment-factors of double-line running vehicles are a bit more than that of single-line running vehicles. Thus, it is reasonable to say that the vibration when vehicles run double-line is more than the one when vehicles run single-line. What's more, all the extreme values computed meet relative requirements of our country for derailment-factor, that is, the value of derailment-factor is required not to go beyond 0.8.

According to Figure 7b, when passenger cars are grouped to run single-line and double-line, the motor cars' the lightening-rations of wheel-loads increase with the increase of the passenger cars' velocities; when passenger cars run single-line, the lightening-rations of wheel-loads vary from 0.07 to 0.23 and when passenger cars run double-line, the lightening-rations of wheel-loads vary from 0.10 to 0.27. When the passenger car runs at the speed of 170 km/h, the lightening-ration of wheel-loads gets the maximal value and when the passenger car runs at the speed of 80 km/h, the lightening-rations of wheel-loads is minimal. Just as the variation of derailment factor, the lightening-rations of wheel-loads of double-line running vehicles

are a bit more than that of singleline running vehicles. Again based on the relative requirements of our country that the lightening-rations of wheel-loads are required not to go beyond 0.6, all the computed extreme values are qualified.

4 CONCLUSION

In the light of what have discussed above, conclusions can be made as follows:

(1) Generally speaking, the vehicle's velocity has remarkable influence on the transverse vibration of the X-style arch bridge. However, the amplitude of transverse vibration does not always increase with the increase of the vehicle's velocity.
(2) The vertical dynamic displacements of the X-style arch bridge are linear with the vehicle's velocity without too much variation.
(3) The dynamic amplification factor of the bridge varies with the variation of a vehicle's velocity. As to the X-style arch bridge, there is a remarkable difference between the maximal dynamic amplification factor and the minimal one.
(4) Within all the computing conditions concerning the X-style arch bridge mentioned in this study, both the derailment-factor and the lightening-rations of wheel-loads increase with the increase of vehicle's velocity, and the values of them meet our country's current requirements which ensure the rigidity of the bridge and the security of the moving vehicles.

REFERENCES

Guo Cheng, "The Analysis on Random Vibration of Vehicle/track Coupling System", A Dissertation submitted to Southwest Jiaotong University in Fulfillment of the Requirement for the Degree of Doctor of Philosophy, 2000, 47–58
Deshan Shan, "Curved-Girder Bridges and Vehicles Coupled Vibration Analysis and Design Study of the Long-span Curved-Girder Bridges in High-speed Railway", A Dissertation submitted to Southwest Jiaotong University in Fulfillment of the Requirement for the Degree of Doctor of Philosophy, 1999a; 32–40
Deshan Shan, Qiao Li, "Numerical Simulation of the Bridges-Vehicles Coupled Vibration and its Software Package", Journal of Southwest Jiaotong University, 1999b; 34: 663–667
Qiao Li, Spatial Analytic Theory of Thin-wall Curved-Girder, A Dissertation submitted to Southwest Jiaotong University in Fulfillment of the Requirement for the Degree of Doctor of Philosophy, 1988, 25–28
Ladislav Fryba, "Vibration of solids and structure under moving loads", Noordhoff International Publishing, Groningen, 1972, 309–324

Ladislav Fryba, "Dynamics of Railway Bridges", Institute of Theoretical and Applied Mechanics, Academy of Science of the Czech Republic, 1996, 162–181

Ray Clough, Joseph Penzien, "Dynamics of Structure", Second Edition (revised), Computer and Structures, Inc., U.S.A., 2003, 116–132

Chen Yunmin, Takemiya Hirokazu, "Environmental Vibration: Prediction, Monitoring and Evaluation", China Communication Press, 2003, 242–266

K.H.Chu, V.K.Grag and A.Wiriyachai, Dynamic Interaction of Railway Train and Bridge, Vehicle System Dynamics, 9 (1980), pp: 207–236

Vijay Kumar Garg & Rao. V. Dukkipati, Dynamic of Railway Vehicle Systems, Academic Press Canada, 1984

Environmental Vibrations – Takemiya (ed.)
© *2005 Taylor & Francis Group, London, ISBN 0 415 39035 4*

Analysis of railway continuous bridges subjected to running trains and non-uniform seismic excitations

Y. Han, H. Xia & N. Zhang
School of Civil Engineering & Architecture, Beijing Jiaotong University, Beijing, China

ABSTRACT: Based on the theory of dynamic interactions between the wheel and rail, the dynamic model of coupled train-bridge system subjected to earthquakes is established, in which the non-uniform characteristics of the seismic waves input from different foundations are considered. A computer simulation program is worked out. The case study is performed to a continuous bridge scheme on the Beijing-Shanghai high-speed railway in China. By inputting different typical seismic waves to the train-bridge system, the whole histories of the trains running through the bridge are simulated and the dynamic responses of the bridge and the vehicles are calculated. The effects of different traveling speeds and phases of the seismic waves on the dynamic responses of the bridge-vehicle system are studied, and many useful results are concluded.

1 INTRODUCTION

The dynamic behavior of train-bridge coupling system and the bridge seismic design problem have been separately studied by many researchers in China and abroad, and many useful results published (Diana 1989, Xu 1993, Wang 1994, Frýba 1999, Xia 2002, Yang 2002, Han 2004). However, the dynamic responses of train-bridge system under earthquakes were not attached importance to until the bridges are brought into intensive vibration with the development of the high-speed railway, where the viaducts stretching tens of kilometers are commonly built to guarantee the smoothness of the track and the safety and stability of the running trains, and thus the probability of the trains on the bridges when an earthquake occurs increases greatly. As a linear spatial structure, the seismic waves input to different foundations of the long-span bridge are not the same. The longitudinal phase differences of the seismic waves may lead to the out-of-phase vibration between the bridge piers and thus increase the dangers of beam dropping. The lateral phase differences of the seismic waves may lead to the out-of-phase displacements of two adjacent piers and thus directly influence the running safety of the train on the bridge. Up to now, there have not been literatures published on the train-bridge vibrations with the seismic traveling wave effect taken into account.

The PC continuous bridges have the merit of great stiffness, convenient construction, low noises and simple maintenance, which make it most commonly used on high-speed railways besides the simply supported bridges. The case study in this paper is made of a PC continuous bridge scheme that will cross the Huaihe River on the Beijing-Shanghai high speed railway in China. By establishing the dynamic model of coupled train-bridge system subjected to earthquakes and inputting typical seismic waves to the train-bridge system, the whole histories of the train running through the bridge are simulated and the dynamic responses of the bridge and the vehicles are calculated. The influences of seismic traveling wave effect on performance of vehicles and bridge undergoing earthquake are studied.

2 DYNAMIC MODEL OF TRAIN-BRIDGE SYSTEM UNDER EARTHQUAKES

The analysis model of train-bridge system subjected to earthquakes can be regarded as a united big spatial dynamic system composed of the vehicle subsystem and the bridge subsystem acting on with each other through the wheel/rail contact interfaces. The track irregularities are considered as the self-excitation source and the seismic loads on the train-bridge system as the external excitation input from different foundations of the bridge, see Figure 1.

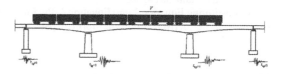

Figure 1. Model of train-bridge system subjected to earthquake.

Theoretically, the seismic movements of the bridge foundations excite the superstructure of the bridge to vibrate through the influence matrix, and act on the vehicles through the dynamic interaction relations between the wheel and the rail, and inversely the vehicle vibration affects the vibrations of the bridge structures.

For bridges stretching very long distances, the seismic waves input to different foundations differ much from one another, and thus can not be considered as the identical excitations in analysis. By separating the whole bridge subsystem into the supporting part and the non-supporting part, the dynamic equations of the bridge subjected to earthquakes can be expressed as (Xu 1993):

$$
\begin{bmatrix} \mathbf{M}_{ss} & \mathbf{M}_{sb} \\ \mathbf{M}_{bs} & \mathbf{M}_{bb} \end{bmatrix} \begin{Bmatrix} \ddot{\mathbf{X}}_s \\ \ddot{\mathbf{X}}_b \end{Bmatrix} + \begin{bmatrix} \mathbf{C}_{ss} & \mathbf{C}_{sb} \\ \mathbf{C}_{bs} & \mathbf{C}_{bb} \end{bmatrix} \begin{Bmatrix} \dot{\mathbf{X}}_s \\ \dot{\mathbf{X}}_b \end{Bmatrix} +
$$

$$
\begin{bmatrix} \mathbf{K}_{ss} & \mathbf{K}_{sb} \\ \mathbf{K}_{bs} & \mathbf{K}_{bb} \end{bmatrix} \begin{Bmatrix} \mathbf{X}_s \\ \mathbf{X}_b \end{Bmatrix} = \begin{Bmatrix} 0 \\ \mathbf{F}_b \end{Bmatrix} \tag{1}
$$

where $\mathbf{M}, \mathbf{C}, \mathbf{K}$ = mass, damping and stiffness matrices of the bridge, $\ddot{\mathbf{X}}, \dot{\mathbf{X}}, \mathbf{X}$ = absolute acceleration, velocity and displacement vector of the bridge, with the subscripts ss, bb and sb standing for the non-supporting part, supporting part and the interacting part, respectively; $\mathbf{F}_b$ = reaction force vector of the supporting part.

Based on the concept of pseudo-static displacement, the total displacement of the bridge under multi-excitations consists of two parts, the pseudo-static displacement and the dynamic response displacement. So the bridge total displacement vector can be expressed as follows:

$$
\{\mathbf{X}\} = \begin{Bmatrix} \mathbf{X}_s \\ \mathbf{X}_b \end{Bmatrix} = \begin{Bmatrix} \mathbf{X}_s^p \\ \mathbf{X}_{bg} \end{Bmatrix} + \begin{Bmatrix} \mathbf{X}_s^d \\ 0 \end{Bmatrix} \tag{2}
$$

where the superscripts p and d represent the pseudo-static response component and the dynamic response component, respectively. $\mathbf{X}_{bg}$ = movement of the bridge supporting foundations, which is a known quantity to a certain earthquake.

By substituting Equation 2 into Equation 1, the following can be derived from the former part of Equation 1:

$$
\mathbf{M}_{ss}(\ddot{\mathbf{X}}_s^d + \ddot{\mathbf{X}}_s^p) + \mathbf{M}_{sb}\ddot{\mathbf{X}}_{bg} + \mathbf{C}_{ss}(\dot{\mathbf{X}}_s^d + \dot{\mathbf{X}}_s^p) +
$$

$$
\mathbf{C}_{sb}\dot{\mathbf{X}}_{bg} + \mathbf{K}_{ss}(\mathbf{X}_s^p + \mathbf{X}_s^d) + \mathbf{K}_{sb}\mathbf{X}_{bg} = 0 \tag{3}
$$

For a structure with only the static displacement at the supporting points and without bearing any other loads, the static equilibrium equations can be written as:

$$
\mathbf{K}_{ss}\mathbf{X}_s^p + \mathbf{K}_{sb}\mathbf{X}_{bg} = 0 \tag{4}
$$

Let $\mathbf{R}_{sb} = -\mathbf{K}^{-1}_{ss}\mathbf{K}_{sb}$ denote the pseudo-static displacement influence matrix. $\mathbf{R}_{sb}$ is the displacement vector of the rest nodes of the bridge induced by one unit displacement of a certain freedom at the structure/soil contacting node. Thus the following expression can be obtained:

$$
\mathbf{X}_s^p = \mathbf{R}_{sb}\mathbf{X}_{bg} \tag{5}
$$

Thus Equation 3 can be expressed as:

$$
\mathbf{M}_{ss}\ddot{\mathbf{X}}_s^d + \mathbf{C}_{ss}\dot{\mathbf{X}}_s^d + \mathbf{K}_{ss}\mathbf{X}_s^d =
$$

$$
-(\mathbf{M}_{ss}\mathbf{R}_{sb} + \mathbf{M}_{sb})\ddot{\mathbf{X}}_{bg} - (\mathbf{C}_{ss}\mathbf{R}_{sb} + \mathbf{C}_{sb})\dot{\mathbf{X}}_{bg} \tag{6}
$$

For common civil engineering structures, the contribution of the damping force caused by the velocity of the supporting node is very small and can be neglected from the right side of Equation 6. Then the dynamic differential equations can be written as:

$$
\mathbf{M}_{ss}\ddot{\mathbf{X}}_s^d + \mathbf{C}_{ss}\dot{\mathbf{X}}_s^d + \mathbf{K}_{ss}\mathbf{X}_s^d = -(\mathbf{M}_{ss}\mathbf{R}_{sb} + \mathbf{M}_{sb})\ddot{\mathbf{X}}_{bg} \tag{7}
$$

When lumped masses are adopted, $\mathbf{M}_{sb}$ is a zero matrix and Equation 7 can be simplified as:

$$
\mathbf{M}_{ss}\ddot{\mathbf{X}}_s^d + \mathbf{C}_{ss}\dot{\mathbf{X}}_s^d + \mathbf{K}_{ss}\mathbf{X}_s^d = -\mathbf{M}_{ss}\mathbf{R}_{sb}\ddot{\mathbf{X}}_{bg} \tag{8}
$$

The seismic traveling wave effects can be considered by differencing the time intervals from the same seismic acceleration record to form the ground acceleration vector $\ddot{\mathbf{X}}_{bg}$ in Equation 8.

The train model is composed of several locomotives and cars. Each vehicle is a complicated MDOF vibration system consisting of car body, bogies, wheel-sets, springs and dashpots. The car body, bogies and wheel-sets in each vehicle are regarded as rigid components. Linear springs and viscous dashpots are used to represent the connections between the car body and bogies, and so as between the bogie and wheel-sets. For each car body or bogie, five degrees of freedom are considered, which are the lateral, rolling, yawing, vertical and pitching movements. Each wheel-set has four degrees of freedom, in lateral, rolling, yawing and vertical directions. Thus for each 2-bogie 4-axle vehicle, the total degrees of freedom are 31 in the study.

By arranging the 31 freedoms of a vehicle in sequence, the dynamic equation of the whole train can be expressed in the matrix form:

$$[\mathbf{M}]\{\ddot{\mathbf{V}}\}+[\mathbf{C}]\{\dot{\mathbf{V}}\}+[\mathbf{K}]\{\mathbf{V}\}=\{\mathbf{F}\} \qquad (9)$$

where $\mathbf{M}$, $\mathbf{C}$, $\mathbf{K}$ = mass, damping and stiffness matrices of the train; $\mathbf{V}$, $\mathbf{F}$ = the displacement and force vector of the train, respectively.

The dynamic interaction between the wheel and rail is the tache of the train-bridge coupled system. In the determination of the wheel/rail forces, the normal interaction is treated according to the nonlinear elastic Hertz contact theory, while the tangent interaction is treated firstly by the Kalker linear theory under small creep ratios and then corrected by the Shen-Hedrick-Elkins nonlinear theory, to make it applicable to the conditions with large creep ratios (Zhai 2002).

According to the nonlinear elastic Hertz contact theory, the normal interaction force between the wheel and rail can be expressed as:

$$N_z(t)=\left[\frac{1}{G}\delta Z(t)\right]^{\frac{3}{2}} \qquad (10)$$

where $\delta Z(t)$ = the compressed quantity between the wheel and rail, G = wheel/rail contact constant.

According to the Kalker theory, the tangent interaction force between the wheel and rail can be expressed as:

$$\begin{cases} F_x = -f_{11}\xi_x \\ F_y = -f_{22}\xi_y - f_{23}\xi_{sp} \\ M_z = f_{23}\xi_y - f_{33}\xi_{sp} \end{cases} \qquad (11)$$

where F_x, F_y = creep forces in longitudinal and lateral direction, M_z = rolling creep movement; f_{11}, f_{12} = longitudinal and lateral creep factor, f_{23} = rotation/lateral-displacement creep factor; f_{33} = rolling creep factor.

The final creep force/movement corrected by the Shen-Hedrick-Elkins theory is expressed as:

$$\begin{cases} F_x' = \varepsilon F_x \\ F_y' = \varepsilon F_y \\ M_z' = \varepsilon M_z \end{cases} \qquad (12)$$

where ε = correcting factor.

When Equation 10 and Equation 12 are applied to calculate the wheel/rail interaction forces, the wheel/rail contact point and the corresponding contact geometry parameters must be determined firstly, which can be obtained by adopting the spatial dynamic interaction relationship model. By introducing the measured

wheel/rail section profile data and figuring them in the fixed coordinate systems with the three-order spline approximation, the coordinates of the wheel/rail contact point can be fount out by computing the minimum perpendicular distance of the left or right wheel/rail at certain time, and then the necessary wheel/rail contact geometry parameters can be determined. The wheel/rail contact force can be worked out in this way. The details of the calculating process can be found in References (Han 2005, Zhai 2002).

By combining the bridge subsystem Equation 1 and the vehicle subsystem Equation 2, together with the wheel/rail contact forces (Equations 10 and 12), the dynamic equations of the coupled train-bridge system subjected to earthquakes can be obtained.

The deformation of the high-speed railway bridge during frequent earthquakes can be approximately considered within the elastic limitation. So the modal comprehension analysis method is adopted for modeling the bridge subsystem. First, the free vibration frequencies and modes of the system are solved. Upon the orthogonality of the modes, the FEM equations coupled with each other can then be decoupled, which makes the bridge model become the superposition of independent modal equations. Owing to the fact that the dynamic response of a structure is dominantly influenced by its several lowest modes, this method has a very great advantage that an adequate estimation on the dynamic response can be obtained by considering only a few modes of vibration, and then the computational effort can be significantly reduced (Xia 2000).

Therefore, the vehicle equations are combined with the modal equations, instead of the direct finite element equations of the bridge.

When the cross section deformations of the bridge deck are taken into account in its vibration modes, its movement at any section can be determined by the lateral movement Y_s, torsional movement θ_s and vertical movement Z_s, and can be expressed by the superposition of the modes as:

$$Y_s(x_{ijk}) = \sum_{n=1}^{N_q} q_n \phi_h^n(x_{ijk})$$

$$\theta_s(x_{ijk}) = \sum_{n=1}^{N_q} q_n \phi_\theta^n(x_{ijk}) \qquad (13)$$

$$Z_s(x_{ijk}) = \sum_{n=1}^{N_q} q_n \phi_v^n(x_{ijk})$$

where $\phi_h^n(x_{ijk})$, $\phi_\theta^n(x_{ijk})$, $\phi_v^n(x_{ijk})$ = normalized lateral, rotational and vertical modal functions of the nth modal at the position of the kth wheel set of the jth bogie of the ith vehicle on the bridge. q_n is the modal coordinate and N_q = the number of modes concerned in the analysis.

Based on the assumptions that there is no relative displacement between the track and bridge deck, the dynamic equations of the coupled train-bridge system subjected to earthquakes can be expressed as:

$$\begin{bmatrix} \mathbf{M}_v & \mathbf{0} \\ \mathbf{0} & \mathbf{M}_b \end{bmatrix} \begin{Bmatrix} \ddot{\mathbf{X}}_v \\ \ddot{\mathbf{X}}_b \end{Bmatrix} + \begin{bmatrix} \mathbf{C}_v & \mathbf{0} \\ \mathbf{0} & \mathbf{C}_b \end{bmatrix} \begin{Bmatrix} \dot{\mathbf{X}}_v \\ \dot{\mathbf{X}}_b \end{Bmatrix} + \begin{bmatrix} \mathbf{K}_v & \mathbf{0} \\ \mathbf{0} & \mathbf{K}_b \end{bmatrix} \begin{Bmatrix} \mathbf{X}_v \\ \mathbf{X}_b \end{Bmatrix} = \begin{Bmatrix} \mathbf{F}_v \\ \mathbf{F}_b \end{Bmatrix} \qquad (14)$$

where $\mathbf{M, C, K}$ = mass, damping and stiffness matrices of the train and the bridge, with the subscripts v and b representing vehicles and bridge, respectively. The details of the mass, damping, stiffness matrices and displacement vector are the same as the corresponding ones without earthquake excitations, and can be found in Reference (Han 2005). $\mathbf{F}_v$ = wheel/rail contact forces acting on the wheel set, determined by the relative displacement between the wheel and the rail. $\mathbf{F}_b$ = generalized seismic force vector acting on the non-supporting parts of bridge, consisting of the generalized wheel/rail contact forces vector $\mathbf{F}_{bv}$ and the generalized seismic forces vector $\mathbf{F}_{bs}$ transmitted by the influence matrix from the supporting parts of bridge.

Suppose there are N_v vehicles traveling on the bridge when an earthquake takes place, the vector of seismic forces acting on the vehicle $\mathbf{F}_v$ can be expressed as:

$$\mathbf{F}_v = [\mathbf{F}_{v1} \quad \mathbf{F}_{v2} \quad \cdots \quad \mathbf{F}_{vN_v}]^T \qquad (15)$$

where $\mathbf{F}_{vi}$ = seismic force vector of the ith vehicle. It can be expressed as:

$$\mathbf{F}_{vi} = [0 \quad 0 \quad 0 \quad \mathbf{F}_{w_{11}ci} \quad \mathbf{F}_{w_{12}ci} \quad \mathbf{F}_{w_{21}ci} \quad \mathbf{F}_{w_{22}ci}]^T \qquad (16)$$

where $\mathbf{F}_{wjkci}$ = the wheel/rail force vector of the kth wheel set of the jth bogie in the ith vehicle on the bridge, which can be expressed as:

$$\mathbf{F}_{w_{jk}ci} = \begin{Bmatrix} -F_{Lyijk} - F_{Ryijk} - N_{Lyijk} - N_{Ryijk} \\ a_{0iL}(F_{Lzijk} + N_{Lzijk}) - a_{0iR}(F_{Rzijk} + N_{Rzijk}) - \\ r_{Lijk}(F_{Lyijk} + N_{Lyijk}) - r_{Rijk}(F_{Ryijk} + N_{Ryijk}) \\ a_{0iL}(F_{Lxijk} + N_{Lxijk}) - a_{0iR}(F_{Rxijk} + N_{Rxijk}) + \\ M_{Lzijk} + M_{Rzijk} + \psi_{wijk}[a_{0iL}(F_{Lyijk} + N_{Lyijk}) \\ -a_{0iR}(F_{Ryijk} + N_{Ryijk})] \\ -F_{Lzijk} - F_{Rzijk} - N_{Lzijk} - N_{Rzijk} \end{Bmatrix}$$
$$(j = 1 \sim 2, k = 1 \sim 2) \qquad (17)$$

where N, F = normal and tangent interaction forces between the kth wheel set of the jth bogie in the ith vehicle respectively; L, R = interaction forces on the left or right side of the corresponding wheel set; y, z denote the coordinate axis of the fixed coordinate system; a_{0iL}, a_{0iR} = lateral projection of the distance from the ith left/right wheel/rail contact point to the cancroids of the wheel set, respectively.

Let N_q denotes the mode number of the finite element model of the bridge concerned in the analysis, the seismic force vector of bridge mode can be expressed as:

$$\mathbf{F}_b = [F_{b1} \quad F_{b2} \quad \cdots \quad F_{bN_q}]^T \qquad (18)$$

where $F_{bn}(n = 1, 2, ..., N_q)$ = seismic force sub-vector of the nth mode, consisting of the generalized wheel/rail contact forces F_{nbv} and the generalized seismic forces F_{nbs} transmitted by the influence matrix from the supporting parts of the bridge, which can be expressed as:

$$F_{bn} = F_{nbs} + F_{nbv} \qquad (19)$$

where:

$$F_{nbs} = \mathbf{\Phi}_n^T \mathbf{M}_{ss} \mathbf{R}_{sb} \ddot{\mathbf{X}}_{bg} \qquad (20)$$

where $\mathbf{\Phi}_n^T$ = transposition of the nth mode shape of all non-supporting nodes of the bridge; $\mathbf{M}_{ss}$ = mass matrix of the corresponding nodes; $\mathbf{R}_{sb}$ = pseudo-static displacement influence matrix of the bridge supporting nodes on non-supporting nodes; $\mathbf{X}_{bg}$ = ground acceleration vector input from the bridge supporting points.

The lateral and vertical seismic excitations are considered in the analysis. When the lumped mass matrix is adopted in the bridge model and the multi-point excitations are considered, the following expression can be found:

$$F_{nbs} = \sum_{i=1}^{N_s} \sum_{j=1}^{N_b} m_{ssi} r_{sbij} (\phi_{hi}^n \ddot{X}_{bghj} + \phi_{vi}^n \ddot{X}_{bgvj}) \qquad (21)$$
$$(n = 1, 2, \cdots, N_q)$$

where N_s, N_b = node number of the non-supporting parts and supporting parts in the bridge finite element model, respectively. m_{ssi} = mass of the ith node of the bridge non-supporting parts. ϕ_{hi}^n, ϕ_{vi}^n = values of the nth bridge mode at the position of the ith node of the bridge non-supporting parts. $\ddot{X}_{bghj}, \ddot{X}_{bgvj}$ = lateral and vertical accelerations component of the earthquake input from the jth bridge supporting node, respectively: r_{sbij} = the effect matrix element of the jth supporting node on ith non-supporting nodes.

The generalized wheel/rail force acting on bridge can be written as:

$$F_{nbv} = \sum_{i=1}^{N_v} \sum_{j=1}^{2} \sum_{k=1}^{2} F_{nijk} \qquad (22)$$

where F_{nijk} = generalized wheel/rail force of the kth wheel set and the jth bogie in the ith vehicle on the bridge deck. When the wheel deviates from the rail, F_{nijk} equals to zero; and when the wheel keep contact with the rail, F_{nijk} can be calculated as:

$$F_{nijk} = \phi_h^n(x_{ijk})(N_{Lyijk} + F_{Lyijk} + N_{Ryijk} + F_{Ryijk})$$
$$+ \left[\phi_v^n(x_{ijk}) + e\phi_\theta^n(x_{yk})\right]\left(N_{Lzijk} + F_{Lzijk} +\right.$$
$$N_{Rzijk} + F_{Rzijk}) + \phi_\theta^n(x_{yk})[-a_{0iL}(N_{Lzijk} +$$
$$F_{Lzijk}) + a_{0iR}(N_{Rzijk} + F_{Rzijk}) + (N_{Lyijk} +$$
$$F_{Lyijk} + N_{Ryijk} + F_{Ryijk})h_s] \qquad (23)$$

where e = distance from the vehicle mass center to the bridge center of gravity; h_s = the distance from the bridge center of gravity to the bridge deck.

When the loads on the vehicle subsystem and the bridge subsystem become known quantities, Equation 14 can be solved by the step-by-step integration method. In each time step, the response of the vehicle and the bridge are calculated separately, iterative processes are adopted to meet the convergence of all generalized displacement of the two subsystems. At each step, a displacement vector of the bridge and the vehicle at the wheel/rail contact point is assumed first and the wheel/rail contact forces are calculated, from which the new displacement vector can be obtained by solving the equations. When the error between the solved displacement and the assumed one is within the given allowance, the solved displacement in this time step is taken as true and the next iterative step goes on.

A computer program of the train-bridge coupled system subjected to earthquakes is worked out based on the formulation described above and is used to perform a case study.

3 CASE STUDY

3.1 *Bridge description and calculation conditions*

The case study concerns a scheme of the bridge across the Huaihe River on the Beijing-Shanghai High-speed Railway in China. The bridge consists of the continuous PC box girders with seven spans of $48 + 5 \times 80 + 48$ m as a unit. The girders are of one-box-one-chamber varying sections, which is 6.1 m

high at the support and 3.8 m at the mid span. The piers are $16\sim22$ m high, with round-end sections and boring piles. On the third pier are mounted with fixed-basin type rubber bearings, the others are moveable bearings.

The software ANSYS is used in establishing the finite element model of the bridge, where the girders and piers are all dispersed by Beam 4 element, the secondary loads of the bridge are distributed on the girders as the additional mass. The connections between the girders and the piers are dealt with main-and-appended freedoms, where the lateral displacement of the girder is assumed as the same as the pier top. The rotational angle of the girder about vertical direction varied with the bearing types, the same at the fixed bearings and different at the moveable bearings. The interactions between the structure foundations and the ground soils are modeled by Matrix 27 element.

Altogether 60 natural frequencies and the mode shapes of the bridge are analyzed. The first vertical mode of the girders is 1.24 Hz and the first lateral mode is 2.33 Hz.

Suppose the bridge site belongs to the second soil field. Two representative seismic acceleration records are selected for study, with their properties shown in Table 1. Three traveling speeds of seismic wave in the ground soil, 400 m/s, 500 m/s and 600 m/s, are used in the simulation. By normalizing the maximum acceleration as 0.1 g in lateral direction and 0.05 g in vertical, the seismic waves are input along the lateral and vertical directions from the bridge foundations at the same time. For the calculated $48 + 5 \times 80 + 48$ m continuous bridge, the maximum time delays of the seismic waves arriving at all piers are 2.76 s, 1.98 s and 1.38 s, respectively.

In this study, the high-speed train made in China called China-Star was used, which is composed of 1 locomotive +9 passenger cars +1 locomotive, with the axle load being 195 kN each for the locomotive and 142.5 kN each for the car. The calculation train speeds are 120, 160, 200, 240, 280 and 320 km/h. The track irregularities measured from the Qin-Shen Special Passenger Railway in China are taken in the calculation.

Suppose the earthquake takes place just when the train moves onto the bridge. The starting time of the

Table 1. Properties of the measured seismic wave records.

Name	Component	Magnitude	Max. acc. cm/s^2	Time
Kern	TAF111	7.4	177.8	1952.7.21
County	TAF-UP		108.7	
North-	UCL360	6.7	473.8	1994.1.17
ridge	UCL-UP		265.0	

/

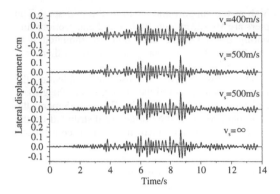

Figure 2. Lateral displacements of bridge at middle side-span.

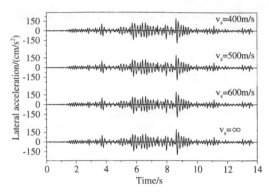

Figure 3. Lateral accelerations of bridge at middle side-span.

input seismic waves are controlled in the calculation to assure that the maximum seismic acceleration occurs during the train passing on the bridge. The integral step is taken as 0.00005 s and the damping ratio as 4%.

The running safety of the train is the most important index. The derail factors and the offload factors giving in this study are calculated in the condition that the wheels do not deviate from the rail, i.e., the deviating time of the wheel is less than 0.015 s and the elevation height less than 0.1 cm.

3.2 Dynamic responses of train-bridge system under seismic loads

The vibration histories of the bridge at the side-span and the first vehicle when the train runs on the bridge at a constant speed of 200 km/h under the second seismic wave of different traveling speeds are presented herein, see Figure 2~Figure 7. In all of the figures, v_s is the traveling speed of seismic waves, and $v_s=\infty$ denotes the uniform excitation.

The results show that the vibration patterns of the vehicles and the bridge are not the same under the different seismic waves at the same ground soil field. For the dynamic response of the bridge at middle side-span under the same excitation, the time histories of vertical deflection are almost the same under different seismic traveling wave excitations, and so do the lateral displacement. The bridge vibrates more intense with the increase of the seismic traveling speed. The time history shapes of the lateral and vertical accelerations of the bridge change with the seismic traveling speed, reaching the maximum value under the uniform excitations. Under the same seismic excitations but with different traveling speeds, the time history shapes of the lateral displacements and accelerations of the bridge are similar, the maximum values are almost the same.

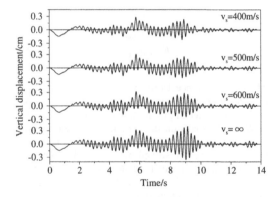

Figure 4. Vertical displacements of bridge at middle side-span.

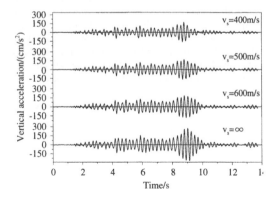

Figure 5. Vertical accelerations of bridge at middle side-span.

The maximum lateral response of the continuous bridge at middle side-spans, both the lateral displacement and lateral acceleration are much greater than the corresponding ones at middle center-spans. Under the same seismic excitation, as the response of the

216

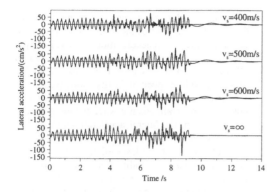

Figure 6.　Lateral accelerations of the first car body.

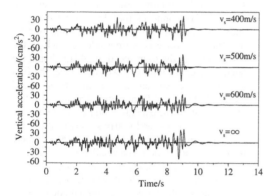

Figure 7.　Vertical accelerations of the first car body.

vehicles such as the lateral and vertical acceleration concerned, the time history shapes varied with the seismic traveling speeds, and differed most from the uniform excitations.

Figure 8～Figure 13 are the distributions of the maximum lateral responses of the bridge and the vehicles with respect to the train speed and the traveling speed of the first seismic wave.

It is clear from Figure 8～Figure 13 that the traveling speed of seismic wave has great influences on the dynamic responses of the bridge and the vehicles. These influences are different on different responses of the bridge and the vehicles.

(1) For the dynamic responses of the bridge, the lateral displacements and accelerations slightly change with the seismic wave traveling speed, lower speed tends to induce greater displacements but smaller accelerations. The dynamic responses of the bridge such as the lateral displacement and acceleration changed versus the train speed and all appear peak values at the train speed of 280 km/h under different seismic wave traveling speeds.

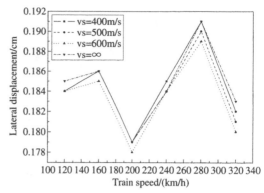

Figure 8.　Maximum bridge displacements versus train speed.

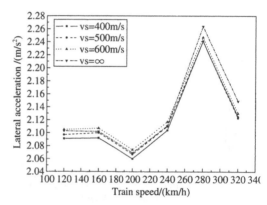

Figure 9.　Maximum bridge accelerations versus train speed.

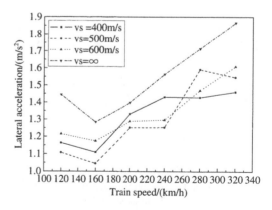

Figure 10.　Maximum vehicle accelerations versus train speed.

(2) For the vehicle vibrations, the accelerations of the car body are greatly influenced by the seismic traveling speed, but with no obvious regulation. For example, when the train runs at a speed of

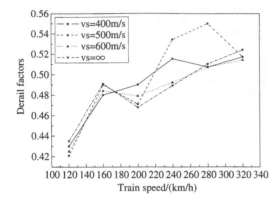

Figure 11. Maximum derail factors versus train speed.

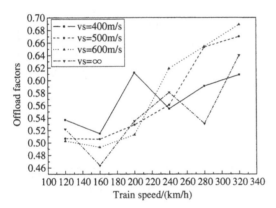

Figure 12. Maximum offload factors versus train speed.

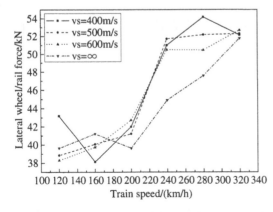

Figure 13. Maximum wheel/rail forces versus train speed.

200 km/h, the lateral acceleration of the car body is 1.289 m/s² under the seismic traveling speed of 600 m/s, which is greater than 1.252 m/s² under the seismic traveling speed of 500 m/s and less than

1.331 m/s² under the seismic traveling speed of 400 m/s; however, when the train runs at a speed of 280 km/h, the lateral acceleration of the car body is 1.469 m/s² under the seismic traveling speed of 600 m/s, which is less than 1.591 m/s² under the seismic traveling speed of 500 m/s and greater than 1.428 m/s² under the seismic traveling speed of 400 m/s. Under all seismic wave traveling speeds in calculation, the lateral accelerations of the car body tend to increase versus the train speed.

(3) The running safety indexes of the train vehicles under earthquakes, including derail factors, offload factors and lateral wheel/rail forces, all increase with the train speed. Under the certain train speed, no obvious laws can be discovered in the influences of the same seismic excitation of different traveling speeds on the running safety indexes of vehicles. The calculated indexes may be estimated apparently smaller than the real ones if the effect of seismic traveling wave is not considered, such as the derail factor at the train speed of 240 km/h and the lateral wheel/rail forces at 280 km/h under the first seismic wave excitation. At the train speed 240 km/h, the derail factors are calculated as 0.581 under the uniform excitation and 0.619 under the seismic traveling speed 600 m/s. Obviously, if the allowance in the Chinese Code GB5599-85 where the offload factor should be smaller than 0.6 is used to assess the train's safety, an error conclusion may be made in the case without considering the effect of seismic traveling wave, and thus a hidden trouble be left over the running safety of trains.

Therefore, the effect of seismic traveling wave should be considered in evaluating the running safety of trains on the bridge during earthquakes.

4 CONCLUSIONS

In this study, the dynamic model of coupled train-bridge system subjected to earthquakes is established by connecting the motion equations of the bridge under earthquakes and those of the vehicles through the nonlinear contact relations between the wheels and rails. The seismic loads are imposed on the bridge by the influence matrix and act on the vehicles through the dynamic wheel/rail interaction relations. By taking the scheme of a continuous PC box girder bridge on the Beijing-Shanghai High-speed Railway as the case study, the whole histories of the train running on the bridge under earthquakes of different traveling speeds are simulated, and the dynamic responses of the train and the bridge are calculated. Some conclusions can be obtained from the study:

(1) The effect of seismic traveling waves have great influences on the dynamic responses of the

218

train-bridge coupled system. The lack of considering the traveling wave effect in calculation may result in unsafe conclusions. It is not always true that the faster the seismic wave traveling speed is, the closer the dynamic response of the train-bridge coupled system is to the corresponding ones under the uniform seismic excitations. Therefore, the seismic traveling wave speed should be considered according to the property of the soil at the bridge site in the calculation of the dynamic response of train-bridge system subjected to earthquakes.

(2) For the dynamic response of train vehicles, the lateral car body accelerations, derail factors, offload factors and lateral wheel/rail forces all increase with the train speed. Therefore the influences of train speed must be taken into account in evaluating the runnability of vehicles on the bridge during earthquakes.

(3) The dynamic analysis of the coupled train-bridge system subjected to earthquakes is quite complicated. The analysis model and the calculation results proposed in this paper may provide a reference for the dynamic design of railway bridges.

ACKNOWLEDGEMENTS

This study is sponsored by the National Natural Scientific Foundation of China (No. 50478059) and the Key Research Foundation of Beijing Jiaotong University (No. 2004SZ005).

REFERENCES

Diana, G. & Cheli, F. 1989. Dynamic interaction of railway systems with large bridges. *Vehicle System Dynamics* 18: 71–106.

Fr"ba, L. 1999. *Vibration of Solids and Structures under Moving Loads.* London: Thomas Telford.

Han, Y. & Xia, H. 2004. Running safety of trains on half-through arch bridge during earthquakes. *Progress in Safety Science & Technology*, Vol. IV: 1945–1951.

Han, Y. 2005. *Dynamic Response of High-speed Railway Bridges and Running Safety of Vehicles during Earthquakes.* Ph.D Thesis, Beijing Jiaotong University.

Wang, F.T. 1994. *Dynamics of Vehicle System.* Beijing: China Railway Press: 106–123.

Xia, H. & Xu, Y.L. 2000. Dynamic interaction of long suspension bridges with running trains. *Sound & Vibration* 237(2): 263–280.

Xia, H. 2002. *Dynamic Interaction between Vehicles and Structures.* Beijing: Science Press: 118–134.

Xu, Zh.X. & Hu, Z.L. 1993. *Analysis of Structures under Earthquakes.* Beijing: Superior Education Press: 80–84.

Yang, Y.B. 2002. *Theory of Vehicle-Bridge Interaction for High Speed Railway.* Taibei: DNE Publisher.

Zhai, W.M. 2002. *Vehicle-track Coupling Dynamics.* Beijing: China Railway Press: 62–78.

Environmental Vibrations – Takemiya (ed.)
© *2005 Taylor & Francis Group, London, ISBN 0 415 39035 4*

Study on dynamic characteristic of the EMS maglev guideway

J. Shi, Q.C. Wei & Y.W. Feng
Beijing Jiaotong University, Beijing, China

ABSTRACT: Maglev train is a new type of ground tracked transport tool, which advantages are attractive and competitive in the 21st century. The maglev demonstration and operation line in Pudong Shanghai has been built successfully with the cooperation between China and Germany. Dynamic simulation is one of the key links in maglev technology research, has a very important reference meaning on promoting maglev technology. Firstly, this paper investigates the dynamic characteristics of magnet/rail relationship under the active control considering the mechanical-electromagnetic coupling field and suspension control system. Subsequently an integrated dynamic model of maglev vehicle-guideway coupling system is established based on TR maglev vehicle and the elevated-guideway technology. The response basic laws of different guideway structure are analyzed; Additionally, The response characteristics of random excitation of guideway are simulated. It is on this basis that the suggestions about the design of magnetic levitation guideway and the measure to reduce vibration are made for safe operation.

1 INTRODUCTION

A high-speed ground transportation system, based on maglev train propelled by a linear electric motor, has been proposed to meet future intercity transportation requirements. The maglev system will offer the advantages of lower noise and emissions and better ride quality relative to conventional rail systems. Research and development in maglev technology is being actively pursued in Europe, Japan, and the U.S.A. Nowadays, Transrapid in German, MLX and HSST system in Japan are ready to the revenue application, moreover, Shanghai high-speed maglev railway adopting the German Transrapid technology has been successfully running, it is therefore reasonable to expect that maglev systems may indeed be a key transportation mode in the 21st century.

The dynamic responses of maglev ground transportation system have important consequents for system safety and stability. As maglev train speeds increase to 300–500 km/h, or as guideways, especially those made of steel, become lighter and more flexible, maglev system may be susceptible to dynamic instability and unacceptable vibration.

Firstly, this paper investigates the dynamic characteristics of magnet/rail relationship under the active control considering the mechanical-electromagnetic coupling field and suspension control system. Subsequently an integrated dynamic model of maglev vehicle-guideway coupling system is established based on TR maglev vehicle and the elevated-guideway technology. The response basic laws of different

structure guideway are analyzed, the limit values of natural frequencies of guideway are determined; Additionally, The response characteristics of random excitation of guideway are simulated. It is on this basis that the suggestions about the design of magnetic levitation guideway and the measure to reduce vibration are made for safe operation.

2 DYNAMIC MODEL

2.1 *The electromagnetic force model*

For the electromagnetic suspension system in Figure 1. $z(t)$ may be taken as the absolute position of the suspended magnet with $\ddot{z}(t)$, being its absolute vertical acceleration at any instance of time t. $\delta(t)$ represent magnetic air gap. $v(t)$ is the displacement of

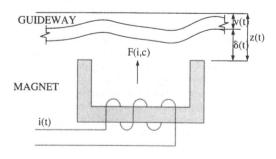

Figure 1. Electromagnetic suspension system.

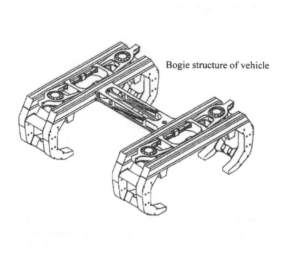

Bogie structure of vehicle

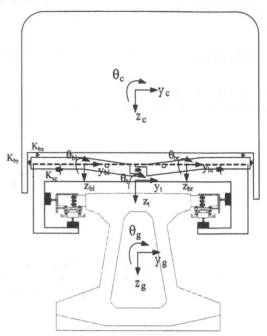

guideway, with these assumptions, the dynamics of the suspension system is described by

$$u(t) = Ri(t) + \frac{d}{dt}\left[L(x,i) \cdot i(t)\right]$$

$$= Ri(t) + \frac{\mu_0 N^2 A}{2c(t)}\frac{di(t)}{dt} - \frac{\mu_0 N^2 Ai(t)}{2\left[c(t)\right]^2}\frac{dc(t)}{dt} \quad (1)$$

$$m\frac{d^2 z(t)}{dt^2} = mg + f_d(t) - F(i,c) \quad (2)$$

$$F(i,\delta) = \Gamma\left[\frac{i(t)}{\delta(t)}\right]^2 \quad (3)$$

Where $\Gamma = \mu_o N^2 A/4$ is a characteristic feature of the suspension magnet, N with the number of turns in the magnet winding and A as the pole face area; R is the total resistance of electrical circuit, $f_d(t)$ is vertical disturbance force, $F(i,\delta)$ is electromagnetic force, mg is the total weight of suspend object, and $u(t)$ is the input voltage to magnet amplifier.

Although various levitation and control strategies and methods have been put forward, the decentralized hierarchical control concept is widely adopted in practice. The current control laws can be given as:

$$\Delta i = k_p \Delta\delta + k_v \Delta\dot{\delta} + k_p \ddot{z} \quad (4)$$

where $\Delta\delta$ represents the air gap change. kp, kv and ka are the feedback gains corresponding to of the change, the velocity and the acceleration of the magnet.

2.2 Vehicle model

The dynamic model of train with secondary spring multi-body and multi-DOF is considered when analysis of vehicle-guideway coupling vibration is performed. The car body is rigid and has a uniform mass. The center of mass is consistent with that of the moment of the inertia. Figure 2 illustrates the vehicle model adopted in this study. It consists of a vehicle body, eight C-shaped frames, eight pieces of detached bolsters, and seven magnets. The spring and the shock absorber in the primary suspension are characterized by spring sti_ness and damping coe_cient respectively. Likewise, the secondary suspension is characterized by spring sti_ness and damping coe_cient respectively.

A 117-DOF spatial model of maglev vehicle is established considering vertical, lateral, sway, pitch and roll motions. The equation of motion can be expressed as follows:

$$[M]_v\{\ddot{a}\}_v + [C]_v\{\dot{a}\}_v + [K]_v\{a\}_v = [Q]_v \quad (5)$$

where $[M]_v$, $[C]_v$ and $[K]_y$ are the mass, damping and sti_ness matrices of the vehicle respectively and $[Q]_y$ denote coupling force between vehicle and guideway.

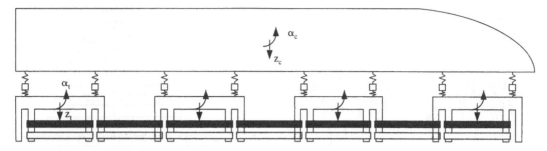

Figure 2. Vehicle-guideway coupling dynamic model.

2.3 The guideway model

For typical guideway systems, span-length-to-width ratios are large enough so that individual spans may be considered as beams rather than as plates. Using dynamic finite element method, the equation of motion for the guideway spans can be expressed as:

$$[M]_g\{\ddot{a}\}_g + [C]_g\{\dot{a}\}_g + [K]_g\{a\}_g = \{Q\}_g \tag{6}$$

The displacements of the guideway at any section in the finite element analysis are identified in terms of vertical, lateral and twist displacement. The displacement vector can be expressed as:

$$\{a\}_g = \{u_i, w_{yi}, w_{zi}, \phi_i, \phi_{yi}, \phi_{zi}; u_j, w_{yj}, w_{zj}, \phi_j, \phi_{yj}, \phi_{zj}\}^T$$

The mass matrix and the stiffness matrix of the guideway can be achieved by simply superposition method to finite elements; the stiffness matrix of the guideway element can be expressed as:

$$[k]^e = \begin{bmatrix} \frac{EA}{l} & 0 & 0 & 0 & 0 & 0 & -\frac{EA}{l} & 0 & 0 & 0 & 0 & 0 \\ & \frac{12EI_z}{l^3(1+b_z)} & 0 & 0 & 0 & \frac{6EI_z}{l^2(1+b_z)} & 0 & \frac{-12EI_z}{l^3(1+b_z)} & 0 & 0 & 0 & \frac{6EI_z}{l^2(1+b_z)} \\ & & \frac{12EI_y}{l^3(1+b_y)} & 0 & \frac{-6EI_y}{l^2(1+b_y)} & 0 & 0 & 0 & \frac{-12EI_y}{l^3(1+b_y)} & 0 & \frac{-6EI_y}{l^2(1+b_y)} & 0 \\ & & & \frac{GI_x}{l} & 0 & 0 & 0 & 0 & 0 & -\frac{GI_x}{l} & 0 & 0 \\ & & & & \frac{EI_y(4+b_y)}{l(1+b_y)} & 0 & 0 & 0 & \frac{6EI_y}{l^2(1+b_y)} & 0 & \frac{EI_y(2-b_y)}{l(1+b_y)} & 0 \\ & & & & & \frac{EI_z(4+b_z)}{l(1+b_z)} & 0 & \frac{-6EI_z}{l^2(1+b_z)} & 0 & 0 & 0 & \frac{EI_z(2-b_z)}{l(1+b_z)} \\ & & & & & & \frac{EA}{l} & 0 & 0 & 0 & 0 & 0 \\ & & & & & & & \frac{12EI_z}{l^3(1+b_z)} & 0 & 0 & 0 & \frac{-6EI_z}{l^2(1+b_z)} \\ & & & & & & & & \frac{12EI_y}{l^3(1+b_y)} & 0 & \frac{6EI_y}{l^2(1+b_y)} & 0 \\ & & & & & & & & & \frac{GI_x}{l} & 0 & 0 \\ & & & & & & & & & & \frac{EI_y(4+b_y)}{l(1+b_y)} & 0 \\ & & & & & & & & & & & \frac{EI_z(4+b_z)}{l(1+b_z)} \end{bmatrix}$$

where EI_y, EI_z and EA are the lateral, vertical, and twist bending rigidity of the guideway, l represent per element length, G is shear modulus, b_y and b_z are coefficient of correct. The mass matrix of the guideway element can be expressed as:

$$[m]^e = \frac{\rho l}{420} \begin{bmatrix} 140 & 0 & 0 & 0 & 0 & 0 & 70 & 0 & 0 & 0 & 0 & 0 \\ & 156 & 0 & 0 & 0 & 22l & 0 & 54 & 0 & 0 & 0 & -13l \\ & & 156 & 0 & -22l & 0 & 0 & 0 & 54 & 0 & 13l & 0 \\ & & & \frac{140I_p}{A} & 0 & 0 & 0 & 0 & 0 & \frac{70I_p}{A} & 0 & 0 \\ & & & & 4l^2 & 0 & 0 & 0 & -13l & 0 & -3l^2 & 0 \\ & & & & & 4l^2 & 0 & 13l & 0 & 0 & 0 & -3l^2 \\ & & & & & & 140 & 0 & 0 & 0 & 0 & 0 \\ & & & & & & & 156 & 0 & 0 & 0 & -22l \\ & & & & & & & & 156 & 0 & 22l & 0 \\ & & & & & & & & & \frac{140I_p}{A} & 0 & 0 \\ & & & & & & & & & & 4l^2 & 0 \\ & & & & & & & & & & & 4l^2 \end{bmatrix}$$

where ρ is the guideway mass per unit length, I_p is polar moment of inertia the damping matrix is formed by the assumption of Rayleigh damping $[C]_g = \alpha[M]_g + \beta[K]_g$; $[Q]_g$; is the exciting force of the guideway due to dead load and electric magnetic force.

The dynamic model of the guideway-vehicle coupling system can be established by the composition of the vehicles and the guideway model.

2.4 Guideway irregularity

Stationary and ergodic random irregularities are described by a power spectral density function (PSD) (see equation (7)). The function can be expressed as:

$$S(\Omega) = \frac{A}{\Omega^2} \ (\text{m}^2.\text{m/rad}) \tag{7}$$

Ω is the spatial frequency and A is a constant which is indicative of the guideway roughness. It was assumed that guideways of interest in maglev would have vertical roughness profile equivalent to present day good quality railroad weld rails. References reveal that $A = 1.5 \times 10^{-7}$ $rad - ft$ is a representative value for high speed track spectrum. The appropriate value of A for these guideways can therefore be estimated from existing experimental data on welded rails. Considering needs of guideway-vehicle dynamic model input, local irregularities are describes as mathematical functions,

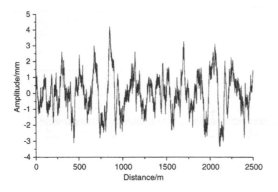

Figure 3. Vertical guideway random irregularities.

such as cosine, sinusoidal, exponent function, random guideway irregularities simulate from PSD by use of trigonometric series method, Figure 3 shows the vertical profile of the guideway generated by this method.

3 NUMERICAL SIMULATIONS

Numerical simulations of dynamic interactions of guideway/vehicle systems were carried out on the basis of the composition equations for the vehicles and guideway, taking guideway irregularities. Because of the coupled dynamic interaction between the vehicle and guideway, a numerical integration method is required to calculate dynamic responses. The separated iterative procedure and its corresponding numeric algorithm are applied in the simulations. A digital power FORTRAN program has been written to perform dynamic analysis, whose flow chart is shown in Figure 4.

4 CALCULATION RESULTS

The whole histories of maglev vehicles passing on the guideway were simulated on computer, the dynamic responses of the guideway and the vehicle were obtained. The 50 m-length two-span guideway are considered in the following simulations. The stiffness of existing guideway is 3.8×10^{10} N · m^{-2}, and the running speed is 400 km/h.

Figure 5 shows midspan deflections of the 50 m-length two-span guideway, which indicates that the maximum deflection of the guideway is 2.5 mm; deformation-to-span ratio is smaller than that (1/4800) of the guideway designed. Figure 6 shows midspan acceleration of the guideway, which indicates that the maximum value is 0.6 m/s^{-2}. Figure 7 shows the change of vertical air gap, which indicates that the change fluctuation is less than 2 mm. Those numerical results show that the 50 m-length two-span guideway parameters are within the safety range.

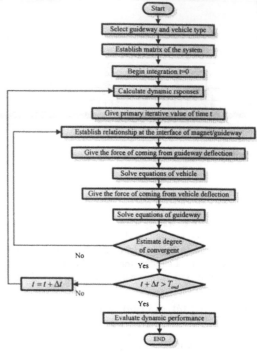

Figure 4. Flow chart of simulation program.

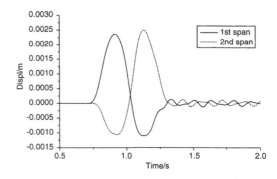

Figure 5. Midspan deflections of the guideway.

Figures 8 and 9 shows the influence of different stiffness and first order frequency on the maximum deflection of the guideway. Those numerical results indicate that the stiffness of existing guideway can be reduce 80%, first order frequency have an obviously influence on deflection of the guideway, and it should greater than 6 HZ.

The random responses of maglev vehicle are simulated and analyzed to evaluate the ride quality of maglev vehicle. Figure 10 shows PSD of car body acceleration, the dominant frequencies of maglev vehicle

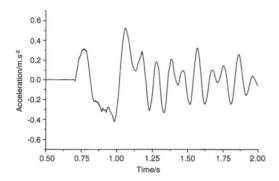

Figure 6. Midspan acceleration of the guideway.

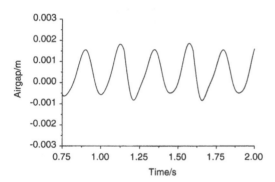

Figure7. Change of vertical air gap.

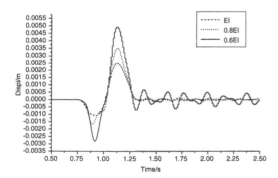

Figure 8. Influence of different stiffness on the maximum deflection of the guideway.

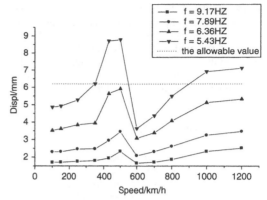

Figure 9. Influence of first order frequency on the maximum deflection of the guideway.

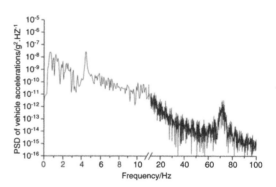

Figure 10. PSD of vehicle.

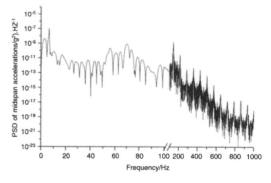

Figure 11. PSD of midspan accelerations.

vibration range from 1 HZ to 2 HZ. Compared to the ride quality guides of some countries, the comfort performance of the vehicle is quite good. Figure 11 shows PSD of midspan acceleration of free beam, the dominant frequencies are less than that of rail of the wheel-rail systems.

5 CONCLUSIONS

An integrated dynamic model of maglev vehicle-guideway coupling system has been established based

225

on TR maglev vehicle and the elevated-guideway technology. The computer simulation method proposed in this paper can well reflect the main vibration characteristic of the guideway and the vehicle. The results prove that the special program can be used to study the dynamic characteristic of the EMS Maglev Guideway and optimize system parameters in the future.

ACKNOWLEDGMENTS

The work described in this paper has been supported by The National High Technology.

Research and Development Program of China (2004AA505210) and Technology Development Foundation of Beijing Jiaotong University (BJTU48008). The authors thank the referees of this paper for their valuable comments and suggestions.

REFERENCES

Grove, A.T. 1980. Geomorphic evolution of the Sahara and the Nile. In M.A.J. Williams & H. Faure (eds), *The Sahara and the Nile*: 21–35. Rotterdam: Balkema

Xia He. 2002. Dynamic Interaction of Bridges and Vehicles, *Science Press*. China: Beijing

Zeng Youwen, Wang Shao-hua, Zhang Kunlun. 1999. A Study of Vertical Coupling Dynamic of EMS Maglev Train and Guideway Systems, *Journal of the China Railway Society:* 21(2): 22–25. China: Beijing

Zhao Chunfa. 2002. Maglev Vehicles System Dynamics, *Southwest Jiaotong University.* China: Chengdu

Y.Cai, S.S. Chen. 1997. Dynamic characteristics of magnetically levitated vehicle systems. *Applied Mechanics Reviews*, 50(11): 647–670

Corresponding author: School of Civil Engineering & Architecture, Beijing Jiaotong University, Beijing 100044, People's Republic of China
E-mail:sjnjtu@163.com

Environmental Vibrations – Takemiya (ed.)
© 2005 Taylor & Francis Group, London, ISBN 0 415 39035 4

Modal identification from free vibration test using a neural network

C.H. Chen
Department of Civil and Environmental Engineering, National University of Kaohsiung, Kaohsiung, Taiwan, R. O. C.

M.C. Huang
Department of Aircraft Engineering, Air Force Institute of Technology, Kaohsiung, Taiwan, R. O. C.

ABSTRACT: The present proposes a neural network-based method for determining the dynamic characteristic parameters of structures from free vibration data. The method employs the observed dynamic responses to train an appropriate neural network. Conventional back-propagation is used to train the artificial neural network. Then, the modal parameters are directly identified by applying weight matrices in the neural network approach. The modal parameters, including frequencies, damping and vibration shapes of a structure can then be determined accurately. The proposed identification procedure is used to determine the dynamic characteristics of a three-span highway bridge, whose characteristics are determined from its free vibration responses by conducting a field impact test.

1 INTRODUCTION

Over the last two decades, artificial neural networks (ANN) have gradually become powerful tools for pattern recognition, signal processing, control, and complex mapping, because of their excellent capacity for learning and high tolerance to partially inaccurate data. An artificial neural network model is a system with its input and output data based on biological nerves. The system may comprise numerous computational elements that operate in parallel and are arranged in patterns that resemble biological neural nets. A neural network is usually characterized by its computational elements, its network topology and the learning algorithm applied (Rumelnart *et al* 1986). Of the various ANNs, including the feedforward, the multilayered and the supervised neural network with the error back-propagation algorithm – the backpropagation network (BPN) is by far the most commonly used neural network learning model. Barai and Pandey (1995) applied an artificial neural network to process the vibration signature to damage detected to a steel bridge. Chen *et al* (1990) explicated the identification of a discrete-time nonlinear system using neural network with a single hidden layer. Masri *et al* (1993) investigated the potential of using neural network method to determine the internal forces of structure-unknown nonlinear dynamic system. Many advanced studies have applied the neural network techniques for damage

detection (Ko *et al* 2002, Ni *et al* 2002, Lee *et al* 2005), structural system identification, (Wu *et al* 2002) and structural health monitoring (Zhao *et al* 1998) in civil engineering.

This investigation presents an efficient identification method, a neural network approach, to analyze data collected in a field measurement test. The technique is employed to determine the dynamic characteristics of the structure. The natural frequencies, the damping ratio and the modal shapes identified from the field vibration test will be compared with those used in finite element analysis.

2 NEURAL NETWORK ALGORITHM

2.1 *Backpropagation neural network*

A BP networks, depicted in Fig. 1, consists of an input layer, one or more hidden layers and an output layer. Every node in each layer is connected to every node in the adjacent layer. Importantly, Hecht-Nielsen (1989) proved that one hidden layer of neurons suffices to model any solution surface of practical interest. Therefore, this study addresses a network with only one hidden layer. Before an ANN can be applied, it must be trained using an existing training set of pairs of input-output elements. The training of a supervised neural network using a BP learning algorithm typically involves three stages. The first stage is to feed-forward the data. The computed output of the ith node in the output layer is defined as follows;

$$y_i = g(\sum_{j=1}^{N_h} (w_{ij} g(\sum_{k=1}^{N_i} v_{jk} x_k + \theta_{vj}) + \theta_{wi}))$$

$$i = 1, 2, \cdots N_o,$$ (1)

where w_{ij} is the connective weight between nodes in the hidden layer and those in the output layer; v_{jk} represents the connective weight between the nodes in the input layer and those in the hidden layer; θ_{wi} (or θ_{vj}) are bias terms that represent the threshold of the transfer function g, and x_k is the input of the kth node in the input layer. Terms N_i, N_h, and N_o are the numbers of nodes in the input, hidden and output layers, respectively. The transfer function can be linear or nonlinear. This work applies a linear transfer function. Typically, the input of each node in the input layer is normalized to be between 1 and -1 in the training of ANN.

The second stage is error back-propagation through the neural network. A system error function can be used to monitor the performance of the network. The system error function is commonly regarded as a function of the desired (or measured) value and the computed value of each node in the output layer.

The final stage adjusts the weightings. A gradient descent method with a learning ratio in the standard BP algorithm is used to train the neural network, and is defined as follows;

$$\mathbf{W}^{(r+1)} = \mathbf{W}^{(r)} + \Delta\mathbf{W}^{(r)}$$ (2)

where

$$\mathbf{W} = (v_{11} v_{12} \cdots v_{jk} \cdots v_{N_h N_i} \theta_{v1} \theta_{v2} \cdots \theta_{vN_h} w_{11} w_{12} \cdots w_{ij} \cdots$$
$$w_{N_o N_h} \theta_{w1} \theta_{w2} \cdots \theta_{wN_o})$$ (3)

$\mathbf{W}$ is a vector of parameters to be determined in an ANN. The weight is adjusted as follows;

$$\Delta\mathbf{W}^{(r)} = -\eta \frac{\partial E(\mathbf{W})}{\partial \mathbf{W}^{(r)}}$$ (4)

where η is the constant learning ratio; the superscript index (r) refers to the rth learning iteration, and $E(\mathbf{W})$ is the system error function. Normally, BP network learning models take an extended period to learn the network. Additionally, the convergence speed of a BP neural network depends on the learning ratio. For computational efficiency, the Marquardt–Levenberg algorithm is applied herein to yield w_{ij}, v_{jk}, θ_{wi} and θ_{vj} by minimizing the system error function in this work.

2.2 Procedure for identifying dynamic parameters

The structural responses are obtained by determining $x_i(t)$ from a numerical simulation or, experimentally, from a field vibration test. Therefore, a neural network to be established, as shown in Fig. 2, to predict $x_i(t)$ from previous responses, $x_i(t-j)$, where $i = 1, 2, 3, \ldots, n; j = 1, 2, \ldots m$. Variables n and m represent the lags in the output and input, respectively. $x_i(t-j)$ describes the observed historical responses of the ith measured degree of freedom, in relation to the basic at the $(t-j)$th time step. For a linear structural system, if the transfer function used is linear, Eq. (1) can be easily applied to yield the neural network in Fig. 2, which can be mathematically described as,

$$\{Y\} = [W][V]\{X\} + ([W]\{\theta_v\} + \{\theta_w\})$$ (5)

where $\{Y\} = (x_1(t), x_2(t) \cdots, x_n(t))^T$,
$$\{X\} = (x_1(t-1)\ x_2(t-1) \cdots x_n(t-1)\ x_1(t-2)$$
$$x_2(t-2) \cdots x_n(t-2) \cdots x_{n-1}(t-m)\ x_n(t-m))$$

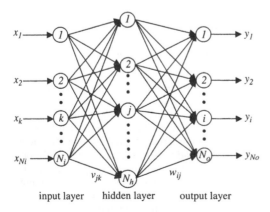

input layer   hidden layer   output layer

Figure 1.   A typical three-layer neural network.

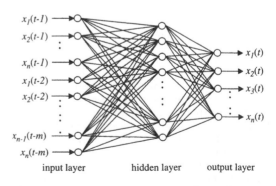

input layer   hidden layer   output layer

Figure 2.   Topology of a three-layer neural network.

The elements of $[W]$ and $[V]$ are w_{ij} and v_{ij}, respectively, and the elements of $\{\theta_w\}$ and $\{\theta_v\}$ are θ_{wi} and θ_{vi}. Carefully expanding equation (5) yields,

$$\begin{Bmatrix} x_1(t) \\ x_2(t) \\ \vdots \\ \vdots \\ x_n(t) \end{Bmatrix} = \sum_{j=1}^{m} \hat{\mathbf{W}}^{(j)} \begin{Bmatrix} x_1(t-j) \\ x_2(t-j) \\ \vdots \\ \vdots \\ x_n(t-j) \end{Bmatrix} + \{C\} \qquad (6)$$

where

$$[\hat{\mathbf{W}}] = [W][V], \qquad (7)$$

$$\hat{\mathbf{W}} = [\hat{\mathbf{W}}^{(1)} \ \hat{\mathbf{W}}^{(2)} \cdots \hat{\mathbf{W}}^{(m)}], \qquad (8)$$

$$\{C\} = [W]\{\theta_v\} + \{\theta_w\}. \qquad (9)$$

The elements in $[W]$ and $[V]$ are w_{ij} and v_{ij}, respectively, and the elements of $\{\theta_w\}$ and $\{\theta_v\}$ are θ_{wi} and θ_{vi}. It has to be mentioned that Eq. (6) is only valid for free vibration response.

Equation (6) is similar to the time series model, AR. The AR model equates the equations of motion. Hence, the dynamic characteristics of the structural system can be obtained from the coefficient matrices of AR. The modal parameters can be determined from the eigenvalues and eigenvectors of $[G]$, by the procedure of Huang (2001), and establishing the following matrix.

$$[G] = \begin{bmatrix} 0 & \mathbf{I} & 0 & 0 & 0 \\ 0 & 0 & \mathbf{I} & 0 & 0 \\ \vdots & \vdots & \vdots & \vdots & \vdots \\ 0 & 0 & 0 & 0 & \mathbf{I} \\ \hat{\mathbf{W}}_1^{(m)} & \hat{\mathbf{W}}_1^{(m-1)} & \cdots & \hat{\mathbf{W}}_1^{(2)} & \hat{\mathbf{W}}_1^{(1)} \end{bmatrix} \qquad (10)$$

The eigenvectors of $[G]$ correspond to the mode shapes of the structural system of interest in the form of state variables, whereas the eigenvalues of $[G]$ relate to the natural frequencies and damping ratios. Let λ_k and $\{\psi_k\}$ represent the kth eigenvalue and eigenvector of $[G]$, respectively. The eigenvalue, λ_k, is a complex number, and can thus be expressed as $a_k + ib_k$. The complex conjugates of λ_k and $\{\psi_k\}$ are also an eigenvalue and eigenvector, respectively. The natural frequency and modal damping of the system, as in Eq. (10) are given by

$$\tilde{\beta}_k = \sqrt{\alpha_k^2 + \beta_k^2} \qquad (11)$$

$$\xi_k = -\alpha_k / \tilde{\beta}_k \qquad (12)$$

where $\tilde{\beta}_k$ is the pseudo-undamped circular natural frequency; ξ_k is the modal damping ratio;

$$\beta_k = \frac{1}{\Delta t} \tan^{-1}(\frac{b_k}{a_k}) \qquad (13)$$

$$\alpha_k = \frac{1}{2\Delta t} \ln(a_k^2 + b_k^2) \qquad (14)$$

and $1/\Delta t$ is the sampling rate of measurement.

The particular composition of $[G]$ in Eq. (9) is such that its eigenvectors exhibit the following property.

$$\{\psi_k\} = (\{\psi_k\}_1^T, \lambda_k \{\psi_k\}_1^T, \lambda_k^2 \{\psi_k\}_1^T, \dots, \lambda_k^{m-1} \{\psi_k\}_1^T)^T \qquad (15)$$

where $\{\psi_k\}_1$ is the modal shape of the system that corresponds to the natural frequency, $\tilde{\beta}_k$.

Notably, the number of eigenvalues of $[G]$ in Eq. (10) commonly exceeds the number of natural frequencies of the structural system. Accordingly, as well as real mechanical modes, extra spurious modes are generated from the constructed ANN. However, the eigenvalues and eigenvectors that correspond to real mechanical modes are observed regularly as m increases (Huang et al 2003, Chen 2003).

Hence, the responses measured in the structure in the field vibration test show that a neural network can be established, as indicated in Fig. 2. The modal parameters of such a structural system can be determined using the foregoing method.

3 THREE-SPAN HIGHWAY BRIDGE

3.1 Description of bridge

The other tested bridge is a unit of the elevated highway bridge system in the northbound line of constructed along the No. 1 National Freeway, which crosses Taipei City in Taiwan, which is completed in 1996 and is called the Yuan-Shan bridge. As depicted in Fig. 3, the Yuan-Shan bridge is a continuous three-span bridge, which consists of pre-stressed concrete box-girders with various cross-sections. The total length of the bridge is 360 m, and the lengths for three spans are 95, 155 and 110 m. The height of the box-girder ranges from 3.5 m to 8 m, and the width of the deck is approximately 12.6 m, supporting two lanes of vehicles. The girder of the bridge is fixed in the transverse and vertical directions but is movable in the longitudinal direction at the two ends; it is rigidly supported by the two central columns.

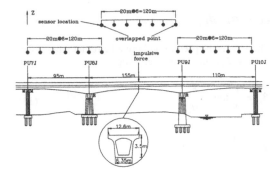

Figure 3. The tested three-span highway bridge and the sensor layout.

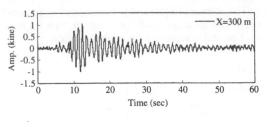

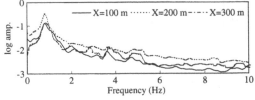

Figure 4. A typical set of recorded data for the impact responses with the corresponding auto-spectra.

3.2 Instruments using in measuring free vibration in the impulse test

The impulse test was performed on the Yuan-Shan bridge immediately before it was opened to the public. An exciting force was generated on the bridge in the transverse (y) direction (Fig. 3), by the passing of a loaded truck with single wheel axle that weighed around 14 tons. The impulsive force in the transverse direction was applied to the center point of the middle span. The transverse force was applied by suddenly breaking the truck at this point on the deck as the truck traveled 30 degrees to the centerline of the deck at approximately 15 km/hr. The transverse impulsive force is thus generated as the friction force between the vehicle's wheels and the deck. The impulsive forces were applied to the top deck of the box-girder, and did not excite any local modes of the deck because of the deck is relatively rigid (Chen et al 1997).

The number of sensors that could be simultaneously used during testing was limited, so the bridge was divided into three segments, each of 120 m. Seven sensors were placed at equal intervals in each segment, along the centerline of the deck (Fig. 3). The free vibration response of the three segments of the bridge was measured when the impulsive force was applied to the center point of the middle span, as presented in Fig. 3. The impulsive force was then applied in the transverse direction and the sensors were deployed in the direction of this impulsive force, in each of the three segments, to record the response of the bridge in that direction. All responses were recorded for a duration of one minute at a sampling rate of 50 Hz, so 3,000 data were recorded in each direction at every measurement station. Notably, the overlapped points at $x = 120$ m and 240 m were used to link the data recorded from the three segments.

Table 1. Identified natural frequencies and modal damping ratios in the transverse direction.

	Impulse test		Ambient test	
Mode	Frequency (Hz)	Damping (%)	Frequency (Hz)	Damping (%)
1	0.80	1.5	0.81	1.3
2	1.02	2.9	1.01	2.0
3	1.13	5.0	1.19	5.3
4	1.88	5.3	1.84	4.6
5	2.70	4.5	2.77	4.7
6	3.72	2.2	3.65	1.6

3.3 Processing and analyzing data

Figure 4 plots a typical set of recorded data concerning the responses in the transverse direction at various locations, when the measurement segments were subjected to an impulsive force in that direction. The figures reveal that the free vibration response can indeed be generated by the proposed testing procedure. The impact responses to large impacts after 10 sec were used to train an ANN, to reduce the effect of noise and thus yield better identified results. Figure 2 presents the architecture of the ANN with $n = 7$, $m = 12$ and 16 nodes in the hidden layer. The velocity responses in each measurement channel in the transverse direction were used to train an ANN. Table 1 lists the natural frequencies and the modal damping ratios identified from the impulse test by the trained ANN, while Fig. 5 shows the modal shapes.

3.4 Results and discussions

Table 1 indicates that a total of six transverse modes were identified. A comparison of the identified

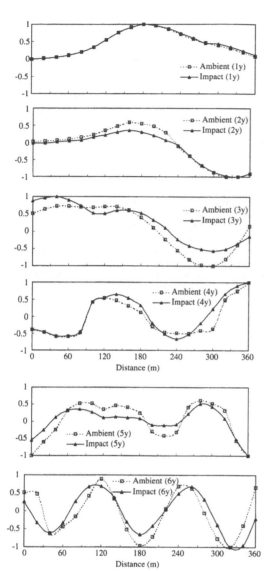

Figure 5. Comparison of identified mode shapes with ambient vibration results.

Table 1 also shows that the consistency between the modal damping ratios identified from the two tests is not as strong as that between the natural frequencies, but is acceptable. The damping ratios identified obtained the tests are smaller than the design value of 5%, expect in the fourth mode. These ratios do not increase with the natural frequency.

Table 2 presents the MAC values for the modal shapes shown in Fig. 5 obtained from the free vibration and ambient vibration tests, which are used to evaluate the consistency between the mode shapes obtained by the two tests. As can be seen, the two sets of modal shapes in the transverse direction are good correlation, especially in the first five modes.

4 CONCLUSIONS

This work uses artificial neural networks to determine the dynamic characteristics of a structural system from its dynamic responses. The approach is based on constructing a neural network to match the observed responses. The modal parameters of the structural system are directly estimated from the weighting matrices associated with the neural network. Then, the modal parameters are determined through matrix operations. A shear-building model was numerically simulated to validate the proposed method. The proposed procedure was demonstrated to be able accurately to determine the dynamic characteristics of the structural system.

This investigation also proposed a method of structural identification by ambient vibration testing, applying the neural network method to the arch pylon of a cable-stayed bridge. The first five modes of the arch pylon were identified from the results of the ambient vibration test. The identified results are reasonably consistent with the results of the finite element analysis for the first five modes; the resulting vibration shapes are particularity consistent. The proposed procedure was also applied to the free vibration responses by performing an impulse test on a three-span highway bridge. The usefulness of the proposed method for estimating the dynamic characteristics of the bridge in the impulse test was confirmed by the excellent agreement between the obtained results and those in the ambient vibration test.

frequencies with those that correspond to the peak of the auto-spectra presented in Fig. 4 reveals the frequencies in the low modes agree with each other. However, the frequencies and damping ratios in higher modes in Table 1 cannot be directly identified from the auto-spectra. Table 1 also presents the results obtained in an ambient vibration test, using the random decrement system identification technique (Huang *et al* 1999). Table 1 reveals that the frequencies obtained in the impulse test agree very closely with those obtained in the ambient vibration test.

REFERENCES

Barai, S. V. and Pandey, P. C. 1995. Vibration signature analysis using artificial neural networks. *Journal of Computing in Civil Engineering,* **9**(4), 259–265.

Chen, C. H., Lu, L. Y. and Yang, Y. B. 1997. Dynamic testing and system identification of a highway bridge. *Structural Engineering,* Taipei, Taiwan, **12**(3), 3–22 (in Chinese).

Chen, C. H. 2003. Determination of flutter derivatives via a neural network approach. *Journal of Sound and Vibration,* **263**(4), 797–813.

Chen, S., Billing, S. A. and Grant, P. M. 1990. Non-linear system identification using neural networks. *Int. J. Control,* **51**(6), 1191–1214.

Hecht-Nielsen, R. 1989. Theory of the back propagation neural network. *Proceedings of International Joint Conference on Neural Networks,* IEEE **1**, 593–605.

Huang, C. S., Yang, Y. B., Lu, L. Y. and Chen, C. H. 1999. Dynamic testing and system identification of a highway bridge. *Earthquake Eng. & Structural Dynamics,* **28**, 857–878.

Huang, C. S. 2001. Structural identification from ambient vibration measurement using the multivariate AR model. *Journal of Sound and Vibration,* **241**(3), 337–359.

Ko, J. M., Sun, Z. G. and Ni, Y. Q. 2002. Multistage identification scheme for detecting damage in cable-stayed Kap Shui Mun Bridge, *Engineering Structures,* **24**(7), 857–868.

Lee, J. J., Lee, J. W., Yi, J. H., Yun, C. B. and Jung, H. Y. 2005. Neural networks-based damage detection for bridges considering errors in baseline finite element models, *Journal of Sound and Vibration,* **280**(3–5), 555–578.

Masri, S. F., Chassiakos, A. G. and Caughey, T. K. 1993. Identification of nonlinear dynamic systems: using neural networks. *Journal of Applied Mechanics.* March, **60**, 123–133.

Ni, Y. Q., Wang, B. S. and Ko, J. M. 2002. Constructing input vectors to neural networks for structural damage identification, *Journal of Smart Materials and Structures,* **11**, 825–833.

Runelnart, D. E., Hinton, G. E. and Williams, R. J. 1986. Learning international representation by error propagation. *Parallel Distributed Processing,* D. E. Rumelnart *et al.,* Eds, The MIT Press, Cambridge, MA 318–362.

Wu, Z. S., Xu, B. and Yokoyama, K. 2002. Decentralized parametric damage based on neural networks, *Computer Aided Civil and Infrastructure Engineering,* **17**, 175–184.

Zhao, J., Ivan, J. N. and Dewolf, T. 1998. Structural health monitoring using neural networks, *Journal of Infrastructure Engineering,* **4**(3), 93–101.

Dynamics of soil-structure interaction

Environmental Vibrations – Takemiya (ed.)
© 2005 Taylor & Francis Group, London, ISBN 0 415 39035 4

Soil dynamics considerations in the automobile industry

R.D. Woods
University of Michigan, Ann Arbor, Michigan, USA

ABSTRACT: From concept to manufacture and beyond, design, production and disposal of automobiles depends on structures, machines, and instruments supported on the ground. Proper design of foundations for facilities, machines and instruments depends on the principles associated with foundation and soil dynamics. Examples of applications of soil/foundation dynamics in the automobile industry from concept through production and to after life are presented.

1 BACKGROUND

The production of modern automobiles depends on careful control of dimensions, tolerances, and material quality. Often the demands of manufacturing are incompatible with refined measurement processes, or neighboring space applications. However, both aspects must be accommodated to produce an acceptable vehicle and to co-exist with the surroundings. The current state of knowledge in foundation dynamics and soil dynamics helps engineers provide adequately stable platforms for many automobile manufacturing applications and to minimize the environmental effects of the manufacturing process to the neighborhood. Some of these applications appear before the production of a specific vehicle while others extend beyond the useful lifetime of the automobile.

2 COMPONENTS OF DYNAMIC DESIGN

2.1 *What is failure?*

Before an appropriate foundation design can be accomplished, it is necessary to know what will be used to judge success or failure of the system. Often the owner/manufacturer will not know what criteria need to be applied in design and the engineer must find an adequate parameter to guide the design. In some instances criteria will be known and provided to the engineer, for example the tolerable motion of the machine base for a contour measuring machine (CMM) may be specified by the CMM maker.

2.2 *Model connecting applied loads, soil conditions and significant foundation dimensions*

The engineer must use some model to connect the effects of unbalanced forces from the manufacturing operation with properties of the ground on which the foundation will be supported and the size, shape and mass of the machine and foundation. There are a few models available to accomplish this task and these will be described below as examples of dynamic design are presented.

2.3 *Methods for measuring critical soil parameters*

A key parameter in the dynamic design of foundations subject to unbalanced loads is the shear modulus of the soil, G. This parameter can be measured either in the laboratory or in the field. Field measurements are preferred for low strain situations, but laboratory measurements may be needed in some high vibration amplitude situations.

2.4 *Outline of presentation*

Examples of foundation dynamics in automobile production will be presented. For each application, the three components of dynamic design presented above will be discussed.

3 CRASH TESTING FACILITY

Before an automobile goes into production, crash tests are performed to determine the crash worthiness of

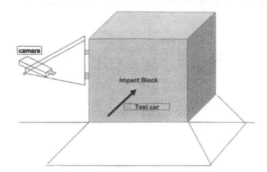

Figure 1. Crash test impact block.

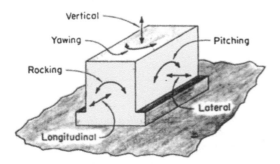

Figure 2. Rigid body modes of motion.

the vehicle. Movement of the inertia block that absorbs the impact of the vehicle is not very critical to the success of a test and the manufacturer did not provide any limiting motion of the block. The inertia block shown in Figure 1 was modeled as a single-degree-of-freedom (SDOF) system for analysis and was analyzed for three of the six modes of motion shown in Figure 2, horizontal translation in the direction of impact, rocking about an axis perpendicular to the impact direction and twisting about a central vertical axis.

The analysis was performed using the Lysmer Analogs, Richart et al (1970). Lysmer developed constant coefficients for single degree of freedom models where the constant coefficients were functions of shear modulus (G), Poisson's ratio (ν) and physical dimensions of the foundation. For vertical translation of a rigid block, Lysmer's SDOF analog is

- $m\ddot{z} + c_z\dot{z} + k_z z = Q(t)$

- $k_z = 4Gr_0 / (1-\mu)$

- $c_z = 3.4\, r_0^2\, (\rho G)^{0.5} / (1-\mu)$

- **G = shear modulus**

- **μ = Poisson's ratio**

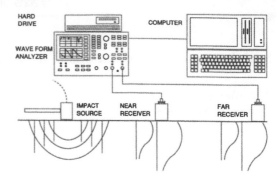

Figure 3. Basic configuration for SASW test (from Gucunski and Woods, 1991).

Shear modulus of the soil was measured using the Spectral-Analysis-of-Surface-Waves (SASW) method described by Nazarian and Stokoe (1984). The basics of the SASW method are shown in Figure 3.

With this technique, the shear modulus profile at a site can be determined from the surface, i.e. in a non-destructive and non-invasive manner.

The unbalanced forces in the crash test case were the impacts of a known weight vehicle traveling at a known velocity and being stopped instantly by the inertia block. The manufacturer provided impulse diagrams for the impact.

The successful function of the crash test facility depended on the ability to photograph the crash with several cameras, one of which was mounted on a bracket attached to the inertia block, see Figure 1. It was the tolerance of that camera that controlled the criteria for motion of the crash block. The design of the crash block was accomplished limiting motion of the camera to 5 mm p-p over a frequency range of 0 to 200 Hz using the Lysmer Analogs and the SASW technique to measure soil properties.

4 LARGE FORCES IN AUTOMOBILE MANUFACTURING

Forging, stamping or pressing machines are common in the manufacture of parts for automobiles. The force required to deform or forge hot metal billets into useful parts may be caused by dropping a large weight, forcing a weight or ram to move by compressed air or steam or through mechanical principles as in the case of a connecting rod and die being displaced by an eccentric on a rotating shaft. This latter mechanical forge is more common, currently, than the other types of forges. Figure 4 is an example of a mechanical forge.

A forge of this configuration was the subject of the design of a foundation to minimize vibrations. Neither the manufacturer nor forge supplier provided

236

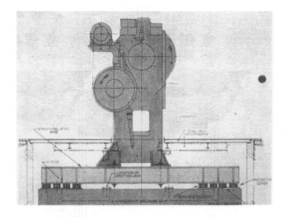

Figure 4. Example of mechanical forge.

Figure 5. Metal forming press in auto plant.

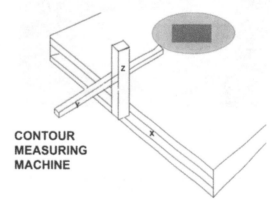

CONTOUR
MEASURING
MACHINE

Figure 6. Schematic of CMM.

any guidance on criteria for satisfactory operation. However, an elementary school was situated about 400 m from the site of the mechanical forge and the students were experiencing troublesome vibrations. The criteria, therefore, became limiting the vibrations exported by the forge to the school to an imperceptible level. That level was taken as 0.25 mm/second particle velocity at all frequencies as suggested in Richart et al (1970).

During observation of the operation of the forge, it was seen that the operator could change the level of vibrations by the way he operated the forge. He stopped the motion of the upper die with a foot activated compressed air brake. When the die was stopped at the top of the motion, large vibrations occurred, but when the die was stopped at the mid-point of the motion much lower vibrations occurred. The major motion of the forge was rocking about the transverse axis and the amplitude of that vibration was influenced by the momentum of the drive and die system at any time. Choosing a minimum momentum point helped reduce motion of the forge.

By altering the stopping point and later automating the stopping point of the forge the vibrations at the elementary school were imperceptible, about 0.3 mm p-p yet satisfactory parts were still produced. In this case the solution did not depend on foundation design, but rather machine operation.

5 QUALITY CONTROL IN THE PLANT

For manufacturing efficiency, it is often necessary to place sensitive measuring equipment in proximity of a highly energetic machine. One such example is the siting of a contour measuring machine (CMM) in the vicinity of a large metal forming press. Figure 5 shows a Komatsu Press in an automobile plant in which the owner wanted to establish a CMM with its closest point

located at about the building column in the foreground in Figure 5, i.e. about 15 m away.

A schematic of the CMM is shown in Figure 6 where the oval represents a pressed metal auto part. The XYZ coordinates are traced and recorded by the stylist and then compared with the design coordinates. For the car parts stamped out here, the criterion was that the measured coordinate must not vary by more than $+/-0.5$ mm from the design coordinate.

The CMM supplier guaranteed that the CMM could discern the 0.5 mm tolerance providing that motion of the foundation of the CMM did not exceed the levels described by the solid line limits on Figure 7. The vibration tolerance varied with frequency because the CMM coordinate Y & Z axis beams had varying natural frequencies.

The foundation shown in Figure 8 was constructed to support the CMM and the solid dots in Figure 7 indicate that measured vibrations satisfied the criteria.

Dynamic response of this foundation was analyzed using impedance functions in a manner described by Gazetas (1991). Impedance functions account for

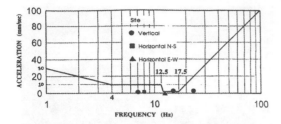

Figure 7. CMM vibration limits.

Figure 8. Foundation for CMM.

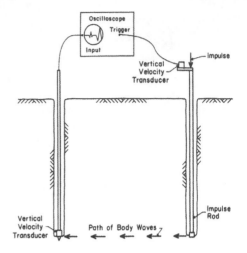

Figure 9. Two-borehole crosshole seismic test (from Stokoe and Woods, 1972).

Figure 10. Basic MAST facility.

frequency dependant stiffness and damping in the form of expressions like the following:

$$K = (Kk + i\omega Cc) \qquad (1)$$

where,
K is impedance
K is static stiffness
k is a dynamic stiffness coefficient
i is the imaginary coefficient
ω is circular frequency
C is a radiation dashpot coefficient
c is a frequency dependant dashpot modifier.

The static components in this equation are dependant on foundation dimensions and shear modulus of the soil. Because the location of the CMM foundation was inside an existing plant, the SASW method was ruled out. It is difficult to perform SASW through a 200 mm thick concrete floor. The shear modulus of the soil was measured by the two-borehole, crosshole seismic method. This approach minimizes the disturbance inside a plant by requiring only two holes through the concrete floor see Figure 9.

In this example, floor vibration was caused by the pressing machine and transmitted through the floor and ground to the location of the CMM. Design involved designing such that the foundation and CMM system did not respond to frequencies generated by the press.

6 SHAKE TESTING CARS AND COMPONENTS

Automobile manufacturers and their parts suppliers routinely test their products for noise, harshness and vibrations. This is done on shaking platforms of three basic types: MASTs, 4-Posters, and Cube Shakers. MASTs (multi-axis shake tables) are the most versatile and robust of the four methods and the design of a foundation for this testing facility will be described. Figure 10 shows a basic MAST facility for testing components.

The supplier of the MAST system provided vibration limits beyond which the MAST could not operate within specifications. A concrete block foundation 6 m × 7 m × 2.5 m thick was designed using the impedance method.

The shear wave velocity at the site for use in design was measured by the three-borehole crosshole method, Figure 11. Analysis showed that the concrete block foundation did not satisfy the criteria provided by the

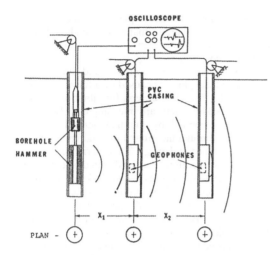

OSCILLOSCOPE

PYC CASING

BOREHOLE HAMMER

GEOPHONES

X₁ — X₂

PLAN –

Figure 11. Schematic of 3-hole crosshole system (Woods and Stokoe, 1985).

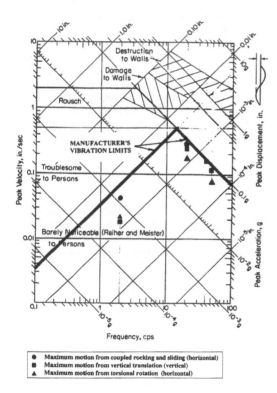

Figure 12. Calculated foundation motion for pile supported MAST.

MAST manufacturer at 50 Hz and above. Therefore, the foundation was redesigned using 500 mm diameter cast-in-place concrete piles 13 m long and analyzed by the PILAY approach, Novak and Aboul-Ella (1977).

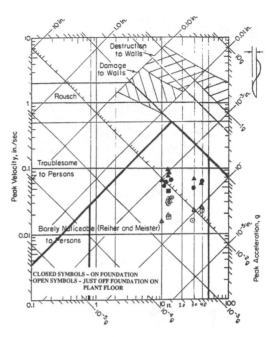

Figure 13. Measured vibrations on MAST foundation.

Frequency dependant dynamic stiffness and damping of the piles was calculated by this method. Shear modulus in PILAY can be modeled layer by layer along the length of the piles. Figure 12 shows calculated vibration amplitudes plotted against the manufacturer's criteria for vibrations showing that the pile supported MAST foundation will satisfy the criteria.

While the MAST manufacturer supplied motion criteria, they did not provide unbalanced force levels to use in design of the foundation for the MAST. It is possible to estimate "maximum" unbalanced forces based on force levels created by the electro-hydraulic actuators that create the vibrations and the mass of the MAST fixtures and test parts. For design calculations the maximum forces assuming worst case phase relationships that produced maximum forces and moments were used. Foundation motion was measured during the performance of the MAST under "extreme" conditions where bolts and other parts were being thrown off the tested component after several hours of operation. Figure 13 shows the measured vibrations relative to the manufacturers criteria.

7 END OF LIFE CYCLE

After the useful life of automobiles, recycling is possible, transforming about 80% of the vehicle to reusable products. The principle machine in recycling automobiles is the hammermill, Figure 14. This machine

239

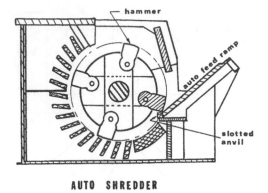

AUTO SHREDDER

Figure 14. Schematic of a hamermill.

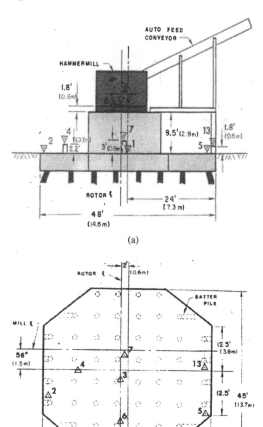

(a)

(b)

Figure 15. (a) Elevation view, (b) plan view of shredder foundation.

shreds the auto into small pieces on the order of 75 mm × 75 mm. Various materials like glass, fabric, aluminum, and steel are then separated by density and magnetic separation techniques.

The shredder itself is very robust and can even tolerate periodic explosions caused by left over gasoline in fuel tanks. Therefore criteria for vibration damage to the shredder itself are meaningless. However, vibrations generated by the hammermill may be transmitted through the ground to neighboring areas, causing nuisance or damage. The criteria for successful operation, therefore, may be dictated by the export of vibrations to the neighborhood. To keep vibrations below the threshold of perception for humans, the criteria for vibrations beyond the perimeter of the hammermill property are often set at about 0.25 mm/sec. particle velocity for all frequencies.

Unbalanced forces needed in calculation of dynamic motion of hammermills are generated in at least two ways: unbalanced centrifugal forces due to uneven hammer wear at the basic rotational speed of the machine and hammer impacts on the auto as it passes over the slotted anvil at four times per cycle, Figure 14.

Using these as unbalanced forces, soil properties measured by crosshole or SASW methods (depending of site conditions), and the dynamic analysis methods using impedance functions or the PILAY method, hammermill foundations can be designed to restrict vibrations exported to the neighborhood to acceptable levels. One example of a pile supported Hammermill foundation is shown in Figures 15 a & b.

8 SUMMARY

Five examples of dynamic foundation design incorporating three components of design have been presented. These examples include situations where criteria for design success have not been provided, situations where unbalanced forces were not available or provided and situations where design criteria were dictated by ground motion response outside the property of the owner. Three mathematical models were described for calculating dynamic response of foundations and two methods of determining soil properties for dynamic design were presented. These examples demonstrate that it is possible now to calculate dynamic response of foundation machine systems and to measure appropriate soil properties in situ.

REFERENCES

Gazetas, G. (1991). Foundation Vibrations, Chapter 15 of Foundation Engineering Handbook, edited by H-Y. Fang. New York, Van Nostrand Reinhold.

Gucunski, N. and Woods, R.D. (1991). Inversion of Rayleigh Wave Dispersion Curve For SASW Test. Proc. of 5th International Conference on Soil Dynamics and Earthquake Engineering. Univ. of Karlsruhe, Sept. 23–26.

Nazarian, S. and Stokoe, K.H. II (1984). In Situ Shear Wave Velocities From Spectral Analysis of Surface Waves. Proceedings of the Eighth World Conference on Earthquake Engineering, San Francisco, California, Vol III, pp. 31–38.

Novak, M. and Aboul-Ella, F. (1977). PILAY, A Computer Program for Calculation of Stiffness and Damping of Piles in Layered Media. University of Western Ontario, London, Ontario.

Richart, F.E., Jr., Hall, J.R., Jr. and Woods, R.D. (1970). Vibrations of Soils and Foundations. Englewood Cliffs, N. J.: Prentice-Hall.

Stokoe, K.H. II and Woods, R.D. (1972). In Situ Shear Wave Velocity by Cross-Hole Method. Journal of the Soil Mechanics and Foundations Division, Proc. ASCE, Vol. 98, No. SM5, pp. 443–460.

Woods, R.D. and Stokoe, K.H. II (1985). Field and Laboratory Determination of Soil Properties at Low and High Strains. Richart Commemorative Lectures, Proc. of session sponsored by Geotechnical Engineering Division, ASCE at ASCE annual Convention, Detroit, Michigan, Oct. 23.

Environmental Vibrations – Takemiya (ed.)
© *2005 Taylor & Francis Group, London, ISBN 0 415 39035 4*

Effect of foundation rigidity on train-bridge coupling vibration response

M.M. Gao, J.Y. Pan & Y.Q. Yang
China Academy of Railway Sciences, Beijing, China

ABSTRACT: Along with the escalation of train speed, effect of foundation on train-bridge coupling vibration response becomes more and more significant. As for a 24 m single-track double T beam on the 3-span bridge of an existing line, it is measured that 7.85 mm lateral amplitude at pier top and 8.21 mm at mid-span when a train passing through within speed of 50–70 km/h. Such violent vibration will endanger running safety. In order to find out main cause of such big displacement, a train-bridge-foundation coupling dynamic analysis model is built. In this model, the effect of foundation is simplified as the equivalent moveable and rotational spring stiffnesses at pier bottom. Result of simulation analysis shows that relative weak foundation rigidity is the major factor. On the basis of this result, several kinds of reinforcement plans are studied respectively, one is main beam reinforcement plan, another is pier reinforcement plan and the third is foundation reinforcement plan. Finally, suggestions are put forward according to dynamic simulation analyses.

1 INTRODUCTION

Increasing of train speed requires adequate rigidity of railway bridge to guarantee running safety and riding comfort[1]. With regard to the single track double T beam and the single cylindrical pier which are largely used on China's existing lines, their relatively severe lateral dynamic performances become emphatical when train speed increasing.

As for a 24 m single-track simple supported beam on the 3-span bridge of an existing line, the superstructure is made of two pre-stressed concrete T girders connected by 7 cross slabs with 24 m span, substructure is single columnar pier with spread foundation, and 9.5 m height between platform top and pier top. The typical cross section of beam is shown in figure 1, and actually measured vibration displacements under wagon effect are shown in table 1.

In order to investigate the effect of foundation rigidity and select an appropriate reinforcement plan, a train-bridge-foundation coupling dynamic model is built, and dynamic response of train and bridge is analyzed under passing of train at a speed of 50–80 km/h in empty and loaded conditions respectively. Finally, various effects of different reinforcement plans are compared.

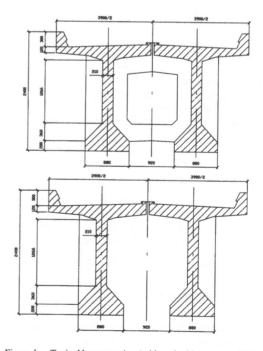

Figure 1. Typical beam section (with and without cross slab).

Table 1. Actually measured lateral vibration values of 24 m single-track simple supported beam on a 3-span bridge with single columnar pier.

Train condition	Velocity (km/h)	Lateral amplitude (mm)			
		Pier 1	Pier 2	Span 2	Span 3
Down-direction wagon	49.2	5.46	4.30	7.22	5.24
Down-direction wagon	55.8	5.32	3.76	5.26	4.89
Up-direction wagon	55.8	7.26	6.39	7.90	8.21
Up-direction wagon	70.1	6.90	4.05	5.95	6.40
Up-direction wagon	71.3	7.03	3.37	4.55	5.59
Up-direction wagon	60.2	7.85	3.82	6.47	4.35
Normal lateral amplitude limit in "Railway Bridge Inspection Specification"	$0.675(v \leqslant 60)$ $0.776(v > 60)$	2.667			
Actually measured natural frequency (Hz)		1.783	1.788	/	/
Normal limit in "Railway Bridge Inspection Specification" (Hz)		2.786		3.75	

2 TRAIN-BRIDGE-FOUNDATION DYNAMIC ANALYSIS MODEL

2.1 Bridge model

Bridge adopts multi-degree-of-freedom finite element model. In order to avoid excessive scale of equations caused by too much degree-of-freedom, space uniform cross-section beam is used. And elastic modulus of concrete material in the beam is reduced by 10% according to Chinese design specification. All pier structures apply uniform cross-section circular pier, with diameter as 1.9 m and concrete grade as C25. Elastic modulus of concrete material in pier is reduced by 20% according to design specification. Constraints of bearing between piers and beams are disposed through master/slave joint, and effect of foundation is simplified as equivalent moveable and rotational spring stiffnesses at pier bottom.

Bridge structure adopts Rayleign damp, with damp ratio as 2%.

2.2 Calculation condition

Formations of calculated train:

(1) DF-4 type diesel locomotive +40 loaded C62 wagons (shown as No.1 formation)
(2) DF-4 type diesel locomotive +40 empty C62 wagons (shown as No.2 formation)

Calculated train speed: 50,60,70,80 km/h.

2.3 Adopted track irregularity spectra

Due to the fact that the track spectra of three main lines in China match American 5-grade track spectra in the main, therefore, track irregularity time domain sample generated from American 5-grade track spectra is used in calculation[2]. And the maximum amplitudes in alignment and longitudinal level of track irregularity sample reach 10.69 mm and 14.87 mm respectively.

3 EVALUATION STANDARD FOR TRAIN-BRIDGE DYNAMIC ANALYSIS

3.1 Evaluation index for running safety and riding comfort[3][4]

3.1.1 Evaluation index for safety

Derailment coefficient $Q/P \leqslant 0.8$
Rate of wheel load reduction $\Delta P/P_0 \leqslant 0.6$
Wheel/rail lateral force $H \leqslant 0.85(10 + P_{st}/3)$

where P_{st} = static wheel load.

3.1.2 Evaluation index of acceleration of carbody

Concerning about vibration intensity, The Limits of wagon's accelerations are:

Vertical 0.7 g
Lateral 0.5 g

3.1.3 Evaluation index for riding comfort

In the light of "Test & Identification Methods and Evaluation Standard For Dynamic Performance of Railway Locomotive TB/T 2360-96", evaluation standard for riding level of locomotive defined by riding index W (namely Sperling index) is:

$W \leqslant 2.75$ $acc_V \leqslant 2.45 \, m/s^2$ $acc_H \leqslant 1.47 \, m/s^2$
Riding level: excellent

$W \leqslant 3.10 \quad acc_V \leqslant 2.95\,m/s^2 \quad acc_H \leqslant 1.96\,m/s^2$
Riding level: good

$W \leqslant 3.45 \quad acc_V \leqslant 3.63\,m/s^2 \quad acc_H \leqslant 2.45\,m/s^2$
Riding level: qualified

Riding level for wagon is evaluated according to riding index W and average maximum vibration acceleration of carbody. In the case of running speed 80 km/h, W and acceleration of carbody shall meet the following requirements:

$W \leqslant 3.5 \quad acc_V \leqslant 2.32\,m/s^2 \quad acc_H \leqslant 1.88\,m/s^2$
Riding level: excellent

$W \leqslant 4.0 \quad acc_V \leqslant 2.82\,m/s^2 \quad acc_H \leqslant 2.38\,m/s^2$
Riding level: good

$W \leqslant 4.25 \quad acc_V \leqslant 3.32\,m/s^2 \quad acc_H \leqslant 2.88\,m/s^2$
Riding level: qualified

3.2 Dynamic response limit of bridge

Referring to "Railway Bridge and Culvert Inspection Specification"(RBCIS) (Tie Yun Han [2004] No.120) and relevant specifications at home and abroad, dynamic response limits of bridge are as follows:

3.2.1 Acceleration limits of bridge
Ballasted deck , $acc_v \leqslant 0.35\,g \quad acc_H \leqslant 0.14\,g$
Ballastless deck, $acc_v \leqslant 0.50\,g \quad acc_H \leqslant 0.14\,g$

3.2.2 Lateral amplitude of pre-stressed concrete beam
According to RBCIS, referring to an 24 m-span beam:
General lateral amplitude at mid-span is 1.905 mm, for running safety, corresponding lateral limit is 2.667 mm.

3.2.3 General lateral amplitude of pier top
According to large numbers of actual measured data, Chinese specification gives general lateral amplitude limit of pier top is: 0.776 mm($V \leqslant 60\,km/h$) and 0.776 mm ($V > 60\,km/h$).

4 NATURAL FREQUENCES AND MODE OF BRIDGE

See Table 2 for natural frequences calculated on the basis of single span beam.
Actually measured natural frequency is 1.783 Hz. The table shows that when considering effect of pier structure and ignoring foundation rigidity, the first order lateral bending frequency is 3.521 Hz, while considering foundation rigidity, frequency falls to 1.700 Hz, a relatively large drop.

Table 2. Summary chart of natural frequences for simple supported 24 m beam with single columnar piers.

Constraints	No.	Frequency (Hz)	Mode
Without consideration of pier and foundation	1	5.013	Symmetrical lateral bending
	2	5.690	Symmetrical vertical bending
With consideration of pier structure, without consideration of foundation	1	3.282	Longitudinal floatation
	2	3.521	Common lateral bending of pier and beam
	3	5.658	Symmetrical vertical bending
	4	6.805	Asymmetrical lateral bending of pier and beam
With consideration of pier structure and foundation rigidity	1	1.700	Common lateral bending of pier and beam
	2	2.314	Longitudinal floatation
	3	2.681	Asymmetrical lateral bending of pier and beam
	4	5.658	Symmetrical vertical bending

5 TRAIN-BRIDGE-FOUNDATION DYNAMIC ANALYSIS RESULT

In order to study effect of foundation rigidity on vibration of whole bridge, train-bridge dynamic analysis has been made on the basis of considering and ignoring effect of foundation rigidity respectively.

5.1 Dynamic response of bridge

Because the vertical and lateral acceleration results satisfy the limit in all working conditions, acceleration results are not listed here. See Tables 3 and 4.

5.2 Dynamic response of locomotive & rolling stock

See Tables 5 and 6.

5.3 Comparison of reinforcement plans

Selected plan 1 is to reinforce pier structure, and plan 2 is to reinforce main beam, both disregard foundation treatment. Plan 3 is to reinforce foundation. See Tables 7 and 8 for dynamic response results of bridge.

Table 3. Dynamic response of bridge (considering effect of foundation rigidity).

| Train formation | Speed (km/h) | Coefficient of impact | Mid-span deflection (mm) | Lateral displacement (mm) | |
				Mid span	Pier top
DF4 + 40*C62	50	1.152	7.21	5.18	6.98
loaded wagon	60	1.156	7.23	5.30	7.06
	70	1.162	7.27	5.40	7.20
	80	1.157	7.23	5.37	7.39
DF4 + 40*C62	50	1.17	6.98	5.46	6.63
empty wagon	60	1.18	7.05	5.58	6.69
	70	1.185	7.09	5.68	6.84
	80	1.186	7.09	5.66	7.02

Table 4. Dynamic response of bridge (ignoring effect of foundation rigidity).

| Train formation | Speed (km/h) | Coefficient of impact | Mid-span deflection (mm) | Lateral displacement (mm) | |
				Mid span	Pier top
DF4 + 40*C62	50	1.152	7.21	2.06	1.31
loaded wagon	60	1.156	7.23	2.68	1.52
	70	1.162	7.27	4.11	2.19
	80	1.157	7.23	4.53	4.56
DF4 + 40*C62	50	1.167	6.98	2.11	1.25
empty wagon	60	1.180	7.05	2.74	1.49
	70	1.185	7.09	4.20	2.25
	80	1.186	7.09	3.74	3.84

Table 5. Dynamic responses of locomotive & rolling stock (considering effect of foundation rigidity).

| Formation and velocity (km/h) | | Q/P | $\Delta P/P$ | F_H (kN) | Acc_V (m/s^2) | Acc_H (m/s^2) | Sperling index | |
							V	H
DF4 + 40*C62	locomotive							
loaded wagon	50	0.59	0.348	36.5	1.177	1.878	3.154	3.305
	60	0.60	0.353	37.2	1.203	1.888	2.949	3.317
	70	0.61	0.358	37.4	1.239	1.919	2.987	3.387
	80	0.64	0.364	38.6	1.255	1.903	3.030	3.402
	trailer							
	50	0.46	0.374	32.4	1.87	2.388	3.883	4.227
	60	0.56	0.380	39.5	1.86	2.417	3.840	4.219
	70	0.55	0.385	39.2	1.91	2.424	3.856	4.229
	80	0.66	0.384	47.1	1.96	2.408	3.869	4.223
DF4 + 40*C62	locomotive							
empty wagon	50	0.584	0.346	36.400	1.168	1.873	3.154	3.304
	60	0.603	0.350	37.100	1.194	1.884	3.075	3.317
	70	0.611	0.355	37.320	1.239	1.916	3.089	3.387
	80	0.642	0.361	38.500	1.255	1.901	3.113	3.402
	trailer							
	50	0.561	0.541	27.833	1.851	2.817	3.963	4.218
	60	0.716	0.542	35.287	1.911	2.844	3.965	4.250
	70	0.715	0.553	35.513	1.917	2.868	3.937	4.221
	80	0.704	0.571	35.073	1.918	2.856	4.175	4.257

Table 6. Dynamic responses of locomotive & rolling stock (ignoring effect of foundation rigidity).

Formation and velocity (km/h)		Q/P	$\Delta P/P$	F_H (kN)	Acc_V (m/s²)	Acc_H (m/s²)	Sperling index	
							V	H
DF4 + 40*C62	*locomotive*							
loaded wagon	50	0.587	0.353	34.720	1.173	1.935	3.144	3.380
	60	0.612	0.359	35.960	1.165	1.962	3.025	3.361
	70	0.615	0.365	37.240	1.155	1.950	3.120	3.423
	80	0.658	0.368	39.160	1.164	1.963	3.108	3.467
	trailer							
	50	0.356	0.393	25.160	1.289	1.978	3.838	4.184
	60	0.383	0.395	27.087	1.271	1.981	3.842	4.201
	70	0.527	0.398	37.240	1.405	2.115	3.900	4.202
	80	0.701	0.397	49.800	1.814	2.426	4.041	4.258
DF4 + 40*C62	*locomotive*							
empty wagon	50	0.587	0.350	34.800	1.173	1.893	3.144	3.380
	60	0.614	0.356	36.020	1.161	1.921	3.180	3.361
	70	0.619	0.362	37.300	1.155	1.909	3.185	3.423
	80	0.659	0.365	39.240	1.164	1.922	3.209	3.467
	trailer							
	50	0.210	0.546	23.653	1.745	2.103	3.980	3.708
	60	0.235	0.538	26.380	1.763	2.097	3.989	3.764
	70	0.375	0.548	42.193	1.860	2.246	3.965	3.947
	80	0.728	0.563	41.633	1.928	2.917	4.223	4.332

Table 7. Bridge dynamic response under reinforcement plan 1 (considering foundation rigidity).

Train formation	Speed (km/h)	Coefficient of impact	Mid-span deflection (mm)	Lateral displacement (mm)	
				Mid span	Pier top
DF4 + 40*C62	50	1.144	7.15	4.06	4.92
loaded wagon	60	1.148	7.18	4.14	5.02
	70	1.154	7.22	4.26	5.11
	80	1.146	7.17	4.38	5.20
DF4 + 40*C62	50	1.162	6.95	4.21	4.67
empty wagon	60	1.175	7.03	4.29	4.77
	70	1.181	7.06	4.41	4.86
	80	1.182	7.07	4.52	4.94

Table 8. Bridge dynamic response under reinforcement plan 2 (considering foundation rigidity).

Train formation	Speed (km/h)	Coefficient of impact	Mid-span deflection (mm)	Lateral displacement (mm)	
				Mid span	Pier top
DF4 + 40*C62	50	1.086	6.80	4.05	5.11
loaded wagon	60	1.096	6.85	4.27	5.19
	70	1.104	6.90	4.26	5.31
	80	1.102	6.89	4.52	5.55
DF4 + 40*C62	50	1.103	6.59	4.26	4.86
empty wagon	60	1.112	6.65	4.49	4.94
	70	1.120	6.69	4.48	5.04
	80	1.118	6.69	4.76	5.27

6 CONCLUSION

1. Lateral displacements of pier top and mid-span resulted from two reinforcement plans are close to each other. After comparison with calculation results with the one of rigid foundation, it is concluded that weak rigidity of substructure is the major cause resulting in excessive lateral displacement.
2. When taking effect of pier structure into account with rigid foundation, the 1st order lateral bending natural frequency is 3.521 Hz, while after taking foundation effect into account, natural frequency reduces to 1.700 Hz, with a relatively big drop, which indicates also the foundation rigidity of this bridge is relatively weak.
3. Although vibration accelerations of bridge can satisfy requirements under various calculation conditions, the lateral amplitude of the bridge fails to meet requirements according to "Railway Bridge Inspection Specification" (Tie Yun Han (2004) No.120), in addition, riding index of locomotive & rolling stock exceeds qualified limit under train speed of 80 km/h, therefore reinforcement treatment is necessary to take.
4. All together 3 reinforcement plans are calculated, of which one is with rigid foundation namely extreme condition of reinforcing foundation, another aims at pier reinforcement, and the last focus on main beam reinforcement. After comparison, it can be found that if pier reinforcement or main beam reinforcement is carried out solely without treatment to foundation, then lateral vibration of bridge remains too large to meet requirements of "Railway Bridge Inspection Specification". If foundation reinforcement is done, significant improvement will be seen in lateral vibration of bridge, and lateral displacement will meet requirements of "Railway Bridge Inspection Specification" under train speed of 60 km, thus, foundation reinforcement is recommended.

REFERENCES

Mangmang Gao, 2001. *Study on Train-Track-Bridge Coupling Vibration and Train Running on High-Speed Railway.* (Doctor Thesis), Beijing, China Academy of Railway Science

Wanming Zai, 1997. *Train-Track Coupling Dynamics,* beijing, China Railway Press

State standard of P.R.C., 1985, *Dynamic Performance Evaluation and Test Identification Specification for Railway Vehicle GB 5599-85*, Beijing, China Railway Press

Ministry-issued standard of P.R.C., 1993, *Dynamic Performance Test Identification Methods and Evaluation Standard TB/T 2360-93 of Railway Locomotive*, Beijing, China Railway Press

He Xia, & Yingjun Chen. 1992, *Interaction Analysis on Train-Beam-Pier System*, Beijing, China Civil Engineering Journal, 25(2) 3–12

Environmental Vibrations – Takemiya (ed.)
© *2005 Taylor & Francis Group, London, ISBN 0 415 39035 4*

System identification of a high-speed railway bridge with consideration of bridge-soil interaction

M.C. Huang
Department of Aircraft Engineering, Air Force Institute of Technology, Kaohsiung, Taiwan, R. O. C.

C.H. Chen
Department of Civil and Environmental Engineering, National University of Kaohsiung, Kaohsiung, Taiwan R.O.C.

T.C. Chen
Department of Aircraft Engineering, Air Force Institute of Technology, Kaohsiung, Taiwan, R. O. C.

ABSTRACT: This research intends to investigate an identification process of a railway bridge subjected to high-speed trains loading using a bridge-soil coupled system. The proposed algorithm can extract the actual system parameters rather than the mode parameters or equivalent ones. The actual system parameters are believed to be more meaningful and useful. The dynamic interaction of the bridge-foundation with the underlying soil is taken into account. The bridge foundation is simulated with an equivalent foundation-spring model. This equivalent spring for the consideration of dynamic interaction between the bridge foundation and the underlying soil, as well as with consideration of the combing radiation damping and material damping simultaneously. With the analysis of system identification, the actual stiffness and the damping coefficients can be obtained. Consequently, the calculations of the actual shears and moments at the bottom of the piers caused by the non-constant velocity motion or sudden brake are possible, which is more helpful for the safety evaluation of a high-speed railway bridge.

1   INTRODUCTION

The research of railway bridges subjected to moving vehicles can be tracked back to as the mid-nineteenth century. In the early research works, the dynamic response of railway bridges subjected to high-speed train loading has been an interested for many engineers. The recent studies focus on numerical simulations, possibly including the effects of train mass inertia, coupling with the train cars suspension systems, tracks stiffness, damping and roughness, especially for ballasted tracks, or rail-wheel contact. [Savin 2001, Klasztorny *et al* 1990, Lin *et al* 1990, Henchi *et al* 1997, Moreno *et al* 1997, Chen *et al* 2000, Yau 1996] Delgado and dos Santos (1997) carried out an analysis to investigate the effect of some parameters such as characteristics of stiffness and mass of a bridge, stiffness of train, bridge span, and track irregularities. Yang proposed the dynamic condensation method for vehicle-bridge interaction analysis and investigated the key parameters that govern the dynamic responses of simple beams using moving load assumption. [Yang *et al* 1994, Yang *et al* 1997]

Much progress has been made recently in the development of analytical procedures for evaluate the dynamic response of railway bridges under moving loading. Successful application of such procedure is essentially dependent on the incorporation of representative soil behavior in the analysis. Whenever a foundation element moves against the surrounding soil, stress wave originate at the contact surface and spread outward. These waves carry away some of the energy transmitted by the foundation into the soil. The magnitude of this damping depends on several factors, i.e. the frequency of excitation, the geometry of soil-foundation system, and the stress-strain characteristics of the soil. As a matter of fact, Knowledge on the subject of dynamic soil-structure interaction has derived primarily from studies for buildings and nuclear containment structures on mat foundation [Luco 1982]. Although the dynamic response of pile supported bridge or similar structures has been the issue of some research efforts, much is yet to be learned on the crucial problem parameters on the response of an isolated bridge foundation on piles and pile groups in multi-layered soil [Takemiya *et al* 1981, Mylonakis *et al* 1997].

In short, the usefulness of the theoretical analysis is limited by the degree of realistic representation of the formulated mathematical models. Obviously, a logical prelude to investigate the dynamic behavior of high-speed railway bridges subjected to trains loading is through the application of system identification techniques. This study intends to develop an identification algorithm to investigate the dynamic properties of a high-speed railway bridge under trains loading. The bridge piers are founded on pile. The dynamic bridge-soil interaction is taken into account. A linear model is used for the pier while a Winkler beam model is chosen to describe the foundational behavior of the bridge piers are founded on piles. The bridge foundation is simulated with an equivalent foundation-spring model. This equivalent spring for the consideration of dynamic interaction between the bridge foundation and the underlying soil, as well as with consideration of the combing radiation damping and material damping simultaneously. With the analysis of system identification, the actual stiffness and the damping coefficients can be obtained. Consequently, the calculations of the actual shears and moments at the bottom of the piers caused by the non-constant velocity motion or sudden brake are possible, which is more helpful for the safety evaluation of a high-speed railway bridge.

2 MOTION EQUATION

A typical 3-span bridge built on soft soil, as shown in Fig.1, is considered. The deck is rigid and the piers under high-speed trains loading are assumed to be linear. Meanwhile, the bridge-pier system involves a one-column bent founded on a group of 5 rigidly-capped piles. A pile is deemed as a beam on Winkler foundation. The role of pile-soil interaction is played through a set of continuously distributed springs and dashpots [Luco 1982]. Those springs and dashpots connect the pile to the free-field soil. The motion of soil serves as the support excitation of the pile-soil system.

Accordingly, the system equation of motion can be expressed as

$$-\left(\frac{m_{pj}}{2}+m_{fj}\right)\ddot{x}_{fj}-R_{fj}\left(\dot{x}_{fj},x_{fj}\right)+$$
$$R_{pj}\left(\dot{x}_{pj}-\dot{x}_{fj},x_i-x_{fj}\right)=\{0\} \qquad j=1\sim4 \qquad (1)$$

$$-\frac{m_{pj}}{2}\ddot{x}_d-R_{pj}\left(\dot{x}_d-\dot{x}_{fj},x_d-x_{fj}\right)+f_j=\{0\}$$
$$j=1\sim4 \qquad (2)$$

$$m_d\ddot{x}_d+\sum_{j=1}^{4}f_j=m_v(\ddot{x}_{vx}-\ddot{x}_d)\delta(x-Vt) \qquad (3)$$

in which m_{fj}, x_{fj} and R_{fj} are the mass, displacement and restoring force of the j-th foundation, respectively; m_{pj}, x_{pj} and R_{pj} are the lumped mass, displacement and restoring force of the j-th pier head; m_d and x_d are the mass and displacement of the bridge deck; m_v, $\ddot{x}_{vx}$ and V are the mass, acceleration and velocity of the center of trains; f_j is the restoring force of the j-th hinge. There storing forces of foundation and substructure are respectively written as

$$R_{fj}\left(\dot{x}_{fj},x_{fj}\right)=C_{fj}\dot{x}_{fj}+K_{fj}x_{fj} \qquad (4)$$

$$R_{pj}\left(\dot{x}_d-\dot{x}_{fj},x_d-x_{fj}\right)=C_{pj}\left(\dot{x}_d-\dot{x}_{fj}\right)+K_{pj}\left(x_d-x_{fj}\right) \qquad (5)$$

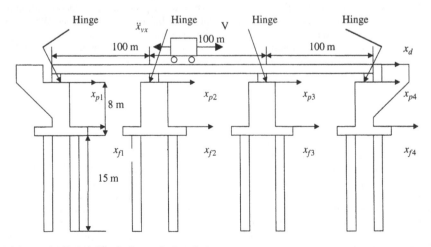

Figure 1. High-speed railway bridge built on pile foundations.

where C_{pj} and K_{pj} represent the damping coefficient and stiffness of the j-th column pier, respectively, while C_{fj} and K_{fj} are the equivalent "dashpot" and "spring" coefficients corresponding to the j-th foundation. Physically, K_{fj} reflects the stiffness and inertia characteristics of the pile-soil system, while C_{fj} expresses the energy loss in the system.

3 EQUIVALENT FORCES AT FOUNDATION

To determine the equivalent spring coefficient at foundation, K_{fj}, one may obtain the horizontal-displacement profile, $y_s(0)$, of the pile subjected to a statically horizontal load of magnitude P_0. This coefficient is then directly computed as $P_0/y_s(0)$. The estimation of such a displacement profile can be done through the application of finite element method. At a particular node i, the reactive force from the surrounding soil was $K_h y_s(0)$. Note that K_h is stiffness constant of soil spring and is given as

$$K_h = k_h A_h \qquad (6)$$

in which k_h is local subgrade reaction coefficient and A_h indicates the effective area. In the analysis, this area is computed by

$$A_h = b\left(\frac{l_{i-1}}{2} + \frac{l_i}{2}\right) = \frac{1}{2}b\left(l_{i-1} + l_i\right) \qquad (7)$$

where l_{i-1} and l_{i+1} are length of elements adjacent to node i and b is pile diameter.

The equivalent dashpot coefficient at pile head is computed from the values of radiation damping and material damping distributed along piles. Radiation or geometric damping represents the radiation of energy by wave spreading geometrically away from the pile-soil interface. Its coefficient c_m can be obtained from appropriate wave propagation problem [Gazetas et al 1984]. On the other hand, material damping refers to the hysteretic dissipation of energy in the soil. Its coefficient c_m is a function of the effective shear strain. The estimation of such strain was proposed by Kagawa and Kraft [1980]. Once it is known, the hysteretic damping ratio, $\beta = \beta(z)$, in the soil can be assessed by widely available experimental data in the form of damping-versus-strain curve [Richart et al 1977].

The radiation-damping coefficient associated with a circular cross section of diameter being b is expressed as [Gazetas et al 1984]

$$c_r = 2b\rho V_s\left\{1 + \left[\frac{3.4}{\pi(1-\upsilon)}\right]^{5/4}\right\}\left(\frac{\pi}{4}\right)^{3/4}\left(\frac{\pi f b}{V_s}\right)^{-1/4} \qquad (8)$$

in which ρ = soil mass density, V_s = S-wave velocity, ω = excitation frequency, and υ = Poisson's ratio. Due to the generation of surface type waves instead of plane-strain body wave at shallow depths, i.e. z less than $2.5b$, the above equation should be modified as

$$c_r = 4b\rho V_s\left(\frac{\pi}{4}\right)^{3/4}\left(\frac{\pi f b}{V_s}\right)^{-1/4} \qquad (9)$$

The material damping coefficient at pile depth z is related to the hysteretic damping ratio, β, by [16]

$$c_m \approx 2k(z)\frac{\beta}{\omega} \qquad (10)$$

in which $k(z)$ is a secant modulus defined as the ratio of the static local soil reaction against a unit length of pile over the corresponding pile deflection, and can be obtained from the local soil Young's modulus E_s

$$k(z) = \delta E_s(z) \qquad (11)$$

The coefficient δ is a function of the type of soil profile, the type of head loading and the ratio of pile stiffness to soil stiffness, E_p/E_s. Guidance for the selection of δ is provided in the literature [Richart et al 1977]. It is also noted that

$$E_s = 2\rho V_s^2(1+\upsilon)^2 \qquad (12)$$

Therefore, the dashpot coefficient at the pile head is evaluated by

$$C_f = \int_0^l (c_m + c_r)\gamma_s^2(z)dz \qquad (13)$$

in which $\gamma_s(z) = y_s(z)/y_s(0)$.

Finally, it is also noted that the equivalent damping ratio of pile head is given by

$$\xi_f = \frac{\omega}{2K(f)} = \frac{\pi f C_f}{K(f)} \qquad (14)$$

4 IDENTIFICATION PROCESSES

When the displacements of deck and foundations are measured, the system identification of the structure can be proceeded. Many system identification techniques are classified as output-error methods. The system parameters are obtained by minimizing the discrepancy between the recorded responses and the theoretical responses of the system. The parameter value so evaluated is called an optimal.

The performance of identification starts with a fixed initial value of C_{p2}. Through several times of minimization process, an optimal K_{p2} is obtained. Then, K_{p2} is fixed at this value and the minimization process is proceeded to again an optimal C_{p2}. Meanwhile, the corresponding system parameters C_{p1}, K_{p1}, C_{f1} and K_{f1} are yielded. This constitutes one cycle of identification. A few cycles may be needed to ensure the convergence of an error index.

The above minimization process is further illustrated as follows. We define the global measure-of-fit as

$$e = \sum_{i=1}^{2} \left[\left(\frac{m_1 + m_2 + m_d}{2} \right) \ddot{x}_d - \frac{m_v}{2} \ddot{x}_{vx} \delta(x - Vt) + \sum_{j=1}^{2} C_{pj} (\dot{x}_d - \dot{x}_{fj}) + \sum_{j=1}^{2} K_{pj} (x_d - x_{fj}) \right]^2 \quad (15)$$

The values of C_{p1} and K_{p1} can be obtained by solving

$$\frac{\partial e}{\partial (C_{p1})} = 0 \qquad \frac{\partial e}{\partial (K_{p1})} = 0 \qquad (16)$$

Similarly, C_{f1} and K_{f1} are assessed from performing identification on Eqs.(1),(4),(5), in which the update values of C_{pj} and K_{pj} are used.

To compare the convergence and accuracy of the identification process, the error index is defined as

$$EI = \left\{ \frac{\int_0^t [(\ddot{x}_{fj})_r - (\ddot{x}_{fj})_t]^2 dt}{\int_0^t [(\ddot{x}_{fj})_r]^2 dt} \right\}^{1/2} \quad (17)$$

where $(\ddot{x}_{fj})_r$ is the recorded or measured acceleration response and $(\ddot{x}_{fj})_t$ is the theoretical or identified response. The latter is calculated from the identified parameter values and the recorded input. We noted that fj refers to the j-th foundation. When fj is replaced by d, Eq.(17) may yield the error indices at deck.

5 NUMERICAL EXAMPLE

The pile profile analyzed in this paper is used five piles of the same diameter (0.5 m) and lengths (15 m) are rigidly capped. The length, width, and height of the

pile head are taken to be 6 m, 6 m, and 0.5 m, respectively. In the finite element analysis for piles, the length of each element between 0~1 m, 1~5 m, and under 5 m are 0.2 m, 0.5 m, and 1.0 m, respectively.

Based on data at different sites, the constant of soil spring is calculated and listed in Table1. Other used numerical data include (1) piers and deck: $m_1/2 = m_4/2 = 76.8 \times 10^3 kg$, $m_{p2}/2 = m_{p3}/2 = 38.4 \times 10^3 kg$, $m_d = 1377.9 \times 10^3 kg$, $K_{p1} = K_{p4} = 431.8 MN/m$, $K_{p2} = K_{p3} = 215.9 MN/m$, (2) foundations: $m_{f1} = m_{f2} = m_{f3} = m_{f4} = 43.2 \times 10^3 kg$, $K_{f1} = K_{f2} = K_{f3} = K_{f4} = 236.97 MN/m$, and (3) Trains: $m_v = 20.0 \times 10^3 kg$, $\ddot{x}_{vx} = -5.0 m/sec^2$, $V_0 = 50.0 m/sec$.

The damping ration of reinforced concrete is assumed to be 0.07. Meanwhile, the equivalent damping ratio of pile head is given by Eqs.(8)~(14) and is equal to $\zeta_f = 0.121 + 0.064f^{3/4}$, Table 2 presents its calculation procedure. By utilizing the concept of compound modal damping and Rayleigh damping, all damping coefficients are estimated. Their values turn out to be $C_{p1} = C_{p4} = 345 kN.s/m$, $C_{p2} = C_{p3} = 237 kN.s/m$, $C_{f1} = C_{f2} = C_{f3} = C_{f4} = 451 kN.s/m$.

The displacements of the deck and foundations under trains loading are calculated by Newmark's linear acceleration method and are treated as the measured responses.

The third identification cycle is then proceeded. The initial value of C_{p2} is equal to 242 kN.s/m. Minimizing the global measure-of-fit, we have $K_{p2} = 216.5 MN/m$, is presented in Fig. 2, which reveals that the least squares estimate is performed. As illustrated in Fig. 3, the optimal estimate of is $C_{p2} = 243 kN.s/m$. In the meantime, the identified parameter values of pier1 are $C_{p1} = 469.0 kN.s/m$ $K_{p1} = 430.9 MN/m$.

Table 1. Constant of soil spring on a pile.

Pile depth z (m)	$K_{h1,2}$ (MN/m)
0 (pile head)	3
0.2~0.8	6
1	10.5
1.5~4.5	15
5	37.5
6~14	60
15	30

Table 2. Equivalent damping ratios of pile heads.

Pile heads	$\Sigma \omega C_m$ $\gamma_s^2 \Delta_z$	$\dfrac{\Sigma C_r}{f^{1/4} \gamma_s^2}$ Δz	K_{fj} (kN/m)	ξ_{fj}
$j = 1~4$	57475*	4810*	236965	$0.121 + 0.064f^{3/4}$

* Five piles.

252

Those related with foundation1 are $C_{f1} = 454\,kN.s/m$, $K_{f1} = 237.57\,MN/m$.

Numerical calculation suggests that three cycles of identification are sufficient. Column 2~3, column 4~5, and column 6~7 in Table 3, respectively summarize the iterative identified parameter values of pier1, pier2, and foundation1.

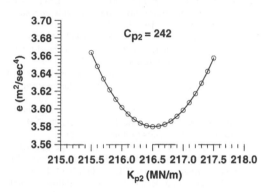

Figure 2. Global measure-of-fit in the third cycle setting $C_{p2} = 242.0$.

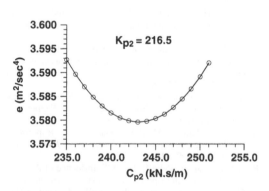

Figure 3. Global measure-of-fit in the third cycle setting $K_{p2} = 216.5$.

The comparison between the identified and measured response is made. Figs 4 and 5 show the displacement responses at deck and foundation1, respectively. The identified and measured ones are virtually identical. They also promise an excellent identification.

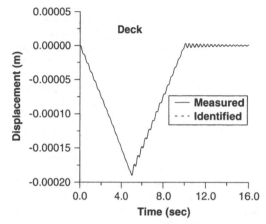

Figure 4. Comparison between identified and measured displacements of deck.

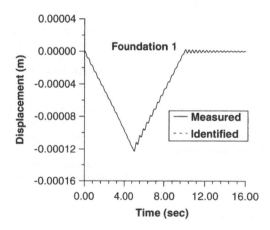

Figure 5. Comparison between identified and measured displacements of foundation 1.

Table 3. Identified parameters of piers and Foundation1.

Number of cycle	C_{p1} kN.s/m	K_{p1} MN/m	C_{p2} kN.s/m	K_{p2} MN/m	C_{f1} kN.s/m	K_{f1} MN/m
1	482	440.5	230.0	210.0	436.0	224.1
2	470	431.7	242.0	216.0	455.0	237.1
3	469	430.9	243.0	216.5	454.0	237.6
True value	475.0	431.8	237.0	215.9	451.0	237.0
Error index (EI)	Deck 0.047				Foundation1 0.072	

6 CONCLUSIONS

This paper develops an identification procedure to examine the dynamic characteristics of a high-speed highway bridge built on pile foundation. A linear model is used for the pier. To be more realistic, piles are founded on soft sits. The pile-soil interaction is taken into account by using a series of distributed springs and dashpots. The ordinary identification process can be performed with only minimal computational effort beyond that required for the identification of a linear system.

The contribution of this paper can be evaluated from several ways. Unlike many other identification methods that generally identify the equivalent system parameters or modal parameters, the proposed method is able to identify the physical system parameters. Those parameters are more meaningful and useful. For instance, if degradation of pier stiffness is found, the appropriate retrofit can be executed. In this regard, the physical parameters obtained through the proposed identification process may provide a basis by which various warning system for a high-speed railway bridge can be established. Above all, the proposed method also provides the stiffness and damping coefficients at foundation. This can be employed to check many currently available methods in calculating those coefficients.

In conclusion, the comparison between the identified and measured responses at deck and foundation1 shows that the developed technique is feasible for identifying a high-speed railway bridge under high-speed trains loading and built on pile foundation. With the analysis of system identification, the actual stiffness and the damping coefficients can be obtained. Consequently, the calculations of the actual shears and moments at the bottom of the piers caused by the non-constant velocity motion or sudden brake are possible, which are more useful and helpful for the safety evaluation of a high-speed railway bridge.

REFERENCES

Savin E. 2001, Dynamic Amplification Factor and Response Spectrum for The Evaluation of Vibrations of Beams under Successive Moving Loads, *Journal of Sound and vibration*, **248(2)**, 267–288

Klasztorny M. & langer J. 1990, Dynamic response of single-span beam bridges to a series of moving loads, *Earthquake Engineering and structural Dynamics*, **19**, 1107–1124

Lin Y.-H. & Trethewey M. W. 1990, Finite element analysis of elastic beams subjected to moving dynamic loads, *Journal of Sound and vibration*, **136**, 323–342

Fafard K. M., Dhatt G. & Talbot M. 1997, Dynamic behaviour of multi-span beams under moving loads, *Journal of Sound and vibration*, **199**, 33–50

Moreno Delgado R. & Dos Santos S. M. 1997, Modelling of railway bridge-vehicle interaction on high speed tracks, *Computer and Structures*, **63**, 511–523

Chen Y. H. & Li C. Y. 2000, Dynamic response of elevated high-speed railway, *Journal of Bridge Engineering*, **5**, 124–130

Yau J. D. 1996, Dynamic response of bridges traveled by trains-analytical and numerical approaches, Ph. D. thesis, national Taiwan University, Taipei, Taiwan

Yang Y. B. & Lin B. H. 1994, Vehicle-bridge interaction analysis by dynamic condensation method, *Journal of Structural Engineering*, ASCE, **12(11)**, 1634–1643

Yang Y. B., Yau J. D. & Hsu L. C. 1997, Vibration of simple beams due to trains moving at high speeds, *Engineering of structure*, **19(11)**, 936–944

Luco J. E. 1982, Linear soil-structure interaction: a review, *Earthquake ground motion effects structures*, ASME, **AMD 53**, 41–57

Takemiya H. and Yamada Y. 1981, Layered soil-pile-structure dynamic interaction, *Earthquake Engineering and Structure Dynamics*, **9**, 437–458

Mylonakis G., Nikolaou A. & Gazetas G. 1997, Soil-pile-bridge seismic interaction: kinematic and inertia effects. Part I: Soft soil, *Earthquake Engineering and Structural Dynamics*, **26**, 337–359

Gazetas G. & Dobry R. 1984, Simple radiation damping model for piles and footings, *Journal of Engineering Mechanics*, ASCE, **110**, 937–956

Kagawa T. & Kraft L. M. 1980, Seismic $p \sim y$ response of flexible piles, *Journal of the Geotechnical Engineering*, ASCE, **106**, 899–918

Richart F. E. Jr. & Wylie E. B. 1977, Influence of dynamic soil properties on response of soil masses, *Structural and Geotechnical Mechanics*, W. J. Hall, ed, Prentice-Hall Inc., Englewood Cliffs, N. J., 141–162

Gazetas G. & Dobry R. 1984, Horizontal response of piles in layered soils, *Journal of Geotechnical Engineering*, ASCE, **110**, 20–40

Environmental Vibrations – Takemiya (ed.)
© *2005 Taylor & Francis Group, London, ISBN 0 415 39035 4*

Vertical impedance for stiff and flexible embedded foundations

M. Liingaard, L. Andersen & L.B. Ibsen
Department of Civil Engineering, Aalborg University, Aalborg, Denmark

ABSTRACT: The dynamic response of offshore wind turbines e.g. eigen frequencies and eigen modes, are affected by the properties of the foundation. The purpose of this paper is to evaluate the dynamic soil-structure interaction of embedded foundations for offshore wind turbines, with the intention that the dynamic properties of the embedded foundation can be properly included in a composite structure-foundation system. The investigation concerns the determination of the vertical dynamic stiffness for surface footings, skirted foundations and solid embedded foundations. The soil surrounding the foundation is homogenous with linear viscoelastic properties. The dynamic stiffness of the foundations is expressed by dimensionless frequency dependent dynamic stiffness coefficients corresponding to the vertical degree of freedom. The dynamic stiffness coefficients for the foundations are evaluated by means of a dynamic three-dimensional coupled Boundary Element Method/Finite Element Method model. Benchmark tests have been performed for analytical half-space solutions for rigid surface footings and there is good agreement between the analytical solution and the BEM/FEM model. The frequency dependent dynamic vertical stiffness of skirted foundations has been investigated and compared with the solutions for the rigid surface footing and the solid embedded foundation.

1 INTRODUCTION

Wind energy has proven to become a significant and powerful global energy resource. The international developments in wind energy technology has been governed by the European market for the last 25 years and the wind energy technologies have been well established in most western European countries. In 2004 Germany achieved 7% (16,629 MW), Spain 6.5% (8,263 MW), and Denmark 20% (3,117 MW) of the total national electrical consumption from wind energy resources, according to the Global Wind Energy Council (www.gwec.net). The general aim of the Renewable Energy Policies for most countries with significant wind energy potential is to increase the percentage of total electrical consumption produced by wind power. The wind power has until recently been based on onshore wind turbines, but the newly developed megawatt sized wind turbines and new knowledge about offshore wind conditions are improving the economics of offshore wind power. Hence, offshore wind energy is rapidly becoming competitive with other power generating technologies.

The continuous improvement in wind turbine technology means that the wind turbines have increased tremendously in both size and performance during the last 25 years. The general output of the wind turbines is improved by lager rotors and more powerful generators. In order to reduce the costs, the overall weight

of the wind turbine components is minimized, which means that the wind turbine structures become more flexible and thus more sensitive to dynamic excitation.

A modern offshore wind turbine (1.5 to 2 MW) is typically installed with a variable speed system so the rotational speed of the rotor varies from, for example, 10–20 RPM. This means that the excitation frequency of the rotor system varies. The first excitation frequency interval then becomes 0.17–0.33 Hz (for 10–20 RPM) and is referred to as the 1Ω frequency interval. The second excitation frequency interval corresponds to the rotor blade frequency that depends on the number of blades. For a three-bladed wind turbine the 3Ω frequency interval is equal to 0.5–1.0 Hz (for 10–20 RPM). Since the first resonance frequency ω_1 of the modern offshore wind turbines is placed between 1Ω and 3Ω, it is of outmost importance to be able to evaluate the resonance frequencies of the wind turbine structure accurately as the wind turbines increase in size.

At present, the wind turbine foundations are modeled simply by beam elements or static soil springs, which means that the foundation stiffness is frequency independent. The purpose of this paper is to investigate the dynamic frequency dependent stiffness for offshore wind turbine foundation concepts. The dynamic stiffness coefficients are investigated with a dynamic three-dimensional coupled Boundary Element Method/Finite Element Method model for a homogeneous elastic half-space. Prior to the analysis of the skirted foundations the BEM/FEM model has been compared with known analytical half-space solutions for rigid surface footings by Veletsos & Tang (1987). The paper is restricted to an examination of the vertical dynamic stiffness for circular skirted foundations. The behavior of the skirted foundation is compared with the dynamic characteristics of surface and solid embedded foundations.

2 SKIRTED FOUNDATION

The skirted foundation (also known as bucket foundation) is an innovative foundation solution that has been developed over the past 5 years and the concept has been utilized for a Vestas 3.0 MW offshore wind turbine in the northern part of Denmark. The concept is sketched in Figure 1.

In the initial phase of the installation process the skirt penetrates into the seabed due to the weight of the structure. In the second phase suction is applied to penetrate the skirt to the design depth. After installation the foundation acts a hybrid of a pile and a gravity based foundation. The stability of the foundation is ensured by a combination of earth pressures on the skirt and the vertical bearing capacity of the bucket. This foundation type is a welded steel structure and the fabrication/ material costs are comparable to those of the monopile foundation concept. The installation phase does not require heavy pile hammers and the decommissioning is a relatively simple process where the foundation can be raised by applying pressure to the bucket structure. See Ibsen & Liingaard (2005) for further details.

3 BEM/FEM MODEL

The dynamic stiffness for the foundations is evaluated by means of the dynamic three-dimensional coupled Boundary Element Method/Finite Element Method program BEASTS by Andersen & Jones (2001). The model used in BEASTS is based on a combined finite element (FE) and boundary element (BE) formulation where the foundation is modeled by finite elements and the soil surrounding the foundation is described by boundary elements. The reason for modeling the surrounding soil with boundary elements is that the boundary element method is superior to the finite element method due to its inherent ability to model radiating waves. The finite element method will not be described in further detail, see e.g. Petyt (1998) for further reading. The boundary element part of BEASTS is an extension of the theory presented by Domínguez (1993), which has been modified to account for open domains and to allow coupling with finite elements.

3.1 Boundary elements in the frequency domain

The governing equation of motion for a three dimensional body Ω in the frequency domain is given by

$$\frac{\partial \sigma_{ij}(\mathbf{x},\omega)}{\partial x_j} + \rho B_i(\mathbf{x},\omega) + \omega^2 \rho U_i(\mathbf{x},\omega),$$
(1)

where $U_i(\mathbf{x},\omega)$ ($i = 1, 2, 3$) are the complex amplitudes of the displacement field, $\rho B_i(\mathbf{x}, \omega)$ are the body forces and $\sigma_{ij}(\mathbf{x}, \omega)$ ($j = 1, 2, 3$) are the stresses that may be computed from the displacements by the constitutive relation. The boundary conditions on the surface Γ of the body Ω are:

$$\left. \begin{aligned} U_i(\mathbf{x},\omega) &= \hat{U}_i(\mathbf{x},\omega) \quad \text{for} \quad \mathbf{x} \in \Gamma_U \\ P_i(\mathbf{x},\omega) &= \hat{P}_i(\mathbf{x},\omega) \quad \text{for} \quad \mathbf{x} \in \Gamma_P \end{aligned} \right\},$$

$$\Gamma = \Gamma_U \cup \Gamma_P, \quad \Gamma_U \cap \Gamma_P = \varnothing,$$
(2)

where the displacement amplitude $U_i(\mathbf{x}, \omega)$ is given on one part of the boundary, Γ_U, and the surface traction $P_i(\mathbf{x}, \omega) = \sigma_{ij}(\mathbf{x}, \omega) n_j(\mathbf{x})$ is given on the remaining part of the boundary, Γ_P. Here $n_j(\mathbf{x})$ are the components of the outward unit normal to the surface.

To obtain the boundary element formulation of equation (1), a second state $U_{il}^*(\mathbf{x}, \omega; \xi)$ is identified as the fundamental solution to the equation of motion

$$\frac{\partial \sigma_{ijl}^*(\mathbf{x},\omega;\xi)}{\partial x_j} + \rho \delta(\mathbf{x}-\xi)\delta_{il} + \omega^2 \rho U_{il}^*(\mathbf{x},\omega;\xi),$$
(3)

Figure 1. Illustration of skirted foundation (bucket foundation).

where $\delta(\mathbf{x} - \xi)$ is the Dirac delta function and δ_{il} is the Kronecker delta. The fundamental solution, or Green's function, for the displacement field is given as $U_{il}^*(\mathbf{x}, \omega; \xi)$. $U_{il}^*(\mathbf{x}, \omega; \xi)$ is the solution to the equation of motion in the frequency domain for a harmonic concentrated load with a unit amplitude (with a circular frequency ω) at the source point ξ in the direction of l. It should be noted that $U_{il}^*(\mathbf{x}, \omega; \xi)$ is a 3×3 matrix, i.e. there are three displacement components for each direction l at the source point ξ.

The fundamental solution is based on wave propagation in the full space and therefore only represents body waves emanating from the source, i.e. dilatation and shear waves with phase velocities c_P and c_S, respectively. In three dimensions, $U_{il}^*(\mathbf{x}, \omega; \xi)$ has a singularity of the order $1/r$.

The fundamental solution is applied as a weight function in the weak formulation of the equation of motion (1) for the physical field and vice versa. After some manipulations, and disregarding body forces in the interior of the domain, Somigliana's identity is derived:

$$C_{il}(\mathbf{x})U_l(\mathbf{x},\omega) + \int_\Gamma P_{il}^*(\mathbf{x},\omega;\xi)U_l(\xi,\omega)d\Gamma(\xi)$$

$$= \int_\Gamma U_{il}^*(\mathbf{x},\omega;\xi)P_l(\xi,\omega)d\Gamma(\xi). \qquad (4)$$

Here $P_{il}^*(\mathbf{x}, \omega; \xi)$ is the surface traction related to the Green's function $U_{il}^*(\mathbf{x}, \omega; \xi)$. $U_{il}^*(\mathbf{x}, \omega; \xi)$ and $P_{il}^*(\mathbf{x}, \omega; \xi)$ are the fundamental solutions, for the displacement and the traction, respectively. $C_{il}(\mathbf{x})$ depends only on the geometry of the surface Γ. $C_{il}(\mathbf{x}) = 0.5\delta_{il}$ on a smooth surface on the boundary Γ and $C_{il}(\mathbf{x}) = \delta_{il}$ inside the body Ω. The detailed derivation of (4) and properties of $C_{il}(\mathbf{x})$ are given in Domínguez (1993) and Andersen (2002).

In order to evaluate the Boundary Integral Equations in (4) for a point i on the boundary the boundary surface is discretized into a finite number of boundary elements. The Boundary Integral Equation can then be solved numerically for any point i on the boundary. The boundary surface can be discretized by different types of elements with varying order of integration. The elements used in the present study are second order elements with quadratic interpolation, due to the fact that 9-noded boundary elements are superior in performance and convergence compared to elements with constant or linear interpolation. By using quadratic elements it is sufficient to use 3 elements per wave length (compared with 10–20 elements for constant or linear elements).

To obtain the BE formulation, the state variable fields on the boundary are discretized. $\mathbf{P}_j(\omega)$ and $\mathbf{U}_j(\omega)$ be the vectors storing the displacements and tractions at the N_j nodes in element j. The displacement and traction fields over the element surface Γ_j then become

$$\mathbf{U}(\mathbf{x},\omega) = \mathbf{\Phi}_j(\mathbf{x})\mathbf{U}_j(\omega), \qquad \mathbf{P}(\mathbf{x},\omega) = \mathbf{\Phi}_j(\mathbf{x})\mathbf{P}_j(\omega),$$
$$(5)$$

where $\mathbf{\Phi}_j(\mathbf{x})$ is a matrix storing the interpolation, or shape, functions for the element. This allows the unknown values of the state variables to be taken outside the integrals in Eq. (4). Finally, the two- or three-row matrices originating from Eq. (4) for each of the observation points may be assembled into a single matrix equation for the entire BE domain,

$$\mathbf{HU} = \mathbf{GP}. \qquad (6)$$

Component (i, k) of the matrices $\mathbf{H}$ and $\mathbf{G}$ stores the influence from degree-of-freedom k to degree-of-freedom i for the traction and the displacement, respectively, i.e. the integral terms on the left- and right-hand side of Eq. (4). The $C_{il}(\mathbf{x})$ terms are absorbed into the diagonal of $\mathbf{H}$.

3.2 Coupling of FE and BE schemes

In some of the analyses the foundation consists of thin structures (skirt) and the use of boundary elements in this region is not appropriate. In these sections of the model finite elements are used. In order to couple a BE domain formulated in terms of surface tractions with an FE region with loads applied in terms of nodal forces, a transformation matrix $\mathbf{T}$ is defined, such that $\mathbf{F} = \mathbf{TP}$. Here $\mathbf{F}$ is the vector of nodal forces equivalent to the tractions $\mathbf{P}$ applied on the domain. The transformation matrix only depends on the spatial interpolation functions, i.e. the shape functions, for the elements along the interaction boundary. Hence, $\mathbf{T}$ may be determined once and for all and applied in all analyses with a given model geometry. Subsequently, for each frequency the matrix

$$\mathbf{TG}^{-1}\mathbf{H} = \mathbf{K}_{BE} \qquad (7)$$

defines an *equivalent dynamic stiffness matrix* for the boundary element domain. This operation turns the BE domain into a macro finite element. It should be noted that $\mathbf{K}_{BE}$ is a fully populated and asymmetrical matrix.

4 DYNAMIC STIFFNESS OF FOUNDATIONS

A generalized mass less axisymmetric rigid foundation has 6 degrees of freedom: one vertical, two horizontal, two rocking and one torsional. The six degrees of freedom and the corresponding forces and moments are shown in Figure 2.

For a harmonic excitation with the circular frequency ω, the dynamic stiffness matrix $\mathbf{K}$ is related to the vector of forces and moments $\mathbf{R}$ and the vector of displacements and rotations $\mathbf{U}$ as follows:

$$\mathbf{R} = \mathbf{KU}, \tag{8}$$

where the components of $\mathbf{R}$, $\mathbf{K}$ and $\mathbf{U}$ are defined as

$$\mathbf{K} = \begin{bmatrix} K_{VV}^{\bullet} & 0 & 0 & 0 & 0 & 0 \\ 0 & K_{HH}^{\bullet} & 0 & 0 & 0 & -K_{MH}^{\bullet} \\ 0 & 0 & K_{HH}^{\bullet} & 0 & K_{MH}^{\bullet} & 0 \\ 0 & 0 & 0 & K_{TT}^{\bullet} & 0 & 0 \\ 0 & 0 & K_{MH}^{\bullet} & 0 & K_{MM}^{\bullet} & 0 \\ 0 & -K_{MH}^{\bullet} & 0 & 0 & 0 & K_{MM}^{\bullet} \end{bmatrix} \tag{9}$$

and

$$\mathbf{R} = \begin{bmatrix} V/GR^2 \\ H_1/GR^2 \\ H_2/GR^2 \\ T/GR^3 \\ M_1/GR^3 \\ M_2/GR^3 \end{bmatrix}, \quad \mathbf{U} = \begin{bmatrix} w/R \\ u_1/R \\ u_2/R \\ \theta_T \\ \theta_{M1} \\ \theta_{M2} \end{bmatrix} \tag{10}$$

R is the radius of the foundation and G is the shear modulus of the soil. The terms in $\mathbf{K}$ are functions of the circular frequency ω and $\mathbf{K}$ reflects the dynamic stiffness or impedance of the soil for a given shape of the foundation. The components of $\mathbf{K}$ can be written as:

$$K_{ij}^{\bullet} = K_{ij}^0 \left(k_{ij} + ia_0 c_{ij} \right), \quad (i,j = H, M, T, V), \tag{11}$$

where K^0_{ij} is the static value of ijth stiffness component, k_{ij} and c_{ij} are the dynamic stiffness and damping coefficients, respectively. $a_0 = \omega R/c_S$ is the dimensionless frequency where R is the radius of the foundation and c_S is the shear wave velocity of the soil. The real part of (11) is related to the stiffness and inertia properties of the soil, whereas the imaginary part describes the damping of the system. For a soil without material damping c_{ij} reflects the damping effects of the energy dissipated by radiation of waves into the subsoil.

5 BENCHMARK TESTS

The boundary element formulation in the BEM/FEM program has been tested and compared with known analytical half-space solutions for rigid surface footings by Veletsos & Tang (1987). Similar analytical solutions are given in Luco & Westmann (1971) for vertical vibrations and in Veletsos & Wei (1971) for lateral and rocking vibrations of rigid circular footings.

The analytical solution given by Veletsos & Tang (1987) is based on a perfect elastic half space with Poisson's ratio $\nu = 1/3$, and relaxed boundary conditions under the foundation are assumed, corresponding to the condition of 'smooth' contact.

The boundary element model contains a mass-less circular foundation with a radius $R = 5$ m. Due to symmetry only half of the foundation is included. The foundation is modeled by 40 finite elements (quadratic interpolation) of varying size. The soil is discretized by a total of 152 boundary elements with quadratic interpolation. The soil below the foundation consists of 40 boundary elements, and the free surface is modeled by 112 boundary elements with a relatively high mesh density close the edge of the foundation. The mesh

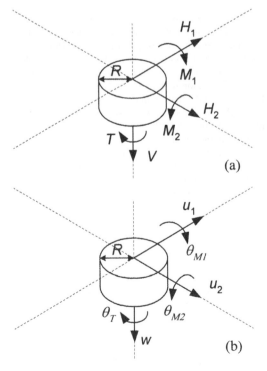

Figure 2. Sign conventions for a) applied force and moment resultants and b) corresponding displacements and rotations.

of the free surface is truncated at 15 m (3 times radius R). The soil is linear elastic with a shear modulus $G = 1 \cdot 10^6$ Pa, $\nu = 1/3$ and $p = 1000\,\text{kg/m}^3$. The connection between soil and foundation corresponds to the condition of 'rough' contact.

The boundary element model has been utilized for 13 different excitation frequencies in the range $a_0 \epsilon [0;\ 6]$. For a given excitation frequency a vertical load equal to 1 N is applied in the centre on top of the foundation and the complex displacements are computed. The vertical components of the impedance matrix (k_{VV} and c_{VV}) are then determined from the load and the displacement response.

The static stiffness $K^0{}_{VV}$ is used to normalize the components k_{VV} and c_{VV}. $K^0{}_{VV}$ is estimated from the lowest excitation frequency ($a_0 = 0.01$). The static value $K^0{}_{VV}$ for rough foundations is given as (Spence, 1968):

$$K_{VV,static} = 4\frac{\ln(3-4\nu)}{(1-2\nu)}. \tag{12}$$

$K^0{}_{VV}$ for $a_0 = 0.01$ is equal to 6.24 and the static value from (12) equal to 6.13, the difference is 1.7% k_{VV} and c_{VV} are shown in Figure 3. There is good agreement,

when it is considered that the analytical solution by Veletsos & Tang (1987) is based on relaxed boundary conditions and the boundary element solution corresponds to welded or rough contacts. The same type of results has been reported by Alarcon et al. (1989).

6 ANALYSIS OF VERTICAL DYNAMIC STIFFNESS

The vertical impedance is investigated for skirted foundations and then compared with solid foundations with the same embedment depth as the skirted foundations in consideration. The skirted foundations have been investigated for skirt depths equal to 1/4, 1/2 and 1 times the diameter of the foundation.

6.1 BE/FE model of the skirted foundations

The BE/FE model of the skirted foundation is shown in Figure 4 for an embedment depth of 1/2 times the foundation diameter.

The model consists of four sections; a mass less finite element section that forms the top of the foundation where the load is applied, a finite element section of the skirts, a boundary element domain inside the skirts and finally, a boundary element domain outside the skirts that also forms the free surface. All the elements are with quadratic interpolation. The models of the skirted foundation contain approx. 100 finite elements and 350 boundary elements. The main model properties are shown in Table 1.

The boundary element mesh size has been investigated by convergence of the vertical static stiffness. The mesh sizes used in the calculations are given in Table 1.

6.2 Vertical dynamic stiffness of skirted foundation

The impedance for the skirted foundations has been investigated for a rigid skirt and a skirt with steel properties (Table 1). There is no significant difference in the impedance. As long as the stiffness of the skirt material is more than 10,000 times larger than the soil, there is no drastic change in the dynamic stiffness properties.

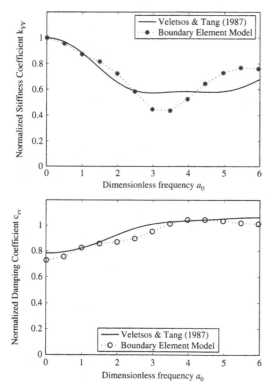

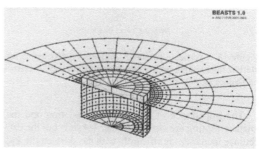

Figure 3. Vertical dynamic stiffness coefficients for a circular mass less foundation on an elastic half-space ($\nu = 1/3$).

Figure 4. Discretization of one half of the skirted foundation.

259

Table 1. Model properties for skirted foundation.

	1/4 skirt	1/2 skirt	1/1 skirt
Radius R	5.0 m	5.0 m	5.0 m
Skirt depth H	2.5 m	5.0 m	10.0 m
BE mesh size	$4 \cdot R$	$5 \cdot R$	$6 \cdot R$
G_{soil}	$1 \cdot 10^6$ Pa		
ν_{soil}	1/3		
ρ_{soil}	1000 kg/m^3		
E_{skirt}	$2 \cdot 10^{11}$ Pa/$1 \cdot 10^{15}$ Pa		
ν_{skirt}	0.25		
ρ_{skirt}	1000 kg/m^3		
t_{skirt}*	25 mm		

* Thickness of skirt.

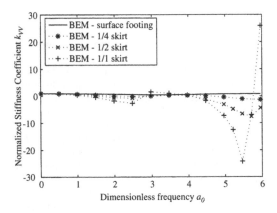

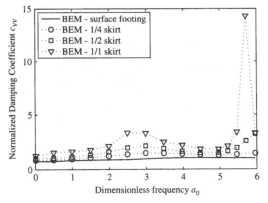

Figure 5. Vertical dynamic stiffness coefficients for skirted foundations with varying skirt depths on an elastic half-space ($\nu = 1/3$).

The impedance of the skirted foundation and the surface footing are similar for the 1/4 skirt but the difference increases significantly with the skirt length. k_{VV} in Figure 5 are characterized by high amplitude oscillations that increase with skirt length around $a_0 = 3$ and

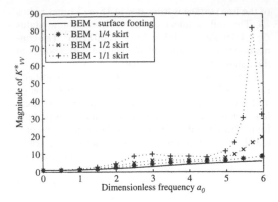

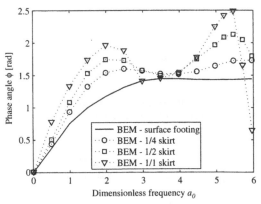

Figure 6. Magnitude and phase angle of the vertical dynamic stiffness for skirted foundations with varying skirt depths on an elastic half-space ($\nu = 1/3$).

the tendency is repeated close to $a_0 = 6$. The variation of c_{VV} with skirt length shows the same pattern, however the oscillations are not as evident as in the case of k_{VV}.

Instead of describing the characteristics of K^*_{VV} by k_{VV} and c_{VV} the complex dynamic stiffness can be represented by the complex modulus (or magnitude) and phase angle of K^*_{VV}.

The magnitude of K^*_{VV} corresponds to the amplitude of the total force necessary to create a unit displacement of the foundation. The phase angle simply describes the phase shift between the total force and the displacement.

In Figure 6 the amplitude of K^*_{VV} increases with a_0 for the surface foundation the foundation with 1/4 skirt. For increasing skirt lengths the magnitude oscillates around $a_0 = 3$ and 6. Furthermore it is evident that the amplitude of K^*_{VV} is dominated by the contribution from the damping, when Figure 5 and 6 are compared.

The sudden change in the phase angle ϕ at $a_0 = 3$–4 is likely to be due to a resonance phenomenon of

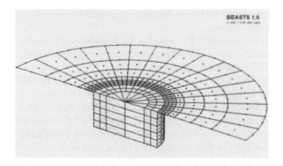

Figure 7.   Discretization of one half of the solid foundation.

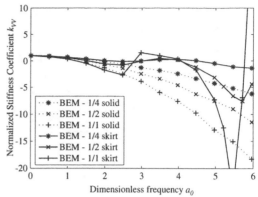

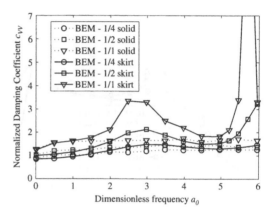

Figure 8.   Comparison of the vertical dynamic stiffness coefficients for the skirted and the solid foundations on an elastic half-space ($\nu = 1/3$).

Table 2.   Model properties for solid foundation.

	1/4 solid	1/2 solid	1/1 solid
Radius R	5.0 m	5.0 m	5.0 m
Embedment depth H	2.5 m	5.0 m	10.0 m
BE mesh size	$4 \cdot R$	$5 \cdot R$	$6 \cdot R$
G_{soil}/E_{soil}	$1 \cdot 10^6$ Pa$/2.667 \cdot 10^6$ Pa		
ν_{soil}	1/3		
ρ_{soil}	1000 kg/m^3		
E_{solid}	$2.667 \cdot 10^{10}$ Pa		
ν_{solid}	0.25		
ρ_{solid}	1000 kg/m^3		

the soil inside the skirted foundation. The tendency seems to be existent again at $a_0 = 6$.

6.3   BE/FE model of the solid foundations

The BE/FE model of the solid foundation is shown in Figure 7 for an embedment depth of 1/2 times the foundation diameter.

The model of the solid foundation consists of two sections; a finite element section of the solid foundation and a boundary element domain for discretizing the surrounding soil.

All the elements are with quadratic interpolation. The models of the skirted foundation contain approx. 200 finite elements and 250 boundary elements. The main model properties are shown in Table 2.

The impedance for the solid foundations has been investigated for a stiff foundation where the foundation stiffness is 10,000 times higher than the soil (Table 2). The density of the solid foundation is equal to the density of the surrounding soil.

6.4   Skirted foundation versus solid foundation

The vertical impedance of the skirted foundation is compared with the impedance of the solid foundation.

The vertical dynamic stiffness coefficients k_{VV} and c_{VV} are presented for the skirted and solid foundation in Figure 8.

The impedance of the solid foundation does not show the high amplitude oscillations as in the case of the skirted foundation. The curvature of k_{VV} for the solid foundations in Figure 8 corresponds to the shape of the classical dynamic stiffness $k - \omega^2 m$ where the inertia term $\omega^2 m$ increases with mass and circular frequency (k and m are some characteristic stiffness and mass of the system, respectively). The variation of c_{VV} with embedment depth shows that the radiational damping increases but the shape of the curves are similar for the three embedment depths.

The impedance of the skirted foundations seem to follow the same path as the solid foundation for a given skirt length until $a_0 = 2.5$ where k_{VV} of the skirted foundations deviates from the path of the solids. The reason may be that there is some sort of interference of P or S-waves in the soil inside the skirt.

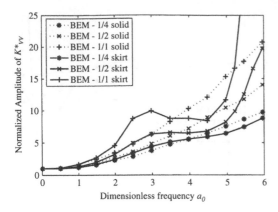

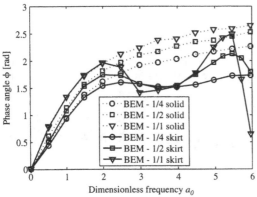

Figure 9. Comparison of the magnitude and phase angle of the vertical dynamic stiffness for the skirted and the solid foundations on an elastic half-space ($\nu = 1/3$).

In Figure 9 the amplitude of K^*_{VV} and the phase angle ϕ is shown for both foundation types. As expected the amplitude and phase angle plots do not show any significant oscillations for the solid foundation. However, it should be noted that the phase angle ϕ for the solid foundation tends towards higher levels as the embedment depth increases, whereas the limiting value for the surface foundation was close to $\pi/2$.

The comparison of the phase angles shows that the dynamic behavior of skirted and solid foundations is similar until a sudden frequency. For a_0 from 0 to 2 the phase angle is almost identical for the two types, after which the phase angle shifts drastically for the skirted foundations. This supports the idea of some sort of resonance or local dynamic mutual reaction inside the skirted foundation.

7 CONCLUSION

A coupled boundary element and finite element method has been applied for analyzing the vertical impedance of a new foundation concept for offshore wind turbines.

The dynamic stiffness for the foundations is evaluated by means of the three-dimensional coupled Boundary Element Method/Finite Element Method program BEASTS by Andersen & Jones (2001). The reason for using the coupled formulation is that the inherent radiotional damping phenomena of an unbounded soil domain can be modeled accurately by means of boundary elements, whereas the complex foundation geometry is best modeled by finite elements.

The boundary element formulation has been tested and compared with known analytical half-space solutions for rigid circular surface footings by Veletsos & Tang (1987). There is good agreement, when it is considered that the analytical solution by Veletsos & Tang (1987) is based on relaxed boundary conditions and the boundary element solution corresponds to welded or rough contacts. Furthermore, the static stiffness K^0_{VV} has been determined by the boundary element model from the lowest excitation frequency ($a_0 = 0.01$) and compared with the analytical expression for rough foundations given by Spence (1968). The difference is 1.7%.

The vertical impedance of skirted foundations and solid foundations has been determined for different excitation frequencies in the range $a_0 \epsilon [0; 6]$. The solid foundation behaves more or less as expected, where the dynamic stiffness is governed by the inertial forces as a_0 increases. The skirted foundation behaves as the solid foundation until $a_0 = 2$–2.5. At this frequency level the skirted foundation deviates from the path of the solid foundation and exhibit relatively high oscillations in the interval of $a_0 = 2.5$–4. The sudden change in the stiffness and phase angle ϕ at $a_0 = 3$–4 is likely to be due to a resonance phenomenon or local dynamic mutual reaction of the soil inside the skirted foundation. The tendency seems to be existent again at $a_0 = 6$

The present study clearly shows that the dynamic properties of a wind turbine foundation have to be accounted for. In particular, a good model of the geometry is required in order to achieve reliable responses of the structure.

REFERENCES

Alarcon, E. & Cano, J.J. & Dominguez, J. (1989), Boundary Element approach to the dynamic stiffness functions of circular foundations, *International Journal for Numerical and Analytical Methods in Geomechanics*, 13, 645–664.

Andersen, L. & Jones, C.J.C. (2001), BEASTS – a computer program for Boundary Element Analysis of Soil and Three-dimensional Structures, *ISVR Technical Memorandum 868* Institute of Sound and Vibration Research, University of Southampton.

Andersen, L. (2002), *Wave propagation in infinite structures and media*. Ph.D. thesis, Structural Dynamics Group, Department of Civil Engineering: Aalborg University.

Domínguez, J. (1993), *Boundary elements in dynamics*, Southampton: Computational Mechanics Publications.

Ibsen, L.B. & Liingaard, M. (2005), Output-only modal analysis used on new foundation concept for offshore wind turbine. *Proceedings of the 1st International Operational Modal Analysis Conference (IOMAC)*, April 26–27, 2005, Copenhagen, Denmark.

Luco, J.E. & Westmann, R.A. (1971), Dynamic response of circular footings, *Journal of Engineering Mechanics*, ASCE, 97(5), 1381–1395.

Petyt, M. (1998), *Introduction to Finite Element Vibration Analysis,* Cambridge: Cambridge University Press.

Spence, D. A. (1968), Self similar solutions to adhesive contact problems with incremental loading, *Proc. Roy. Soc. A.*, 305, 55–80.

Veletsos, A.S. & Tang, Y. (1987), Vertical Vibration of Ring Foundations, *Earthquake Engineering and Structural Dynamics*, 15, 1–21.

Veletsos, A.S. & Wei, Y.T. (1971), Lateral and Rocking Vibration of Footings, Journal of the Soil Mechanics and Foundation Division, 97 (SM 9), 1227–1248.

Environmental Vibrations – Takemiya (ed.)
© *2005 Taylor & Francis Group, London, ISBN 0 415 39035 4*

Dynamic response of culvert – embankment transitions in high speed railways

Z.Q. Li, H.R. Zhang, J.K. Liu, J. Zhang & Y.F. Hou
School of Civil Engineering and Architecture, Beijing Jiaotong University, Beijing, China

ABSTRACT: The objective of this paper is to present the theoretical and experimental results on the dynamic responses – the stress (earth pressure), displacement and acceleration, of the transition part built up by graded crushed stone between a reinforced concrete box culvert and the adjacent embankment in high speed railway. The transition is in a shape of upside-down trapezoid, where the embankment is embanked with granular soil and graded stone. In-situ experiments have been conducted to investigate the dynamic behavior of the transition and to verify the design criteria. A series results have been obtained through the in-situ measurements for the train speed from 180 km/h to 250 km/h, including the maximum stress, displacement and acceleration at the surface of the roadbed (transition), and the relationship between the dynamic responses and the train speeds. Furthermore, a numerical model is proposed to simulate the train running across the box culver – transition – embankment system to obtain the dynamic responses of the transition.

1 INTRODUCTION

The structures composed of a railway line, such as embankment, culvert, bridge and tunnel, are built up by different methods with special materials. Therefore, their dynamic response behaviors to running trains are distinguished. The steady state response of a rigidly supported, granular layer to a moving load is discussed in the paper (Suiker et al. 2000). The datum about the Measurement of stress histories using a test train on short span railway bridges in Victoria are provided by Chitty et al (1990). The paper (C. S. Pan & G. N. Pande 1985) presents an analysis of dynamic response to passing trains of a tunnel built in loess soil. The most remarkable difference is the vertical stiffness between the embankment and bridge or culvert, which will affect running smoothness or the safety of train when across these two parts. To overcome this problem, a transition part between two kinds of structures is necessary, where the stiffness changes gradually from the rigid part (culvert, bridge or tunnel) to the soft part (embankment). There are some methods in many papers (e.g., P. Chen & C. C. Zheng 1998; Q. Luo & Y. Cai 1999; L. Gao & H. Peng 2001) to build the transition, such as approach slabs, graded crushed stone and reinforced earth. Some publications also contribute to the dynamic behaviors of the transition. The influence of the irregularities of the bridge-subgrade transition on high speed running vehicle was analyzed by Q. Luo et al (1999). It had been shown that the track

bending resulting from the settlement difference between bridge and subgrade was the main factor, which affected the running safety and comfort of the high speed train. The computational results in the paper (L. J. Mao et al. 2001) had also showed that abrupt changes in track trasition vertical stiffness didn't lead to the increase in dynamic coefficient of action between wheel and rail directly, while it will increase sharply once existing angle in track trasition area.

Among these methods, graded crushed stone embanked transition is widely used in the high speed passenger railway line from Qinhuangdao, Hebei province, to Shenyang, Liaoning Province. The transitions are built by graded crushed stone with a shape of upside – down trapezoid, whereas the embankment is built with granular soil and graded stone.

To investigate the dynamic behavior of the transition segment and to verify the design criteria, in-site experiments have been conducted on the railway line. A large number of transducers, including accelerometers, earth pressure cells, have been embedded in the process of construction of the roadbed and the transition. A series results have been obtained through the in-site measurements for the train speed from 180 km/h to 250 km/h, including the maximum stress, displacement and acceleration at the surface of the roadbed (transition), the dynamic stiffness of transition segment, and the relationship between the dynamic responses and the train speeds.

Furthermore, to analyze the dynamic responses of the transition, a numerical model is proposed to simulate the train running across the box culver – transition – embankment system to obtain the responses.

In this paper, the theoretical and experimental results on the dynamic responses – the stress (earth pressure), displacement and acceleration, of the transition segment built up by graded crushed stone between a reinforced concrete box culvert and the adjacent roadbed in high speed railway, are presented.

2 IN-SITU EXPERIMENT

2.1 The structure of the embankment and transition

The embankment consists of three layers:

(1) 0.6 m thick of the top layer of the roadbed embanked with graded stone;
(2) 1.9 m thick of the bottom layer of the roadbed embanked with grain soil;
(3) Normal embankment layer down to the subsoil.

Ballast track system is used in this part of the railway line. The thickness of the ballast is 0.35 m.

There is only 0.6 m top layer of roadbed laid on the top of the culvert. Fig. 1 shows the detail of the transition, culvert, embankment and the installation of the transducers. The finished embankment and culvert (without ballast) is shown in Fig. 2.

2.2 Installation of transducers

The distribution of the embedded earth pressure cells and accelerometers is shown in Fig. 1. Fig. 3 shows the

Figure 2. Overview of the culvert and embankment.

Figure 3. Installation of earth pressure transducer.

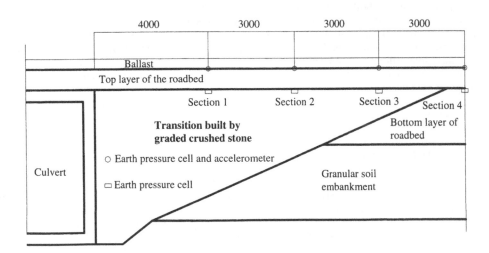

Figure 1. Details of the culvert, embankment, transition and the installation of transducers.

266

method to embed the earth pressure cell. Fig. 4 shows the containers applied to protect the accelerometer from the water. The containers are sealed up with silicon rubber.

2.3 The train applied in the measurement

The train used in the measurement is China Star, as shown in Fig. 5. The maximum speed in the test is over 300 km/h. The train passing this measurement site consists of 1 locomotive and 4 passenger coaches. The running speed is from 180 km/h to 250 km/h.

2.4 Measured results

The results obtained from the running speed of 180 km/h to 250 km/h in this measurement site, with over 10 running passes. Fig. 6 to Fig. 8 show the sample figures of the measured dynamic stress, acceleration and dynamic displacement at the top surface of the transition.

Figure 4. Protection of accelerometers.

Figure 5. Train used in the test.

The maximum values measured for the dynamic stress, acceleration and dynamic displacement at the surface of the transition are 34.07 kPa, 11.5 m/s^2 and 0.87 mm, respectively.

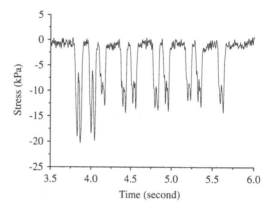

Figure 6. Measured dynamic stress.

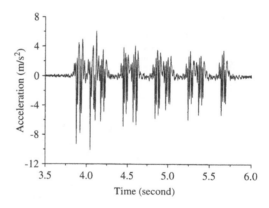

Figure 7. Measured acceleration.

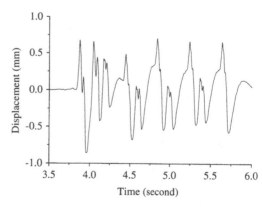

Figure 8. Measured dynamic displacement.

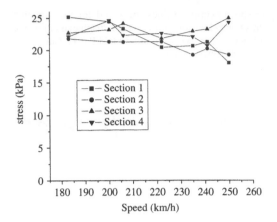

Figure 9. Relationship between dynamic stress and train speed.

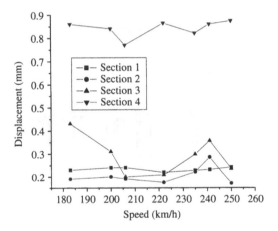

Figure 10. Relationship between displacement and train speed.

From the data, we can also find the relationship between the train speed and the measured dynamic stress, displacement and acceleration at the surface of the transition, as shown in Fig. 9 to Fig. 11.

It can be found from Fig. 9 to Fig. 11 that the measured responses have little change with the train speed in the measured speed range.

3 NUMERICAL MODEL AND SIMULATION

3.1 *Principal assumption*

Between the wheeltrack system of train and roadbed of embankment are complex three-dimensional space problem. But, the calculation work is very sophisticated

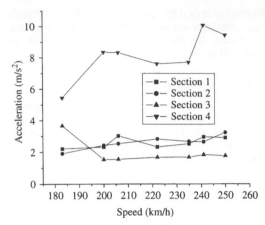

Figure 11. Relationship between acceleration and train Speed.

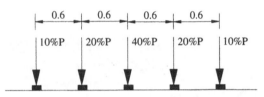

Figure 12. Distribution of the wheel load.

if the three-dimensional FEM model was used. Some studies had showed that the influence to the calculation hadn't significant difference if neglecting the cross direction action of track, but, the calculation work will be obviously decreased.

So, in this paper, a two-dimensional FEM model was created to simulated the real situations. In order to simplify the numerical model, the following assumptions have been adopted:

(1) The deformation of all the structures, including the culvert, embankment and the transition part are small and therefore, in linear elastic state;
(2) The train loads are simulated with moving load of the weight of the wheels;
(3) The wheel load is distributed in 5 adjacent sleepers, as shown in Fig. 12, where P is the wheel load.

3.2 *Finite element model*

Base on the above assumption, a finite element model is built up as shown in Fig. 13, where the transition and embankment are simulated by solid elements, and the culvert is modeled as beam elements.

To simplify the numerical model, half of the culvert length which equals to 2.5 m is used, the length

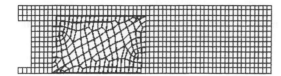

Figure 13. Finite element model of culvert-embankment system.

Table 1. Parameter of material.

Situation (material)	Young's modulus (Kpa)	Density (kg/m³)	Poisson's ratio
Embankment (Soil)	$0.8 \times E4$	1800	0.35
Culvert (Reinforced concrete)	$2.6 \times E10$	2500	0.25
Transition (Graded crushed stone)	$1.2 \times E4$	1950	0.30

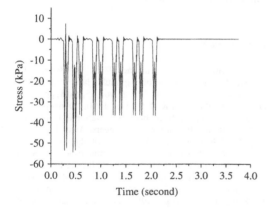

Figure 14. Calculated stress.

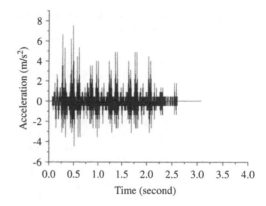

Figure 15. Calculated acceleration.

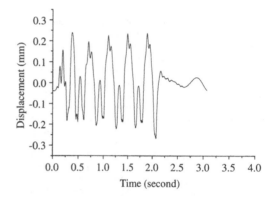

Figure 16. Calculated displacement.

It can be seen that the shapes of the dynamic response curves are similar to the measured responses. The maximum dynamic stress and displacement at the top of the transition obtained from the numerical simulation is 55.3 kPa and 0.72 mm.

of transition is 13 m, at the same time, the size at the back of transition is 15 m. The thickness of the top and bottom layer of the roadbed are 0.6 m and 1.9 m respectively, normal embankment layer down to the subsoil is 3.1 m. Some details were shown in Fig.1. All parameters of material which used in this model sees Table 1.

3.3 Numerical results

Some numerical results including dynamic stress and displacement were obtained from the running speed of 180 km/h to 250 km/h . Fig. 14 to Fig. 16 show the samples of the dynamic response of stress, acceleration and displacement at the top of the transition.

4 COMPARISION OF IN-SITU EXPERIMENT AND NUMERICAL SIMULATION

It can be found from Table 2 to Table 3 that the stress obtained from the numerical simulation is larger than that obtained from the in-situ measurement, the maximum measured and calculated dynamic stress responses at the top surface of the transition are 34.07 kPa and 55.3 kPa. Whereas the displacement is smaller that the measured one. The maximum dynamic displacement obtained from the experiment and numerical simulation are 0.87 mm and 0.72 mm.

The reason of this result maybe the assumption of the wheel load distribution and the accuracy of the

Table 2. Dynamic stress value of calculation and experiment at the top of the transition (KPa).

Speed (km/h)	Value	Situation			
		Section 1	Section 2	Section 3	Section 4
180	Calculation	22.04	21.97	24.33	22.68
	Experiment	25.19	32.65	22.70	22.12
200	Calculation	22.50	22.41	29.37	21.31
	Experiment	24.43	34.07	23.14	24.60
220	Calculation	29.49	27.25	31.39	21.43
	Experiment	20.44	32.29	32.59	22.60
240	Calculation	32.37	28.36	34.49	21.85
	Experiment	21.32	31.54	23.30	20.74
250	Calculation	37.21	33.76	55.30	22.57
	Experiment	18.07	32.32	25.06	24.37

Table 3. Dynamic displacement value of calculation and experiment at the top of the transition (mm).

Speed (km/h)	Value	Situation			
		Section 1	Section 2	Section 3	Section 4
180	Calculation	0.22	0.21	0.36	0.54
	Experiment	0.23	0.19	0.43	0.86
200	Calculation	0.25	0.23	0.38	0.56
	Experiment	0.24	0.20	0.31	0.84
220	Calculation	0.27	0.25	0.41	0.59
	Experiment	0.22	0.17	0.24	0.86
240	Calculation	0.32	0.28	0.45	0.63
	Experiment	0.23	0.29	0.36	0.86
250	Calculation	0.36	0.31	0.49	0.72
	Experiment	0.20	0.17	0.24	0.87

parameters used in the simulation, especially the modulus of the materials. It may be also caused by the measurement accuracy.

5 CONCLUSIONS AND DISCUSSIONS

From the results by both experiments and the numerical simulation, the following conclusions can be made:

(1) The measured and calculated dynamic stress responses at the top surface of the transition are 34.07 kPa and 55.3 kPa, which are much smaller than the bearing capacity of the embankment of 100 kPa.
(2) The dynamic displacement responses are very small and therefore have no influence to the running of the train.
(3) The dynamic responses change very little with the train speed.

It also should be mentioned that improvements are necessary both in the measurement and the numerical simulation such as:

(1) The dynamic response at lower frequency of the accelerometers used in the measurement is not satisfactory. The lower cut off frequency is only about 1 Hz. Therefore, the response to the weight of the train cannot be measured, as shown in Fig. 8.
(2) The soil (or stone) parameters used in the numerical simulation strongly affects the dynamic displacement response. The accuracy method to obtain these parameters is remains a challenge to engineers and researchers.
(3) In the numerical simulation, the non-linear contact between the wheels and the rail was neglected. However, this effect should be included in rigorous analysis.

ACKNOWLEDGEMENTS

The research of this paper is financially supported by the research project of The Ministry of Railway of China, Grant No. 2000G47.

REFERENCES

A. S. J. Suiker, A. V. Metrikines & R. De Borst, 2000. "Steady state response of a granular layer to a moving load – a discrete model." Heron, Vol. 45, No 1, pp. 75–87
G. B. Chitty, P. Grundy & H. McTier, 1990. "Dynamic stress in short span railway bridges." National Conference Publication – Institution of Engineers, Australia, No. 90 pt 10, 1990, pp 213–218, Adelaide, Australia, Oct 3–5, 1990
C. S. Pan & G. N. Pande. 1985. "Dynamic response of a railway tunnel due to passing trains." Proceedings of the Fifth International Conference on Numerical Methods in Geomechanics., Vol. 2, pp 1149–1160, Nagoya, Japan, 1985
P. Chen & C. C. Zheng. 1998. "Structural Analysis on the Approach Slabs between Pavements and Abutment." (in Chinese), Journal of Xi' an Highway University, Vol. 18, No. 3 (B), pp 244–251
Q. Luo & Y. Cai.1999. "Study on technological treatment methods of high speed railway bridge-subgrade transition." (in Chinese), Journal of Railway Engineering Society, No. 3, pp 30–33
L. Gao & H. Peng. 2001. "Setting mode of sleeper on HSR bridge-roadbed transition section for ballast track." Proc. International Symposium on Traffic Induced Vibrations & Controls, pp 229–234, Beijing, China, Nov. 6–7, 2001
Q. Luo, Y. Cai & W. M. Zhai.1999. "Dynamic performance analysis on high speed railway bridge-subgrade transition." (in Chinese), Engineering Mechanics, vol. 16, No. 5, pp 65–70
L. J. Mao, H. Z. Du & X. Y. Lei. 2001. "Dynamic Response Analyses on Track Transition in High Speed Railway." (in Chinese), Journal of East China Jiaotong University, Vol. 18, No. 1, pp 35–40

Environmental Vibrations – Takemiya (ed.)
© *2005 Taylor & Francis Group, London, ISBN 0 415 39035 4*

Damage evaluation of bridge foundations considering subsoil properties

J.W. Zhan, H. Xia & J.B. Yao
School of Civil Engineering & Architecture, Beijing Jiaotong University, Beijing, China

ABSTRACT: In this paper, the natural frequencies are selected as the indicators of damages of bridge foundations, which are mainly influenced by the rigidities of foundation bodies and the subsoil properties. The quantitative influences of the two factors are analyzed by some numerical examples in which damage scenarios are simulated by degrading either the body rigidity coefficients or the subsoil spring coefficients, and some laws are obtained. The concept of soundness index is put forward and the index is used to evaluate the damages of bridge foundations. For the familiar spread foundations, pile foundations and caisson foundations in railway bridges, according to the provisions on the safety coefficient of foundation design in Chinese Code for Bridge Foundation Design, the safety limitation values of soundness indexes are determined on the basis of numerical examples and real engineering disaster cases.

1 INTRODUCTION

In China, the damage evaluation of bridge foundations is mainly based on the Code for Rating Existing Railway Bridges in which the vibration amplitudes of foundations under train loads are restricted to the allowable ranges (China 2004). According to some studies, however, it is not easy to use the measured amplitudes to evaluate the damages of foundations because these amplitudes are related to train types, vehicle weights, vehicle lengths, train speeds, rail conditions, etc. It is often difficult to consider all these factors in processing the measured amplitudes (Nishimura 1992, 2001).

On the other hand, the structural damages of bridge foundations will lead to the changes of their vibration properties, such as natural frequencies, mode shapes and damping ratios (Pandey 1991, Salawu 1997, Hassiotis 2000, Palacz 2002, Wang 2001, Ren 2001). Among these vibration properties, the natural frequencies of foundations are the most sensitive indicators of damages and can be obtained easily and reliably (Nishimura 1987, Zhan & Xia 2005), they are thus selected for the damage evaluation of bridge foundations.

In this paper, various possible damage scenarios of bridge foundations are studied, which are divided into two groups: (1) For the foundation bodies, the changing of damage locations with the same damage extent and changing of damages with different extents at the same location; (2) For the subsoil, the changing of covering depth and degrading of subsoil stiffness. Two groups of damages are simulated, respectively by

reducing the body rigidity coefficients and the subsoil spring coefficients. The influences of these damages on the natural frequencies of bridge foundations are analysed quantitatively and some laws are gotten. If the natural frequencies in sound conditions are utilized as baseline values and the ratios of natural frequencies after damage to them are defined as soundness indexes, the indexes can be used to evaluate the damages of bridge foundations. Based on the safety coefficient provisions on foundation design, the safety limitation values of soundness indexes can be determined by numerical computations and analysis of real engineering disaster cases. When the soundness index of a foundation is below the safety limitation value, the foundation is in unsafe condition resulting from damages, and in contrary, the foundation is in critical condition or safe condition. The damage evaluation rule can be constituted on this safety limitation value.

2 NUMERICAL SIMULATION OF DAMAGES

In general, there are two common types of damages that cause the changes in the states of the foundations: the strength decrease in materials such as concrete that constitute the foundation members and the bearing capacity decrease of the ground that supports the foundations. The decrease of the material strengths induces structural damage which is caused by weathering with aging or cracking. While the decrease in

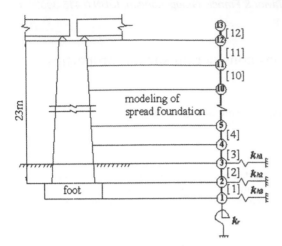

Figure 1. Outline of the spread foundation.

Table 1. Damage scenarios for spread foundation.

Scenarios	Objects	Degrading ratio (%)
	Element 2-11 alternatively	50
A*:	All rigidity coefficients	10
		25
		50
		75
B*:	All spring coefficients	10
		25
		50
		75

* Scenario A: Degrading of body rigidity coefficients.
* Scenario B: Degrading of subsoil spring coefficients.

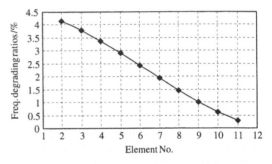

Figure 2. Frequency degrading ratios *vs* damage element.

bearing capacity of the ground results in supporting damage, which is caused by the scouring of subsoil around the foundation due to the water flow or the gaps made between the foundation and the ground due to subsidence of ground (Nishimura 1992). The two types of damages are unified as the damages of foundations. The numerical examples are given below to simulate these damages.

In the analysis, the pier-foundation-subsoil system and the foundation are of the same meaning, and the natural frequency and the first-order lateral natural frequency refer to the same item.

2.1 Damage simulation of foundation bodies

This type of damage is simulated by degrading the rigidity coefficients of foundation bodies. A high spread foundation and a pile foundation are analyzed herein, respectively.

2.1.1 Spread foundations

Usually most spread foundations are very rigid, and their lateral vibrations behave in rigid swing mode, so the degrading of body rigidity coefficients may hardly influences the natural frequencies of foundations, the rigidity of the beam-foundation-subsoil system is mainly provided by subsoil. But for flexible spread foundations with great heights, their lateral vibrations are predominated by the first-order bending vibration of the foundation bodies. At this time, the rigidities of foundation bodies may have a comparatively greater influence on their natural frequencies.

A 23 m high pier with spread foundation shown in Figure 1 is used to analyze the influence of body rigidity on the natural frequencies, which is divided in to

12 elements. Two damage scenarios shown in Table 1 are investigated.

Scenario A denotes the deterioration of the foundation body, where two sub-cases are analyzed: firstly, the ten elements [2] to [11] of the foundation alternatively suffer the damages with 50% rigidity reduction, and the analysis result is shown in Figure 2; Secondly, the rigidity of the whole foundation decreases at different levels which simulate the body deterioration caused by original faults in construction or later environmental erosion. Figure 3 shows the analysis result.

Scenario B denotes the relaxation of supporting subsoil caused by scouring or earthquake loads. The influence curve is shown in Figure 3.

From Figure 2, it can be found that with the distance increase of the elements to the foot, they have less and less influences (at last, less than 0.5%) on the natural frequency, and element 2 has the biggest influence. From FEM analysis, it is seen that the element 2 has higher stress than the others, which proves that the damages at the regions of comparatively high stress or high curvature result in greater reduction in the natural frequencies of the foundations. As is shown in Figure 3, the degrading of body rigidities

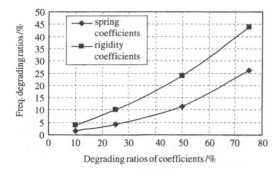

Figure 3. Frequency degrading ratios *vs* coefficients.

has greater influence on natural frequencies than the deterioration of the subsoil when the degrading ratios are the same. Therefore, for flexible spread foundations, more attention should be paid to the monitoring and maintenance of the foundation bodies; while for rigid spread foundations, it is more important to inspect the subsoil conditions.

2.1.2 *Pile foundations*

According to the relative location between the top surface of pile cap and the ground surface, pile foundations are classified into two categories: the low-capped pile foundation when the top surface of pile cap is below the ground surface, and the elevated pile foundation when the bottom surface of pile cap is above the ground surface. The main difference of the two types of pile foundations is that the low-capped pile foundations are supported by subsoil, while elevated pile foundations are not. For this reason, the two types of foundations have different vibration properties and stress distributions (Zhou 2000). The same characteristics of them is that their vibration properties are all greatly influenced by pile properties.

The usual disease of pile foundations is the breaking of piles caused by original bad construction or some exceptional external forces in service. This type of damage greatly reduces the natural frequencies of the foundations.

In this section, a low-capped pile foundation, as shown in Figure 4, is selected to quantitatively analyze the influence of pile parameters on the natural frequencies of foundations. In this example, it is assumed that pile 1 and pile 2 simultaneously break at the same position. This extreme state is simulated by a 3-Dimensional pile foundation analysis software. In total, there are ten analytical cases where the pile elements of the aforementioned two piles, from No.1 to No.10, break alternately. The analysis result is shown in Figure 5. It can be seen that the breaking of the pile at any location reduces the natural frequencies by more than 38%, and

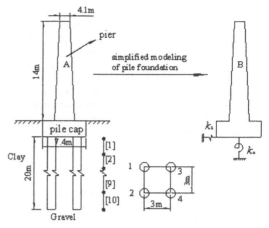

Figure 4. Outline of the pile foundation.

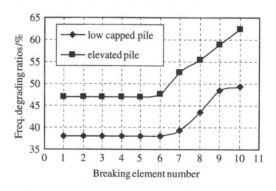

Figure 5. Frequency degrading ratios *vs* damage element.

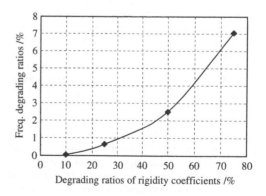

Figure 6. Frequency degrading ratios *vs* pier rigidity.

the upper parts of piles have greater influences than other parts. However, Figure 6 shows that the deterioration of pier body lightly influences the natural frequency: the degrading ratio of natural frequency is less

273

than 8% even when the pier rigidity degrades by 75%. The vibration modes indicate that the vibration of pier is in rigid swing, while the vibration of pile is predominated by bending and axial deformation. This is why the deterioration of piles has a greater influence on the natural frequency. Suggestion is given that rigid regulations should be executed during the construction of foundations to guarantee the construction quality of piles, so as to guarantee the bearing capacity and vibration properties of pile foundations.

2.2 Damage simulation of subsoil scouring

For most bridges, their main functions are to span rivers, lakes and seas, so their foundations will be under water. As is known to all, the flow of water always causes the decrease of supporting subsoil around the foundations. Especially when a flood happens, the serious scouring of the subsoil may decrease the bearing capacity of the ground, sometimes may even disable the functions of the foundation. Some disasters caused by scouring have been reported. When the bearing capacity of the ground decreases, the natural frequencies of the foundation will decrease accordingly. Therefore, it is meaningful to analyze the influence of scouring on the natural frequencies of foundations. In this section, a pile foundation and a caisson foundation are studied.

2.2.1 Pile foundations

For low-capped pile foundations, the scouring of subsoil can turn them into elevated pile foundations, and will greatly change their dynamic properties. The pile foundation shown in Figure 4 is again taken as the example to analyze the influence, but it is assumed that the subsoil around the pile cap is completely scoured. According to the numerical calculation, the natural frequencies before and after scouring are 1.832 Hz and 1.56 Hz, respectively, and thus the degrading ratio is about 15%.

When the same damages of pile bodies for this elevated pile foundation are simulated, similar results as those for the low-capped pile foundation in 2.1.2 are obtained (see Fig. 5). It can be seen that the influence is greater than that of the low-capped pile foundation. The explanation is that the supporting subsoil around the pile caps degrades the sensitiveness of the natural frequencies to pile damages. Similar principle is embodied in the difference between the lateral natural frequencies and the longitudinal ones: the constraint of the beams and the tracks on the lateral vibrations of foundations is very small, so the lateral frequencies are very sensitive to the damages of foundations; while in contrast, the longitudinal vibrations of foundations are strongly constrained by the beams and the tracks, so the longitudinal natural frequencies are less sensitive to the damages of foundations (Cao 2003).

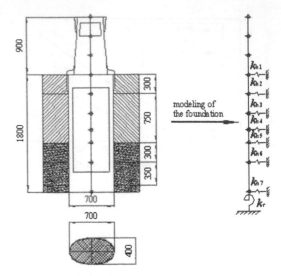

Figure 7. Outline of the caisson foundation.

Table 2. Scenarios for scouring of subsoil.

Scouring depth (m)	Degrading ratio of embedding depth (%)
2	11
4	22
6	33
9	50

2.2.2 Caisson foundations

The bearing capacities of caisson foundations are mainly provided by the surrounded subsoil, so their natural frequencies are more sensitive to the scouring of subsoil. A caisson foundation shown in Figure 7 is investigated. According to the Chinese Code for Bridge Foundation Design, the interaction between the foundation and the subsoil is simulated by the subsoil springs, and the m-method is used to form the spring coefficients (Chow 1987, China 1999).

The soil outside the caisson is clay with $m = 10,000 \, kPa/m^2$, and the soil under the caisson is gravel with $m = 30,000 \, kPa/m^2$. The damage scenarios are shown in Table 2. The ratios of the scouring depths to the original depth of subsoil are defined as degrading ratios of embedding depth. The solid line in Figure 8 shows the influence quantitatively. The results indicate that the scouring of subsoil influences the natural frequencies greatly: when the degrading ratio of embedding depth is 50%, the degrading of natural frequency is up to 60%. The slope rate of the dashed line in Figure 8 is bigger than that of the solid

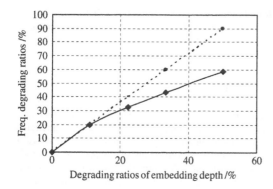

Figure 8. Frequency degrading ratios *vs* embedding depths.

Table 3. Influence of spring coefficient on *SI*.

Foundation type	Objects	β (%)	Soundness index (*SI*)
Spread	All	50	0.7
Pile	Torsional	50	0.91
	Spring	75	0.97
	Horizontal	50	0.77
	Spring	75	0.90
	All	50	0.71
	Spring	75	0.86
Caisson	Torsional	50	0.83
	Spring	75	0.93
	Horizontal	50	0.96
	Spring	75	0.98
	All	50	0.73
	Spring	75	0.90

line, which proves that the upper parts of the subsoil influence the natural frequencies of foundations more than the other parts.

3 DAMAGE EVALUATION RULES

In this paper, the natural frequency of the foundation before damage, namely in sound condition is utilized as baseline value and the ratio of natural frequency after damage to it is defined as the soundness index (*SI*) which can be written as equation (1). This index is used for damage evaluation.

$$SI = \frac{\text{Natural frequency after damage}}{\text{Baseline value of natural frequency}} \quad (1)$$

The Chinese Code prescribes that the safety coefficient for foundation design is 2.0. If a damage happens, the soundness index will decrease, and also will the safety coefficient decrease. When the safety coefficient decreases to half of the original value, i.e. 1.0, the foundation is in critical condition. The soundness index at this time is called the safety limitation values. When the soundness index is below the safety limitation value, the foundation is in unsafe condition resulting from various damages, and *vice versa*. The bigger the soundness index *SI* is, the safer the foundation is. This is the principle of damage diagnosis using the natural frequencies. The following work is to determine the safety limitation value for every kind of foundation.

3.1 Spread foundations

For most spread foundations concerned, the body rigidities are much bigger than those of the subsoil, so their lateral vibrations behave as rigid swing. When the beam-foundation-subsoil system is simplified as a single-freedom vibration system, the stiffness coefficient *k* of the vibration system is mainly provided by the subsoil (Li 2004). The natural frequency *f* of the vibration system can be written as:

$$f = \frac{1}{2\pi}\sqrt{\frac{k}{m}} \quad (2)$$

where *m* is the equivalent mass of the vibration system. When *k* decreases to the half of the original value, the foundation is in critical condition. According to equation (1), the soundness index *SI* will decreases to 0.7. Hence for spread foundations, the safety limitation value of soundness index can be determined as 0.7, which is listed in Table 3.

3.2 Pile foundations

It is noted that the piers of pile foundations vibrate in rigid swing modes, and the bearing capacities of pile foundations are provided by the subsoil and the pile bodies. The Chinese Code for Bridge Foundation Design simplifies the pile foundation from form A to form B, as is shown in Figure 4, where the bearing capacities of the pile foundation are simulated by torsional spring k_r and horizontal spring k_1. The damages are simulated by degrading the spring coefficients at different levels. The ratio of the computational value of the spring coefficient to the original value is defined as β. The influence law of the springs on the natural frequencies is shown in Table 3. According to the results, it is concluded that the safety

275

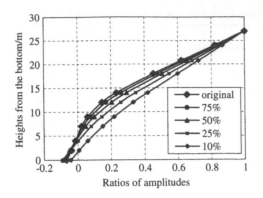

Figure 9. Vibration modes *vs* horizontal spring coefficients.

limitation value of soundness index for pile foundations is about 0.7.

3.3 *Caisson foundations*

Subsoil plays a much more important role in the bearing capacity of caisson foundation than the foundation body. For this reason, the damages of caisson foundations can be simulated by degrading subsoil spring coefficients at different levels. Table 3 lists the analysis results. Taking the modeling errors into account, the safety limitation value of soundness index for caisson foundation is approximately determined as 0.7. Figure 9 shows the influence of horizontal spring coefficients on the vibration modes of foundations.

3.4 *Safety limitation value of soundness index*

Since natural frequencies are sensitive to the damages of foundations, the soundness indexes based on natural frequencies can be used to evaluate the damages of foundations. The key of this method is to get the natural frequencies in sound conditions. If the dynamic experiment was executed immediately after construction, the measured natural frequencies can be the baseline values. When no such data obtained, it is suggested to gather the natural frequencies of sound foundations and establish the standard values of natural frequencies by statistical regressive method on the basis of these values for each category of foundation. In Japan, the regressive formulae of standard values for all types of foundations have been established, and the standard values computed by them are used as the baseline values for damage evaluation (Nishimura 2001). Above numerical examples proves that 0.7 is the critical value of the soundness index. Many disaster cases showed that the soundness indexes of these damaged

foundations were less than 0.7 (Nishimura 1992). Therefore, 0.7 is the critical value of the soundness index in real sense.

4 CONCLUSIONS

In this paper, the quantitative influences of body damages and subsoil damages on the natural frequencies of spread foundations, pile foundations and caisson foundations are studied. The results prove that soundness index is an index sensitive to all kinds of damages and can be used for the damage evaluation of bridge foundations. The safety limitation values of soundness indexes are determined as 0.7 on the basis of numerical examples and real engineering disaster cases. When the soundness index decreases to below 0.7, the foundation is considered as unsafe, and detailed inspection or reinforcement work should be done. When this method is used for damage evaluation of foundations in China, the urgent and important work is to establish the standard values of the natural frequencies for all types of existing foundations. And it is also suggested to grasp the natural frequencies of foundations for new-built bridges as the baseline values for future damage evaluation.

ACKNOWLEDGEMENTS

This study is sponsored by the National Natural Scientific Foundation of China (No. 50478059).

REFERENCES

Pandey, A.K. & Biswas, M.S. 1991. Damage detection from changes in curvature mode shapes. *Sound & Vibration* 145: 321–332.
Salawu, O.S. 1997. Detection of structure damage through changes in frequency: a review. *Engineering Structures* 19: 718–723.
Hassiotis, S. 2000. Identification of damage using natural frequencies and Markov parameters. *Computers & Structures* 74: 365–373.
Palacz, M. & Krawczuk, M. 2002. Vibration parameters for damage detection in structures. *Sound & Vibration* 249(5): 999–1010.
Wang, X. 2001. Structural damage identification using static test data and changes in frequencies. *Engineering Structures* 23: 610–621.
Ren, W.X. & De Roeck, G. 2001. Structural damage identification using modal data. I: Simulation Verification. *Structural Engineering* 128(1): 81–95.
Zhan, J.W., Xia, H. & Yao, J.B. 2004. Safety evaluation for foundations of existing railway bridges. *Proc. ISSST'2004*, Shang'hai: 1952–1958.

Zhan, J.W., Xia, H. & Yao, J.B. 2005. An impact vibration test method for measuring natural frequencies of existing piers. *Journal of Beijing Jiaotong University* 29(1): 14–17.

Cao, J.A. & Leng, W.M. 2003. Disease diagnosis and evaluation method of railway bridge pier supported by pile foundation. *Journal of Changsha Railway University* 21(4): 24–29.

Xia, H. 2002. *Dynamic Interaction of Vehicles and Structures.* Beijing: Science Press.

Nishimura, A. & Nakano. 1987. Grasping the vibration characteristics of structures by impact vibration test. *Proc. the 19th Japan national conference on soil mechanics and foundation engineering.* Tokyo: 563–569.

Nishimura, A. 1992. *Research on The Diagnosis Method of Integrity for Existing Bridge Foundations.* Tokyo: Railway Technical Research Institute.

Nishimura, A. 2001. Examination of bridge substructure for integrity. *Proc. TIVC'2001*, Beijing: 131–142.

Zhou, H.L. 2000. Experimental and finite element analysis of bridge pier with high-rising pile cap under dynamic train loading. *Railway Engineering Society* 3: 36–40.

Li, Y.S. 2004. *Theory and Experimental Studies on Lateral Vibration of Railway Bridge Piers.* Ph.D Thesis, Beijing Jiaotong University.

Chow, Y.K. 1987. Three-dimensional analysis of pile groups. *Geotechnology Engineering* 113(6): 637–651.

Ministry of Railways P.R. China. 1999. *Code for Design on Subsoil and Foundation of Railway Bridge and Culvert.* Beijing: China Railway Publishing House.

Ministry of Railways P.R. China. 2004. *Code for Rating Existing Railway Bridges.* Beijing: China Railway Publishing House.

Environmental Vibrations – Takemiya (ed.)
© *2005 Taylor & Francis Group, London, ISBN 0 415 39035 4*

Impact analysis and absorber design for collision protection of part-buried structures

B. Zhu

Department of Civil Engineering, Zhejiang University, Hangzhou, China
Department of Building and Construction, City University of Hong Kong, Hong Kong, China

A.Y.T. Leung

Department of Building and Construction, City University of Hong Kong, Hong Kong, China

ABSTRACT: Based on the *p*-version finite element method (FEM), an elasto-plastic impact model abandoning the Hertzian contact law for the collision protection of part-buried structures is presented. With the model, one can predict the whole impact process including the impact force and responses of the impactor and the structure automatically and straightforwardly. An impact experiment of a three-dimensional column-plate structure attached with elastic and elasto-plastic absorbers is carried out to verify the proposed model. For the engineering application, some Cellular Reinforced Concrete Blocks (CRCBs) with different strength are designed as impact absorbers. Unlike steel and Fiber Reinforced Polymer (FRP), the proposed blocks would not be easily rusted and aged. The quasi-static tests prove that they have the excellent energy absorbing characteristic. These blocks are used for protection of the T structure resisting the famous Qiantang tide and protecting the dams at both sides of the Qiantang River.

1 INTRODUCTION

Most of studies used the Hertzian contact law (Zukas 1982) to calculate the impact force and analyze beams and shells subjected to impact loadings. Other alternative methods such as the spring-dashpot and momentum balance methods utilized the coefficient of restitution as an input to the dynamic analysis (Khulief & Shabana 1986, Palas et al. 1992). In some problems, the local deformation absorbs a significant portion of the impact energy so that it must be modeled adequately in the analysis. As early as in 1940, a simple elastic impact model to simulate the impact of a mass striking a beam was presented (Lee 1940). Recently, some elasto-plastic impact models using the mass-spring system were proposed (Wu & Yu 2001). The authors have also attempted several simple methods to analyze the structural impact (Chen et al. 2002, Zhu et al. 2003). Some of these methods are for the elastic impact and the others are only for the analysis of simple structures. Generally, the impact process can be predicted by the conventional plastic finite elements accurately, even for some complex structures. But by this way, much computational time is used for the iteration of the plastic analysis and many degrees of freedoms (DOFs) should be considered to obtain satisfied solutions. It needs a simple and efficient

elasto-plastic impact model to analyze the whole impactor-absorber-structure system to design the optimal absorber for the structural collision protection in civil engineering.

Some materials were used to smooth the impact force and absorb the impact energy. An overview of rate effects in cellular material was presented in the monograph of Gibson & Ashby (1997). Reid & Peng (1997) studied the dynamic crushing of cylindrical specimens of five different kinds of woods for impact velocities up to 300 m/s. They showed that the substantial enhancement of the initial crushing stress for wood loaded along the grain is due to micro-inertia inhibiting cell wall buckling modes. Many interests focused on the analysis of high strain rate compressive honeycombs (Zhao & Gary 1998), as well as the polyurethane foam and aluminum alloy foam subjected to the impact loading (Deshpande & Felck 2000, Shim et al. 2000), but the impact force they can resist is very limited, or their volume is very large. Davalos et al. (2001) and Qiao et al. (2004) designed and modeled the fiber-reinforced plastic honeycomb sandwich panels and I-Lam sandwich system for the protection of highway bridges. Commercial finite element package was used in the analysis, and some applications were described.

An elasto-plastic impact model based on the *p*-version FEM is presented in this paper. The impact force and responses of the impactor-absorber-structure system can be predicted automatically. An impact experiment of a three-dimensional column-plate structure attached with elastic and elasto-plastic absorbers is installed to verify the proposed model. Concrete blocks (CRCBs) are designed as the impact absorbers used in engineering application. The optimal block is selected for the T structure in the Qiantang River to resist the collision of moving ship on the river. This CRCB can also be attached on lampposts to minimize impact casualty and to save lives, and be applied to the protection of other ocean and offshore structures.

2 ELASTO-PLASTIC IMPACT MODEL

Since many iterations are involved in the time-domain impact simulation, much computational time can be saved when using the p-version finite elements. A hierarchical element for beams resting on two-parameter foundations based on the Kirchhoff hypothesis is presented in Figure 1(a). The development of the element can refer to Leung & Zhu (2004) and Zhu & Leung (2005). Based on the Mindlin plate theory, a quadrilateral hierarchical element for plates resting on two-parameter foundations (Leung & Zhu 2005) is shown in Figure 1(b). The internal DOFs of the both elements are represented by the additional hierarchical terms in the shape functions. With the enriching DOFs in these p-version elements, the accuracy of the vibration analysis is greatly improved, and it is a good choice to use these p-version elements in the simulation of structural impact analysis.

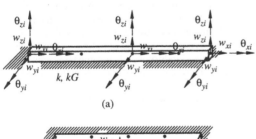

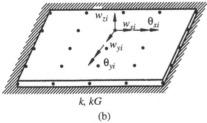

Figure 1. Geometry and DOF system for p-version elements of beam and plate resting on Pasternak foundations (a. beam element; b. plate element).

The governing equations for the impactor-absorber-structure system are

$$\mathbf{M}\ddot{\mathbf{X}}(t)+\mathbf{C}\dot{\mathbf{X}}(t)+\mathbf{K}\mathbf{X}(t)=\mathbf{F}(t) \tag{1a}$$

$$m\ddot{w}(t)+f(t)=0 \tag{1b}$$

where $\mathbf{K}$, $\mathbf{M}$, $\mathbf{C}$ are the stiffness, mass and Rayleigh damping matrices of the structure, respectively; $\mathbf{F}(t)$ is the force vector applied on the structure; $f(t)$ is the impact force; m is the mass of the impactor; $w(t)$ and $\mathbf{X}(t)$ are the displacement of the impactor and the vector of displacements of the structure. The indentation of the absorber can be expressed as

$$\alpha(t)=w(t)-x(t) \tag{2}$$

where $x(t)$ is the displacement where the absorber is located on the structure. Before the iteration in each time step, it is assumed that

$$w_n^{(0)}(t)=w_{n-1}(t)+\dot{w}_{n-1}(t)\Delta t+\frac{1}{2}\ddot{w}_{n-1}(t)\Delta t^2 \tag{3}$$

$$x_n^{(0)}(t)=x_{n-1}(t)+\dot{x}_{n-1}(t)\Delta t+\frac{1}{2}\ddot{x}_{n-1}(t)\Delta t^2 \tag{4}$$

where n and the superscript (i) indict the i-th iteration in the n-th time step. Then the iteration starts to obtain the impact force and responses of the impactor and the structure as follows:

(a) Calculation of the indentation of the absorber from Equation (2).
(b) Obtaining the impact force $f_n^{(i)}(t)$ from the quasi-static load-deformation relationship of the absorber.
(c) Substituting the impact force $f_n^{(i)}(t)$ into Equation (1a) and computing the responses of the structure $\mathbf{X}_n^{(i)}(t)$, $\dot{\mathbf{X}}_n^{(i)}(t)$ and $\ddot{\mathbf{X}}_n^{(i)}(t)$ according to the Newmark linear acceleration method (Clough & Penzien 1993). A new value of $x_n^{(i)}(t)$ is also obtained.
(d) Substituting $f_n^{(i)}(t)$ into Equation (1b) and computing the acceleration of the impactor $\ddot{w}_n^{(i)}(t)$.
(e) Obtaining the new values of $\dot{w}_n(t)$ and $\ddot{w}_n(t)$ by

$$w_n^{(i+1)}(t)=w_{n-1}(t)+\dot{w}_{n-1}(t)\Delta t$$
$$+\frac{1}{3}\ddot{w}_{n-1}(t)\Delta t^2+\frac{1}{6}\ddot{w}_n^{(i)}(t)\Delta t^2 \tag{5}$$

$$\dot{w}_n^{(i+1)}(t)=\dot{w}_{n-1}(t)+\frac{\Delta t}{2}\left(\ddot{w}_{n-1}(t)+\ddot{w}_n^{(i)}(t)\right) \tag{6}$$

where Δt is the time step length. The above procedure is iterated until acceptable and stable responses of the impactor and the structure are obtained. By this way, one can simulate responses of the whole system when the mass and the initial velocity of the impactor are given.

3 VERIFICATION OF MODEL BY EXPERIMENT

3.1 Experiment installation

A three-dimensional column-plate structure attaching elastic and elasto-plastic absorbers subjected to impact deformation is shown in Figure 2. The configuration of elasto-plastic aluminum absorbers is shown in Figure 3. The weight of the impactor can vary as 9.622 kg, 11.622 kg, 13.622 kg, 15.622 kg, 17.622 kg and 19.622 kg. Some associated filters, amplifiers and the data collector used in the experiment are also shown in Figure 2(a).

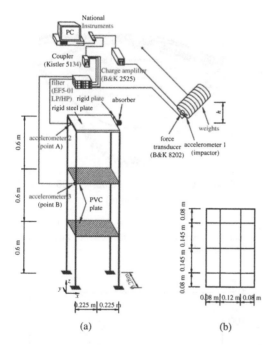

(a)   (b)

Figure 2. Experimental setup and analysis model (a. experimental setup; b. p-version mesh of the PVC plate).

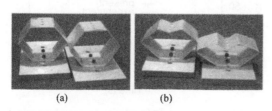

(a)   (b)

Figure 3. Elasto-plastic aluminum absorbers (a. before impact; b. after impact with plastic deformation).

3.2 Impact simulation by proposed model

One p-version beam element and twelve p-version plate elements shown in Figure 2(b) are used for each column and each PVC plate respectively. The number of additional hierarchical terms used in the beam element and the plate element are 2 and 3, respectively. Since the top steel plate is rigid, DOFs at the top of the four columns connecting the steel plate are same. The steel plate is simulated as four masses attached at the top of these four columns. Some materials properties used in the simulation are listed in Table 1. Damping ratios of the structure are identified by the continuous wavelet transform (CWT) (Ruzzene et al. 1997, Lardies & Gouttebroze 2002, Leung et al. 2005), and the first two modes in the x direction are 0.53% and 0.22% respectively, which are used for the Rayleigh damping matrix in the computation.

3.3 Quasi-static test

The quasi-static test of the aluminum absorber is performed by a LLOYD (50 kN). In the test the loading speed is set as 3 mm/min while the loading routes are set up as $0 \rightarrow 10$ mm $\rightarrow 0$, $0 \rightarrow 15$ mm $\rightarrow 0$, $0 \rightarrow 20$ mm $\rightarrow 0$ and $0 \rightarrow 30$ mm, respectively. The force is chosen as the control variable in the test. The experimental results and the numerical simulation are plotted in Figure 4.

3.4 Impact with elasto-plastic and elastic absorbers

For the elasto-plastic impact, the simulated and experimental impact forces with $m = 11.622$ kg and different initial velocities of the impactor are compared in Figure 5. For the case of $m = 11.622$ kg and $h = 0.02$ m, where h is the initial height of the impactor, comparison of values of the three accelerometers between the simulation and the experiment is carried out in Figure 6. From these plots, it can be found that the simulated results are in good agreement with the experimental

Table 1. Properties of materials used for the column-plate structure.

Steel columns	Young's modulus E	210×10^9 Pa
	Material density ρ	7800 kg/m
	Length L	0.6 m
	Width b	0.02532 m
	Thickness t	0.00470 m
PVC plates	Young's modulus E	3.7×10^9 Pa
	Material density ρ	1400 kg/m
	Area A	0.45×0.28 m^2
	Thickness t	0.004 m
	Shear correction factor κ	5/6
	Poisson's ratio ν	0.3
Steel plate	Mass	25.26 kg

281

ones. Good agreement between the simulated and experimental results can also be observed for the elastic impact as shown in Figures 7 and 8. The elastic spring with the stiffness $k = 7691.5\,\text{N/m}$ is used herein.

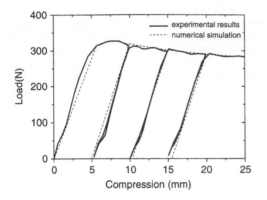

Figure 4.   Quasi-static experimental results and numerical simulation of the absorber.

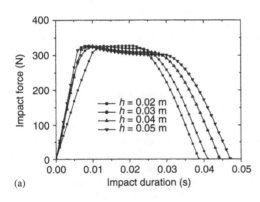

(a)

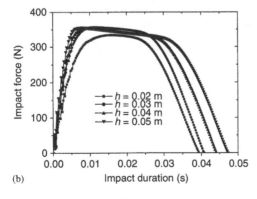

(b)

Figure 5.   Comparison of the impact force for the case of $m = 11.622\,\text{kg}$ between the experiment and the simulation (a. simulated results; b. tested results).

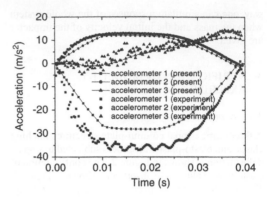

Figure 6.   Comparison of the value of three accelerometers between the simulation and the experiment for the case of $m = 11.622\,\text{kg}$ and $h = 0.02\,\text{m}$.

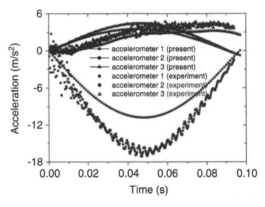

Figure 7.   Comparison of impact force between the simulation and the experiment for the elastic impact with $m = 9.622\,\text{kg}$ and $h = 0.01\,\text{m}$.

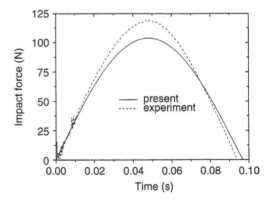

Figure 8.   Comparison of the value of three accelerometers between the simulation and the experiment for the elastic impact with $m = 9.622\,\text{kg}$ and $h = 0.01\,\text{m}$.

4 CHARACTERISTIC OF ABSORBER CRCB

Some small CRCBs for the plane surface (CRCB-P) and for the circular surface (CRCB-C) are made in the laboratory as shown in Figure 9. The quasi-static tests of CRCBs are performed by a LLOYD (50 kN) and a MTS 815 Rock Mechanics Test System (1600 kN). The force is chose as the control variable in tests.

The tested result and the numerical simulation for a typical CRCB are plotted in Figure 10. The simulated bilinear relationship between the load and the compression is: $F = F_0 x/D_0$ $(0 \leqslant x \leqslant D_0)$ and $F = \alpha x + F_0$ $(x > D_0)$ where F(N) and x(m) are the applied load and the compression of the block, respectively. $F_0 = 8000$ N, $D_0 = 0.73377 \times 10^{-3}$ m and $\alpha = 78533$ are shown in the figure for this CRCB. In this study, for different strength and size of the block, it is assumed that the simulated relationship between the load and the compression depends only on the values of F_0, D_0 and α.

Two types of quasi-static tests are carried out for CRCB-Ps. For test A as shown in Figure 11, the CRCB-P is pressed by a plane surface; while it is pressed by a sharp surface for test B as shown in Figure 12. The tests are used to study the performance of CRCBs for the two main impact cases. The tested

results for the two test cases are presented in Figures 13 and 14, respectively. Three types of CRCB-Ps with water cement ratios of 0.85, 0.7 and 0.55 respectively are used in the tests. In these tests, cellular holes of

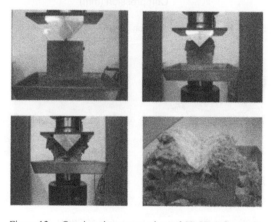

Figure 11. Quasi-static test procedure of CRCB-P for test A.

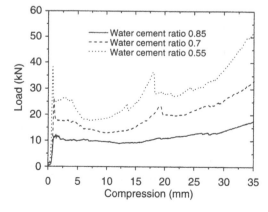

Figure 12. Quasi-static test procedure of CRCB-P for test B.

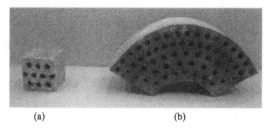

(a)   (b)

Figure 9. Configuration of CRCBs (a. CRCB-P; b. CRCB-C).

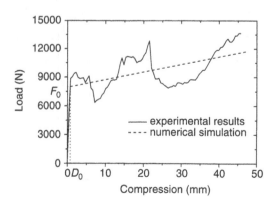

Figure 10. Typical load-compression relationship of CRCB-P.

Figure 13. Load-compression plot of CRCB-Ps for test A.

the block are crushed layer by layer for test A but one by one for test B, and reinforcements in blocks constrain the development of vertical cracks. While the normal concrete cube is shear failure when the load approaches its failure strength and then the load will be decreased abruptly. So the CRCB has much more excellent characteristic of plastic deformation than the normal concrete cube does. For the CRCB-C, it is very difficult to overcome the stress concentration on the middle vertical surface, but from Figures 15 and 16, the excellent plastic deformation is observed just similar to that of the CRCB-P.

The CRCB can vary in strength by using different proportion of cement, water, and aggregates, and it can also vary in size and shape. So it is convenient to be used for engineering application. For a real structure subjected to the impact loading, the present impact model is an alternative choice to compute the optimal size and strength of the CRCB.

5 COLLISION PROTECTION OF T STRUCTURE ON QUIANTANG RIVER

A T structure resisting the famous Qiantang tide and protecting Qiantang dams is shown in Figure 17. There are two rows of reinforced concrete piles with

Figure 14. Load-compression plot of CRCB-Ps for test B.

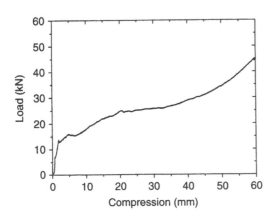

Figure 15. Quasi-static test procedure of CRCB-C.

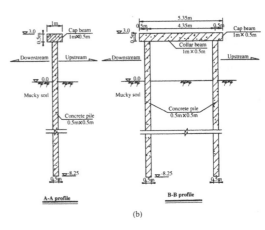

Figure 17. T structure on the Qiantang River (a. plane view; b. side view).

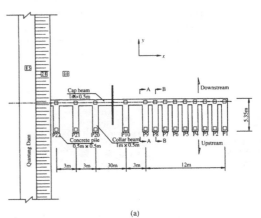

Figure 16. Load-compression relationship of CRCB-C.

length of 11 m buried in the mucky soil, which are connected to each other (see Figure 17(a)) by cap beams and collar beams. The row of piles toward the upstream of the river are numbered as P1-P22.

Since the structure is transversely extended into the river, impact by moving ships on the Qiantang River is common. Some physical properties of the structure and the moving ship are shown in Table 2. Just parts of DOFs of the structure are taken into account to simplify the computation (see Figure 18). There is one element for each cap beam and each collar beam,

Table 2. Physical properties of impact analysis of T structure.

T structure	Young's modulus E	30×10^9 Pa
	Material density ρ	2500 kg/m
	Shear correction factor κ	1
	Poisson's ratio ν	0.3
Moving ship	Moving speed	3 m/s
Foundation	Winkler foundation modulus k	2×10^6 Pa
	Shear foundation modulus k_G	2.5×10^7 N

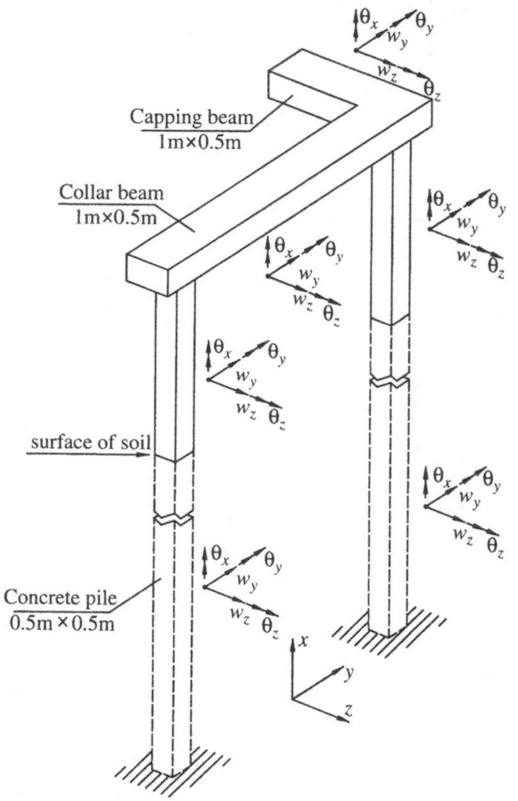

but two elements for each pile. The number of additional hierarchical terms in each element is 4. The first five transverse vibration mode shapes for the top of piles P1 ~ P22 are plotted in Figure 19.

It is assumed that $D_0 = 7.3377 \times 10^{-3}$ m and various F_0 are used for simulating the relationship between the load and the compression of different CRCBs. The right part of the structure (see Figure 17(a)) tends to be impacted more than the left one, and it is clear that responses of the structure will be largest if the top of the pile P1 is impacted. Thus, only the transverse impact at the top of the pile P1 is analyzed herein, and the appropriate CRCB is selected for the protection of this location. It is assumed that the moving ship comes from the upstream of the river. As shown in Figures 20 and 21, time duration of the impact and indentation of the absorber decrease rapidly with the increasing strength of CRCBs. It noted that the compression stress of the absorber during the impact should be smaller than that of concrete of the structure can endure. The characteristic of the CRCB determines responses of the impactor as shown in Figure 22. For the

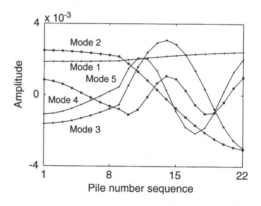

Figure 19. Transverse vibration mode shapes at the top of piles P1 ~ P22 for the first five modes of the T structure.

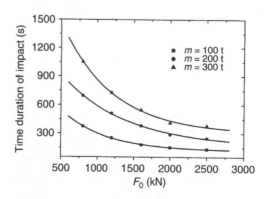

Figure 20. Time duration of impact vs. F_0.

Figure 18. *p*-version finite element model of the substructure.

285

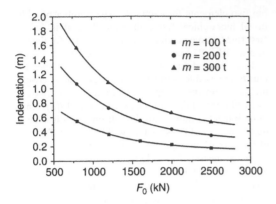

Figure 21. Indentation of the absorber vs. F_0.

impactor with mass 200 t, the responses at the impacted position on the structure are shown in Figure 22. The responses of the structure are also heavily influenced by the strength of the CRCB. With the obtained displacement responses of the structure, the strength of the structure can be checked according to the corresponding structural design codes, and the optimal CRCB will be selected.

6 CONCLUSIONS AND DISCUSSION

An elasto-plastic impact model is presented. The simulated responses of the structure are analyzed by the p-version elements. Then, one can obtain the same accurate solutions using less DOFs compared with the conventional finite elements. An experiment of a three-dimensional column-plate structure attached with different absorbers subjected to impact deformation is carried out. The simulated results are in good agreement with the experimental ones for the elastic impact as well as the elasto-plastic impact. To design a suitable small elasto-plastic absorber generating small responses of the impactor-structure system for a structure subjected to a certain impact is required in an engineering application.

As mention previously, the analysis will be complex and many DOFs should be involved in the computation if the conventional plastic finite elements are used in the impact analysis. For the impact analysis of the structure attached with the absorber, usually, only the absorber will deform plastically. With the present simple impact model and fast convergent p-version elements, the analysis of this type of impact is reasonable and much computational time can be saved.

CRCBs with excellent characteristic of plastic deformation are presented. With the load-compression relationship from the quasi-static test and the proposed impact model, responses of the impactor-absorber-structure system can be calculated without

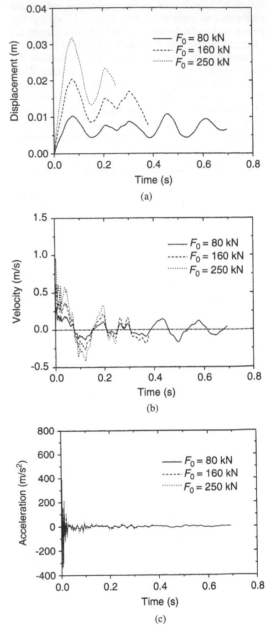

Figure 22. Responses at impacted position on structure for impactor with mass of 200 t and different CRCBs (a. displacement responses; b. velocity responses; c. acceleration responses).

difficulty. The strength of CRCB heavily affects responses of the structure. Explicit design procedure of CRCB is desired for engineering applications.

REFERENCES

Chen, Y.M. et al. 2002. Dynamic response of ocean trestle to horizontal of moving mass. *China Ocean Engineering* 16(1): 51–60.

Clough, R.W. & Penzien, J. 1993. *Dynamic of structures*. 2nd edn, New York: McGraw-Hill.

Davalos, J.F. et al. 2001. Modeling and characterization of fiber-reinforced plastic honeycomb sandwich panels for highway bridge applications. *Composite Structures* 52: 441–452.

Deshpande, V.S. & Felck, N.A. 2000. High strain rate compressive behaviour of aluminium alloy foams. *International Journal of Impact Engineering* 24: 277–298.

Gibson, L.J. & Ashby, M.F. 1997. *Cellular solids: structure and properties*. 2nd Edition, Cambridge: Cambridge University Press.

Khulief, Y.A. & Shabana, A.A. 1986. Dynamic analysis of constrained system of rigid and flexible bodies with intermittent motion. *Journal of Mechanisms, Transmissions, and Automation in Design* (ASME) 108: 38–45.

Lardies, J. & Gouttebroze, S. 2002. Identification of modal parameters using the wavelet transform. *Mechanical Sciences* 44: 2263–2283.

Lee, E.H. 1940. The impact of a mass striking a beam. *Journal of Applied Mechanics* A129-38.

Leung, A.Y.T. & Zhu, B. 2004. Fourier *p*-elements for curved beam vibrations. Thin-Walled Structures 42(1): 39–57.

Leung, A.Y.T. & Zhu, B. 2005. Transverse vibration of Mindlin plates on two-parameter foundations by analytic trapezoidal *p*-elements. *Journal of Engineering Mechanics*, ASCE, accepted.

Leung, A.Y.T. et al. 2005. Experimental study on damage detection of a cantilever beam attached with spring. *Engineering Structures*, submitted.

Palas, H. et al. 1992. On the use of momentum balance method in transverse impact problems. *Journal of Vibration and Acoustics* (ASME) 114: 364–373.

Qiao, P. et al. 2004. Impact analysis of I-Lam sandwich system for over-height collision protection of highway bridges. *Engineering Structures* 26: 1003–1012.

Reid, S.R. & Peng, C. 1997. Dynamic uniaxial crushing of wood. *International Journal of Impact Engineering* 19: 531–570.

Ruzzene, M. et al. 1997. Natural frequencies and damping identification using wavelet transform: application to real data. *Mechanical Systems and Signal processing* 11(2): 207–218.

Shim, V.P.W. et al. 2000. Two-dimensional response of crushable polyurethane foam to low velocity impact. *International Journal of Impact Engineering* 24: 703–731.

Wu, K.Q. & Yu, T.X. 2001. Simple dynamic models of elastic-plastic structures under impact. *International Journal of Impact Engineering* 25: 735–754.

Zhao, H. & Gary, G. 1998. Crushing behaviour of aluminium honeycombs under impact loading. *International Journal of Impact Engineering* 21(10): 827–836.

Zhu, B. et al. 2003. Transient response of piles-bridge under horizontal excitation. *Journal of Zhejiang University: Science* 4(1): 28–34.

Zhu, B. & Leung, A.Y.T. 2005. Linear and nonlinear vibration of non-uniform beams on two-parameter foundations using *p*-element. *Computers & Structures*, submitted.

Zukas, J.A. et al. 1982. Impact Dynamics. New York: Wiley.

*Simulation of ground vibrations due to
traffic and other sources*

Environmental Vibrations – Takemiya (ed.)
© 2005 Taylor & Francis Group, London, ISBN 0 415 39035 4

A unified approach to numerical modelling of traffic induced vibrations

G. Degrande & G. Lombaert
Department of Civil Engineering, K.U. Leuven, Kasteelpark Arenberg, Leuven, Belgium

ABSTRACT: This paper presents a unified approach for the prediction of free field vibrations due to road or rail traffic. It is assumed that the road or track and the (horizontally layered) soil are invariant in the longitudinal direction so that the dynamic road/track-soil interaction equations can be formulated in the frequency-wavenumber domain. Dynamic train axle loads are computed accounting for the track compliance in the moving frame of reference; in the case of road traffic, the compliance of the tyres is much higher than the compliance of the road-soil system so that the latter can be disregarded. The free field response due to moving loads is computed applying the dynamic reciprocity theorem, which can be simplified when the vehicle speed is low with respect to the body wave velocities in the soil. A first numerical example illustrates how free field vibrations due to road traffic can efficiently be predicted on a large three-dimensional mesh of receiver points. A second example considers the vibration isolation efficiency of a floating slab track. For the unisolated and the floating slab track, the free field response is compared for the passage of a single tram axle on uneven rails.

1 INTRODUCTION

Road traffic induced vibrations are due to the passage of vehicles on an uneven road surface or on discrete unevenness as speed bumps, traffic plateaus and man-holes. Train induced vibrations are characterized by different excitation mechanisms, such as quasi-static excitation, parametric excitation due to the discrete supports of the rails on sleepers, transient excitation due to rail joints and wheel flats and the excitation due to wheel and rail roughness and track unevenness.

A lot of attention has recently been paid to the development of numerical models for traffic induced vibrations that account for the interaction between the vehicle, the road or track and the soil. Krylov (1996, 1997) has developed an analytical prediction model that only includes quasi-static excitation and aims to predict ground vibrations at train speeds below and above the critical phase velocity of the track-soil system. Degrande and Lombaert (2001) have reformulated Krylov's prediction model in the frequency-wavenumber domain, resulting in a more efficient calculation of the response.

Sheng et al. (1999a, 1999b) have coupled a layered beam model of infinite extension, representing the track, to a layered halfspace model of the soil and consider the case of a fixed and harmonic point load; they also have coupled a train model to the track (Sheng et al. 2003, 2004) and indicate that, when the train speed is low compared to the wave velocities in the soil, the dynamic component of the axle loads,

rather than quasi-static excitation, determines the free field vibration levels. Auersch (2005) couples a finite element model of the track to a boundary element model of the soil to calculate the track compliance that is used to solve the vehicle-track interaction problem. Metrikine et al. (2005) study the stability of a moving train bogie, modelled as a 2DOF system, which is coupled to an infinite beam track model and a homogeneous half space model. Kaynia et al. (2000) and Madshus and Kaynia (2000) model the track with coupled beam elements, while the disc Green's functions for a horizontally layered halfspace are employed to compute the soil impedance. Andersen and Nielsen (2003) apply a boundary element method for the steady-state response of an elastic medium in a moving frame of reference. Ekevid and Wiberg (2002) combine the finite element method and the scaled boundary element method for the quasi-static response of the coupled track-soil system.

Within the frame of this paper, a unified formulation for both road and train traffic induced vibrations is presented; the model is initially developed for road traffic induced vibrations (Lombaert et al. 2000, 2001, 2003, 2004) and solves the dynamic road-soil interaction problem in the frequency-wavenumber domain, exploiting the invariance of the geometry in the direction along the road's longitudinal axis. The model is extended to railway induced vibrations (Lombaert et al. 2004, 2004,) accounting for quasi-static, transient and roughness excitation. Dynamic axle loads are computed accounting for the interaction between the train, the track and the soil. As in many of the aforementioned models, the track is modelled by means of infinite beams, coupled by springs and dampers, while the impedance of the soil is modelled with a boundary element method, using the Green's functions of a layered halfspace. The model can also be applied to classical ballasted tracks, neglecting parametric excitation due to the discrete support of the rails, which is only valid in the low frequency range of interest. Similar developments exist for moving dynamic axle loads on periodic media (Metrikine and Popp 1999; Clouteau et al. 2001; Vostroukhov and Metrikine 2003; Clouteau et al. 2005), but are not considered in the present paper.

The terminology for train-track interaction is prioritized throughout the paper. It is shown how the calculation of the dynamic axle loads accounts for the interaction between the vehicle and the track-soil system. The prediction of free field vibrations is discussed and several approximations for small vehicle speeds are proposed. A first numerical example illustrates how road traffic induced vibrations can efficiently be predicted on a large three-dimensional mesh of receiver points. A second example considers the vibration isolation efficiency of a floating slab track.

2 TRAIN-TRACK INTERACTION

The dynamic axle loads applied by the train on the track are determined by the dynamic interaction between the train, the track and the soil (Figure 1).

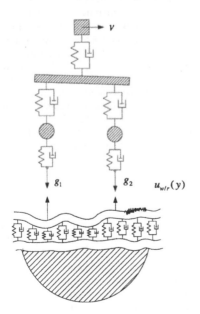

Figure 1. The train–track–soil interaction problem.

A point contact is assumed between the axles and the rails and only the vertical interaction is considered. The distribution of n vertical axle loads on the coupled track-soil system is written as:

$$\mathbf{F}(x, y, z, t) = \sum_{k=1}^{n} \delta(\mathbf{x} - \mathbf{x}_S) g_k(t) \mathbf{e}_z \qquad (1)$$

where $\mathbf{x}_S = \{x_S, y_k + vt, z_S\}^{\mathsf{T}}$ denotes the instantaneous position of the k-th axle load $g_k(t)$ with initial position $\{0, y_k, 0\}^{\mathsf{T}}$, that moves with the train speed v along the $\mathbf{e}_y$ direction.

In the frequency domain, the vehicle's equations of motion are written as:

$$\begin{bmatrix} \mathbf{S}_{bb} & \mathbf{S}_{ba} & \mathbf{0} \\ \mathbf{S}_{ab} & \mathbf{S}_{aa} + \mathbf{S}_{aa}^W & -\mathbf{S}_{aa}^W \\ \mathbf{0} & -\mathbf{S}_{aa}^W & \mathbf{S}_{aa}^W \end{bmatrix} \left\{ \begin{array}{c} \hat{\mathbf{u}}_b \\ \hat{\mathbf{u}}_a \\ \hat{\mathbf{u}}_c \end{array} \right\} = \left\{ \begin{array}{c} \mathbf{0} \\ \mathbf{0} \\ -\hat{\mathbf{g}} \end{array} \right\}$$

$$(2)$$

where the vectors $\hat{\mathbf{u}}_b$, $\hat{\mathbf{u}}_a$ and $\hat{\mathbf{u}}_c$ collect the displacements of the body, the axles and the contact points between the axles and the rail, respectively. The co-efficient matrix contains block submatrices of the dynamic stiffness matrix of the vehicle (above the wheels), while the matrix $\mathbf{S}_{aa}^W$ denotes the dynamic stiffness matrix corresponding to the Hertzian springs between the wheels and the rails (or the tyres of a car). Equation (2) can alternatively be written as:

$$\begin{bmatrix} \bar{\mathbf{S}}_{aa} + \mathbf{S}_{aa}^W & -\mathbf{S}_{aa}^W \\ -\mathbf{S}_{aa}^W & \mathbf{S}_{aa}^W \end{bmatrix} \left\{ \begin{array}{c} \hat{\mathbf{u}}_a \\ \hat{\mathbf{u}}_c \end{array} \right\} = \left\{ \begin{array}{c} \mathbf{0} \\ -\hat{\mathbf{g}} \end{array} \right\} \quad (3)$$

where $\bar{\mathbf{S}}_{aa} = \mathbf{S}_{aa} - \mathbf{S}_{ab} \mathbf{S}_{bb}^{-1} \mathbf{S}_{ba}$. The displacement vector $\hat{\mathbf{u}}_c$ is equal to the sum of the rail displacement $\hat{\mathbf{u}}_r$ and the rail unevenness $\hat{\mathbf{u}}_{w/r}$:

$$\hat{\mathbf{u}}_c = \hat{\mathbf{u}}_r + \hat{\mathbf{u}}_{w/r} \qquad (4)$$

It follows immediately from equations (3) and (4) that the interaction forces $\hat{\mathbf{g}}$ can be written as:

$$\hat{\mathbf{g}} = \mathbf{S}_{aa}^W \left[\hat{\mathbf{u}}_a - \left(\hat{\mathbf{u}}_r + \hat{\mathbf{u}}_{w/r} \right) \right] \qquad (5)$$

In the case of a rigid contact between the wheels and the rails, the stiffness $\mathbf{S}_{aa}^W$ is infinitely large and the axle displacements $\hat{\mathbf{u}}_a$ are equal to the displacements of the contact points $\hat{\mathbf{u}}_c$, or, according to equation (4):

$$\hat{\mathbf{u}}_a = \hat{\mathbf{u}}_r + \hat{\mathbf{u}}_{w/r} \qquad (6)$$

Elimination of the displacements $\hat{\mathbf{u}}_c$ of the contact points from equation (3) results into:

$$\bar{\mathbf{S}}_{aa} \hat{\mathbf{u}}_a = -\hat{\mathbf{g}} \qquad (7)$$

Upon elimination of the axle displacements $\hat{\mathbf{u}}_a$ from equation (3), and accounting for equation (4), the following alternative equation is obtained:

$$\hat{\mathbf{C}}^v \hat{\mathbf{g}} = -\left(\hat{\mathbf{u}}_r + \hat{\mathbf{u}}_{w/r}\right) \tag{8}$$

where $\hat{\mathbf{C}}^v = \overline{\mathbf{S}}_{aa}^{-1} + \mathbf{S}_{aa}^{W-1}$ is the compliance matrix of the vehicle in the contact points between the axles and the rails. This vehicle compliance matrix reduces to $\hat{\mathbf{C}}^v = \overline{\mathbf{S}}_{aa}^{-1}$ in the case of a rigid contact between the axles and the rails.

It will be shown further how the track displacement $\hat{\mathbf{u}}_r$ can be written in a similar way as a function of the interaction forces $\hat{\mathbf{g}}$:

$$\hat{\mathbf{u}}_r = \hat{\mathbf{C}}^t \hat{\mathbf{g}} \tag{9}$$

where each element $\hat{C}_{kl}^t$ of the track compliance matrix $\hat{\mathbf{C}}^t$ represents the track displacement at the time-dependent position of the k-th axle due to an impulsive load at the time-dependent position of the l-th axle.

Introduction of equation (9) into equation (8) results in the following equilibrium equation:

$$\left[\hat{\mathbf{C}}^v + \hat{\mathbf{C}}^t\right] \hat{\mathbf{g}} = -\hat{\mathbf{u}}_{w/r} \tag{10}$$

from which the interaction forces $\hat{\mathbf{g}}(\omega)$ can be computed. For road traffic induced vibrations, the road compliance can be neglected with respect to the vehicle compliance in equation (10). This is not the case for train axle loads, as the compliance of the combined track-soil subsystem has the same order of magnitude as the compliance of the train; resonance of the wheel or vehicle masses against the stiffness of the track becomes important.

The frequency content $\hat{\mathbf{u}}_{w/r}(\omega)$ of the track unevenness is calculated from the wavenumber domain representation $\tilde{u}_{w/r}(k_y)$ of the unevenness $u_{w/r}(y)$:

$$\hat{\mathbf{u}}_{w/r}(\omega) = \frac{1}{v}\tilde{u}_{w/r}\left(-\frac{\omega}{v}\right) \exp\left(i\omega\frac{\mathbf{y}_a}{v}\right) \tag{11}$$

where the vector $\mathbf{y}_a$ contains the initial positions y_k of all axles k. The vector $\exp\left(i\omega \mathbf{y}_a/v\right)$ includes the phase difference between the excitation and the axles.

3   TRACK(ROAD)-SOIL INTERACTION

The equations of motion of the coupled track-soil system are solved for a vertical impulse load at a fixed position $\{x_S, 0, z_S\}^T$ on the track, which is located at the surface of a horizontally layered halfspace. The geometry of the track-soil system is assumed not to vary in the longitudinal direction $\mathbf{e}_y$ of the track. This

assumption is also valid for discretely supported tracks as the discrete support of the rails does not importantly influence the track receptance in the range of low frequencies relevant for railway induced vibrations (Knothe and Grassie 1993).

The invariance of the geometry with respect to the longitudinal coordinate y allows to perform a Fourier transformation of y to the horizontal wavenumber k_y. This results in a solution procedure in the frequency–wavenumber domain, where the following equations of motion are formulated for the coupled track-soil system (Aubry et al. 1994; Clouteau et al. 2001):

$$\left[\tilde{\mathbf{K}}_{tr}(k_y, \omega) + \tilde{\mathbf{K}}_s(k_y, \omega)\right] \tilde{\mathbf{u}}_{tr}(k_y, \omega) = \tilde{\mathbf{f}}_{tr}(k_y, \omega) \tag{12}$$

$\tilde{\mathbf{K}}_{tr}$ and $\tilde{\mathbf{K}}_s$ represent the track's and the soil's impedance matrices, respectively, $\tilde{\mathbf{u}}_{tr}$ is the track displacement vector and $\tilde{\mathbf{f}}_{tr}$ is the force vector applied to the track. Equation (12) is generic and can be generally applied to compute the response of a road or a track due to a vertical impulse load.

3.1   Road-soil interaction

In the case of road traffic, the road is represented by a beam with a rigid cross section, for which the longitudinal bending and torsional deformations are accounted for (figure 2). The displacement vector equals $\tilde{\mathbf{u}}_{tr} = \{\tilde{u}_{sl}, \tilde{\beta}_{sl}\}^T$, with $\tilde{u}_{sl}$ the vertical displacement of the section's centre of gravity and $\tilde{\beta}_{sl}$ the rotation about this centre. The road-soil interface is also assumed to be rigid in the plane of the road's cross section as continuity of displacements must hold. The vertical displacement $\tilde{u}_s$ and the rotation $\tilde{\beta}_s$ of the centre of the road-soil interface are therefore equal to $\tilde{u}_{sl}$ and $\tilde{\beta}_{sl}$, respectively. The vertical displacements of the road at an arbitrary position are written as follows in the frequency-wavenumber domain:

$$\tilde{u}_{rz}(x, k_y, z, \omega) = \mathbf{\Phi}_{tr}(x)\tilde{\mathbf{u}}_{tr} \tag{13}$$

The vector $\mathbf{\Phi}_{tr} = \{1, x\}^T$ collects the displacement modes of the rigid cross section. The displacements $\tilde{\mathbf{u}}_{tr}$ can also be considered as the modal coordinates of the road's deformation modes.

Figure 2.   The road model (cross section).

The road impedance matrix $\tilde{K}_{tr}$ is equal to

$$\begin{bmatrix} \tilde{K}_{sl,b} & 0 \\ 0 & \tilde{K}_{sl,t} \end{bmatrix}$$

where $\tilde{K}_{sl,b} = E_{sl}I_{sl}k_y^4 - \rho_{sl}A_{sl}\omega^2$ and $\tilde{K}_{sl,t} = G_{sl}C_{sl}k_y^2 - \rho_{sl}I_{sl,t}\omega^2$ are the slab's bending and torsional impedance, respectively.

The formulation of the vertical equilibrium at the road-soil interface Σ allows to calculate the elements of the soil impedance matrix $\tilde{K}_s$:

$$\left[\tilde{K}_s \right]_{ij} = \int_\Sigma \tilde{\phi}_{ti} \tilde{t}_{sz}(\tilde{\phi}_{sj}) \, d\Gamma \tag{14}$$

where $\tilde{t}_{sz}(\tilde{\phi}_{sj})$ is the frequency-wavenumber domain representation of the vertical component of the soil tractions $\tilde{t}_s = \tilde{\sigma}_s \mathbf{n}$ on a boundary with a unit outward normal $\mathbf{n}$ due to the scattered wavefield $\tilde{\phi}_{sj}$ in the soil, while $\tilde{\phi}_{ti}$ is a vertical road displacement mode. A boundary element method is used to calculate the soil tractions $\tilde{t}_{sz}(\tilde{\phi}_{sj})$ at the track-soil interface (Aubry et al. 1994; Lombaert et al. 2000), based on the boundary integral equations in the frequency-wavenumber domain and using the Green's functions of a horizontally layered elastic halfspace (Kausel and Roësset 1981).

The force vector $\tilde{f}_r$ in equation (12) contains the contribution of the impulse load applied in the point $(x_S, 0, z_S)$ of the road-soil interface Σ and is equal to $\{1, x_S\}^T$ in the present case.

As both the road and soil impedance matrices are diagonal, the bending and torsional modes of the road are uncoupled.

3.2 Track-soil interaction

In the case of an unisolated slab track (figure 3), the track displacement vector $\hat{\mathbf{u}}_{tr}$ is equal to $\{\tilde{u}_{r1}, \tilde{u}_{r2}, \tilde{u}_{sl}, \tilde{\beta}_{sl}\}^T$, where $\tilde{u}_{r1}$ and $\tilde{u}_{r2}$ are the vertical displacements of the left and right rail, and $\tilde{u}_{sl}$, and $\tilde{\beta}_{sl}$ are the vertical displacement and rotation of the centre of gravity of the slab, which are equal to the displacement and rotation at the interface between the track and the soil. The track force vector $\tilde{\mathbf{f}}_{tr}$ contains the forces applied at both rails and equals $\{\tilde{f}_{r1}, \tilde{f}_{r2}, 0, 0\}^T$.

The track impedance $\tilde{\mathbf{K}}_{tr}$ matrix in the equations of motion (12) of the coupled track-soil system is:

$$\begin{bmatrix} \tilde{K}_r + \tilde{K}_{rp} & 0 & -\tilde{K}_{rp} & -\tilde{K}_{rp}l_1 \\ 0 & \tilde{K}_r + \tilde{K}_{rp} & -\tilde{K}_{rp} & -\tilde{K}_{rp}l_2 \\ -\tilde{K}_{rp} & -\tilde{K}_{rp} & \tilde{K}_{sl,b} + 2\tilde{K}_{rp} & \tilde{K}_{rp}(l_1 + l_2) \\ -\tilde{K}_{rp}l_1 & -\tilde{K}_{rp}l_2 & \tilde{K}_{rp}(l_1 + l_2) & \tilde{K}_{sl,t} + \tilde{K}_{rp}(l_1^2 + l_2^2) \end{bmatrix}$$

The rails are modelled as Euler beams; the rail impedance $\tilde{K}_r = E_r I_r k_y^4 - \rho_r A_r w^2$ is determined by the rail bending stiffness $E_r I_r$ and the mass $\rho_r A_r$ per unit

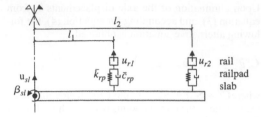

Figure 3. The unisolated slab track model.

length. l_1 and l_2 are the locations of both rails with respect to the centre of the track. For a single symmetric track, $l_1 = -l_2$. The rail pads are modelled as continuous spring-damper connections with a dynamic stiffness defined as $\tilde{K}_{rp} = \bar{k}_{rp} + i\omega\bar{c}_{rp}$. The sleeper distance d is used to calculate a smeared value $\bar{k}_{rp} = k_{rp}/d$ and $\bar{c}_{rp} = c_{rp}/d$ from the rail pad stiffness k_{rp} and damping c_{rp}, respectively, as the rail is assumed to be continuously supported. $\tilde{K}_{sl,b} = E_{sl}I_{sl}k_y^4 - \rho_{sl}A_{sl}w^2$ and $\tilde{K}_{sl,t} = G_{sl}C_{sl}k_y^2 - \rho_{sl}I_{sl,t}\omega^2$ are the slab's bending and torsional impedance.

The elements of the soil impedance matrix $\tilde{\mathbf{K}}_s$ are formulated analogously as in equations (14).

More complicated track models can be formulated analogously. In the case of the isolated slab track (figure 4), a resilient mat is introduced below the slab, which is modelled as a continuous spring-damper connection. The dynamic response of the slab and the soil now decouple and a distinction is made between the degrees of freedom $\tilde{u}_{sl}$ and $\tilde{\beta}_{sl}$ that describe the vertical slab response and the degrees of freedom $\tilde{u}_s$ and $\tilde{\beta}_s$ that describe the vertical response at the interface between the track and the soil. The track displacement vector $\hat{\mathbf{u}}_{tr}$ is $\{\tilde{u}_{r1}, \tilde{u}_{r2}, \tilde{u}_{sl}, \tilde{\beta}_{sl}, \tilde{u}_s, \tilde{\beta}_s\}^T$ and the track force vector $\tilde{\mathbf{f}}_{tr}$ equals $\{\tilde{f}_{r1}, \tilde{f}_{r2}, 0, 0, 0, 0\}^T$.

The track impedance matrix $\tilde{\mathbf{K}}_{tr}$ in the equations of motion (12) is now equal to:

$$\begin{bmatrix} \tilde{K}_r + \tilde{K}_{rp} & 0 & -\tilde{K}_{rp} \\ 0 & \tilde{K}_r + \tilde{K}_{rp} & -\tilde{K}_{rp} \\ -\tilde{K}_{rp} & -\tilde{K}_{rp} & \tilde{K}_{sl,b} + 2\tilde{K}_{rp} + \tilde{K}_{sm} \\ -\tilde{K}_{rp}l_1 & -\tilde{K}_{rp}l_2 & \tilde{K}_{rp}(l_1 + l_2) \\ 0 & 0 & -\tilde{K}_{sm} \\ 0 & 0 & 0 \end{bmatrix} \cdots$$

$$\cdots \begin{bmatrix} -\tilde{K}_{rp}l_1 & 0 & 0 \\ -\tilde{K}_{rp}l_2 & 0 & 0 \\ \tilde{K}_{rp}(l_1 + l_2) & -\tilde{K}_{sm} & 0 \\ \tilde{K}_{sl,t} + \tilde{K}_{rp}(l_1^2 + l_2^2) + \tilde{K}_{sm}\frac{B^2}{3} & 0 & -\tilde{K}_{sm}\frac{B^2}{3} \\ 0 & \tilde{K}_{sm} & 0 \\ -\tilde{K}_{sm}\frac{B^2}{3} & 0 & \tilde{K}_{sm}\frac{B^2}{3} \end{bmatrix}$$

where $\tilde{K}_{sm} = \bar{k}_{sm} + i\omega\bar{c}_{sm}$ is the impedance of the slab mat that couples the slab to the soil. The soil

294

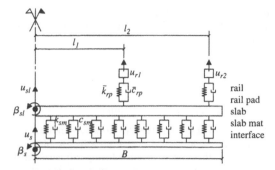

Figure 4. The isolated slab track model.

impedance matrix $\check{K}_s$ is similar as in the case of the unisolated slab track.

The solution of the track-soil interaction equation (12) provides the track displacement vector $\tilde{\mathbf{u}}_{tr}$. The track receptance functions (Fourier transform of the track impulse response functions) are equal to the inverse Fourier transformation of the track displacement vector:

$$\hat{\mathbf{u}}_{tr} = \frac{1}{2\pi} \int_{-\infty}^{+\infty} \tilde{\mathbf{u}}_{tr} \exp\left(-ik_y y\right) dk_y \qquad (15)$$

4 THE FREE FIELD RESPONSE

4.1 *The track-soil transfer functions*

The soil tractions $\tilde{t}_{sz}(x, k_y, z = 0, \omega)$ at the track-soil interface Σ are equal to:

$$\tilde{t}_{sz}(x, k_y, z = 0, \omega) = \tilde{t}_{sz}(\tilde{\phi}_s)\tilde{\alpha} \qquad (16)$$

where $\tilde{t}_{sz}(\tilde{\phi}_s)$ collects the vertical component of the soil tractions due to the scattered wavefields $\tilde{\phi}_s$ in the soil and $\tilde{\alpha} = \{\tilde{u}_s, \tilde{\beta}_s\}^{\mathrm{T}}$ collects the coefficient of these modes, which are equal to the displacements at the track-soil interface.

The dynamic reciprocity theorem is used to compute the track-soil transfer function $\tilde{h}_{zi}(x, k_y, z, w)$:

$$\tilde{h}_{zi}(x, k_y, z, \omega)$$

$$= \int_{-B}^{+B} \tilde{u}_{zi}^G(x - x', k_y, z, \omega)\tilde{t}_{sz}(x', k_y, z' = 0, \omega)\, dx' \qquad (17)$$

where $\tilde{u}_{zi}^G(x, k_y, z, \omega)$ is the Green's function of the supporting horizontally layered halfspace that represents the displacement in the direction $\mathbf{e}_i$ due to an impulse load in the vertical direction $\mathbf{e}_z$.

The frequency content of the free field displacements to an impulse load on the track is calculated as the inverse Fourier transform of the transfer functions:

$$\hat{h}_{zi}(x, y, z, \omega)$$

$$= \frac{1}{2\pi} \int_{-\infty}^{+\infty} \tilde{h}_{zi}(x, k_y, z, \omega) \exp\left(-ik_y y\right) dk_y \qquad (18)$$

4.2 *Response to moving loads*

Accounting for the invariance of the track-soil system in the longitudinal y-direction, the Betti-Rayleigh reciprocal theorem allows to derive the response in a point $\{x, y, z\}^{\mathrm{T}}$ due to the k-th axle load:

$$u_{si}(x, y, z, t)$$

$$= \sum_{k=1}^{n} \int_{-\infty}^{t} g_k(\tau)h_{zi}(x, y - y_k - v\tau, z, t - \tau)\, d\tau \qquad (19)$$

The response due to a moving load can therefore be calculated from the response for a concentrated impulse load at a fixed position $\{x_S, 0, z_S\}^{\mathrm{T}}$ on the track. A double forward Fourier transformation allows to derive the following expression in the frequency–wavenumber domain:

$$\tilde{u}_{si}(x, k_y, z, \omega)$$

$$= \tilde{h}_{zi}(x, k_y, z, \omega) \sum_{k=1}^{n} \hat{g}_k(\omega - k_y v) \exp\left(+ik_y y_k\right) \qquad (20)$$

The response in the frequency domain is written as:

$$\hat{u}_{si}(x, y, z, \omega) = \frac{1}{2\pi} \int_{-\infty}^{+\infty} \tilde{h}_{zi}(x, k_y, z, \omega)$$

$$\sum_{k=1}^{n} \hat{g}_k(\omega - k_y v) \exp\left[-ik_y(y - y_k)\right] dk_y \qquad (21)$$

A change of variables according to $k_y = (w - \tilde{w})/v$ moves the frequency shift from the axle loads to the transfer function:

$$\hat{u}_{si}(x, y, z, \omega) = \frac{1}{2\pi v} \int_{-\infty}^{+\infty} \tilde{h}_{zi}\left(x, \frac{\omega - \tilde{\omega}}{v}, z, \omega\right)$$

$$\sum_{k=1}^{n} \hat{g}_k(\tilde{\omega}) \exp\left[-i\left(\frac{\omega - \tilde{\omega}}{v}\right)(y - y_k)\right] d\tilde{\omega} \qquad (22)$$

which clearly illustrates the frequency shift between the source and the receiver known as the Doppler effect. When it can be assumed that the vehicle speed v is small with respect to the wave velocities in the soil, the motion of the vehicle and, consequently, the shift $k_y v$ in the argument of the axle force $\hat{g}_k(w - k_y v)$ in equations (21) and (22) can be neglected, resulting in the following expression of the free field response in the spatial domain:

$$
\hat{u}_{si}(x, y, z, \omega) = \sum_{k=1}^{n} \hat{g}_k(\omega) \left[\frac{1}{2\pi} \int_{-\infty}^{+\infty} \right.
$$

$$
\left. \tilde{h}_{zi}(x, k_y, z, \omega) \exp\left[-ik_y(y - y_k)\right] dk_y \right]
$$

(23)

5 THE TRACK COMPLIANCE

In order to derive the expression for the track compliance, the double inverse Fourier transform of equation (21) is used to calculate the response in a moving frame of reference $(x, \hat{y}, z, t)$, with $\hat{y} = y - vt$:

$$
u_{si}(x, \hat{y}, z, t) = \frac{1}{4\pi^2} \int_{-\infty}^{+\infty} \int_{-\infty}^{+\infty}
$$

$$
\hat{g}_k(\omega - k_y v) \tilde{h}_{zi}(x, k_y, z, \omega)
$$

$$
\exp\left[-ik_y(\hat{y} + vt - y_k)\right] \exp\left(+i\omega t\right) dk_y \, d\omega
$$

(24)

The circular frequency w is replaced by $\tilde{w} + k_y v$:

$$
u_{si}(x, \hat{y}, z, t) = \frac{1}{2\pi} \int_{-\infty}^{+\infty} \left[\frac{1}{2\pi} \int_{-\infty}^{+\infty} \right.
$$

$$
\hat{g}_k(\tilde{\omega}) \tilde{h}_{zi}(x, k_y, z, \tilde{\omega} + k_y v)
$$

(25)

$$
\left. \exp\left[-ik_y(\hat{y} - y_k)\right] dk_y \right] \exp\left(+i\tilde{\omega}t\right) d\tilde{\omega}
$$

The bracketed term represents the Fourier transform $\hat{u}_{si}(x, \hat{y}, z, \tilde{w})$ of the response in the moving frame of reference. This equation allows to derive the element $\hat{C}_{lk}^t(\tilde{w})$ of the track compliance matrix, representing the track response at the time–dependent position y_l of axle l due to a unit impulse $(\hat{g}_k(\tilde{\omega}) = 1)$ at axle k.

$$
\hat{C}_{lk}^t(\tilde{\omega}) = \frac{1}{2\pi} \int_{-\infty}^{+\infty} \tilde{h}_{zz}(x, k_y, z, \tilde{\omega} + k_y v)
$$

(26)

$$
\exp\left[-ik_y(y_l - y_k)\right] dk_y
$$

In the limit for small train speeds, and for a single axle, the track compliance corresponds to the track receptance in equation (15).

6 EXAMPLES

6.1 Road traffic induced vibrations

The first example considers the vibrations induced by the passage of a two-axle truck on a traffic plateau, installed on the road as speed reducing infrastructure. The road has a width $2B = 4\,\text{m}$ and consists of an asphalt top layer, a layer of crushed stone and a crushed concrete subbase layer (table 1). The road is supported by a homogeneous halfspace with a shear wave velocity $C_s = 200\,\text{m/s}$, a longitudinal wave velocity $C_p = 400\,\text{m/s}$, a density $\rho = 1750\,\text{kg/m}^3$ and a material damping ratio $\beta = 0.025$ in deviatoric and volumetric deformation.

A two-axle truck with a wheel base of 5.2 m passes at a vehicle speed $v = 50\,\text{km/h}$ on the traffic plateau which has a top length $L = 10\,\text{m}$, a height $H = 0.12\,\text{m}$ and sine-shaped ramps with a length $l = 1.2\,\text{m}$. The model of Lombaert et al. (2000, 2001, 2003) is used to compute the free field vibrations. The axle loads are derived from a two-dimensional 4DOF vehicle model, neglecting the compliance of the road-soil system in equation (10). The time history of the front axle load in figure 5 shows the passage on both ramps; the frequency content is dominated by the pitch and bounce mode at 1.9 Hz and the axle hop mode at 10.8 Hz.

Chebyshev based shape functions are used with a Galerkin boundary element formulation to compute the soil's impedance, as well as the soil tractions, allowing for an accurate and efficient calculation of the free field response in a large number of output points (François et al. 2005). Figure 6 shows the free field vertical velocity in a receiver a 6 m from the centre of the road. The four peaks in the time history correspond to the passage of the vehicle axles on both ramps of the traffic plateau. The free field response is dominated by the axle hop modes.

Figure 7 shows the norm of the displacement vector in the free field during the passage of the rear axle on the first ramp. Cylindrical Rayleigh wave fronts are observed, as the bending stiffness of the road is negligible with respect to the stiffness of the soil.

Table 1. The parameters of the road model.

Layer	d [m]	E [$\times 10^6\,\text{N/m}^2$]	v [−]	ρ [kg/m^3]
1	0.15	9150	1/3	2100
2	0.20	500	1/2	2000
3	0.25	200	1/2	1800

6.2 Performance of a floating slab track

The second example considers the vibration isolation efficiency of a floating slab track, consisting of a concrete slab with a width $2B = 2.5$ m, a height $h_{sl} = 0.55$ m, a density $\rho_{sl} = 2500$ kg/m^3 and a Young's modulus $E_{sl} = 30000$ MPa. A single track is installed on the slab consisting of UIC60 rails, discretely supported at a distance $d = 0.6$ m by rail pads with a stiffness $k_{rp} = 213 \times 10^6$ N/m and a damping $c_{rp} = 14.8 \times 10^3$ Ns/m. The rail is assumed to be continuously supported and smeared values $\bar{k}_{rp} = k_{rp}/d$ and $\bar{c}_{rp} = c_{rp}/d$ of the rail pad stiffness and damping are used. The soil is modelled as a homogeneous half-space, with a shear wave velocity $C_s = 150$ m/s, a longitudinal wave velocity $C_p = 300$ m/s, a density $\rho = 1800$ kg/m^3 and a material damping ratio $\beta = 0.025$.

The floating slab track has a resilient mat below the concrete slab with a stiffness $K_{sm} = 15 \times 10^6$ N/m^3 and a damping $C_{sm} = 30 \times 10^3$ Ns/m^3. The corresponding stiffness and damping value per unit length in the longitudinal direction of the track are equal to $\bar{k}_{sm} = 2BK_{sm}$ and $\bar{c}_{sm} = 2BC_{sm}$.

Figure 8 shows the response of the track-soil interface of the unisolated and isolated track as a function of the frequency w and the dimensionless wavenumber $\bar{k}_y = k_y C_s/w$. Superimposed on this figure are the dispersion curve of the non-dispersive Rayleigh wave at $\bar{k}_y = 1.073$ and the dispersion curve of the first free track bending wave; the dispersion curve of the first fixed track bending wave of the floating slab track is

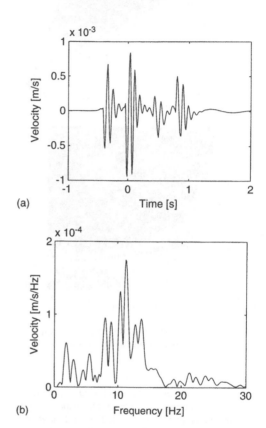

(a)

(b)

Figure 6. (a) Time history and (b) frequency content of the free field vertical velocity at 6 m from the centre of the road due to the passage of a truck on a traffic plateau.

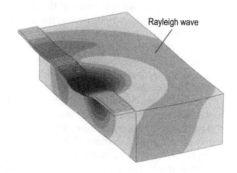

Figure 7. Displacements in the free field at $t = 0.08$ s due to the passage of a truck on a traffic plateau.

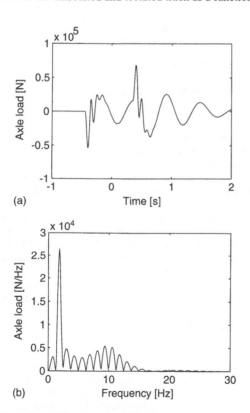

(a)

(b)

Figure 5. (a) Time history and (b) frequency content of the front axle load.

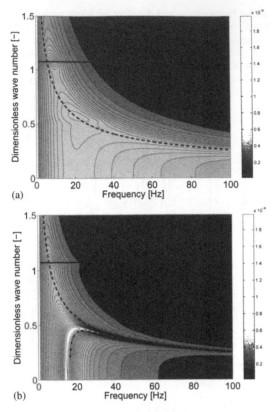

(a)

(b)

Figure 8. Response of the track-soil interface as a function of ω and $\bar{k}_y$ for the (a) unisolated and (b) isolated track.

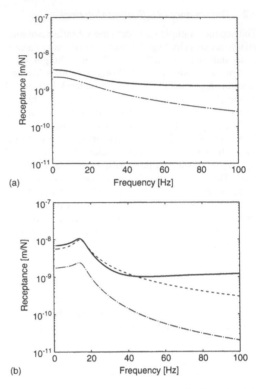

(a)

(b)

Figure 9. The displacement of the rail (solid line), the slab (dashed line) and the track-soil interface (dashed-dotted line) for the (a) unisolated and (b) isolated track.

also shown for frequencies beyond the cut-on frequency at 16.3 Hz. In the unisolated case, the response at the track-soil interface is concentrated around the dispersion curve of the free track wave and has maxima near 20 Hz and near the intersection point of the dispersion curves of the Rayleigh wave and the first free track wave. For the floating slab track, the response at the track-soil interface shows a maximum at the slab resonance frequency. At higher frequencies, the maxima follow the dispersion curve of the slab bending waves.

Figure 9 shows the track receptance for the unisolated and the isolated track. In the first case, the displacement of the slab and the track-soil interface coincide. At low frequencies, the rail and slab response are higher in the case of the isolated track, due to the flexibility of the mat, whereas the response at the track-soil interface has a similar magnitude. At the slab resonance frequency, the displacement of the rail, the slab and the track-soil interface is much higher in the case of the isolated track. The presence of the slab mat reduces the response at the track-soil interface at higher frequencies. The response of the rail and the slab

decouple and tend to a similar order of magnitude as in the case of the unisolated track.

The free field velocity at a perpendicular distance of 24 m to the track and three lateral distances along the track are compared in figure 10 for the unisolated and isolated track. For the floating slab track, the response increases around the resonance frequency of the slab on the resilient mat, and is substantially reduced at higher frequencies. Whereas for the unisolated track, the response decreases with increasing coordinate y along the track, this is not the case for the isolated track at frequencies higher than the slab resonance frequency, where the radiation of waves is clearly modified. This can better be appreciated on a plot of the free field displacements in a large number of receiver points on the track and in the free field. Figure 11 compares the track and free field displacements due to a harmonic load at 8 Hz for the unisolated and isolated track. The displacements of the rail and the slab are larger in the latter case due to the flexibility of the resilient mat. In a direction perpendicular to the track, the free field displacements are also larger, due to the dynamic amplification near the slab resonance

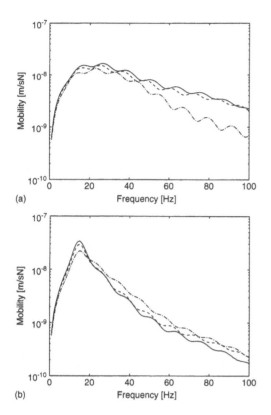

Figure 10. The free field mobility at $x = 24$ m and $y = 0$ m (solid line), $y = 8$ m (dashed line) and $y = 16$ m (dashed-dotted line) for the (a) unisolated and (b) isolated track.

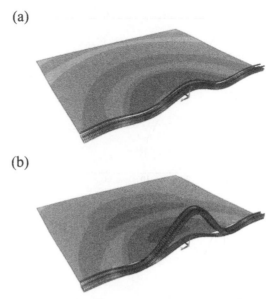

Figure 11. The free field displacements due to a harmonic load on the track at 8 Hz for the (a) unisolated and (b) isolated track.

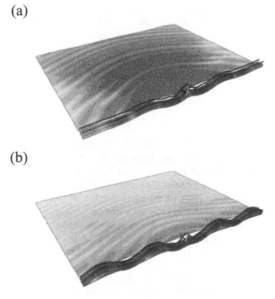

Figure 12. The free field displacements due to a harmonic load on the track at 32 Hz for the (a) unisolated and (b) isolated track.

frequency. Figure 12 shows similar results at a frequency of 32 Hz, well beyond the slab resonance frequency. The presence of the resilient mat completely changes the radiation of waves from the track. The slab is uncoupled from the soil and the bending waves in the uncoupled slab lead to larger displacements along the track and an effective reduction of waves radiated perpendicular to the track.

The response due to a single axle of a T2000 tram on an uneven rail is subsequently considered. A 2DOF model for a drive wheel is used, consisting of an unsprung mass $m_u = 80$ kg connected to the central part of the wheel that includes the motor and represents a sprung mass $m_v = 1890$ kg. Both masses are connected by a spring-damper system with $k_v = 145.0$ MN/m and $c_v = 3.0$ kNs/m, while the Hertzian contact spring stiffness $k_H = 1$ GN/m. The primary suspension linking the central part of the wheel to the bogie and the secondary suspension that links the bogie to the car are soft, so that the dynamics of the superstructure does not influence the frequency content of the axle loads.

Figure 13 illustrates the importance of the track compliance (it is assumed that the tram speed is low) in the total compliance, especially near the resonance frequency of the sprung mass. The coupling of the

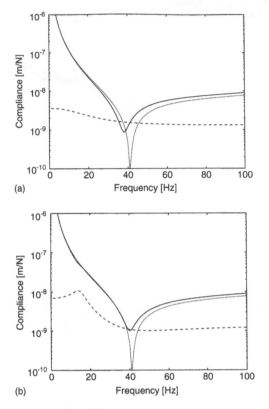

(a)

(b)

Figure 13. The vehicle compliance $\hat{C}^v$ (dotted line), the track compliance $\hat{C}^t$ (dashed line) and the total compliance $\hat{C}^v + \hat{C}^t$ (solid line) for the (a) unisolated and (b) isolated track.

track and the vehicle shifts this resonance frequency to a lower value, and increases the damping as the vehicle can radiate energy into the track-soil system. As the slab resonance frequency is well below the resonance frequency of the sprung mass, the total compliance of both tracks is very similar.

The tram's axle is excited by an uneven rail of medium quality, characterized by a single-sided PSD function $S_{wr}(n) = A/(1 + n/n_0)^3$, with $n_0 = 0.0489$ cycles/m and $A = 308 \times 1.15$ mm²/cycli/m. The PSD of the vehicle-track interaction force is calculated from the PSD of the unevenness and the total compliance of the vehicle-track system:

$$\hat{S}_g(\omega) = \left|\hat{C}_t + \hat{C}_v\right|^{-2} \frac{1}{v} \tilde{S}_{w/r}\left(-\frac{\omega}{v}\right) \quad (27)$$

Figure 14 illustrates that the vehicle-track interaction force is only weakly affected by the presence

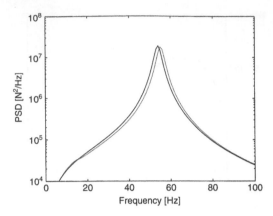

Figure 14. PSD of the vehicle-track interaction force for the unisolated (solid line) and the isolated (dotted line) track.

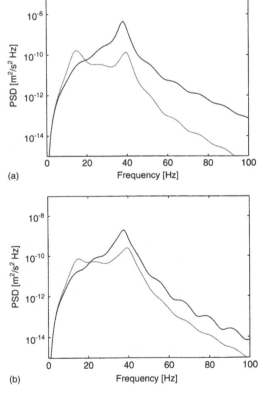

(a)

(b)

Figure 15. PSD of the free field velocity at $x = 24$ m and (a) $y = 0$ m and (b) $y = 16$ m for the unisolated (solid line) and isolated (dotted line) track.

of the resilient mat. The free field response will therefore mainly be determined by the transfer functions. When the train speed is low, the PSD of the free field velocity is computed from the free field mobility and the PSD of the vehicle-track interaction forces:

$$\hat{S}_v(\omega) = \left| i\omega \hat{h}_{zi}(x,y,z,\omega) \right|^2 \hat{S}_g(\omega) \qquad (28)$$

Figure 15 compares the PSD of the free field velocity at 24 m from the track and two lateral distances for the unisolated and isolated track. The response is maximum near the resonance frequency of the sprung mass, which is attenuated by material damping in the soil. At low frequencies, the PSD of the free field velocity is similar for both track structures, while at the slab resonance frequency, the response in the isolated case is much larger. At higher frequencies, a reduction of vibrations is achieved. The efficiency of the isolation decreases with the distance y parallel to the track.

7 CONCLUSION

An efficient methodology has been presented for the prediction of road or railway induced vibrations, where the dynamic road/track-soil interaction problem is solved in the frequency-wavenumber domain. The model is flexible and easily allows to incorporate simplifying assumptions for the calculation of the dynamic axle loads and for low vehicle speeds. The formulation is also efficient from a computational point of view as it allows to compute the free field response in a large number of output points, as needed for example on the interaction horizon of the soil and the foundation of a nearby building when a subsequent dynamic soil-structure interaction analysis is envisaged. The formulation has also been validated by a lot of in situ vibration measurements for road and train induced vibrations, the discussion of which is beyond the scope of the present paper.

REFERENCES

Andersen, L. and S. Nielsen (2003). Boundary element analysis of the steady-state response of an elastic half-space to a moving force on its surface. *Engineering Analysis with Boundary Elements 27*, 23–38.

Aubry, D., D. Clouteau and G. Bonnet (1994, December). Modelling of wave propagation due to fixed or mobile dynamic sources. In N. Chouw and G. Schmid (Eds.), *Workshop Wave '94, Wave propagation and Reduction of Vibrations*, Ruhr Universität Bochum, Germany, pp. 109–121.

Auersch, L. (2005). The excitation of ground vibration by rail traffic: theory of vehicle-track-soil interaction and measurements on high-speed lines. *Journal of Sound and Vibration 284*(1–2), 103–132. Accepted for publication. In press.

Clouteau, D., M. Arnst, T. Al-Hussaini, and G. Degrande (2005). Free field vibrations due to dynamic loading on a tunnel embedded in a stratified medium. *Journal of Sound and Vibration 283*(1–2), 173–199.

Clouteau, D., G. Degrande, and G. Lombaert (2001). Numerical modelling of traffic induced vibrations. *Mecca-nica 36*(4), 401–420.

Degrande, G. and G. Lombaert (2001). An efficient formulation of Krylov's prediction model for train induced vibrations based on the dynamic reciprocity theorem. *Journal of the Acoustical Society of America 110*(3), 1379–1390.

Ekevid, T. and N.-E. Wiberg (2002). Wave propagation related to high-speed train. A scaled boundary FE-approach for unbounded domains. *Computer Methods in Applied Mechanics and Engineering 191*, 3947–3964.

François, S., G. Lombaert and G. Degrande (2005). Comparison between local and global shape functions in a boundary element method for the calculation of traffic induced vibrations. *Soil Dynamics and Earthquake Engineering*. Accepted for publication.

Kausel, E. and J. Roësset (1981). Stiffness matrices for layered soils. *Bulletin of the Seismological Society of America 71*(6), 1743–1761.

Kaynia, A., C. Madshus and P. Zackrisson (2000). Ground vibration from high speed trains: prediction and countermeasure. *Journal of Geotechnical and Geoenvironmental Engineering, Proceedings of the ASCE 126*(6), 531–537.

Knothe, K. and S. Grassie (1993). Modelling of railway track and vehicle/track interaction at high frequencies. *Vehicle Systems Dynamics 22*, 209–262.

Krylov, V. (1996). Vibrational impact of high-speed trains. I. Effect of track dynamics. *Journal of the Acoustical Society of America 100*(5), 3121–3134.

Krylov, V. (1997). Vibrational impact of high-speed trains. I. Effect of track dynamics. *Journal of the Acoustical Society of America 101*(6), 3810. Erratum.

Lombaert, G. and G. Degrande (2001). Experimental validation of a numerical prediction model for free field traffic induced vibrations by in situ experiments. *Soil Dynamics and Earthquake Engineering 21*(6), 485–497.

Lombaert, G. and G. Degrande (2003). The experimental validation of a numerical model for the prediction of the vibrations in the free field produced by road traffic. *Journal of Sound and Vibration 262*, 309–331.

Lombaert, G., G. Degrande and D. Clouteau (2000). Numerical modelling of free field traffic induced vibrations. *Soil Dynamics and Earthquake Engineering 19*(7), 473–488.

Lombaert, G., G. Degrande and D. Clouteau (2004). The non-stationary free field response for a random moving load. *Journal of Sound and Vibration 278*, 611–635.

Lombaert, G., G. Degrande, S. François, J. Kogut and L. Pyl (2004, January). Validation of a numerical model for the prediction of railway induced vibrations in the free field. In *11th International Conference on Soil Dynamics and Earthquake Engineering and 3rd International Conference on Earthquake Geotechnical Engineering*, Berkeley, CA, USA.

Lombaert, G., G. Degrande, B. Vanhauwere and B. Vandeborght (2004, September). A numerical study of vibration isolation for rail traffic. In *ISMA2004 International Conference on Noise and Vibration Engineering*, Leuven, Belgium.

Madshus, C. and A. Kaynia (2000). High-speed railway lines on soft ground: dynamic behaviour at critical train speed. *Journal of Sound and Vibration 231*(3), 689–701.

Metrikine, A. and K. Popp (1999). Vibration of a periodically supported beam on an elastic half-space. *European Journal of Mechanics, A/Solids 18*(4), 679–701.

Metrikine, A., S. Verichev and J. Blauwendraad (2005). Stability of a two-mass oscillator moving on a beam supported by a visco-elastic half-space. *International Journal of Solids and Structures 42*, 1187–1207.

Sheng, X., C. Jones and M. Petyt (1999a). Ground vibration generated by a harmonic load acting on a railway track. *Journal of Sound and Vibration 225*(1), 3–28.

Sheng, X., C. Jones and M. Petyt (1999b). Ground vibration generated by a load moving along a railway track. *Journal of Sound and Vibration 228*(1), 129–156.

Sheng, X., C. Jones and D. Thompson (2003). A comparison of a theoretical model for quasi-statically and dynamically induced environmental vibration from trains with measurements. *Journal of Sound and Vibration 267*(3), 621–635.

Sheng, X., C. Jones and D. Thompson (2004). A theoretical model for ground vibration from trains generated by vertical track irregularities. *Journal of Sound and Vibration 272*(3–5), 937–965.

Vostroukhov, A. and A. Metrikine (2003). Periodically supported beam on a visco-elastic layer as a model for dynamic analysis of a high-speed railway track. *International Journal of Solids and Structures 40*(21), 5723–5752.

Environmental Vibrations – Takemiya (ed.)
© *2005 Taylor & Francis Group, London, ISBN 0 415 39035 4*

An extended analytical model including a layered embankment to simulate ground vibrations from railway traffic

A. Karlström & A. Boström

Division of Dynamics, Department of Applied Mechanics, Chalmers University of Technology, Göteborg, SWEDEN

ABSTRACT: Ground vibrations are investigated with an analytical approach. The ground is modelled as a layered half-space on which a layered embankment is placed. Situated on top of the embankment, the sleepers are modelled with an anisotropic Kirchhoff plate and the rails as Euler-Bernoulli beams. In the equation for the rails, propagating wheel loads with constant velocity are accounted for. The analyzing method is based on Fourier transforms in time and along the space coordinate parallel with the track. In the transverse direction Fourier transforms are adopted in the ground, whereas Fourier cosine and Fourier sine series are used in the embankment layers. The numerical scheme is very efficient. In particular the beam-like rectangular elastic region (embankment) is compared with the Euler-Bernoulli beam, both supported on a layered ground. With the same bending stiffness, the two solutions are seen to differ already from 0 Hz.

1 INTRODUCTION

At certain ground conditions, excessive ground vibrations from high speed trains cause environmental problems. A well documented case with large vibrations due to the Swedish high speed train X2 occurred at the site Ledsgård in Sweden (Banverket 1999). The problems arise when the train velocity passes the surface wave velocity in the ground, which is typical for grounds with various types of soft clays where the shear wave velocity may be as low as 30–40 m/s.

The interest in modelling ground vibrations caused by high speed trains has increased during the past 10 years. A common analytical approach is to represent the whole track with an Euler-Bernoulli beam, supported on an elastic ground. Dieterman & Metrikine (1996) determine the critical speeds of a uniformly moving point load on an Euler-Bernoulli beam and give further references. Recent contributions are made by Kaynia et al. (2000), Madshus & Kaynia (2000) and Takemiya (2003). They model the track with a beam resting on a layered viscoelastic halfspace and favourably compare the results with the measurements at Ledsgård.

An alternative to the analytical method is to employ FEM or some other discretization method. The advantage is the lack of restrictions on the geometries and the possibility to introduce nonlinear effects. Ekevid & Wiberg (2002) presented a model where the scaled boundary finite element method is used to treat the in-finite domains. A disadvantage, however, is that many degrees of freedom must be adopted, which results in relatively small discretized regions. In this case the model is 40 m long and less than that in the transverse direction (with a 107 m long train).

In this paper, the refined semi-analytical model presented by Karlström & Boström (in press) is developed further to include a layered embankment, where the layers are modelled as rectangular elastic regions with hysteretic damping. On top of the embankment, the rails and the sleepers are supported. The rails are governed by Euler-Bernoulli's equation for flexural vibrations in the vertical and transverse directions and by the rod equation for longitudinal movement. The sleepers are introduced as an anisotropic Kirchhoff plate. The ground is modelled as a stratified half-space

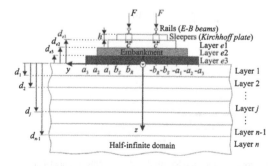

Figure 1. A cross section of the model showing geometrical properties and the applied wheel loads. The rails and the sleepers are introduced as boundary conditions on top of the layered embankment.

with elastic layers and hysteretic damping. Only the moving mass of the train is considered, i.e. influence of rail roughness etc. is not studied.

Finally, the beam-like rectangular elastic region (embankment) is compared with the Euler-Bernoulli beam, both supported on a layered ground. With the same bending stiffness, the solutions are seen to differ. Differences in the results are discussed.

2 PROBLEM FORMULATION

The model developed by Karlström & Boström (in press) is here extended to include a layered elastic embankment. The geometry of the model is given in Figure 2, where the cartesian coordinate system is given with x along the track, y in the transverse direction and z directed downwards. The embankment layer ei is defined by the width $2a_i$ and the height $d_{ei+1} - d_{ei}$. The rails and the sleepers are both modelled on top of the upper embankment layer. The rails are governed by the Euler-Bernoulli equation for the coupling in the vertical and transverse directions and by the rod equation for coupling in the longitudinal direction. They are positioned at $\pm b_R$ and the width in contact with the embankment is c. The sleepers are modelled as an anisotropic Kirchhoff plate and are given by the dimensions $2b_S \times h$. The ground consists of n layers, which are positioned with the coordinates $d_1, ..., d_{n-1}$. In the Euler-Bernoulli equation for the vertical coupling between the rails and the embankment, the wheel loads are represented with propagating point loads given by $F = F_0 \delta(x - V_0 t)$. F_0 is the amplitude, V_0 the constant velocity and t the time. The displacement components are $\mathbf{u}_j = \{u_j, v_j, w_j\}$, where $j = e1, e2, e3$ refers to the embankment layers and $j = 1, 2, ..., n$ refers to the ground layers. The material in each layer is defined by the density ρ_j and the complex Lamé constants λ_j and μ_j. In this way hysteretic damping (constant damping for all frequencies) is simulated. Provided the Fourier transform pair:

$$\tilde{f}(\omega) = \int_{-\infty}^{\infty} f(t)\, e^{i\omega t} dt, \tag{1}$$

$$f(t) = \frac{1}{2\pi} \int_{-\infty}^{\infty} \tilde{f}(\omega)\, e^{-i\omega t} d\omega, \tag{2}$$

for the time and frequency, the elastodynamic wave equation is given in the frequency domain as:

$$c_j^{P\,2}\nabla(\nabla \cdot \tilde{\mathbf{u}}_j) - c_j^{S\,2}\nabla \times (\nabla \times \tilde{\mathbf{u}}_j) = -\omega^2 \tilde{\mathbf{u}}_j, \tag{3}$$

where $\tilde{\mathbf{u}}_j$ is the displacement vector in layer j given in the frequency domain, $c_j^P = ((\lambda_j + 2\mu_j)/\rho_j)^{1/2}$ is the pressure wave velocity and $c_j^S = (\mu_j/\rho_j)^{1/2}$ is the shear wave velocity.

All interfacial boundaries in the model are assumed to be in welded contact. The traction on a plane with the normal direction $\mathbf{e}_z$ is denoted by $\tilde{\mathbf{t}}^{(\mathbf{e}_z)}$. Starting with the ground this gives:

$$\tilde{\mathbf{u}}_j = \tilde{\mathbf{u}}_{j+1}, \tag{4}$$

$$\tilde{\mathbf{t}}_j^{(\mathbf{e}_z)} = \tilde{\mathbf{t}}_{j+1}^{(\mathbf{e}_z)}, \tag{5}$$

for $z = d_j$ and $j = 1, 2, ..., n - 1$. On the ground surface, $z = 0$, the traction vanishes, except within the region below the embankment where the displacement and traction vectors are continuous:

$$\tilde{\mathbf{u}}_1 = \tilde{\mathbf{u}}_{e3}, \qquad |y| \leq a_3, \tag{6}$$

$$\tilde{\mathbf{t}}_1^{(\mathbf{e}_z)} = \begin{cases} \tilde{\mathbf{t}}_{e3}^{(\mathbf{e}_z)}, & |y| \leq a_3, \\ 0, & |y| > a_3, \end{cases} \tag{7}$$

As the width of the embankment layers varies, the boundary conditions are given as:

$$\tilde{\mathbf{u}}_{ei} = \tilde{\mathbf{u}}_{ei-1}, \qquad |y| \leq a_{i-1}, \tag{8}$$

$$\tilde{\mathbf{t}}_{ei}^{(\mathbf{e}_z)} = \begin{cases} \tilde{\mathbf{t}}_{ei-1}^{(\mathbf{e}_z)}, & |y| \leq a_{i-1}, \\ 0, & a_{i-1} < |y| \leq a_i, \end{cases} \tag{9}$$

at $z = -d_{ei}$ for $i = 2, 3$. To enable series expansions of the displacement fields in the embankment, the boundary conditions on the sides of the embankment layers must be designed in a special way:

$$\begin{cases} \tilde{v}_{ei} = 0, & -d_{ei} < z < -d_{ei+1}, \\ \partial_y \tilde{u}_{ei} = 0, & -d_{ei} < z < -d_{ei+1}, \\ \partial_y \tilde{w}_{ei} = 0, & -d_{ei} < z < -d_{ei+1}, \end{cases} \tag{10}$$

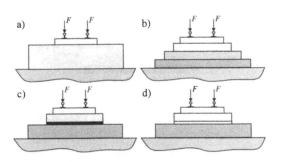

Figure 2. Track models used in the example, Figures 4 and 4. a) Rectangular embankment. b) Layered embankment with same materials as in a) in all layers. c) Same as in b), but with the middle layer replaced with a thin concrete layer. Same as in c), but with the concrete layer replaced by a rubber layer.

which is achieved by choosing to restrict the normal displacement component $\tilde{v}_{ei}$ to zero and using the natural conditions on the shear stresses:

$$\begin{cases} \tilde{v}_{ei} = 0, & -d_{ei} < z < -d_{ei+1}, \\ \tilde{\sigma}_{xyei} = 0, & -d_{ei} < z < -d_{ei+1}, \\ \tilde{\sigma}_{zyei} = 0, & -d_{ei} < z < -d_{ei+1}, \end{cases} \quad (11)$$

where $|y| = a_i$ for $i = 1, 2, 3$ and $d_{e4} = 0$ by definition. This approach was already used by Karlström & Boström (in press) for their rectangular embankment. They showed that the constraint on the normal displacement gives a very good approximation of the vertical displacement on top of both the embankment and the ground provided that the loads are applied vertically.

On top of the embankment the two rails are placed. They are characterized by the cross-sectional area A_b, the modulus of elasticity E_b, the mass density ρ_b and the area moment of inertia about the y and z axis I_{yb} and I_{zb} (subscript b denotes beam). In the transverse direction they are governed by Euler-Bernoulli's equation and in the longitudinal direction by the rod equation.

An anisotropic Kirchhoff plate is used to represent the sleepers. Vostroukhov & Metrikine (2003) showed that the vertical displacement due to train passage over a track with discrete sleeper positions is almost identical to the results when the sleepers are uniformly distributed along the track. Hence the mass and stiffness from the sleepers are accounted for by introducing a transversely isotropic Kirchhoff plate (Ambartsumyan 1970). The shear stiffness and the Young's modulus in the x direction should both be equal to zero if the rails are supported by sleepers. However, a slab track carries forces in the longitudinal direction via the sleepers. A transversally isotropic plate model is able to describe both situations. The plate material has a mass density ρ_s, a modulus of elasticity E_{sk} and Poisson's ratio v_{syz} in the $y - z$ plane and E_{sx} in the x direction and v_{sxk} in the $x - k$ plane, where k represents all directions perpendicular to the x direction. In the x direction the shear modulus is G_{sx} and in the isotropic $y - z$ plane it is simply $G_{sk} = E_{sk}/2(1 + v_{syz})$. Note that $v_{sxk}E_{sx} = v_{skx}E_{sk}$ due to symmetry of the stiffness tensor. On the free surface next to the plate, i.e. at $b_S \leqslant y < a_1$, the traction vanishes. The equations are given by Karlström & Boström (in press).

3 TRANSFORM SOLUTION

To get analytical solutions for the displacement fields, three scalar potentials in the embankment layers ($j = e1, e2, e3$) and in the ground layers ($j = 1, 2, ..., n$) are employed so that $\tilde{\mathbf{u}}_j = \nabla \tilde{\varphi}_j + \nabla \times (\mathbf{e}_z \tilde{\psi}_{SHj}) + \nabla \times \nabla \times (\mathbf{e}_z \tilde{\psi}_{SVj})$. Here $\tilde{\varphi}_j, \tilde{\psi}_{SHj}$ and $\tilde{\psi}_{SVj}$ are potentials for longitudinal, horizontal transverse and vertical transverse waves which satisfy scalar wave equations.

The problem formulation implies that it is convenient to apply a Fourier transform also with respect to x, where the corresponding transform variable is q. The Fourier transform pair for the space coordinate is:

$$\bar{g}(q) = \int_{-\infty}^{\infty} g(x)\, e^{-iqx}\, dx, \quad (12)$$

$$g(x) = \frac{1}{2\pi} \int_{-\infty}^{\infty} \bar{g}(q)\, e^{iqx}\, dq. \quad (13)$$

The doubly transformed fields (with respect to t and x) are denoted by a hat.

3.1 Solution in the ground ($j = 1, 2, ..., n$)

In the layered ground a Fourier transform in y makes it possible to represent the displacement fields by employing the potentials:

$$\hat{\varphi}_j = \frac{1}{2\pi} \int_{-\infty}^{\infty} \left(A_{jd} e^{ih_j^P(z-d_{j-1})} + A_{ju} e^{-ih_j^P(z-d_j)} \right) e^{ipy}\, dp, \quad (14)$$

$$\hat{\psi}_{SHj} = \frac{1}{2\pi} \int_{-\infty}^{\infty} \left(B_{jd} e^{ih_j^S(z-d_{j-1})} + B_{ju} e^{-ih_j^S(z-d_j)} \right) e^{ipy}\, dp, \quad (15)$$

$$\hat{\psi}_{SVj} = \frac{1}{2\pi} \int_{-\infty}^{\infty} \left(C_{jd} e^{ih_j^S(z-d_{j-1})} + C_{ju} e^{-ih_j^S(z-d_j)} \right) e^{ipy}\, dp, \quad (16)$$

where p is the transform variable to y. The wave numbers are $k_j^P = \omega/c_j^P$ and $k_j^S = \omega/c_j^S$. Corresponding wave numbers in the z direction are $h_j^P = (k_j^{P2} - q^2 - p^2)^{1/2}$ and $h_j^S = (k_j^{S2} - q^2 - p^2)^{1/2}$, where the roots are defined so that $\text{Im} h_j^P \geqslant 0$ and $\text{Im} h_j^S \geqslant 0$. With this choice of the wave numbers, $A_{jd} = A_{jd}(p)$, $A_{ju} = A_{ju}(p)$, $B_{jd} = B_{jd}(p)$, $B_{ju} = B_{ju}(p)$, $C_{jd} = C_{jd}(p)$ and $C_{ju} = C_{ju}(p)$ are the amplitudes of the down-going (subscript d) and up-going (subscript u) P, SH and SV waves, respectively. In the last halfinfinite layer ($j = n$) there are no reflected waves, i.e. $A_{nu} = 0$, $B_{nu} = 0$ and $C_{nu} = 0$, and the radiation condition with down-going or evanescent waves has been applied. The d_{j-1} and d_j in the exponents are crucial as they prevent exponential growth. With this choice the absolute value of the exponential functions never exceeds one. When $j = 1$ by definition: $d_0 = 0$.

3.2 Solution in the embankment ($j = ei$ and $i = 1, 2, 3$)

The boundary conditions (10) were designed to enable series expansions of the displacements in the

305

embankment. Due to the symmetric loading about $y = 0$, the displacement vector $\tilde{\mathbf{u}}_{ei}$ becomes symmetric. This means that $\tilde{u}_{ei}$ and $\tilde{w}_{ei}$ are even and $\tilde{v}_{ei}$ is odd. In view of Equation (10) the displacement components can be developed in Fourier cosine and Fourier sine series, with the wave numbers $p_{mi} = m\pi/a_i$ in the y direction. This gives the following choice for the potentials:

$$\hat{\varphi}_{ei} = \sum_{m=0}^{\infty} (D_{1m}^{i} \sin h_{mi}^{P} z + E_{1m}^{i} \cos h_{mi}^{P} z) \cos p_{mi} y, \tag{17}$$

$$\hat{\psi}_{SHei} = \sum_{m=1}^{\infty} (D_{2m}^{i} \sin h_{mi}^{S} z + E_{2m}^{i} \cos h_{mi}^{S} z) \sin p_{mi} y, \tag{18}$$

$$\hat{\psi}_{SVei} = \sum_{m=0}^{\infty} (E_{3m}^{i} \sin h_{mi}^{S} z - D_{3m}^{i} \cos h_{mi}^{S} z) \cos p_{mi} y. \tag{19}$$

D_{nm}^{i} and E_{nm}^{i} are unknown amplitudes in the embankment layer ei, where $n = 1, 2$ and 3 give the amplitudes for the P, SH and SV waves, respectively. The wave numbers are similar to those in the half-space, but with subscript $j = e1, e2, e3$ to denote the layers in the embankment: $k_{ei}^{P} = \omega/c_{ei}^{P}$ and $k_{ei}^{S} = \omega/c_{ei}^{S}$. The wave numbers in the z direction are $h_{mi}^{P} = (k_{ei}^{P2} - q^2 - p_{mi}^2)^{1/2}$ and $h_{mi}^{S} = (k_{ei}^{S2} - q^2 - p_{mi}^2)^{1/2}$, where the roots are defined so that $\mathrm{Im} h_{mi}^{P} \geqslant 0$ and $\mathrm{Im} h_{mi}^{S} \geqslant 0$.

3.3 General solution procedure

There are two possible methods to adopt to obtain the relation between the constants in the embankment and the coefficients in the ground, one of which is the socalled Thomson-Haskell approach or transfer matrix technique (Mal 1988). The idea is to obtain a linear relation between the coefficients by recursive elimination, starting from the bottom layer. However, with increasing frequency the calculations will eventually suffer from precision problems and the algorithm becomes unstable (Mal 1988). Sheng et al. (1999) modify the method to overcome the problem, whereas the global matrix approach used by e.g. Mal (1988) is adopted in this model. The idea is to obtain the coefficients in the ground expressed in the constants in the third layer in the embankment simultaneously for each p using the interfacial conditions (4)–(5) and the boundary condition (7). In this way all the unknowns $A_{jd}, A_{ju}, B_{jd}, B_{ju}, C_{jd}$ and C_{ju} that depend on the continuous Fourier variable p can be expressed in D_{nm}^{3} and E_{nm}^{3} (which do not depend on p). The equations are outlined in the work by Karlström & Boström (in press).

When the unknowns in the ground are eliminated, the constants in the embankment remain to be solved. The equation system is obtained with the interfacial conditions (8) and (9) together with the boundary condition on top of the embankment, which includes the rails, sleepers and wheel loads, and the condition for the coupling to the ground, which includes the expression for the p dependent unknowns in the ground.

Employing a reverse Fourier series with respect to y over the width of the embankment ($-a_3 < y < a_3$) of the boundary condition (6) together with similar reverse Fourier series of the boundary conditions on top of the embankment ($-a_1 < y < a_1$) and reverse Fourier series with respect to y in Equation (8) in layer $ei - 1$ and in Equation (9) in layer ei for $i = 2, 3$ at the interface between layer $ei - 1$ and ei, the remaining unknowns D_{nm}^{i} and E_{nm}^{i} are solved.

4 EXAMPLE

In the paper by Karlström & Boström (in press), the rectangular embankment supported on a layered ground was developed and demonstrated. The welldocumented site Ledsgård, with material parameters and measurements of the ground displacements due to train passage of an X2-train (Banverket 1999), was used to verify the model. The simulations are seen to be very good approximations of the measurements at both 70 km/h and 200 km/h. The bending stiffness used by Kaynia et al. (2000) and Takemiya (2003) in their beam models was used to extract the material properties in the rectangular embankment. With the documented embankment height 1.4 m it was however seen that the embankment became too stiff no matter the width of the embankment. Hence the arbitrary geometry $0.5 \times 8\,\mathrm{m}^2$ was used with great success.

In this paper, the site Ledsgård is used again for the configuration of the ground, which contains five layers of surface crust, organic clay and marine clays. The track is exited with only one wheel pair to exclude the interference between the wheels obtained with the full train. The axle load is taken as 200 kN, propagating at the supercritical velocity 200 km/h. The rectangular embankment model from the work by Karlström & Boström (in press) is used in a comparison with three variations of the layered embankment model from this paper, see Figure 3.2. In case a) the rectangular embankment is depicted, where the geometry is taken as $1.4 \times 8\,\mathrm{m}^2$. The material is made softer compared to the preceding paper to obtain as realistic results as possible, see next section. In case b) the layered embankment is depicted where the geometry is taken as $a_1 = 2\,\mathrm{m}$, $a_2 = 3\,\mathrm{m}$ and $a_3 = 4\,\mathrm{m}$ and $d_1 = 1.4\,\mathrm{m}$, $d_2 = 1.0\,\mathrm{m}$ and $d_3 = 0.5\,\mathrm{m}$ (Fig. 2). The same material parameters as for case a) is adopted in all three layers. In case c) and d), the model in case b) is modified to

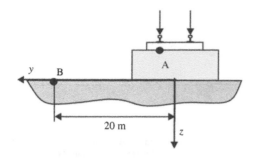

Figure 3. The positions A and B from where the frequency responses in Figures 4 and 4 are plotted, respectively.

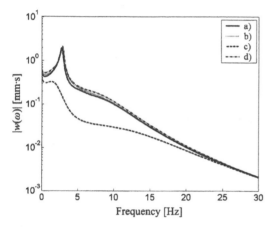

Figure 4. Frequency response at position A in Figure 3.3 due to the axle load 200 kN moving at 200 km/h at Ledsgård. The curves a)–d) refer to 4 different embankments depicted in Figure 3.2.

include a 0.2 m thick middle layer representing a concrete layer and a rubber layer, respectively. The geometries are $a_1 = a_2 = 2$ m and $a_3 = 4$ m and $d_1 = 1.4$ m, $d_2 = 1.0$ m and $d_3 = 0.8$ m. The concrete layer is modelled with the material parameters for a standard monoblock sleeper NS90 with the elasticity modulus $E = 38450$ MPa, the Poisson's ratio $\nu = 0.2$ and the density $\rho = 635$ kg/m^3, whereas the rubber is modelled with $E = 5$ MPa, $\nu = 0.499$ and $\rho = 1000$ kg/m^3.

The layered embankment model provides new possibilities to investigate the effects of embankment geometry and, in particular, the material properties. The frequency responses on top of the embankment and on the ground at $y = 20$ m, position A and B shown in Figure 3, are plotted in Figures 4 and 4 for the cases a)–d) in Figure 3.2 due to the propagating axle load. Comparing case a) (the solid curve) and b) (the dotted curve) at the position A, i.e. directly under the rails, the geometry seem to have some small influence on the outcome at supercritical velocity. The discrepancies

are evident around 1 Hz and between 3 and 10 Hz. In general the differences are very small. However, at subcritical speed the geometry does not affect the result no matter how the widths of the upper two embankment layers are varied. Exchanging the middle layer to rubber, case d), the similar observations are made and instead of damping the vibrations, the rubber layer contribute to somewhat increased levels in the range 0–30 Hz. However, replacing the rubber layer with a concrete layer, case c), the vibration levels are strongly reduced for the frequency range 0–25 Hz, whereas for higher frequencies the reduction vanish.

Considering position B, i.e. on the ground at $y = 20$ m, similar behaviour as for position A is observed with some differences. The low frequency response up to about 2 Hz is similar for all models, which was not the case for the concrete layer model at position A. Also, the high frequency response differs for the concrete layer model, compared to the others. Neither this was the situation at position A. It should be mentioned that the concrete layer have been placed on top of and at the bottom of the embankment, with similar results. This suggests that a slab track could be used to reduce high vibration levels at the sides of the track due to soft ground conditions where the X-2 train has to operate at supercritical velocities. As for position A, the rubber layer (case d) gives a somewhat higher response than the other models. However, it should be noted that the excitation is made with a moving mass and that dynamic effects due to train-track interaction is not accounted for. The rubber material is known to be an efficient damper of high frequency vibrations and the rubber layer would certainly contribute to damping if the excitation mechanism included dynamic effects. Also, it should be noted that the simulations are only carried out for frequencies less than 30 Hz.

5 COMPARISON BETWEEN RECTANGULAR ELASTIC REGION AND EULER-BERNOULLI BEAM

The rectangular embankment model developed by Karlström & Boström (in press) was compared with measurements and it was seen that the arbitrary geometry 0.5×8 m^2 instead of 1.4×3 m^2 (Takemiya 2003) gave results that agreed very well. A comparison between inter alia the Euler-Bernoulli beam representation and the proposed track model (rails and sleepers modelled as Euler-Bernoulli beams and a Kirchhoff plate, supported on a rectangular region) was made. The beam representation was seen to deviate from the more refined elastic model already at 1–2 Hz at 70 km/h and at 2–3 Hz at 200 km/h on top of the embankment at $y = 0$ m. Also, the low frequency displacements were somewhat higher than those in the proposed model. And when the time domain displacement is investigated

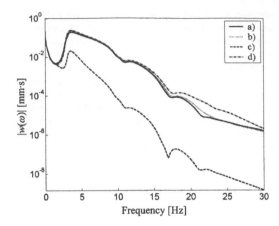

Figure 5. Frequency response at position B in Figure 3.3 due to the axle load 200 kN moving at 200 km/h at Ledsgård. The curves a)–d) refer to 4 different embankments depicted in Figure 3.2.

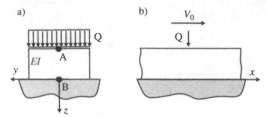

Figure 6. The model used to investigate the elastic rectangular region with the Euler-Bernoulli beam, both with the same bending stiffness. A line-load of 160 kN propagate at the site Ledsgård with the velocity 70 km/h. The positions A and B are shown, from where the frequency response in Figures 5 and 5 are calculated.

it is in general observed that for the same bending stiffness the elastic embankment is remarkably stiffer than the corresponding Euler-Bernoulli beam representation. To find explanations to the differences, the track response due to a line load of 160 kN is investigated for an elastic rectangular region and an Euler-Bernoulli beam with the same bending stiffness with respect to the centre of gravity with the geometry $1.4 \times 3 \, m^2$ for the cross section, see Figure 5. The bending stiffness is taken as 200 MNm² and the ground properties are adopted from Ledsgård.

In Figure 5, the frequency response (logarithmic scale) at position A and B in Figure 5 for the elastic region is compared to the response at position B for the Euler-Bernoulli beam. The line load propagate at the speed 70 km/h and the frequency range is 0–20 Hz. The Euler-Bernoulli equation is an approximation based on assumptions, which roughly spoken differs from the exact theory for wavelengths smaller than 5 times the diameter for a circular cross section. The limit gives a hint of the validity for the rectangular beam. Provided the width 3.0 m instead of the diameter, the beam is a valid approximation for wavelengths larger than 15 m. The surface Rayleigh wave velocity is about 47 m/s, which gives a limit frequency of 3 Hz. Thereafter the Euler-Bernoulli beam theory differs from the exact solution. This is confirmed in Figure 5. The displacement at position A for the elastic region deviates at 3 Hz, whereas the displacement at position B for the elastic region and the Euler-Bernoulli beam follow each other up to about 13 Hz. As the Euler-Bernoulli beam provides the same solution in A and B, something in between the response at A and B for the elastic region ought to be the appropriate curve to compare with. The result in the frequency range 0–3 Hz

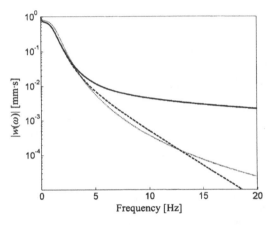

Figure 7. Frequency response plot due to a line-load of 160 kN propagating at 70 km/h at Ledsgård. The curves are for the elastic region at position A (solid curve) and B (dashed curve), plus the Euler-Bernoulli beam at position B (dotted curve) from Figure 5. The frequency range is 0–20 and the vertical scale is logarithmic.

is enlarged in Figure 5, here shown with a linear scale. Evidently the displacement differs up to 10% within this region, which show the discrepancy that has been observed in the simulations and confused the authors for a while.

The mathematical assumptions made to obtain the simple Euler-Bernoulli equation disregard important features when it couples to the ground. Bövik (1996) made similar findings for a Tiersten plate model. He developed surface boundary conditions for elastic surface waves guided by thin films by means of series expansions, starting from the three-dimensional elasticity theory. The results for surface Rayleigh waves (flexural vibrations) were then compared with the results where the boundary was derived with the

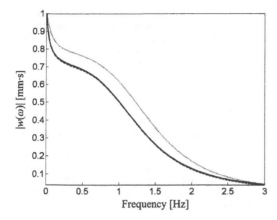

Figure 8. The result in the frequency range 0–3 Hz from Figure 5 is here enlarged, given with a linear scale. Frequency response plotted for an elastic region at position A (solid curve) and B (dashed curve), plus an Euler-Bernoulli beam at position B (dotted curve).

Tiersten plate model (obtained by approximate equations for low frequencies) including the flexural stiffness. It was found that with higher frequency the boundary condition developed by Bövik (1996) was a better approximation of the exact solution. This emphasize the validity only at lower frequencies.

As the Euler-Bernoulli beam only accounts for vertical loads, an attempt to explain the discrepancy between 0 and 3 Hz was made by disregarding the longitudinal and transverse shear traction components σ_{xz} and σ_{yz} at the interface between the elastic embankment and the ground. Small effects were found, but still not enough to explain the differences. Moreover, the equation for the Euler-Bernoulli beam was extended adding the longitudinal shear traction and the rotation about the y-axis, ending up with two coupled differential equations. Neither this approach made the differences disappear. Actually the differences were even smaller than if just the shear traction was excluded for the elastic embankment.

The Euler-Bernoulli equation is derived with the neutral layer (provided that only bending is considered) defined in the centre of gravity for, in this case, the rectangular cross section. The loading on the beam is given as a distributed load on the surface. When the beam is used as a track representation, it is used with a loading given by the normal traction from the interface between the ground and the embankment (as in the derivation), but it is also kinematically constrained to the ground motion. Similarly to a sandwich beam, the axis for calculating the area moment of inertia is probably moved from the centre of gravity towards the ground surface.

The relation between the stiffness in the ground and in the beam is also relevant. The stiffer the beam is compared to the ground, the more accurate the beam equation is. Kaynia et al. (2000) and Takemiya (2003) use a beam that has an elasticity modulus that is about 2 or 5 times higher than in the surface ground layers, depending on train speed (70 km/h or 200 km/h). Hence the beam equation becomes a tolerable approximation.

In order to obtain material properties that works for the elastic embankment model, it is obviously pointless to try to adopt the material property for the Euler-Bernoulli beam. In Chapter 4 the material properties in the embankment was simply adjusted to comply with the measurements from Ledsgård performed by Banverket (1999), ending up with the shear-wave velocity $c_{ei}^S = 55$ m/s and the pressure wave velocity $c_{ei}^P = 470$ m/s in all embankment layers at 70 km/h, and $c_{ei}^S = 60$ m/s and $c_{ei}^P = 220$ m/s in all embankment layers at 200 km/h.

6 CONCLUSIONS

A layered embankment model is developed from the rectangular model made by Karlström & Boström (in press). It is seen to provide similar results as the rectangular embankment model, which was verified with measurements. The layered embankment opens for new possibilities to investigate effects of varying embankment materials. Also effects due to geometry can be studied.

A comparison was made to investigate the difference in outcome due to a rectangular elastic region and an Euler-Bernoulli beam supported on a ground, with the same load and bending stiffness. The conclusion is that the beam theory results in a larger low frequency response than the elastic region, and it is therefore not possible to adopt the same material parameters from the beam to the elastic material.

ACKNOWLEDGEMENTS

This work is part of the activities of the Center of Excellence CHARMEC (CHAlmers Railway MEChanics, www.charmec.chalmers.se).

REFERENCES

Ambartsumyan, S.A. 1970. *Theory of anisotropic plates* (Volume II). Stamford, USA: Technomic Publication.
Banverket, 1999. *High speed lines on soft ground: evaluation and analyses of measurements from the West Coast line.* Borlänge, Sweden: Swedish National Rail Administration.
Bövik, P. 1996. A comparison between the Tiersten model and O(H) boundary conditions for elastic surface waves

guided by thin layers. *Journal of Applied Mechanics – Transactions of the ASME* 63 (1): 162–167.

Dieterman, H.A. & Metrikine, A. 1996. The equivalent stiffness of a half-space interacting with a beam. Critical velocities of a moving load along the beam. *European Journal of Mechanics – A/Solids* 15 (1): 67–90.

Ekevid, T. & Wiberg, N.-E. 2002. Wave propagation related to high-speed train: A scaled boundary FE-approach for unbounded domains. *Computer Methods in Applied Mechanics and Engineering* 191 (36): 3947–3964.

Karlström, A. & Boström, A. in press. An analytical model for train induced ground vibrations from railways. *Journal of Sound and Vibration*.

Kaynia, A.M., Madshus, C. & Zackrisson, P. 2000. Ground vibration from high-speed trains: prediction and counter-measure. *Journal of Geotechnical and Geoenvironmental Engineering* 126 (6): 531–537.

Madshus, C. & Kaynia, A.M. 2000. High-speed railway lines on soft ground: dynamic behaviour at critical train speed. *Journal of Sound and Vibration* 231 (3): 689–701.

Mal, A.K. 1988. Wave propagation in layered composite laminates under periodic surface loads. *Wave Motion* 10 (3): 257–266.

Sheng, X., Jones, C.J.C. & Petyt, M. 1999. Ground vibration generated by a load moving along a railway track. *Journal of Sound and Vibration* 228 (1): 129–156.

Takemiya, H. 2003. Simulation of track-ground vibrations due to a high-speed train: the case of X2000 at Ledsgard. *Journal of Sound and Vibration* 261 (3): 503–526.

Vostroukhov, A.V. & Metrikine, A.V. 2003. Periodically supported beam on a visco-elastic layer as a model for dynamic analysis of a high-speed railway track. *International Journal of Solids and Structures* 40 (21): 5723–5752.

Environmental Vibrations – Takemiya (ed.)
© *2005 Taylor & Francis Group, London, ISBN 0 415 39035 4*

Propagation of ground vibrations induced by moving trains

J. Lu, H. Xia & Y.M. Cao
School of Civil Engineering & Architecture, Beijing Jiaotong University, Beijing, China

ABSTRACT: The propagation of ground vibrations induced by moving trains is a basic subject in the research field of environmental vibrations. And it is a complex problem dependent on a lot of factors. Although many researches have been done on this subject, there are still many problems remain unclear. In this paper, the propagation of train-induced vibrations in free field is studied using numerical method. The effects of distance from the railway track, the axle loads of vehicles, the train speed, the characteristics of ground soil, and the depth of bedrock are analyzed. Different trains used in China are considered in the analyses. Two kinds of ground soils and their combinations are used in the calculations.

1 INTRODUCTION

With the development of social economy and improvement of science and technology, the vibration nuisance becomes an issue of great concern. The propagation of ground vibrations induced by moving trains is a basic subject in the research field of environmental vibrations. Many researches have been done on this subject, in which both theoretical analysis and numerical method are used to solve this problem (Xia 2002, Zhai 1997, Krylov 1995, Yang 1996, Xia 2001). And many in-situ experiments have also been done to find out the laws of train-induced ground vibrations (Xia 2005). From these studies, it has been already known that the vibration levels of ground decay with the distance between the receiver points and the vibration sources, and it is even found that there exist some amplified zones where the vibration levels increase at the certain distances from the vibration sources (Xia 2002).

Because of its complexity, there are always many simplifications in the analysis models. So for the train-induced ground vibrations only a few influencing factors are considered and there are still many problems unclear on this subject. In this paper, the analysis results obtained by using the numerical method are introduced.

2 NUMERICAL MODEL

The numerical method used in this paper is a spatial train-track-roadbed-soil system model, which is composed of two two-dimensional subsystem models including a X–Z train-track system model and a Y–Z ground model.

2.1 Train-track system model

The train-track system model consists of a vehicle model and a track model connected by assumed wheel-track relations (see Fig. 1). Because the vibrations in vertical direction are much more significant than that in other directions, only the vertical vibration is considered in this model.

The vehicle model is a train composed of many carriages. Each one is a multiple degree of freedom vibration system composed by one car, two bogies, four wheels, and two groups of spring-damper devices. Two degrees of freedom (Z and RX) are considered for each car or each bogie. Only one degree of freedom (Z) is considered for each wheel.

The track is simplified to a 3-level mass-spring-damper system, in which the influence of rail, ballast, sleeper and roadbed is considered.

2.2 Ground model

An FEM model shown in Figure 2 is used as the ground model. The following assumptions are used in

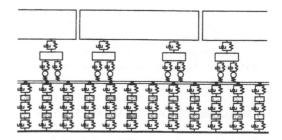

Figure 1. Train-track system model.

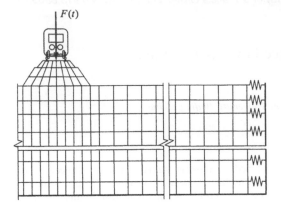

Figure 2. Ground model.

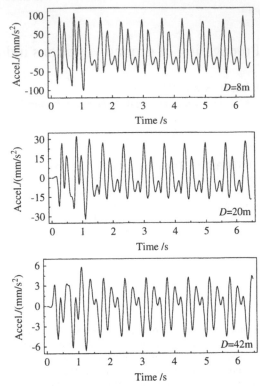

Figure 3. Time-history of acceleration.

the analysis: (1) The soil is regarded as linear elastic material; (2) The bottom of soil is fixed to simulate the influence of bedrock.

The reaction forces from the train-track system model are taken as the excitations on the ground.

3 CALCULATION PARAMETERS

The analysis is performed by using the above numerical model to investigate the propagation laws of ground vibrations induced by moving trains.

In the calculation, four types of trains and two kinds of ground soils are considered. The trains include the DF4 + C62 freight trains, the SS8 passenger trains, the Hong Kong light railway trains, and the China Star high-speed trains, with their running speeds from 30 km/h to 300 km/h. Two types of ground soils are considered: a hard soil with elastic modular $E = 1.642 \times 10^8 \, N/m^2$, Poisson ratio $\nu = 0.45$, density $\rho = 2.033 \times 10^3 \, kg/m^3$ and damping ratio $\xi = 0.05$; An other is soft soil with $E = 2.52 \times 10^7 N/m^2$, $\nu = 0.25$, $\rho = 1.80 \times 10^3 \, kg/m^3$ and $\xi = 0.05$.

The time histories of displacement, velocity and acceleration of different positions of ground surface are obtained by means of numerical calculations. As an example, the time-history curves of ground acceleration are given in Figure 3 when the DF4 + C62 freight train moves on hard soil with the speed of 90 km/h. In the graphs, D is the distance between the receiver point and the central line of railway track.

4 ANALYSES OF CALCULATION RESULTS

Further analyses of calculations results are done to study the influence laws of different parameters.

4.1 Effect of distance from railway track (D)

To consider the effect of the distance from railway track (D) systemically, further analyses were made on all four types of trains with different speed. The results are shown in Figure 4 and Figure 5.

Through the 32 curves in these figures, it is easy to draw the following conclusions:

(1) The ground vibration levels attenuate rapidly with the distance from railway track (D).
(2) The speed of vibration attenuation decreases with D, namely, the vibration level reduces to a fairly low value in a small area near the railway track. The larger the distance from railway track is, the smoother curves become.
(3) There is an amplified zone on each curve. It shows that when D is in a special range, the ground vibration level will increase with D. And the trend is more obvious on ground with hard soil than with soft soil by comparing of Figure 3 and Figure 4.

4.2 Effect of wheel loads (W)

As the main parameter of a train, the wheel load of train plays an important role on train loads. It is reasonable

312

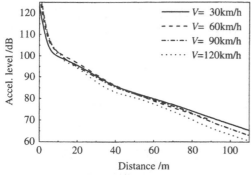

(a) Hong Kong light railway train

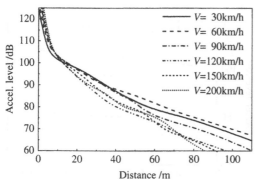

(b) SS8 passenger train

Figure 4.   Effect of D on soft soil (V-train speed).

to imagine that the vibrations induced by trains with different wheel load must be different. For the trains considered in this paper, the wheel loads are 323.5 kN for DF4 + C62 freight train, 210.0 kN for SS8 passenger train, 194.9 kN for China Star high-speed train and 142.0 kN for Hong Kong light railway train, respectively.

The curves of peak values of the acceleration of ground vibration versus the distance from railway track when four types of trains pass by are shown in Figure 6, in which the train speeds are 30 km/h and 90 km/h, respectively. By comparing Figures 4–6, it is easy to draw the conclusion that the larger the wheel loads of the train (W) are, the higher the vibration levels are. But the trend of vibration level decrease with D does not vary with W.

4.3   Effect of train speeds (V)

According to common understanding, the higher the train speed V is, the more obvious the impact effect of the train is, and then the stronger the train-induced vibration will be, which has already been shown

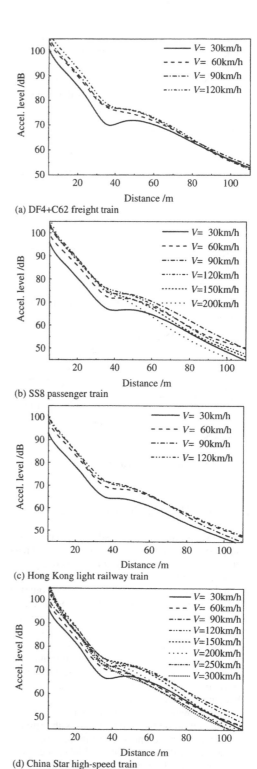

(a) DF4+C62 freight train

(b) SS8 passenger train

(c) Hong Kong light railway train

(d) China Star high-speed train

Figure 5.   Effect of D on hard soil (V-train speed).

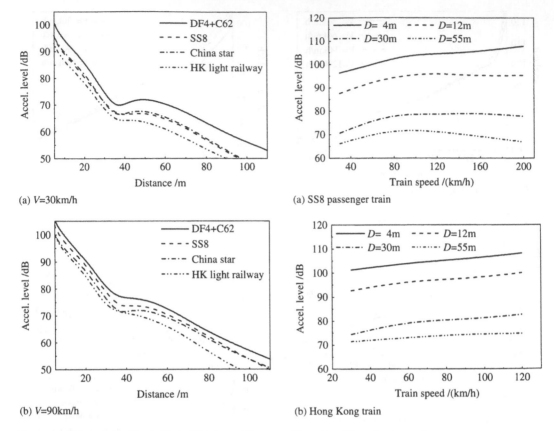

(a) V=30km/h

(b) V=90km/h

Figure 6.   Effect of wheel load of train (V-train speed).

(a) SS8 passenger train

(b) Hong Kong train

Figure 7.   Effect of train speed.

in Figure 4–6. However, different laws are found in further analysis.

Figure 7 shows the curves of peak value of the acceleration of ground vibration vary with V when SS8 passenger trains pass by, in which D is respectively 4 m, 12 m, 30 m and 55 m.

From these curves, it is not difficult to find out that when V is fairly low, the peak value of the acceleration of ground vibration has the tendency to increase with the improvement of V, and when V is beyond certain value, the peak value of the acceleration of ground vibration will be reduced with improvement of V. The larger D is, the lower the critical speed is. When the train speed is still higher, the variety trend will become more complicated.

This phenomenon can be explained by using the fact that the vibration induced by trains with different speeds has different composition of frequency.

It is shown in the study that the higher the speed is, the higher the frequency of train loads is (see Figure 8), and the larger the frequency range is. In the process of their propagation in the ground, vibrations

with different frequencies decay in different speeds. The higher the frequency is, the quicker the vibration decays.

To express the trend more intensively, harmony analyses are carried out on the original model. In the analysis, simple harmonic loads of 100 kN on amplitude applied in the position of train loads. The result is shown in Figure 9.

It is obvious that the vibrations with higher frequency decrease more rapidly with the distance from the track, and this trend is more obviously in soft soil. In the zone near the source, vibrations induced by loads with higher frequencies are stronger than those induced by lower frequency loads, which is similar with the common understanding. Then the trend that vibrations vary with V can be explained. Trains with higher speed induce stronger vibrations near the tracks, and the frequencies are higher, the frequency range of vibration is larger.

With the increase of D, those with highest frequencies decay first. So the vibrations induced by high

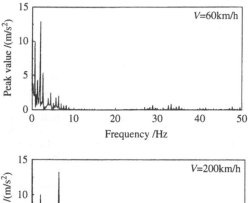

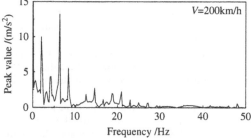

Figure 8. Spectrum curves of train loads.

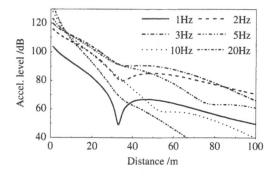

Figure 9. Effect of frequency (hard soil).

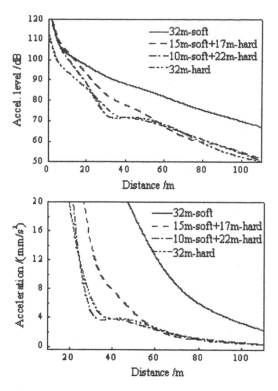

Figure 10. Effect of soil properties.

In addition, Figure 9 also indicates that there are more significant amplified zones in the propagation path of lower frequency vibrations.

4.4 Effect of soil properties

It is in soil layers that the train-induced vibration propagates, so the properties of soil have great influence on the vibration levels. Calculations have been done on both hard soil and soft soil above mentioned.

These calculations are aimed at single layer soil with the thickness of 32 m. In order to study the effect of soil properties, analyses of vibration induced by SS8 passenger trains on layered ground are carried out. The result is gathered in Figure 10, from which the following conclusions can be found:

(1) The harder the ground soil is, the higher the vibration levels are.
(2) When there exist soft layers above the hard soil, the surface vibration acceleration peak value curve is close to soft soil in the area near the track, while it is close to hard soil in the farther zone. The thicker the soft layer is, the larger the peak value of vibration acceleration is.

speed trains decrease with D faster than others. The difference between the two is diminished thereupon. The trend continues with the increase of D. When D is beyond a critical value, the phenomenon occurs that vibrations induced by high speed trains are lower than those induced by low speed trains.

For a point whose distance from the track is a decided value, it appears that there is a critical train speed at which the peak value of the train-induced ground vibrations occur, and either the train speed is lower or higher than the critical train speed, the ground vibrations decrease. With the further increase of the train speed, the frequency range of train loads is further enlarged, and the distribution law versus train speed will be more complicated.

315

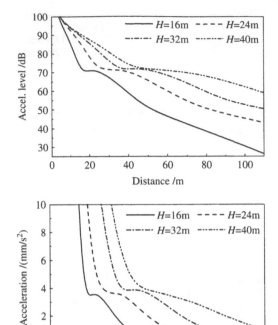

Figure 11.   Effect of the bedrock depth.

5   CONCLUSIONS

The following conclusions can be drawn from the analyses presented in this paper:

(1) The ground vibrations decrease with the distance from the railway track significantly, and the farther the distance is, the less the effect of the distance varying is.
(2) There exist certain zones where the ground vibrations are amplified. The distances from the amplified zones of the ground vibration to the railway track increase with the depths of the bedrock.
(3) Different train loads induce different ground vibrations: heavier trains induce greater vibrations than lighter trains.
(4) The ground vibrations increase with the train speed when the speed is fairly low. But the distribution trend becomes complex when the train speed is higher.
(5) The properties of ground soils have great influences on the ground vibrations: the harder the ground soil is, the higher the vibration levels are.

ACKNOWLEDGMENTS

This study is sponsored by the Natural Scientific Foundation of Beijing (No. 8042017) and the Key Research Foundation of Beijing Jiaotong University (No. 2004SZ005).

4.5   Effect of bedrock depths

Furthermore, the special phenomena of the existence of vibration-amplified zones are analyzed. Many factors affect the position of the amplified zones, including train speeds, propagation velocity of waves, soil properties and bedrock depths. To certain extent, the laws of their influence reflect the results of analyses presented above. It is found from the analyses that the depths of bedrock influence the position of the amplified zone significantly.

Figure 11 gives the vibration variety curves of ground with different depths of bedrock when the SS8 passenger trains pass by with the speed of 60 km/h. It is not difficult to drawn a conclusion from these curves that the distance from the railway track to the amplified zone increase with the depth of the bedrock.

Through the comparison of the curves shown in Figure 11, it can also be found that due to the restraint function on the soil layer of the bedrock, the shallower the bedrock is, the more rapidly the peak value of vibration acceleration decrease with the distance from the railway track.

REFERENCES

Xia, H. 2002. *Dynamic Interaction of Vehicles and Structures.* Beijing: Science Press.
Zhai, W.M. 1997. *Dynamics of Coupled Train-Track.* Beijing: China Railway Press.
Krylov, V.V. 1995. Generation of ground vibration by super-fast trains. *Applied Acoustics* 44: 149–164.
Yang, Y.B. & Kuo, S.R. 1996. Frequency-independent infinite elements for analyzing semi-infinite problems. *Numer. Meth. Engng.* 39: 3553–3569.
Xia, H. & Wu, X. 2001. Environmental vibration induced by urban rail transit system. *Journal of Northern Jiaotong University* 23(4): 1–7.
Xia, H. & Zhang, N. 2005. Experimental study of train-induced vibrations of environments and buildings. *Sound & Vibration* 2005 (280): 1017–1029.
Hung, H.H. & Jenny, K. 2001. Reduction of train-induced vibrations on adjacent buildings. *Structural Engineering & Mechanics* 11(5): 503–518.
Woods, R.D. 1968. Screening of surface waves in soils. *Soil Mech. Found. Div. ASCE* 94: 951–979.
Clough, R.W. & Penzien, J. 2003. *Dynamics of Structures.* New York: McGraw Hill Inc.
Wang, M.C. 2003. *Finite Element Method.* Beijing: Tsinghua University Press.

Environmental Vibrations – Takemiya (ed.)
© *2005 Taylor & Francis Group, London, ISBN 0 415 39035 4*

Evaluation of site vibration around Shinkansen viaducts under bullet train

X. He & S. Yamaguchi
Graduate School of Science and Technology, Kobe University, Kobe, Japan

M. Kawatani
Dept. of Civil Eng., Kobe University, Kobe, Japan

S. Nishiyama
Nikken Sekkei Civil Engineering, Ltd., Tokyo, Japan

ABSTRACT: Dynamic characteristics of the train-bridge interaction system are clarified by means of three-dimensional dynamic analysis using a 15 DOF bullet train model and the reaction forces at the pier bottoms of the viaducts are calculated. Employing the reaction forces as input excitations, site vibration around the viaducts is analyzed using a general program. Analytical results are compared with experimental ones to demonstrate the validity of the analytical procedure. Therefore, a reliable and effective approach to simulate the site vibration around Shinkansen viaducts is established. Especially in this analytical procedure, with a 15 DOF train model that can express sway and yawing motions of the bullet train, horizontal dynamic responses of the bridge as well as of the ground can be simulated.

1 INTRODUCTION

The main lines of Japan's high-speed railway usually pass directly over densely populated urban areas, where the railway structures mainly comprise elevated bridges of reinforced concrete in the form of a portal rigid frame. The vibrations caused by bullet trains on elevated bridges are propagated to the ambient ground via footings and pile structures, thereby causing some environmental problems related to site vibration around the viaducts. Those vibrations can influence precision instruments installed in hospitals and laboratories or people who are studying and resting in schools and residences. Along with further urbanization and more rapid transportation facilities, there is rising public concern about the environmental problems in modern Japan.

Although the importance and urgency of environmental problems have been recognized, the development and propagation mechanism of site vibration caused particularly by running vehicles on viaducts remains unclear because of its complicated nature. Without a clear grasp of the site-vibration mechanism through analytical procedures, environmental vibration problems are traditionally evaluated and predicted based on field test data. The efficiency of such a process is limited to particular cases. For more general cases, an effective analytical approach to simulate the environmental vibration problems is anticipated. In recent years, a great deal of effort has been devoted to studies of site vibrations induced by trains moving on the ground surface. Fujikake (1986) proposed a predictive method for vibration levels of the surrounding environment. Takemiya (2003) conducted a simulation of track-ground vibrations caused by a high-speed train for predicting train-track and nearby ground-borne vibrations. Yang et al. (2003) also examined train-induced wave propagation in layered soils using a 2.5D finite/infinite element approach. Nonetheless, little is known about the ground vibration caused by trains moving over viaducts because of its complicated nature: vibrations are transmitted to the ground via piers, footings and piles. Recently, Xia et al. (2002) evaluated the vibration related effects of light-rail train-viaduct system on the surrounding environment using a 2D interaction model of a "train-bridge" system for obtaining the dynamic loads of moving trains on bridge piers and a 2D dynamic model of "pier-foundation-ground" system for analyzing vibration responses of the ground. Wu et al. (2002) attempted to establish a semi-analytical approach to deal with ground vibration induced by trains moving over viaducts. In that approach, the train-bridge system is assumed as elastically supported beams under a series of moving loads.

In these studies, some assumptions and approximations were made either by simplifying the dimensions of the interaction system or by neglecting the dynamic interaction of the train-track system. An approach that can more accurately model the train-bridge interaction system and the foundation-ground interaction system is desirable. In this study, an effective approach to evaluate site vibration around viaducts of Shinkansen will be established, using three-dimensional (3D) dynamic analysis, in which the dynamic interactions of the train and track as well as of the foundation and ground can be simulated accurately. Because of the complicated system and limited computational capacity, the whole interaction system is divided into two subsystems: one describing train-bridge interaction and another showing foundation-ground interaction.

2 TRAIN-BRIDGE INTERACTION

Dynamic wheel loads of high-speed bullet trains directly contact the tracks and are transmitted to the foundation structures via the deck slabs and pier structures of the elevated bridges. To perform exact analysis of site vibration around the viaducts, the accurate dynamic reaction forces at the bottoms of the piers are demanded for use as input external excitations to the foundation-ground interaction system. First, it is necessary to simulate the accurate dynamic responses of elevated bridges to obtain the dynamic reaction forces at the bottoms of the piers. That simulation must take account of the coupled vibration of the train-bridge interaction system, which is a subject of many researchers' interest in structural dynamics.

Toward clarification of the vibration characteristics of elevated bridges, theoretical studies of the vibration of elevated railway bridges caused by traveling trains have been carried out since the middle 1960s (Tanabe et al. 1993). Recently in Kawatani's researches (Kawatani & Kim 2001, Kim & Kawatani 2001, Kawatani et al. 2004), a 3D vehicle model with eight-DOF and a 3D train model with nine-DOF were developed for dynamic response analyses of various bridge-vehicle interaction systems.

Considering the structural features of elevated bridges and the high-speed of bullet trains, the high-frequency components are predicted to be important for dynamic response; the influence of train model on bridge responses is expected. In this study, for the 3D dynamic response analysis of a train-bridge interaction system, the nine-DOF train model mentioned above is developed into a 15 DOF train model that can appropriately express the horizontal dynamic responses of the bridge.

2.1 Analytical procedure of train-bridge interaction

Dynamic responses of the high-speed railway elevated bridges under moving bullet trains are analyzed

in consideration of the wheel-track interaction with the rail surface roughness. The elevated bridges are modeled as 3D beam elements and simultaneous dynamic differential equations of the bridge are derived using modal analysis (Kawatani et al. 2000). The direct numerical integration of Newmark's β method is applied to solving the dynamic differential equations. Adopting 1/4 as the value of β, the responses can be obtained accurately, with less than 1/1000 error for acceleration at each time step.

For the limit of pages, dynamic differential equations of the developed 15 DOF bullet train model (see Fig. 1) are briefly described below. More detailed theoretical information of train idealization is obtainable from (He et al. 2005).

2.1.1 Dynamic differential equations of train body

Lateral translation of train body:

$$m_1 \ddot{y}_{j1} - \sum_{l=1}^{2} \sum_{m=1}^{2} (-1)^m v_{jylm}(t) = 0 \qquad (1)$$

Bouncing of train body:

$$m_1 \ddot{z}_{j1} + \sum_{l=1}^{2} \sum_{m=1}^{2} v_{jzlm}(t) = 0 \qquad (2)$$

Rolling of train body:

$$I_{x1} \ddot{\theta}_{jx1} - \sum_{l=1}^{2} \sum_{m=1}^{2} (-1)^m \lambda_{y3} v_{jzlm}(t)$$
$$- \sum_{l=1}^{2} \sum_{m=1}^{2} (-1)^m \lambda_{z1} v_{jylm}(t) = 0 \qquad (3)$$

Pitching of train body:

$$I_{y1} \ddot{\theta}_{jy1} + \sum_{l=1}^{2} \sum_{m=1}^{2} (-1)^l \lambda_{x1} v_{jzlm}(t) = 0 \qquad (4)$$

Yawing of train body:

$$I_{z1} \ddot{\theta}_{jz1} + \sum_{l=1}^{2} \sum_{m=1}^{2} (-1)^{l+m} \lambda_{x1} v_{jylm}(t)$$
$$+ \sum_{l=1}^{2} \sum_{m=1}^{2} (-1)^m \lambda_{y4} v_{jxlm}(t) = 0 \qquad (5)$$

Herein, the subscript j indicates the sequence number of the car. The subscripts relative to the motion of the train body are described as: $l = 1, 2$ indicates the front and rear bogies; $m = 1, 2$ indicates the left and right sides of the train, respectively. $v_{jxlm}(t)$, $v_{jylm}(t)$ and $v_{jzlm}(t)$ denote the forces due to the expansion quantities of the upper springs of relative directions, respectively.

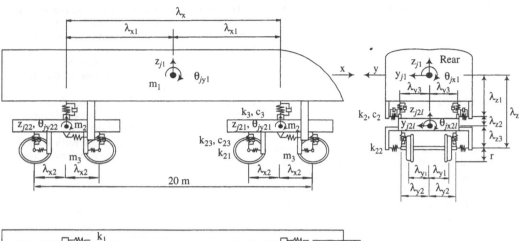

Figure 1. 15 DOF bullet train model.

2.1.2 *Dynamic differential equations of bogies*

Sway of front or rear bogie:

$$m_2 \ddot{y}_{j2l} + \sum_{m=1}^{2} (-1)^m v_{jylm}(t) - \sum_{k=1}^{2}\sum_{m=1}^{2} (-1)^m v_{jylkm}(t) = 0$$

(6)

Parallel hop of front or rear bogie:

$$m_2 \ddot{z}_{j2l} - \sum_{m=1}^{2} v_{jzlm}(t) + \sum_{k=1}^{2}\sum_{m=1}^{2} v_{jzlkm}(t) = 0$$

(7)

Axle tramp of front or rear bogie:

$$I_{x2}\ddot{\theta}_{jx2l} - \sum_{m=1}^{2} (-1)^m \lambda_{z2} v_{jylm}(t) + \sum_{m=1}^{2} (-1)^m \lambda_{y3} v_{jzlm}(t)$$

$$-\sum_{k=1}^{2}\sum_{m=1}^{2}(-1)^m \lambda_{z3} v_{jylkm}(t) - \sum_{k=1}^{2}\sum_{m=1}^{2}(-1)^m \lambda_{y2} v_{jzlkm}(t) = 0$$

(8)

Windup motion of front or rear bogie:

$$I_{y2}\ddot{\theta}_{jy2l} - \sum_{k=1}^{2}\sum_{m=1}^{2} (-1)^k \lambda_{z3} v_{jxlkm}(t)$$

$$+\sum_{k=1}^{2}\sum_{m=1}^{2} (-1)^k \lambda_{x2} v_{jzlkm}(t) = 0$$

(9)

Yawing of front or rear bogie:

$$I_{z2}\ddot{\theta}_{jz2l} - \sum_{m=1}^{2} (-1)^m \lambda_{y4} v_{jxlm}(t)$$

$$+\sum_{k=1}^{2}\sum_{m=1}^{2} (-1)^{k+m} \lambda_{y2} v_{jxlkm}(t)$$

$$+\sum_{k=1}^{2}\sum_{m=1}^{2} (-1)^{k+m} \lambda_{x2} v_{jylkm}(t) = 0$$

(10)

Herein, the subscripts relative to the motion of the bogies are described as: $k = 1, 2$ indicates the front and rear axles of the bogie, $m = 1, 2$ indicates the left and right sides of the bogie, respectively. $v_{jxlkm}(t)$, $v_{jylkm}(t)$ and $v_{jzlkm}(t)$ denote the forces due to the expansion quantities of the lower springs of relative directions, respectively.

2.2 *Simulation of reaction forces at bottoms of piers*

Dynamic reaction forces at the bottoms of piers cannot be obtained accurately through modal analysis by means of calculating the shear forces at the ends of piers because of the Gibbs phenomenon. Therefore in this study, the dynamic reaction forces are calculated with Eq. (11) using the influence value matrix of the reaction force (Kawatani et al. 2004):

$$R(t) = K_R(P_{vst} + P_{vdy}) + K_R P_{sdy}$$

(11)

319

where $R(t)$ denotes the reaction force vector and K_R indicates the influence value matrix of the reaction force. Vectors P_{vst}, P_{vdy}, and P_{sdy} respectively denote the static and dynamic components of the wheel loads and the inertia forces of the bridge nodes.

3 ANALYTICAL MODELS

3.1 Elevated bridge model

A typical Shinkansen elevated bridge of reinforced concrete in the form of a rigid portal frame is adopted for the dynamic response analysis. Three blocks (72 m) of bridges with 24 m length of each block are modeled as 3D beam elements with six-DOF at each node as shown in Figure 2. The lumped mass system and Rayleigh damping (Agabein 1971) are applied for beam elements of the bridge. A damping constant of 0.03 is assumed for the first and second natural modes of vibration of the bridge system. The mass of the ballast and the rail is also taken into account. Double nodes defined as two nodes of independence sharing the same coordinate are adopted at the bottoms of the piers to simulate the effect of a ground spring (Kobori & Kubo 1985). That spring includes the effects of the footing and the piles. Table 1 shows the ground spring constants.

The rail is also modeled as 3D beam elements with six-DOF at each node. Double nodes are also defined here to simulate the elastic effect of sleepers and ballast at the positions of sleepers. Table 2 shows the properties of the rail and the spring constant of the track. The vertical spring constant of the track is derived from the ratio of the wheel load to the vertical displacement of the rail. The horizontal spring constant of the track is assumed from experimental and experiential values to be 1/3 of the value in the vertical direction. The roughnesses in the vertical and horizontal directions of the rail are taken into account.

3.2 Train model

Bullet trains composed of 16 cars, modeled as 15 DOF system for each car (Fig. 1), are employed for analysis. Table 3 shows dynamic properties of the moving trains. As shown by the parameters of the train model in Table 3, the natural frequency of the bogies is higher than that of the train body, which can engender resonance in a higher-frequency field and contribute to high-frequency components of dynamic responses of the bridge. The train velocity is assumed to be 270 km/h, referring to the actual Shinkansen operation speed.

4 DYNAMIC RESPONSES OF BRIDGE

4.1 Dynamic responses of elevated bridge

The eigenvalue analysis of the bridge model is performed. The predominant frequency of the horizontal natural mode is observed as 2.20 Hz, showing good agreement with the value obtained from the field test,

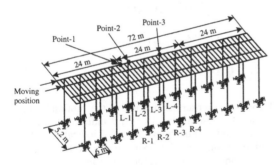

Figure 2. 3-block elevated bridge model.

Table 1. Ground spring constant.

Explanation	Longitudinal	Transverse
Vertical spring of pile top (kN/m)	3.86×10^6	
Rotating spring of pile top (kN · m/rad)	3.64×10^6	2.42×10^6
Horizontal spring of footing (kN/m)	4.84×10^3	4.72×10^3
Horizontal spring of pile top (kN/m)	8.22×10^4	8.08×10^4

Table 2. Property of railway.

Area (m^2)	7.75×10^{-3}
Mass (t/m)	0.0608
Moment of inertia (m^4)	3.09×10^{-5}
Spring constant of track (MN/m)	70

Table 3. Dynamic properties of moving cars.

Mass (t)	m_1 (Body)	32.818
	m_2 (Bogies)	2.639
	m_3 (Wheels)	0.9025
Spring constant k (N/m)	k_u (Upper)	8.86×10^5
	k_l (Lower)	2.42×10^6
Damping coefficient c (N · s/m)	c_u (Upper)	4.32×10^4
	c_l (Lower)	3.92×10^4
Natural frequency (Hz)	f_u (Upper)	1.07
	f_l (Lower)	7.41

which is 2.19 Hz. Therefore, the bridge model validation can be confirmed. More detailed results are available in (Kawatani et al. 2004).

The analytical acceleration responses and the experimental ones in vertical direction, of point-1 through point-3 of elevated bridges indicated in Figure 2, are shown respectively in Figure 3, and those in horizontal direction of point-3 are shown in Figure 4. Herein, point-1, point-2, and point-3 respectively indicate the point of hanging part, the top of the first pier and the top of the third pier of the elevated bridge, with respect to the direction that the train runs towards. As shown in Figure 3 and Figure 4, analytical results using the 15 DOF train model indicate good agreement with experimental results, thereby validating this analytical procedure. The hanging parts of the elevated bridge are connected with neighboring ones by rails and ballast in the actual structure, but only the rails' connecting effect can be incorporated into analysis. Presumably for that reason, the vibrations are predominant at lower frequencies and analytical acceleration responses display larger amplitudes than do experimental ones at point-1 in Figure 3.

4.2 Dynamic reaction forces at bottoms of piers

Reaction forces at the bottoms of the piers in vertical and horizontal directions, as respectively shown in Figure 5, are calculated using the influence value matrix of the reaction forces. As shown in Figure 2, L-1 to L-4 and R-1 to R-4 respectively indicate the piers on the left and right sides of the middle block of the bridge, with respect to the train's direction.

The vertical reaction forces of the piers on the left side are much stronger than those on the right side because the trains are assumed to run along the left sides of the bridges. On the other hand, the reaction forces on the left and right sides in horizontal direction display similar amplitudes. In particular, for both directions in Figure 5, the amplitude at L-1 is somewhat larger than that of L-2. Figure 3 shows the probable reason: the maximum acceleration response that engenders a larger inertia force appears at the hanging part of the bridge. Dynamic reaction forces obtained here are useful as input external excitations in further analyses of site vibration problems.

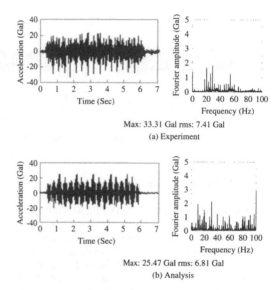

Max: 33.31 Gal rms: 7.41 Gal

(a) Experiment

Max: 25.47 Gal rms: 6.81 Gal

(b) Analysis

Figure 4. Acceleration of bridge (Point-3, horizontal).

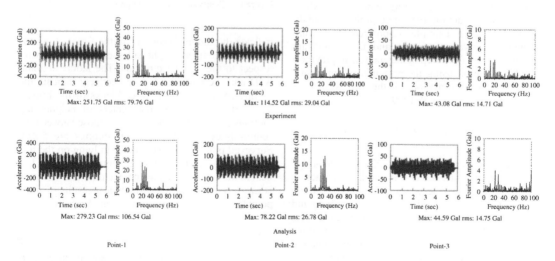

Max: 251.75 Gal rms: 79.76 Gal   Max: 114.52 Gal rms: 29.04 Gal   Max: 43.08 Gal rms: 14.71 Gal

Experiment

Max: 279.23 Gal rms: 106.54 Gal   Max: 78.22 Gal rms: 26.78 Gal   Max: 44.59 Gal rms: 14.75 Gal

Analysis

Point-1   Point-2   Point-3

Figure 3. Acceleration of bridge (vertical).

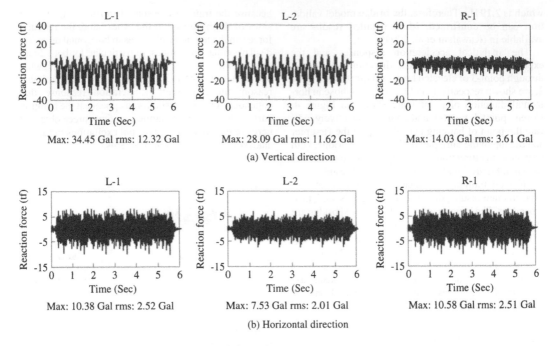

(a) Vertical direction

Max: 34.45 Gal rms: 12.32 Gal   Max: 28.09 Gal rms: 11.62 Gal   Max: 14.03 Gal rms: 3.61 Gal

Max: 10.38 Gal rms: 2.52 Gal   Max: 7.53 Gal rms: 2.01 Gal   Max: 10.58 Gal rms: 2.51 Gal

(b) Horizontal direction

Figure 5. Dynamic reaction forces at bottoms of piers.

5 FOUNDATION-GROUND INTERACTION

For the foundation-ground interaction system in this study, employing the previously obtained reaction forces at the bottoms of piers of the bridges as input dynamic forces, site vibrations around the viaducts of the high-speed railway are simulated using a general computer program named SASSI2000. The soil-structure interaction (SSI) problem is analyzed conveniently using a substructuring approach by which the linear soil-structure interaction problem is subdivided into a series of simple sub-problems. Each sub-problem is solved separately and the results are combined in the final step of the analysis to provide a complete solution using the principle of superposition. Detailed information of the SSI system is described in (Lysmer et al. 1999).

5.1 Description of soil-structure interaction model

Properties of the actual site around the elevated bridge used in the previous bullet train-bridge inter-action system are employed to establish the analytical site model. The structural model contains the footing and piles of the previous elevated bridges. Figure 6 depicts the surveyed points. In all, 24 footings of the three blocks of bridges used in previous bridge vibration analyses are adopted to be excited here and thereby simulate site vibration. Black rectangles in the figure

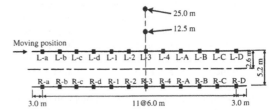

Figure 6. Positions of surveyed points.

indicate the footing positions. In the figure, L and R denote the left and right sides of the bridge with respect to the moving direction. The letters a–d and A–D and the numbers 1–4 respectively indicate the footing sequences in the three blocks of the bridge. The distances between the centers of neighboring footings on the same side are 6.0 m; those between the central lines of left and right footings are 5.2 m. Surveyed points of 12.5 m and 25 m lying on the line passing through the centers of footings R-3 and L-3 denote the surveyed points at which the site vibration is measured in field tests. Analytical results of these points are compared with experimental results.

5.2 Site model

Table 4 shows surveyed values of actual site properties. The site mainly comprises three strata separated

322

Table 4. Ground properties.

Depth of stratum (m)	0–6.8	6.8–17.2	17.2–$$
Unit mass (t/m³)	1.6	1.8	2.0
Shear modulus G (kN/m²)	10 400	66 300	250 000
Poisson's ratio ν	0.49	0.49	0.49
S wave velocity Vs (m/s)	80	190	350
Damping constant	0.05	0.05	0.05

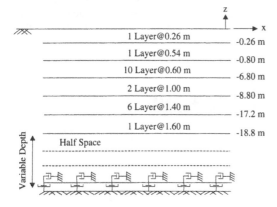

Figure 7. Profile of site model.

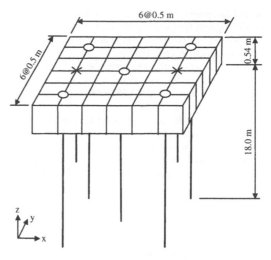

Figure 8. Structural model.

Table 5. Properties of footing.

Unit mass (t/m³)	Young's modulus E (kN/m²)	Poisson's ratio ν	Damping constant
2.50E+06	25	0.2	0.05

Table 6. Properties of piles.

Type	1	2
Unit mass (t/m³)	2.50E+06	2.50E+06
Cross-section area A (m²)	0.058	0.045
Young's modulus E (kN/m²)	3.50E+07	3.50E+07
Moment of inertia I (m⁴)	6.22E−04	3.50E−04
Poisson's ratio ν	0.2	0.2
Damping constant	0.05	0.05

at depths of 6.8 m and 17.2 m. The velocity of an S-wave in the first stratum is 80 m/s, from which the site condition can be considered as relatively inferior. The damping constant is assumed as 5%, determined from experiential values. For analysis, the site model is divided further into 21 thin layer elements, whose profiles are shown in Figure 7. The maximum thickness of each layer is determined in compliance with the criterion that it does not exceed 1/5 λs, where λs is the shortest S wavelength in that layer (Lysmer et al. 1999). Layer elements are established down to the depth of 18.8 m, to which the structural model is embedded. The program then automatically adds some extra layer elements and the viscous boundary at the base to simulate the effect of half space.

5.3 Structure model

One structural set consisting of one footing and seven piles is modeled as Figure 8. Properties of the footing and the piles are shown respectively in Table 5 and Table 6. The actual footing structure is in the shape of rectangular parallelepiped at the base and a trapezoid at the top. To simplify the analyses, the footing is approximated as a rectangular parallelepiped divided into 36 solid elements according to the conversion of volume. The sizes of the solid elements also meet the criterion that they be less than 1/5 of the shortest S

wavelength in the corresponding layer (Lysmer et al. 1999). The footing surface is assumed to extend 0.26 m under the ground surface. The piles are divided into two types according to their length: Type 1 is 7 m long and Type 2 is 18 m. The $\bigcirc$ and $\times$ marks indicate the positions at which the piles are connected vertically to the footing. Herein, $\bigcirc$ represents 18-m-long piles and $\times$ represents 7-m-long piles. The piles are modeled as 3D beam elements. The ends of the beam elements are established at the soil layer interfaces.

6 ANALYTICAL RESULTS OF SITE VIBRATION

Considering the predominant frequency components of the external forces that are confirmed within 15 Hz, the soil damping effect, and the efficiency of the analysis, the highest frequency addressed in the analysis is determined as 25 Hz. By applying the dynamic reaction forces obtained in elevated bridge vibration analyses at 24 total footings, the site response analyses using the analytical models described in previous sections is carried out using the SASSI2000 computer program (Lysmer et al. 1999).

Analytical results and experimental values in vertical and horizontal directions simultaneously with the maximum and rms values, of the points of 12.5 m and 25 m indicated in Figure 6, are shown in Figure 9 and Figure 10. For both vertical and horizontal directions,

the amplitudes of analytical results show good agreement with the experimental ones. On the other hand, the predominant frequency components, particularly in the horizontal direction, indicate somewhat disagreement with those of the experimental ones. The reasons can be considered as follows. First, since the wheel sets of the train are not modeled, the components of the analytical results are inevitable to have some differences with the actual responses especially for the horizontal direction. Furthermore, the discrepancies of the results arise from the difference between the actual site properties and the idealized model. However, considering the complicated nature of the

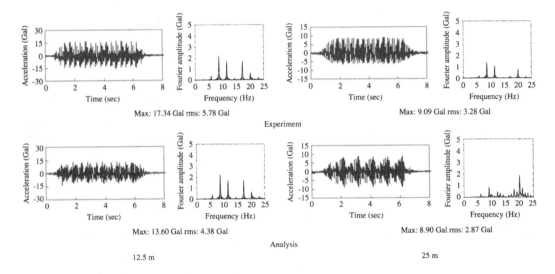

Figure 9.   Acceleration of ground (vertical).

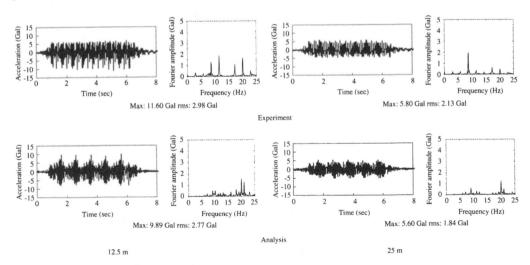

Figure 10.   Acceleration of ground (horizontal).

whole train-bridge-ground interaction system, the analytical results obtained here are considered good enough to evaluate site vibration around viaducts in the further studies.

7 CONCLUSION

In this study, a 15 DOF bullet train model was developed and the dynamic responses of Shinkansen viaducts were simulated, considering the train-bridge interaction. Using the reaction forces at the pier bottoms obtained in the bridge vibration analysis as input excitations, site vibration around the viaducts of the high-speed railway was simulated using a general computer program named SASSI2000. Analytical results were validated through comparison with experimental ones. Employing the analytical procedure established in this study, it is possible not only to simulate and evaluate site vibration around viaducts that is caused by running trains, but also to investigate the effectiveness of presumed countermeasures against bridge and site vibration by reinforcing the bridge structure, improving the ground conditions, or employing other means.

REFERENCES

Agabein, M.E. 1971. The Effect of Various Damping Assumptions on the Dynamic Response of Structure. *Bulletin of International Institute of Seismology and Earthquake Eng.*, Vol. 8, pp.217–236.

Fujikake, T.A. 1986. A Prediction Method for the Propagation of Ground Vibration from Railway Trains. *Journal of Sound and Vibration*, 111(2), pp.289–297.

He, X., Kawatani, M., Sobukawa, R. & Nishiyama, S. 2005. Dynamic Response Analysis of Shinkansen Train-Bridge Interaction System Subjected to Seismic Load. *Fourth International Conference on Current and Future Trends in Bridge Design, Construction and Maintenance, Kuala Lumpur, Malaysia*. (In press)

Kawatani, M., He, X., Sobukawa, R. & Nishiyama, S. 2004. Traffic-Induced Dynamic Response Analysis of High-speed Railway Bridges. *Proceedings of the Third Asian-Pacific Symposium on Structural Reliability and Its Applications, Seoul, Korea.*

Kawatani, M. & Kim, C.W. 2001. Computer Simulation for Dynamic Wheel Loads of Heavy Vehicles. *Structural Engineering and Mechanics*, Vol. 12, No. 4, pp.409–428.

Kawatani, M., Kobayashi, Y. & Kawaki, H. 2000. Influence of Elastomeric Bearings on Traffic-Induced Vibration of Highway Bridges. *Transportation Research Record*, No. 1696, Vol. 2, pp.76–82.

Kim, C.W. & Kawatani, M. 2001. A Comparative Study on Dynamic Wheel Loads of Multi-Axle Vehicle and Bridge Responses. *Proc. ASME International 2001 Design Engineering Technical Conferences, Symposium on Dynamics and Control of Moving Load Problems, Pittsburgh, USA.*

Kobori, T. & Kubo, M. 1985. A Practical Dynamic Analysis Method of Continuous Girder Bridge with Spring Connections on Spring Supports. *Proceedings of JSCE/ Japan Society of Civil Engineers*, No. 356/I-3, pp.395–403. (In Japanese)

Lysmer, J., Ostadan, F. & Chin, C.C. 1999. SASSI2000 Theoretical Manual & User's Manual – A System for Analysis of Soil-Structure Interaction. Academic Version, University of California, Berkeley.

Takemiya, H. 2003. Simulation of Track-Ground Vibrations due to a High-Speed Train: the Case of X-2000 at Ledsgard. *Journal of Sound and Vibration*, 261, pp.503–526.

Tanabe, M., Wakui, H. & Matsumoto, N. 1993. The finite element analysis of dynamic interactions of high-speed Shinkansen, rail, and bridge. *ASME Computers in Engineering*, Book No. G0813A, pp.17–22.

Wu, Y.S., Hsu, L.C. & Yang, Y.B. 2002. Ground Vibrations Induced by Trains Moving over Series of Elevated Bridge. *Proc. of the 10th Sound and Vibration Conference, Taipei, Taiwan*, pp. 1–7.

Xia, H., Cao, Y.M., Zhang, N. & Qu, J.J. 2002. Vibration Effects of Light-Rail Train-Viaduct System on Surrounding Environment. *International Journal of Structural Stability and Dynamics*, Vol. 2, No. 2, pp.227–240.

Yang, Y.B., Hung, H.H. & Chang, D.W. 2003. Train-induced wave propagation in layered soils using finite/infinite element simulation. *Soil Dynamics and Earthquake Engineering*, 23(4), pp.263–278.

Environmental Vibrations – Takemiya (ed.)
© *2005 Taylor & Francis Group, London, ISBN 0 415 39035 4*

Ground vibration around adjacent buildings due to a nearby source

H. Pezeshki

Department of Civil Engineering, Kobe University, Kobe, Japan

Y. Kitamura

Construction Engineering Research Institute Foundation, Kobe University, Kobe, Japan

ABSTRACT: This paper provides a research on dynamic analysis for multiple buildings on layered media, where the lumped masses are used to model the structures, a thin layer method approach is used to model the layered soil, and the mixed boundary value problem is solved using the discretized Green's function method in frequency domain. Coupling effects of multiple structures are evaluated with an iterative approach. The problem is solved for different types of soil deposits in urban areas in Japan, where the results indicate the effects of the layering on ground vibration around the buildings with different heights and separate distances.

1 INTRODUCTION

Vibration pollution in urban areas is one of the significant factors that may cause damages to adjacent structures as well as disturbance to neighbors. The predominant frequencies and amplitude of the ground vibration due to road traffics, railways and construction works depend on many factors, which make these vibration-related problems complex and difficult to identify the causes. It is therefore necessary to consider the different effects of characteristics of vibration sources and sites on ground vibration and response of structures. The dynamic behavior of a ground surface and its influence on soil coupling between structures has been discussed in many researches. One of the first analytical studies of the interaction of two rigid circular foundations was done by *Warburton et al (1971)[1]*, and this was followed by other researches using numerical techniques. *Lin et al (1987)[2]* used the finite element method and *Yoshida et al (1984)[3]* applied the boundary element method to solve the problem, while *Wong & Luco (1976)[4]* introduced the discretized Green's function method to study the coupling effects. The boundary element method was used by *Qian & Beskos (1995)[5]* to study the cross interaction between two rigid foundations subjected to obliquely incident seismic waves. Meanwhile, the Thin Layer Method (TLM) is employed when the mechanical systems on the ground under the structure have material properties that changes at most in one coordinate direction. TLM was first introduced by *Lysmer (1970)[6]* to study the propagation of seismic Rayleigh waves in layered earth strata. Since then, the method has found application in other studies; for

instance, *Waas (1980)[7]* and *Kausel (1981)[8]* used it to obtain the Green's functions for point force acting on a layered medium.

This research involves the study of the dynamic responses of multi-buildings with rigid bases (modeled as lumped masses) on layered soil and the ground surface around them due to waves radiated from an adjacent source. The elastic medium consists of layers of constant thickness with discontinuities in material properties in the vertical direction resting on elastic half-space. The discretized Green's function method in frequency domain, which is used in solving a mixed boundary value problem in the calculation of driving force and dynamic stiffness, is applied in this study. In calculating the stiffness of the layered soil, the Thin Layer Method (TLM) approach in which each layer is laminated in the direction of layering is used. The displacements are formulated in the TLM using the solution of a complex quadratic eigen value problem. Finally, the coupling effects of structures are evaluated using an iteration method to reduce the computation time and computer memory storage.

2 ANALYSIS PROCEDURE

2.1 *Outline of the analysis*

Consider a group of buildings with rigid bases resting on the surface of a visco-elastic layered media excited by waves $Pe^{i\omega t}$ radiated from an adjacent harmonic source. The model is illustrated in Figures 1 and 2. Three translational and three rotational degrees of freedom describe the response of each structure.

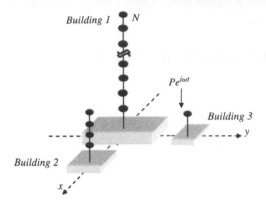

Building 1

N

$Pe^{i\omega t}$

Building 3

y

Building 2

x

Figure 1.   Model and coordinates.

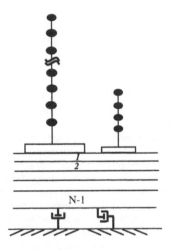

2

N-1

Figure 2.   Modeling of ground.

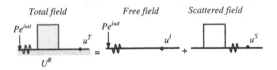

Total field   Free field   Scattered field

$Pe^{i\omega t}$ $Pe^{i\omega t}$

u^T $=$ u^I $+$ u^S

U^R

Figure 3.   Outline of analysis.

In Figure 3, u^T is the displacement vector at any arbitrary response point in the total wave field. It is the sum of the displacement vector u^I in the free wave field in the absence of any structures, and the displacement vector u^S in the scattered wave field. The motions of the structures, which are assumed to be perfectly bonded to the ground, generate the displacement vector u^S. The following boundary condition must be satisfied in the contact area[9].

$$(u^T)_S + U^R = 0 \qquad (1)$$

where S denotes the contact area.

2.2   Basic equations

2.2.1   Discretized green function

In soil-rigid base interaction the force displacement relationships at the contact area can be formulated as an integral representation:

$$u_i(x, y) = \sum_{j=x,y,z} \iint_S G_{ij}(x, y / \xi, \eta) q_j(\xi, \eta) d\xi d\eta \qquad (2)$$

where u_i represents the displacement vector, q_j is the pressure on the contact area S and G_{ij} is the Green's function for visco-elastic layered media. The equation above is solved numerically by discretizing the contact area, where we assumed the contact pressure to be uniformly distributed over each element. After discretization the equation (2) can be expressed as:

$$\bar{u}_i(k) = \sum_{j=x,y,z} \sum_{m=1}^{M} \hat{G}_{ij}(k / m) \bar{q}_j(m) \quad (k=1\sim M) \qquad (3)$$

where $\hat{G}_{ij}(k/m)$ is the influence coefficient, which is defined as the displacement of element k due to the uniformly distributed load of element m with unit amplitude, using TLM. The resulting influence coefficient $\hat{G}_{ij}(k/m)$ is computed as the displacement at the center of element k (x_k, y_k) due to unit disk load of element m.

2.2.2   Formulation of layer stiffness matrices

The global stiffness is constructed by overlapping the contribution of layer matrices at each interface. The stiffness is an algebraic function on the wave number. The layer stiffness matrix of Nth layer, $[K_N]$, in the discrete case were obtained as (Kausel (1974)[10]):

$$[K_N] = [A_N]k^2 + [B_N]k + [G_N] - \omega^2 [M_N] \qquad (4)$$

where k is the wave number. The detail of each matrix is given in reference 11.

2.2.3   Formulation of soil structure system

With reference to the soil structure system illustrated in Figure 4, the structures were assumed to rest on elastic layered half-space. The buildings have different N lumped floor masses, and are subjected to a harmonic wave radiated from an adjacent source. In Figure (5) m_b is the base mass, $I_t = I_b + I_I + \cdots + I_N$, and $P(t)$ and $Q(t)$ are the base shear and moment which can be derived from the elasto-dynamics solution of the harmonic forced vibration (horizontal and rocking) of a massless, rigid base resting on the elastic layered half-space. Both $P(t)$ and $Q(t)$ are called the driving force.

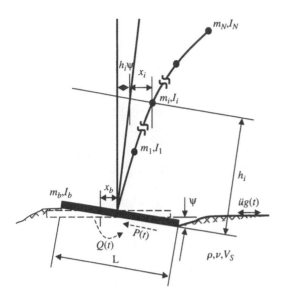

Figure 4.   Building-foundation system.

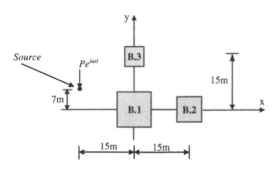

Figure 5.   Arrangement of the buildings.

The relation for the harmonic vibration expressing the equilibrium of the building will be:

$$[\overline{M}]\{\ddot{\overline{u}}\} + [\overline{C}]\{\dot{\overline{u}}\} + [\overline{K}]\{\overline{u}\} = -\ddot{u}_g \{\overline{f}\} \quad (5)$$

where

$$\{\overline{u}\} = \begin{Bmatrix} \{x_i\} \\ \cdots \\ x_b \\ \Psi \end{Bmatrix}, \quad \{\overline{f}\} = \begin{Bmatrix} m_i \\ \cdots \\ m_b + \sum_{i=1}^{N} m_i \\ \sum_{i=1}^{N} m_i h_i \end{Bmatrix} \quad (6)$$

and

$$[\overline{M}] = \begin{bmatrix} [M] & \{m_i\} & \{m_i h_i\} \\ & m_b + \sum_{i=1}^{N} m_i & \sum_{i=1}^{N} m_i h_i \\ SYM & & I_t + \sum_{i=1}^{N} m_i h_i^{2} \end{bmatrix} \quad (7)$$

In the calculation of driving force and dynamic stiffness, an iteration method is used to solve a mixed boundary value problem. This iteration method is based on the discretized Green's function method in frequency domain and the unknown contact pressures under each rigid base[12].

3   NUMERICAL ANALYSIS

3.1   Model description

In the numerical model, it is assumed that there are three multi-lumped mass buildings with the arrangement shown in Figure 5.

The position of building 1 in the model is fixed at (0,0) that corresponds to the center of building 1. In all cases, the buildings are assumed to have rigid bases on elastic layered ground. In the numerical analysis, three cases base on different soil properties are studied Figures 6 to 8[13].

In the numerical analysis, three buildings with the following specification are studied (Table 1).

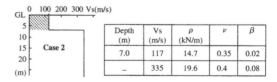

Depth (m)	Vs (m/s)	ρ (kN/m)	ν	β
7.0	203	17.6	0.4	0.04
–	335	19.6	0.4	0.08

Figure 6.   Soil properties (Case 1).

Depth (m)	Vs (m/s)	ρ (kN/m)	ν	β
7.0	117	14.7	0.35	0.02
–	335	19.6	0.4	0.08

Figure 7.   Soil properties (Case 2).

Depth (m)	Vs (m/s)	ρ (kN/m)	ν	β
5.5	215	17.6	0.4	0.04
11.0	117	14.7	0.35	0.02
–	352	19.6	0.4	0.08

Figure 8.   Soil properties (Case 3).

3.2 *Multi-lumped mass buildings in layered media*

Figures 9, 10 and 11 show the contours of equal ground vibration for selected frequencies of three cases of layered soil. The unit of the vertical displacement is in μm. Regarding the scale of these figures, the response of the free field in low frequency shows a similarity between cases 2 and 3 and in the high frequency between cases 1 and 3.

When vibrations are transmitted from the source through the soil into the building foundations, the soil has a strong influence on the intensity of the vibrations received in the building. Under similar excitations at the vibration source, grounds with softer surface layer show the trend to have larger amplitude vibrations than grounds with hard materials.

At 2 Hz, the presence of the structures does not affect the ground response, as the wavelength is long.

Table 1. Foundation size.

| Buildings | Foundation size | | | |
	Length (m)	Width (m)	Height (m)	Floor (m)
Building 1	14.0	14.0	8.0	15
Building 2	10.0	10.0	2.0	4
Building 3	8.0	8.0	1.0	1

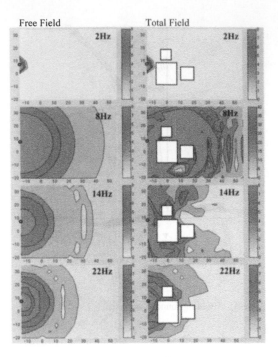

Figure 10. Contour of vertical displacement (Case 2).

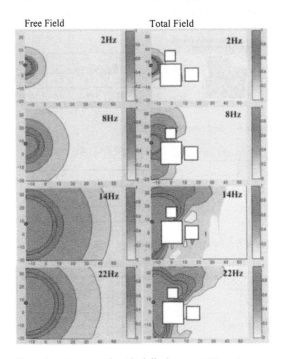

Figure 9. Contour of vertical displacement (Case 1).

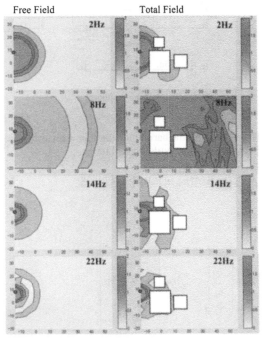

Figure 11. Contour of vertical displacement (Case 3).

In case 1, the responses of the ground surface are large in 14 Hz and 22 Hz because these frequencies are near the natural frequency 17.8 Hz for the longitudinal vibration of the surface layer. It is noticed that the area of large ground response is deformed and reduced at these frequencies. This seems to be due to the input loss by the presence of the structure in high frequency. Another peculiarity is that the reduced zone of the ground response is spread in the side of the structure 1 and 2, while the ground response in the side of structure 3 does not decrease.

In case 2 the response of the ground surface is large at 8 Hz because this frequency is almost the same as the natural frequency of 8.7 Hz for the longitudinal vibration of the surface layer. At higher frequencies, for the deformed area of ground response the phenomenon is similar to case 1, although the contour scale is different.

In case 3 at 8 Hz, the change of the contour line for the ground response around the buildings, where it decreases behind building 2 and spreads behind building 3, should be due to the interference between the free field and the scattered field and a small effect of input loss. The contours in the higher frequency do not show a clear effect, but the phenomenon of the ground vibration around the structures, which is mentioned for case1 and 2, has occurred in a smaller scale.

For reference, the contours of the ground vibration for the half-space with the same material properties of the first layer in case 2 are shown in Figure 12. The shape of the contours is different because of the ground amplification. On the other hand, the shape of the contours is affected by the interaction between the buildings and ground at 22 Hz because the shapes of the contours are similar. From these phenomena, it is understood that the contribution of the ground response around the buildings must be considered due to the ground layer and soil-structure interaction in the high frequency.

3.3 Comparison of buildings responses

It is observed that in Case 2 the longitudinal natural frequency of the soil is around 8 Hz and has very large amplitude. The same phenomenon can also been observed in case 3 at 8 Hz, although the natural frequency of soil for this case is not computed. To find out the reason for this, the ground response in the half space, the horizontal and vertical response of building 1 (shown in Figure 13 for the frequencies 2,8,14 and 22 Hz) are drawn. At 8 Hz, building 1 has a high response for both vertical and horizontal directions.

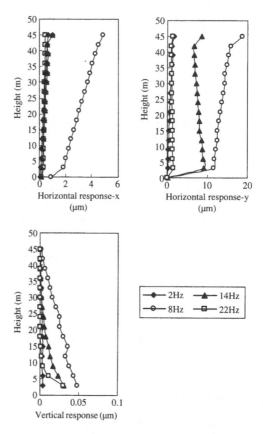

Figure 13. Comparison of horizontal and vertical response in each frequency (Case 2).

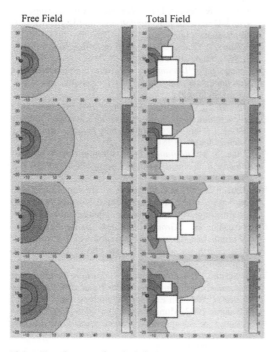

Figure 12. Contour of vertical displacement of the half-space.

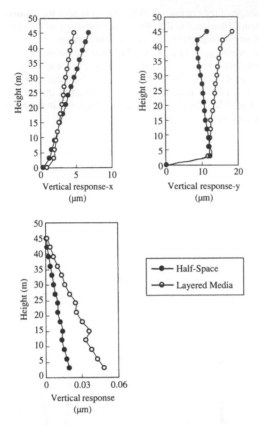

Figure 14. Comparison of horizontal and vertical response at 8 Hz for building 1 between half-space and layered-media (Case 2).

Figure 14 shows the comparison of the response of building 1 in half-space and layered media at 8 Hz. The ground in the half-space has the identical properties of the first layer of case 2. From this figure, we can see that building 1 in half-space has the same high response as in the layered media. The responses of the layered media at 8 Hz include the effect of both resonance of the layered soil and the structures.

We can conclude that the reason for the amplification at 8 Hz in structure 1 is the coincidence of the natural frequency of the ground and the building. Also, the layering effects which cause amplifications can be seen in different frequencies that are related to the ground's natural frequency.

4 CONCLUSION

Soil-structure interaction is an important phenomenon when studying the response of buildings to dynamic forces applied in the neighborhoods of these buildings.

In any soil-structure interaction problem, the dynamic characteristics of the soil have to be determined, especially in the case of structures obstructed by high-rise buildings. Structures located near a high-rise building are more affected by wave radiated from a nearby source.

The response of adjacent buildings to a vibration source is also governed by the following factors: (a) the arrangement of the buildings in the vicinity; (b) the differences of the height; and (c) the vibration characteristic on layered soil.

REFERENCES

1. Warburton, G.B., Richardson, J.D. & Webster, J.J.: "Forced vibrations of two masses on an elastic half space", Journal of Applied Mechanics, ASME, Vol. 38, pp. 148–156, (1971).
2. Lin, H.T., Roesset, J.M. & Tassoulas, J.T.: "Dynamic interaction between adjacent foundations", Earthquake Engineering Structural Dynamics, Vol. 15, pp. 323–343, (1987).
3. Yoshida, K., Sato, T. & Kawase, H.: "Dynamic response of rigid foundations subjected to various type of seismic waves", Proc. 8th World Conf. Earthquake Eng., San fransisco. Vol.3 (745–751), (1984).
4. Wong, H.L. & Luco, J.E.: "Dynamic interaction between rigid foundations in a layered half space", Soil dynamics, Earthquake Engineering, Vol. 5, pp. 149–158, (1986).
5. Qian, J. & Beskos, D.E.: "Harmonic wave response of two 3-D rigid surface foundations". Soil Dynamics and Earthquake Engineering, Vol. 15, pp. 95–110, (1995).
6. Lysmer, J.: "Lumped mass method for Rayleigh waves", BSSA, 60, 89–104, (1970).
7. Waas, G.: "Dynamisch belastete fundamente auf geschichtetem baugrund", VDI Berichte, Nr. 381, 185–189, (1980).
8. Kausel, E.: "An explicit solution for the Green functions for dynamic loads in layered media", Research Report R81-13, Dept. Civil Engineering, M.I.T. Cambridge, Massachusetts, (1981).
9. Thau, S.A.: "Radiation and scattering from rigid conclusion in an elastic medium", Appl. Mech. ASME, 34, 509–511, (1967).
10. Kausel, E.: "Forced vibrations of circular foundations on layered media", Research Report R74-11, Dept. Civil Engineering, M.I.T., Cambridge, Massachusetts, (1974).
11. Pezeshki, H.: "Ground Vibration in the Vicinity of Multiple Structures on a Layered media due to Waves Radiated by an Adjacent Source", Ph.D. Thesis, Kobe University, Japan, (2003).
12. Pezeshki, H. & Kitamura, Y.: "Vibration Transmission Characteristics through Multi-Rigid Structures on Layered Soil", 5th European Conference on Structural Dynamic 2002, Munich, Vol.2, pp.1321–1326, (2002).
13. Osaki, Y. & Sakaguchi, O.: "Major types of soil deposits in urban areas in Japan", Soils and Foundations, 13(2), 49–65, (1973).

Environmental Vibrations – Takemiya (ed.)
© 2005 Taylor & Francis Group, London, ISBN 0 415 39035 4

Effect of track parameters to the ground dynamic response due to metro train

Y.Q. Zhang & W. Chen
College of Civil Engineering, Shijiazhuang Railway Institute, Shijiazhuang, China

W.N. Liu
College of Civil Engineering and Architecture, Beijing Jiaotong University, Beijing, China

ABSTRACT: With the development of metro, the vibration and noise induced by metro trains have been paid more and more attention. In this paper, an algorithm is tried to study the transmission rule of the vibration. Firstly, a space due to underground moving loads is adopted as the model of the metro. The dynamic response of a point ξ is expressed as the periodic analyze solution. Secondly, combined with numerical computation, the dynamic response of a point ξ in the ground surface is given. Finally, the effect of sub-rail parameters to the dynamic response of ground due to metro train are calculated and analyzed in time domain and frequency domain.

1 INTRODUCTION

With the continuous expansion of central cities, metro has become a primary choice to ease the heavy traffic, as its convenience, capacity and safety. However, the vibration and noise induced by metro trains effects structures and people nearby. Being a mass transition system for a city, some parts of the metro lines have to be built under dense populated residential areas with a lot of masonry buildings, or even fragile historic buildings are located. In dense populated residential areas, the vibration induced by metro trains has received many complainants as it causes shaking of windows, or pictures hanging on walls and rumbling noise more than 16 hours a day. In extreme cases, the vibration even damages the structures of the surrounding structures. According to the test results in Beijing, the vibration degree of the area above the metro railway approaches to a level higher than 85 dB when the train is moving at the speed of 15~20 km/h. In rush hours, as there are 60 trains passing by the same line in an hour, the vibration is much more serious. Thus, the vibration induced by the metro trains has brought great influence to the surrounding environment. Up to now, many measures have been developed in China and abroad to solve this problem, many new track structures are development.

In relative, the theory in this aspect is far from sufficient. In this paper, an algorithm is tried to study the vibration due to metro train. Based on the general dynamical response expression of a point in the spatial object due to train loads, combined with transfer matrix method and numerical computation, the dynamic response of a point in the ground surface of tunnel is given; and the effect of sub-rail parameters to the dynamic response of ground are calculated and analyzed in time domain and frequency domain.

2 GROUND DYNAMIC RESPONSE DUE TO METRO TRAINS

2.1 Generalized Duhamel integral of ground dynamic response due to metro trains

Firstly, a space due to underground moving loads is adopted as the model of the train moving in the subway. The dynamic response of a point ξ in a elastic half space model subjected to underground moving loads is expressed as the periodic analyze solution:

$$\hat{u}_i(\xi_x, \widetilde{\xi}_y + n_{\xi_y}L, \xi_z, \omega) = \sum_{n_y=-\infty}^{+\infty} \frac{1}{v} \int_{\widetilde{y}_k}^{\widetilde{y}_k+L} g(\frac{\widetilde{y} - \widetilde{y}_k + (n_y - n_k)L}{v})$$

$$\times \hat{h}_n(\widetilde{y}, \widetilde{\xi}_y + (n_{\xi_y} - n_y)L, \omega) \exp[-i\omega(\frac{\widetilde{y} - \widetilde{y}_k + (n_y - n_k)L}{v})]d\widetilde{y}$$

(1)

Where y is the position of load in time t: $y = y_k + vt$, its projection in the element local coordinate $\widetilde{y} = y - n_yL$, the n_y is the number of sleeper space between from the load position to origin. The $\widetilde{\xi}_y$ is the

333

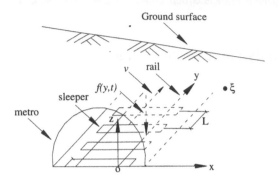

Figure 1. Model of a metro.

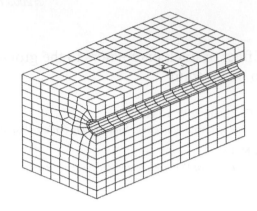

Figure 2. Finite element mesh for metro structure.

projection in the element local coordinate of the y coordinate of the point $\xi : \tilde{\xi}_y = \xi_y - n_{\xi y}L$, $n_{\xi y}$ is the number of sleeper space between from the point ξ to the origin. The y_k is the initial position of the load, its projection in the element local coordinate is: $\tilde{y}_k = y_k - n_k L$, n_k is the number of sleeper space between from the number k load position to the origin.

Formula 1 indicates that the train moving on the rail can be transformed to a combined moving of the load moving only in a period L and the vibration point jump L distance to the opposite position along with n_y, and then, via n_y changing from $-\infty$ to $+\infty$, the dynamic response of point ξ in frequency domain is got. And the response in time domain can be got through Fourier transform. To the formula 1, the key question is how to get the transfer function $\hat{h}()$.

2.2 Dynamic response of ground due to metro train

Zhang,Y.Q. 2004. Development the loading on the ballast. In order to get the dynamic response of the ground, the first question is to get the transfer function. In this paper, a hole circle tunnel is as a example. Tunnel's depth is 9 m, diameter is 6 m, the inner diameter is 2.7 m, the outer diameter is 3 m. The structure of line is reinforced concrete. The parameters of tunnel stratums are: Ground: density is 19 kN/m³, cohesion is 22 kPa, Φ is 25°, elastic module is 57 MPa, μ is 0.35; Concrete: density is 23 kN/m³, elastic module is 30 GPa, μ is 0.2.

Using the above model, we can get the transfer function of a point (distance 5 m from the tunnel axle) in the ground surface of tunnel due to a inner moving load. The impulse dynamic response function is fitted as the function as plane coordinate x, y and time t

$$h(x, y, t) = Ae^{-\lambda_1 x}e^{-\lambda_2 y}(B + e^{-\lambda_3 t}(C \sin \omega_1 t + D \cos \omega_1 t))$$

(2)

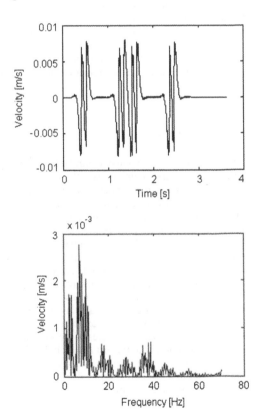

Figure 3. Vibration velocity curve of ground surface with initial parameters.

Where $\lambda_1, \lambda_2, \lambda_3, A, B, C, D, \omega_1$ is the fitted coefficient. $\lambda_1 = 0.2821$, $\lambda_2 = 0.1902$, $\lambda_3 = 0.84$, $A = 0.176e{-}6$, $B = -0.32$, $C = -0.33$, $D = -0.23$, $\omega_1 = 9.0$. Then the corresponding transfer function is obtained through Fourier transform. The dynamic response of the ground surface due to metro trains can be got.

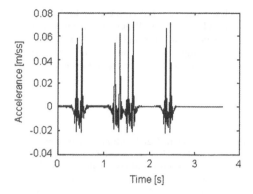

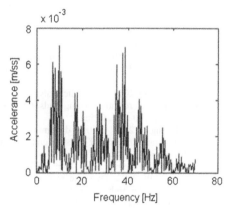

Figure 4. Vibration accelerance curve of ground surface with initial parameters.

2.3 Track parameter effect to the dynamic response of ground

Based on the above theory, the dynamic response of a point on the ground surface is calculated. In the calculation, the train axle height is $14\,t$, the length is $22.732\,m$, and the wheelbase is $2.5\,m$. The sub-rail parameters are as below: the stiffness of railpad is $80e6\,N/m$; the damp coefficient of railpad is $50e3\,Ns/m$; the mass of sleeper is $50\,kg$; the stiffness of sleeperpad is $100e6\,N/m$; the damp coefficient of sleeperpad is $50e3\,Ns/m$; the space between sleeper is $0.6\,m$. The velocity of train is $100\,km/h$. Firstly, with the above sub-rail parameters, the dynamic response of the point is given out in time domain and frequency domain; Secondly, through changing the different sub-rail parameter, the effects of the sub-rail parameter are analyzed. Figure 3 and Figure 4 give the dynamic response of the point (distance $5\,m$ from the tunnel axle) in time domain and frequency domain with the initial sub-rail parameters.

The changing of the sub-rail parameters is as below:
(1) Kr40: changing the stiffness of railpad to $40e6\,N/m$;
(2) Ks60: changing the stiffness of sleeperpad to

Table 1. First sixth amplitude of the point vibration velocity in different sub-rail parameter (m/s)

Sub-rail	1	2	3	4	5	6
Initial	0.0013	0.0028	0.0006	0.0004	0.0006	0.0002
Kr40	0.0010	0.0023	0.0006	0.0003	0.0005	0.0002
Ks60	0.0013	0.0025	0.0006	0.0004	0.0006	0.0002
Ms70	0.0012	0.0024	0.0006	0.0003	0.0006	0.0002
Prid5	0.0013	0.0025	0.0005	0.0003	0.0005	0.0002

Table 2. First sixth amplitude of the point vibration acceleration in different sub-rail parameter (m/s²).

Sub-rail	1	2	3	4	5	6
Initial	0.0018	0.0094	0.0059	0.0053	0.0098	0.0057
Kr40	0.0015	0.0070	0.0044	0.0038	0.0069	0.0040
Ks60	0.0025	0.0082	0.0047	0.0037	0.0074	0.0033
Ms70	0.0016	0.0063	0.0038	0.0031	0.0065	0.0034
Prid5	0.0014	0.0061	0.0037	0.0025	0.0044	0.0063

$40e6\,N/m$; (3) Ms70: changing the mass of sleeper to $70\,kg$; (4) Prid5: changing the distance between sleepers to $0.5\,m$. Through calculating, the first six frequency of the metro are: $3.55\,Hz$, $8.5\,Hz$, $17.5\,Hz$, $27.8\,Hz$, $37.96\,Hz$ and $45.2\,Hz$. The frequency amplitude analysis with different sub-rail parameters are listed in table 1 and table 2.

From the frequency amplitude analysis, we can see that the vibration acceleration range is larger than the vibration velocity range. Along with the decrease of the stiffness of railpad and sleeperpad, the amplitude of vibration velocity and vibration acceleration are all decrease, especially the amplitude in 0~20 Hz, the amplitude in high frequency is less. Along with the increase of the sleeper mass, the amplitude increase; When the sleeper distance decrease from 0.6 m to 0.5 m, the amplitude in 0~40 Hz is decrease, the amplitude in 40~50 Hz is increase.

In the following, the vibration virtual value with different parameters is listed in Table 3 and Table 4.

From the vibration history curve, we can see that along with the decrease of the stiffness of railpad and sleeperpad, the vibrations of the ground are all decrease. When the stiffness of railpad decreases from 80e6 N/m to 40e6 N/m, the vibration velocity decreases 11.5%, the vibration acceleration decrease 29.6%. When the stiffness of sleeperpad changes from 100e6 N/m to 60e6 N/m, the vibration velocity decreases 8.2%, the vibration acceleration decreases 25.3%. When the mass of sleeper increases from 50 kg to 70 kg, the vibration velocity decreases 18.0%, the vibration acceleration decrease 36.6%. When the distance between sleeper changes from 0.6 m to 0.5 m, the vibration velocity decreases 16.4%, the vibration acceleration decreases 32.4%.

Table 3. Virtual value of the point vibration velocity in different sub-rail parameter (m/s).

Sub-rail	Initial	Kr40	Ks60	Ms70	Prid5
	0.0061	0.0054	0.0056	0.0051	0.0051

Table 4. Virtual value of the point vibration acceleration in different sub-rail parameter (m/s^2).

Sub-rail	Initial	Kr40	Ks60	Ms70	Prid5
	0.071	0.050	0.053	0.045	0.048

3 CONCLUSIONS

(1) Based on the dynamic response of sub-rail in literature, half space due to inner moving loads is adopted as the model of metro, the transfer function of the ground surface is fitted. And the dynamic response of the ground point (distance 5 m from the tunnel axle) due the metro train is given in time domain and frequency domain.

(2) By compare the dynamic response with different sub-rail parameters, the conclusion can be summarized:

   (a) Decreasing the stiffness of railpad and sleeper-pad, the distance between sleepers and increasing the sleeper mass, the vibration of the point on the tunnel surface decrease;

   (b) From frequency analysis, along with the decreasing of the stiffness of railpad and sleeperpad, the increasing the sleeper mass, the decrease of vibration amplitude is mostly in low frequency, the vibration in high frequency is relatively less; When the distance between sleepers minish from 0.6 m to 0.5 m, the vibration amplitude in 40~50 Hz increase, the amplitude in the other frequency decrease.

REFERENCES

Huang, Z.P. & Tong, K.Z. 2001. Monitoring on vibration and crack induced by blasting for daping mucking tunnel of chongqing city express-railway. *Chinese Journal of Rock Mechanics and Engineering* 20(addition): 1838–1841.

Lei ,X.Y. & Wang, Q.J. 2003. Study on environmental vibration and vibration noise induced by the urban rail transit system. *Journal of the China Railway Society* 25(5): 109–112. China.

Lei, X.Y. & Mao, L.J. 2001. Analyses of dynamic response of vehicle and track coupling system with random irregularity of rail vertical profile. *China Railway Science* 22(6): 38–43. China.

Liu, W.N. & Zhang, Y.Q. 2004. A periodic analytical solution of railway track structure under moving loads. *Engineering Mechanics* 21(5): 100–102. China.

Tang, Y.Q. & Huang, Y. 2003. Critical dynamic stress ratio and dynamic strain analysis of soils around the tunnel under subway train loading. *Chinese Journal of Rock Mechanics and Engineering* 22(9): 1566–1570.

Xie, W.P. & Hu, J.W. 2002. Dynamic response of track-ground systems under high speed moving load. *Chinese Journal of Rock Mechanics and Engineering* 21(7): 1075–1078. China.

Zhang, Y.Q. & Liu, W.N. 2003. Dynamic response of surrounding object due to running train load. *Journal of the China Railway Society* 25(4): 84–88. China.

Zhang, Y.Q. & Wang, X.Y. 2003. Study reality and prospect on the environment effect due to metro train. *Journal of Northern JiaoTong University* 27(4): 48–51. China.

Environmental Vibrations – Takemiya (ed.)
© *2005 Taylor & Francis Group, London, ISBN 0 415 39035 4*

The subway-train-induced vibration effects on surrounding buildings

W.Q. Liu & J.Q. Hong
College of Civil Engineering, Nanjing University of Technology, Jiangsu, China

ABSTRACT: In this paper, the vibration responses of the buildings near the subway are calculated by the finite element program, ANSYS. A series of different frequencies' harmonic loads are used to simulate the vibration loads induced by subway trains. The analysis involves 2D finite element dynamic models of soil—structure or tunnel—soil—structure. During the analysis, some factors are considered, including different floors in the same building, the load point acting on the surface of soil or the bottom of tunnel, the different distances between load and building, the different frequencies' loads, and the different types of soil.

1 INTRODUCTIONS

In recent years, vibration effects on environment and surrounding buildings induced by subway trains have been focused on by researchers and engineers. Many researchers in many countries have discussed this problem. Based on a lot of data collected from the on site experiments, Mao Yuquan (1987) researched the characteristic and attenuation of vibration at ground surface, which caused by traffic transportations by means of the statistics method of complex regressive analysis. Chua et al. (1992) presented a finite element model for determining ground borne vibrations in buildings due to subway trains. Yang Yinghao and Wang Jiexian (1995) obtained the attenuation formulas of vertical and horizontal displacements with the model, an infinite line model of vertical disturbed harmonic force acting along a semi-space surface. Liu Weining et al. (1996) studied the simulations of the train-track system vibration, which provides the excitation loads of moving trains onto the tunnel structure, and obtained the transmitting properties of ground acceleration and velocity with the model of track-tunnel-ground system. Hunt (1996) presented a methodology for the calculation of vibration transmission from railways into buildings in frequency domain. The method permits existing models of railway vehicles and track to be incorporated. Zhang Yu' e et al. (1997) evaluated the vibration levels of environment induced by rail vehicles in time domain. Nonlinear soil model and Mohr-Coulomb rule are used in that procedure. Wang Fengchao et al. (1999) studied the propagation laws of vibration in soil and the effects on surrounding buildings, which used 2D dynamic model analysis of train-foundation-soil-structure. Tao Lianjin et al. (2003) presented a linear-elastic plane subway-soil system

model to evaluate the ground propagation attenuation law. This analysis is studied with the program, FLAC (Fast Lagrangian Analysis of Continua).

In this paper, the vibration responses of the buildings near the subway are calculated by the finite element program, ANSYS. A series of different frequencies' harmonic loads are used to simulate the vibration loads induced by subway trains. The analysis involves 2D finite element dynamic models of soil—structure or substructure—soil—structure. During the analysis, some factors are considered, including different floors in the same building, the load point acting on the surface of soil or the bottom of tunnel, the different distances between load and building, the different frequencies' loads, and the different types of soil.

2 ILLUSTRATION OF LOADS AND FINITE ELEMENT MODEL

2.1 *Illustration of loads*

The train-induced vibration shows the characteristic of harmonic vibration according to the data measured from the on site test. So it is rational to reduce the vibration load induced by subway train to the harmonic vibration load to some degree, shown in Figure 1. Based on many calculations about the vibration load induced by train vehicles by Chen Xuefeng, the frequencies of the vibration load acting on the surface of the tunnel bedding distribute from 2 to 20 HZ mostly.

In this paper, the subway-induced vibration load is reduced to the harmonic load along the track, which acts on the surface of the tunnel bedding and vibrates with the same phase and different frequencies, as shown in equation (1)

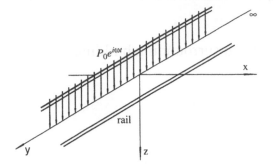

Figure 1. The sketch map of train-induced load.

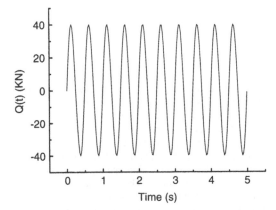

Figure 2. 2 HZ sine-wave load.

$$P(t) = 80 \sin(2\pi f t) \text{ KN} \quad (1)$$

where t is the continuous time of load action, unit is s; f is the frequency of the load, unit is HZ, varying from 2 to 20 HZ. The finite element model is assumed to be symmetrical in this paper, so $Q(t) = P(t)/2$, as shown in equation (2)

$$Q(t) = 40 \sin(2\pi f t) \text{ KN} \quad (2)$$

Figure 2 shows the sketch map of $Q(t)$, where $t = 5$s and $f = 2$ HZ.

2.2 Finite element model

Based on the on site test and the FEM analysis, Pan C.S and Liu W.N presented that 2D FEM model can provide the enough accurate results contrasting with 3D FEM model during the calculation of the vibration induced by train. The 2D finite element dynamic model of soil—structure or tunnel—soil—structure is used in this paper. The soil is simulated by the plane strain element and the building structure is

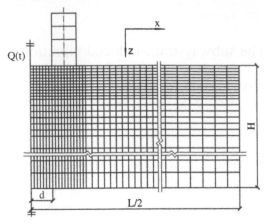

Figure 3. Ground line model.

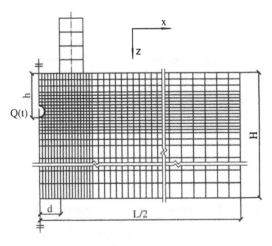

Figure 4. Underground line model.

simulated by the plane beam element. On the assumption that the 2D FEM model is symmetrical to the center line of the rail, half of the model is used to the calculation, shown in Figure 3 and Figure 4.

The width and depth of soil model, L and H, is 300 m and 60 m respectively. The depth of tunnel, h, is 15 m; the diameter is 6 m; and the thickness of concrete lining is 0.3 m. The material parameters of soil are shown in Table 1.

The building structure model is a four-floor frame of reinforced concrete, which the height of each floor is 3.6 m and the column spacing is 5 m. The beam size is 200×500 mm (width $\times$ depth), and the column size is 450×450 mm. The material parameters of the frame and lining are the Young's Elastic Modulus of concrete Ec = 28000 N/mm^2, Poisson ratio $\mu = 0.2$, Mass density $\rho = 25$ KN/m^3.

Table 1. Soil parameters.

No.	Soil type	Density ρ (t/m³)	Poisson ratio μ	Cohesion c (Kpa)	Friction angle φ (°)	Shear-wave velocity Vs (m/s)	Elastic modulus E (Mpa)
1	Soft soil	1.9	0.46	20.7	21	128	93
2	Medium soft soil	1.9	0.37	7.1	34.7	173	160
3	Medium hard soil	2.0	0.33	4.3	34.2	263	380

*The classifications of soil in Table 1 are referred to GB50011-2001: Soft soil, Vs < 140 m/s; Medium soft soil, 140 < Vs ⩽ 250 m/s; Medium hard soil, 250 < Vs ⩽ 500 m/s, Vs is the shear-wave velocity.

2.3 Boundary conditions

From the point of view of the vibration source, the vibration induced by train is different from that by earthquake. The train-induced vibration source, an interior vibration source, is contained in the calculation model, and the vibration wave propagates from the interior to the exterior. But the earthquake-induced vibration source, an exterior vibration source, is not contained in the calculation model, and the vibration wave propagates adversely contrasting with the former. So the two types of boundary conditions are different. Pan C.S presented that when the vibration direction is vertical mostly, horizontal restrictions can be added onto two side directions of the model, and vertical restrictions can be added on the bottom of the model; when the vibration direction is horizontal mostly, vertical restrictions can be added onto two side directions, and the bottom of model should be placed on the foundation rock. In this paper, the surface of soil is free, horizontal restrictions are added onto two side directions of the model, and vertical restrictions are added on the bottom of the model, because the subway-train-induced vibration is vertical mostly.

2.4 Illustration of calculation method

The finite element program, ANSYS, and Newmark hidden integration on time are used in this paper, adopting: the integration constants, $\delta = 0.5$, $\gamma = 0.25$; Rayleigh damping coefficient $C = \alpha[M] + \beta[K]$, Rayleigh damping factor, $\alpha = 0.04$, $\beta = 0.01$; analysis time, t = 5s; time step, ⩾t = 0.004.

3 ANALYSIS OF CALCULATION RESULTS

Zhang Nan et al. presented that the influence bound of vibration induced by subway train into the surrounding buildings are not usually beyond 100 m from the railway. d, the distance between the center line of the left column of the frame and the center line of the railway is −2.5 m (The building is just over the subway) ~100 m. AY1, AY2, AY3, AY4 denotes the vertical vibration acceleration of 1st, 2nd, 3rd, 4th floor respectively at the midpoint of the floors. The

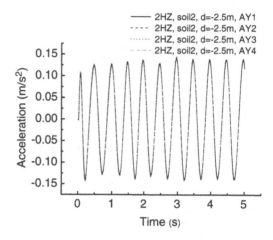

Figure 5. The acceleration curves at the different floors in the same building, d = 5 m.

vibration decibel (dB) of the acceleration is calculated by equation (3),

$$dB = 20\lg|AY| + 70 \tag{3}$$

AY—vertical acceleration, m/s².
Soil 1~3 in figures is referred to Table 1.

3.1 Different floors in the same building

Figure 5 shows the acceleration time curves of different floors in the same building over the subway in the field of soft soil, excited by 2 HZ vertical harmonic load. Figure 6 and Table 2 show the ratios of the maximum absolute acceleration among the different floors in the same building in the field of soft soil, where d = −2.5 m~100 m.

The calculation results show that with the floor number increasing, the vibration of the floor increases a little. The maximum vibration happens at the 4th floor. Table 2 shows that the maximum ratio is 1.06 at the building (d = 60 m). Generally, the vertical vibration levels are same at the different floors in the same building, and the above floors' vibrations are a little

339

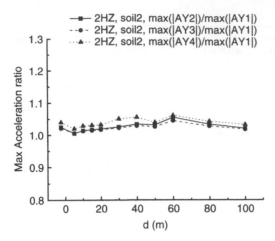

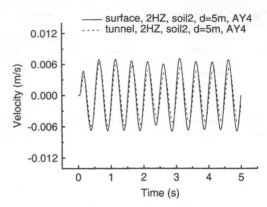

Figure 7. The acceleration curves of the 4th floor excited by the ground and underground line loads, d = 5 m.

Figure 6. The vertical absolute acceleration ratios among the different floors in the same building, d = −2.5∼100 m.

Table 2. The vertical absolute acceleration ratios among the different floors, d = −2.5∼100 m.

d (m)	$\dfrac{\text{Max}(|AY2|)}{\text{Max}(|AY2|)}$	$\dfrac{\text{Max}(|AY3|)}{\text{Max}(|AY1|)}$	$\dfrac{\text{Max}(|AY4|)}{\text{Max}(|AY1|)}$
−2.5	1.03	1.02	1.04
5	1.00	1.02	1.02
10	1.01	1.01	1.03
15	1.02	1.01	1.03
20	1.02	1.02	1.03
30	1.02	1.03	1.05
40	1.03	1.03	1.05
50	1.03	1.03	1.04
60	1.05	1.04	**1.06**
80	1.03	1.03	1.04
100	1.02	1.02	1.03

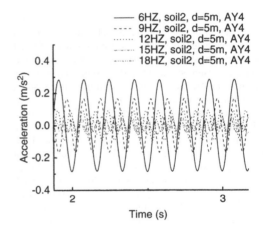

Figure 8. The acceleration curves of the 4th floor excited by 6∼8 HZ loads, d = 5 m.

stronger than the below. During the subsequent analysis, the 4th-floor vertical acceleration is supposed to be the analysis quantity.

But, a question should be noticed that the concrete ascending value of the vibration at the above floor relates to the properties of load and building. This question is not discussed in details in this paper.

3.2 The ground and underground line of subway

Figure 7 shows that the vertical acceleration time curves of the 1st floor in the building (d = 5 m) under the loads of the ground and underground lines of the subway. Wang Chaofeng presented that the vibration of the building decreases with the depth of the tunnel increasing. The calculation results in this paper validate the Wang's conclusion to some extent and show

simultaneously that the vibration due to the ground line is larger than that due to the underground line.

3.3 The different frequencies loads

Figure 8 shows the vertical acceleration time curves of the 4th floor in the building (d = 5 m) in the field of medium soft soil, excited by 6∼18 HZ vertical harmonic loads, which time varies from 2∼3s. Figure 9 shows the vertical acceleration time curves of the 4th floor in the building (d = 5 m) in the field of medium soft soil, excited by 2∼7 HZ vertical harmonic loads, which time varies from 2∼3s. Figure 10 shows the maximum vertical accelerations (dB) of the 4th floor in the building (d = 5 m) in the field of medium soft soil, excited by 2∼20 HZ vertical harmonic loads.

Figure 8 presents the vibration effects on surrounding buildings due to low frequency loads are larger

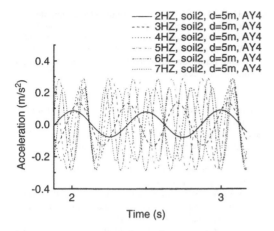

Figure 9. The acceleration curves of the 4th floor excited by 2~7 HZ loads, d = 5 m.

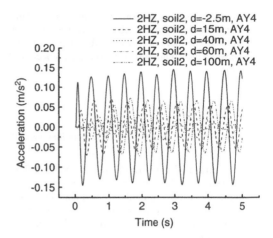

Figure 11. The acceleration curves of the 4th floor excited by the different-distance loads.

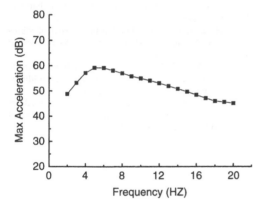

Figure 10. The vertical absolute accelerations (dB) excited by different frequencies' loads.

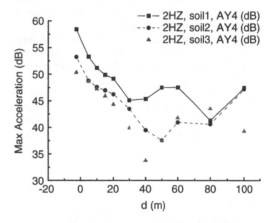

Figure 12. The vertical absolute accelerations (dB) excited by the different-distance loads in different types of soil.

than those due to high frequency loads. In other words, the vibration of the surrounding building decreases with the frequency of the load increasing. Figure 9 and Figure 10 present that the vibration of the building ascends firstly, and then descends when the frequency of the load is within low frequency band. The maximum vibration happens when the frequency is about 6 HZ. So during the analysis about the vibration effects on surrounding buildings caused by subway trains, the low-frequency-band loads should be the key. In addition, the band width of the low-frequency load is wider than that considered in earthquake engineering, which should be attached importance to.

3.4 Different distances between load and building

Figure 11 shows the vertical accelerations of the 4th floor in the building (d = 5 m) in the field of medium soft soil, which the distances are different between the load and the building. Figure 12 presents the accelerations in decibel calculated in equation (3). The calculation results present that the vibration of the building decreases in fluctuation with the distance, d, increasing. The site of the maximum response happens just over the subway. The trough is appears at 30~50 m from the subway, and the rebound crest appears at 60~80 m, where the vibration level is lower that over the subway greatly.

3.5 Different types of soil

Figure 12 shows that the vibration laws are same when the building is on the different types of soil. The responses of vibration are largest when the building sitting on soft soil, and smallest on medium hard soil.

That is, the responses decrease with the soil hardening. The distance from the rebound crest to the load on hard soil lags behind that on soft soil. The difference between the trough and the rebound crest becomes larger too.

4 CONCLUSIONS

From the results of this study, some conclusions can be obtained:

(1) The vertical vibration levels are same at the different floors in the same building, and the above floors' vibrations are a little stronger than the below.
(2) The vibration effects on surrounding buildings induced by ground line is stronger than that by underground line.
(3) The vibration effects on surrounding buildings induced by low frequency load are stronger than that by high frequency load. And the width of the low frequency band is wider than that considered in earthquake engineering.
(4) The vibration effects on surrounding buildings induced by subway train decreases in fluctuation with the distance between the load and the building increasing.
(5) The vibration laws are same when the buildings is on the different types of soil. But the response of vibration is biggest when the building sitting on soft soil, and decreases with soil hardening.

REFERENCES

Code for seismic design of buildings (GB5011-2001). Beijing. China Architecture & Building Press.
Chua, K. H. 1992. Ground borne vibrations due to trains in tunnels. Earthquake Engineering & Structural Dynamics 21(5): 445–460.
Hunt, H. E. M. 1996. Modeling of rail vehicles and track for calculation of ground-vibration transmission into buildings. Journal of Sound and Vibration 193(1): 185–194.
Liu, W. N. et al. 1996. Study of vibration effects of underground trains on surrounding environments. Chinese Journal of Rock Mechanics and Engineering 15(supp): 586–593.
Mao, Y. Q. 1987. Characteristics and attenuation of ground vibration caused by traffic vehicle. Journal of Building Structure 8(1): 67–77.
Pan, C. S. Numerical analysis in tunnel mechanics. 1995. Beijing. China Railway Publishing House.
Wang, F. C. et al. 1999. Vibration effects of subway trains on surrounding buildings. Journal of Northern Jiaotong University 23(5): 45–48.
Shen Yanfeng. 2000. Study on the decay laws of the ground vibration induced by Beijing underground Diameter Line. Dissertation of Master's Degree.
Tao, L. J. et al. 2003. Study on attenuation law of ground motion induced by subway shock. World Earthquake Engineering 19(1): 83–87.
Xia, H. et al. 1999. Environmental vibration induced by urban rail transit system. Journal of Northern Jiaotong University 23(5): 1–7.
Yang, Y. H. & Wang, J. X. 1995. On the transfer of vibrowave in soil caused by passing trains. J. Xi'an Univ. of Arch. & Tech 27(3): 329–334.
Zhang, N. & Xia, H. 2001. Study on the vibration effects induced by subway trains on surrounding buildings. Engineering Mechanics (supp): 199–203.
Zhang, Y. E. et al. 1997. Evaluation on vibration effects induced by subway trains on environment. Journal of Noise and Vibration Control 2: 37–41.

Empirical prediction of ground vibrations due to traffic and construction work

Environmental Vibrations – Takemiya (ed.)
© 2005 Taylor & Francis Group, London, ISBN 0 415 39035 4

An evaluation method for train-speed dependency of Shinkansen-induced vibration

H. Yokoyama, K. Ashiya & N. Iwata
Railway Technical Research Institute, Tokyo, Japan

ABSTRACT: Speedups of Shinkansen-trains are in progress on several Shinkansen lines. To preserve the environmental conditions along the Shinkansen track, it is necessary to assess the effect of increased speeds on ground vibration. An average increment of Shinkansen-induced ground vibration has already been estimated for each Shinkansen line with the previous measurement results; however, train-speed dependency varies much within one line. Therefore, a method for the train-speed dependency at each point along the line is required to estimate the cost and the length of needed countermeasures. In this paper, the characteristics of Shinkansen-induced ground vibration are discussed first. Then, an evaluation method for the train-speed dependency of train-induced vibration is proposed with simple models of dynamic characteristics of the ground, excitation force, and the periodic effect of axel loads.

1 INTRODUCTION

Speedups of Shinkansen-trains have been in progress on several Shinkansen lines. To preserve the environmental condition along the Shinkansen, it is needed to assess the effect of ground vibrations caused by increased speed.

An average increment of Shinkansen-induced ground vibration has already been estimated for each Shinkansen line with the previous measurement results; however, the train-speed dependency varies much within one line. Therefore, a method for train-speed dependency at each point along the line is required to estimate the cost and the length of needed countermeasures.

This paper discusses the evaluation method for the speed power exponent with existing data such as a standard penetration test result.

2 TRAIN-SPEED DEPENDENCY OF SHINKANSEN-INDUCED GROUND VIBRATION

The relation between train speed and the vibration level of Shinkansen-induced ground vibration is approximated to the power of velocity law,

$$VL = 10\log_{10}V^n + VL_0 = 10n \log_{10}V + VL_0. \quad (1)$$

Where VL is the vibration level, n is the speed power exponent, V is velocity, and VL_0 is a constant.

According to the result of previous high-speed test up to 300 km/h, the speed power exponent n differs much between lines; it also varies much even on the same line. Cumulative frequency distribution curves of n for each Shinkansen-line are shown in Figure 1. The average values of the speed power exponent for each line are shown in Table 1.

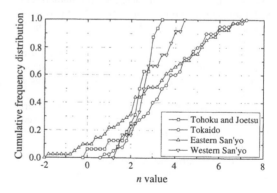

Figure 1. Cumulative frequency of n value.

Table 1. Average n for each Shinkansen line.

Line	average n
Tokaido	3.4
Eastern San'yo	3.1
Western San'yo	2.7
Tohoku and Joetsu	2.2

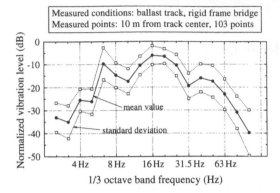

Measured conditions: ballast track, rigid frame bridge
Measured points: 10 m from track center, 103 points

mean value

standard deviation

Figure 2. A typical third-octave band spectrum of ground vibration induced by Shinkansen running at a speed of around 200 km/h.

Figure 2 shows a typical third-octave band spectrum of ground vibration induced by Shinkansen train running at a speed of around 200 km/h. Dominant frequency is 16~20 Hz band, which corresponds to 2.5 m, the rigid wheel base of standard Shinkansen train.

3 CHARACTERISTICS OF GROUND VIBRATION INDUCED BY A VERY HIGH SPEED TRAIN

Figure 3 shows the results of recent high-speed tests up to 360 km/h. Track, structures, and soil conditions of each point are shown in Table 2. Note that the type of cars used for the high-speed test at Line A and Line B are different.

At point a, the average speed power exponent n is 5.6, which is much bigger than the average value of each line as shown in Table 1, while the average n observed at point b is 2.7.

Third-octave band spectrums of each point are shown in Figure 4. Due to the difference of cars used, the shape of spectrums differs between the points a and b.

The most significant difference is the dominant frequency in a very-high-speed range (i.e. over 300 km/h).

At point a, the dominant frequency changes from a middle-frequency band (20~40 Hz) to a very-low-frequency band (2~4 Hz) as the train speed increases. The increment of ground vibration is quite significant at this range, and it determines the train speed dependency of all-pass vibration level in the very-high-speed range.

On the other hand, at point b, the dominant frequency changes from a middle range to a low-frequency band (6.3~12.5 Hz).

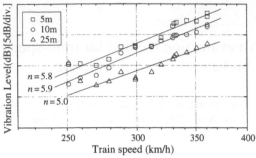

(a) Measured point a.

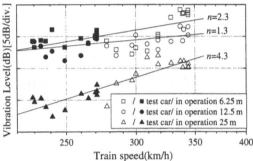

(b) Measured point b.

Figure 3. Train speed dependency of ground vibration measured in high-speed test.

Table 2. Conditions of each measure point.

Measured point	a	b
Line	A	B
Type of cars	Series A	Series B
Track	Slab track	Slab track
Structure	Rigid frame bridge	Rigid frame bridge
Depth to bearing layer	50 m<	≈20 m
Average N value of soil above bearing layer	≈5	≈26
Condition of ground surface	Paved with asphalt (5 m, 10 m), soil(25 m)	Soil

According to the previous measurement, the dominant frequency bands of ground vibration induced by Shinkansen-train running at approximately 200 to 300 km/h are a middle-frequency band and a low-frequency band. However, the result mentioned as above shows that a very-low-frequency band, which was not considered in the previous study (Ashiya & Yoshioka 1994), could be dominant in some ground condition for Shinkansen-train running over a speed of 300 km/h.

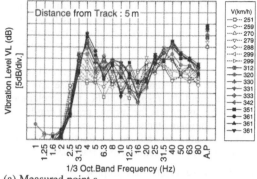

(a) Measured point a.

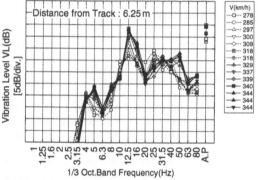

(b) Measured point b.

Figure 4. Third-octave band spectrums of measured ground vibrations.

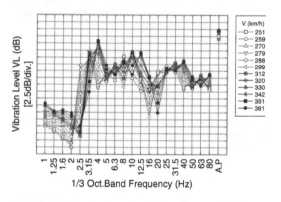

Figure 5. Theoretic spectrum of periodic effect of axle load.

4 EVALUATION METHOD FOR TRAIN-SPEED DEPENDENCY OF GROUND VIBRATION

4.1 *Periodic effect of axle load*

A periodic effect of axle load strongly affects the frequency characteristics of Shinkansen-induced ground vibration (Yoshioka & Ashiya 1995). Figure 5 shows

the theoretic spectrum (Yoshioka & Ashiya 1995, Yoshioka 2000) of periodic effect of axle load for the test car measured at point a.

Peak frequencies at each train speed of measured spectrum (Figure 4(a)) and theoretic spectrum (Figure 5) are in good agreement. Figure 5 also shows that the train-speed dependency of very-low frequency band (2~4 Hz) is bigger than other frequency band. This also agrees with the measured spectrum.

If a standard Shinkansen train runs at 200~360 km/h, three peaks will appear by the periodic effect of axle load: very-low-frequency band (2~4 Hz), low-frequency band (6.3~12.5 Hz), and middle-frequency band (20~40 Hz).

Peak frequency of each band can be determined as below:

$$F_l = v/25,\tag{2}$$

$$F_m = v/8.54,\tag{3}$$

$$F_h = v/2.5,\tag{4}$$

Where F_l, F_m, F_h, are the peak frequency of very-low, low, and middle frequency band, and v is train speed in m/s. F_l is determined by car length (25 m), and F_h is determined by rigid wheel base (2.5 m). F_m depends on both the car length and the bogie-centers distance (17.5 m).

4.2 *Scheme of the evaluation method*

Figure 6 shows the model to represent the spectrum change of ground vibration due to speedup of train.

The change of vibration level due to speedup from V_0 to V is denoted as dVL, and the dominant frequencies of very-low, low, and middle frequency band at the speed of V_0 and V are denoted as (F_{l0}, F_{m0}, F_{h0}), (F_l, F_m, F_h), respectively. Each frequency can be evaluated using equation (2)~(4).

It is assumed that the speed power exponent n and dVL can be approximated from these equations:

$$n = dVL/(10\log(V/V_0)),\tag{5}$$

$$dVL = -10\log(1+10^a+10^b)+10\log(10^c+10^d+10^e),\tag{6}$$

$$a = -dS_1/10,\tag{7}$$

$$b = -dS_0/10,\tag{8}$$

$$c = dVL_2/10,\tag{9}$$

$$d = -dS_1/10+dVL_1/10,\tag{10}$$

$$e = -dS_0/10+dVL_0/10.\tag{11}$$

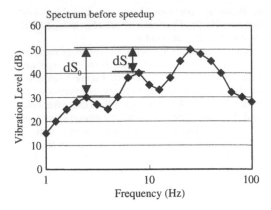

Spectrum before speedup

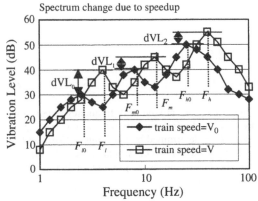

Figure 6. Spectrum change model of speedup.

The all-pass vibration level is assumed representative by the power sum of the vibration level of three peaks (i.e. the peak of very-low-frequency band, low-frequency band, and middle-frequency band).

Vibration level change of each band, dL_0, dL_1, and dL_2, can be roughly evaluated using the theoretic model of periodic effect of axle load.

The other two variables, dS_0 and dS_1, represent the relative characteristics of spectrum before speedup, which determine the dominant band. These variables depend on the conditions of soil, structures, and other local conditions.

4.3 A simple model for the dynamic characteristics of ground

Ashiya and Yoshioka (1994) proposed a simple model to predict the dynamic characteristics of ground from existing ground condition data. From the results of two-dimensional FEM analysis for various ground conditions, they concluded that:

1 The acceleration response spectrum of ground with surface layer at the ground surface has one characteristic frequency, Fc. In the frequency range under Fc, as the frequency goes up, the response at the ground surface increases much. However, the response change is small in the higher frequency range.

2 The Fc value itself varies due to the ground condition, but by normalizing the response spectrum with Fc and the response at Fc, the acceleration response spectrums for various ground conditions can be approximated with one curve.

3 The Fc value depends on the fraction of S-wave velocity and the thickness of surface layer.

Based on these results, they proposed a model shown below:

$$A(F) = 18.8 \tan^{-1}[6\{\log(F / F_c) + 0.3\}] - 20, \quad (12)$$

$$F_c = (Vs / H)^{0.766}, \quad (13)$$

$$Vs = 89.8 N^{0.341}, \quad (14)$$

Where $A(F)$ is the relative level (in dB) of acceleration response spectrum, Vs is the S-wave velocity of surface layer (in m/s), H is the thickness of surface layer, and N is the average N-value of surface layer. In this model, the ground from the surface to the bearing layer (load bearing layer of the structure or the layer with N > 50) is treated as one layer, and called "surface layer".

With this model, the dynamic characteristics of ground can be roughly estimated with two parameters, N and H, which can be obtained from the result of standard penetration test.

Note that the function appears in equation (12) is introduced only to fit the result of two-dimensional FEM analysis, and has no physical meaning.

4.4 Determination of model parameters

The model parameters, which appear in Figure 6, are determined both theoretically and statistically.

At first, the determination of three variables, dVL_0, dVL_1, and dVL_2 are discussed. These variables represent the change of peak value of three bands due to speedup, and can be roughly estimated by the theoretic model of periodic effect of axle load. In this model, the effect of frequency shift by the speedup is also considered, and formulated as below:

$$dVL_{0,1,2} = \alpha_{l,m,h} + dG_{l,m,h}, \quad (15)$$

$$\alpha_{l,m,h} = 10(\gamma_0 + \gamma_{l,m,h}) \log(V / V_0), \quad (16)$$

$$dG_{l,m,h} = A(F_{l,m,h}) - A(F_{l0,m0,h0}) \cdot \quad (17)$$

Table 3. Measured speed power exponent and the ground condition.

No.	Measured points	n	Ground condition	
			Average N of surface layer	Thickness of surface layer (m)
1	12.5 m,25 m	2.9	5	24
2	12.5 m,25 m	2.1	24	28
3	12.5 m,25 m	2.8	9	23
4	12.5 m,25 m	2.0	19	5
5	12.5 m,25 m	4.7	8	22
6	12.5 m,25 m	1.3	30	13
7	12.5 m,25 m	1.5	21	7
8	12.5 m,25 m	1.2	33	16
9	5 m,10 m,25 m	5.6	9	35
10	6.25 m,12.5 m,25 m	2.7	18	10

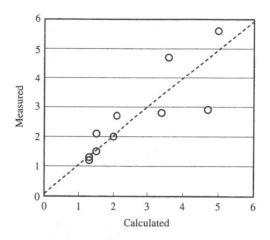

Figure 7. Calculated and Measured n.

In these equations, α_l, α_m, and α_h are the train-speed dependencies of each frequency band, and the parameters γ_0 and $\gamma_{l,m,h}$ correspond to the speed power exponent. The $\gamma_{l,m,h}$ are the speed power exponents of each band, and evaluated from the model of periodic effect of axle load. γ_0 represents the common train-speed dependency of all band, and is determined statistically using past data of the line concerned. The frequency shift effect is represented by $dG_{l,m,h}$, where $A(F)$ is the relative acceleration response level of ground as shown in equation (12).

The other two parameters, dS_0 and dS_1, are determined by the relative acceleration response level of ground, $A(F)$, and excitation model for the ground. The latter is empirically modeled using past measurements. The results are:

$$dS_{0,1} = \beta - \varepsilon_0(F_{h0} - F_{l0,m0})\eta + dA_{l,m},\tag{18}$$

$$\eta = 0.598 \times 0.945^N,\tag{19}$$

$$dA_{l,m} = A(F_{h0}) - A(F_{l0,m0}).\tag{20}$$

The first two terms in equation (17) are the factors for excitation model. β and ε_0 are determined statistically from past data.

In addition to the ground conditions, six parameters, γ_0, $\gamma_{l,m,h}$, β, and ε_0, are still undecided. By employing the theoretical calculation results, $\gamma_{l,m,h}$ are determined to (4.1,1.5,0), respectively.

Three parameters left depend on the conditions of lines, so they are determined from the result of past speedup test at the same line up to 300 km/h.

Measured speed power exponent and the ground conditions are shown in Table 3. Note that the n value shown here is the average of each point. For example, the n value of No.1 section is 2.9, which is the average

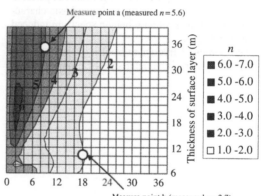

Measure point a (measured $n = 5.6$)

Measure point b (measured $n = 2.7$)

Average N value of surface layer

Figure 8. A trial calculation of train-speed dependency.

of n value of 12.5 m point and 25 m point. Using the data in Table 2, γ_0, β and ε_0 are determined to 0.5, 10, and 2.6, respectively.

Figure 7 shows the relation between calculated and measured (shown in Table 3) n value. The error of the calculated n is within one for most of the measured sections, and is acceptable for rough estimation.

5 A TRIAL CALCULATION OF TRAIN-SPEED DEPENDENCY

The method proposed in this paper evaluates the speed power exponent n from the average N value and the thickness of surface layer. Figure 8 shows the result of a trial calculation of the expected speed power exponent for various ground conditions along with a certain Shinkansen line.

The speed power exponent increases as the ground become softer (i.e. the average N become smaller) in Figure 8. This figure also shows that the results of the high-speed test discussed in Section 3 are well represented with this method.

6 CONCLUSIONS

To assess the effect of speed improvements on ground vibration, an evaluation method for the train-speed dependency of train-induced vibration is discussed. The followings are the conclusions obtained by the study.

1. In some ground condition, a very-low-frequency band around 4 Hz, which was not considered in the previous study, could be dominant for Shinkansen-train running over 300 km/h.
2. A previously proposed theoretic model of periodic effect of axel load can approximate basic characteristics of the ground vibration spectrum induced by Shinkansen.
3. The dominant frequency band is mainly decided by the dynamic characteristics of ground and the excitation force. Therefore, a simple model is discussed to represent the dynamic characteristics of these factors.
4. An evaluation method is proposed to assess the speed power exponent from the average N value and the thickness of surface layer. With this method, ground conditions that may cause bigger increment can be approximately estimated.

REFERENCES

Ashiya, K. & Yoshioka, O. 1994. A method to estimate ground vibrations induced by high-speed trains (in Japanese). *RTRI Report* 8(6): 37–42.

Yoshioka, A. 2000. Basic characteristics of Shinkansen-induced ground vibration and its reduction measures. In Chouw & Schmid (eds), *Wave 2000*: 219–237. Rotterdam: Balkema.

Yoshioka, A. & Ashiya, K. 1995. A dynamic model on excitation and propagation of Shinkansen-induced ground vibrations (in Japanese). *Butsuri-tansa* 48(5): 299–315.

Environmental Vibrations – Takemiya (ed.)
© *2005 Taylor & Francis Group, London, ISBN 0 415 39035 4*

Field test of a semi-empirical model for prediction of train-induced ground vibration

C. With & A. Bodare

Department of Civil and Architectural Engineering Royal Institute of Technology (KTH), Sweden

ABSTRACT: This paper presents a field test of the semi-empirical model ENVIB-01 in predicting train-induced ground vibrations. The model is meant to be used in an early design phase when the major concern is to locate areas with heightening risk of excessive ground motion or before allowing operators to increase speed and/or load. ENVIB-01 is easy to use and can quickly produce prediction of ground motion at the railway embankment and in the vicinity of the track. Measurements at one site are here presented and the results from the embankment and at distances of up to 30 m from the track are compared to a prediction achieved with the model.

1 INTRODUCTION

Train-induced ground vibrations are of concern. Excessive vibrations can cause malfunctioning to sensitive equipment and discomfort to people. Damages to buildings are less common. Large vibrations at the embankment do also increase the maintenance cost and is therefore also in rural areas of concern.

A great number of countermeasures has been developed and implemented, Jones (1994), Woods (1968), Ahmad and Al-Hussaini (1991). However, preferable is of course if areas where excessive ground vibrations can be expected is acknowledged in an early stage and taken into consideration, either by avoiding the area or by taking mitigating actions. The largest cost in a project is if it has to be redone with the consequence that the operation on the line being disrupted.

Several models, empirical and numerical, have been developed to support engineers in predicting expected magnitudes of vibrations on the track and in the vicinity. Some are sophisticated while other is rather simple in the design, Jones et al. (2000), Madshus et al. (1996), Paolucci et al. (2003), Sheng et al. (2003), Takemiya (2003) and US DOT (1998). They all have advantages and disadvantages. A simple model that does not require large amount of high-quality data can with advantage be used in an early stage for screening and reducing the amount of work that then has to be done with a more sophisticated model. One that quite likely demand site data that might be costly to collect and is more time-consuming to use.

This paper presents ENVIB-01, a semi-empirical model developed by Bahrekazemi (2004) to be used in

an early design phase. The target group is geotechnical engineers with basic knowledge in soil dynamic.

2 ENVIB-01

The model assumes a linear relationship between the particle velocity v_{rms} calculated as one-second-root-mean-square and the speed of the train:

$$v_{rms} = aV + b \tag{1}$$

The equations describing the gradient, $a = a_1 F_{rms} + a_2$, and the value for the intercept, $b = b_1 F_{rms} + b_2$, for any given line can be derived by plotting these in a separate graph vs. the wheel force. F_{rms} is the r.m.s.-wheel force applied on the rail.

The model is based on the assumption that the railway embankment can be seen as a Bernoulli-Euler beam on a Winkler foundation:

$$EI \frac{\partial^4 w}{\partial x^4} = -\kappa w - \rho \frac{\partial^2 w}{\partial t^2} + p(x,t) \tag{2}$$

EI is the bending resistance, κ is the distributed spring constant, ρ is the line density, w is the displacement and p is the distributed load. The vertical particle velocity at the embankment can be written as:

$$v_{max} = -V \left(\frac{dw}{d\xi} \right)_{min} = VF \frac{k_c^2}{\sqrt{2\kappa}} \frac{e^{-\frac{\beta}{\alpha} arctg \frac{\alpha}{\beta}}}{\beta}; \quad 0 < m < 1 \tag{3}$$

Where F is a point load and $\alpha = (1 + m^2)^{0.5}$, $\beta = (1 - m^2)^{0.5}$ and $m = V/c_c$, c_c being the characteristic velocity, often called the critical velocity. $k_c = \sqrt[4]{\kappa/4EI}$.

The term b in Equation 1 is a result of assuming a linear relationship between the particle velocity and the speed of the load. This assumption is acceptable within a certain range, Figure 1. Extrapolation of the assumed linear relationship gives a non-zero value on the abscissa. This is due to the limitation of the model and not a physical phenomenon. There is of course no possibility for vibrations to originate when the load is standing still.

The particle velocity at an arbitrary distance is calculated with the result from the predicated motion on the embankment and the quote of the distance between the point of interest and a reference point to the power of $-n$:

$$v_{rms1} = v_{rms2}\left(\frac{r_1}{r_2}\right)^{-n} \tag{4}$$

Figures 2 and 3 presents the measured the wheel-load and the particle velocity of a high-speed train (X2000) on the embankment. The calculated 1s-r.m.s. is also presented.

3 MEASUREMENTS

One site on the railway line named *Godsstråket* between Hallsberg-Örebro in the middle of Sweden was investigated to test the model. The measurements were conducted in September 2004 at Säbylund.

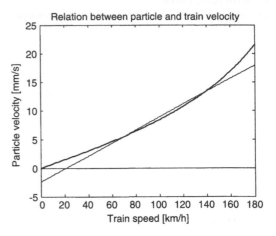

Figure 1. Particle velocity on the railway embankment as a function of the speed of the train.

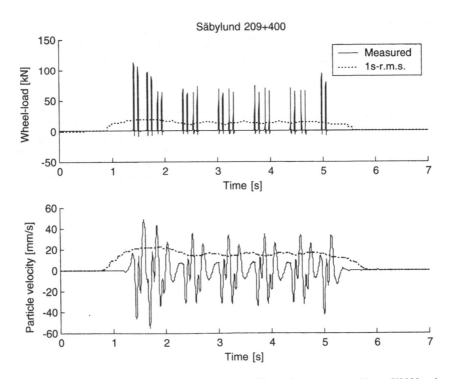

Figure 2. Wheel-force on the rail and the particle velocity on top of the embankment caused by an X2000-train.

Vibrations from passages of freight trains and normal passenger trains together with passages of some high-speed trains of model X2000 have been recorded. Results from the measurements were taken at the upper surface of the railway embankment and at positions of up to 30 m perpendicular to the track. The vertical ground motion was measured by using geophones (SM 1A-4.5 Hz), seismometers (Mark 4A-2 Hz), and accelerometers (B&K 8318). Strain gauges (KFW-5-120-D16-11L1M2S) were used on the rail to measure the wheel-load.

The geotechnical profile in Figure 4 shows that the soil consists of about 0.8 m peat above 2 m of gyttja and clay/gyttja, followed by 5 m clay and a thin layer of silt, approximately 1.5 m. The density is around 1500 kg/m^3. The undrained shear strength of the soil is about 24 kPa at the top one meter and then approximately 18 kPa down to 6 m below the ground surface.

4 RESULTS

Measured and predicted vertical particle velocity on the embankment as well as at a distance of 15 and 30 m from the track for Säbylund are presented in Figures 5, 6 and 7. The prediction values used the model parameters given from a section (km 209 + 850) 450 m north of the one now measured (km 209 + 400). Bahrekazemi (2004):

$a_1 = 0.028$ (mm/s)/(kN*km/h)
$a_2 = 0$ (mm/s)/(km/h)
$b_1 = -1.423$ (mm/s)/(kN)
$b_2 = 0.001$ mm/s
$n = 0.941$ (freight train)
$n = 1.285$ (passenger train)

Five freight trains and twelve passenger trains were recorded, including both commuters and high-speed

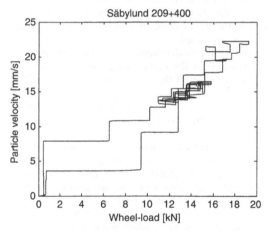

Figure 3. Particle velocity r.m.s. vs. the wheel-force r.m.s. on the rail on top of the embankment caused by an X2000-train.

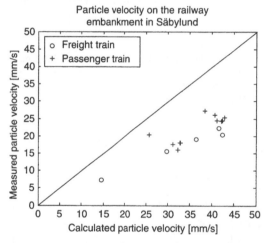

Figure 5. Measured particle velocity at Säbylund site at the top of the railway embankment vs. predicted ground motion.

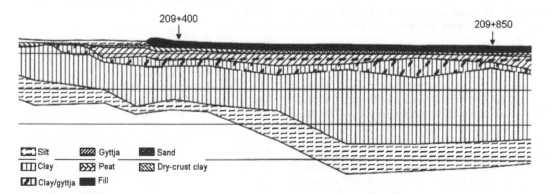

Figure 4. Interpreted soil profile at Säbylund. Measurements in Section km 209 + 850 was by Bahrekazemi (2004) analyzed. Parameters from this site were then used to predict the ground vibrations at section km 209 + 400.

353

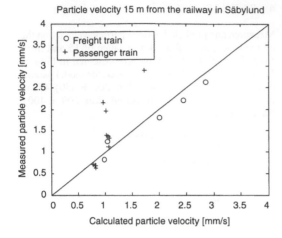

Figure 6. Measured particle velocity at Säbylund site at a distance of 15 m from the railway vs. predicted ground motion.

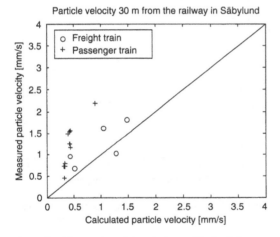

Figure 7. Measured particle velocity at Säbylund site at a distance of 30 m from the railway vs. predicted ground motion.

models. The speed of the trains varied between 75–133 km/h with an average of 123 km/h for freight trains and 101 km/h for passenger trains. The wheel loads were in the range of 15–31 kN with a mean value of 24 kN and 18 kN respectively for freight and passenger trains.

The calculated 1s-r.m.s. particle velocity on top of the railway embankment in Säbylund gave an average error of 49% and 40% for respectively freight and passenger trains, overestimating the vibrations by 16 mm/s respectively 15 mm/s.

The calculated 1s-r.m.s. particle velocity at a distance of 15 m from the railway embankment in Säbylund gave an average error of 5% and −25% for respectively freight and passenger trains, overestimating the vibrations by 0.1 mm/s respectively underestimating by 0.3 mm/s.

The calculated 1s-r.m.s. particle velocity at a distance of 30 m from the railway embankment in Säbylund gave an average error of −43% and −168% for respectively freight and passenger trains, underestimating the vibrations by 0.3 mm/s respectively 0.7 mm/s.

5 DISCUSSION AND CONCLUSION

The purpose of ENVIB-01 is to screen the position of a future railway line and reduce the number sites where excessive vibrations might be expected, or to predict the increase of ground motion if the velocity and/or the wheel-force of which the trains operate with are increased.

Though the model overestimated the vertical particle velocity of the embankment; it gave relatively good prediction of the motion in the vicinity of the track. However, knowing that the model overestimated the vibrations on the embankment, it can be concluded, that the attenuation factor was assumed too large.

The surveys were conducted on the western track and the trains where north bound but the boundary condition, i.e. the profile of the bedrock is different. This and local variation of the material parameters of the soil can have caused the imperfection in the comparison between prediction and measurements. This needs to be addressed in a more in depth analysis.

Considering the crude tool that ENVIB-01 is, it has shown promising results but need further investigation before it can be considered of use in practice.

ACKNOWLEDGEMENTS

This paper was prepared as part of a research study on how to predict train-induced ground vibration, sponsored by the Swedish National Rail Administration (Banverket). Participation in the conference was made possible after contribution of the Erik Philip's foundation. Dr. K. Lindgren is acknowledged for his valuable assistance and advice during preparation of the test. Mr. P. Dahlin's participation in the field tests is highly appreciated. The Lagerfelt family is acknowledged for allowing measurements to be carried out on there property.

REFERENCES

Ahmad, S. & Al-Hussaini, T.M. 1991. Simplified design for vibration screening by open and in-filled trenches. *Journal of Geotechnical Engineering*; 117:67–88.

Bahrekazemi, M. 2004. Train-Induces Ground Vibration and Its Prediction. *Royal Institute of Technology*: PhD Thesis.

Jones, C.J.C. 1994. Use of numerical models to determine the effectiveness of anti-vibration systems for railways. *Proc. of the Institute of Civil Engineers, Transportation*: 43–51.

Jones, C.J.C., Sheng, X. & Petyt, M. 2000. Simulation of ground vibration from a moving harmonic load on a railway track. *Journal of Sound and Vibration*; 231(3): 739–751.

Madshus, C., Bessason, B. & Hårvik, L. 1996. Prediction model for low frequency vibration from high speed railways on soft ground. *Journal of Sound and Vibration*; 193(1): 195–203.

Sheng, X., Jones, C.J.C. & Thompson, D.J. 2003. A comparison of a theoretical model for quasi-statically and dynamically induced environmental vibration from trains with measurements. *Journal of Sound and Vibration*; 267(2003): 621–635.

Paolucci, R., Maffeis, A., Scandella, L., Stupazzini, M. & Vanini, M. 2003. Numerical prediction of low-frequency ground vibrations induced by high-speed trains at Ledsgaard, Sweden. *Soil Dynamics and Earthquake Engineering*; 23: 425–433.

Takemiya, H. 2003. Simulation of track-ground vibrations due to a high-speed train: the case of X-2000 at Ledsgard. *Journal of Sound and Vibration*; 261(2003): 503–526.

U.S. Department of Transportation. 1998. High-speed ground transportation noise and vibration impact assessment. *Office of Railroad Development*, report 293630-1.

Woods, R.D. 1968. Screening of surface waves in soils. *Journal of the soil and mechanics and foundations division*; 951–979.

Environmental Vibrations – Takemiya (ed.)
© 2005 Taylor & Francis Group, London, ISBN 0 415 39035 4

Environmental ground-borne vibrations from train operations in the US 36 Corridor (Denver-Boulder, CO)

D.A. Towers

Harris Miller, Miller & Hanson Inc., Burlington, MA, USA

ABSTRACT: To characterize the existing environmental vibration conditions and to help predict and assess future commuter rail vibration levels along the US 36 Corridor, ground vibration measurements were carried out during freight train operations in representative areas at various distances from the existing track. The measurement results were normalized to an average train speed of 40 km/h, and the existing ground vibration levels as a function of distance from the track were generalized into curves for three geographical areas along the corridor. The results indicated that the prediction curve for locomotive-hauled trains from the U.S. Federal Transit Administration (FTA) General Vibration Assessment Method falls within the range of the generalized curves. However, the results also showed that the vibration levels measured at a site in downtown Louisville, CO were about 10 decibels (VdB) higher than at the other sites. A review of the geology along the corridor suggested that this is most likely due to the existence of abandoned coal mines that are located at depths ranging between 15 meters and 60 meters below the ground surface throughout downtown Louisville.

1 INTRODUCTION

The U.S. Federal Highway Administration (FHWA) and Federal Transit Administration (FTA), in cooperation with the Colorado Department of Transportation (CDOT) and the Regional Transportation District (RTD) for the Denver Metropolitan Region, have jointly initiated a project to identify multi-modal transportation improvements between Denver and Boulder, CO. The alternatives under consideration include the implementation of commuter rail service along an existing 48-km long Burlington Northern Santa Fe (BNSF) freight rail alignment that traverses a number of residential areas in the cities of Denver, Westminster, Broomfield, Louisville and Boulder. With many residences located as close as 15–30 meters from the track, the effect of environmental ground-borne vibrations from train operations was a significant concern that was investigated as part of the Environmental Impact Statement (EIS) study for the project.

This paper summarizes the results of ground vibration measurements that were made to characterize the existing environmental vibration conditions along the project corridor. The paper also describes how the results were used to predict future ground-borne vibration levels from the proposed commuter rail operations and discusses the implications of the results with regard to vibration impact assessment and mitigation.

2 MEASUREMENTS AND DATA ANALYSIS

Ground vibration measurements were carried out at five representative locations, designated as Sites V-1 through V-5. As shown on the map in Figure 1, these sites were geographically distributed over the rail corridor. Ground vibrations from between one and four train passages were measured at each site, comprising a total of eleven train events.

The ground vibration measurements were made using high-sensitivity accelerometers mounted in a vertical orientation on either paved surfaces or on top of steel stakes driven into the soil. At each site, the measurement positions were located at distances ranging from 7.5 m to 45 m from the track centerline. The acceleration signals were amplified and recorded on a multi-channel digital audio tape (DAT) recorder and were analyzed in the laboratory to obtain data in terms of maximum root-mean square (rms) vibration velocity level (in VdB re 1 micro-inch per second with a one-second averaging time). A hand-held radar speed detector was used in the field to determine the train speeds.

To generalize the existing maximum vibration levels from train operations along the rail corridor, the measurement results were first normalized to an average train speed of 40 km/h. The normalization assumed that vibration levels are approximately proportional

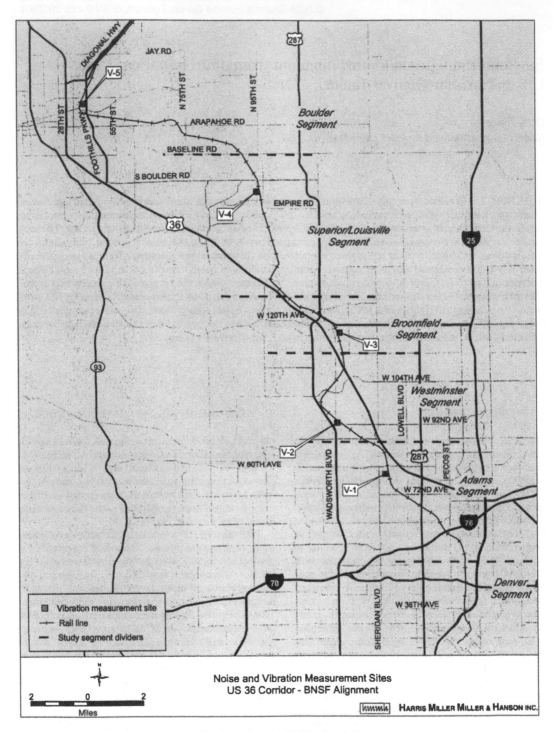

Figure 1. Ground vibration measurement locations along the BNSF railroad alignment.

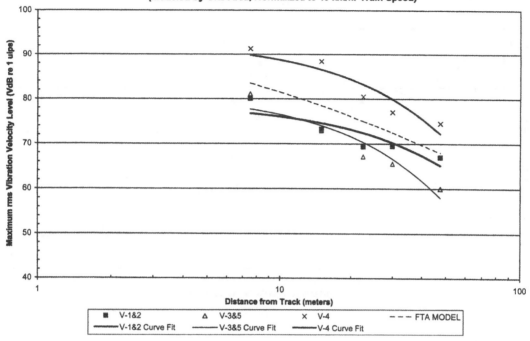

Ground Vibration from Freight Trains on BNSF Corridor
(Modeled by Site Area, Normalized to 40 km/hr Train Speed)

Figure 2. Ground vibration from freight trains on BNSF corridor.

to $20 * \log_{10}[\text{speed/speed}_{\text{ref}}]$ (U.S. Federal Transit Administration, 1995).

3 MEASUREMENT RESULTS

The results of the vibration measurements suggested that the existing ground vibration levels from train operations along the rail corridor, as function of distance from the track, could be generalized into three curves representing different geographical areas. These curves are shown in Figure 2, along with the prediction curve from the U.S. Federal Transit Administration (FTA) General Vibration Assessment method (U.S. Federal Transit Administration, 1995).

The first curve in Figure 2, representing an average of the data at Site V-1 and Site V-2, was used to characterize the existing ground vibration levels from trains along the track in Denver, Adams County and Westminster. The second curve, representing and average of the data at Site V-3 and V-5, was used to characterize the existing ground vibration levels from trains along the track in Broomfield, in Boulder and in portions of Louisville outside the downtown area.

The third curve, representing the data at Site V-4, was used to characterize the existing ground vibration

levels from trains along the track in downtown Louisville. In this area, the normalized ground vibration levels from train operations were found to be about 10 decibels (VdB) higher than at all other locations. A review of the geology along the corridor suggested that this is most likely due to the existence of abandoned coal mines that are located at depths ranging between 15 meters and 60 meters below the ground surface throughout downtown Louisville.

The results in Figure 2 indicate that the general FTA method predicts ground vibration levels for locomotive-hauled train operations that are within the range modeled for the BNSF rail corridor, with roughly similar propagation characteristics. However, the three generalized curves also indicate significant site-specific variations along the corridor.

Further insight into the site-specific variations can be obtained from Figure 3, which provides average one-third octave band ground-borne train vibration spectra at 22.5 m from the track for the three general areas. While the overall vibration level for the areas represented by Site V-1 and Site V-2 is similar to that for the areas represented by Site V-3 and Site V-5, the spectrum for the former areas is dominated by very low frequency energy (with a spectrum peak at 8 Hz) whereas the spectrum for the latter areas is dominated

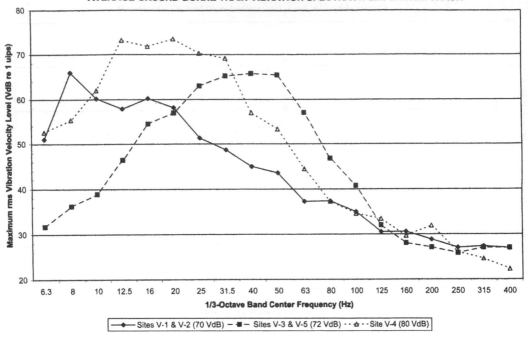

Figure 3.   Average ground-borne train vibration spectra.

by energy in the 25 Hz to 50 Hz frequency range. The overall vibration level for the area represented by Site V-4 is much higher than for the other areas, and is dominated by energy in the 12.5 Hz to 31.5 Hz frequency range.

4   VIBRATION PROJECTION & MITIGATION

The projection of ground-borne vibration from the proposed commuter rail operations on the BNSF corridor was based on the measurements and generalizations of existing freight train vibration described above, assuming that maximum ground-borne vibration levels are due to the locomotive. Depending on location along the corridor, the appropriate curve from Figure 2 was used to determine the ground-borne vibration level based on distance from the track for a train speed of 40 km/h, and the vibration level was adjusted for the actual train speed according to FTA methodology. The projections assume a ground-to-building coupling loss of 0 VdB, except for a few large masonry buildings where the coupling loss was taken to be 5 VdB.

The projected maximum ground vibration levels from commuter rail operations for various areas of the corridor are shown in Figure 4 as a function of distance for a typical average train speed of 80 km/h.

Based on the 72 VdB FTA criterion for residences, the curves in Figure 4 suggest that ground-borne vibration impact is likely at residential buildings at distances within 30 m to 60 m from the track, depending on geographical location.

With regard to mitigation of the potential impacts, the results of the spectral analysis imply that standard track vibration isolation treatments may only be practical for the areas represented by Site V-3 and Site V-5. In these areas, the overall vibration level is dominated by energy in a higher frequency range where mitigation measures are most effective.

5   CONCLUSIONS

The results described in this paper indicate that the projected ground vibration levels from train operations obtained using the FTA General Vibration Assessment method were representative of the average vibration conditions along the US 36 rail corridor. However, the measurement results demonstrated that there are significant variations in ground-borne vibration propagation depending on area-specific ground characteristics. In particular, it appears that the existence of abandoned coal mines not very far below the surface resulted in elevated vibration levels in a localized area.

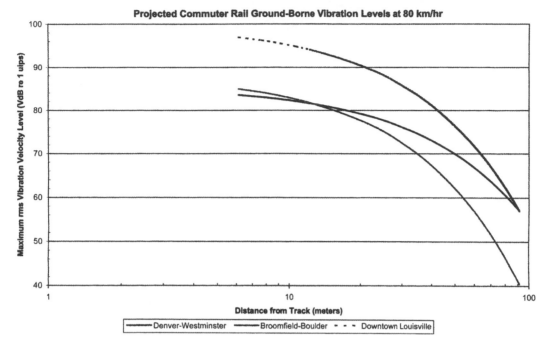

Figure 4.   Projected commuter rail ground-borne vibration levels.

The implications of the results are that while the general FTA method can be useful for estimating the overall extent of environmental vibration impact, a refined analysis is necessary prior to determining the types and locations of vibration mitigation measures for a project. Such an analysis should include site-specific measurements and spectral analysis, along with a review of local geological conditions.

REFERENCE

U.S. Federal Transit Administration, 1995. Transit Noise and Vibration Impact Assessment, Final Report DOT-T-95-16.

Environmental Vibrations – Takemiya (ed.)
© *2005 Taylor & Francis Group, London, ISBN 0 415 39035 4*

Ground vibration caused by magnetic-levitating train and its effects on surroundings

G.Y. Gao & Z.Y. Li
Department of Geotechnical Engineering, Tongji University, Shanghai, China

Z.Q. Yue
Department of Civil Engineering, The University of Hong Kong, Hong Kong, China

ABSTRACT: Magnetic-levitating train is a new type of ground transportation means. The first commercial magnetic-levitation line in the world has been built in Pu Dong of Shanghai, China. However, to the authors' knowledge, few tests and analyses about ground vibration and the influences on environment induced by actual magnetic-levitating train have been reported in the open literature. In this paper, the large-scale test results about the magnetic-train in Shanghai are reported. The characters of ground vibration and wave propagation induced by magnetic-train at different speeds are studied. The results indicate that the ground vibration by magnetic-levitating train is less than that caused by wheel-track train. The dramatic increase of ground vibration, which is existed in wheel-track train when the speed exceeds Rayleigh wave, was not found.

1 INSTRUCTIONS

Train induced vibrations are an environmental concern, as waves propagate through the soil and interact with nearby buildings, where they may cause malfunctioning of sensitive equipment and discomfort to people, as well as re-radiated noise in properties. The rapid extension of the high-speed rail network throughout the world has initiated a lot of research on field test and analytical prediction models for train induced vibrations [H. Xia, N. Zhang & Y.M. Cao, 2005; Degrande G., 2001; Takemiya, H. & Yuasa, S., 1999; Krylov, V.V., 1998; Sheng, X., Jones, C. & Petyt, M., 1999]. However, Magnetic-levitating train, as the newest type of ground transportation means, has not attracted enough attention from the researchers. To the authors' knowledge, few tests and analyses about ground vibration and the influences on environment induced by actual magnetic-levitating train have been reported in the open literature.

In this paper, the magnetic-train in Pu Dong of Shanghai is briefly introduced. The large-scale test results are reported. The characters of ground vibration and wave propagation induced by magnetic-train at different speeds are studied. The results indicate that the ground vibration by magnetic-levitating train is less than that caused by wheel-track train. The dramatic increase of ground vibration, which is existed in wheel-track train when the speed exceed Rayleigh wave, was not found. The case studies reported in this paper will be useful for the further utilization of magnetic-levitating trains and the protection for the environment around the line.

2 INTRODUCTION TO MAGNETIC-LEVITATING TRAIN AND THE TEST SITE

The line of magnetic-levitating train located at Pudong of shanghai China. The ground soil of the site is a classical soft soil, which is very general in Shanghai. Stratification of the soil and the dynamic soil characteristic are shown by Table 1, which summarizes the layer thickness h, compression modulus E, shear modulus G, the density γ, and shear wave velocity c_s for the soil layers.

The major structure for the track is composed by elevated bridges and has a full length of 35.0 km. Each span consists of a piece of beam of 25 m in length, and two square columns of a section-length of 1.8 m to support the beam. A pile foundation composed by 16 drilled piles with length of 36 m and diameter of 0.6 m, and a cap with cross-section of 10 m × 8 m and height of 1.5 m, to support a column.

3 EXPERIMENT OF VIBRATION CHARACTERS INDUCED BY MAGNETIC-LEVITATING TRAIN

In the experiment, the test train is composed by three carriages. The measured contents include column vibration, which can be regarded as the source for the ground vibration; and ground vibration as well as attenuation along and perpendicular to the line, respectively.

In January 2003, Accelerations and velocities of the column and the surrounding ground at some points were measured during magnetic-levitating train passage for two times, one is carried out when the train running at straight-line with speed of 430 km/h, and the other is done when the train running at curved line with speed of 210 km/h, respectively.

3.1 Column vibration (source vibration)

For this purpose, there are two points located on the two lateral centerlines of the column at a distance of 0.6 m from the root, respectively. The vibration characters of one point can be regarded as a source to induce ground vibration along the line and the other induce ground vibration perpendicular to the line.

Table 1. Stratification of the soil and the dynamic soil characteristic.

Stratification	h (m)	E (MPa)	γ (kN/m^3)	c_s (m/s)	G (MPa)
Mixed backfill soil	0.8	2.0	18.9		
Brown-yellow clay	0.8	5.6	18.5	89.76	14.9
Gray-yellow soft clay	6.5	2.9	18.5	124.6	28.7
Saturated soft clay	12.4	2.3	17.9	141.4	35.8
Gay saturated soft clay	2.7	3.1	20.4	154.9	48.9
Brown-gray clay	1.7	4.3	20.4	158.0	50.9
Gray-green clay	1.7	8.0	20.4	160.3	52.4
Gray fine sand	5.7	11.2	19.2	289.7	161.1
Gray silt fine sand	10	14.8	19.2	310.9	185.6
Gray clay	30	6.5	20.1	222.7	99.7

The time histories of vertical and horizontal accelerations and velocities of the column are shown by Figs 1 and 6 and Table 2, and frequency content of velocity and acceleration illustrated by Figs 7 and 8.

From Fig. 1 and Fig. 2, it can be observed that the time histories of vertical and horizontal accelerations and velocities along and perpendicular to the line have slightly difference when the train running along the curved line with the speed of 210 km/h.

Figure 3, Fig. 4 and Table 2 represent that horizontal velocity is slightly higher than the vertical one and horizontal acceleration is lower than the vertical one along the line during the train passage at the speed of 430 km/h. The velocity spectrum along the line in Fig. 7 shows that dominant frequency in horizontal and vertical direction has on difference and nearly equal to 115 Hz.

Figure 5, Fig. 6 and Table 2 indicate that horizontal velocity is nearly equal to the vertical one and horizontal acceleration is slightly higher than the vertical

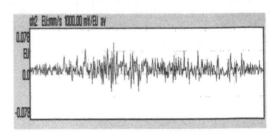

(a) Perpendicular to the line

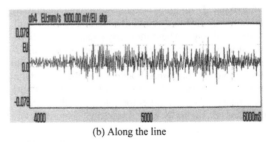

(b) Along the line

Figure 1. Time history of horizontal velocity during train passage at speed of 210 km/h. (4000–6000 ms).

Table 2. Velocities and accretions of the source.

Measurement location		Velocity (mm/s)		Acceleration (m/s^2)	
		Horizontal	Vertical	Horizontal	Vertical
Curved line (210 km/h)	Along line	0.014		0.079	
	Perpendicular to line	0.012		0.088	
Straight-line (430 km/h)	Along line	0.036	0.035	0.089	0.125
	Perpendicular to line	0.076	0.071	0.257	0.194

one in the perpendicular direction to the line during the train passage at the speed of 430 km/h. Frequency content of velocity shown by Fig. 8 indicate the dominant frequencies of horizontal and vertical velocity are identical and nearly equal to 320 Hz in the perpendicular direction to the line when the train running at the straight line with the speed of 430 km/h. Figure 4, Fig. 6 and Table 2 show that the velocities and accelerations perpendicular to the line are obviously larger

than that along the track when the train running at the straight line.

In addition to, it can be observed that velocities and acceleration increase with the increasing speed of the train. However, velocities and acceleration perpendicular to the track are always higher than that along the track when the train running at the straight line with the speed of 430 km/h. Frequency content of velocity during the train passage at the speed of 210 km/h, which is not listed in the paper, indicated there is no dominant frequencies along as well as perpendicular to the line.

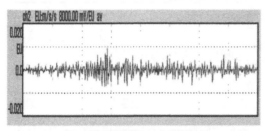

(a) Perpendicular to the line

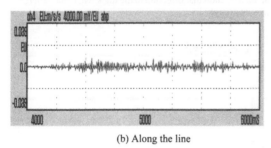

(b) Along the line

Figure 2. Time history of horizontal acceleration during train passage at speed of 210 km/h (4000–6000 ms).

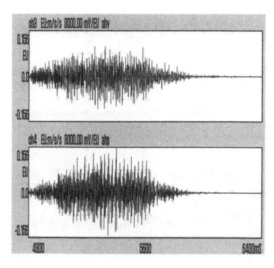

Figure 4. The time history of horizontal and vertical acceleration along the track when the train running at straight line (4800–6400 ms).

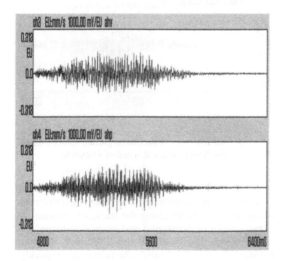

Figure 3. Time history of horizontal velocity perpendicular to and along the line during train passage at speed of 430 km/h (4800–6400 ms).

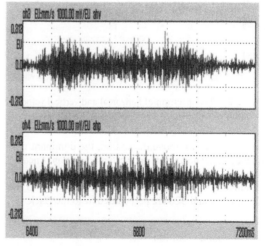

Figure 5. The time history of horizontal and vertical velocity at the perpendicular direction to the track when the train running straight line (6400–7200 ms).

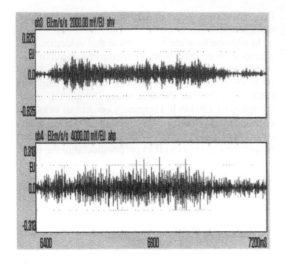

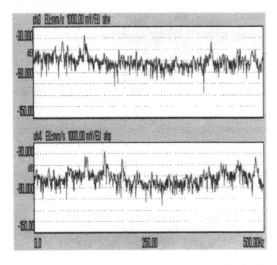

Figure 6. The time history of horizontal and vertical acceleration at the perpendicular direction to the track when the train running at straight line (6400–7200 ms).

Figure 7. The spectrum of horizontal and vertical velocity along the track when the train running at straight line.

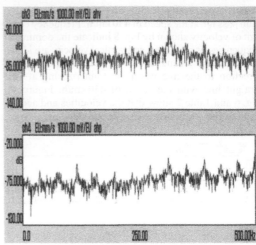

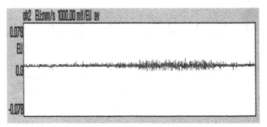

Figure 8. The spectrum of horizontal and vertical velocity at the perpendicular direction to the track when the train running at straight line.

(a) Point at a distance of 6m to the column

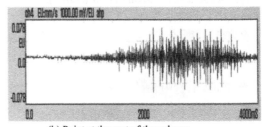

(b) Point at the root of the column

Figure 9. Time history of horizontal ground velocity along the line during train passage at speed of 430 km/h.

However, just as above discussion, the dominant frequency in the horizontal and vertical direction is nearly identical, and the dominant frequency along and perpendicular to the track different greatly with 115 Hz along track and 320 Hz perpendicular to the track.

3.2 Ground vibration

When the train running with the speed of 430 km/h at the straight line, ground vibration was measured along and perpendicular to the track, respectively.

For this purpose, there are two points, one locating at the root of the column and the other at the distance of 6m to the column along the track, are measured. Time history of horizontal and vertical velocity are illustrated by Fig. 9 indicate that the velocity near the column is equal 0.0057 mm/s, then attenuating quickly to 0.0017 mm/s at distance of 6 m from the column. However, the vibration near the column is much lower than that of the column (reference Table 2), which has

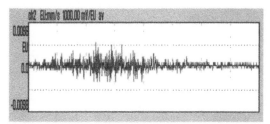

(a) Point at a distance of 10m to the column

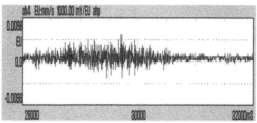

(b) Point at a distance of 5m to the column

Figure 10. Time history of vertical ground velocity perpendicular to the line during train passage at speed of 430 km/h.

the horizontal velocity of 0.036 mm/s. This is because the raft with the thick of 2 m and piles under the raft can isolation vibration as a barrier.

There are two points perpendicular to column located at the distance of 5 m and 10 m, respectively. The results shown by Fig. 10 indicate that the vertical velocities for the two points different slightly with the values of 0.0015 mm/s for the first point at distance of 5 m to the column and 0.0013 mm/s for the second point at distance of 10 m to the column perpendicular to the line. However, just as Table 2 suggested the vertical velocity of the column is 0.071 mm/s, which is nearly 47 times of 0.0015 mm/s. Therefore, it can be concluded that vibration attenuate quickly near the column and slowly after certain distance of 5m from the column, i.e., the nearer the distance, the greater the attenuation.

4 CONCLUSIONS

From the measured data, it can be observed that there are obvious differences about the ground vibration induced by wheel track vehicle and magnetic-levitating train. The main points can be obtained as follows:

Ground vibration by magnetic-levitating train, mainly induced by the magnetic force between the train and the track, propagating from track to column, then to the pile, at last transmitting to the surrounding ground through the basement, is completely different with vibration transimition induced by wheel track vehicle. In addition to, comparing with wheel track vehicle, the time histories of vertical velocity induced by magnetic-levitating train has no obvious periodic peak value.

Ground vibration by magnetic-levitating train is less than that caused by wheel-track train because there is no bump between the wheel and rail for magnetic-levitating train. The peak velocity of the ground induced by the two means of communication even different tens or hundreds of times. Therefore, environment vibration by magnetic-levitating train is less than that caused by wheel-track train.

Attenuation trend of the ground vibration with the distance from the line by the two means of communication is nearly identical. The peak velocity attenuates rapidly with the distance from the track due to the material damping and radiation damping. However, for magnetic-levitating train, only a small increase in the amplitude of the vibration is observed with the increasing speed, which is different with that of wheel-track train. The dramatic increase of ground vibration, which is existed in wheel-track train when the speed exceed Rayleigh wave, was not found in the test. The reason is worthy of further study.

Just as wheel track train, the ground vibration by magnetic-levitating train attenuates rapidly with the increase of the distance, while at certain distances to the track, the attenuation becomes slowly for different train speeds. For this test, this certain distance is 5 m, which is much smaller than that of wheel track train.

The case studies reported in this paper will be useful for the further utilization of magnetic-levitating trains and the protection for the environment around the line.

ACKNOWLEDGEMENTS

The financial aid from the Natural Science Foundation of China under grant 50178056 is gratefully acknowledged.

REFERENCES

Degrande G. 2001. Free field vibrations during the passage of a THALYS high-speed train at variable speed. Journal of sound and vibration, 247(1): 131–144.

H. Xia, N. Zhang, Y.M. Cao. 2005. Experimental study of train-induced vibrations environments and buildings. Journal of Sound and Vibration. 280: 1017–1029.

Krylov, V.V. 1998. Effects of Track Properties on Ground Vibrations Generated by High-Speed Trains. Acustica-Acta Acoustic. 84(1): 78–90.

Sheng, X., Jones, C. and Petyt, M. 1999. Ground Vibration Generated by a Harmonic Load Acting on a Railway Track. Journal of Sound and Vibration. 225(1): 3–28.

Takemiya, H. and Yuasa, S. 1999. Lineside Ground Vibrations Induced by High speed Train and Mitigation Measure WIB. In L. Frýba and J. Nápstrek, eds, Proceedings of the 4th European Conference on Structural Dynamics: Eurodyn 99, A.A. Balkema, Rotterdam, Prague, Czech Republic, 821–826.

Environmental Vibrations – Takemiya (ed.)
© *2005 Taylor & Francis Group, London, ISBN 0 415 39035 4*

Ground vibrations caused by impact pile driving

K.R. Massarsch
Vibisol International AB, Bromma, Sweden

ABSTRACT: The limitations of energy-based vibration attenuation relationships are reviewed. The large scatter of reported results can be explained by the fact that such correlations do not take into account important dynamic pile and soil parameters. The propagation of driving energy from the hammer into the pile and from the pile to the soil along the shaft and at the base is discussed, using wave propagation theory. The importance of the impact velocity, the pile and soil impedance and of the distribution of the dynamic soil resistance are highlighted. The dynamic soil resistance, which depends on the strain-dependent impedance, is the source of ground vibrations. Guidance is given regarding the selection of soil impedance values, taking into account the strain-softening effect. Vibration propagation from the base and along the shaft can be calculated using analytical methods, which take into account the transfer of driving energy.

1 INTRODUCTION

Driven piles are among the most common and cost-effective foundation solutions in variable ground conditions. In recent years, equipment manufacturers have developed new, more powerful pile installation equipment, such as hydraulic impact hammers and sophisticated vibratory hammers. Electronic data acquisition systems have made it possible to monitor and document pile penetration and to determine the dynamic response of the ground and of buildings. With the introduction of wave equation computer programs, the prediction of the static pile capacity based on dynamic measurements during driving has become more reliable. Significant progress has also been made with respect to the determination of dynamic soil properties from seismic field and laboratory measurements.

Today, the efficient use of driven piles and sheet piles is many times limited due to concerns regarding negative environmental effects, such as excessive noise and ground vibrations or ground movements. In many cases, design engineers choose therefore alternative foundation solutions as they are not confident how to assess the risk of ground vibrations during pile driving.

At present, and in contrast to other aspects of pile dynamics, the prediction of ground vibrations is still based on crude empirical rules, developed about 30 years ago. While energy-based predictions of pile bearing capacity have been discarded due to their inaccuracy, these concepts are still being used to predict ground vibrations due to pile driving. These prediction methods neglect fundamental aspects of

dynamic pile-soil interaction. Surprisingly few publications have addressed the problem of dynamic pile-soil interaction in a rational way (e. g. Martin, 1980; Massarsch, 1992; Selby, 1991). In the present paper, an attempt is made to describe in a rational way the parameters which govern the propagation of driving energy from the source to the surrounding soil layers, cf. Figure 1.

The following main aspects of vibration propagation during the driving of piles with an impact hammer can be identified: (A) – Wave propagation in the pile: the energy generated by the drop hammer (1) impacts

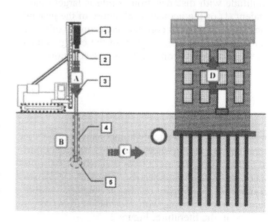

Figure 1. Transfer of pile driving energy from the hammer through the pile and into the surrounding soil and adjacent buildings.

the pile cap and the pile head (2) and the vibration energy is transmitted through the pile (3); (B) – Pile-soil interaction: along the pile shaft (4) and at the pile base (5); (C) – Wave propagation in the ground: transmission of vibrations through soil layers and the ground water; (D) – Dynamic soil structure interaction: dynamic response of buildings and installations in the building.

The dynamic soil-structure interaction of piled foundations at small strains is described extensively in the literature, such as the dynamic response of machine foundations (Nowak & Janes, 1989). However, little information is available regarding the dynamic pile-soil interaction during driving at large strain levels (Massarsch, 1992).

Most investigations, which address vibrations caused by pile driving, discuss the propagation of vibrations in the ground and the response of buildings subjected to vibrations but do not consider the conditions at the source, i.e. the transfer of driving energy from the pile to the soil. Therefore, in this paper emphasis is placed on the discussion of parameters which govern the propagation of waves in the pile (A) and the dynamic pile-soil interaction (B), Figure 1.

1.1 *Energy-based vibration attenuation*

Most investigations reporting measurements of ground vibrations from pile driving have adopted concepts which were developed for the case of vibrations due to rock blasting. The vibration intensity at a distance from the source is assumed to be a function of the energy released at the source.

Attewell & Farmer (1973) analysed results of vibration measurements in different ground conditions and for different pile types. They came to the conclusion that the attenuation of ground vibration amplitude with distance from a pile is largely independent (!) of the geotechnical nature of the ground. They suggested that a conservative estimate of vibration velocity v at distance r from the energy source (pile) can be made from the following equation

$$v = k \frac{\sqrt{E}}{r} \qquad (1)$$

where E is the input energy at the source and k is an empirical constant. It should be noted that the empirical factor k depends on the units used to define the distance and energy. Brenner & Viranuvut (1975) used equation (1) to compare results of vibration measurements from pile driving with information from projects reported in the literature, Figure 2.

In spite of the double-logarithmic scale, the scatter is large and would be unacceptable for most conventional geotechnical design applications. In spite of the

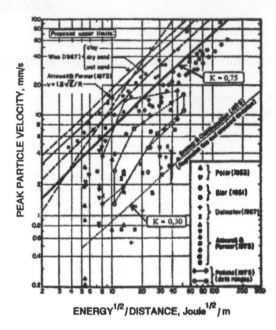

Figure 2. Attenuation of peak particle velocity versus scaled energy (Brenner & Viranuvut, 1975).

uncertainties associated with this concept, the energy-based prediction of ground vibrations is still frequently used (New, 1986, Woods & Jedele, 1985, Jedele, 2005).

One of the most widely reference publication on this subject is the State-of-the-Art paper by Wiss (1981) on construction vibrations, which uses a similar relationship as the one proposed by Attewell & Farmer (1973). The peak particle velocity v is expressed as a function of the scaled distance

$$v = K \left(\frac{D}{\sqrt{E}} \right)^{-n} \qquad (2)$$

where K and n are empirical constant and is D the distance from the energy source E. However, guidance is not given in the literature how the distance should be chosen when a pile penetrates into the ground. This aspect will therefore be discussed in the following sections.

1.2 *Influence of pile penetration depth*

The energy-base relationships require the assessment of the distance from the vibration source to the observation point. With the exception of the initial phase of driving when the pile penetrates dense soil layers close to the ground surface, the depth of the energy source will change as the pile penetrates into the ground. In most cases which have been reported in the literature,

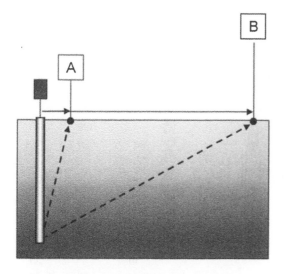

Figure 3. Influence of pile penetration depth on the distance to measuring points at two locations, A and B.

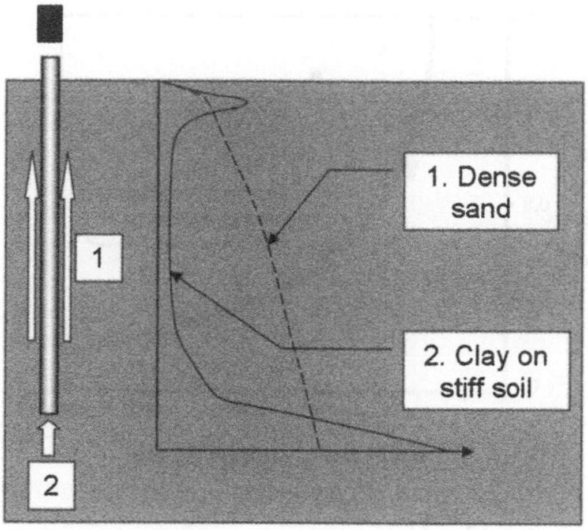

Figure 4. Conceptual picture of soil resistance along the shaft and base during driving of a pile into dense sand and soft clay on a stiff layer, respectively.

the horizontal distance at the ground surface is used when correlating results from vibration measurements. Figure 3 illustrates the problem for two measuring points, located at different distances from the pile. The distance to the energy source becomes particularly important in the case of measuring point A, which is located in the vicinity of a long pile. If the horizontal distance is used for all vibration measurements at location A during the penetration of the pile, it is not surprising that the scatter in reported values becomes large. On the other hand, if the distance to measuring point B corresponds to several pile length, the effect of pile penetration becomes less important. However, in most cases, problems associated with vibrations from pile driving occur at distances less than one pile length.

The location of the energy source during driving of a pile depends also on the soil resistance along the shaft and at the base. This case is illustrated in Figure 4, where two penetration resistance curves (e.g. cone penetration test) are shown. In the first case, it is assumed that the pile is driven into a sand deposit with gradually increasing stiffness. A significant amount of the driving resistance will be generated along the shaft of the pile. Waves will propagate mainly as cylindrical or conical shear waves, with wave attenuation similar to surface waves (Selby, 1991, Massarsch, 1992).

In the second case, it is assumed that a pile is driven into a clay layer with a dry crust (or surface fill) close to the ground surface. During the initial phase, vibrations will propagate along the ground surface in the form of surface waves. When the pile is driven into the soft clay layer, ground vibrations will be negligible.

However, when the tip of the pile is driven into the dense bottom layer, most of the soil resistance will be generated at the base of the pile.

At the base of the pile, vibrations will propagate mainly in the form of body waves (compression and shear waves).

During impact driving, vibrations can be transmitted along the pile shaft and/or at the pile base. The source of vibration emission (origin of dynamic soil resistance) can change during pile penetration (shaft and/or base) and depends strongly on the geotechnical conditions. Thus, pile driving can give rise to vibrations emitted in the form of body waves at the pile base but at the same time also due to shear waves and surface waves along the shaft.

1.3 Influence of pile impedance

In addition to the geotechnical conditions, also the material properties of the pile are of importance. Hackman & Hagerty (1978) showed that the intensity of ground vibrations is affected by the dynamic properties of the pile material. In Figure 5 the K-value as defined in equation (2) is shown as a function of the pile impedance. From the reported measurements it can be concluded that ground vibrations increase markedly when the impedance of the pile decreases. Ground vibrations can be ten times larger in the case of a pile with low impedance compared to a pile with high impedance (Massarsch, 1992). This aspect will be discussed in more detail in the following sections.

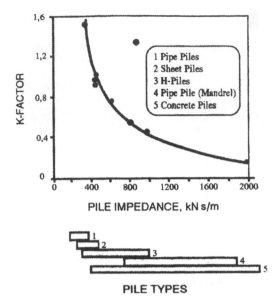

PILE TYPES

Figure 5. Influence of pile impedance on the vibration factor K, cf. equation (2), adapted from Hackman & Hagerty (1978).

2 VIBRATION PROPAGATION IN PILE

The propagation of the driving energy in a pile is a complex problem. During the past thirty years, major progress has been made in the area of pile dynamics. Today the use of dynamic testing methods is generally accepted and the assessment of pile capacity during driving is undertaken routinely.

The accuracy of dynamic methods for estimating the static pile capacity has been improved significantly. With the aid of modern measuring and data acquisition systems it is possible to monitor the dynamic force and wave propagation in the pile during driving. However, these concepts have not yet been used to analyse ground vibration problems associated with pile driving.

One of the main limitations of dynamic testing is that the total pile resistance, R_{tot} is composed of a static component, R_{stat} and a dynamic component, R_{dyn}. As will be shown in the following sections, ground vibrations are caused by the dynamic (velocity-dependent) soil resistance. Therefore, it is important to determine R_{dyn} when predicting ground vibrations during pile driving. At first, the propagation of stress waves in the pile will be addressed. Thereafter, the dynamic interaction between the pile and the adjacent soil layers will be discussed.

2.1 Stress waves in piles

The vibration attenuation relationships given in equations (1) and (2) have major shortcomings as they are

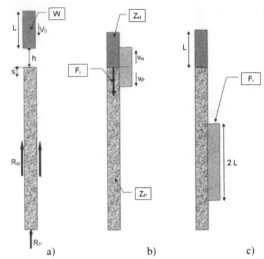

Figure 6. Definition of parameters governing stress wave propagation in piles.

based on the assumption that energy in the pile driving system is conserved. The driving energy is determined from the potential energy of the hammer at the top of the stroke with reference to the pile top. A simplified model of a pile driving arrangement using an impact hammer with mass W and drop height h is shown in Figure 6.

The work that is done when the pile penetrates a distance s can be calculated if the assumption is made that the soil resistance acts along the pile shaft, R_M and at the pile point, R_P. This aspect can be analysed using stress wave propagation theory, which can be used to study the pile penetration problem.

Detailed descriptions of stress wave theory have been presented in the literature and derivations are therefore not presented in this paper, (Broms & Bredenberg, 1982, Goble et al. 1980). The relationship between the axial force F and the particle velocity in the pile v_P is defined by

$$F = Z v_P \qquad (3)$$

where Z is the impedance of a pile with area A. The pile impedance can be determined if the modulus of elasticity E is known. Alternatively, the impedance can be determined from the product of the cross-sectional area A, the wave propagation velocity c and the material density of the pile ρ

$$Z = \frac{AE}{c} = Ac\rho \qquad (4)$$

Inserting equation (4) into equation (3) yields the well-known relationship

$$F = \frac{AE}{c} v_P \tag{5}$$

which can be used to calculate the axial force in a pile based on measurement of the particle velocity during driving. When the hammer strikes the pile with the velocity v_0, a compression wave will be generated simultaneously in the pile and in the hammer. The hammer and the pile will remain in contact only for a short time, the impact time. Since the force between the hammer and the pile must be equal

$$Z_H v_H = Z_P v_P \tag{6}$$

where Z_H and Z_P are the impedances of the hammer and of the pile, respectively. The corresponding particle velocities are v_P and v_H, respectively. Since the particle velocities in the hammer and in the pile are the same at the contact surface

$$v_0 - v_H = v_P \tag{7}$$

Combining equations (6) and (7) and rearranging the terms, the particle velocity v_P in the pile can be calculated from

$$v_P = \frac{v_0}{1 + \dfrac{Z_P}{Z_H}} \tag{8}$$

In the case that $Z_H = Z_P = Z$ the vibration velocity in the pile v_P will be half the hammer impact velocity

$$v_P = 0.5 v_0 \tag{9}$$

Thus the particle velocity in the pile will correspond to half the initial velocity v_0 of the impacting hammer. The force F_i during the impact depends thus on the striking velocity of the hammer v_0 and on the impedance of the pile Z_P and can be calculated from

$$F_i = 0.5 v_0 Z_P \tag{10}$$

immediately after the hammer strikes the pile the particle velocity in the pile behind the wave front will be $v_0/2$, cf. Figure 6b. The duration of the impact, when the pile and the hammer are in contact, t_0 will be equal to

$$t_0 = \frac{2L_H}{c} \tag{11}$$

where L_H is the length of the hammer. The length of the stress wave in the pile will thus be $2LH$, cf. Figure 6c. If the material properties in the hammer are different to that in the pile but the impedances are the same due

to difference in cross section, the duration of the impact will then be

$$t_0 = \frac{2L_H}{c_H} \tag{12}$$

where c_H is the velocity of the stress wave in the hammer. The length of the stress wave in the pile will in this case be

$$L = 2L_H \frac{c_P}{c_H} \tag{13}$$

If an infinitely rigid hammer impacts an elastic pile, the top of the pile will be set in motion at the velocity of the impacting hammer. The force generated in the pile slows down the motion of the hammer and a stress wave is generated in the pile (Goble, 1995). In the next instant the hammer will be moving slower and the generated particle velocity will be smaller. The force at the top of the pile F_i will decay exponentially according to the relationship

$$F = F_i e^{-\frac{Z_P}{M_H} t} \tag{14}$$

where M_H is the mass of the hammer and F_i is the force at impact given by equation (5). With some simple algebraic modifications equation (14) can be modified to the form

$$F = F_i e^{-\frac{M_P}{M_H} \alpha} \tag{15}$$

where M_P is the mass of the pile and α is a variable expressing the time in L/c units. It is thus apparent that the force in the pile is also affected by the ratio of the pile and hammer mass.

2.1.1 Maximum force at pile base

When the initial wave $F_i(t)$ reaches the pile base it starts to move. The force $F_p(t)$ at the pile base will increase with increasing displacement. At equilibrium

$$F_p(t) = F_i(t) + F_r(t) \tag{16}$$

where $F_r(t)$ is the reflected wave. For the case that the material below the pile is infinitely rigid, which is of interest in the case of pile vibrations, then $F_r = F_i$ and $F_p = 2F_i$. In this case, the downward directed compression wave will be reflected and the stress will increase by up to 100%. Thus, it is possible to estimate with rather simple theory the upper limit of the force F_p that can occur at the pile base during hard driving.

2.1.2 Impact velocity vs. energy

To illustrate the limitation of the energy concept, two pile driving cases are compared. It is assumed that both piles are driven with the same energy. At first a hammer with a mass of 4 tons strikes the pile from a height of 1 m, yielding an energy of 40 kJ. The impact velocity v is obtained from

$$v = \sqrt{2gh} \qquad (17)$$

It should be pointed out that the impact velocity is independent of the mass of the hammer. At a drop height of 1 m, the velocity at hammer impact is 4.3 m/s. In the second case, a hammer with a mass of 2 tons strikes the pile from a height of 2 m, thus generating the same energy (40 kJ). However, the impact velocity is 6.3 m/s. The stress in the pile σ_P can be calculated, based on equation (5), from the following relationship

$$\sigma_P = \frac{E_P}{c_P} v_P \qquad (18)$$

The stresses in the respective piles are calculated, using the material properties given in Table 1. In the case of a concrete pile and assuming the same driving energy, but increasing the drop heights (from 1 to 2 m), the stress in the pile increases from 44 MPa to 63 MPa.

2.1.3 Maximum stress in pile
It may be of interest to determine maximum force that can be propagated in the pile. The maximum stress in a pile can be calculated by inserting equation (17) in equation (18)

$$\sigma_P = \frac{E_P}{c_P} \sqrt{2gh_{cr}} \qquad (19)$$

from which the critical drop height h_{cr} can be calculate for different pile materials

$$h_{crit} = \frac{\sigma_{max}^2}{2g\rho E} \qquad (20)$$

In the case of a concrete pile with cylinder strength ranging between 30 and 60 MPa and the material properties listed in Table 1, the critical (effective) drop height varies between 0.47 m and 1.9 m. In the

case of a steel pile with a dynamic strength of 450 MPa, the critical drop height increases to 6.3 m.

The above brief discussion demonstrates that stress wave propagation during pile driving is affected by several factors, such as hammer weight, hammer impact velocity and pile impedance. It is thus not surprising that a single parameter, driving energy, cannot describe the pile driving operation correctly.

2.2 Pile-soil interaction

The total soil resistance R_{tot} during pile driving is composed of a displacement-dependent (static) component R_{stat} and a velocity-dependent (dynamic) component R_{dyn}

$$R_{tot} = R_{stat} + R_{dyn} \qquad (21)$$

A fundamental aspect of predicting ground vibrations due to pile driving is that ground vibrations are caused by the dynamic soil resistance. Thus, if a pile is pushed slowly into the ground, the dynamic resistance does not exist and ground vibrations will be negligible. If the penetration velocity increases, the dynamic soil resistance (R_{dyn}) increases, and consequently giving rise to ground vibrations. Thus, the dynamic soil resistance and ground vibrations are closely associated, as will be discussed in the following sections.

2.2.1 Dynamic base resistance
The soil resistance below the pile can be modelled conceptually as a spring with stiffness k and a dashpot with viscous damping c ($J_c Z_P$) Figure 7. When the pile base is moved a distance u, the total force at the pile base, R_{tot}

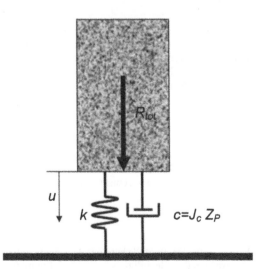

Figure 7. Model of pile base interaction according to Smith model.

Table 1. Typical material properties for driven piles.

Material	Density, ρ (kN/m^3)	Modulus E_P (MPa)	Wave velocity, C_P (m/s)
Steel	78.5	210 000	5120
Concrete	24.5	40 000	4000
Timber	10	16 000	3300

is resisted by a static component, which depends on spring stiffness k, and on a dynamic component, which depends on damping, c ($J_c Z_p$).

Goble et al. (1980) suggested that the dynamic resistance at the tip of the pile can be expressed by the following relationship

$$R_{dyn} = J_c Z_P v_P \qquad (22)$$

where J_c is a dimensionless damping factor. It is generally assumed that J_c depends only on the dynamic soil properties. Typical values of J_c were determined empirically and typical values are given in Table 2.

The damping factor J_c is generally considered to depend only on the dynamic soil properties. However, Iwanowsky & Bodare (1988) derived the damping factor J_c analytically, using the model of a vibrating circular plate in an infinite elastic body. In this way it was possible to describe the interaction between the pile base and the surrounding soil. They showed that the damping factor J_c depends not only on the soil but also on the impedance of the pile at the tip. They arrived at the following relationship

$$J_c = 2 \frac{Z_s}{Z_P} \qquad (23)$$

which implies that the damping factor is a function of the ratio of the soil impedance and the pile impedance. The dynamic component of the driving resistance at the tip can thus be readily calculated from

$$R_{dyn} = 2Z_s v_P \qquad (24)$$

Table 2. Damping factor J_c for different soils (Rausche et al. 1985).

Soil type	J_c
Clay	0.60–1.10
Silty clay and clayey silt	0.40–0.70
Silt	0.20–0.45
Silty sand and sandy silt	0.15–0.30
Sand	0.05–0.20

The damping factor J_c does not appear in equation (24), cf. equation (22). Instead, the soil impedance Z_s is sufficient to determine the dynamic behaviour at the pile-soil interface. In Table 3, typical J_c damping values are calculate according to equation (23) for pile with material properties, as stated in Table 1. Thus, it is possible to determine J_c factors for different pile types, geometries and material properties.

2.3 Dynamic soil resistance at base of pile

The dynamic soil resistance below the pile base can be modelled using the theoretical concept developed by Herlitz (1984). He presented a closed-form solution of the displacement of the centre of a circular plate for arbitrary time-dependent forces acting in an elastic body. He also gave the expression of the displacement for a step-load, which was used by Bodare & Orrje (1985) to describe the response of a plate subjected to a quadratic sinusoidal force. The displacement u_0 at the centre of a circular plate inside an elastic medium, when subjected to a uniformly distributed step load F_0 can be described by the following three terms

$$
\begin{aligned}
u_0 &= F_0 \frac{s}{2\pi a^2 z} t & 0 \le t \le t_p \\
u_0 &= F_0 \frac{1}{3m} (t^2 + t_p^2) & t_p \le t \le t_s \\
u_0 &= F_0 \frac{1+s^2}{4\pi Ga} & t_s \le t
\end{aligned}
\qquad (25)
$$

using the following definitions

$$s = \frac{c_s}{c_p} = \sqrt{\frac{1-2v}{2-2v}} \qquad (26)$$

where v is Poisson's ratio. The specific impedance of the material z is defined as

$$z = c_s \rho = \sqrt{G \rho} \qquad (27)$$

where ρ is the material density and G is the shear modulus. The mass of the sphere with a radius a in the elastic material is defined as follows

Table 3. Values of the J_c damping factor for different pile materials. An average soil density of $\rho = 1.8\,t/m^3$ was chosen. For pile material properties cf. Table 1.

	Compression wave velocity at pile base, c_{Psoil} (m/s)					
Material	250	500	750	1000	1250	1500
Steel	0.02	0.04	0.07	0.09	0.11	0.13
Concrete	0.09	0.18	0.28	0.37	0.46	0.55
Wood	0.27	0.55	0.82	1.09	1.36	1.64

$$m = \frac{4}{3}\pi a^3 \rho \tag{28}$$

The parameters t_s and t_p are the times for a transverse or longitudinal wave to travel one radius a

$$t_s = \frac{a}{c_s}; \quad t_p = \frac{a}{c_p} \tag{29}$$

By differentiation of the first part of equation (25) with respect to time, it can be seen that the velocity at the centre of the plate is constant

$$v_0 = F_0 \frac{s}{2\pi a^2 z} \tag{30}$$

In the second part, the acceleration a_0 at the centre is constant

$$a_0 = F_0 \frac{2}{3m} \tag{31}$$

In the third part the displacement is constant and is the same as the static displacement

$$u_0 = F_0 \frac{1+s^2}{4\pi Ga} \tag{32}$$

The displacement u_0 at the centre and the average displacement u_{av} is illustrated in Figure 8.

For the case of a plate inside an elastic medium, the relationship between force F_0 and the particle velocity v_0 is given by equation (30). This relationship can be rewritten to express the stress as a function of the velocity

$$\sigma_0 = \frac{2z}{s}v_0 = 2\rho c_p v_0 \tag{33}$$

This stress-particle velocity relationship according to equation (33) is shown in Figure 9. Orrje (1996) used this relationship to analyse the dynamic response of a plate impacting on an elastic half space.

Figure 9 demonstrates that the specific impedance z can be determined from field tests if the mobilized dynamic stress is plotted versus the velocity during impact of the plate. By rearranging the terms of equation (30) it is possible to determine the particle velocity that is transmitted from the plate to the underlying soil. The vibration velocity v_{max} which is generated due to the mobilized stress, σ_{mob} can be determined from the following equation

$$v_{max} = \frac{\sigma_{mob}}{\rho c_s^*} s \tag{34}$$

where c_s^* is the shear strain-dependent shear wave velocity.

The shear wave velocity decreases with strain level and can be calculated from

$$c_s^* = R_c c_s \tag{35}$$

where c_s is the shear wave velocity at small strain and R_c is the wave velocity reduction factor. The effect of shear strain on the deformation properties of fine-grained soils was discussed in detail by Massarsch (2004). Semi-empirical solutions are available to estimate the shear wave velocity at small strain and the effect of

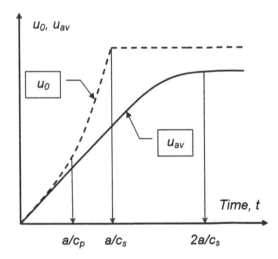

Figure 8. Schematic illustration of the displacement at the centre, u_0 and the average displacement u_{av} due to a step load uniformly distributed across a circular area of radius a, Bodare & Orrje (1985), cf. equation (25).

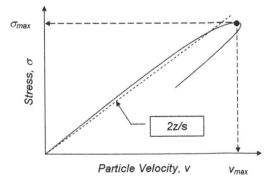

Figure 9. Relationship between dynamic stress s and velocity v during dynamic loading of a plate in an elastic material, cf. equation (32).

shear strains. Massarsch (2004) showed that in the case of fine-grained soils the shear wave reduction factor, R_c depends on the plasticity index, I_P of the soil. In Figure 10, the influence of shear strain on the shear wave reduction factor R_c is given.

In soils with low plasticity (sandy soils), the shear wave velocity at large strain (failure) can decrease by over 70%. For example, in a sandy silt, the shear wave velocity can decrease from its maximum value of around 100 to 150 m/s at small strains ($<10^{-4}$%) to about 30 to 50 m/s at failure (strains exceeding 1%). In soils with higher plasticity, the reduction of the shear wave velocity will be considerably smaller. In case of a plastic clay ($I_P = 50$%) the shear wave velocity will decrease by about 45% of its maximum value. The effect of shear strain on wave propagation velocity and thus also on the soil impedance is generally neglected, but is of great importance for accurate predictions of pile-soil interaction and wave propagation in the near-field.

2.4 Dynamic soil resistance along shaft

The dynamic force transmitted to the soil F_{Dm} can be calculated based on the following relationship

$$F_{Dm} = v_{max} z^* = v_{max} \rho c_s^* \qquad (36)$$

where v_{max} is the maximum velocity of the pile and z^* is the strain-dependent impedance of the soil at the interface with the pile. As mentioned previously, the shear wave velocity – and thus the impedance of the soil – decreases with strain level, cf. Figure 10. Thus, there is an upper limit to the vibration energy which can be transmitted along the pile shaft (and at the pile base). The maximum vibration velocity, v_{max} which is transmitted from the pile shaft to the soil can be estimated from

$$v_{max} = \frac{F_{Dm}}{\rho c_s^*} \qquad (37)$$

3 PREDICTION OF GROUND VIBRATIONS

3.1 Dynamic pile-soil interaction

In order to make rational predictions of ground vibrations during pile driving, it is important to consider the most important factors, which control the transfer of driving energy. As has been shown above, these factors are: hammer impact velocity and mass, pile impedance and dynamic soil resistance along the shaft and at the base of the pile. It is possible to estimate – based on the above given relationships – the upper limits of the vibration velocity which can be transmitted along the pile shaft and at the base.

The significance of different parameters can be assessed by relatively simple analytical methods. However, a more reliable prediction of vibration propagation in the pile can be made based on dynamic measurements during pile driving. Stress wave measurements have gained widespread acceptance and details of performing such measurements are not presented in this paper. Figure 11 shows a principle drawing of the results of stress wave measurements in a pile during driving. From the velocity and strain measurements, the force in the pile and the shaft resistance can be readily determined.

Based on such measurements it is possible to obtain a rather clear picture of the location (source) and extent of vibration emission from the pile shaft and at the base. It is also possible to obtain the input parameters (mobilized resistance) for assessing the

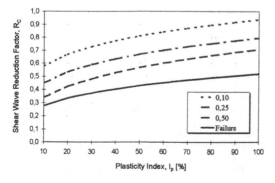

Figure 10. Reduction of shear wave velocity as a function of Plasticity Index, I_P for different shear strain levels, cf. equation (35).

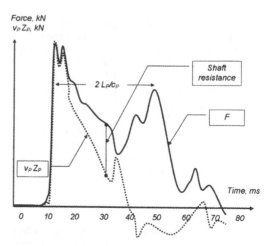

Figure 11. Principle sketch of results from stress wave measurements in a pile.

Table 4. Exponent n for different wave types.

Wave type	Exponent n
Body wave	1.0
Body wave at surface	2.0
Surface wave	0.5

dynamic interaction between the pile and the soil along the shaft and at the base.

3.2 Vibration propagation in elastic materials

Sophisticated computational methods are available for studying vibration propagation in complex ground conditions. However, the propagation of vibrations in the soil can then be assessed using relatively simple, well-established methods (Massarsch, 1992).

The vibration amplitude A_2 at distance R_2 can be calculated if the vibration amplitude A_1 is known at distance R_1

$$\frac{A_2}{A_1} = (\frac{R_2}{R_1})^{-n} e^{-\alpha(R_2-R_1)} \tag{38}$$

The exponent n depends on the wave type as shown in Table 4.

The attenuation of ground vibrations is strongly affected by the absorption coefficient α, which depends on the soil damping coefficient D, the vibration frequency f and the shear wave velocity c_S (for the case of compression waves, c_p should be used)

$$\alpha = \frac{2\pi D f}{c_s} \tag{39}$$

For the case of vibration propagation in an elastic medium (far-field problem), the soil damping coefficient D is on the order of 3–6%. However, at large strain in the near field of the vibration source, soil damping can increase significantly.

Values of the absorption coefficient α, given in the literature vary within a wide range and make a rational analysis difficult. Comparison of measured and predicted vibration velocity values based on equation (38) and (39) show good agreement. However, it is important that the effect of shear strain level on the wave velocity is taken into consideration.

4 CONCLUSIONS

Empirical vibration attenuation relationships, which are based on the energy transmitted to the pile using the concept of scaled distance, are not reliable, as they do not take into account basic pile dynamic considerations.

Two major shortcomings are the definition of the scaled distance using hammer impact energy and the fact that dynamic pile-soil interaction is neglected. Thus, it is not surprising that the scatter of reported results is large and not acceptable for reliable prediction of ground vibrations.

The fundamental aspects of hammer-pile-soil interaction are presented, which make it possible to estimate the particle velocity and thus the dynamic forces in the pile. Instead of the hammer energy, the impact velocity at the pile head is a more suitable parameter on which to base prediction of ground vibrations.

The distribution of the dynamic soil resistance along the pile shaft and at the pile base depends on the dynamic pile and soil properties. A fundamental parameter required for the analysis of pile vibration problems is the impedance. The main advantage of stress wave theory is that the relative importance of different parameter can be evaluated in a rational way, thereby giving a better understanding of this complex problem.

The most important aspect of dynamic pile-soil interaction is the fact that the intensity of ground vibrations depends on the impedance of the soil and on the particle velocity. It is possible to estimate the dynamic response based on theoretical considerations. This information can be used to select appropriate driving equipment and pile types.

The most reliable method of predicting ground vibrations is by stress wave measurements during pile driving. In this way it is possible to determine the source(s) of vibration during different phases of pile installation.

ACKNOWLEDGEMENT

Part of this research was conducted in connection with, and funded by, the European Research Project, SIPDIS. The author is indebted to Prof. Anders Bodare for valuable suggestions and discussions.

REFERENCES

Attewell, P. B. & Farmer, I. W., 1973. Attenuation of ground vibrations from pile driving. *Ground Engineering*, Vol. 6, Nr 4, pp. 26–29.

Brenner, RP, Chittikuladilok, B., 1975. Vibrations from pile driving in the Bangkok area. Geotechnical Engineering, Vol. 6, nr 2, pp. 167–197.

Bodare, A. & Orrje, O., 1988. Impulse load in a circular surface in an infinite elastic medium. (Closed solution according o the theory of elasticity). Extended version. KTH jord och bergmekanik, JoB Report No. 23, Stockholm.

Brenner, R. P. & Viranuvut, S., 1977. Measurement and prediction of vibrations generated by drop hammer piling in Bangkok subsoils. Proceedings of the 5th Southeast

Asian Conference on Soil Engineering, Bangkok, July 1977, pp. 105–119.

Broms, B. B. & Bredenberg, H., 1982. Applications of stress-wave theory on pile driving – a State-of-the-Art Report. Southeast Asian geotechnical conference, 7, Hong Kong, Nov. 1982. Proceedings, Vol. 2, pp. 195–238.

Goble, G. G., Rausche, F. & Likins, G. E., 1980. The analysis of pile driving – A State-of-the-Art Report. Int. Sem. On the Application of Stress-Wave theory on Piles, June 4–5, 1980, Stockholm, pp. 131–162.

Goble, G., 1995. What causes piles to penetrate. Deep Foundations Institute 20th Annual Members Conference and Meeting October 16–18, 1995, Charleston, South Carolina, Proceedings, pp. 95–101.

Herlitz, S., 1984. Oscillatory loads in an infinite elastic body. Institute of Technology, Uppsala University, Report UPTEC 8479R, 28 p.

Iwanowski, T. & Bodare, A., 1985. On soil damping factor used in wave analysis of pile driving. International conference on Application of Stress-wave Theory to Piles, Ottawa, May, 1988. pp. 343–352.

Jedele, L. P., 2005. Energy-Attenuation Relationships from Vibrations Revisited. Geo-Frontiers 2005, Soil Dynamics Symposium in Honour of Prof. R. D. Woods. Austin, Texas, January 24–26, 2005, 14 p.

Martin, D. J., 1980. Ground Vibrations from Impact Pile Driving during Road Construction. Transport and Road Research Laboratory, TRRL Supplementary Report 544, 16 p.

Massarsch, K. R., 1992. Static and Dynamic Soil Displacements Caused by Pile Driving, Keynote Lecture, Fourth International Conference on the Application of Stress-Wave Theory to Piles, the Hague, the Netherlands, September 21–24, 1992, pp. 15–24.

Massarsch, K. R., 2004. Deformation properties of fine-grained soils from seismic tests. Keynote lecture, International Conference on Site Characterization, ISC'2, 19–22 Sept. 2004, Porto, pp. 133–146.

New, B. M., 1986. Ground Vibrations caused by Civil Engineering Works. Transport and Road Research Laboratory, TRRL Research Report 53, 19 p.

Novak, M. & Janes, M., 1989. Dynamic and static response of pile groups. International Conference on Soil Mechanics and Foundation Engineering, 12, Rio de Janeiro, Aug. 1989. Proceedings, Vol. 2, pp. 1175–1178.

Orrje, O., 1996. The use of dynamic plate load tests in determining deformation properties of soil. Doctoral thesis, Royal Institute of Technology (KTH). TRITA-AMI PHD 1007, ISSN 140–1284, 111 p.

Rausche, F., Goble, G. G. & Likins, Jr., G. E., 1985. Dynamic determination of pile capacity. ASCE Journal of Geotechnical Engineering. Vol. 111, No. 3, pp. 367–383.

Selby, A. R., 1991. Ground Vibrations Caused by Pile Installation. Proceedings 4th International Conference on Piling and Deep Foundations, Stresa, Italy, pp. 497–502.

Smith, E. A. L., 1960. Pile-driving analysis by the wave equation. Journal of the Soil Mechanics and Foundation Engineering Division, Proceedings ASCE. No. 86 SM4, pp. 35–61.

Wiss, J. F., 1981. Construction vibrations: State-of-the-Art. ASCE Geotechnical Engineering Division. Journal, 1981. Vol. 107, No. GT2, pp. 167–181.

Woods, R. D. & Jedele, L. P., 1985. Energy-Attenuation Relationships from Construction Vibrations. Proceedings, ASCE Convention in Detroit sponsored by the Geotechnical Engineering Division, pp. 229–246.

Environmental Vibrations – Takemiya (ed.)
© 2005 Taylor & Francis Group, London, ISBN 0 415 39035 4

Three-dimensional predictive analysis of ground vibrations produced by construction work

T. Hanazato, N. Taguchi & Y. Nagataki
Taisei Corporation, Technology Center, Yokohama, Japan

Y. Ikeda
Taisei Corporation, Tokyo, Japan

ABSTRACT: Construction of high-tech facilities that must be protected against vibrations caused by machine, traffics and construction vibrations has been in great demand. In order to satisfy the requirement of performance of these facilities, it is needed to control the vibrations within the allowable limit being strict for precision instruments in buildings. This implies necessity for development of 3-dimensional dynamic soil-structure analysis that makes it possible to accurately predict the vibrations transmitting from sources to structures via soils, as well as, to employ it to develop the most suitable measures for reduction of vibrations. Therefore, we have developed the analysis technique that combines 3-D FEM with thin layer method to predict the ground vibrations produced by traffics, machines and construction operations. In the present technique, 3-D finite element and thin layer models represent near-field including structures and far-field, respectively.

1 INTRODUCTION

Construction of high-tech facilities that must be protected against vibrations caused by machines, traffics and construction operations has been in great demand. In order to satisfy the requirement of performance of these facilities, it is necessary to control the vibrations within the allowable limit being strict for precision instruments in buildings. This implies necessity for development of 3-dimensional dynamic soil-structure analysis that makes it possible to accurately predict the vibrations transmitting from sources to structures via soils, and to apply it to the most suitable measures for reduction of vibrations. However, there is a problem in the capacity and computing speed of the computer to model the whole soil-structure with 3-D FEM. Therefore, we have developed the analysis technique, shown in Figure 1, which combines 3-D FEM with thin layer method to predict the ground vibrations produced by traffics, machines and construction operations. Hence, 3-D finite element and thin layer models represent near-field including structures and far-field, respectively.

This paper outlines the analysis system developed together with its application to the analysis of the ground vibrations produced by pile driving test.

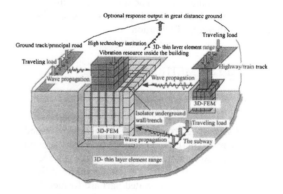

Figure 1. Concept of modeling.

2 CONCEPT OF ANALYSIS

The analysis system developed in the present research consists of the post processor that displays the analysis results, the solver that performs the analysis and the pre processor that makes the analysis model.

The constitution of the present system is shown in Figure 2. The analysis procedure is explained as;

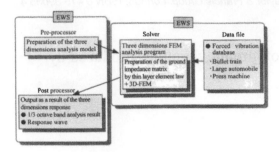

Figure 2.   Outline of system.

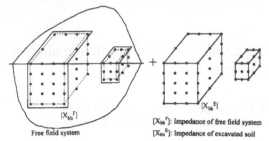

[X_{bb}^F]: Impedance of free field system
[X_{bb}^E]: Impedance of excavated soil

Figure 3.   Concept of calculating the impedance.

STEP 1 (Modeling): Foundations of buildings and their near-fields (surrounding soils) are modeled with 3-dimensional finite elements in the pre-processor. On the other hand, free field is represented by thin layered semi-infinite region. Furthermore, the points for calculation of response and for exciting force are given in the finite element region or/and in the thin-layered region.

STEP 2 (Analysis): Impedance matrix of a semi-infinite region of soils is calculated by using the thin layer technique in the solver. This matrix is combined with the stiffness matrix of the finite element region. Next, response analysis is carried out.

STEP 3 (Display of results): Numerical results are displayed by using the post processor. Here, the wave propagation in the field can be simulated. Furthermore, the 1/3 octave band filtered spectra can be displayed for the evaluation of vibration.

3   ANALYSIS METHODOLOGY

3.1   Combination method of 3-dimensional finite element and thin layered region

Figure 3 shows the procedure for calculation of impedance. Excavated parts (foundations with their surrounding soils of structures on which source and receipt are) are discretized into finite elements. The stiffness and inertia clause of excavated part are formed from finite element technique. The impedance [X_{bb}^E] of the excavated part is defined on the outer boundary of the excavated part. On the otherhand, the impedance [X_{bb}^F] on the boundary of excavated part of a free field system is calculated by using the 3-dimensional point exciting thin-layer method. These two impedances [X_{bb}^E], [X_{bb}^F] and the dynamic stiffness matrix [$S_{aa} \sim S_{bb}$] of structures are combined. The equilibrium equation in frequency domain is derived on the basis of flexible volume method.

$$[S]\{U\}=\{P\} \tag{1}$$

3.2   Response in far field

Response {Ur} at far field idealized by thin-layered region can be derived as;

$$\{Ur\}=[A_{bb}^F]\{Ps\} \tag{2}$$

where,

{Ur}:   displacement vector at far field
[A_{bb}^F]:   flexible matrix defined between points at the boundary and those in the far field
{Ps}:   force vector induced on the boundary

3.3   Response to exciting force produced in thin layer region

The sub-equation in the thin layer region and the sub-equation in the finite element regions are formulated as,

$$\{P_b\}=[X_{bb}^G]\{U_b\} \tag{3}$$

$$\{P_F\}=[K_F]\{U_F\} \tag{4}$$

where,

{P_b}:   force vector of soil
{U_b}:   displacement vector of soil
[X_{bb}^G] = [X_{bb}^F] − [X_{bb}^E]: impedance of excavated part of soil
{P_F}:   force vector of FE region.
{U_F}:   displacement vector of FE region.
[K_F]:   impedance of FE region

The equation (5) of the whole system are derived from formula (3), (4).

$$\{P_T\}=([K_F]+[X_{bb}^G])\{U_T\} \tag{5}$$

Suppose that the external force act as a series of forces {Pa}, response of the combined model can be calculated by the external force {P_T}.

$$\begin{aligned}\{P_T\}&=[\{0\},\{0\},\{Pa\}]\\ \{U_T\}&=([K_F]+[X_{bb}^G])^{-1}\{P_T\}\end{aligned} \tag{6}$$

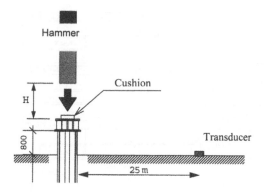

Figure 4.   Rapid load test.

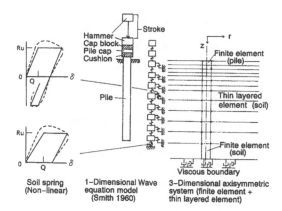

Figure 5.   Analysis model of pile driving.

The responses can be calculated from equation (6) in the finite element region, on the boundary and in the far field of thin layered region.

3.4   *1/3 octave band analysis functions*

Since the permissible limit of vibrations both for human bodies and for high-tech equipment are usually given by the amplitude in the center frequency of 1/3 octave bands, 1/3 octave analysis functions is installed in the present tool of analysis.

4   ANALYSIS OF VIBRATIONS PRODUCED BY PILE DRIVING

4.1   *Outline of field test*

Rapid load tests of piles were conducted to evaluate the baring capacity of existing cast-incase concrete piles. In the rapid load test, when the pile head was hit by a hammer, the dynamic displacement of the pile head was recorded to obtain the ultimate baring capacity of the pile toe. Since ground vibrations were produced during this integrity test, we measured them not only to assess the vibrations produced by the rapid load test but also to analyze them by employing the analysis technique introduced in the present paper. Figure 4 shows the schematic drawing of the rapid load test using the cushion of forming urethane. A drop hammer pile driver was utilized with a hammer of 39.2 kN.

The piles were cast-in-case ones constructed about 20 years ago, of which overall diameter and length were 750 mm and 8 m, respectively. Ground vibrations were measured at a distance of 25, 45 and 65 m from the pile.

4.2   *Analysis procedure*

For impact pile driving operations utilizing a drop hammer under horizontally deposited soil conditions, the dynamic analysis method combines the non-linear pile

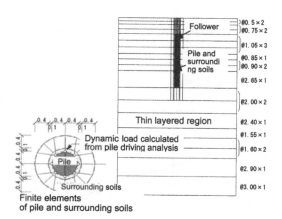

Figure 6.   Analysis model of ground vibration using proposed method.

driving analysis of a hammer-pile-soil model (See Figure 5) with the linear analysis of three-dimensionally propagating waves in multi-layered media. Although such analysis procedure was introduced by us in ISEV2003 (T. Hanazato), the analysis technique proposed in the present paper was employed to conduct the latter analysis. The pile and the surrounding soils were represented by the 3-dimensional finite elements, shown in Figure 6. The dynamic loads acting on the interface between the finite element region including the pile and the thin-layered model was calculated by the former pile driving analysis. Figure 7 shows the soil model used in the present analysis.

4.3   *Analysis results*

Figures 8 and 9 compare the analysis results with the measurements. In these figures, X and Z denote horizontal (Radial) and vertical components, respectively.

GL−(m)	地質	N (杭用)	Vs (m/s)	ρ	E (kN/m²)	G (kN/m²)	h
−1.00	埋土		120	2.10	88500	30200	0.04
−2.50	埋土		170	2.10	176100	60700	0.04
−5.65	砂質粘土	9	260	1.60	321100	108200	0.04
−8.30	砂質粘土	3	170	1.60	138100	46200	0.04
−10.95	砂礫	24	470	2.10	1356000	463900	0.04
−14.95	砂		410	1.90	937700	319400	0.04
−17.35	細砂		430	1.95	1056000	360600	0.04
−22.10	粘土		340	1.75	596100	202300	0.04
−28.00	細砂		490	1.95	1332000	465200	0.04

Figure 7.   Soil model.

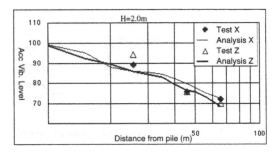

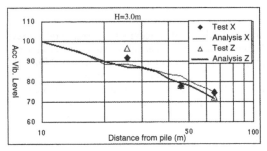

Figure 8.   Calculated attenuation with distance of acceleration vibration level, compared with measurements.

The calculated attenuation with distance of the acceleration vibration level was in good agreement with the measurements, shown in Figure 8. The simulated waveforms also show good correlation between the analysis and the measurements, shown in Figure 9. It should be noticed that not only the vertical component but also

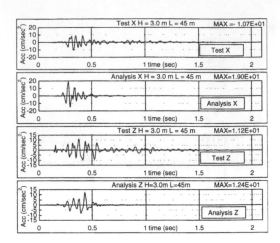

Figure 9.   Calculated waveforms of acceleration, compared with measurements.

the horizontal components show good comparison. Furthermore, the Fourier Spectra of the numerical results were in excellent agreement with the measurements. Those results demonstrate that the analysis method presented in this paper is useful in predicting ground vibrations produced by pile driving, indicating that it can be employed for predicting ground vibrations produced by machines, traffics and construction operations.

5   CONCLUSIONS

The analysis method presented in this paper can be employed for predicting ground vibrations produced by machines, traffics and construction operations. In order to predict ground vibrations produced those sources, it is essential to evaluate exciting forces as accurately as possible.

REFERENCES

Tajimi, H. 1980. A Contribution to theoretical prediction of dynamic stiffness of surface foundation, 7WCEE, Vol. 5, pp 105–112.
Hanazato, T. 2003. Ground vibrations produced during construction of pile foundations, Proc. of ISEV2003, 186–201.

Environmental Vibrations – Takemiya (ed.)
© *2005 Taylor & Francis Group, London, ISBN 0 415 39035 4*

The measured vibration data of construction sites

H. Yoshinaga, A. Hayashi, K. Yoshida & H. Yamamoto
The Public Works Research Institute, an Incorporated Administrative Agency, Tsukuba, Ibaraki-ken, Japan

ABSTRACT: Multivariate analysis was applied to vibration data measured at various construction sites using the Bornitz formula, and the measured data were put in order on the following basis : 1) Geometric attenuation coefficient n is adopted as 0.75, because approximate values are nearly equal in the two cases of geometric attenuation coefficient $n = 0.5$ and $n = 0.75$; 2) The attenuation coefficient α should be constant except for super high-frequency vibration hammer, because the number of measurements is insufficient to permit the subdivision of α by the construction categories and types of ground.

1 INTRODUCTION

Japanese law requires that project whose scale exceeds a certain size conduct an environmental impact assessment during the planning stage. Many environmental factors are part of the environmental impact assessment. To date, the Public Works Research Institute (2000) has published assessment techniques in Environmental Impact Assessment Techniques for Road Projects (2) etc. The authors were responsible for the construction vibration assessment techniques in these publications. There is insufficient data to develop prediction techniques for construction vibration for use with the various construction methods, so on-site measurements have continued since 2000.

Here, we will apply the Bornitz formula, analyze the on-site measurement data, compile typical values for each type of construction, and present the results.

2 THE BASIC METHOD FOR PREDICTING CONSTRUCTION VIBRATION

In regard to the vibrations transmitted in the earth, it is known that P-waves, S-waves, and Rayleigh waves exist as solutions to wave equations that are applied to isotropic and homogeneous semi-infinite elastic solids. Also, the Bornitz formula that experientially incorporates attenuation is used for actual forecasts. The Public Works Research Institute (2000) and others directed to forecast the vibration level of construction vibration using a formula based on the Bornitz formula (1). The institute designates $r_0 = 5\,\mathrm{m}$ and $n = 0.75$ of the parameters as the uniform values. The

value $n = 0.75$ is an intermediate value of the body wave $n = 1$ that disperses in three dimensions and the Rayleigh wave $n = 0.5$ that disperses in two dimensions. Also, based on the on-site measurements, L_{va0} for each type of construction, $\alpha = 0.001$(for hard ground, including rock and stone) and $\alpha = 0.019$ (for un-hard ground, including clay and sand) are fixed.

$$L_{va} = L_{va0} - 20n\log_{10}(r/r_0) - 8.68\alpha(r - r_0) \quad (1)$$

where L_{va}[dB] is the vibration level at the distance of r[m] from the vibration source; L_{va0} is the vibration level at the distance of r_0[m] from the vibration source; n is the geometric attenuation coefficient, and α is the attenuation coefficient.

The site of origin for vibration at an actual construction site is not a point. It spreads along a horizontal plane or underground, and its force is different in each direction. Also, the earth is not uniform, and there are cases when the ground structure is layered, there are rocks or buried structures in the ground, or there is variation in the water content. The vibration will be curved, reflected, or obstructed. Therefore, the influence of these factors must be rigorously considered for measurement and analysis.

Due to the large diversity in construction methods in civil engineering projects, however, it is difficult to conduct these measurements and analyses. Also, as the early release of the data is sought, the formula in (1) is used initially for the actual measured values on site, and representative statistical values are set by least squares method.

3 EXAMINATION OF THE METHODS FOR ANALYZING MEASUREMENT DATA

3.1 *Measurement values*

Vibration level is measured on the basis of the distribution in Figure 1 for construction work using construction machinery. If there are multiple machines, the machine with the greatest effect is regarded as the source of the vibrations, and as a rule is placed at measurement points that minimize to the greatest extent possible the effect of the other machines. The distance from the source of the vibrations to the measurement points is, as a basis, 7 to 30 meters horizontally along the earth's surface.

The vibration level differs with the construction process, so the measurements are made in those processes with the highest vibration levels. Also, the vibration level is measured in a plumb bob vertical direction using measurement methods determined in accordance with the Japanese Vibration Regulation Law. The 90th percentile is used as the basis for measurement values when the vibration level varies over time. They are the largest values when there is an impact vibration.

Those data less than 40 dB are eliminated from analysis, because they are thought to be influenced by background vibration. Also eliminated are vibration measurement values too close to the source of origin. To determine whether the distance is too close, we used as a reference examples by the Research Committee of Construction Noise Prediction in the Acoustical Society of Japan (2002) in which the distance that can be regarded as the point sound source for the noise measurement is more than 1.5 times the sound source size.

3.2 *Examination of methods of analysis*

The measurement values for formula (1) are L_{va}, r, and the parameters are L_{va0}, n, α. Method A and Method B are compared as methods of analysis for the parameters based on measurement values.

Method A: Formula (1) is the primary formula for the parameters L_{va0}, n and α so a solution is sought approximating the line using the least square method. A general purpose multiple regression analysis program is used.

Additionally, there is a relation between the explanatory variables $20 \log_{10}(r/r_0)$ and $8.68(r - r_0)$, but there is no linear relationship. Therefore, no multicollinearity is considered to exist.

Method B: Formula (1) is converted to decibels, so the vibration level returns to the antilogarithm and is approximated using the least square method. In this instance, the parameters are non-linear and existing programs cannot be used, so a program was created for the analysis.

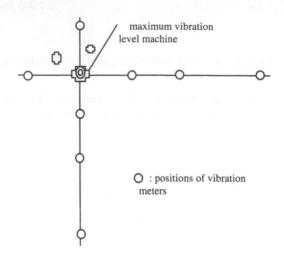

Figure 1. Position of measuring points.

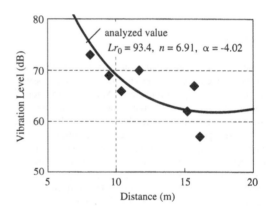

Figure 2. Example of analysis by the method of least squares of dB.

Figure 2 is an example of analysis using Method A, and Figure 3 is an example of analysis using Method B with the same data. The true values are thought to be completely different – for example, the analytic value of the attenuation coefficient α is a negative value, and the geometric attenuation coefficient n is an aberrant value. This is thought to be caused by a large variation in the data for analysis using three parameters, and the insufficient amount of data. In addition, analysis using antilogarithms is greatly influenced by measurement values at a close distance, and there will be a large fluctuation in measurement values for forecast distances of greater than 10 meters in assessment work. With other on-site measurement values, the analytic values were aberrant.

Therefore the geometric attenuation coefficient n is fixed at 0.5 or 0.75. There are two parameters, L_{va0},

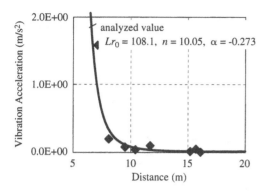

Figure 3. Example of analysis by the method of least squares of anti-logarithm.

α, and the decibels are approximated as is using the least square method.

4 ANALYSIS RESULT

4.1 Geometric attenuation coefficient n

For the measurement values in each type of construction, Table 1 presents an example of the analysis of the parameters for work whose ground conditions are dealt with as un-hard ground, including clay and stone. As the geometric attenuation increases, optimization occurs with the decrease in the internal attenuation and other factors. The average values for the multiple correlation efficient are 0.686 when $n = 0.5$ and 0.691 when $n = 0.75$. Thus, it is apparent that there is no differential as the real values are approached. That's because for distances ranging from 7 to 30 meters, there is little difference in approach accuracy in formula (1) between the curved line resulting from geometric attenuation and the straight line resulting from internal attenuation.

In this case, there are no grounds for changing the geometric attenuation coefficient and there is no difference in the approach accuracy. Therefore, as there will be no problem in conducting the work, this will be maintained at the conventional 0.75.

4.2 Attenuation coefficient α

Figure 4 shows the creation of a histogram for the attenuation coefficient in Table 1. While there is a negative value, the average value is 0.014, close to the value of 0.019 in Public Works Research Institute (2000). Formula (2) comparing the vibration frequency is generally used for attenuation coefficient α as in INCE/JRTV-MODEL 2003.

$$\alpha = \frac{\omega\eta}{V} \qquad (2)$$

where ω is circular frequency of the wave motion ($\omega = 2\pi f$); η is ground attenuation constant.

The super high-frequency vibration hammer in Table 1 applies this principle as a measure against vibration. Figure 5 shows the frequency characteristics for each type of construction in Table 1. To facilitate a comparison, all pass levels are set at 0 dB, with sensory compensation. It can be seen that compared to other types of construction, the super high-frequency vibration hammer has a high proportion of high frequency elements, centered on 50 Hz. In Figure 6, this analytic value is showed in the curve of $\alpha = 0.135$. Attenuation curves 0, 0.019 and 0.04 are also recorded. It is thought appropriate to adopt an individual value for attenuation coefficient α different from other construction methods.

A contradiction occurs with other construction methods, however, when the attenuation coefficient is set by maintaining the analytic value for each method as is. This is shown in Figure 4. The attenuation coefficient, except that for the super high-frequency vibration hammer, is distributed in a range of ± 0.03 from the center, and this deviation is thought to be the primary cause. In addition, the range of $\alpha = 0$ and $\alpha = 0.04$ in Figure 6 is roughly at the same range of measurement values for this construction method. Therefore, the difference in the results of the analysis for α is thought to be primarily due to the variation in the data caused by factors outside the hypotheses of formula (1), rather than a difference in real values.

Based on the foregoing, it is appropriate that the attenuation coefficient α is a specific value when the cause is clear as the super high-frequency vibration hammer. But the cause of the difference of α is unclear except for that construction category in conditions with little measurement data. So, it is appropriate to fix a uniform value rather than subdivide.

As is shown in Table 1, the average value for α is roughly the same when the site of origin are on land as the cases they are underground. For cases in which the source of the vibration moves from the surface to underground, as with pile driving work, the differences in distances from the vertical location of origin and the differences in wave motion should be considered. In this research, we did not conduct a detailed investigation or analysis. Therefore we used the uniform values for now.

4.3 Summation of analysis results

The extreme right-hand column in Table 1 shows the vibration level at the reference distance when the significant values for the attenuation coefficient other than with the super high-frequency vibration hammer are a single digit, and they are made uniform at an average of 0.01. Before making Table 1, distinctions had been made in the types of ground. For the vibration

Table 1. Example of analysis of construction vibration parameter[1].

Construction category	Positions of vibration sources[2]	Types of ground surface[3]	Soil types[4]	Vibration levels[5]	Analyzed as n = 0.75		Analyzed as n = 0.5		n = 0.75, α fixed	
					α	$R^{[6]}$	α	$R^{[6]}$	Lr_0	α
Demolition work by giant breaker	1	2	1,2	L_{10}	0.034	0.82	0.045	0.82	73	0.01
Mobile recycler	1	2	2	L_{10}	0.015	0.91	0.026	0.90	69	0.01
Dismantlement of old bridge	1	3,5	1,2	L_{10}	0.010	0.84	0.028	0.82	76	0.01
Stabilization for subgrade	1	5	1,2	L_{10}	0.017	0.56	0.029	0.56	66	0.01
Sand mat work	1	3,5	1,2	L_{10}	0.018	0.83	0.028	0.82	71	0.01
Slope spraying work	1	2	1	L_{10}	−0.033	0.72	−0.013	0.78	48	0.01
Concrete bridge erection	1	2	1	L_{10}	−0.008	0.35	0.006	0.35	55	0.01
Pavement-base of lower layer	1	2	2	L_{10}	0.005	0.68	0.017	0.68	59	0.01
Vibration hammer method	2	2,5	2	L_{10}	0.035	0.82	0.047	0.82	77	0.01
Vibration hammer and w.j. method	2	2,3	2	L_{10}	0.019	0.62	0.029	0.62	75	0.01
Steel sheet pile with earth auger	2	3	2	L_{10}	0.020	0.68	0.031	0.68	59	0.01
Pile jacking machine	2	4	1	L_{10}	−0.008	0.39	0.004	0.41	62	0.01
Diesel pile hammer	2	3	2	L_{MAX}	−0.011	0.35	−0.003	0.43	81	0.01
Hydraulic pile hammer	2	5	1,2	L_{MAX}	0.021	0.59	0.031	0.59	81	0.01
Auger caison piling method	2	5	1,2	L_{10}	0.017	0.80	0.025	0.80	63	0.01
Reverse circulation drilling method	2	5	2	L_{10}	−0.009	0.80	0.001	0.79	54	0.01
All casing	2	2,4,5	1,2	L_{10}	0.013	0.71	0.024	0.70	63	0.01
All casing for hard ground	2	2	3	L_{10}	0.034	0.84	0.048	0.85	61	0.01
Down-the-hole hammer	2	2,3	1,2	L_{10}	0.029	0.74	0.039	0.73	67	0.01
Earth drill	2	4	2	L_{10}	−0.006	0.47	0.004	0.46	56	0.01
Pile installation by excavation	2	2	2	L_{10}	−0.006	0.35	0.009	0.38	62	0.01
Open caisson work	2	2,5	1,2	L_{10}	0.012	0.94	0.027	0.94	55	0.01
Chemical grouting	2	2	1	L_{10}	−0.020	0.30	−0.008	0.33	53	0.01
Sand compaction pile	2	3,5	1	L_{10}	0.011	0.89	0.022	0.88	81	0.01
Acked sand drain	2	5	1	L_{10}	0.014	0.53	0.024	0.54	83	0.01
Dry jet mixing	2	5	1	L_{10}	−0.002	0.25	0.007	0.26	62	0.01
Super high-frequency vibration hammer	2	3	1	L_{10}	0.135	0.86	0.157	0.86	81	0.14
Average of all construction category					0.014	0.68	0.026	0.69		
Except super high-frequency vibration hammer					0.011	0.68	0.022	0.69		
Positions of vibration sources are on the surface of the ground					0.008	0.72	0.021	0.74		
Positions of vibration sources are underground (except super high-frequency vibration hammer)					0.012	0.66	0.023	0.66		

[1] The distance range of the measurement are from 7 to 30 meters.
[2] 1:on the surface of the ground, 2:underground.
[3] 1:concrete/asphalt, 2:hard surface, 3:grass/rice paddy, 4:farm, 5:no information.
[4] 1:cray, 2:sand, 3:gravel.
[5] L_{10} is 90th percentile and L_{max} is largest value.
[6] multiple correlation coefficient.

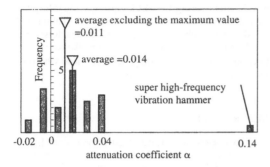

Figure 4. Analyzed value of α (un-hard ground).

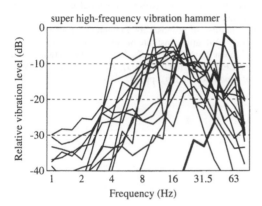

Figure 5. Construction vibration spectrum.

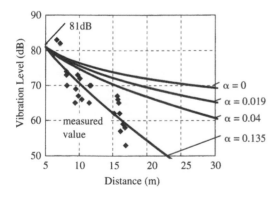

Figure 6. Attenuation of vibration by a super high-frequency vibration hammer.

level at the reference distance, the average values for each construction type had been compiled.

For reference, in Figure 7 there are the measurement values for demolition work by giant breakers and approximated curved lines from the analytic values of Table 1. There is also the attenuation curve with the

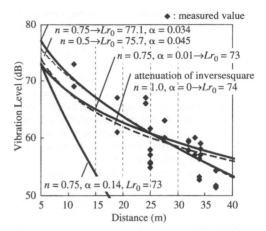

Figure 7. Attenuation of vibration caused by demolition works by giant breakers.

Table 2. Comparison between this report and previous report.

	Previous report	This report
Geometric attenuation coefficient n	0.75	0.75
Attenuation coefficient α un-hard ground	0.019	0.01
super high-frequency vibration hammer	0.019	0.14

super high-frequency vibration hammer and the attenuation curve through the simple inverse-square law. When approximating the measurement value with two parameters, there is no significant difference when the value for the geometric coefficient n is 0.5 and when the approximate curve is 0.75. Further, the attenuation curve of $n = 0.75$, $\alpha = 0.01$ has a rough correspondence with the curve $n = 1.0$, $\alpha = 0$ using the simple inverse square law. Unless a special construction method is used, such as the super high-frequency vibration hammer, construction vibration can be easily forecast when the distance attenuation using the inverse square law is identical to that for noise.

When using Table 1 for forecasts, the higher value should be set at 30 meters for the applicable range for the distance from the site of origin to the forecast point (horizontal distance). The lower value should be either 7 meters or the figure 1.5 times the size of the vibration source, whichever is larger.

5 CONCLUSION

We analyzed construction vibration measurement values applying the Bornitz formula. Table 2 shows a result of the comparison with previous reports. Geometric coefficient n is the conventional value and there are two values for internal attenuation – one for the super high-frequency vibration hammer and one for other construction methods.

Other than increasing the number of measurement values during on-site measurements as in this case, other methods for improving the accuracy of the geometric attenuation coefficient and the attenuation coefficient would be through experimentation under known conditions for the location, expansion, frequency, and ground conditions of the site of origin. It is also thought that the vibration level at the reference distance differs with each type of ground, so in the future it will be necessary to formulate a theory and conduct an analysis.

The measurement data is from surveys of construction sites conducted by persons affiliated with the Japanese Ministry of Land, Infrastructure, and Transport. The analyses were conducted with the cooperation of the contractors M. Sano of the Japan Construction Method & Machinery Research Institute and H. Aoi

of CTI Engineering Co., Ltd. Opinions regarding the analytic methods were received from Y. Tokita and J. Kaku of the Kobayashi Institute of Physical Research, and M. Shioda of Kogakuin University. We would like to express our deepest gratitude to all those who provided their advice and assistance.

REFERENCES

Public Works Research Institute 2000. Environmental Impact Assessment Technique for Road Project (2), *Technical Note of Public Works Research Institute* No.3743.

Research Committee of Construction Noise Prediction in Acoustical Society of Japan 2002. ASJ Prediction Model 2002 for Construction Noise, *The Journal of the Acoustical Society* Vol.58, No.11.

Subcommittee of proposal for Prediction formula of Road Traffic Vibration, Institute of Noise Control Engineering of Japan 2004. Road Traffic Vibration Prediction Model (INCE/J RTV-MODEL 2003), *The Journal of the INCE of Japan* Vol.28, No.3 pp.207–216.

Mitigation theories and applications against traffic induced ground vibrations: numerical analyses

Environmental Vibrations – Takemiya (ed.)
© 2005 Taylor & Francis Group, London, ISBN 0 415 39035 4

Scattering of Rayleigh waves by heavy masses as method of protection against traffic-induced ground vibrations

V.V. Krylov

Department of Aeronautical and Automotive Engineering, Loughborough University, Loughborough, UK

ABSTRACT: A promising and cost effective method of screening ground vibrations from rail and road traffic can be based on placing heavy masses on the ground surface alongside roads (e.g. concrete or stone blocks of 400–800 kg). The principle of their operation relies on the fact that natural frequencies of vibration for such masses, which depend on the mass values and on the ground stiffness, can be chosen within the frequency range of railway- or road-generated ground vibrations. When the masses are shaken at resonance in vertical and horizontal directions under the impact of incident Rayleigh waves, they scatter the incident energy very efficiently at different directions on the surface and into the ground depth, thus resulting in noticeable attenuation of transmitted ground vibrations. The aim of the present paper is to give a brief introduction to the resonant mass scatterers and to discuss some problems that still need to be considered.

1   INTRODUCTION

Ground vibrations generated by rail and road traffic are major sources of environmental noise and vibration pollution. They cause significant disturbance for local residents and sensitive community building such as schools, hospitals, etc. They are also very disruptive for precise manufacturing, which is so important in our time of rapid development of high technology.

Theoretical investigations of ground vibrations from rail and road traffic undertaken during the last decade (see e.g. Krylov 1995, 1996, 1998, 2001a,b; Krylov et al. 2000; Sheng et al. 1999, 2003; Grundmann et al. 1999; Watts et al. 2000; Takemiya 2001; Degrande et al. 2001) contributed to understanding the reasons why different levels of vibrations are generated at different conditions and parameters of vehicles and the infrastructure. In particular, it has been first predicted theoretically by the present author (Krylov 1994, 1995, 1998) that especially large increase in railway-generated vibrations can occur if train speeds v exceed the velocity of Rayleigh surface waves in the ground c_R. When this happens, *a ground vibration boom* takes place, similar to a sonic boom from supersonic aircraft.

In October 1997, a ground vibration boom has been observed for the first time on the newly opened West-coast Main Line from Gothenburg to Malmö in Sweden (see Madshus & Kaynia 1998). In particular, at the location near Ledsgård the Rayleigh wave velocity in the ground was only 45 m/s, so that an increase in train speed from 140 to 180 km/h lead to about 10 times increase in generated ground vibrations. This was in good agreement with the above-mentioned theoretical papers.

Although the theory of ground vibrations from rail and road traffic is now well understood and some progress in their suppression at source has been achieved, there are still relatively few investigations aimed to protect the affected areas by influencing propagation of ground vibrations from a source to a receiver. The ability to suppress ground vibration on the propagation path would be especially important in the cases where it is difficult or impossible to reduce the intensity of generated ground vibrations at source.

Note that, in contrast to the case of ground vibrations, the progress in controlling air-borne noise generated by rail and road traffic by influencing wave propagation is much more noticeable. The most popular way of protecting built up areas in this case is acoustic screening, i.e. erecting anti-noise barriers close to the transport routes concerned. However, for ground vibrations induced by rail and road traffic, the complex nature of elastic fields propagating in the ground, which may contain bulk and surface Rayleigh waves, does not permit direct analogies of acoustic screening. Nevertheless, some measures resembling simple acoustic screening have been designed in the past for protection of built up areas against ground vibrations as well.

Among such measures one can mention specifically constructed protective trenches that can screen sensitive buildings from Rayleigh surface waves, that are carrying most of the energy of railway- or traffic-generated ground vibrations from surface sources (see e.g. Degrande et al. 1996). Unfortunately, the above-mentioned protective trenches are very expensive to build and to maintain. Moreover, due to the specific features of Rayleigh wave scattering, they can not provide a full screening of Rayleigh waves even theoretically, regardless of how deep they are.

A promising and cost effective alternative to trenches as means of screening ground vibrations on the propagation path can be placing heavy masses on the ground surface, such as concrete or stone blocks. The principle of their work is based on the fact that any surface disturbance, including stones or topographic irregularities, affects Rayleigh wave propagation, causing scattering of the incident Rayleigh waves into bulk waves and into other surface Rayleigh waves propagating in different directions. For heavy masses placed on the ground, the scattering can be especially strong at the natural frequencies of vibration of such masses resting on the elastic ground. These frequencies can be chosen within the frequency range of railway- or road-generated ground vibrations. Then, when such resonant masses are shaken in vertical and horizontal directions under the impact of incident Rayleigh waves, they scatter the incident Rayleigh waves into the depth of the ground and at different directions on the surface, thus resulting in significant resonant attenuation of transmitted Rayleigh waves.

While some initial efforts have been made to investigate the above-mentioned mass scatterers, mainly by means of numerical calculations, very little progress in understanding their properties has been made so far. Therefore, this promising method of protection against ground vibrations from rail and road traffic remains largely unexplored.

The aim of the present paper is to give a brief introduction to the resonant mass scatterers and to discuss some problems that still need to be considered to achieve a fuller understanding of their operation as means of damping of ground vibrations generated by rail and road traffic.

2 GENERAL DESCRIPTION OF THE METHOD

The method of screening ground vibrations by heavy masses (e.g. by concrete or stone blocks of 400–800 kg) placed on the ground surface has been first proposed about 20 years ago (see Jones et al. 1986; Ford 1990). Although the term 'resonant scattering' was not mentioned in the above-mentioned two papers, that employed numerical calculations in their studies, it was noted that natural vibration frequencies of such resting masses play an important role in the behaviour of such masses as means of protection against propagating ground vibrations.

The natural frequencies $f_0 = \omega_0/2\pi$ of heavy masses resting on the ground can be defined by the well-known relationship $f_0 = (K/m)^{1/2}/2\pi$, where K is the equivalent concentrated stiffness of the ground associated with the resting mass of finite dimensions, and m is the mass value.

If f_0 is chosen, by selecting the appropriate m, within the frequency range of rail- and road-generated ground vibrations (normally from 5 to 50 Hz), then, under the impact of incident Rayleigh surface waves, the mass will be shaken at its natural frequency and the amplitudes of its vibrations will be strongly amplified. Because of this amplification, the masses will scatter incident surface waves very efficiently both into the ground depth (as longitudinal and shear bulk elastic waves) and at different directions on the surface (into scattered Rayleigh surface waves). As a result of these processes, the energy of transmitted Rayleigh waves will be reduced, thus resulting in strong attenuation of transmitted ground vibrations.

The scattering properties of individual masses can be further enhanced to achieve the required protection of the built up areas by using their suitable combinations (arrays).

The important aspect of using heavy masses as means of protection against rail- and road-generated vibrations is that it may have an additional benefit for society. Namely, in addition to their scattering properties, the masses, such as natural stones or specifically designed concrete blocks, can be used also to enhance visual appearance of the area. For example, in some cities in the UK, architectural compositions of heavy natural stones are already used for this purpose without any thought of their possible complementary application for protection against traffic-induced vibrations (see, for example, Figure 1 showing the composition of decorative stones off Clifton Boulevard (A52) in Nottingham). Using such decorative stones with the additional purpose of protecting built up areas against traffic-induced vibrations would make life in the locations concerned not only aesthetically more pleasant, but also much quieter.

As was mentioned above, the initial research into using heavy masses for protection against ground vibrations has been carried out by applying numerical techniques (Jones et al. 1986; Ford 1990). It has been shown that large concrete blocks placed on the ground surface can act as vibration dampers. However, the physics of this process was not clarified, and a little progress in understanding the properties of such devices as resonant scatterers of Rayleigh surface waves has been made.

Figure 1. Composition of decorative stones off Clifton Boulevard in Nottingham.

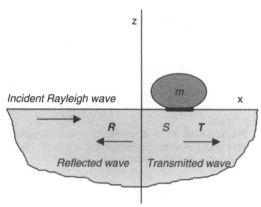

Figure 2. Rayleigh wave scattering by a heavy mass placed on the ground.

In the same time, the theory of scattering of Rayleigh waves by different surface inhomogeneities is now well developed in respect of ultrasonic non-destructive testing applications and signal processing devices using surface acoustic waves (see e.g. the book of Biryukov et al. 1995 and references there). Moreover, during the last two decades some important advances have been made in understanding resonant scattering of surface Rayleigh waves, in particular for applications to solid state physics (see e.g. Maradudin et al. 1988; Plessky et al. 1991; Mayer (Garova) 1994). Therefore, it is natural to apply and develop the existing knowledge of Rayleigh wave resonant scattering to the specific case of heavy masses placed on the ground, with the purpose of protecting built environment against rail- and road-generated ground vibrations.

3 THEORETICAL APPROACH

In this section we briefly consider only the outline of the theoretical approach to the problem of resonant scattering by heavy masses, without any detailed derivations and discussions of the existing results.

Let us assume that a harmonic Rayleigh wave (factor $exp(-i\omega t)$ assumed) which is characterised by the amplitudes of displacement $u_i^0(r)$ is incident upon a mass placed on a flat ground surface $z = 0$ (see Figure 2). Due to the interaction with the mass (also called 'scatterer'), the total displacement field u_i in the elastic half space representing the ground differs from the field of the incident wave, so that the scattered field is $u_i^{sc} = u_i - u_i^0$.

One of the possible ways to approach this self-consistent problem can be based on the solution of the integral equation describing elastic wave scattering by surface inhomogeneities. It can be shown (see Biryukov et al. 1995, p. 282–287) that in the general case of Rayleigh wave scattering the total field u can be related to u^0 by the integral expression

$$u_m(\mathbf{r}) = u_m^0(\mathbf{r}) + \int_C n_j \sigma_{ij}(\mathbf{r}')G_{im}(\mathbf{r},\mathbf{r}')d\mathbf{r} \ . \tag{1}$$

Here $n_j = (0,0,1)$ is a unit vector normal to the surface, $n_j\sigma_{ij}(r') = n_z\sigma_{iz}(r')$ are non-zero stress components applied to the surface over the contact area C due to the presence of a scattering mass, and $G_{im}(r,r')$ is the dynamic Green's tensor that satisfies the free boundary conditions everywhere at the ground surface $z = 0$:

$$n_j c_{ijkl}G_{im,k} = 0. \tag{2}$$

In the latter equation c_{ijkl} are elastic constants of the medium, which for simplicity is assumed isotropic.

If the scatterer-related stress components $n_j\sigma_{ij}(r')$ can be expressed in terms of u_m, which generally is a difficult problem that should be considered separately for each specific type of scatterer, then by placing the point of observation r in Eqn (1) on the surface within the area of its contact with the scatterer, one can obtain the integral equation versus u. Solving this equation analytically or numerically and substituting the obtained values of $n_j\sigma_{ij}(r')$ again into the integral of Eqn (1), one can calculate $u_i^{sc} = u_i - u_i^0$ at any point on the surface or in the bulk of the ground.

As was mentioned above, the solution of the integral Eqn (1) is a difficult problem that depends on the dynamic properties of a scatterer, its size and geometry. However, in certain important cases the problem can be simplified and the solution can be obtained analytically. In particular, in the case of heavy masses placed on the ground one can normally assume that the size of the mass is much smaller than the Rayleigh wave lengths of interest, so that one can ignore elastic deformations inside the mass and consider its movement as a whole. In the case of three dimensions such a simplified model is often called a 'point' model, and in the two-dimensional case – a 'line' model.

The non-zero stress components $n_z\sigma_{iz}(r')$ applied to the surface over the contact area C in Eqn (1) then can be defined by the 2nd Newton's law:

$$n_z\sigma_{iz}(r') = -(m/S)\omega^2 u_i, \qquad (3)$$

where S is the area of contact. Substituting Eqn (3) into Eqn (1), one can obtain the integral equation versus u_i in the area of contact. Solving Eqn (1) versus u_i and using the found value of u_I in Eqn (1) again, now for calculation of the scattered Rayleigh waves in backward and forward directions, u_{back}^{sc} and u_{forw}^{sc} respectively, one can obtain relative values of the amplitudes of reflected and transmitted waves.

In the case of two-dimensional geometry (a line obstacle) one can introduce the reflection and transmission coefficients of Rayleigh waves as $R = u_{back}^{sc}/u_0$ and $T = 1 + u_{forw}^{sc}/u_0$.

From the point of view of mathematical solution, it is often convenient to rewrite Eqn (1) in the wave number domain, as the expressions for the Green's tensor are simpler in this case. The discussion of the corresponding mathematical aspects, however, is beyond the scope of this paper.

The important point to note is that all scattered fields as functions of frequency, and hence the expressions for R and T, show a resonant behaviour, i.e. they have broad maxima around a certain value of frequency that can be associated with the above-mentioned resonant frequency $f_0 = (K/m)^{1/2}/2\pi$. Note, however, that the value of the equivalent stiffness K associated with the ground elasticity is derived automatically as the result of the solution of the above-considered self-consistent problem of Rayleigh wave scattering. Apparently, the value of K can be estimated also from a simpler model of a vibrating mass resting on Winkler foundation, using the known relationships between stiffness of Winkler foundation and real elastic parameters of the ground.

So far, detailed calculations of the reflection and transmission coefficients have been carried out for a two-dimensional case (see Plessky et al. 1991; Mayer (Garova) 1994). According to these calculations, for typical parameters of the mass scatterers and of the elastic half space the resulting absolute values of the reflection coefficient $|R|$ can be as high as 0.3–0.5 in a rather wide frequency range of about 40% around

the resonant frequency of the mass scatterer ω_0. In the same frequency range the values of the transmission coefficient $|T|$ can be as low as 0.2–0.4, thus describing a desirable significant resonant attenuation of transmitted Rayleigh waves. The remainder of the incident Rayleigh wave energy is scattered into bulk waves, according to the well-known relationship $|R|^2 + |T|^2 + |B|^2 = 1$. Here $|B|^2$ is the relative amount of energy scattered into bulk longitudinal and shear waves (here we do not take into account part of the energy transformed into heat).

4 SUGGESTIONS FOR FUTURE RESEARCH

In spite of the basic understanding of the phenomenon of resonant scattering of Rayleigh waves by masses resting on the ground, much work remains to be done. First of all, there is a need to develop a more comprehensive theoretical model of the phenomenon in question, especially in the light of the current understanding of field distributions of traffic-induced ground vibrations, including those from high-speed trains and heavy lorries.

Essential part of the work should be concentrated around developing the theory for real large stones and concrete blocks resting on the ground, in contrast to highly idealised models of point or line masses used in the previous theoretical works (see, e.g. Plessky et al. 1991; Mayer (Garova) 1994). Basically, one needs to know how to calculate the values of real resonant masses for given elastic parameters of the ground, expected levels and frequency contents of traffic-induced vibrations, building locations, etc. Another important topic to be considered is Rayleigh wave scattering by arrays of masses placed on the ground surface. Finally, on the basis of this knowledge, one should be able to work out practical recommendations on protection of built up areas against rail-and road-generated ground vibrations using suitable combinations of heavy masses. The more detailed explanation of the above-mentioned topics is given below.

To fully understand resonant scattering of Rayleigh waves by heavy masses one should analyse the integral Eqn (1) in the general case of finite dimensions of a scatterer, taking into account its elastic deformations. The links should be established between the obtained more general solutions and the results following from the existing simplified point and line models. In addition to this, the attention should be paid to analysing the possibility of using simplified elastic models of the ground, such as Winkler foundation model, for predictions of resonant frequencies of Rayleigh wave scattering by heavy masses. The obtained solutions should be validated by numerical calculations, e.g. using finite element method (FEM) or boundary element method (BEM), and by direct measurements of Rayleigh wave scattering in real outdoor conditions.

The basic solutions for single-mass scatterers discussed above then can be used for analysing the combined effects of groups (arrays) of several scattering masses distributed over the ground surface to enhance their wave attenuation performance. It is well known that arrays of scatterers can multiply their individual action even if the scatterers are positioned in random. And this is especially so when the scatterers are distributed in such a way that the individual scattered fields are combined in phase. As the simplest type of such an array, one can consider a combination of stones of similar size placed in a straight line very close to each other. Obviously, such a combination is equivalent to a two-dimensional mass scatterer considered above. If we combine several two-dimensional scatterers formed by individual stones placed in lines, then for simple incident waves, such as plane waves, the in-phase amplification of the resulting scattering can take place if the lines of scatterers are placed periodically (the so-called Bragg's reflective array).

One should remember, however, that in the case of ground vibrations generated by rail and road traffic, the incident waves may have much more complex distributions in space and time. Therefore, special investigation would be needed to find out optimal positions and mass distribution of individual scatterers in such complex cases.

The important exemption is the case of ground vibration boom generated by trans-Rayleigh trains mentioned in the Introduction to this paper. In this case the incident waves are quasi-plane waves propagating at the angles $\Theta = \cos^{-1}(c_R/v)$ relative to the track, so that Bragg's reflective arrays can be directly applied for efficient damping of generated ground vibrations.

Since resonant mass scatterers under consideration are strong enough, the attention should be paid also to the effects of multiple scattering of Rayleigh waves by their combinations.

The obtained knowledge on scattering of Rayleigh waves by single resonant masses and by arrays of masses should then be applied to the calculations of different practical situations to protect built up areas against traffic-induced ground vibrations. Real parameters of trains, lorries, rail tracks, roads, ground and mass scatterers should be taken into account. Optimal solutions should be sought and the resulting practical recommendations made.

The expected outcomes of the above-mentioned outline program of further research into Rayleigh wave resonant scattering by heavy masses placed on the ground would assist in better understanding of this promising method of protection of the built environment against railway- and road-generated ground vibrations. They will also give the ideas on practical aspects of combined usage of specially designed compositions of stones and concrete blocks for protection against traffic-induced vibrations and for decorative purposes.

5 CONCLUSIONS

According to the existing theoretical works, suitable combinations of heavy masses placed on the ground, such as large stones or concrete blocks, can be used as efficient resonant dampers of Rayleigh waves generated by rail and road traffic. The main advantage of using such dampers is the expected low cost of their construction and maintenance.

The additional advantage of such scatterers is their ability to enhance visual appearance of the area. In some cities, architectural compositions of heavy natural stones are already used for this purpose without any thought of their possible complementary application for protection against traffic-induced vibrations. Using such decorative stones with the additional purpose of protecting built up areas against traffic-induced vibrations would make life in the locations concerned not only aesthetically more pleasant, but also much quieter.

In spite of the above-mentioned basic understanding of the phenomenon of resonant scattering of Rayleigh waves by heavy masses, much work remains to be done to develop more comprehensive theoretical models of the phenomenon. This is especially topical in the light of the current understanding of field distributions of traffic-induced ground vibrations, including those from high-speed trains and heavy lorries.

REFERENCES

Biryukov, S.V., Gulyaev, Yu.V., Krylov, V.V. and Plessky, V.P. (1995) *Surface Acoustic Waves in Inhomogeneous Media*, Springer-Verlag, Berlin, Heidelberg 1995. – 390 p.

Degrande, G., De Roeck, G., Dewulf, W., Van den Broeck, P. and Verlinden, M. (1996) Design of a vibration isolating screen. *Proc of the International Conf. On Noise and Vibration Engineering*, Catholic University of Leuven, Belgium, 18–20 September 1996, p. 823–834.

Degrande, G. and Lombaert, G. (2001) An efficient formulation of Krylov's prediction model for train induced vibrations based on the dynamic reciprocity theorem. *Journ. Acoust. Soc. Amer.*, **110**(3), Pt. 1, 1379–1390.

Ford, R.A.J. (1990) Inhibiting the transmission of ground-borne vibrations by placing masses on the surface of the ground. *Proc. Institution of Engineers Australia Vibration and Noise Conference*, Melbourne, 18–20 September 1990, p. 227–231.

Grundmann, H., Lieb, M. and Trommer, E. (1999) The response of a layered half-space to traffic loads moving along its surface. Archive of Applied Mechanics, **69**(1), 55–67.

Jones, D.V. and Petyt, M. (1986) Ground borne vibrations from passing trains: the effect of masses placed on the ground's surface. ISVR Technical Memorandum, No. 671, University of Southampton, 1986.

Krylov, V.V. (1995) Generation of ground vibrations by superfast trains. *Applied Acoustics*, **44**, 149–164.

Krylov, V.V. (1996) Generation of ground vibrations by accelerating and braking road vehicles. *Acustica-acta acustica*, **82**, No 4, 642–649.

Krylov, V.V. (1998) Effect of track properties on ground vibrations generated by high-speed trains. *Acustica-acta acustica*, **84**, No 1, 78–90.

Krylov, V.V. (2001a) Generation of ground elastic waves by road vehicles, *Journal of Computational Acoustics*, **9**(3), 919–933.

Krylov, V.V. (2001b) Generation of ground vibration boom by high-speed trains. In: *Noise and Vibration from High-Speed Trains*, Ed. V.V. Krylov, Thomas Telford Publishing, London, p. 251–283.

Krylov, V.V., Dawson, A.R., Heelis, M.E. and Collop, A.C. (2000) Rail movement and ground waves caused by high-speed trains approaching track-soil critical velocities. *Proc. Instn. Mech. Engrs., Part F: Journ. Rail and Rapid Transit*, **214**, 107–116.

Madshus, C. and Kaynia, A.M. (1998) High speed railway lines on soft ground: dynamic behaviour at critical train speed, *Proc. 6th International Workshop on Railway and Tracked Transit System Noise*, Ile des Embiez, France, 108–119.

Maradudin, A.A., Ryan, P. and McGurn, A.R. (1988) Shear horizontal acoustic shape resonances. *Physical Review B*, **38**, 3068–3074.

Mayer (Garova), E.A. (1994) Reflection of surface acoustic waves from strong single inhomogeneities and periodic structures, PhD Thesis, Moscow (in Russian).

Plessky, V.P. and Simonian, A.W. (1991) Rayleigh wave reflection and scattering on a resonator. *Physics Letters A*, **155**, 281–284.

Sheng, X., Jones, C.J.C. and Petyt, M. (1999) Ground vibrations generated by a load moving along a railway track. *Journal of Sound and Vibration*, **228**(1), 129–156.

Sheng, X., Jones, C.J.C. and Thompson, D.J. (2003) A comparison of a theoretical model for quasi-statically and dynamically induced environmental vibration from trains with measurements. *Journal of Sound and Vibration*, **267**(3), 621–635.

Takemiya, H. (2001) Ground vibration alongside tracks induced by high-speed trains: prediction and mitigation. In: *Noise and Vibration from High-Speed Trains*, Ed. V.V. Krylov, Thomas Telford Publishing, London, p. 347–393.

Watts, G.R. and Krylov, V.V. (2000) Ground-borne vibration generated by vehicles crossing road humps and speed control cushions. *Applied Acoustics*, **59**, No 3, 221–236.

Environmental Vibrations – Takemiya (ed.)
© *2005 Taylor & Francis Group, London, ISBN 0 415 39035 4*

Prediction of train/traffic induced ground vibrations and mitigation by WIB

H. Takemiya

Department of Environmental and Civil Engineering, Okayama University, Okayama

ABSTRACT: This paper, based on the authors publications, describes the prediction methods of traffic induced vibrations and the mitigation procedures. Special interest is focused on the so-called low frequency vibrations. First, the prediction method is presented in view of the wave propagation in ground from traffic. The dispersive nature in the layered soils is detected for identifying the wave predominance. Then, by using thus obtained knowledge, the wave impeding barriers (WIB) are developed. The basic idea is to shift the cut-off frequency for wave propagation in field. Herein, illustrations are demonstrated for the honeycomb shaped WIB when considered around the viaduct pier or on field off the viaduct. An innovative active control procedure is also briefly introduced.

1 INTRODUCTION

In the last decades, the ground vibrations in the vicinity of the train track grow significantly with increasing train speed, bringing disturbances to the residence area and high-tech production facilities alongside the train/traffic lines. The environmental problems caused by the emitted vibrations from high-speed train or highway traffic have aroused the more concern of people.

Viaducts are widely used for supporting road/railway structures, especially for those that run through urban area or soft soil deposited area. The Japan Shinkansen viaduct in Fig. 1 and a highway viaduct in Fig. 2 are mainly focused in this work. Both structures are supported by substructures, Caisson or grouped piles. For vibration assessment of the alongside ground vibrations, the vehicle-structure, the structure-ground interaction and the wave propagation in ground from the foundation positions should be taken into account. The resultant wave field should be interpreted according to the vehicle movement. This is in very contrast to a continuous loading along vehicle motion for the direct on-ground tracks.

In order to predict traffic induced vibrations, the information on the source properties, wave propagation characteristics are essential. The structural and ground borne vibrations are crucial as well, since they are mostly in low frequencies that affects. The evaluation should be made not only in terms of response amplitude but also in terms of frequency contents. Especially, in the case of low frequency vibrations, there are vibration claims, even though the levels are below the specified levels. This low frequency vibration problem is therefore targeted in this study.

In what follows, the author discusses the train/traffic induced vibrations from the prediction method to mitigation procedures, with the main model for analysis of the viaducts illustrated in Fig. 1 and Fig. 2. First, regarding the wave propagation in ground, the site characteristics are mainly referred to with respect to the soil layering. This information is essential not only for predicting the train/traffic induced vibrations and for developing the mitigation measures. Applications of

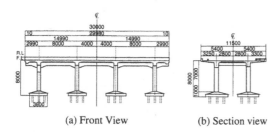

Figure 1. Geometry of Japan Shinkansen viaduct unit.

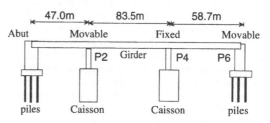

Figure 2. Geometry of highway viaduct.

the basic theories lead to the effective solutions to the practical problems.

The details of the individual issues are referred to the respective publications listed in References.

2 SITE CHARACTERISTICS

2.1 Wave propagation theory

Actual grounds constitute layered soils. Under train/traffic loads, the shear wave velocity contrast of the softer layer lying on the stiffer layers generates the surface waves: the generalized Rayleigh wave or the generalized Love wave whose wavelengths depend on frequency. The wavelength λ is expressed by

$$c = f\lambda \qquad (1)$$

where c = the wave velocity and f = the frequency. The group velocity is a more appropriate indicator for the dispersive wave propagation. Given a driving frequency f_0, then the frequency vs. wavenumber relationship is described by

$$f = ck + f_0 \qquad (2)$$

The minimum group velocity defines the Airy phase that corresponds to the gravity center of wave energy transmission. Therefore, by this velocity, the most expected wave is predictable. If the frequency-wavenumber relation is available, then the group velocity is assessed as

$$c_g = 2\pi \frac{df}{dk} \simeq 2\pi \frac{\Delta f}{\Delta k} \qquad (3)$$

where Δf, Δk denote respectively increments of frequency and wavenumber.

2.2 Interpretation of field data

In order to investigate the nearby wave field of the viaduct under the train passage, the spectral analysis is carried out by the well-known SASW method. (Takemiya, 2004a, Takemiya, et al., 2004b). The results are drawn in Fig. 3 by triangle symbols. The trend that the phase velocity strongly depends on the focused frequency in the low frequency range below 30 Hz gives an evidence of the dispersive wave nature in this range. Supplemental field vibration measurements have been conducted also. The results by the guided hamper impact tests for higher frequency range information are indicated by solid circles. The ambient measurements are plotted to supplement the very low frequency range information by solid squares in the same figure. In

the very low frequency range below several Hz, the associated wavelength becomes several ten meters and more. Here noteworthy is the trend that the Shinkansen train induced ground vibrations, showing a transition of wave speed by the frequency, is different from that of the ambient motions where only low wave speed dominates in the frequency range below 15 Hz. The impact-induced vibrations, on the other hand, show consistency with those of the Shinkansen induced vibrations in the range above 10 Hz.

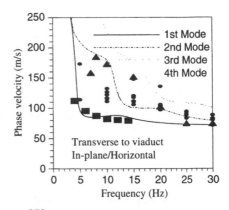

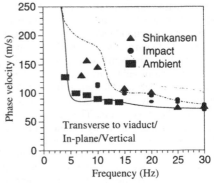

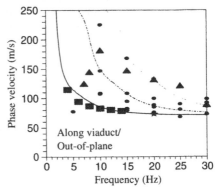

Figure 3. Wave dispersion characteristics with phase velocities vs. frequencies.

2.3 Thin layer method prediction

For the mathematical prediction the thin layer method, in which the finite element discretization is introduced along depth is applied for the wave field. In view of the long wavelengths in low frequency range, an extended base layer should mostly be taken into account below the surface layers for the numerical computation model. The eigenvalue solution for wavenumbers derives the wave velocities for given frequencies. The resultant wave modes curves are depicted in Fig. 3 for the in-plane as well as for the out-of-plane motions. They are termed the 1st, 2nd, 3rd and the 4th mode in the order of wavenumbers from small to large values. The graphs indicated as "In-plane/Horizontal" is obtained from the horizontal response component in the transversal direction to the viaduct and those as "In-plane/Vertical" is from the associated vertical response component. The graphs indicated as "Out-of-plane" is obtained from the horizontal response component along the viaduct axis. The in-plane motion is herein related largely to the Rayleigh wave propagation, whereas the out-of-plane motion to the Love wave propagation. Under prescribed frequencies the wave speeds are estimated from the phase difference of the measured points. With reference to these modes characteristics, it is interesting to note that in the frequency range below 10 Hz the measured in-plane data from the Shinkansen train fall on the 1st mode to the 2nd mode curves with increasing frequency whereas those from the ambient motions stay on the 1st mode only. The out-of-plane motions indicate the higher modes are concerned more.

In order to examine the wave characteristics at the site more in detail, we computed the phase-velocity vs. wavenumber curves, and the group-velocity vs. frequency relationship. They are shown in Fig. 4. The quotient of the phase velocity against the wavenumber gives the velocity unit so that the train speed lines are drawn in the same figure, as starting from the origin

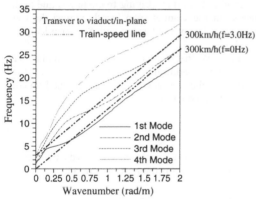

(a) In-plane motion in the transverse section to viaduct

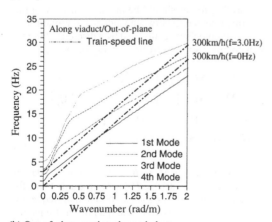

(b) Out-of-plane motion along viaduct

Figure 4.   Phase velocity.

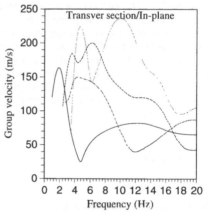

(a) In-plane motion in the transverse section to viaduct

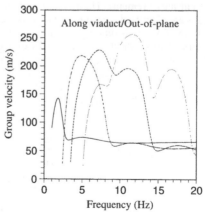

(b) Out-of-plane motion along viaduct

Figure 5.   Phase velocity and group velocity.

or from 3 Hz at zero wavenumber. The first speed line corresponds to a smooth moving load condition and the latter speed line to a moving load accompanied by a harmonic oscillation of 3 Hz. This frequency is chosen to represent a predominant driving frequency from the finding for the driving frequency under the train passage. Crossings of these train speed lines with the modal characteristic curves indicate the significantly amplified frequencies of the emitted vibration from the viaduct foundations. These are read off in Fig. 5 as 7 Hz for the 1st mode and 13 Hz for the 2nd mode in the in-plane motions while 10 Hz for the 1st mode and 15 Hz for the 2nd mode in the out-of-plane motions. The Airy phase, given as the local minimum in the group velocity vs. wavenumber relationship, on the other hand, provides the information of the ground-borne vibration while propagating. The 1st mode for the in-plane motion indicates a clear-cut Airy phase at 4.5 Hz while the corresponding one at 3 Hz for the out-of-plane motion. For the 2nd mode they are respectively 12 Hz and 9 Hz. Hence, we derive the following prediction. For the in-plane motion the 1st mode dominates the response in the frequency range below several Hz but the 2nd mode does for the higher frequency than it. The same trend holds for the out-of-plane motion. The critical frequency however is lower for the out-of-plane motion than the in-plane motion. The above statements are evidenced in Fig. 3 from the measurements.

3 PREDICTION METHOD

3.1 *Moving source loads*

Moving train induces vibrations according to the train geometry and sleepers distance against the speed (Takemiya, 2003). This is verified for high-speed trains by field measurements. The predominant frequency peaks are identified accordingly. For instance, suppose a train of N cars is running with the speed c. The geometry of axle wheels in individual cars are described by the distance a and b for the axle load span length of a bogie and the shortest distance of axle loads of the neighboring bodies in one car. Then, the loading function by these axle loads are formulated in the time domain as

$$f_n(x-a) = P_{n1}\delta(x-a+\sum_{s=1}^{n-1}L_s+L_0) + P_{n1}\delta(x-a+a_n+\sum_{s=1}^{n-1}L_s+L_0)$$
$$+P_{n2}\delta(x-a+a_n+b_n+\sum_{s=1}^{n-1}L_s+L_0) + P_{n2}\delta(x-a+2a_n+b_n+\sum_{s=1}^{n-1}L_s+L_0)$$

$$(4)$$

Where L_0 = the distance from the first axle to a focused position, L_n = individual car length, and $\delta(\cdot)$

denotes the Dirac's delta function. The Fourier transform in the frequency domain is given as

$$f'_N(\omega,x) = \sum_{n=1}^{N} f'_n(\omega) =$$

$$\sum_{n=1}^{N} \frac{1}{c} e^{i(\sum_{s=1}^{n-1}L_s + L_0 + x)} \left\{ P_{n1}(1+e^{i a_n \omega/c}) + P_{n2}(e^{i(a_n+b_n)\omega/c} + e^{i(2a_n+b_n)\omega/c}) \right\}$$

$$(5)$$

The Eq.(5) is a deterministic description for the corresponding source forces.

The Japan Shinkansen train, Nozomi, comprises 16 cars whose axle forces distribution in space is specified by the multiplies of 2.5 m for the axle distance in a bogy. Each car has same axle weights $P = 150$ kN. The moving speed of $c = 300$ km/h is assumed for the present analysis. The car length is 25 m, which gives the main driving frequency 3.3 Hz and the frequencies of its multiplies. It is very clear that the train geometry, especially the axle arrangement, concerns the predominant peaks.

Another important driving frequency is due to the inverse of the viaduct span length against the train speed. For Japan Shinkansen with 8 m span length is 10.4 Hz.

In the case that the viaduct is curved so that the lateral loading condition associated with the centrifuge action for the vertical axle loads are additionally taken into account by

$$F_h = F_v c^2 / gR \qquad (6)$$

in which F_v is the vertical axle load, c is the train speed, R is a radius and g is the gravity constant.

3.2 *Driving forces on foundation*

The driving forces should be obtained in view of the dynamic interaction between viaduct and ground. The viaduct should first be modeled by a discretized model with complex valued stiffness spring supports.

$$-\omega^2 \begin{bmatrix} \mathbf{M}_{ss} & \mathbf{M}_{si} \\ \mathbf{M}_{is} & \mathbf{M}_{ii} \end{bmatrix} \begin{Bmatrix} \mathbf{U}'_s \\ \mathbf{U}'_i \end{Bmatrix} + i\omega \begin{bmatrix} \mathbf{C}_{ss} & \mathbf{C}_{si} \\ \mathbf{C}_{is} & \mathbf{C}_{ii} \end{bmatrix} \begin{Bmatrix} \mathbf{U}'_s \\ \mathbf{U}'_i \end{Bmatrix}$$
$$+ \begin{bmatrix} \mathbf{K}_{ss} & \mathbf{K}_{si} \\ \mathbf{K}_{is} & \mathbf{K}_{ii} + \mathbf{K}_f(\omega) \end{bmatrix} \begin{Bmatrix} \mathbf{U}'_s \\ \mathbf{U}'_i \end{Bmatrix} = \begin{Bmatrix} \mathbf{F}'_s \\ 0 \end{Bmatrix} \qquad (7)$$

where the notations M, C, K refer to the mass, damping and stiffness matrices; F defines the axle loads for a moving trains, and ω stands for the circular frequency. The superscript $(\ )'$ denotes the quantity in frequency transformed domain. By solving the governing of such interaction system, we can get internal forces at the foundation top directly.

402

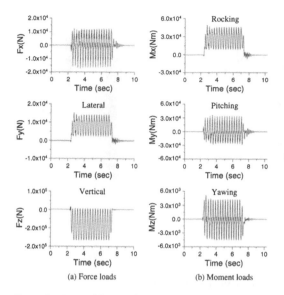

(a) Force loads   (b) Moment loads

Figure 6.   Internal force actions on the footing top.

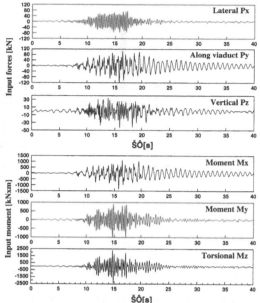

Figure 7.   Driving forces at foundation top for a test run by a track (20 tf); P4: Caisson.

Figure 6 is the thus obtained internal forces for the Shinkansen viaduct in Fig. 1. Since the viaduct has a curvature 1/4000 so that the corresponding centrifuge lateral force is additionally considered to the vertical axle loads. The associated moments are accordingly taken into account.

3.3   Semi-empirical evaluation

On the other hand, traffic loads on highways are very random in vehicle weights and flows. The source loadings should be interpreted in random theory.

Alternatively, by using the foundation response (displacement), the internal forces at the foundations are obtained indirectly (Takemiya, 2004). In the case of road traffic, the loading condition has a very random nature so that the mathematical prediction should be made in the random vibration theory. For one event, a half analytical method can be taken, if the measurement data are available at a specific point of the foundation in the 6-degrees of freedom provided that the pier is taken as a rigid body (Takemiya, et al., 2005). The associated internal forces motions in the frequency domain are obtained as the product of the displacement and the soil stiffness. Namely,

$$P_0 e^{i\omega t} = K(\omega) U_0 e^{i\omega t} \tag{8}$$

Thus obtained forces are emitted from the viaduct foundation into ground. Since the measurement data are obtained in time domain for a specific loading condition, Eq.(8) can be converted into the time domain by Fourier transform. Fig. 7 represents the results for a test run by using a track (20 tf) run for the important

foundations P4 in Fig. 2. The internal force actions reflect the vibration characteristic of the viaduct. The actions along the viaduct axis have the lower frequencies that than those in the transverse section of the viaduct. The horizontal forces are almost the same amplitude along the viaduct and the transverse direction to it and they are important for the consequent ground motion prediction.

The low frequency vibration emission into the ground has a high possibility of generating the low frequency ground vibrations in the neighborhood.

3.4   Induced ground vibrations

The ground response due to the emitting vibration from foundations are computed for individual foundations concerned. Since a multiple of foundations support a viaduct structure, the wave emission from them should be taken into account. The wave superposition should therefore be conducted with due consideration for their phase effect. For this computation, the FEM is used to advantage for the complex boundary condition involved. In order to take into account of the extending far field, the transmitting boundary is employed effectively. The ground response is expressed as

$$U_0 e^{i\omega t} = F(\omega) P_0 e^{i\omega t} \tag{9}$$

where $F(\omega)$ denotes the flexibility matrix of the surrounding ground.

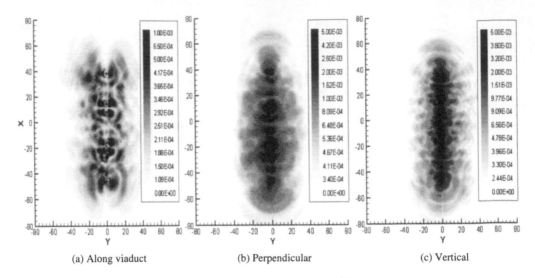

| (a) Along viaduct | (b) Perpendicular | (c) Vertical |

Figure 8.   Contour maps for the maximum velocity responses on ground surface for two unit viaducts.

An illustrative example is demonstrated for the Shinkansen viaduct in Fig. 1. Since the train speed is assumed a constant, the consequent wave field is stationary. Fig. 8 is the contour maps for the maximum velocity responses on the nearby ground surface for two unit viaducts. It is noted that the substantial intensity response is limited in the narrow area along the viaduct. The motion along viaduct indicates the wave interferences among those emitted from involved foundations in a certainly determined space and time coordinates along the viaduct. The present two viaduct units proved sufficient for the ground motion prediction.

The maximum velocity responses from the inverse Fourier transform are depicted in Fig. 9. In order to validate the analysis method and to choose appropriate involved parameters, the field test data of maximum velocities at the measurement site are plotted. The train passages are considered either on near-side or on far-side track to the measurement points, as described separately in Fig. 9. For both cases, the horizontal loads acting on the viaduct girder exist because of viaduct's curvature. It is clearly noted that when the Shinkansen train runs on the near-side track to the focused points there are more intensified motions than those on the far track. The matching is good between the computation and the field measurement for the maximum vibration velocities, especially regarding the vertical component that shows a monotonically decreasing trend with the distance from the viaduct.

The maximum acceleration level (VAL) is commonly used in Japan for the environmental legislation. It is defined as

$$VAL = 20 \, log\left(A / A_0\right) \text{ with } A_0 = 10^{-5} m/^2 \qquad (10)$$

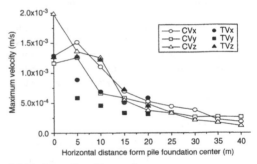

(a) Train on far-side track

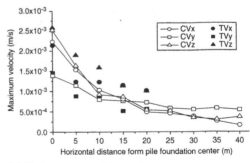

(b) Train on near-side track

Figure 9.   Maximum velocity responses of ground surface. (Prefix 'T' for field test data, 'C' for computation results.)

The VAL presentation is indicated in Fig. 10 for a traffic flow on the viaduct in Fig. 2 by using the force action in Fig. 7. The vibration sources are assumed at the foundations P4 and P6. The vibrations are simply

404

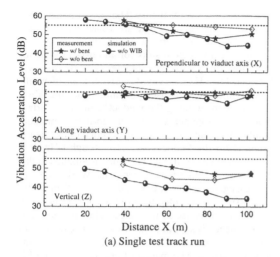

(a) Single test track run

Figure 10. Comparison between prediction and measurement.

summed up vectorically from these foundation-ground systems for both P4 and P6. The comparison between the field measurement and prediction has better agreement for the component along the viaduct direction than other components. The former response is larger than others. Because of the significant environmental vibration problem to the neighborhood, the viaduct is now supported by the bents softly at the middle section of each span. The measurements have been conducted therefore with and without these bents. The computer simulation has been carried out for the latter situation.

4 VIBRATION MITIGATION MEASURE WIB

4.1 Concept

The author has been developing a variety of Wave Impeding Barriers (WIB). (Takemiya et al., 1994; 1998; 1996; 2004; 2005). Although these WIB are different in appearance, the basic idea is to shift the cut-off frequency of the wave field higher range that reduces the vibration amplitudes in a targeted frequency range. This is illustrated in Fig. 11.

4.2 Passive type WIBs

A stratum model underlain by a rigid bottom has a cut-off frequency for wave propagation (Takemiya & Fujiwara, 1994, Takemiya & Goda, 1998). The vibration transmission in such a layer is interpreted in the normalized quantities as indicated for the SH wave propagation, in which f = frequency, D = depth and Vs = shear wave velocity. The wave impeding effect is read off from Fig. 12 in the normalized frequency range of

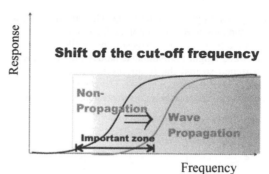

Figure 11. Cut off frequency against wave propagation.

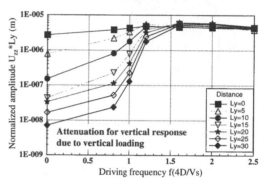

Figure 12. Normalized ground response of a soil stratum.

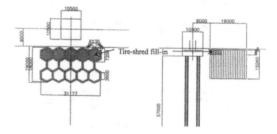

Figure 13. Honeycomb-cell WIB on wave field.

$$f \frac{4D}{V_s} < 1.5 \qquad (11)$$

Given a site of the shear velocity and the focused frequency, the wave impediment is put into effect as the bottom depth is shortened.

Another approach is to use a cluster of cells, among them honeycomb-shapes are most effective stable structure. These cell-type WIBs as illustrated on the wave propagation field in Fig. 13 (Takemiya, 2004) or in Fig. 14 positioned around the foundation that emits vibrations (Takemiya and Shimabuku, 2005). The basic idea lies in the stiffness contrast between cell-walls

405

and ground dynamic interaction. The fill-in stiffness and damping have also important factors. In the course of wave propagation across stiffened cells, the original wave length is modulated, namely, disturbed and shortened so that the frequency is shifted to higher range, as illustrated in Fig. 11. The criteria for honeycomb cell size and their construction area are derived

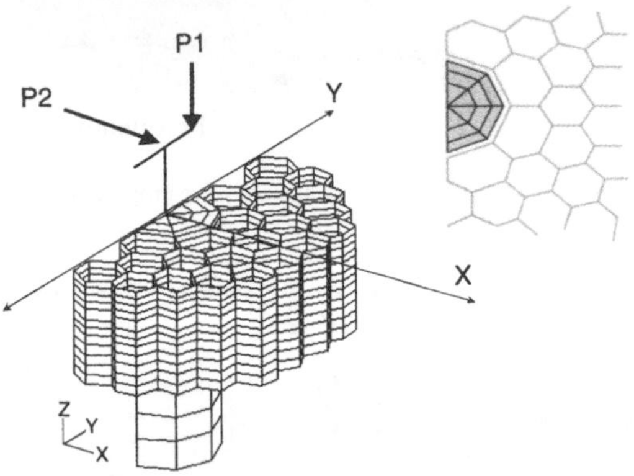

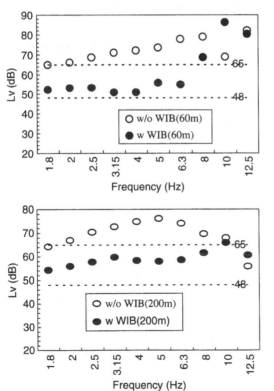

Figure 14. Honeycomb-cell WIB around vibration emitting foundation.

from the wave cut-off theory. The size of honeycomb cells and construction area are based on the wave field knowledge (Takemiya, et al., 2004).

The Fig. 15 is the results for an on-field WIB for vibration mitigation for targeted low frequencies between 3 to 8 Hz. The significant reduction of amplitudes is attained at focused positions. Since the vibration component along the viaduct is more important than other components, so that the former component is intensively investigated. The design parameters are oriented for the frequencies 3 Hz through 6 Hz. It is clearly noted a significant reduction is attained in this range for the distances beyond the WIB zone. An impressive reduction rate is read off more than 10 dB. The WIB effect is noted to shift the cut-off frequency so that the response in the range below the new cut-off frequency is subdued significantly. The degrees of reduction depend on the focused position on the ground surface. The average trend indicates the SH wave propagation is more subdued than the P and SV waves. This trend is interpreted physically from the wave theory in Section 2.

The Fig. 16 is the results for the WIB in Fig. 14. In this model, the different wave field is analyzed for the SH wave field and the P and SV wave field. In order

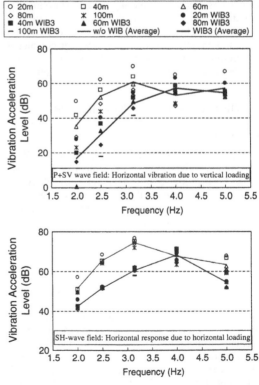

Figure 15. Response reduction by WIB on wave field.

Figure 16. Response reduction by surrounding WIB.

to substantiate the above change of wave field, Fig.16(a) and (b) are prepared by plotting the velocity amplitudes at different distances from the loaded foundation for the indicated frequencies. It is clearly seen the shift of the cut-off frequency toward higher frequency and the response reduction is more gained for the SH wave field than for the P and SV wave field. From these figures, the zone for wave propagation is clearly noted different by different loading conditions. The wave field is generated according to the loading direction. For the eccentric vertical loading, due to the rocking of the foundation, the wave emits and propagates toward the bridge axis direction supposedly by the SH wave. The most significant intensity appears around the 3 Hz driving frequency, which corresponds to the SH wave cut-off frequency from Fig. 5. The concentric 3-layer honeycomb WIB works significantly for response reduction in the indicated frequencies, most effectively below 5 Hz. The wave field due to the horizontal loading, the wave emits and propagated toward the normal direction to the loading direction, most significantly at the 3.15 Hz. This is also supposed to be the SH wave field. The honeycomb WIB subdues this wave field greatly.

4.3 Active type WIB

Just recently, the author proposed an innovative idea that use additional vibration source for producing tranquilized area for the wave filed against induced by source loading (Takemiya et al., 2005). This is an active vibration control procedure. Preliminary theoretical investigation is conducted for the harmonic wave field due to a vertical surface loading. An effective wave canceling procedure is to use a pair of shear force action on the wave field. The 3-dimensional computation result is demonstrated in Fig. 17 in which an apparent shift of the cut-off frequency is noted significantly. The reduction mechanism is proved by the field experiment.

A field test has been conducted in order to verify the shear loading by a buried rigid plate. A source shaker is used for the source loading while a RC plate is loaded by an actuator on top of it. Harmonic force action is produced at fixed frequencies. The same frequency harmonic load is applied but with a different phase that is controlled to minimize the total response at a targeted point on the ground surface behind the plate. Fig. 18 is the result for a preliminary simulation by using 3D FEM. The predominant fundamental

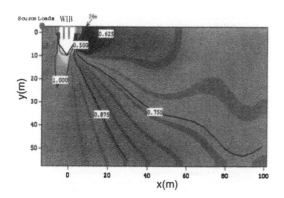

Figure 17.   3-D simulation for active controlled field.

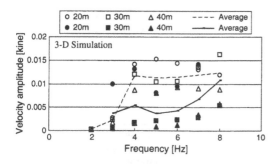

Figure 18.   Response reduction by active control.
Open symbols: before active control
Solid symbols: with active control

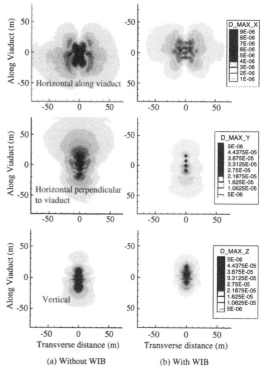

(a) Without WIB   (b) With WIB

Figure 19.   Response reduction by WIB of ground surface velocities for one-unit viaduct (only vertical axle loading is assumed).

wave mode is subdued by the wave interference, resulting in an apparent shift of the cut-off frequency to the second wave mode. This is validated from the field experiment with detailed interpretation (Takemiya, et al., 2005d) from the wave theory.

5 FEASIBLE PRACTICAL APPLICATIONS

5.1 *High-speed Shinkansen viaduct case*

For the purpose of vibration mitigation, especially in low frequencies, Takemiya (2005) proposed a cell-type wave-impeding barrier (WIB). An illustrative

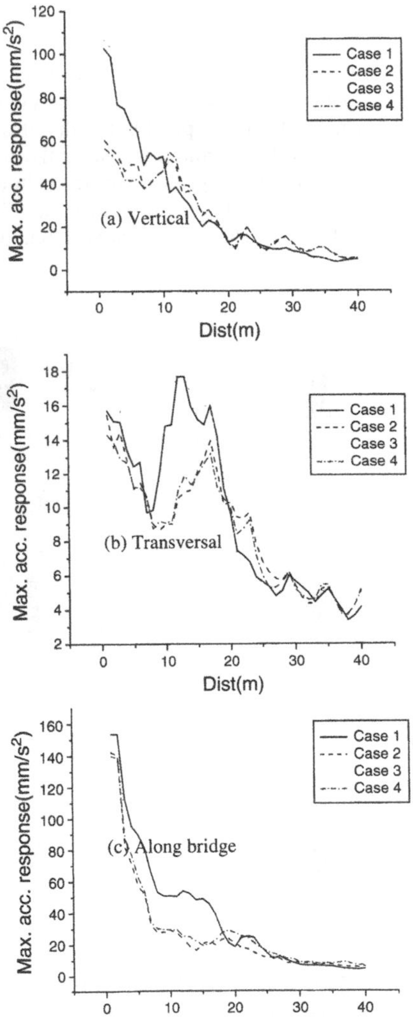

Figure 20. Maximum acceleration, Nozomi, near track 300 km/h.

demonstration is given here by approximation by taking the ring-type model for Fig. 14. The present computation is limited to a single viaduct unit. Fig. 19 is depicted for the ground surface. The comparison with and without WIB indicates a significant reduction by WIB within 25 m horizontal extent from the wave emitting foundations.

For the actual Shinkansen passage on viaduct, a multiple foundations emit the vibrations according to the train system and the speed in which the three-span continuous viaduct units concern dynamically. Then, the wave superposition occurs as the distance for observation is increased. Under a constant speed, however, a stationary assumption holds in space and time. In order to prove the vibration mitigation by the honeycomb-WIB, herein three units are taken into consideration by connecting them by 1/10 stiffness of the transverse section. The results are shown in Fig. 20 for the ground acceleration. The Case 1 indicates the situation with no measures, the Case 2 with the WIB but approximated by a concentric rings around individuation foundations. Also the model connecting adjacent viaduct units by a cross beam is considered as the Case 3 and the combined use of WIB as the Case 4.

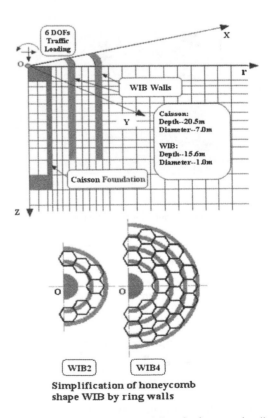

Figure 21. Axisymmetric modeling for honeycomb-cell WIB.

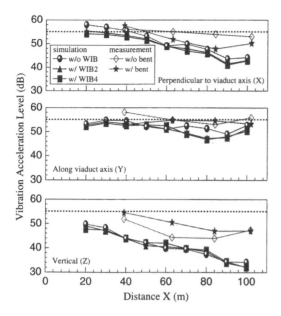

Figure 22. Maximum acceleration response for normal traffic.

5.2 *Highway traffic viaduct case*

A highway traffic viaduct is analyzed (Takemiya, Chen & Ida, 2005c). The measurement data at the foundations are utilized for semi-analytical prediction for the foundation driving forces. The computation results are validated from field measurements of the test run by using a heavy track. Fig. 21 illustrates an approximate model for Fig. 14 by the axisymmetric walls around the caisson foundation at P4 pier. Fig. 22 shows the computation results that demonstrates the mitigation effect for the major horizontal motions. However, this model results in less effective results when compared with the original honeycomb WIB in previous section.

6. CONCLUSION

A prediction method for the soil-foundation-superstructure dynamic interaction has been developed to simulate the viaduct and ground vibrations due to the train running on viaduct structures. From the illustrative case studies for the Shinkansen passage, we derived that:

(1) Among the responses on ground surface the vertical component is most predominant in near field, while the horizontal components become more important than the vertical one with the distance increases from the viaduct, namely the one along the viaduct and also the one perpendicular to it in the curved viaduct.

(2) The soil-pile foundation-structure interaction (DSSI) effect is proved to be crucial for accurately determining the viaduct vibrations and the nearby ground vibrations.

(3) The ring-type WIB for the vibration mitigation is proved to be effective for low frequency vibrations.

REFERENCES

Takemiya, H. 1985. Three-dimensional seismic analysis for soil-foundation superstructure based on dynamic substructure method. Proc. JSCE; Vol.3, No.1:139–149.

Takemiya, H. 1986. Ring pile analysis for grouped piles subjected base motion, Structural Eng./Earthquake Eng; Vol.3, No.1 JSCE:195–202.

Takemiya, H. and Fujiwara, A. 1994. Wave propagation/impediment in a stratum and wave impeding block(WIB) measured for SSI response reduction, Soil Dynamics and Earthquake Engineering, 13, 49–61.

Takemiya, H., Goda, K. and Sato, N., Response reduction effect by passive use of w wave impeding block (WIB), Proc. Japan Society of Civil Engineers, 549/I-37, 221–230.

Takemiya, H. Kellezi, L. 1998. Paraseismic behavior of wave impeding barrier (WIB) measured for ground vibration reduction, 10. Japan Earthquake Engineering Symposium, E3-13, 1879–1884.

Takemiya, H. 2002. Controlling of track-ground vibration due to highspeed train by WIB Eurodyn 2002, Lisse: Swets & Zeitlinger: 497–502.

Takemiya, H. 2003. Simulation of track-ground vibrations due to a high-speed train: the case of X-2000 at Ledsgard, Journal of Sound and Vibration 261, 503–526.

Takemiya, H. 2004a. Vibration transmission in ground of Shinkansen train vibration from viaduct, Proc. Railway Mechanics, JSCE, 8, 53–58.

Takemiya, H. Bian, X.C. Yamamoto, K. and Asayama, T. 2004b. High-speed train induced ground vibration: Transmission and mitigation for viaduct case., IWRN8:107–118.

Takemiya, H. et al. 2004. Patent application.

Takemiya, H. 2005a. Field vibration mitigation by honeycomb WIB for pile foundations of a high-speed train viaduct, Soil Dynamics and Earthquake Engineering, 24, 69–87.

Takemiya, H. and Shimabuku, J. 2005b. Honeycomb-WIB for mitigation of traffic-induced ground vibrations, Proc. ISEV2005.

Takemiya, H. Chen, F. and Ida, K. 2005c (Submitted). Traffic-induced vibration from viaduct: Hybrid procedure of measurement and FEM analysis for prediction and mitigation of ground motions. JSCE.

Takemiya, H. et al. 2005d. Environmental vibration control by active piezo-actuator system, Proc. ISEV2005.

Environmental Vibrations – Takemiya (ed.)
© 2005 Taylor & Francis Group, London, ISBN 0 415 39035 4

Honeycomb-WIB for mitigation of traffic-induced ground vibrations

H. Takemiya
Department of Environmental and Civil Engineering, Okayama University, Okayama, Japan

J. Shimabuku
Kozokekaku Kenkyuusho, Co. Ltd. Tokyo, Japan

ABSTRACT: The authors demonstrated that a honeycomb-WIB (Wave Impeding Barrier of honeycomb cells) works for vibration mitigation for the traffic induced vibrations from viaduct when constructed to surround the foundation, especially in the low frequency range. The thin layered method is taken to analyze the involved wave field to determine the size of the WIB cells and their area for construction. The interaction of structure-foundation-soil is analyzed by the 3D substructure method to show the WIB effect for the ground response reduction. The comparison with the buried wall for an alternative measure demonstrates the superiority of the honeycomb-WIB, clamming the reduction more than several dB at low frequencies.

1 INTRODUCTION

The traffic induced vibrations are increasingly affecting the built-in area alongside, causing disturbances and annoyances. These vibrations include mostly the low frequency nature from 3 to 8 Hz frequencies. The vibration legislation in Japan limits of vibration acceleration level (VAL) in the range of 60–65 dB during daytime while 55-d0 dB during nighttime. Protecting built-in areas from exposures to detrimental environment vibrations or vibration prolusion is strongly desired.

Some measures have been attempted for avoiding exposures to vibration sources in the built-in area. They are a row(s) of concrete piles or steel sheet-piles, open trenches of in-fill trenches with concrete or bentonite. These barriers aim to protect structures and facilities by creating a shadow zone. Recently, Takemiya (2004) developed a special cell-type WIB that can modulate the propagating waves and impeding them across their zone. They are classified all as passive type mechanism for reduction. The detail review is referred to the above paper.

In the previous paper (Takemiya, 2004), a honeycomb-type WIB is considered on the wave field off the structural foundations. During the wave propagation across the WIB zone, the frequency contents are modulated and because of the energy scattering, the wave amplitudes are reduces beyond that zone. The frequencies for mitigation can be determined priori by the knowledge of the free field waves.

This paper demonstrates the mitigation effect by use of the honeycomb-WIB when constructed around viaduct foundations. The size of the cells is based on the wavelength at site. The 3-dimensional wave field is taken into account with due consideration of the stiffness contrast of WIB walls and the fill-in and the shape effect for the vibration mitigation through wave modulation.

2 FINDINGS FROM FIELD MEASUREMENTS

2.1 *Time histories and the Fourier spectral densities*

The Fig. 1 is a targeted viaduct for investigation. The structural design is the three-span continuous girder

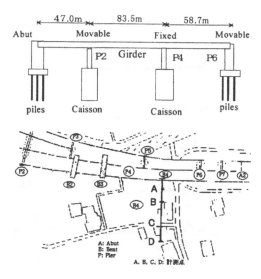

Figure 1. A traffic viaduct on soft ground.

bridge in which the connection to the pier P4 is rigid while the rest of the connections to piers are movable. Therefore, the motion of the viaduct along the axis is only resisted by the P4. The viaduct has a curvature of 1/800 so that beside of the vertical loading by traffic, the lateral loading results due to their centrifuge force action.

The measurements have been conducted while there is traffic flow. The velocity responses are depicted in Fig. 2 for the time histories for the normal traffic flow and in Fig. 3 for the Fourier spectral densities. They correspond to those at the distances of 7, 15, 35 and 65 m from the viaduct pier P4. They are indicated as A, B, C, and D in Fig. 1. It is noted that the motions are very random from a multiple cars on

the girder and the frequency contents differ at the ground surface with the distance from the viaduct. At the viaduct foundation (0 m) the major peak frequencies are observed around 3 and 5 Hz in every horizontal component. The ground vibration at long distance has the more dominancy at 3 Hz in the longitudinal while closer to 5 Hz in the transversal direction. The vertical vibration has wide spread frequencies below 10 Hz at the ground surface as well as at the foundation.

2.2 *Wave propagation characteristics*

The SASW (Spectral Analysis of Surface Waves) has been applied to get information on traffic induced

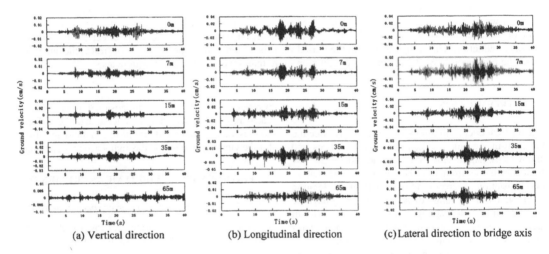

(a) Vertical direction   (b) Longitudinal direction   (c) Lateral direction to bridge axis

Figure 2. Time histories of response velocities.

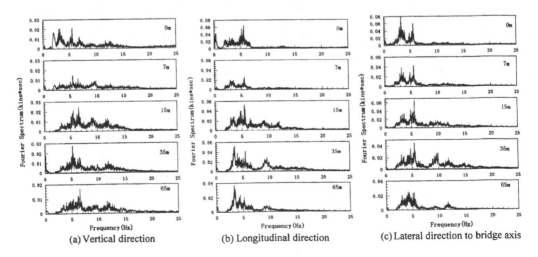

(a) Vertical direction   (b) Longitudinal direction   (c) Lateral direction to bridge axis

Figure 3. Fourier amplitudes envelope.

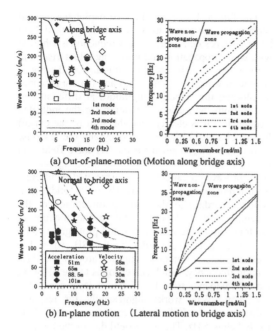

(a) Out-of-plane-motion (Motion along bridge axis)

(b) In-plane motion (Lateral motion to bridge axis)

Figure 4. Wave dispersion characteristics.

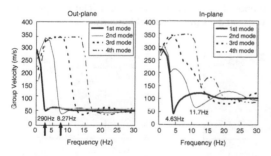

Figure 5. Group velocity va. frequency.

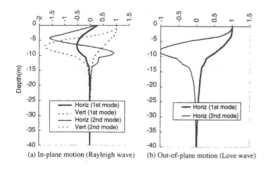

(a) In-plane motion (Rayleigh wave)   (b) Out-of-plane motion (Love wave)

Figure 6. Modal soil profile at Airy phase.

wave propagation. The results (given by open symbols) are depicted in Fig. 4 in which the data from the accelerograms measurement (solid symbols) are also added. The theoretical prediction is also taken by using the thin layer method. They are depicted for the first four wave modes. The dispersive nature of the wave propagation is clearly noted. In view of the predominant peak frequencies in the range of 3 to 6 Hz roughly, the modal contributions are significant from the first and second modes. For the 3 Hz predominance, the phase velocity is read off as around 170 m/s for the in-plane first mode motion and 150 m/s for the out-of-plane corresponding motion. The concerned wavelengths are then estimated respectively as 60 m/s and 50 m/s.

For the layered ground, the wave propagation is better interpreted in terms of the group velocity. This quantity indicates the speed of the wave energy transmission. This is given in Fig. 5. The modal shapes at the Airy phase frequencies are shown in Fig. 6. This confirms that the amplitudes of major ground motions are mostly in the depth 15 m.

3 3-D FEM MODEL ANALYSIS

3.1 Honeycomb-WIB

A cluster of hexagon cells or honeycomb-cells are arranged around the viaduct pier as illustrated in Fig. 7. Based on the field measurement data (Takemiya, Chen and Ida, 2005), assume that the viaduct motion under traffic is replaced by the force action at the pier top horizontally P_1 as well as P_2 vertically. From the field measurement these forces are almost equal in magnitude for the focused viaduct. Therefore, herein $P_1 = P_2 = 20$ tf are taken.

3.2 Computation model and results

The computation is carried out by using the 3-dimensional finite element method. The girder and pier parts are modeled by beam elements and pier and soils are solid elements and the honeycomb cells by shell elements.

The honeycomb-WIB is designed against the low frequencies at 5 Hz or below. This is the main vibration frequency as observed from the traffic in the neighborhood ground of the viaduct. The dimension of the WIB is determined that the depth is 13.6 m in view of the amplitude profile in depth and the soil layering at the site. The representative size of the hexagon cell is 3.5 m. The extent of the WIB is considered by the 3 cells in radial direction in view of one-third of the most important wavelength, as predicted 50 to 60 m at the site, from Fig. 4.

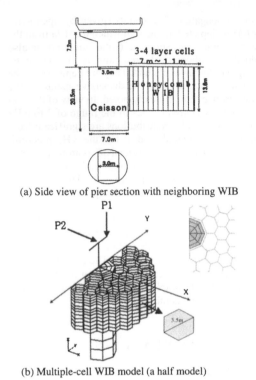

(a) Side view of pier section with neighboring WIB

(b) Multiple-cell WIB model (a half model)

Figure 7.   Pier and honeycomb-WIB.

The Fig. 8, Fig. 9 and Fig. 10 depict the ground surface response due to an eccentric vertical load P1 = 20 tf. A significant response reduction is achieved with wave propagation and impeding effect visible in the contour map. In order to substantiate the above change of wave field, Fig. 12(a) and (b) are prepared by plotting the velocity amplitudes at different distances from the loaded foundation for the indicated frequencies. It is clearly seen the shift of the cut-off frequency toward higher frequency and the response reduction is more gained for the SH wave field than for the P and SV wave field. From these figures, the zone for wave propagation is clearly noted as different by different loading conditions. The wave field is generated according to the loading direction. For the eccentric vertical loading, due to the rocking of the foundation, the wave emits and propagates along the bridge axis, supposedly by the SH wave. The most significant intensity appears around the 3 Hz driving frequency, which corresponds to the SH wave cut-off frequency from Fig. 5. The 3-layer concentric honeycomb-WIB works significantly for response reduction in the indicated frequencies, most effectively below 5 Hz. The wave field due to the horizontal loading along bridge axis, the wave emits and propagated toward the normal direction to the loading direction, most significantly at the 3.15 Hz. This is also supposed to be the SH wave field. The honeycomb-WIB subdues this wave field greatly.

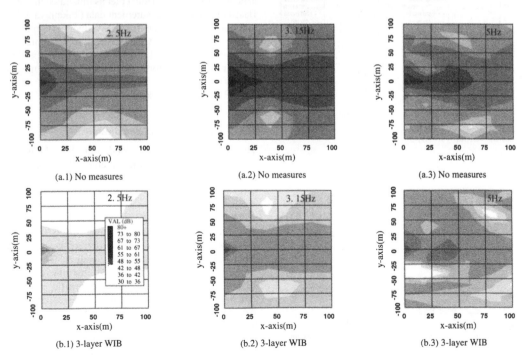

(a.1) No measures　　　　(a.2) No measures　　　　(a.3) No measures

(b.1) 3-layer WIB　　　　(b.2) 3-layer WIB　　　　(b.3) 3-layer WIB

Figure 8.   Ground surface response in lateral direction to bridge axis due to eccentric vertical force (P1 = 20 tf).

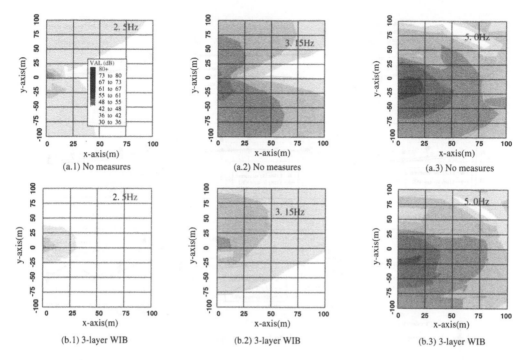

Figure 9. Ground surface response in vertical direction due to harmonic eccentric vertical load (P1 = 20 tf).

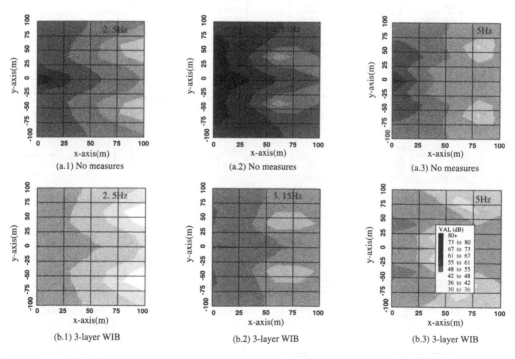

Figure 10. Ground surface response along bridge axis due to harmonic eccentric vertical load (P1 = 20 tf).

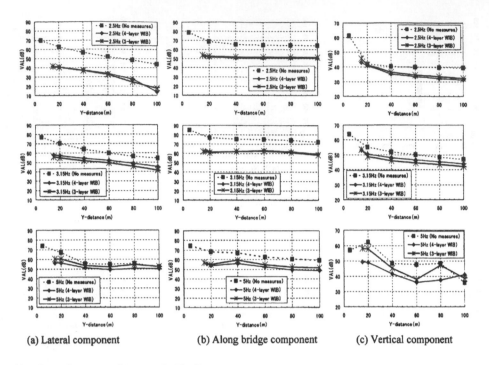

(a) Lateral component   (b) Along bridge component   (c) Vertical component

Figure 11.   Ground responses due to vertical loading.

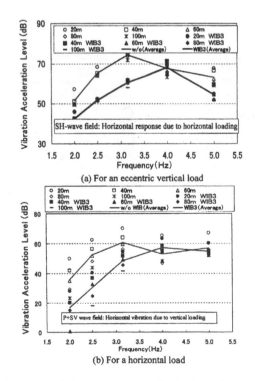

(a) For an eccentric vertical load

(b) For a horizontal load

Figure 12.   Velocity response amplitudes due to harmonic loading.

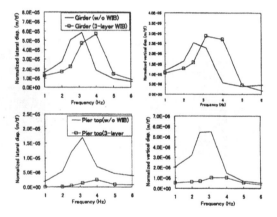

Figure 13.   Effects on viaduct pier response.

Lastly, the effect of installing of the honeycomb-WIB around the foundation on the viaduct pier response is investigated. Fig.13 shows the responses at the girder and at the pier top before and after the measure. Their comparison indicates the shift of the response peak frequency of the pier motions from 3 Hz to 4 Hz with higher response at the girder but greatly reduced response the pier top. This means that the WIB constrains the foundation motion but not to the degree to increase the constraint force leading to

416

greater energy emission into the soil. This is already convinced in Fig. 8 through Fig. 9.

4 CONCLUSIONS

This paper aims at developing an effective measure against the traffic induced low frequency vibrations. A honeycomb-type wave impeding barrier (WIB) is proposed around the vibration emitting foundation. In order to check its vibration reduction effect, a three dimensional soil-structure and wave field analyses have been conducted by the finite element method. The size of the WIB is properly determined based on the knowledge of the concerned wave field. From the computation results, the proposed WIB works for mitigation of the ground response more than 10 dB and most effectively for the SH wave at 3 Hz that is supposedly the most targeted environmental vibration along the viaduct in this study.

REFERENCES

Takemiya, H. 2004. Field vibration mitigation by honeycomb-WIB for pile foundations of a high-speed train viaduct, Soil Dynamics and Earthquake Engineering. 24, 69–87.

Takemiya, H. and Shimabuku, J. 2005. (Submitted). Honeycomb-WIB for mitigation of traffic-induced ground vibrations when constructed around foundation, JSCE.

Takemiya, H., Chen, F. and Ida, K. 2005. (Submitted). Traffic-induced vibration from viaduct. JSCE.

Environmental Vibrations – Takemiya (ed.)
© *2005 Taylor & Francis Group, London, ISBN 0 415 39035 4*

Passive/active WIB for mitigating traffic-induced vibrations

F. Chen & H. Takemiya
Okayama University, Okayama, Japan

ABSTRACT: This paper concerns developing a mitigation method against traffic-induced ground vibrations. First, a passive use of honeycomb shaped wave impeding barrier (WIB) is presented for vibration reduction for a caisson foundation of a traffic viaduct. Secondly, an active use of the WIB is proposed, which is driven to generate another wave. This procedure utilizes wave interference phenomenon to cancel out the source wave by another wave or lessen the total response peak values in the propagation zone. Since the apparatus that produces the counter wave needs a limited space so that it suits the urban built-up situation. However, the precise functioning/manipulation of the apparatus is required to get satisfactory results. The present computation work proves the fundamental information how the effective mitigation can be achieved by this actively controlled WIB.

1 INSTRUCTION

Environmental vibration is one of the modern pollutions that are being given more and more concerns. Traffic induced vibration is a typical kind of this category. In the last couple of decades, the development of transportation systems in most countries has encountered conflicts between the heavier traffic and the more conservation of built-up areas. Heavy traffic lines have annoyed the nearby residents and caused malfunctioning of sensitive instruments or detrimental effects on high-tech manufactures alongside. Effective measures to reduce such traffic-induced vibrations are strongly demanded.

Conventionally, there are some typical measures being widely used for this aim; among others open/filled trenches or concrete walls. The authors have proposed an innovative type of countermeasure for both ground response and seismic response reduction, i.e. the wave impeding barrier (WIB) (Takemiya 2004, Chen et al. 2004). The honeycomb shaped WIB is developed aiming at shifting the cut-off frequency of the wave field to a higher range so that the vibration amplitude can be reduced in a targeted frequency range (Takemiya 2005).

This paper focuses on a highway viaduct founded on soft ground. As shown in Figures 1–2, the viaduct consists of a 3-span girder and another 2-span girder. The major portion of the 3-span girder is founded on foundations A1, P2, P4 and P6, where the joint on P4 is fixed while others are movable along the girder axis. The heavy traffic flow on the viaduct generates vibrations on the girders, which are transferred to the

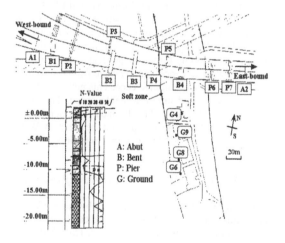

Figure 1. The site of field measurement.

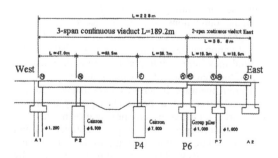

Figure 2. The 3-span viaduct of the study.

419

foundations and emitted to the surrounding built-up residential area, particularly in the soft belt zone as marked by the hatched zone in Figure 1. The interaction between the structures and the soil should be considered to analyze the vibration effect at site. The predominant low frequencies of the vibration are of great importance, which may cause resonance to the houses nearby. Presently, for avoiding it, some bent frames (indicated by B in Fig. 1) are used at several sections to shift the natural frequencies of the viaduct to the higher range.

Field experiments have been conducted for the viaduct by using a heavy truck and normal traffic flow. The test data enable the derivation of the driving force functions on the viaduct foundations, when the combined procedure is taken from the finite element method (FEM). The loading function is thereafter used to predict the ground responses resulting from the traffic.

Two different WIB measures are attempted in this paper for reducing the traffic-induced vibrations. One is the honeycomb shaped WIB that can be regarded as a passive use, termed as P-WIB. Its mitigation effect is first demonstrated by the 3-dimensional FEM simulation for the test-run truck loading and the normal traffic flows. The other is an application of actively controlled WIB, termed as A-WIB that makes use of a pairs of driving forces to generate another out-of-phase wave to cancel out the wave from the source loading by the wave interference. This new type WIB can satisfy limited urban available space. The field experiment of the A-WIB has been carried out recently (Takemiya 2005).The comparison between A-WIB and P-WIB is provided herein for discussion.

2 WAVE PROPAGATION CHARACTERISTICS

The features of the vibrations caused by the traffic on the viaduct and transmitted to ground can be changed as they travel far distance in the field. This shifted in frequency domain is due to the propagation characteristics of the stratified soil. The natural frequencies of the girders may lead to resonance of the vibration with the surrounding field when the waves emitted from the foundations have large frequency contents at the predominant frequencies of the ground. The investigation on the propagation characteristics of the ground may help to interpret the vibration phenomena of the field test and the simulation results.

Herein, the wave dispersion curves that describe the relation between wave speed and frequency are shown in Figure 3 for different wave propagation modes. The top figure corresponds to the in-plane motion of SV and P waves, while the bottom figure to the out-of-plane motion of SH wave. The computation is carried out by the thin layer method procedure. The surface soft layers

mainly determine the wave propagation characteristics. The ground measurement showed the predominant frequency of the vibration was about 3.7 Hz (Takemiya et al. 2005). For the same frequency of 3.7 Hz, the corresponding wave phase velocity of the first mode of the out-of-plane motion is about 130 m/s, and the wavelength is about 35 m. The curve of group velocities of the in-plane motion has a minimum at 4.46 Hz for the first mode, which implies that the ground motion around this frequency dominates the wave field (Airy phase). On the other hand, for the out-of-plane motion, the minimum group velocity of first mode is located at 2.9 Hz. Both the frequencies are close to 3.7 Hz, which may interpret the predominant ground motion.

The information on the wave propagation characteristics is requisite to determine a proper size of WIB for vibration mitigation. It is suggested that the horizontal extent of the WIB zone should be longer than the 1/3 of the important wavelength, with at least three honeycomb cells.

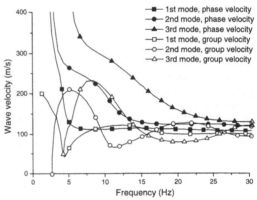

(a) Dispersion curves of in-plane motion

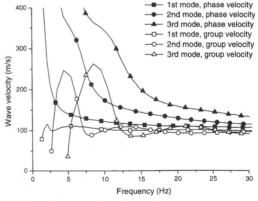

(b) Dispersion curves of out-plane motion

Figure 3.   Wave dispersion characteristics at site.

3 PASSIVELY USED WIB

3.1 *Traffic loading function*

Vibrations in the ground are directly influenced by the loading conditions at the pier bottom while the traffic loads are transferred from the superstructures. Since the analyses of the ground responses are the aim of this paper, we can substitute the effective loading function at the footing top for the traffic loads on the girder. In view of the relatively high rigidity of the pier and the footing, as well as small deformation in traffic situation, the structures are regarded as rigid bodies during the motions. Therefore, the deformation at footing top can be derived from measured data at the pier body.

We utilize the measured records at the top and near the bottom of piers to gain the response at the foundation top. First, we get the displacement by integrating the measured acceleration records by the formulation

$$U = -U(\omega)/\omega^2 \tag{1}$$

Then, from these displacement vectors we inversely obtain the associated response at the footing top by

$$\begin{Bmatrix} u_A(t) \\ u_B(t) \end{Bmatrix} = \begin{bmatrix} A_A \\ A_B \end{bmatrix} \begin{Bmatrix} u_O(t) \\ \theta_O(t) \end{Bmatrix} = A U_O(t) \tag{2}$$

where, $A_{6\times6}$ = geometric matrix of the rigid body condition of the pier.

The finite element formulation is taken to derive the foundation impedance functions with respect to the foundation degrees of freedom (DOF). The axisymmetric modeling for the approximate 3-dimensional behavior that can deal with anti-axisymmetric mode (A3D) is taken as an effective modeling and solution method. The illustration is given in Figure 4. The 6-DOFs behavior is described in the steady state harmonic motions as

$$P_O e^{i\omega t} = K(\omega) U_O e^{i\omega t} \tag{3}$$

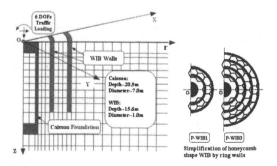

Figure 4.   Foundation-ground FEM model.

where $K(\omega)$ = soil impedance function matrix evaluated at the footing top 'O'; P_O = loading function vector with respect to this point. This defines the equivalent driving forces. Then, by using the compliance functions $F(\omega)$ of the foundation-soil system, we get the ground response by

$$U_g e^{i\omega t} = F(\omega) P_O e^{i\omega t} \tag{4}$$

In the case that the vibration emitted from a multiple foundations are considered, the superimposition should be made accordingly. Herein, only the contributions from foundations P4 and P6 are taken into account. Then,

$$U_g^t = U_g^{P4} + U_g^{P6} \tag{5}$$

The caisson foundation is modeled by solid elements except the cap portion that is replaced by a rigid body, while the grouped piles of P6 are discretized into ring-arranged beam elements.

Figure 5 depicts the obtained frequency dependent driving forces for a westbound truck test-run. It is noted that, as the results of the structural borne vibration at the viaduct, the predominant frequencies are different

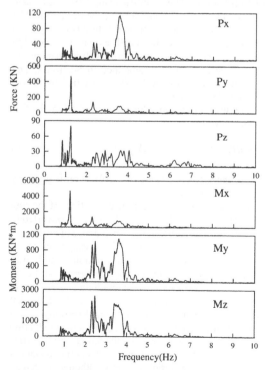

Figure 5.   Fourier spectra of effective driving forces on foundation (P4) for west bound test-run.

according to the motions we refer. The in-plane motion along the viaduct axis indicates the predominant frequency at 1.2 Hz while that in the transverse section of the viaduct at 3.7 Hz.

3.2 Ground response mitigation by P-WIB

The ground responses predicted by the A3D procedure with the loading functions given above were evaluated by the measurement data (Takemiya et al. 2005), which enables the further investigation. The honeycomb shaped WIB is employed as a countermeasure for the traffic-induced vibration mitigation (Takemiya & Shimabuku, 2005). Considering the performance of this measure, especially in comparison with the other procedure studied later, the present WIB is regarded as passively used WIB (P-WIB), for it works by its existence without any additional manipulation.

For the present situation, since the caisson foundation P4 gives the main contribution to the entire vibration in near field, the P-WIB is considered just around it. The present WIB originally consists of a multiple of soil-cement mixed short columns connected among neighboring ones, and they are arranged in honeycomb cells shape. The honeycomb configuration is simplified by the equivalent concentric ring type walls as shown in Figure 4 for the convenience of the axisymmetric FEM modeling. The simplified ring WIB walls have 1 m for the thickness and 15.6 m for the depth. The horizontal extent covers from 6 m to 20 m away from the foundation center for 3 cells extent. These design values are based on the wave propagation characteristics of the layered soil mentioned in Section 2. The mitigation effect of the P-WIB is studied by taking different WIB extent.

The ground response level is evaluated by the following formula

$$Lv[dB] = 20 \log_{10}\left(A_{max} / A_r\right) \qquad (6)$$

where the reference value of acceleration is given by $A_r = 10^{-5}$ (m/s^2).

Figure 6 illustrates the mitigation effect of the ring P-WIB. The ground responses have been converted to the vibration level dB for the convenience of evaluation of mitigation. Since the viaduct for investigation is temporarily supported by the construction bent frames against excessive deformation of girders under heavy traffic. The measurement data with and without bent frames are also provided for reference. A significant reduction is gained in the horizontal responses by 3 cells (denoted by P-WIB3) and satisfactory reduction may be obtained even by one cell (denoted by P-WIB1) over the WIB zone and even substantial distances beyond it. For the vertical response, although the reduction ratio is not as significant as those in horizontal directions, it is not necessary to pay more attention to

subdue it because of its relatively much smaller value. The measurement data show that the bent frames can reduce the ground responses in horizontal directions, while some amplification occurs in vertical direction.

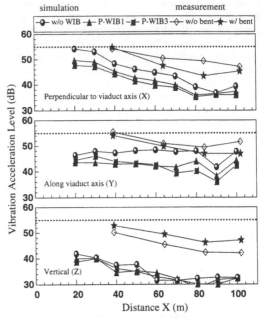

(a) Response of a test-run of a heavy truck

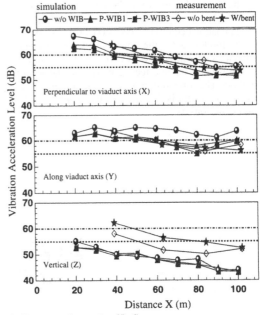

(b) Response of normal traffic flow

Figure 6. Ground response reduction effect of P-WIB.

4 THE ACTIVELY DRIVEN WIB

4.1 Introduction of the reformed device

Due to the environmental problem in built-up situation, where buildings are highly concentrated, it is not easy to find an enough open space for the occupation of the countermeasure. Therefore, other measures with smaller size are strongly demanded.

A new type device is put forth herein which is called actively driven WIB (A-WIB) to substitute the P-WIB. The mechanical model is described in Ref. (Takemiya et al. 2005). The A-WIB is designed on the basis of wave interference phenomenon. Figure 7 illustrates that the A-WIB comprises two parallel WIB walls with an open trench between them, and a pair of opposite driving forces are given on the inner faces of the walls. The open trench between the walls is indispensable to avoid the neutralization of the opposite forces effect.

Suppose that harmonic waves are generated at the source and consequently at the barriers. They are simply transmitted horizontally. Thus, the ground responses at a position X (x, z) can be described as

$$R_p = \alpha(x)A_p \exp\left[i(kx - \omega t)\right] \tag{7}$$

$$R_{q+} = \alpha(s-x)A_{q+} \exp\left\{i\left[-k(x-s)-\omega t\right]\right\} \tag{8}$$

$$R_{q-} = \alpha(x-s-d)A_{q-} \exp\left\{i\left[k(x-s-d)-\omega t+\pi\right]\right\} \tag{9}$$

where R_p, R_{q+} and R_{q-} = ground responses at X caused by loading P, q+ and q− respectively; A = vibration amplitude, $A_{q-} = -A_{q+}$, and we suppose $A_{q+} = \beta A_p$; k = wavenumber; s = position of A-WIB away from the source center; d = measure width; α = amplitude decay factor, which is assumed as a function of the distance away from the source.

It is assumed that the trench inside the barriers cut the wave propagation rightward of q+ at the left barrier. Then, for the area $0 < x < s$ as indicated by subscript '1',

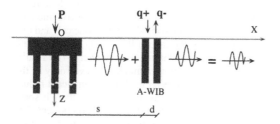

Figure 7. Sketch of the mechanism of A-WIB.

$$R_1 = R_p + R_{q+} = \alpha(x)A_p \exp\left[i(kx - \omega t)\right]$$
$$+\alpha(s-x)A_{q+} \exp\left\{i\left[k(s-x)-\omega t\right]\right\} \tag{10}$$

The phase gap is a function of x,

$$\Delta\varphi_1 = kx - (ks - kx) = k(2x - s) \tag{11}$$

If $\Delta\varphi_1 = 2n\pi$, the response is most amplified; if $\Delta\varphi_1 = (2n - 1)\pi$, the response is most reduced. Here, n is an integer number. Since $\Delta\varphi_1$ is a function of x, the responses here are not stable. For the present study, we focus on the area beyond the measure.

For the area $x > s + d$, as indicated by subscript '2', the response is

$$R_2 = R_p + R_{q-} = \exp\left[i(kx - \omega t)\right]$$
$$\times\left\{\alpha(x)A_p + \alpha(x-s-d)A_{q-} \exp\left[i(ks+kd+\pi)\right]\right\} \tag{12}$$

The phase delay is,

$$\Delta\varphi_2 = k(s+d) + \pi \tag{13}$$

If the wavenumber k is a constant, $\Delta\varphi_2$ is a constant, which implies that the ground vibrations at any positions of this range can be reduced most if the phase of q− is well shifted with respect to that of P.

Finally, the response in the superposition field of the source and the A-WIB is described as follows,

$$R(t) = R_{tr}(t) + R_{aw}(t + \Delta t) \tag{14}$$

where R_{tr} = ground response due to source loading with an open trench; R_{aw} = ground response due to driving force at the A-WIB; Δt = time delay for phase shift.

Herein, the caisson foundation P4 is taken as the example to analyze A-WIB. For comparison, the passively used WIB (P-WIB) is also simulated by the 2 dimensional model as depicted in Figure 8(a). Figure 8(b) illustrates the reformed device with a section size of 2.5 m × 3.4 m, which is quite suitable for urban situation. Mechanically pairs of opposite driving forces are exerted horizontally as well as vertically on the inner face of the walls along depth.

4.2 Effect of the A-WIB for simple loading

The computation is conducted with a 2-dimensional model directly in time domain. The evaluation of the 2D procedure is illustrated in Figure 9 by the ground responses for the case without any measure.

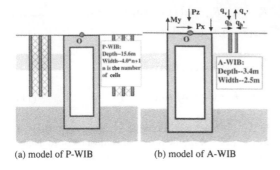

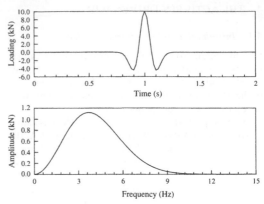

(a) model of P-WIB   (b) model of A-WIB

Figure 8.   Passively and actively used WIB.

Figure 10.   Ricker wavelet form loading for A-WIB.

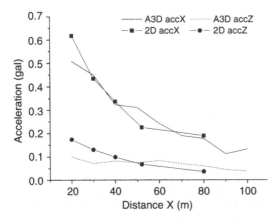

Figure 9.   Computation results comparison between the A3D model and the 2D model.

For simplicity, a Ricker wavelet-form load is employed here for both the source and the artificial loading pattern. The predominant frequency is set as 3.7 Hz in view of the vibration characteristics of the measurement data. The time history and the Fourier spectrum are depicted in Figure 10.

When the additional forces are not given on the A-WIB walls, the A-WIB becomes an open trench for vibration reduction. Therefore, the responses of this open trench are also provided for discussion.

Herein, the out-plane loadings in Figure 5 are considered separately, i.e. P_X, P_Z and M_Y are given respectively for the horizontal and vertical forces and the rocking moment about the axis perpendicular to the model illustration. For the Case 1, where P_X is specified, the corresponding forces in A-WIB are exerted horizontally, denoted as q_h and q_h'. For the other two cases, Case 2 and Case 3, corresponding to the action of P_Z and M_Y, the forces in A-WIB are given vertically, as denoted by q_v and q_v' in the figure.

Relative phase delay and amplitude of the additional vibrations are the two essential parameters to influence

the resultant response. Since the superposed responses are very sensitive to the phase difference of two waves, the manipulation should be carried out precisely. Herein, only the validity of the principle is investigated. The phase delay is substituted by the time delay Δt, which is determined mostly by the distance S (Fig. 7). For amplitude adjustment, it is treated in the post-processing process by determining the loading ratio β in view of the linear modeling. The work to determine the factor β is the pre-study to design a certain A-WIB. The amplitude of forces on A-WIB is determined by the vibration of the targeted area. If it is installed close to the foundation, the additional forces have to be large enough to alleviate the vibration beyond the device. The A-WIB of the present study is installed close to the foundation center, i.e. the distance S is 6 m. Thus the force amplitude ratio β is set about 0.33 and Δt is 0.10 s to get an optimum result. Another problem is that the phases of horizontal and vertical responses may not be changed synchronously, which may cause response reduction in one direction while amplification in the other. The principle of the control of A-WIB is to reduce the predominant vibration component without leading the minor component to be dominant. A mean-square-minimization of two perpendicular components should be taken to keep the balance between them.

Figure 11 shows the normalized responses on the ground surface when the Ricker-type P_X loading is given. The notations P-WIB1 and P-WIB2 indicate respectively the cases of P-WIB with one cell and two cells widths. For the case of P_X loading, the horizontal response is the predominant component. The results show that the reduction cannot be achieved only by the open trench. The A-WIB leads to a significant reduction in the horizontal direction, although there is some amplification in the other directional component. The vertical response is kept in the same level of the horizontal one by following the principle. The reduction

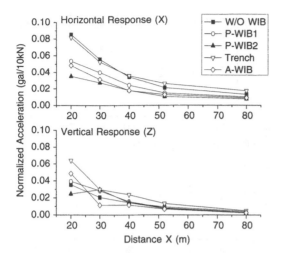

Figure 11. Ground response vs. distance for Ricker-P_X loading.

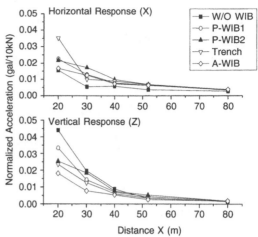

Figure 13. Ground response vs. distance for Ricker-M_Y loading.

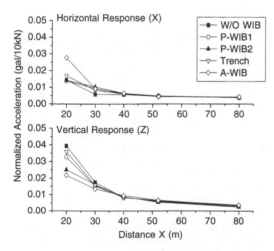

Figure 12. Ground response vs. distance for Ricker-P_Z loading.

Figure 13 illustrates the case of M_Y loading. The moment M_Y is replaced by a pair of vertical forces acting in opposite directions on the foundation footing. It seems that the A-WIB is most effective for the case of the moment action as the vibration mitigation. In comparison with the effect of P-WIB, obviously the A-WIB can give the more reduction. Because of the loading type of moment is similar to the forcing function of the A-WIB, the reduction effect is quite evident even at 100 m distance from the source load position. The moment type loading is of great importance in the view of loading functions from realistic traffic (Fig. 5), which makes the A-WIB measure the more attractive.

The A-WIB seems like a measure to balance the vibration components to an acceptable level. This kind of function of the A-WIB is rather helpful because most realistic vibration differs much in different components, while the environmental pollution is usually only caused by the predominant part.

4.3 Effect of A-WIB for a realistic situation

A realistic situation has been dealt with of the highway viaduct in Figures 1 and 2. The loading patterns of the source and the A-WIB only differ in the phase difference and amplitude. The loading function of Figure 5 is used. Because of the 2-dimensional analyses, only the out-plane force actions that include P_X, P_Z and M_Y are taken into consideration. The time histories of these loading components are shown in Figure 14.

Since the linear elastic behavior is assumed for the traffic-induced vibrations, the entire response can be obtained by superposing the response resulting from individual forces. P_Z has relatively smaller value than the other two components, so the phase adjustment of

effects of P-WIB1 and P-WIB2 are obvious on horizontal component, and also there is some amplification on the other component for both cases.

The ground responses caused by the Ricker-type P_Z are depicted in Figure 12. The case without the WIB shows the vertical component is dominant. The effect of an open trench is not obvious relative to the other measures. The A-WIB is set to lessen the vertical response and keep the horizontal component to the same level. A significant reduction in vertical direction is achieved, showing almost half of the original response is cancelled out in the area of 20–30 m away from the foundation center.

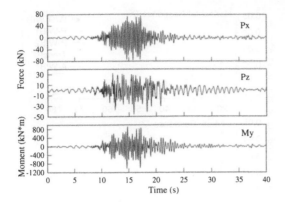

Figure 14.   Time histories of the realistic loadings.

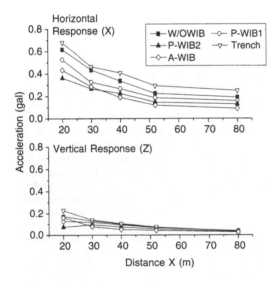

Figure 15.   Ground response reduction vs. distance for realistic loading.

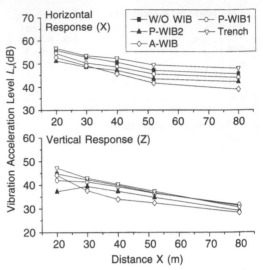

Figure 16.   Ground response level vs. distance for realistic loading.

the A-WIB is targeted to P_X and M_Y. Presume that the apparatus has two independently controlled components to give pair forces in both the horizontal and vertical directions. In the present situation, because the P_X leads to the larger response in the horizontal direction, one component of the A-WIB is well managed to alleviate the horizontal ground response caused by P_X only. While the M_Y loading brings the more deformation vertically, the other component of the apparatus is designed for it. Both components of A-WIB are manipulated on the same principle as mentioned above. Finally, the parameters β and Δt are optimized respectively for P_X and M_Y: the consequence is that β equals 0.45 and 0.15, while Δt equals 0.12 s for the both.

Figure 15 shows the ground responses for the combined out-plane loadings. The case of trench without force action amplifies the responses of both horizontal and vertical components. When an additional pair of forces are given on the trench corresponding to P_X and M_Y, significant response reductions are achieved both in horizontal and vertical responses. It seems that the A-WIB apparatus can get more mitigation effect in the zone along the far distance from the foundation than other measures. It is obvious in a rather large range. From the idea of two components of the A-WIB focusing on different directions, the ideal A-WIB can be designed as ring more obvious when the responses are represented in terms of vibration acceleration level in Figure 16. Although the P-WIB2 gives more reduction in the nearer area, the advantage of A-WIB is still quite walls like P-WIB model in Figure 4, to alleviate horizontal responses in various radiative directions.

Another advantage of A-WIB is its flexibility to generate various kinds of waves you want. Further, the tiny size of A-WIB relative to traditional measures for vibration reduction satisfies most situations in highly-populated area, which expands the application scope of vibration reduction measures.

5   CONCLUSIONS

The application of passively and actively used WIBs has been discussed by analyzing the ground responses caused by traffic on a highway viaduct supported by caisson/pile foundations.

The passively used WIB can result in significant reduction if the horizontal extent and the depth of the

WIB zone are designed appropriately based on the wave propagation characteristics at the focused site.

The actively driven WIB with less land occupation than the passive WIB may suit the highly built-up urban situation where environmental vibration mitigation is strongly demanded. If the active WIB measure is designed properly and operated with high precision with respect to the phase adjustment, a remarkable reduction effect can be gained for the total vibration through the wave interferences with the exiting waves from the source. The reduction effect is more than several dB for the actual situation and extends several tens meters behind of its position.

REFERENCES

Chen, F. Takemiya, H. & Shimabuku, J. 2004. Seismic performance of a WIB-enhanced pile foundation, *13th World Conference on Earthquake Engineering*, Vancouver, Canada, paper No. 1273.

Takemiya, H. 2004. Field vibration mitigation by honeycomb WIB for pile foundations of a high-speed train viaduct. *Soil Dynamics & Earthquake Engineering* 24: 69–87.

Takemiya, H. 2005. Prediction of train/traffic induced ground vibrations and mitigation by WIB. *ISEV2005*, Okayama, Japan.

Takemiya, H. Chen, F. & Ida, K. 2005. Traffic-induced vibrations from viaduct: hybrid procedure of measurement and FEM analysis for prediction and mitigation of ground motions. *Journal of Structural/Earthquake Engineering*, JSCE (submitted).

Takemiya, H. & Shimabuku, J. 2005. Honeycomb-WIB for mitigation of traffic-induced ground vibrations, *ISEV2005*, Okayama, Japan.

Environmental Vibrations – Takemiya (ed.)
© *2005 Taylor & Francis Group, London, ISBN 0 415 39035 4*

Vibration screening with sheet pile walls

Lars Andersen & Morten Liingaard
Department of Civil Engineering, Aalborg University, Aalborg, Denmark

ABSTRACT: The present analysis concerns the utilization of driven sheet pile walls in vibration mitigation. Wave barriers in the form of infilled trenches are a well-known means of vibration mitigation. However, the installation of such barriers may be costly by means of, for example, the cut-and-cover or jet grouting techniques. Driven sheet piles are an alternative to these methods. The soil between two walls may simply be removed, thus leaving an open trench – or the original soil may be replaced with another material such as bentonite. In any case, the goal is to obtain a strong impedance mismatch between this sandwich structure and the surrounding soil. In addition to this, non-vertical sheet piles may be driven, which may lead to a further increase of the efficiency when compared with vertical barriers. The aim of this paper is to investigate whether vertical or non-vertical walls are preferable in order to achieve efficient vibration screening. Further, the optimal dynamic properties of the fill material between two walls are discussed. The computations are carried out in the frequency domain with a two-dimensional coupled boundary element–finite element scheme.

1 INTRODUCTION

In densely populated built-up areas, traffic induced vibration is often a nuisance. Railway tracks and main roads lie close to residential buildings, laboratories, hospitals and other facilities, where only small levels of vibration can be tolerated. Faster passenger transportation and heavier freight trains and trucks only lead to an increase in the vibration levels. This necessitates a means of vibration mitigation. Here it may be of particular interest that traffic typically leads to vibration with a peak amplitude in the range 10 to 80 Hz, which is perceptible as whole-body vibrations and may lead to resonance of, for example, equipment utilized in the nano-technology industry.

A well-known technique in vibration isolation is the establishment of wave barriers along a track. These may take the form of open trenches, which have been studied both numerically (Banerjee et al. 1988; Shrivastava & Kameswara Rao 2002) and experimentally (Woods 1968). Alternatively, the trenches may be infilled with concrete or bentonite (Al-Hussaini et al. 2000), or flexible rubber chip barriers may be applied (Kim et al. 2000). Recently, Andersen and Nielsen (2005) found that the efficiency of wave barriers along a track depend on both the direction of excitation and the speed of a vehicle, using a coupled boundary element–finite element (BE–FE) model formulated in a moving frame of reference. A full review of the research concerning wave barriers is beyond the scope of this paper.

Instead of trenches, wave impeding blocks (WIBs) may be included under a track in order to reduce ground vibration (Takemiya et al. 1995; Chouw & Pflanz 2000), or piles may be driven along the track (Kattis et al. 1999). These solutions may be less expensive than open or infilled trenches, since both WIBs and piles may be installed by, for example, jet grouting; alternatively driven piles may be applied.

To the authors' knowledge, studies of vibration screening has been limited, almost entirely, to horizontal WIBs and vertical wave barriers. However, if sheet piles are utilized for the vibration mitigation, it may be possible to install a barrier with a non-vertical inclination. This produces a hybrid of a WIB and a vertical barrier, which may possibly provide a more efficient isolation than vertical barriers. The aim of the present study is to answer this question. To this end a two-dimensional coupled BE–FE analysis is carried out in the frequency domain. A three-dimensional model may be needed for an accurate evaluation of the vibration levels, but the plane model is better suited for parameter studies due to its low computation cost.

2 THEORY

Wave propagation in a viscoelastic domain with the boundary Γ is considered. The complex amplitudes of the displacements and tractions, $U_i(\mathbf{x}, \omega)$ and $P_i(\mathbf{x}, \omega)$, on Γ are related by the integral identity (Domínguez 1993),

$$C(\mathbf{x})\,U_i(\mathbf{x},\omega) + \int_\Gamma P_{ik}^*(\mathbf{x},\omega;\mathbf{y})U_k(\mathbf{y},\omega)d\Gamma_\mathbf{y}$$

$$= \int_\Gamma U_{ik}^*(\mathbf{x},\omega;\mathbf{y})P_k(\mathbf{y},\omega)d\Gamma_\mathbf{y}. \tag{1}$$

Here it has been assumed that no forces are applied in the interior of the domain. Further, in the present analyses plane strain is assumed. Thus, the indices i, k take the values 1, 2. In Eq. (1) $U_{ik}^*(\mathbf{x}, \omega; \mathbf{y})$ and $P_{ik}^*(\mathbf{x}, \omega; \mathbf{y})$ are the Green's functions for the displacement and surface traction, respectively. These describe the response at point $\mathbf{x}$ in direction i to a point force, or displacement, varying harmonically with the circular frequency ω and unit amplitude, and applied at point $\mathbf{y}$ in direction k. Finally, $C(\mathbf{x})$ is a constant that depends on the geometry of the boundary at point $\mathbf{x}$. In two dimensions, $C(\mathbf{x}) = \phi/2\pi$, where ϕ is the internal angle between the two parts of Γ on either side of the point. For an internal point, $C(\mathbf{x}) = 1$.

The boundary element (BE) form of Eq. (1) is achieved by discretization of the physical fields. In the present analyses quadratic interpolation is applied. Hence, Γ is divided into a number of boundary elements with three nodes. Let $\mathbf{U}_j(\omega)$ and $\mathbf{P}_j(\omega)$ denote the vectors storing the amplitudes of the displacements and tractions, respectively, at the nodes belonging to element j. The corresponding amplitudes on the boundary represented by this element are subsequently found by interpolation,

$$\mathbf{U}(\mathbf{x},\omega) = \Phi_j(\mathbf{x})\mathbf{U}_j(\omega), \;\; \mathbf{P}(\mathbf{x},\omega) = \Phi_j(\mathbf{x})\mathbf{P}_j(\omega), \tag{2}$$

where $\Phi_j(\mathbf{x})$ are vectors storing the quadratic shape functions. Insertion of Eq. (2) into Eq. (1) provides the system of equations for a BE domain,

$$\mathbf{H}(\omega)\,\mathbf{U}(\omega) = \mathbf{G}(\omega)\,\mathbf{P}(\omega), \tag{3}$$

where the geometry constants $C(\mathbf{x})$ are absorbed into the diagonal of $\mathbf{H}(\omega)$. The vectors $\mathbf{U}(\omega)$ and $\mathbf{P}(\omega)$ store the displacements and tractions, respectively, for all nodes on Γ. $\mathbf{H}(\omega)$ and $\mathbf{G}(\omega)$ store the influence from degree-of-freedom k to degree-of-freedom i for the displacement and traction, respectively. Special attention has to be made with regard to the contributions from a node to itself due to the singularities of the Green's functions. The weak singularities of $U_{ik}^*(\mathbf{x}, \omega; \mathbf{y})$ are treated by a coordinate transformation (Domínguez 1993), whereas the diagonal terms of $\mathbf{H}(\omega)$ are evaluated by the local enclosing elements technique proposed by Jones et al. (2000).

The finite element (FE) part of the model is described by the well-known system of equations,

$$(-\mathbf{M}\omega^2 + \mathrm{i}\mathbf{C} + \mathbf{K})\,\mathbf{U} = \mathbf{K}_{FE}\mathbf{U} = \mathbf{F}, \tag{4}$$

where $\mathbf{M}$, $\mathbf{C}$ and $\mathbf{K}$ are the mass, damping and stiffness matrices, respectively, whereas $\mathbf{U}$ and $\mathbf{F}$ are the nodal displacements and forces, respectively. Finally, $\mathrm{i} = \sqrt{-1}$ is the imaginary unit. Hysteretic material dissipation is assumed. Hence, the damping term is independent of the circular frequency, ω.

A coupling of the FE and the BE regions is carried out in a context of nodal forces, i.e. in a finite element manner. This involves that each boundary element domain be transformed into a *macro finite element* as depicted in Figure 1. Here $\mathbf{T}$ is a frequency independent transformation matrix expressing the relationship between the nodal forces and the surface traction applied over the surface of the BE domain.

In a standard BE formulation, discontinuous tractions over an element edge involve that contributions to a collocation node from the elements on either side of the edge are stored in separate columns of $\mathbf{G}(\omega)$. However, in the present method the contributions from all elements adjacent to a node are assembled into the same columns of $\mathbf{G}(\omega)$ before the transformation into equivalent nodal forces.

The coupled BE–FE scheme described above has been implemented in the computer program TEA (Jones et al. 1999), which has been applied for the present analysis.

The present study concerns the mitigation of ground vibration by means of sheet piles. Hence, a proper measure of vibration level reduction is required. As suggested by Jones et al. (2002), the vibration at a point on the surface of the ground is measured in terms of the *pseudo-resultant response*,

$$R = \sqrt{|U_1|^2 + |U_2|^2}, \tag{5}$$

which combines the horizontal and vertical amplitudes of motion. Further, reductions in noise and vibrations

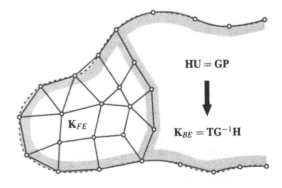

Figure 1.   Coupling of finite elements with a boundary element domain.

are suitably given in dB, and for this purpose, the following quantity is introduced:

$$\Delta_R = 20 \log_{10} \left(\frac{R}{R_0} \right).$$ (6)

Here R_0 is the reference pseudo-resultant response, which in the present analyses is taken as the value of R recorded directly under the load for the excitation frequency 10 Hz. Thus, the results provided in the following analyses indicate the degree of dissipation with distance from the source for different geometries and material properties of the sheet pile barriers; but in addition to this a direct comparison can be made between the vibration levels at different frequencies.

3 PARAMETER STUDY

In this section, the two-dimensional coupled BE–FE model is utilized for a parameter study, in which the efficiency of sheet pile wave barriers with different geometries and material properties are compared.

The geometry of the considered problem is illustrated in Figure 2. The track is modelled as a 2.0 m wide concrete slap with a thickness of 0.2 m. Material properties are given in Table 1, where E is the Young's modulus, ν is Poisson's ratio, ρ is the mass density, and η is the loss factor. As an approximation, the sheet piles are modelled as continuous steel plates with the thickness 0.2 m. Both structures are discretized with quadrilateral solid finite elements employing quadratic interpolation. As indicated by Figure 2, the present study is confined to double-layer walls, and the pile length is $l = 6$ m in all analyses.

The subsoil consists of dense sandy soil, and in order to assure radiation of waves into the semi-infinite half-plane, boundary elements are applied for the discretization. The maximum element length in both the FE and BE part of the model is 1 m. This provides a minimum of approximately three elements per wave length of the Rayleigh wave in all the analyses, which has been found to provide an accurate solution.

The truncation edges of the computation mesh are placed 20 m and 40 m away from the track, respectively. The larger of the two distances is on the side of the track where the barrier has been included in the ground. A convergence study indicates that this produces a fully converged model. Reliable results are achieved even close to the edge due to the small impedance mismatch between the half-space and the full-space forming the basis for the Green's functions.

Figures 4 to 7 show the pseudo-resultant response along the surface of the ground for different backfill materials, wall inclinations, θ, and frequencies, f. In all cases, the distance from the centre of the track to the centre of the barrier is $D = 6$ m, and the horizontal in-plane distance between the centre lines of the piles is $d = 2$ m. In particular, Figure 4 shows the results for the situation in which the original soil is present between the two walls. In Figure 5, the soil has been removed, leaving an open trench, and Figures 6 and 7 show the results for trenches infilled with concrete and rubber, respectively. In all the figures, the thick line indicates the pseudo-resultant response in the reference situation with no vibration screening.

From Figure 4 it is concluded that a reduction of the vibration level is achieved in the case with no soil removal or replacement for all frequencies and $\theta < 30°$. The barriers with $\theta < 0°$ are particularly useful at low frequencies. This was to be expected since these barriers act as a hybrid between a vertical wall and a WIB. In contrast to this, the barriers with $\theta = 45°$ and $\theta = 30°$ lead to an amplification of the vibration in some areas for low and intermediate

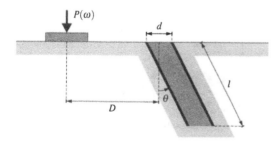

Figure 2. Geometry of the problem in the case of two parallel sheet pile walls.

Table 1. Material properties.

Material	E (MPa)	ν	ρ (kg/m³)	η
Soil	200	0.3333	2000	0.05
Steel	200,000	0.1500	7850	0.01
Concrete	20,000	0.3000	2500	0.03
Rubber	10	0.3333	1000	0.10

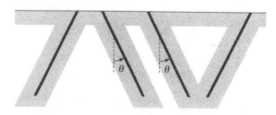

Figure 3. Non-parallel sheet pile walls in A-formation (left) or V-formation (right).

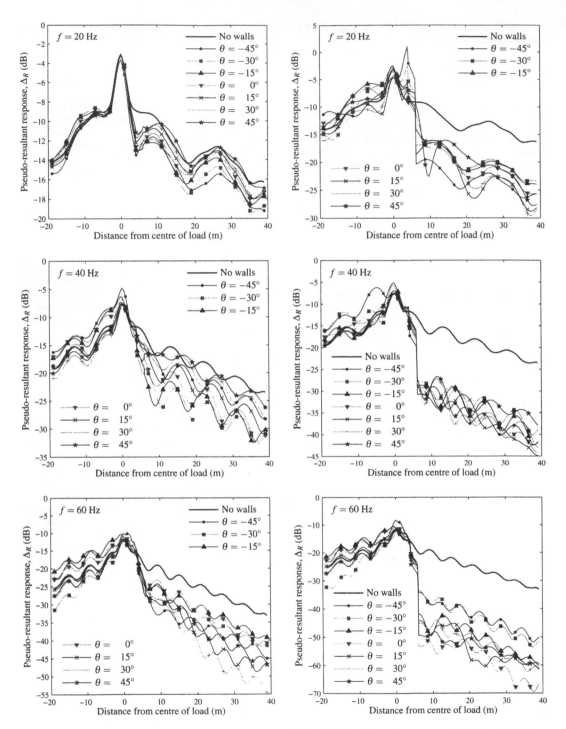

Figure 4.   Pseudo-resultant response in dB for double-layer sheet pile walls with different inclinations, θ, as defined in Figure 3. *Original soil between the walls*.

Figure 5.   Pseudo-resultant response in dB for double-layer sheet pile walls with different inclinations, θ, as defined in Figure 3. *No soil between the walls*.

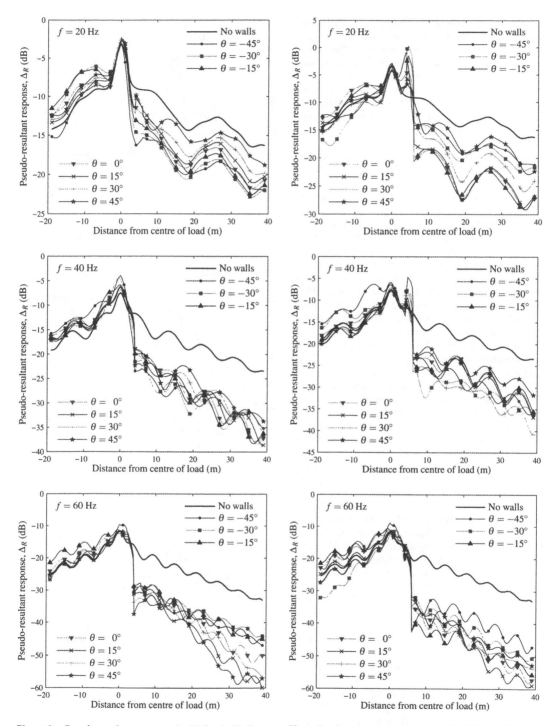

Figure 6. Pseudo-resultant response in dB for double-layer sheet pile walls with different inclinations, θ, as defined in Figure 3. *Concrete between the walls.*

Figure 7. Pseudo-resultant response in dB for double-layer sheet pile walls with different inclinations, θ, as defined in Figure 3. *Rubber between the walls.*

frequencies. However, at the frequency 60 Hz, apparently $\theta = 30°$ provides the better solution. Hence, depending on the frequency spectrum of the excitation, different inclinations may be preferable. Overall, $\theta = 15°$ to $30°$ seems to be a proper choice, since these inclinations provide a great reduction in the vibration level for the frequencies 20 and 40 Hz, and for the frequency 60 Hz all values of θ lead to a reduction of approximately 10 dB or more.

The open and infilled trenches, cf. Figures 5 to 7, provide an efficient shielding for all frequencies and inclinations. Further studies, the results of which are not included in this paper, indicate that similar reductions in the vibration level are achieved for other frequencies in the interval 10 to 80 Hz.

Generally, the open trench is more efficient than the infilled trenches and the barrier with the original soil between the walls. The trench infilled with rubber is slightly better than the concrete barrier in most configurations. This stems from the fact that the sandwich structure composed of steel plates and a rubber core leads to a greater impedance mismatch than does the steel–concrete sandwich when compared with the original half-plane with no barrier.

It turns out that $\theta = 0°$ actually provides an efficient vibration isolation. Other inclinations may lead to a further reduction of the vibration level. However, a vertical barrier may be cheaper to install than a non-vertical barrier. Thus, $\theta = 0°$ may eventually be preferable in the case of an open or infilled trench.

The results achieved for different distances, d, between two vertical walls are plotted in Figure 8. For comparison, the vibration levels achieved with either a 6 or 12 m deep single wall are included. The same amount of material is required for the 12 m deep wall and two 6 m deep walls. It turns out that a small distance, i.e. $d = 1$ to 3 m, is preferable at the frequencies 40 and 60 Hz. However, at 20 Hz a wide barrier provides an efficient vibration isolation. Apparently, in most situations the single wall with double depth leads to a better vibration screening than any of the double-layer walls. A comparison with Figures 6 and 7 indicates that better results are achieved with the double-layer walls if the soil inside the barrier is replaced by a softer or stiffer material. However, this is a more expensive barrier than the single wall.

Figure 9 shows the results for 2 m wide barriers (i.e. $d = 2$ m) placed at different distances from the track. For the frequencies 20 Hz and 60 Hz, it is advantageous to put the barrier close to the source. However, at 40 Hz the optimal distance from the track to the centre of the barrier is apparently $D = 6$ m. The barrier placed 3 m from the track is still more efficient very close to the source, but for all distances beyond 7 m, $D = 6$ m provides the better solution. It is worthwhile to note that the barrier with $D = 15$ m provides poor shielding – even at distances just above 16 m.

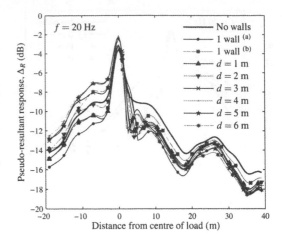

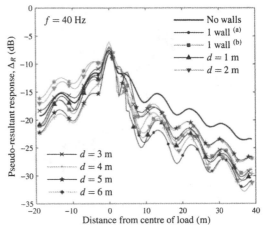

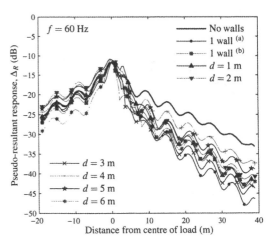

Figure 8. Pseudo-resultant response in dB for double-layer sheet pile walls with *different distances, d, between the walls*. For comparison, single-layer walls with the depth (a) $l = 12$ m and (b) $l = 6$ m are included.

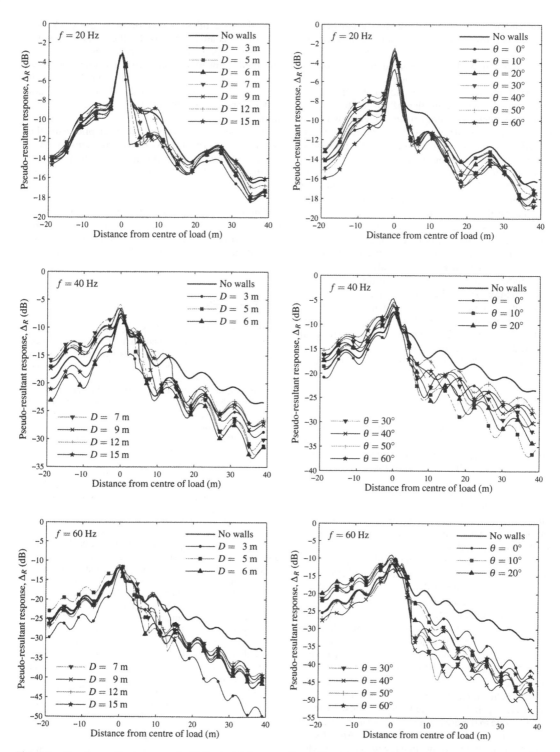

Figure 9. Pseudo-resultant response in dB for double-layer sheet pile walls at *different distances, D, from the track.*

Figure 10. Pseudo-resultant response in dB for double-layer sheet pile walls in *A-formation with different inclinations, θ.*

Instead of two parallel walls, a barrier may be constructed as shown in Figure 3. Results for a barrier in A-formation are given in Figure 10. No soil has been replaced or removed. It is concluded that an A-formation with $\theta = 30°$ or $40°$ provide the better solution for the frequencies 20 and 60 Hz. However, for the frequency 40 Hz an inclination of $10°$ to $20°$ is more efficient. Generally, a comparison of Figures 8 and 10 indicates that the barrier in A-formation is more efficient than the barrier with two vertical sheet pile walls. Further, a comparison of Figures 4 and 10 shows that the barrier in A-formation with $\theta = 20°$ or $30°$ is more efficient than the barrier with two parallel walls for all frequencies. It is noted that the optimal value of θ depends on the frequency in the case of the parallel walls, whereas $\theta = 20°$ or $30°$ provides consistently good results for the barrier in A-formation.

Alternatively, the walls may be driven in V-formation as illustrated in Figure 3. Analyses of walls in V-formation with $d = D = 6$ m and different angles of inclination show that this barrier provides less efficient shielding than may be achieved with the other formations discussed above. Combined with the fact, that this kind of barrier takes up more space near the ground surface than any of the other barriers, it is concluded that the V-formation is generally unsuitable.

4 CONCLUSIONS

The efficiency of wave barriers along a track has been investigated using a two-dimensional coupled boundary element–finite element scheme in the frequency domain. The barriers are made of two sheet pile walls that may, or may not, be parallel. A parameter study is carried out in order to examine the effects of different inclinations of the walls and distances from the source to the barrier. Further, the influence of the material and distance between the walls has been studied.

The effect of the inclination is highly dependent on the excitation frequency. It has been found that a barrier with two parallel walls at an inclination of $45°$ towards, i.e. down under, the track provides relatively good screening at low frequencies for most materials, whereas poor results are achieved for high frequencies. In general an inclination of $15°$ to $30°$ away from the track leads to a great reduction of the vibration level for most excitation frequencies and infill materials. The barrier infilled with rubber is more efficient than trenches infilled with concrete or original soil. However, open trenches provide the most efficient solution due to the great impedance mismatch between the structure and the surrounding soil. An efficient vibration reduction is obtained with open trenches strengthened by vertical walls, which may be preferable from a practical point of view.

A distance of 3 to 6 m between the track and the barrier is advantageous. Further, a barrier width of 1 to 3 m provides a good vibration isolation for intermediate or high frequencies. For lower frequencies a wider barrier may be preferred. However, it turns out that a single wall of double length provides better vibration isolation than any of the double-layer walls.

As an alternative to parallel sheet pile walls the walls may be driven in A- or V-formations as illustrated in Figure 3. Barriers in A-formation with an inclination of $20°$ to $30°$ provide better screening than two parallel walls for all frequencies, whereas barriers in V formations are generally unsuitable.

The present study shows that vibration screening by means of sheet pile walls is an alternative to the costly barriers used today. Future studies may involve three-dimensional analyses and experimental work.

REFERENCES

Al-Hussaini, T.M., Ahmad, S. & Baker, J.M. 2000. Numerical and experimental studies on vibration screening by open and in-filled trench barriers. In N. Chouw & G. Schmid (Eds.), *Wave 2000*, Bochum, Germany, pp. 241–250. Rotterdam: A.A. Balkema.

Andersen, L. & Nielsen, S.R.K. 2005. Reduction of ground vibration by means of barriers or soil improvement along a railway track. *Soil Dynamics and Earthquake Engineering (in press)*.

Banerjee, P.K., Ahmad, S. & Chen, K. 1988. Advanced application of BEM to wave barriers in multi-layered three-dimensional soil media. *Earthquake Engineering & Structural Dynamics 16*(7), 1041–1060.

Chouw, N. & Pflanz, G. 2000. Reduction of structural vibrations due to moving load. In N. Chouw & G. Schmid (Eds.), *Wave 2000*, Bochum, Germany, pp. 251–268. Rotterdam: A.A. Balkema.

Domínguez, J. 1993. *Boundary elements in dynamics*. Southampton: Computational Mechanics Publications.

Jones, C.J.C., Thompson, D.J. & Petyt, M. 1999. TEA – a suite of computer programs for elastodynamic analysis using coupled boundary elements and finite elements. ISVR Technical Memorandum 840, Institute of Sound and Vibration Research, University of Southampton.

Jones, C.J.C., Thompson, D.J. & Petyt, M. 2000. Studies using a combined finite element and boundary element model for vibration propagation from railway tunnels. In *Proceedings the Seventh International Congress on Sound and Vibration*, Garmish-Partenkirchen, pp. 2703–2710.

Jones, C.J.C., Thompson, D.J. & Petyt, M. 2002. A model for ground vibration from railway tunnels. *Proceedings of the Institution of Civil Engineers, Transport 153*(2), 121–129.

Kattis, S.E., Polyzos, D. & Beskos, D.E. 1999. Vibration isolation by a row of piles using a 3-d frequency domain BEM. *International Journal for Numerical Methods in Engineering 46*, 713–728.

Kim, M., Lee, P., Kim, D.-H. & Kwon, H.-O. 2000. Vibration isolation using flexible rubber chip barriers. In N. Chouw & G. Schmid (Eds.), *Wave 2000*, Bochum, Germany, pp. 289–298. Rotterdam: A.A. Balkema.

Shrivastava, R.K. & Kameswara Rao, N.S.V. 2002. Response of soil media due to impulse loads and isolation using trenches. *Soil Dynamics and Earthquake Engineering* 22, 695–702.

Takemiya, H., Shim, K.-S. & Goda, K. 1995. Embankment train track on soil strutum and wave impeding block (WIB) measured for vibration reduction. In A. Çakmak & C. Brebbia (Eds.), *Soil Dynamics and Earthquake Engineering VII*, pp. 103–112. Southampton: Computational Mechanics Publications.

Woods, R.D. 1968. Screening of surface waves in soil. *Journal of Soil Mechanics and Foundations Division ASCE 94*, 951–979.

Environmental Vibrations – Takemiya (ed.)
© *2005 Taylor & Francis Group, London, ISBN 0 415 39035 4*

Study on vibration effects upon precise instruments due to metro train and mitigation measures

X.J. Sun, W.N. Liu, H. Zhai & D.Y. Ding
School of Civil & Architecture Engineering, Northern Jiaotong University, Beijing, China

J.P. Guo
Beijing General Municipal Engineering Design & Research Institute, Beijing, China

ABSTRACT: Vibration effects on sensitive instruments due to metro trains are complicated. According to the study on mitigation measures of one segment of Beijing subway, comprehensive measures are proposed to reduce the vibration effects on instruments from several aspects including design, construction, operation and vibration isolation of instruments. The design philosophy is put forward in the paper. Four kinds of floating slab track frames are set out tentatively, whose effects are compared by numerical analysis using ANSYS program. The normal and T form floating slab tracks are recommended with special measures adopted to mitigate the basic frequency vibration. The horizontal vibration below 4 Hz is mainly reduced by vibration isolation on the ground. Vibration isolation foundations of some special precise instruments are used or improved. Treatment methods of vibration reduction in construction and operation are also presented.

1   INTRODUCTION

Train induced vibration has aroused people's great attention and becomes a hot research subject. People have obtained lots of research productions. Takemiya & Fujiwara (1994) and Takemiya & Shiotsu et al (2000) researched the ground vibration along the lines induced by high-speed train and the counter vibration measure X-WIB. Jones C., Thompson D.J. and Petyt M. used combined finite element and boundary element model to study the ground-borne vibration and noise from trains. Weining liu & Yanfeng Shen studied the ground-borne vibration both in different tunnel depth and for straight link between two railway stations of Beijing. S.Wolf did the site testing and numerical analysis using FLAC^{3D} program, in order to study the potential low frequency ground vibration (<6.3 Hz) impacts from underground LRT operations. C. Madshus, B. Bessason and L. Harvik researched the low frequency-vibration from high speed railways and presented a prediction model. R.Paolucci et al researched the low-frequency ground vibration induced by high-speed trains, using two-dimensional and three-dimensional finite element method. However the research of vibration effects on precise instruments is limited.

In recent years, with the urban rail transit system is developing rapidly and some new research labs of the universities, hospitals and other companies are constructed at the same time, these may result in some cases such as new lines going through existing companies using precise instruments or constructing new companies over the existing subway structure. This problem emerges in several countries. For example, in Beijing the metro project will pass through under the labs of some research companies where installed different instruments; the Taiwan Shinkansen (bullet train) under construction is to pass through Taiwan Industrial Science Park (TNISP), a medical building is planned to be built over the subway structure in Atlanta, and the Sound Transit Link Light Rail LRT system is planned to pass through the University of Washington which may have effects on Physics and Astronomy Building (PAB). The low-frequency vibration due to train load decays slowly and propagates to distant place of 100–200 m, which could affect the work environment of sensitive instruments in the nearby buildings. And it may cause great loss in economy and research without dealt efficiently. Therefore, how to choose suitable mitigation measures to reduce vibration effects has become a critical issue and needs further research, in order to guarantee the transportation network and the city develop smoothly. In the paper, the mitigation measures are discussed focused on the station in Beijing mentioned above.

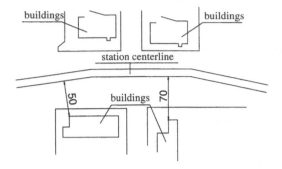

Figure 1. The relation of experiment center and subway.

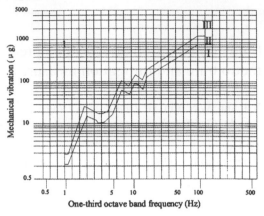

Figure 2. The environmental vibration requested by the central part of one instrument (I, all allowed; II, 1 peak allowed; III, not allowed).

2 VIBRATION EFFECTS ON PRECISE INSTRUMENTS

There are some buildings with research laboratories along the line and the distance is about 70 m, as shown in Figure 1. In this segment, the station length is 266 m, the station center rail surface locates 13 m below the ground surface, the structure bottom locates 17 m below the ground surface and the influence segment length is 724 m. In the buildings, there are 79 precise instruments requiring strict work environment. The environmental vibration required is 0.5~100 μg in the range of 1~10 Hz, 10 μg less-than 2 Hz, 2 μg less-than 1 Hz. The environmental vibration requested by the central part of one instrument is shown in Figure 2.

3 DESIGN PHILOSOPHY

Several philosophies should be taken into account in the design of mitigation measures:

(1) The comprehensive measures should be adopted including the track mitigation technologies, the foundation treatment of structure and ballast, the secondary isolation of the instruments.
(2) Improved and optimum design of existing and mature technique should be considered priorly such as reducing the natural frequency under safe condition, which will not increase difficulties in construction and servicing.
(3) Vibration isolation of precise instruments should be adopted if needed.
(4) It can be considered that the influence vibration range can be divided into several small range, therefore, different measures can be adopted according to each small vibration range.
(5) The design should insure the minimum influence to the instruments but not enlarge the normal vibration effects.
(6) The vibration mitigation structure should need little or no servicing.
(7) The drainage problem should be taken into account.

The stiffness of the vibration isolation platform and the corrective action of platform inclination should be considered in vibration isolation design.

4 COMPREHENSIVE MEASURES

The vibration influence of metro trains on sensitive instruments is a complicated problem as considering the influence not only in operation but also in construction. Therefore, comprehensive measures should be adopted to reduce the vibration effects on instruments from several aspects including design, construction, operation and vibration isolation for some special instruments.

4.1 Design

Design of vibration control includes track mitigation technologies, the tunnel structure, the propagation path and some other aspects.

4.1.1 Track mitigation technologies

The segment is constructed continuous welded rail track with PD3 rail (which is used in railway) to guarantee that the train operation is stable with little vibration. And the steel spring floating slab track with special design is used to control vibration.

4.1.2 Vibration control of tunnel structure

The tunnel structure is a low-frequency harmonic oscillator whose self-vibration frequency is determined by the design scheme. The tunnel structure should be improved by changing the thickness and its mass per meter to reduce the dynamic response.

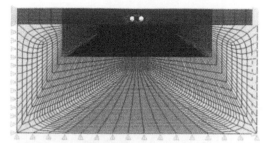

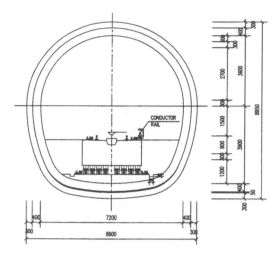

Figure 3. Thickening floating slab track framework.

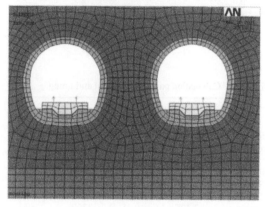

Figure 5. The element in the model.

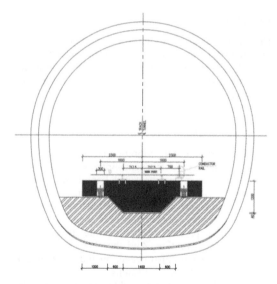

Figure 6. Magnified local element of T form floating slab.

Figure 4. T form floating slab track framework.

4.1.3 *Propagation path*

In order to reduce the vibration, some measures should be adopted to cut off the propagation pat such as vibration insulation ditch and wave impeding barrier (WIB), etc. These measures should be chosen after further researched.

4.1.4 *Special design of track system*

Based on the practical situation, four kinds of floating slab track frames are set out tentatively including the normal floating slab, the thickening floating slab, the thickening and heavy graded concrete floating slab and the T form floating slab.

The structure of normal floating slab is shown in the reference (written by Jing Ren), whose length is 31 m, width is 3.3 m and thickness is 0.66 m. It uses C40 concrete with 17 steel springs each side. The structure of thickening floating slab is shown in Figure 3, whose length is 25 m, width is 3 m and thickness is 1.25 m. It uses C40 concrete with 16 steel springs per meter. The thickening and heavy graded concrete floating slab has the same framework as the thickening floating slab except using heavy graded concrete whose density is 3000 kg/m³. The structure of T form floating slab is shown in Figure 4, whose length is 28.72 m, width is 4.6 m, thickness changes from 0.6 m to 1.2 m and the cross-sectional area is 3.96 m². It uses C40 concrete with 16 steel springs in each side. The same tunnel structure is used, shown in Figure 3.

4.1.4.1 Numerical model

A two-dimension model is used in numerical analysis and the ground is modeled as a layered linear elastic medium, as shown in Figures 5–7. The surrounding soil is divided into three layers according to the sheer wave velocity. The soil and liner is presented by plane elements with four nodes (plan42) and the boundary condition is a combination of elastic damping boundary

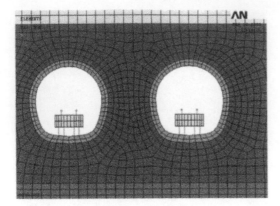

Figure 7. Magnified local element of other floating slabs.

Table 1. Calculation parameters of soil and lining.

Soil layer and lining	Dynamic elastic modulus (MPa)	Dynamic Poisson's ratio	Density (kg/m³)	Damping ratio
First layer	1.17E+08	0.341	1900	0.05
Second layer	2.89E+08	0.313	2023	0.03
Third layer	7.04E+09	0.223	1963	0.01
Lining	2.30E+11	0.25	2300	

and boundary large elements. The elastic stiffness is half of the foundation stiffness and the damping is 10 times of the corresponding ground damping. The calculation parameters are shown in Table 1.

4.1.4.2 Comparison of effectiveness

The acceleration data are obtained of the point on the ground surface where is 20–100 m away from the centerline, through harmonic analysis with the load frequency changed from 1 Hz to 15 Hz. The Y acceleration response spectrum is shown in Figure 8.

So the following conclusions can be obtained:

(a) When the load frequency is in the range of 9–15 Hz, the reduction effectiveness of the four style floating slabs has little differences.

(b) The T form floating slab and the normal floating slab are more efficient during the excitation is below 5 Hz, but they enlarge the vibration during the excitation is in the range of 5–8 Hz because of resonance vibration.

(c) The thickening floating slab has only one peak in the range of 3–4 Hz, and it's same to the thickening and heavy graded concrete floating slab. It indicates that these two style floating slabs are resonant with

the whole system which is unfavorable to reduce the low-frequency vibration. So they shouldn't be adopted.

4.2 Construction methods

(1) To reduce the vibration impact on instruments, some useful methods are taken during construction as follows: selecting the lower-vibration machine, staggering the construction period, adjusting the structural design to meet the requirement of vibration-reduction and using the vibration isolation foundation to decrease the transfer-energy of vibration.

(2) To design the rail of vibration-reduction conveniently, metro station and tunnel are chosen, in which measures of vibration-reduction are simple and reliable. The metro station is constructed by cut method and the tunnel by mining method.

4.3 Operation period

The modern and new-style rail-check vehicle will be adopted to reduce the vibration during operation period. The vehicle can check the status of rail, and eliminate unevenness of rail working in conjunction with rail-polish vehicle.

4.4 Passive vibration-isolation of instruments

It is necessary to use the vibration isolation of instruments to reducing low-frequency vibration which isn't attenuated by the above measures. For vibration isolation design, vibration data of nearby source and some instruments datum are needed such as type, specification, size, mass, centroid position, skeleton of seat, affiliated equipments, even-adjusting requirement and allowable vibration. And the chosen vibration-equipments data are also needed including technical parameter, size, fixing conditions.

Independent foundation could be chosen to isolate the base of instruments from that of buildings. In addition, the isolation system can be installed between the workbench of instruments and the independent base, or between the base of instruments and the central parts. The vibration isolation equipments in common use include steel-spring vibration isolator, rubber vibration isolator, gas vibration isolator and so on. The vibration isolation system of some sensitive instruments in the buildings is shown in Figure 9. These measures could maintain the work environment. During construction and operation, the isolation system should be adjusted according to the actual situation, for example, fixing vibration isolation system for instruments, rectifying the adopted isolation system to get better effectiveness, etc.

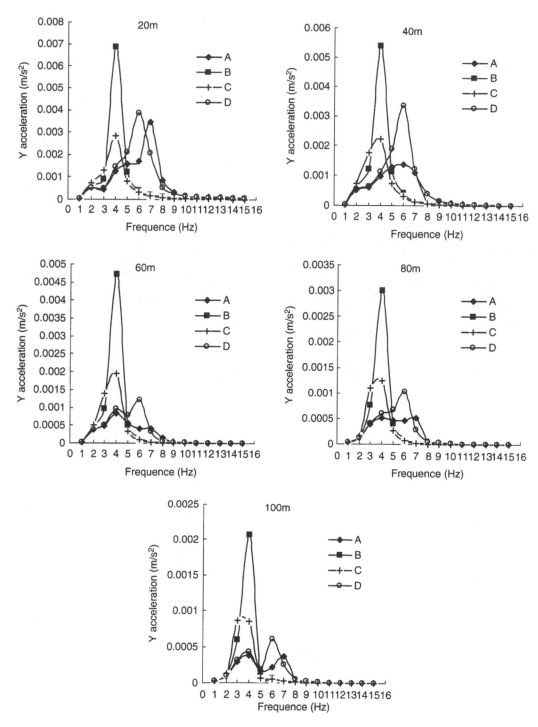

Figure 8.   Y acceleration response spectrum on the ground surface (A, the normal floating slab; B, the thickening floating slab; C, the thickening and heavy graded concrete floating slab; D, the T form floating slab).

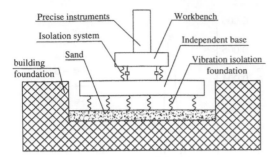

Figure 9.   The isolation system of precise instruments.

5   CONCLUSIONS

Vibration effects on precise instruments due to trains become unavoidable with the development of urban mass transit. Therefore, how to choose the suitable measures to mitigate the effects is important for smoothly operating of the urban mass transit planning with high research value. The study based on this segment is a beginning of further research in China.

The following conclusions could be obtained through research in the paper:

(1) Vibration effects on sensitive instruments due to metro trains are complicated. Therefore, comprehensive measures should be adopted to reduce the vibration effects on instruments from several aspects including design, construction, operation and vibration isolation of instruments.

(2) When the load frequency is in the range of 9–15 Hz, the reduction effectiveness of the four style floating slabs has little differences. The thickening and heavy graded concrete floating slab and the thickening floating slab are resonant with the whole system in the range of 3–4 Hz, which is unfavorable to reduce the low-frequency vibration. The normal and T form floating slab tracks are recommended with special measures adopted to mitigate the basic frequency vibration. The special measures include changing the natural vibration frequency of tunnel structure.

(3) The research of vibration effects on precise instruments is limited, and it needs further study including site testing of vibration acceleration on the ground surface due to metro trains, setting up a uniform standard, strengthening the research of new mitigation measures, etc.

REFERENCES

Hirokazu & Takemiya. 2004. Field vibration mitigation by honeycomb WIB for pile foundations of a high-speed train viaduct. *Soil Dynamics and Earthquake Engineering* 24 (1): 69–87

Jing Ren & Jianba Jiang. 2002. The application of steel spring floating slab track bed to Xizhimen Station of urban rail system. *Railway Standard Design* (9): 15–16

Jones C., Thompson D.J. & Petyt. 1999. Ground-borne vibration and noise from trains: Elasto dynamic analysis using the combined boundary element and finite element. *ISVR Technical Memorandum No.844. University of Southampton*

Jones C., Thompson D.J. & Petyt M. 2000. Studies using combined finite element and boundary element model for vibration propagation from railway tunnels. *Proceedings of Seventh International Congress on Sound and Vibration. Germany*: 2703–2710

Madshus C., Bessason B. & Harvik L. 1996. Prediction model for low frequency vibration from high speed railways in soft ground. *Journal of Sound and Vibration* 193(1): 195–203

Paolucci R., Maffeis A. & Scandella L. et al. 2003. Numerical prediction of low frequency ground vibration induced by high-speed trains at Ledsgaard, Sweden. *Soil Dynamics and Earthquake Engineering* 23: 425–433

Takemiya H. & Fujiwara A. 1994. Wave propagation/impediment in a stratum and wave impeding measured for SSI response reduction. *Soil Dynamics and Earthquake Engin* 13: 49–61

Takemiya H., Shiotsu Y. & Yuasa S. 2000. Features of ground vibration induced by high-speed trains and the counter vibration measure X-WIB. *Proceedings of Japan Society of Civil Engineers*: 33–42

Weining Liu & Yanfeng. 2001. Prediction of ground train borne vibrations for straight link between two railway stations of Beijing. *TIVC'2001, International Symposium on Traffic Induced Vibrations & Controls*: 201–206

Weining Liu & Yanfeng. 2001. Propagation of ground-borne vibrations in different tunnel depth. *TIVC'2001, International Symposium on Traffic Induced Vibrations & Controls:* 207–211

Wilson, Ihrig & Associates. 1996. Recent developments in ground-borne noise and vibration control. *Journal of sound and vibration* 193(1): 367–376

Wolf S. 2003. Potential low frequency ground vibration (<6.3 Hz) impacts from underground LRT operations. *Journal of sound and vibration* 267: 651–661

Yunqing Zhang, Weining Liu & Xiuli Wu. 2004. Control measures for vibration induced by metro trains. *Proceedings of World Tunnel Congress and 13th ITA Assembly, Singapore*: H17

Environmental Vibrations – Takemiya (ed.)
© *2005 Taylor & Francis Group, London, ISBN 0 415 39035 4*

Theoretical study on phase interference method for passive reduction of multiple excitation forces – reduction method of vibration due to rhythmic action of concert audience

R. Inoue & Y. Hashimoto
Takenaka Research & Development Institute, Chiba, Japan

Y. Yokoyama
Tokyo Institute of Technology, Tokyo, Japan

ABSTRACT: This paper presents a proposal of phase interference method for reduction of multiple excitation force. Firstly, conventional vibration reduction methods are discussed, mainly on the merits of general elastic suspension methods and on their demerits of difficulties on reduction of lower frequency vibration components. Secondly, a new method of vibration reduction is presented which is effective on lower frequency vibration case. Theoretical reduction effects of it are studied and differences from conventional isolation methods are discussed. Finally, practical effects and feasibilities of the newly presented method are studied and discussed, considering of various application situations on the building vibration-proof engineering.

1 INTRODUCTION

Large groups of people often move together in time to the music during performances at clubs, concert halls and so on. In recent years, this has become a problem in an increasing number of cases, as it may produce vibrations in buildings in the surrounding area.

The source of these vibrations is the motion of hundreds or even tens of thousands of people in time to the music, which produces (in-phase) excitation force. The combined excitation force applied to structural floors and dirt floors (hereafter "structures") is enormous. The vibrations produced by this excitation force are transmitted through the ground throughout a wide area and appear in surrounding buildings as unpleasant vibrations that are felt by the people in these buildings.

One way of reducing these vibrations is to implement measures for the ground and buildings in the surrounding area. However, during a concert performance or the like, vibrations are transmitted throughout an extremely wide area. Particularly in the case of major concerts that can be expected to be attended by tens of thousands of people, the vibrations produced may affect buildings several hundred meters away or more. For this reason, implementing measures for the ground and buildings in the surrounding area would be impractical. Accordingly, the idea of taking steps for the source of the vibrations itself is being explored. However, reducing the enormous excitation force produced by the motion of large numbers of people with existing technologies would entail many difficulties from the standpoints of scale and cost and so on, and at present no effective measures have yet been devised.

This paper will discuss vibration-proofing, which for many years has been the most commonly used method of dealing with vibration sources. This method involves using rubber springs or rubber pads, metal springs, air cushions and so on to provide elastic support in order to reduce the excitation force transmitted to the structure. The theory of reducing excitation force using this method will be discussed, and the difficulty of reducing excitation force caused by the motion of spectators during a concert or the like using this method will be shown. Subsequently, a new method will be proposed: the phase interference method for passive reduction of multiple excitation forces. A theoretical study of the effectiveness of this method in reducing excitation force will be conducted, and this method will be compared with vibration-proofing. Finally, the suitability of this technology with respect to buildings will be examined from a variety of perspectives.

2 OBJECTIVE AND SCOPE

The objective of this study is to devise an effective and realistic method for eliminating the source of vibrations resulting from the movement of large numbers of people during performances at clubs and concert halls. This paper represents the initial stage of proposing a new method called the phase interference method for passive reduction of multiple excitation forces, and verifying theoretically that this method is capable of reducing the total excitation force applied to structures, as well as examining the applicability of this method to buildings. Subsequent papers will deal with the experimental verification of the effectiveness of this method in reducing vibrations, as well as the detailed studies and so on needed at that time for excitation forces caused by the motion of spectators during concert performances.

3 BASIC THEORY

For many years, vibration-proofing has been the most commonly used method for dealing with sources of ordinary environmental vibration problems. As noted in section 1, vibration-proofing involves the use of rubber springs or rubber cushions, metal springs, air cushions and so on to provide elastic support in order to reduce the excitation force transmitted to the structure. This method is widely used to deal with vibrations produced by the operation of equipment, solid-borne sound, floor impact noise and so on. Specific methods that are adopted include providing elastic support for equipment or its foundation or using rubber insulation, glass wool or other materials to provide elastic support of the entire floor – a method known as the floating floor construction method. However, these methods are generally used for reducing vibrations higher than approximately 8 Hz.

The theory of reducing excitation force through vibration-proofing will be briefly explained using the single degree of freedom spring–mass system model shown in Figure 1. The model shown in the figure is a vibration system in which the mass point for mass, m, is supported by means of a spring with spring constant, k, and a damper with viscous damping coefficient, c. If the mass point displacement is x, the excitation force acting on the mass point is $F_0(t)$, the excitation force applied to the support point for the vibration system of the structure is $F_{TR}(t)$, and the frequency of the excitation force acting on the mass point is ω, the equation of motion for this model can be expressed as follows:

$$F_0(t) \; = F_0 \cdot e^{j\omega t} = m\ddot{x} + c\dot{x} + kx \qquad (1)$$

$$F_{TR}(t) \; = F_{TR} \cdot e^{j\omega t} = c\dot{x} + kx \qquad (2)$$

If $x = C \cdot e^{j\omega t}$ (C: complex amplitude) is used as the formula and this value is substituted in equations (1)

and (2), C is derived as follows:

$$C = \frac{1}{1 - (\omega/\omega_0)^2 + 2h(\omega/\omega_0)j} \cdot \frac{F_0}{k} \qquad (3)$$

$$C = \frac{1}{1 + 2h(\omega/\omega_0)j} \cdot \frac{F_{TR}}{k} \qquad (4)$$

where h and ω_0 are the damping factor and natural frequency, respectively, for the vibration system. From equations (3) and (4), the transfer coefficient, τ, for the transfer of excitation force from the mass points to the vibration system support points can be determined as shown in equation (5).

$$\tau = \frac{F_{TR}}{F_0} = \left| \frac{1 + 2h(\omega/\omega_0)j}{1 - (\omega/\omega_0)^2 + 2h(\omega/\omega_0)j} \right| \qquad (5)$$

In addition, the phase lag ϕ for the excitation force applied to the mass points and the excitation force transmitted to the support points of the vibration system for the structure are as shown in equation (6).

$$\phi = \tan^{-1} \left\{ \frac{2h(\omega/\omega_0)^3}{1 - (\omega/\omega_0)^2 + 4h^2(\omega/\omega_0)^2} \right\} \qquad (6)$$

Figure 2 shows the relationship between the τ value derived from equation (5) and the ratio of the excitation frequency to the natural frequency of the vibration system (ω/ω_0). The figure shows three curves, for $h =$ 0.5%, 5% and 10%. These three h values were selected

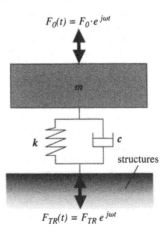

$$F_0(t) = F_0 \cdot e^{j\omega t}$$

$$F_{TR}(t) = F_{TR}\, e^{j\omega t}$$

Figure 1. Single degree of freedom spring – mass system model.

as representative examples from the range of damping factors achievable with existing elastic support materials.

From the figure, it can be seen that, in the range in which $\omega/\omega_0 < \sqrt{2}$, τ is greater than 1, and τ is a particularly large value when $\omega/\omega_0 = 1$. However, in the range in which $\sqrt{2} < \omega/\omega_0$, τ is less than 1, and τ decreases as ω/ω_0 increases. In other words, for the vibration-proofing method, the natural frequency of a vibration system composed of elastic support materials must be set to at least a value lower than $1/\sqrt{2}$ of the excitation frequency. Moreover, with this method, the smaller the value to which the natural frequency of the vibration system is set, the greater will be the effectiveness in reducing excitation force.

Table 1 shows the ω/ω_0 values for each of the three h values such that τ will become 1/1, 1/2, 1/3, 1/4 and 1/5 within the range of $\sqrt{2} < \omega/\omega_0$. As the table shows, in order to, for example, reduce the previously noted excitation force of around 8 Hz or greater to 1/2 or less, for an h value of 0.5% it is necessary for the ω_0 value to be $\omega_0 = 0.58 \times 8\,\text{Hz} = 4.64\,\text{Hz}$ or less. The vibration system with a natural frequency of this degree is achievable – for example, through the method of using metal springs with a damping factor of about 0.5% to support the equipment or its foundation.

However, vibrations are produced during a concert performance when the spectators are standing and moving, mainly during songs with a tempo of 2–3 beats per second.[1),2),3)] In other words, the frequency of excitation force resulting from the motions of people during a concert performance is approximately 2–3 Hz.

To reduce this type of excitation force by means of vibration-proofing, the natural frequency of the vibration system must be reduced to at least 1.41 Hz, 1/2 times the 2 Hz that is the smallest excitation frequency. With existing technologies, this would require the use of air cushions. However, the h value for a vibration system that uses air cushions is generally around 5–10%. Even if h were 5%, in order to reduce the excitation force to 1/2 or less, according to Table 1 ω_0 must be reduced to $\omega_0 = 0.57 \times 2\,\text{Hz} = 1.14\,\text{Hz}$ or less. There are limits to the degree to which a vibration system with a natural frequency this low can be achieved merely by reducing spring rigidity, so the mass of the mass points must be increased. Specifically, the mass of the sections of the equipment foundation and floating floor that are supported with elastic materials must be increased to a degree that is difficult to support in ordinary structures. In addition, as the capacity of the air cushions will also be increased, a great deal of equipment will be needed for their operation, including auxiliary tanks, control valves, piping systems and so on. Moreover, many technical issues can be expected to arise that must be resolved before the basic performance required of the floor can be secured, for example control of horizontal shaking and so on.

For these reasons, the use of vibration-proofing to reduce the excitation force of approximately 2–3 Hz produced by the motion of people during a concert performance will be an enormous undertaking in terms of both scale and cost, and many technical difficulties can be anticipated as well. For this reason, the use of this method is seen as unrealistic, and at present it is not being implemented.

4 PHASE INTERFERENCE THEORY

Vibrations during concert performances are produced mainly by the rhythmic motion of spectators in time to the music. Accordingly, if half of the people moved in the opposite phase, the excitation forces would cancel one another out, and the total excitation force transmitted to the structure would be close to zero.

Figure 3 shows the relationship between ω/ω_0 and the phase lag ϕ between the excitation force applied to the mass points of the single degree of freedom

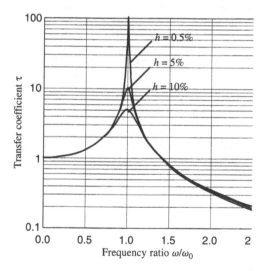

Figure 2.   Relationship between τ and ω/ω_0 (vibration-proofing).

Table 1.   Sample relationship between τ for individual h values and ω/ω_0 (vibration-proofing).

h (%)	0.5%					5%					10%				
τ	1/1	1/2	1/3	1/4	1/5	1/1	1/2	1/3	1/4	1/5	1/1	1/2	1/3	1/4	1/5
ω/ω_0	1.41~	1.73~	2.00~	2.24~	2.45~	1.41~	1.74~	2.01~	2.26~	2.48~	1.41~	1.76~	2.05~	2.32~	2.57~
ω_0 (Hz)	~0.71 ω	~0.58 ω	~0.50 ω	~0.45 ω	~0.41 ω	~0.71 ω	~0.57 ω	~0.50 ω	~0.44 ω	~0.40 ω	~0.71 ω	~0.57 ω	~0.49 ω	~0.43 ω	~0.39 ω

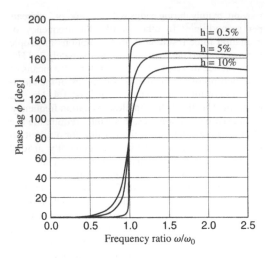

Figure 3. Relationship between ϕ and ω/ω_0 (vibration-proofing).

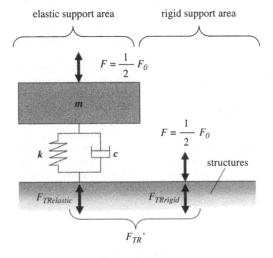

Figure 4. Explanatory model (phase interference theory).

vibration system shown in Figure 1 and the excitation force transmitted to the support points of the vibration system for the structure. As the figure shows, at approximately $\omega/\omega_0 = 1$, ϕ changes around 180 degrees. In other words, when motion occurs on a floor provided with elastic support, the excitation force transmitted to the structure becomes roughly the opposite phase. This means that if approximately half of the spectators are caused to move on a floor provided with elastic support, it is as if they are moving in the opposite phase, and it is theoretically possible to cancel out the excitation force. In this study, the method using this principle is called "the phase interference method for passive reduction of multiple excitation forces."

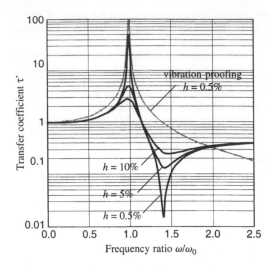

Figure 5. Relationship between τ' and ω/ω_0 (phase interference theory).

The theory of reducing excitation force with this method will be explained using the model shown in Figure 4. This method is characterized by dividing the floor into two areas: one provided with elastic support (elastic support area) and one not provided with elastic support (rigid support area). In the figure, these areas are shown on the left and right, respectively. The elastic support area is modeled using the single degree of freedom spring–mass system in Figure 1.

The excitation force transmitted to the structure from the elastic support area and the rigid support area is denoted by $F_{TRelastic}$ and $F_{TRrigid}$, respectively, and the sum of these values is denoted by F_{TR}'.

$$F_{TR}' = F_{TRelastic} + F_{TRrigid} \tag{7}$$

Here $F_{TRelastic}$ and $F_{TRrigid}$ are expressed by equations (8) and (9) when half of the excitation force is applied to each area.

$$F_{TRelastic} = \frac{1}{2} F_0 \cdot \frac{1 + 2h\left(\omega/\omega_0\right)j}{1 - \left(\omega/\omega_0\right)^2 + 2h\left(\omega/\omega_0\right)j} \tag{8}$$

$$F_{TRrigid} = \frac{1}{2} F_0 \tag{9}$$

From the relationships depicted in equations (7) through (9), the transfer coefficient τ' for excitation force with this method is expressed by equation (10).

$$\tau' = \frac{F_{TR}'}{F_0} = \left| \frac{1 - 1/2\left(\omega/\omega_0\right)^2 + 2h\left(\omega/\omega_0\right)j}{1 - \left(\omega/\omega_0\right)^2 + 2h\left(\omega/\omega_0\right)j} \right| \tag{10}$$

Figure 5 shows the relationship between the τ' derived with equation (10) and the ratio of the excitation

Table 2. Sample relationship between τ' and ω/ω_0 for each h value (phase interference theory).

h (%)	0.5%					5%					10%				
τ	1/1	1/2	1/3	1/4	1/5	1/1	1/2	1/3	1/4	1/5	1/1	1/2	1/3	1/4	1/5
ω/ω_0	1.15	1.22	1.26 / 2.00	1.29 / 1.73	1.31 / 1.63	1.15	1.23	1.28 / 1.97	1.32 / 1.70	1.35 / 1.59	1.15	1.26	1.35 / 1.88	1.47 / 1.53	-
ω_0 (Hz)	0.87ω	0.82ω	0.50ω / 0.79ω	0.58ω / 0.78ω	0.61ω / 0.76ω	0.87ω	0.81ω	0.51ω / 0.78ω	0.59ω / 0.76ω	0.63ω / 0.74ω	0.87ω	0.79ω	0.53ω / 0.74ω	0.65ω / 0.68ω	-

frequency to the natural frequency of the vibration system in the elastic support area (ω/ω_0). For purposes of comparison, the figure also shows the relationship between τ and ω/ω_0 for vibration-proofing (when $h = 0.5\%$).

From this figure, τ' is greater than 1 in the range of $\omega/\omega_0 < \sqrt{4/3}$ and is extremely large when $\omega/\omega_0 = 1$. In the range of $\sqrt{4/3} < \omega/\omega_0$, τ' is less than 1, and the value is particularly small when $\omega/\omega_0 = \sqrt{2}$. It can also be seen that, in the range of $\sqrt{2} < \omega/\omega_0$, τ' converges to 0.5 as ω/ω_0 increases. In other words, with this method, the natural frequency of the vibration system in the elastic support area must be set at least to a value smaller than $1/\sqrt{4/3}$ the excitation frequency. Setting the natural frequency of the vibration system to approximately $1/\sqrt{2}$ the excitation frequency will provide greater effectiveness in reducing excitation force.

Table 2 shows the ω/ω_0 value needed so τ' is 1/1, 1/2, 1/3, 1/4 and 1/5 in the range of $\sqrt{4/3} < \omega/\omega_0$, for each of the three h values. From the table, it can be seen that, to reduce the excitation force of approximately 2–3 Hz produced by the motion of people during a concert performance as noted in section 3, at the very least the natural frequency of the vibration system in the elastic support area must be set to $1/\sqrt{4/3}$ of 2 Hz (1.73 Hz) or less. Moreover, to reduce excitation force to 1/2 or less, in the case of $h = 0.5\%$, ω_0 must be set to $\omega_0 = 0.82 \times 2$ Hz = 1.64 Hz or less; in the case of $h = 0.5\%$, ω_0 must be set to $\omega_0 = 0.81 \times 2$ Hz = 1.62 Hz or less. A vibration system with a natural frequency of this degree is achievable with existing technologies, from the standpoint of the rigidity of the springs used and the mass of the mass points and so on.

Moreover, in contrast to the vibration-proofing method, in which the entire floor must be provided with elastic support, this method achieves greater effectiveness in reducing excitation force than with the vibration-proofing method, even though elastic support is provided for only about half the floor area. Combined with the fact that the natural frequency of the vibration system can be set to a high frequency value, this method can be considered to be more realistic in terms of scale and cost as regards reducing the excitation force produced by the motion of people during a concert performance.

5 APPLICABILITY STUDY TO BUILDING STRUCTURES

5.1 Proportion of excitation force

In section 4, as the first stage of the study of this method, a study was conducted in which approximately half of the excitation force was applied to each of two areas: an area provided with elastic support and an area not provided with elastic support. However, during an actual concert performance, it is thought to be extremely difficult to ensure that there are always equal numbers of spectators in each area, and that half of the excitation force is always applied to each area. For this reason, a study was conducted in which the proportion of excitation force applied to the elastic support area and the rigid support area varied from a 1:1 ratio.

If the proportions of excitation force for the elastic support area and the rigid support area are represented by α and β ($\alpha + \beta = 1$), respectively, then $F_{TRelastic}$ and $F_{TRrigid}$ can be expressed by equations (11) and (12):

$$F_{TRelastic} = \alpha\, F_0 \cdot \frac{1 + 2h(\omega/\omega_0)j}{1 - (\omega/\omega_0)^2 + 2h(\omega/\omega_0)j} \qquad (11)$$

$$F_{TRrigid} = \beta F_0 \qquad (12)$$

From the relationship of equations (7), (11) and (12), the transfer coefficient τ' for excitation force can be expressed by equation (13).

$$\tau' = \frac{F_{TR}'}{F_0} = \left| \frac{1 - (1-\alpha)(\omega/\omega_0)^2 + 2h(\omega/\omega_0)j}{1 - (\omega/\omega_0)^2 + 2h(\omega/\omega_0)j} \right| \qquad (13)$$

Figure 6 shows the relationship between ω/ω_0 and τ' derived by equation (13). The figure shows five curves, for $\alpha = 30\%, 40\%, 50\%, 60\%$ and 70%, for each of the three h values.

From this figure, it can be seen that, as α becomes smaller, the lower limit of ω/ω_0 such that τ' is smaller than 1, as well as the ω/ω_0 value such that τ' becomes extremely small, are reduced. At this point, the lower limit for the value of ω/ω_0 such that τ' becomes less

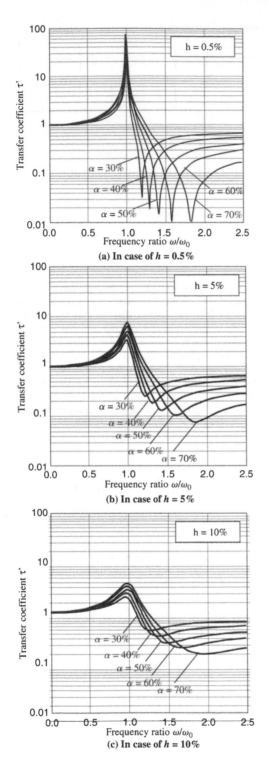

(a) In case of $h = 0.5\%$

$h = 0.5\%$

$\alpha = 30\%$
$\alpha = 40\%$
$\alpha = 60\%$
$\alpha = 50\%$
$\alpha = 70\%$

Transfer coefficient τ'

Frequency ratio ω/ω_0

(b) In case of $h = 5\%$

$h = 5\%$

$\alpha = 30\%$
$\alpha = 40\%$
$\alpha = 50\%$
$\alpha = 60\%$
$\alpha = 70\%$

Transfer coefficient τ'

Frequency ratio ω/ω_0

(c) In case of $h = 10\%$

$h = 10\%$

$\alpha = 30\%$
$\alpha = 40\%$
$\alpha = 50\%$
$\alpha = 60\%$
$\alpha = 70\%$

Transfer coefficient τ'

Frequency ratio ω/ω_0

Figure 6. Relationship between τ' and ω/ω_0 (when the excitation force ratio fluctuates).

than 1 can be determined from equation (13). If, for purposes of simplicity, h is set to $h = 0\%$, this can be expressed by equation (14).

$$\omega/\omega_0 = \sqrt{\frac{1 + 2\alpha}{1 + \alpha}} \tag{14}$$

It is also possible to use equation (13) to determine the value of ω/ω_0 such that τ' becomes extremely small. If $h = 0\%$ in the same manner as before, this can be expressed by equation (15).

$$\omega/\omega_0 = \sqrt{\frac{1}{1 - \alpha}} \tag{15}$$

The value such that τ' will converge when the value for ω/ω_0 is sufficiently large is the same as the value for β. In other words, the transfer coefficient will converge to the same value as the proportion of excitation force applied to the rigid support area. This is because, when ω/ω_0 is sufficiently large, $F_{TRelastic}$ will converge to zero and the equation will become as follows:

$$F_{TR}' = F_{TRrigid} = \beta F_0 \tag{16}$$

Table 3 shows the ω/ω_0 value needed to make τ' equal to 1/1, 1/2, 1/3, 1/4 and 1/5 when $\alpha = 30\%$, 40%, 50%, 60% and 70%, for each of the three h values. From the table, it can be seen that, to reduce the excitation force produced by the motion of spectators during a concert performance, the natural frequency of the vibration system in the elastic support area should be set to 0.81 of 2 Hz (1.62 Hz) or less when α fluctuates in the range of 30–70%, and to 0.83 of 2 Hz (1.66 Hz) or less when α fluctuates in the range of 40–60%. Compared to section 4, in which it was noted that the value should be set to 1.73 Hz or less when $\alpha = 50\%$, the larger the range in which α is predicted to fluctuate, the smaller the value to which the natural frequency of the vibration system in the elastic support area must be set.

As in section 3 and 4, a study was also conducted of the ω_0 value needed to reduce excitation force of 2–3 Hz to 1/2 or less. Here h was set to 5% to enable comparison with the study in section 4.

From the table, it can be seen that, when α is expected to fluctuate within the range of 40–60%, ω_0 must be set in the range of $\omega_0 = 0.45\,\omega - 0.77\,\omega$. Substituting a value of $\omega = 2\,\text{Hz}$ here results in a value of $\omega_0 = 0.9$–1.54 Hz; substituting a value of $\omega = 3\,\text{Hz}$ here results in a value of $\omega_0 = 1.35$–2.31 Hz. Accordingly, in order to reduce by 1/2 the total excitation force in the range of 2–3 Hz, ω_0 must be set within the range of 1.35–1.54 Hz. The vibration system with a natural frequency of this degree is achievable with existing technologies through the use of air cushions, with no need to set the mass of the mass points particularly large.

Table 3. Relationship between τ' and ω/ω_0 for each h value (when the excitation force ratio fluctuates).

h (%)	0.5%					5%					10%				
τ'	1/1	1/2	1/3	1/4	1/5	1/1	1/2	1/3	1/4	1/5	1/1	1/2	1/3	1/4	1/5
$\alpha \approx 30\%$ τ'/ω_0	1.09	1.12	1.14	1.15	1.16	1.09	1.13	1.17	–	–	1.09	1.19	–	–	–
		1.58	1.35	1.29	1.26		1.56	1.31							
ω_τ (Hz)		0.63ω	0.74ω	0.78ω	0.79ω		0.64ω	0.76ω	–	–			–	–	–
	0.92ω	0.89ω	0.88ω	0.87ω	0.86ω	0.92ω	0.88ω	0.85ω			0.92ω	0.84ω			
$\alpha \approx 40\%$ τ'/ω_0	1.12	1.17	1.20	1.21	1.23	1.12	1.18	1.22	1.25	1.28	1.12	1.22	1.34	–	–
		2.24	1.58	1.46	1.41		2.21	1.55	1.42	1.35		2.15	1.41		
ω_τ (Hz)		0.45ω	0.63ω	0.68ω	0.71ω		0.45ω	0.65ω	0.70ω	0.74ω		0.47ω	0.71ω	–	
	0.89ω	0.85ω	0.83ω	0.83ω	0.81ω	0.89ω	0.85ω	0.82ω	0.80ω	0.78ω	0.89ω	0.82ω	0.75ω		
$\alpha \approx 50\%$ τ'/ω_0	1.15	1.22	1.26	1.29	1.31	1.15	1.23	1.28	1.32	1.35	1.15	1.26	1.35	1.47	–
			2.00	1.73	1.63			1.97	1.70	1.59			1.88	1.53	
ω_τ (Hz)			0.50ω	0.58ω	0.61ω			0.51ω	0.59ω	0.63ω			0.53ω	0.65ω	–
	0.87ω	0.82ω	0.79ω	0.78ω	0.76ω	0.87ω	0.81ω	0.78ω	0.76ω	0.74ω	0.87ω	0.79ω	0.74ω	0.68ω	
$\alpha \approx 60\%$ τ'/ω_0	1.20	1.29	1.35	1.39	1.41	1.20	1.30	1.37	1.41	1.45	1.20	1.32	1.42	1.50	1.61
			3.16	2.23	2.00			3.12	2.20	1.95			3.01	2.07	1.76
ω_τ (Hz)			0.32ω	0.45ω	0.50ω			0.32ω	0.45ω	0.51ω			0.33ω	0.48ω	0.57ω
	0.83ω	0.78ω	0.74ω	0.72ω	0.71ω	0.83ω	0.77ω	0.73ω	0.71ω	0.69ω	0.83ω	0.76ω	0.70ω	0.67ω	0.62ω
$\alpha \approx 70\%$ τ'/ω_0	1.24	1.37	1.45	1.51	1.55	1.24	1.38	1.47	1.53	1.58	1.24	1.40	1.51	1.60	1.68
				3.87	2.82				3.82	2.78				3.65	2.61
ω_τ (Hz)				0.26ω	0.35ω				0.26ω	0.36ω				0.27ω	0.38ω
	0.81ω	0.73ω	0.69ω	0.66ω	0.65ω	0.81ω	0.72ω	0.68ω	0.65ω	0.63ω	0.81ω	0.71ω	0.66ω	0.63ω	0.60ω

In the event that α is expected to fluctuate within the range of 30–70%, ω_0 must be set in the range of $\omega_0 = 0.64\omega - 0.72\omega$. Substituting a value of $\omega = 2\,\mathrm{Hz}$ here results in a value of $\omega_0 = 1.28$–$1.44\,\mathrm{Hz}$; substituting a value of $\omega = 3\,\mathrm{Hz}$ here results in a value of $\omega_0 = 1.92$–$2.16\,\mathrm{Hz}$, with no overlap between these two frequency ranges. These results indicate that it is not possible to reduce by 1/2 the total excitation force in the range of 2–3 Hz.

Based on the above results, it is clear that, although there are limits to the fluctuation range for the proportion of excitation force for which this method can be applied, the method has practical applicability. For example, when the fluctuation range is 40–60%, existing technologies can be used to reduce the excitation force to 1/2 or less.

5.2 Habitability to the floor vibrations

Next, let us study the floor vibrations produced in the elastic support area by the motion of spectators, from the standpoint of comfort or habitability.

In the model shown in Figure 4, if the excitation force applied to the elastic support area is represented by $F\ (=F_0/2)$ and the displacement amplitude of the mass points is represented by x, the compliance x/F, meaning the ease with which the mass points vibrate (the displacement amplitude for each unit of excitation force), is expressed by equation (17).

$$\frac{x}{F} = \frac{1}{m\omega^2 \sqrt{\left\{1-\left(\dfrac{\omega}{\omega_0}\right)^2\right\}^2 + \left\{2h\left(\dfrac{\omega}{\omega_0}\right)\right\}^2}} \qquad (17)$$

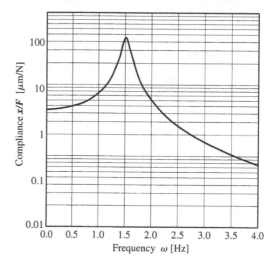

Figure 7. Sample relationship between x/F and ω.

Figure 7 shows an example of the relationship between x/F and ω when spectators move above this mass point. Here ω is set to 1.54 Hz based on the results of the study in section 5.1. As the mass point, a concrete panel ($m = 2400\,\mathrm{kN}$) measuring $10\,\mathrm{m} \times 10\,\mathrm{m} \times 1\,\mathrm{m}$, thought to be within the achievable range judging from the standpoint of the load on the structure, is used. h is set to 5% assuming the use of air cushions.

Of the excitation forces produced by the motion of spectators during a concert performance, the excitation frequency component of approximately 2–3 Hz is thought to be approximately 0.25 times the aggregate weights of all of the spectators.[2),3)] Based on this assumption, the excitation force in the event of movement by 300 spectators (weighing 0.6 N each) is $0.6 \times 300 \times 0.25 = 45$ kN. Accordingly, when $\omega = 2$ Hz, the compliance value is 0.064 mm/kN, and the displacement amplitude is $0.064 \times 45 = 2.9$ mm (acceleration amplitude 45.5 gal). Similarly, when $\omega = 3$ Hz, the displacement amplitude is 0.7 mm (acceleration amplitude 25.4 gal). However, a previous study found that the allowance in cases in which the vibrations produced by aerobics, by people standing and moving, etc. can be felt by those producing the excitation is proportional to acceleration if it the vibrations are approximately 8 Hz or less, and in many cases the allowance is thought to be approximately 100 gal.[3)] If this value is compared to the above calculations, the vibrations produced are much lower than the allowance and pose virtually no problem from the standpoint of comfort. Furthermore, the mass of the mass points in the elastic support area can be reduced and the damping factor can be lowered. The same study was also conducted for harmonic components two to three times larger, and it was confirmed that there was greater latitude than in the case of the excitation frequency, so these components posed no problem.

Based on the above, this method has been shown to be sufficiently applicable from the standpoint of floor vibrations produced in the elastic support area during a concert performance as well.

6 CONCLUSIONS

The conclusions of this study are as follows:

- Vibration-proofing, which for many years has been the most commonly used method for dealing with vibration sources, was examined and the theory by which this method reduces excitation force was discussed. It was also noted that the use of vibration-proofing to reduce excitation force of approximately 2–3 Hz caused by the motion of spectators during a concert performance would be an enormous undertaking in terms of scale and cost, based on the standpoint of the rigidity and mass, etc. of the springs that would need to be used, and moreover that this method would involve technical difficulties as well.
- The authors proposed a new method of dealing with vibration sources called the phase interference method for passive reduction of multiple excitation forces, and presented the theory of how this method reduces excitation force. It was shown that this method allows the natural frequency for the elastic support area to be set to a higher value than in the case of vibration-proofing, and from the standpoint

of the rigidity and mass, etc. of the springs used, it is achievable with existing technologies. Furthermore, whereas with vibration-proofing elastic support must be provided for the entire floor, with this new method only about half of the floor need be provided with elastic support, and the method provides greater effectiveness in reducing excitation force than vibration-proofing. Thus it has many advantages in terms of both scale and cost.

- The application of this method to an actual structure was envisioned, and a study was conducted in which the excitation force applied to an area provided with elastic support and an area not provided with elastic support was varied starting from a 1:1 ratio. The study found that, although there were limits to the fluctuation range for the proportion of excitation force within which this method could be used, the method was capable of reducing by one-half or more the excitation force within a fluctuation range of, for example, 40–60% using existing technologies. This demonstrated that method has sufficient practical applicability.
- A study was also conducted for the habitability to floor vibrations produced within the elastic support area. The study found that if the mass of the mass points and the damping factor for the elastic support area were set so as to be within the achievable range from the standpoint of the load on the structure and the properties of the springs used, the vibrations produced would be far less than the allowable value. Accordingly, it was determined that this method has sufficient practical applicability from the standpoint of the comfort of the persons producing the excitation force during concert performances as well.

ACKNOWLEDGEMENTS

The authors would like to thank Takayuki Abe and Hi-rokazu Yoshioka of the Takenaka Research and Development Institute for invaluable assistance and cooperation regarding various aspects of this study.

REFERENCES

(1) Yutaka Yokoyama, Hiroshi Kushida, Takeru Hiro-matsu and Hidenori Ono, "Study on vibration caused by actions of audience during concert is given: Investigation of vibrational environment and estimation of exciting force" (Papers of the Architectural Institute of Japan (Structure) No. 434, April 1992, pp. 21–30).
(2) Architectural Institute of Japan, "Guidelines for the evaluation of habitability to building vibration" (2004).
(3) Architectural Institute of Japan, "Recommendations for Loads on Buildings" (2004).

Environmental Vibrations – Takemiya (ed.)
© *2005 Taylor & Francis Group, London, ISBN 0 415 39035 4*

Vibration isolation of gas/water-filled cushion wall barrier

T. Takatani & Y. Kato

Department of Civil Engineering, Maizuru National College of Technology, Kyoto, Japan

ABSTRACT: A vibration mitigation countermeasure using gas-filled cushion has been proposed in recent years and several experiments have been conducted in some countries. Because gas cushion material has characteristics of both open trench and concrete wall barrier, it may be the same effective as open trench for vibration mitigation countermeasure in low frequency range. A numerical analysis is carried out in order to investigate the vibration isolation effect of a hybrid model, which consists of a sheet pile and gas/water-filled cushions. The vibration mitigation countermeasure using the hybrid model using gas-filled cushion has a significant vibration isolation effect in comparison with the concrete wall barrier or the sheet pile barrier.

1 INTRODUCTION

Many vibration isolation experiments, numerical simulations and analyses have been carried out in order to investigate not only the vibration isolation mechanism but the vibration isolation effect using many mitigation countermeasures such as open trench, sheet pile, concrete wall, polyethylene foam and so on. In general, both concrete wall and sheet pile barriers are effective for the vibration mitigation countermeasure in high frequency range, while they are not effective in low frequency range. Although open trench is generally effective against vibration obstacles due to propagating waves from artificial oscillations, it has some problems such as its long-term stability and the maintenance after construction. In recent years, a vibration mitigation countermeasure using gas-filled cushion has been proposed (Massarsch, K.R., 2005) and several experiments on gas-filled cushion barrier have been conducted in some countries. Because gas cushion material has characteristics of both open trench and concrete wall barrier, it may be the same effective as open trench for the vibration mitigation countermeasure in low frequency range.

In this paper, a numerical analysis is carried out in order to investigate the vibration isolation effect of a hybrid model using gas/water-filled cushions. The numerical analysis is two-dimensional dynamic nonlinear finite element method based on an effective stress theory. In this finite element analysis, a nonlinear relationship between shear stress and shear strain of soil element is accurately expressed by a multi shear spring model.

The vibration isolation effectiveness for the hybrid model using gas/water-filled cushions are numerically investigated when both gas and water-filled cushions are assumed to be simulated by changing the values of both density and bulk modulus.

In this paper, the vibration isolation model-soil interaction analysis is carried out by using an actual vertical loading wave in order to clarify the differences among some models. The vibration acceleration level on the ground surface is numerically obtained. The vibration isolation effects for open trench, sheet pile, water and gas cushions and hybrid models using both sheet pile and gas/water cushions are investigated from a view point of the vibration acceleration level in this paper.

2 NUMERICAL ANALYSIS

2.1 *Hybrid vibration isolation model*

Recently, both sheet pile and concrete wall barrier have been used as a vibration isolation countermeasure. Although they have a significant vibration isolation effect in high frequency range, they do not have any effect in low frequency range of the vibration pollution. On the other hand, open trench may be more effective in low frequency range if the depth of open trench can be one of four for the length of propagating wave in soil. Open trench has a problem on the maintenance of construction and its long-term stability. In order to solve there problems mentioned above, the vibration isolation countermeasure using gas cushion material have been developed by Massarsch, K.R. (2005).

In Japan, Ohtsuka et al.(2004) conducted a field experiment on a hybrid vibration isolation model, which consists of a sheet pile, soil cement and gas-filled cushion as shown in Figure 1. This hybrid model

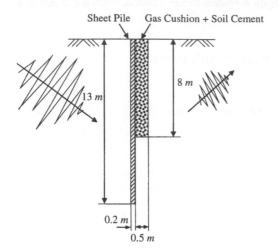

Figure 1. Hybrid vibration isolation model.

Table 1. Soil properties of construction site for hybrid model.

Depth (m)	Soil Type	Unit weight (t/m³)	Shear wave velocity (m/sec)
0.0~1.8	Sandy Gravel with clay	2.661	150
1.8~3.5	Fine sand with Silt	2.665	100
3.5~5.0	Sandy Silt	2.667	115
5.0~8.0	Sandy Silt	2.678	115
8.0~10.5	Fine Sand	2.711	230
10.5~15.0	Fine Sand with Clay	2.694	320
15.0~20.5	Clay Sand	2.693	320

has the same effect as open trench in low frequency range and also can secure its long-term stability. Sheet pile is used for gas cushion material not to rise to the ground surface. Soil cement wall protects gas cushion material from the earth pressure by filling the space between gas cushion material and soil with a bentonite fluid.

Table 1 shows soil properties in the construction site of the hybrid model by Ohtsuka et al.(2004). The soil properties used in numerical analysis are assumed in this paper, based on the soil properties shown in Table 1.

2.2 Numerical analysis for hybrid isolation model

In this paper, the hybrid vibration isolation model-soil interaction analysis is carried out by using two-dimensional non-linear finite element method in order to evaluate the vibration isolation effect of the hybrid model proposed by Ohtsuka et al.

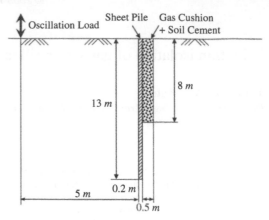

Figure 2. Analytical model for hybrid isolation method.

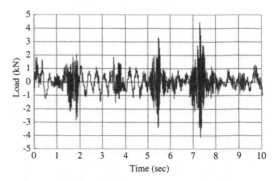

Figure 3. Input vertical loading wave.

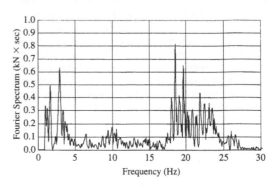

Figure 4. Fourier spectrum of input vertical loading wave.

Figure 2 shows a sketch of the hybrid model and an oscillation point. Because unfortunately the oscillation point was not clearly indicated in the paper by Ohtsuka et al., the oscillation point is assumed to locate as shown in Figure 2 in this paper. The vertical load shown in Figure 3 is employed for an input loading wave acting at the oscillation point, where is 5 m away from the sheet

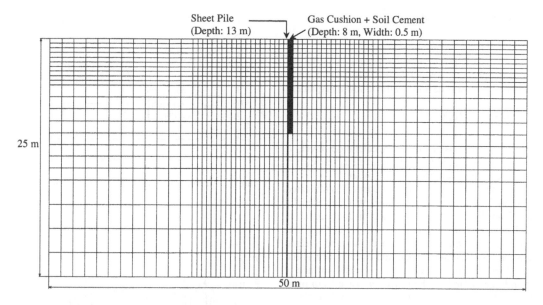

25 m

50 m

Figure 5. Finite element mesh for vibration isolation analysis.

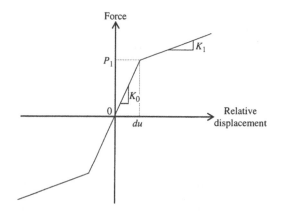

Figure 6. Characteristic of non-linear spring element.

Table 2. Non-linear spring element properties.

	P_1 (kN)	K_0 (kN/m)	K_1 (kN/m)
Water Cushion	10^4	10^5	10^3
Gas Cushion	10	10^2	10

pile of hybrid model. This vertical load was used in the ground vibration analysis carried out by Tokunaga et al.(2000). Fourier spectrum for the vertical input loading wave is indicated in Figure 4. It can be seen from this figure that there exist two significant peak frequencies, 2.83 Hz and 18.85 Hz, in Fourier spectrum.

A numerical analysis used in this paper is two-dimensional dynamic non-linear finite element method based on an effective stress theory. In this finite element analysis, a non-linear relationship between shear stress and shear strain of soil element is accurately expressed by a multi shear spring model (Towhata & Ishihara, 1985) and also the cyclic mobility model (Iai et al.

1990). The significant feature of this model is that the concept of the multiple mechanism, within the framework of plasticity theory defined in strain space, is used a vehicle for decomposing the complex mechanism into a set of one dimensional mechanism.

Figure 5 shows a finite element mesh for the vibration isolation analysis for hybrid model. The sheet pile length is 13 m and the length of gas/water filled-cushion wall barrier is 8 m. The analytical domain is 25 m ×50 m. The transmitting boundary elements are adopted on both sides of finite element mesh to simulate the infinite domain, and also the viscous boundary elements are employed along the bottom of finite element mesh.

In this paper, a bi-linear spring element as shown in Figure 6 is employed to simulate gas/water-filled cushion wall barrier. Generally, the elastic bulk moduli of water and air are 2.2×10^6 and 1.42×10^2 kPa, respectively. Therefore, the properties for non-linear spring element shown in Table 2 are assumed in this paper by considering that gas/water-filled cushion wall barrier is made with soil cement in actual construction site. The relative displacement between two nodal points, du, is assumed to 0.1 m.

455

3 NUMERICAL RESULTS

3.1 Open trench and sheet pile

Figure 7 shows the vibration acceleration levels for "Open trench" and "Sheet Pile" models. The vibration acceleration level is defined by the following equation,

$$L_a = 20 \cdot \log_{10}(a / a_0) \tag{1}$$

where, L_a (dB) is vibration acceleration level, a (m/sec^2) is measured acceleration, and a_0 (m/sec^2) is fundamental acceleration($= 10^{-5}$ m/sec^2).

The vibration level L_a for "No Barrier" is indicated in Figure 7, too. The vibration level of "Open Trench" model on the rear ground surface of open trench is much smaller than "Sheet Pile" model, while the vibration level of "Open Trench" on the ground surface between the oscillation point and open trench is the largest in all. This is because the ground surface is strongly amplified by the reflection wave from open trench. On the other hand, the vibration level of "Sheet Pile" model between the oscillation point and sheet pile is smaller than that of "Open Trench", because the reflection wave from sheet pile is restrained by itself. As the vibration level on the rear ground surface of sheet pile is larger than that of "Open Trench", the vibration isolation effect of "Open Trench" is larger than "Sheet Pile" on the rear ground surface of vibration isolation model.

Figure 8 indicates the vertical acceleration and its Fourier spectrum at the ground surface of the oscillation point in "Open Trench" model. As can be seen from this figure, the vertical acceleration is very large and also there exists a peak frequency at about 2.8 Hz in Fourier spectrum shown in Figure 8(b), and also there is no peak in the frequency range over 10 Hz in comparison with Fourier spectrum of the input vertical loading wave shown in Figure 4. On the other hand, the vertical acceleration and its Fourier spectrum at the ground surface point of 6 m away from oscillation point in "Open Trench" model are illustrated in Figure 9. It can be observed from this figure that the vertical acceleration behind open trench is much smaller than that at the oscillation point shown in Figure 9(a) and also some peak frequencies in Fourier spectrum shown in Figure 9(b) are almost similar to Fourier spectrum at the oscillation point. It should be noted that some peak frequencies in Fourier spectrum of the vertical loading wave disappear in the high frequency range over 5 Hz in Fourier spectrum for "Open Trench" model. In the shear wave velocity $Vs = 150$ m/sec and $f = 5$ Hz, the length of propagating wave is 30 m. Consequently, the depth of open trench is needed about 7.5 m in order to intercept the propagating wave by using "Open Trench" model. This may suggest that "Open Trench" with 8 m depth

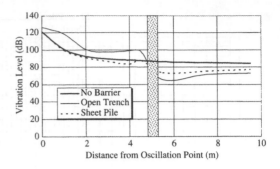

Figure 7. Vibration levels for "Open Trench" and "Sheet Pile".

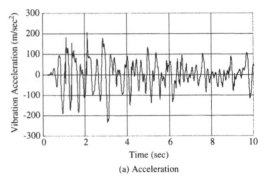

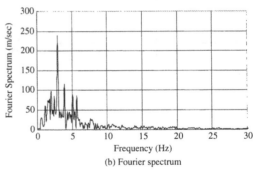

Figure 8. Vertical acceleration and Fourier spectrum at vibration point in "Open Trench".

has a significant effect of intercepting the propagating wave with high frequency component.

3.2 Gas/water-filled cushion model

The vibration acceleration levels for "Sheet Pile", "Water Cushion" and "Gas Cushion" models are shown in Figure 10. As can be seen from this figure, the vibration level between "Gas Cushion" model and the

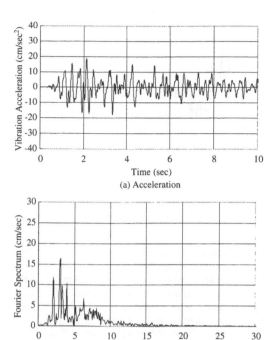

(a) Acceleration

(b) Fourier spectrum

Figure 9. Vertical acceleration and Fourier spectrum behind "Open Trench".

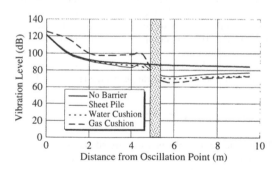

Figure 10. Vibration levels for "Open Trench", "Sheet Pile", "Water Cushion" and "Gas Cushion".

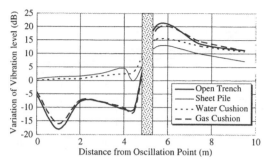

Figure 11. Vibration isolation effects for four models.

non-linear spring constant value for "Water Cushion" model is larger than that for "Gas Cushion" model as shown in Table 2.

The vibration isolation effects for "Open trench", "Sheet Pile", "Water Cushion" and "Gas Cushion" models are indicated in Figure 11 to clarify the difference of vibration isolation effect among four models. The positive value means a significant isolation effect in comparison with "No Barrier" model. Although "Open Trench" and "Gas Cushion" models do not have isolation effect on the ground surface between each model and the oscillation point, they have a significant isolation effect on the rear ground surface of each model in comparison with "Sheet Pile" and "Water Cushion" models. It should be noted that "Gas Cushion" model has the same significant vibration isolation effect as "Open Trench" model on the rear ground surface of vibration isolation barrier, judging from the maintenance of construction and the long-term stability.

3.3 Hybrid vibration isolation model

The vibration acceleration levels for "Sheet Pile + Open Trench", "Sheet Pile + Water Cushion" and "Sheet Pile + Gas Cushion" models are illustrated in Figure 12. The vibration level for "Sheet Pile + Open Trench" model is larger than "Open Trench" model and the vibration isolation effect for "Sheet Pile + Open Trench" is smaller than "Open Trench", because "Sheet Pile + Open Trench" model has an isolation effect of "Sheet Pile" model with a vibration restraining effect because of its large stiffness. On the other hand, "Sheet Pile + Gas Cushion" model has relatively more significant isolation effect in three hybrid models.

The vibration isolation effects for three hybrid models and "Sheet Pile" model are indicated in Figure 13. It can be observed from this figure that "Sheet Pile + Gas Cushion" model has the largest isolation effect in three hybrid models from a view point of the

oscillation point is much larger than those for other models, because the ground surface between "Gas Cushion" model and the oscillation point is strongly amplified by the reflection wave from "Gas Cushion" model the same as "Open Trench" model. However, the vibration level on the rear ground surface of "Gas Cushion" model is the smallest in other models. The vibration level for "Water Cushion" model is slightly larger than "Gas Cushion" model, because the

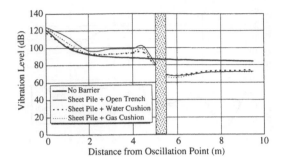

Figure 12. Vibration levels for hybrid models ("S.P. + O.T.", "S.P. + W.C.", and "S.P. + G.C.").

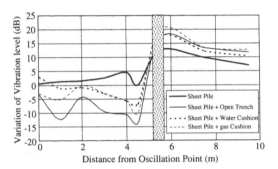

Figure 13. Vibration isolation effects for hybrid models.

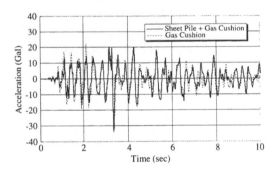

Figure 14. Comparison of acceleration waves for "Sheet Pile + Gas Cushion" and "Gas Cushion" models.

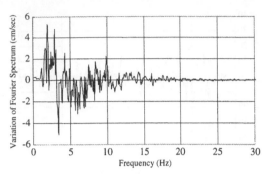

Figure 15. Variation of Fourier spectra for "Sheet Pile + Gas Cushion" and "Gas Cushion" models.

vibration isolation effect on the rear ground surface of each model.

In order to investigate the significant isolation effect of "Sheet Pile + Gas Cushion" model, the comparison of the acceleration waves for "Sheet Pile + Gas Cushion" and "Gas Cushion" model at the same measuring point behind each model is shown in Figure 14. The measuring point locates behind each model and is 6 m away from the oscillation point. As can be seen

from this figure, the difference between both models is not clear. Therefore, the difference in Fourier spectra for both acceleration waves is indicated in Figure 15. This figure means the significance of "Sheet Pile + Gas Cushion" model to "Gas Cushion" model in Fourier spectrum at the measuring point. Fourier spectrum of "Sheet Pile + Gas Cushion" model is larger than that of "Gas Cushion" model within frequency 3 Hz. It is found that "Sheet Pile + Gas Cushion" model cannot intercept the propagating wave with the frequency range within 3 Hz. On the other hand, Fourier spectrum of "Gas Cushion" model is almost larger than "Sheet Pile + Gas Cushion" model in the frequency range from 3 Hz to 10 Hz. "Sheet Pile + Gas Cushion" model can significantly intercept the propagating wave with the frequency range from 3 Hz to 10 Hz in comparison with "Gas Cushion" model. It should be noted that the vibration isolation effect due to the existence of sheet pile is greatly significant in the frequency range from 3 Hz to 10 Hz.

4 CONCLUSIONS

Finite element analysis was carried out in order to evaluate the vibration isolation using gas/water-filled cushion. The numerical analysis in this paper is two-dimensional dynamic non-linear finite element method. The vibration isolation effects for hybrid models were numerically investigated by using an actual vertical loading wave. In this paper a non-linear spring element was employed for the simplest method to simulate gas/water-filled cushion wall barrier model. As a result, it should be noted that "Sheet Pile + Gas Cushion" model has a significant vibration isolation effect in other hybrid models proposed in this paper.

Although the comparison of the analytical result with the experimental data in field test is not included in this paper, it is necessary for an intensive study on the comparison because it is the most important role

458

in the design of vibration mitigation countermeasure. In addition, the vibration isolation effect of hybrid model on the horizontal acceleration waves due to the vertical oscillation may be needed to make some concrete conclusions, although it is not included in this paper. Moreover, a non-linear spring element with bi-linear relationship was used to simulate gas/water-filled cushion material in the hybrid model. A non-linear relationship for gas/water-filled cushion material may be expressed by other mechanism such as a joint element and a non-linear solid element and so on. There seems to be a need for the comparison study concerning the non-linear spring model and the joint element model in future.

REFERENCES

Iai, S., Matsunaga, Y. and Kameoka, T. 1990. Strain space plasticity model for cyclic mobility, Report of Port and Harbour Research Institute, 29(4), 27–56.

Massarsch, K.R. 2005. Vibration isolation using gas-filled cushions, Soil Dynamics Symposium to Honour Prof. Richard D. Woods, Geo-Frontiers 2005, Austin, Texas.

Ohtsuka, M., Tsuboi, H., Isoya, S., Nozu, M., Hioki, K. and Kushihara, S. 2004. Development of hybrid vibration isolation technique using gas-filled cushion, Foundation Engineering & Equipment, Vo.32, No.11, 81–85.(in Japanese)

Tokunaga,N., Morio, S., Iemura, H. and Nishimura, T. 2000. On the participation factor of the surface wave to the ground vibrations propagating from the urban viaduct, J. Struct. Engrg., Vol.46A, 1703–1713.(in Japanese)

Towhata, I. and Ishihara, K. 1985. Modelling soil behaviour under principal stress axes rotation, Proc. 5th Int. Conf. Num Method Geomech., Nagoya, 523–530.

*Mitigation measures against traffic induced
ground vibrations: field measurements*

Environmental Vibrations – Takemiya (ed.)
© *2005 Taylor & Francis Group, London, ISBN 0 415 39035 4*

Some examples about ground vibration isolation using wave barriers

K. Hayakawa

Ritsumeikan University, Shiga, Japan

ABSTRACT: In order to develop an effective and practical method against ground vibration caused by running trains and cars, the authors tried to apply sheet pile walls barrier beside of real railway tracks. And PC wall-piles wave barrier and EPS (Expanded Poly-styrene) wave barrier also were constructed along the real elevated roads. Field tests were carried out to evaluate the effectiveness of these countermeasures for reducing the ground vibrations. An estimation method for vibration reduction by these wave barriers is presented, being based on FEM simulation analysis.

1 INTRODUCTION

Vibration arising form running cars and trains propagates through soil resulting on a ground vibration in their neighborhood. Such ground vibration as well as psychologically adverse effects on the inhabitants around its source. This paper describes a measurement examples conducted on a elevated road and railway.

In first case of the study, field tests were carried out on a wave barrier built of steel sheet-piles to identify its performance in reducing ground vibration arising form a running train. A simulation analysis was subsequently conducted by using tri-dimensional model to qualitatively establish the mechanism through which the wave barrier was capable of reducing such vibration. From the results of conclusion, the analytical method in this study was found to be capable of qualitatively reproducing the performance against vibration of a wave barrier built with steel sheet-pile.

An elevated road or bridge addressed in the second case is located on the west side of the Kuwana Higashi Interchange of the Higashi Meihan(Nagoya-Kobe) Highway. Because the road was built on a relatively soft ground, the passing cars produce strong ground vibration. So, residence in the surrounding areas complained and requested that the disturbing ground vibration should be reduced. PC wall-pile wave barrier was performed around this bridge to isolate the traffic ground vibration.

Numerical analyses and field measurement were carried out to identify the effectiveness of such wave barrier and they could quantitatively simulate and evaluate the effectiveness of PC wall-pile wave barrier.

2 FIELD TESTS

2.1 Test set-up

Shown in Fig.1 is the set-up for the tests, including the relative positions between the railway track, the barrier and the points at which measurement was taken of vibration, including the composition of soil there and the distribution of its N-value. The barrier was 17.2 m long and built of SPll type steel sheet-piles in direction parallel to the railway track. The piles were driven by pressure in two stages, first to a depth of 2 m, and subsequently to 4.5 m, so that data could be taken with the barrier built to a depth above,

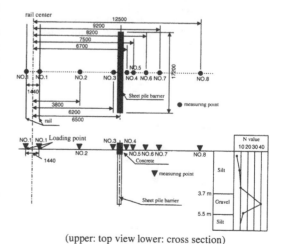

(upper: top view lower: cross section)

Figure 1. The set-up for the tests, including the relative positions.

and also into the gravel stratum having a N-value greater than 30, as identified through a preliminary soil survey, and assumed to be capable of bearing the sheet piles and reflecting the ground vibration.

Vertical acceleration of ground vibration generated by passing train was measured at seven points (No.2 through to No.8) along a line normal to the track and intersecting a joint with a clearance between two rails, the nearest being at 3.8 m, and the farthest at 12.5 m, from the center of the track. As shown in Table 1, data were taken in each of the four phases of the work to built the barrier, offering different conditions.

2.2 Results and analysis

2.2.1 Effect of the barrier

Plotted in Fig. 2 are acceleration levels as recorded during each of four different phases shown in Table 1 as a train was passing by to show how they decreased with distance under four different conditions. The data shown are "all-pass values" (general values without having been filtered through a band-pass) of the averages of several records generated by as many trains passing by. Furthermore, certain correction was made on the data taken at Point No.2 during Phase ll through lV to make them coincide with those recorded at the same point during Phase l with no barrier, so as to facilitate their comparison. It is seen that, with the barrier driven to a depth of 2 m, the level was reduced by about 6 dB at a point just behind it, and with the barrier further

Table 1. Test case.

Case	Countermeasure
STEP1	No barrier
STEP2	Steel Sheet-pile (length 2.0 m)
STEP3	Steel Sheet-pile (length 4.5 m)
STEP4	Steel Sheet-pile (length 4.5 m) + Concrete

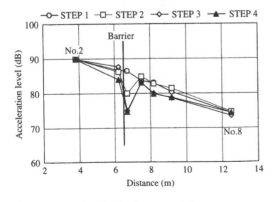

Figure 2. Acceleration levels as recorded.

driven to 4.5 m, by about 12 dB, as compared with that recorded at the same point with no barrier. Under the former condition, however, the level is seen to have grown again to a similar value to that resisted with no barrier, as the distance increased from the barrier, providing no effect at Point No.5 behind the barrier. It can be observed that the positive effect of the barrier as driven to 4.5 m deep was particularly noticeable just behind it, and the effect further increased, albeit slightly, as a concrete crown was built on top of it. With the data taken in front of the barrier some effect is also noted of reflection, which was reduced as the crown was built. At Point No.8, 12.5 m away from the barrier, all the data are seen to converge to the level as recorded with no barrier. These phenomena are as corroborated through a series of experiments conducted on a model by Yoshioka et al[2000].

3 NUMERICAL SIMULATION ANALYSIS

A numerical simulation analysis was conducted by using the data presented above to see whether it was capable of predicting the performance of the barrier as herein addressed in reducing ground vibration.

3.1 The model

The velocity (Vs) at which the S-wave propagates through the soil was calculated by substituting an appropriate N-value as identified through a soil survey into following formula:

$$Vs= 80N^{1/3} \text{ (for sandy soils)}$$
$$Vs=100N^{1/3} \text{ (for clayey soils)} \quad (1)$$

The velocity of the P wave assumed to be 1500 m/s in every case because underground water was present close to the ground surface.

Given in Table 2 are the elements of the model used, and their particulars. The soil was finely divided into eight layers, every one of them thinner than 1/6 of the wave length consistent with velocity Vs at which a S wave propagates through it at frequencies up to 100 Hz. In conformance with the tri-dimensional FEM; the model of the barrier was worked out into shell elements composing a plate having an isotropic stiffness against bending towards its outer surface. The subsequent analysis was conducted with a Super FLUSH/3D program.

3.2 Result and consideration

Figure 3 shows the reduction in acceleration level with distance from the track as calculated. Data are also shown as measured for comparison.

3.2.1 *Reduction on acceleration level with distance*

With no barrier, the result from the analysis are seen to generally agree well with the measured data,

Table 2. Soil property.

Layer	Soil	Unit weight (kN/m³)	S-Wave speed (m/sec)	Damping ratio (%)	Thickness (m)
1	Silt	16.66	170.0	5.00	0.60
2	Gravel	16.66	170.0	5.00	0.80
3	Silt	16.66	170.0	5.00	0.40
4	Silt	16.66	180.0	5.00	1.40
5	Silt	16.66	220.0	5.00	0.50
6	Gravel	16.66	270.0	5.00	1.80
7	Silt	16.66	150.0	5.00	1.00
8	Silt	16.66	180.0	5.00	1.50
Base		16.66	300.0	5.00	

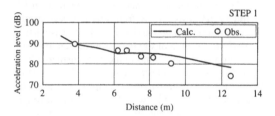

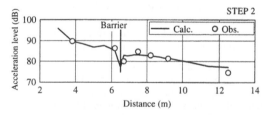

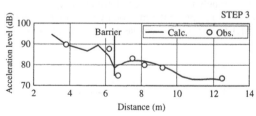

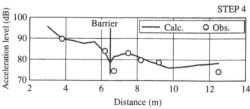

Figure 3. Reduction in acceleration level with distance.

though discrepancy tends to grow to a maximum of about 3 dB as the distance increases from the track. With the barrier in place acceleration level at Point No.4, just behind it, is seen to have been substantially lower than with no barrier. Some effect can be identified of the barrier in the results as calculated as well as those as measured, although the effect with former is seen to be slightly smaller than that by the latter. An increase is seen in the level as calculated at Point No.5, which is particularly noticeable with the barrier driven to 4.5 m deep, and in very good agreement with the measured data. Similar tendencies are observed of the data as calculated as well as those as measured under different conditions. In addition, at every point behind the barrier, the data calculated under any of the conditions are seen to be in excellent agreement with those as measured.

3.2.2 *Effect of possible bending of the barrier*

Shown in the two diagrams in Fig. 4 is the distribution of vertical acceleration as worked out by using the simulation method with a sine wave having a frequency of 60 Hz imparted as input for a period of 8 seconds, and assuming that the barrier was 4.5 m depth. The upper diagram presents a result as worked out by taking into account the bending the barrier may suffer

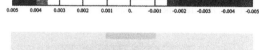

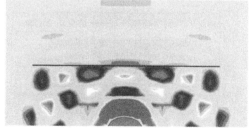

Upper: The result as worked out by taking into account the bending
Lower: The result with no such bending taken into account

Figure 4. Distribution of vertical acceleration.

465

when subject to vibration, while the lower one is the result with no such bending taken into account. In the former the acceleration is seen to have been amplified to some extent while propagating to a point just behind barrier. In the latter, however, no such phenomenon is observed.

Shown in Fig. 5 are two different modes in which the barrier is assumed to bend, and in Fig. 6, the reduction in acceleration level with distance, as worked out for each mode. The data shown indicate that the increase in

acceleration just behind the barrier can be simulated more accurately by taking into account the possible bending the barrier may suffer than otherwise. They further suggest that some additional steps may be needed to limit the deformation the barrier may suffer in longitudinal direction to further improve the performance of this type of vibration barrier.

4 CONCLUSIONS (IN THE FIRST CASE)

In this study, field tests were carried out on a barrier built of steel sheet-piles to identify its performance in reducing ground vibration arising from a running train. A simulation analysis was subsequently conducted by using a tri-dimensional model to quantitatively establish the mechanism through which the barrier was capable of reducing such vibration. From the results of conclusions were drawn as follows:

(1) At point just behind the barrier driven to 2 m and 4.5 m, vibration level was reduced by about 6 dB and 12 dB, respectively from that recorded at the same point with no barrier. The effect was particularly noticeable of the barrier as driven to 4.5 m depth, which was found to increase further, albeit slightly, as a concrete crown was added on top of it.

(2) The effect of the barrier to amplify the level of acceleration behind it was clearly verified through an analysis incorporating the effect of bending the barrier may have suffered. This suggests that steps may need to be taken to keep the barrier from bending in longitudinal direction to further improve its performance against vibration.

The analytical method used in this study was found to be capable of qualitatively reproducing the performance against vibration of a barrier built with steel sheet-piles. Quantitatively, however, the method still leaves much to be desired. Further study is being planned to improve the method by working out a new model to accurately represent both the soil and the barrier.

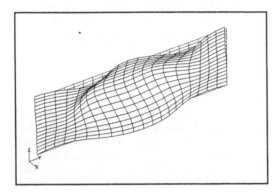

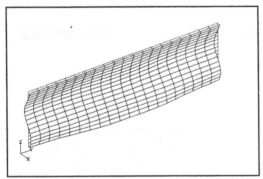

Upper: The result as worked out by taking into account the bending
Lower: The result with no such bending taken into account

Figure 5.   Different modes in which the barrier.

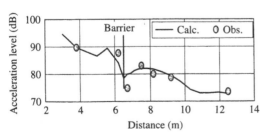

The result as worked out by taking into account the bending

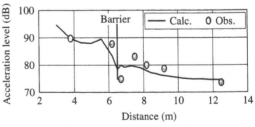

The result with no such bending taken into account

Figure 6.   The reduction in acceleration level with distance.

5 FIELD MEASUREMENT OF GROUND VIBRATION CAUSED BY AN ELEVATED ROAD

5.1 Location and assignment

The source of ground vibration in this study is an elevated road, hereafter simply referred to as the "bridge" with an its over-all span length is 294.5 m with three continuous girders supported by 10 piers at interval of 32.5 m shown in Fig. 7., Fig. 8 shows piers, the ground, PC wall-piles as well as the positions where ground vibration was measured. The boring log shows the composition of the soil. The soil consists of a sandy layer with a N-value between 2 and 20 along a depth of 10.0 m. From a depth of 32.5 m, there is a thick soft-clay stratum having a N-value smaller than 5. Beneath this soft soil, a sandy bearing layer

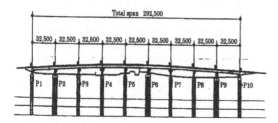

Figure 7.   General view of the bridge.

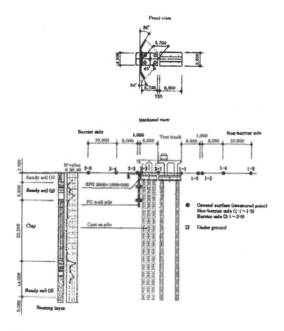

Figure 8.   Relative position of the pier, the barrier and the vibro-meters as set-up.

with an approximate N-value of 20 exists. 12 m-long PC wall-piles were driven one after another to build the continuous wave barrier by using a similar commonly used technique for hollow piles. They were driven with a boring method using an earth auger passing through their hollows. The depth to which the piles were driven were determined in such a way that the barrier can effectively isolate the surface wave and it can also serve as a soil containment wall while work was in progress to build the footing of the bridge. About 300 piles were driven to build the barrier in total at 45 degrees from the ends of the lower work of the bridge.

5.2 Outline of field measurement

An 8-ton sprinkler truck with a weight of 23 tons was used as a test vehicle generating vibration. Vibration was measured on ground surface as well as underground. 10 vibration-level meters were set up on ground surface in an equal pattern both on the barrier side and non-barrier side with respect to the bridge as shown in Fig.8. Underground vibration was measured by using vibro-meters especially designed for the experiment and set up on the inside faces and outside ones at depths of 5 m and 10 m.

5.3 Isolation effect of PC pile-wall barrier

Vibration levels were measured on the soil at four points along each of the two lines. One passing through pier number P1, the other pier number P4. Underground vibration at depths of 5 m and 10 m within the hollow of a pile were also measured. Because the vibration levels measured along the line intersecting P1 were roughly the same as those recorded along the line passing through P4, it was decided to address the former only. These are shown in Fig. 9. It can be observed that on both sides, barrier and non-barrier sides. The levels were reduced as the distance from the footing or from the foot of the pier increases. The reduction was about 10 dB at a distance of 25 m. Furthermore, the vibration level on the barrier side was strongly reduced to the level reached a point immediately in front of the barrier, although it increased again after passing through the barrier. Overall, a reduction of about 5 dB due to the barrier at 25 m away from the footing is observed. Fig. 10 shows the levels of underground vibration as measured in the hollow of the pile in three directions. X-direction is normal to the bridge. Y-direction is parallel to the bridge, and Z is the vertical direction. The vibration levels are higher than those at the further location from the bridge. It is noted that the difference is about 5 dB in the vertical direction, which can be observed as the reduction effect of the barrier. At any rate, effect is seen to have been the largest on Z-component.

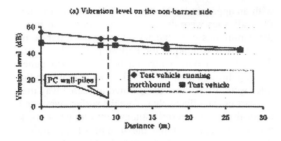

(a) Vibration level on the non-barrier side

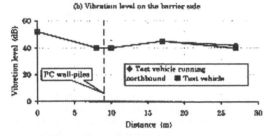

(b) Vibration level on the barrier side

Figure 9. Surface vibration level along the line passing through P1.

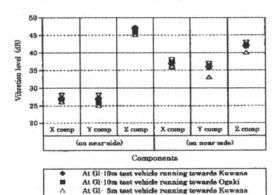

Figure 10. Vibration levels at underground.

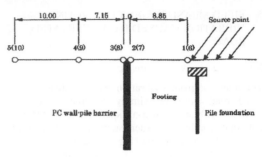

Figure 11. Source point and measured point.

6 NUMERICAL SIMULATION OF FIELD MEASUREMENT

6.1 Modeling and conditions for the analysis

As already mentioned, the ground vibration was measured on the ground surface just above the edges of footing. One measuring point is located just in front of the PC wall-pile barrier and another is located at its behind. Other two points were further away from the barrier. A dynamic response analysis was conducted using a finite element method. The acceleration recorded at the point just above edge of the footing was used as input excitation in order to reproduce the ground vibration at the barrier side and that on the non-barrier side (Kani & Hayakawa, 2002).

Figure 11 shows the observation locations (indicated by the white circle symbol) and the location of applied load (indicated by arrows). Numbers (1) to (5) indicate the observation locations in the area with wave barrier. Numbers (6) to (10) were located in the area without wave barrier. The input acceleration was recorded at ground surface just above the footing. Accelerations at four locations of the ground surface, indicated by the arrows in Fig.11, are used as loading in the numerical calculation. Because many prominent peak are at frequencies around 3 Hz and 10 Hz, unit size of 1 m for the elements over the distance from the accelerating point to the measuring was chosen to improve the accuracy of the response in the high frequency range. The analysis was conducted by using a direct integration technique. The considered response period was 20 sec.

6.2 Discussion and consideration

Figure 12 shows the response on the barrier-side to induced acceleration at the four locations.

The measured data and the calculated response are compared in Fig. 12. The broken line indicates the calculated response on the non-barrier side, while the solid line that on the barrier side. Data shown with white circle symbol are values as measured on the non-barrier side while those with black circle are those on the barrier side. It is seen that the calculated results are in fairly good agreement with the data as measured. It can therefore be assumed that the model used in the analysis is capable of reproducing in a fairly accurate manner albeit with slight errors. Note should be taken that the observed ground vibration was amplified to some extent while it was transmitted from the source to the barrier through the ground. No explanation is possible at this stage on this phenomenon since no data are available. The phenomenon however may be assumed to have arisen from the combination of the ground vibration from source and a secondary one as reflected from the barrier.

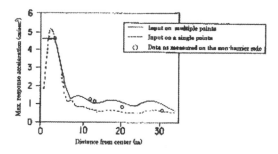

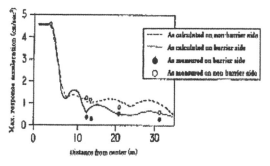

Figure 12. Response of the maximum acceleration.

7 CONCLUSIONS (IN THE SECOND CASE)

In this study, a PC wall-pile barrier was addressed as a means for reducing ground vibration due to road traffics. Both field measurements and an FEM analysis of a elevated road were performed. Also, field model experiments and their simulation are performed. From this study, the following conclusions can be made.

(1) The PC wall-pile barrier was found to be capable of reducing both surface and underground vibration by about 5 dB at a distance 25 m away from the footing of the considered bridge.
(2) The results from the analysis suggested that the footing responded as a rigid body.
(3) The analytical method proved to be capable to fairly accurately reproduce the reduction of the maximum response acceleration on ground surface.

REFERENCES

Yoshioka, O. [2000]. "Basic characteristics of Shinkansen-induced ground vibration and its reduction measures", Proceedings of the International Workshop WAVE 2000, pp. 219–237.
Hayakawa, K., Kani, Y., Matsubara, Y., Matsui, T. and Woods, R.D. [1998a]. "Ground vibration isolation by PC wall-piles", Proc. 4th Intern. Conf. on Case Histories in Geotech. Eng., St. Louis, Missouri, pp. 672–677.
Hayakawa, K., Kani, Y., Matsubara, Y. and Woods, R.D. [1998b]. "The Effectiveness of Pre-Cast Wall-Piles in Reducing Ground Vibration", Building Acoustics, Vol.5, No.3, pp. 185–199.
Hayakawa, K. and Kani, Y. [1999]. "An experimental study on ground vibration isolation using a model", Inter-Noise 1999, pp. 871–874.
Kani, Y., Hayakawa, K. and Maekawa, Y. [2000]. "An experimental study on ground vibration isolation using a model (second report)", Inter-Noise 2000, pp. 2020–2023.
Kani, Y. and Hayakawa, K. [2002]. "Reducing ground vibration around an elevated road with a PC wall-pile barrier", Inter-noise 2002, pp. 1–6.
Matsubara, Y., Kani, Y., Hayakawa, K. and Woods, R.D. [1998]. "Isolation effects of ground vibration due to PC wall-piles on several sites", Earthquake Geotech. Eng., pp. 427–432.

Environmental Vibrations – Takemiya (ed.)
© *2005 Taylor & Francis Group, London, ISBN 0 415 39035 4*

Effect of PC wall-piles weight on ground vibration isolation in the field model tests

Y. Nabeshima
Osaka University, Osaka, Japan

K. Hayakawa
Ritsumeikan University, Shiga, Japan

Y. Kani
Nippon Concrete Co. Ltd., Nagoya, Japan

ABSTRACT: PC wall-piles are practically used as the ground vibration barrier in the field works. Their effectiveness of reducing the ground vibration level was recognized, however, PC wall-piles could not isolate all range of ground vibration and their isolation mechanism was not fully understood. In this study, the isolation mechanism was examined through the field model tests using different weight of PC wall-piles and a numerical simulation. High isolation effect of ground vibration and the transition of isolation mechanism as the pile weight increasing were recognized through the field model tests. The heavy PC wall-piles reduced the ground vibration level by the reflection and light PC wall piles by the transmission of ground vibration.

1 INTRODUCTION

Ground vibration caused from construction works, machineries in factories and various sources of transportation including road and railway through ground is recognized a significant environmental problem. However its transition mechanism through ground is not fully understood because of the complexity of ground formation. Such ground vibration often brings physiologically as well as psychologically adverse effects on the inhabitants around its source, and it also causes mechanical damage to near-by buildings and precision machinery. Many countermeasures for reducing ground vibration were proposed at vibration source, during transition path and at receiving point, however an almighty countermeasure does not exist.

A PC pile means a rectangular pre-stressed concrete pile which has an inner cavity as shown in Figure 1 and its wall barrier installed on the way to transition path of ground vibration is effective to isolate and reduce the level of the ground vibration. A number of reports on the results of field measurements, field model tests, an indoor model test and their theoretical analysis on the effectiveness of PC wall-piles as ground vibration barriers is given by Hayakawa et al. [1998a, 1998b and 1999a], Kani et al. [2000, 2002 and 2003], Matsubara et al. [1998] and Nabeshima et al. [2004a, 2004b].

Figure 1. Appearance of a PC pile.

In this paper, at first, an effectiveness of PC wall-pile barrier for the isolation of ground vibration is summarized on the base of previous works. Then, a series of field model tests of PC wall-pile barrier and its numerical simulation using hybrid three dimensional finite element method are carried out to investigate the effectiveness of PC wall-pile barrier more deeply. Finally, main conclusions are summarized.

2 ISOLATION EFFECT OF A PC WALL PILE BARRIER

Many small size and full size model tests [e.g., Ejima, 1980], field measurements [e.g., Yoshioka et al., 1980],

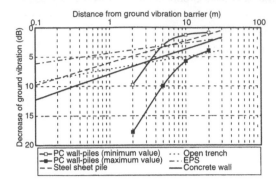

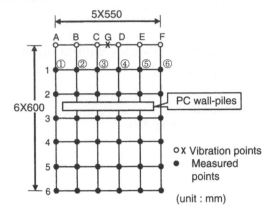

Figure 2. Isolation effect of every ground vibration barriers.

Figure 3. Layout of PC wall-piles and measuring points.

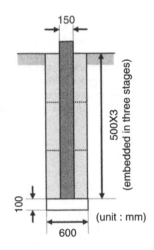

Figure 4. Cross section (side view) of a PC wall.

numerical simulations [e.g., Ohbo & Katayama, 1983] such finite element method and boundary element method were conducted to evaluate the isolation effect of ground vibration barriers. Figure 2 shows a comparison of isolation effect of every ground vibration barriers, which was summarized by Hayakawa [1999b]. Four straight lines in Figure 2 shows regression lines of previous results for steel sheet pile, open trench, expanded polystyrene (EPS) and concrete wall. The embedded depth of the ground vibration barriers were from 1.5 to 2.0 meter for EPS, from 5 to 10 meters for concrete wall and deeper than 10 meters for steel sheet pile.

A PC wall-pile barrier is the most effective in all barriers around 2 meters from a ground vibration barrier. As the distance from the barrier increases, the isolation effect of PC wall-piles rapidly decreases, in contrast, the isolation effect of open trench or EPS is bigger than that of PC wall-piles. Hence, the isolation effect of ground vibration barrier depends on the distance from the vibration source.

3 FIELD MODEL TEST

3.1 Test procedures

A series of field model tests were conducted in a court in front of the Soil Mechanics Laboratory in Ritsumeikan University. Four fields were arranged and each field has 420 cm long and 275 cm wide. The layout of PC wall-piles and measuring points are shown in Figure 3. The soil conditions in the site was as follows; water content of $17.9 \sim 21.8\%$, wet density (or unit weight) of $2.11 \sim 2.31$ g/cm^3, dry density (or unit weight) of $1.79 \sim 1.91$ g/cm^3, and representative shear wave velocity of 149 m/sec. We excavated first a trench of 60 cm wide, 275 cm long and 150 cm deep between the measuring points No. 2 and 3, and constructed a subsurface wall in the trench by placing twelve model PC wall-piles in an array. Each model wall-pile is 150 cm long with 15 cm $\times$ 15 cm square section and contains a cavity of 10 cm in diameter. The

cross section (side view) of the PC wall thus constructed is shown in Figure 4. Three kinds of aggregate were used to change weight of a PC wall pile. The crusher run is used as an aggregate for a normal PC wall-pile, on the other hand, the heavy aggregate (dry density, $3.53 \sim 3.66$ g/cm^3) for heavy PC wall-pile and the light aggregate (dry density, $1.37 \sim 1.84$ g/cm^3) for light PC wall-pile. The weight of a normal PC wall-pile was about 50 kg, a heavy PC wall-pile about 67 kg, a light PC wall-pile about 40 kg. Figure 5 shows an embedding operation of PC wall-piles in an array. Table 1 shows a summary of density and total weight of normal, heavy and light model PC wall-piles.

The ground vibration was measured with six vibration-level meters. As a vibration source, 2.5 or 5 kg weight was dropped from the height of about 0.2 or 0.4 meter, and the propagated waves in vertical acceleration levels in VAL (vibration acceleration level in dB) were measured. The measuring points were set along

472

Figure 5.   Embedding operation of PC wall-piles in an array.

Table 1.   Summary of PC wall-piles.

	Density (ton/m³)	Total weight (kg)
Normal PC wall-pile	2.30	about 50
Heavy PC wall-pile	3.03	about 67
Light PC wall-pile	1.83	about 45

Figure 6.   A testing scene in the field model test.

six lines from ① to ⑥ at 60 cm interval, each interval between lines is 55 cm. Ground vibrations at every measuring points on the lines were measured when the weight was dropped at the impact points A through F, and the acceleration levels on the same lines were compared. A series of measurements were conducted to draw a counter map of ground vibration level when the weight was dropped at the impact point G. Figure 6 shows a testing scene in the field model test.

3.2   Test results

Figure 7 shows the vibration acceleration level measured on the lines of ③ and ④ at every point when

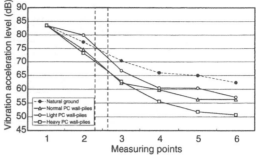

(a) Ground vibration on line ③ by dropping 2.5kg

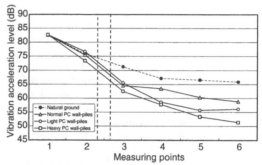

(b) Ground vibration on line ④ by dropping 2.5kg

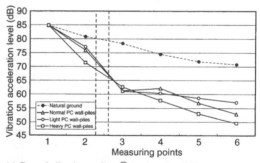

(c) Ground vibration on line ③ by dropping 5.0kg

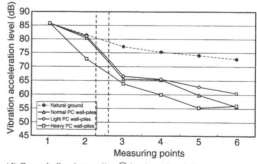

(d) Ground vibration on line ④ by dropping 5.0kg

Figure 7.   Vibration acceleration level with distance from vibration source.

473

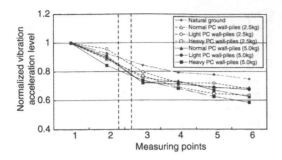

Figure 8. Normalized vibration acceleration level with distance from vibration source.

2.5 kg weight was dropped from 0.2 m and 5.0 kg weight was dropped from 0.4 m. To easily compare the decrease in ground vibration level, the every value at No.1 were adjusted in this figure. As the vibration when 2.5 kg weight was dropped form 0.2 m was different from that 5.0 kg weight was dropped from 0.4 m, the initial value of vibration acceleration level in Figure 7(c) and (d) was about 3 dB bigger than that in Figure 7(a) and (b). The rapid decrease in vibration acceleration level was observed in the improved ground by PC wall-piles regardless of their weight compared to that in natural ground. This shows the high isolation effect of PC wall-pile barrier. The decrease in vibration acceleration level in front of the heavy PC wall-pile barrier was obvious in Figure 7(c) and 7(d). Figure 8 shows the normalized vibration acceleration level in all tests. The decrease in vibration acceleration level was clearly observed behind the PC wall piles compared to the normal ground. The decrease in ground acceleration level of heavy PC wall-pile barrier was biggest among the three PC wall-pile barriers regardless of the dropped weight.

Figure 9 shows counter maps of the ground vibration level in the field model test when the weight was dropped at impact point G. As shown in Figure 9, the clear decrease in vibration acceleration level was observed behind PC wall-piles regardless of their weight. However, the isolation mechanism of heavy PC wall piles seems to be different from that of light PC wall piles. The vibration acceleration level in case of heavy PC wall-pile barrier decreases in front of wall, and the acceleration level in case of light PC wall-pile decreases during transmission of PC wall-piles.

Figure 10 shows a comparison of the isolation effect of every PC wall-piles with different weights. To clarify the effect of wall weight on the isolation mechanism, the relationship between decrease of ground vibration and density of a PC wall-pile was demonstrated. The decrease in front of PC wall-pile barrier and decrease between front and behind the barrier are shown. Although the decrease in front of PC wall pile barrier increases as the density of the PC wall-pile increases,

the decrease between front and behind the barrier increases as the density of the PC wall-pile decreases. Therefore, the reflection of ground vibration is dominant of isolation mechanism in heavy PC wall piles, the transmission is dominant in light PC wall piles.

4 A NUMERICAL SIMULATION

4.1 *Outline of numerical simulation*

A numerical simulation was conducted to evaluate the isolation effect of PC wall-pile barrier. Because a full three-dimensional FEM analysis takes long time, a hybrid analysis was adopted in this study. That is, the soil ground was modeled by 15 laminar elements and the PC wall-pile barrier by 180 three-dimensional shell elements as shown in Figure 10. Impact force by dropping of 5.0 kg weight from 1 meter high was estimated by an inversion analysis based on the above model from actual seismograms measured at the measuring point No.1 from the impact point. Parameters for soil ground and a PC wall-pile for three-dimensional FEM analysis are given in Table 2. The parameters for soil ground were decided from field measurements or previous data and the parameters for a PC wall-pile were decided from those of normal concrete. When the model soil ground was divided into 15 laminas, the thickness was decided to be less than 1/6 of wavelength with 250 Hz, which was the maximum frequency of the numerical simulation program.

4.2 *Discussions*

Measured and calculated acceleration levels along the measuring line of ③ are shown in Figure 11. Calculated results are compensated to make the computed value at measuring point No.1 in the natural ground equal to the measured value at the same point in the improved ground by PC wall-piles. The computed and measured values agree well for the case without a countermeasure of PC wall-piles, implying that the assumed parameters are appropriate. Although the computed acceleration level rapidly dropped at the PC wall-pile barrier in the numerical simulation, it was confirmed in the field model test that the PC wall-piles hardly vibrate. In case of the improved ground by PC wall-piles, although a computed value closely behind the PC wall-piles is different from the measured value, the other points agree appreciably well with the measured values. The numerical simulation can evaluate the behavior of PC wall-pile barrier and its screening effect.

Figure 11 shows a result of frequency analysis in the numerical simulations with and without PC wall-piles. The acceleration levels with higher frequency than 50 Hz obviously decreases in the numerical simulation with PC wall-piles compared with that

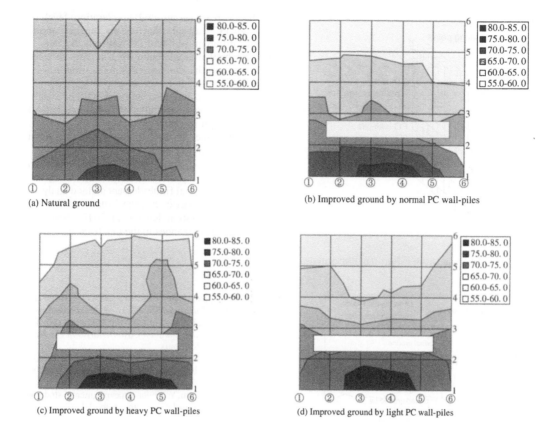

(a) Natural ground   (b) Improved ground by normal PC wall-piles

(c) Improved ground by heavy PC wall-piles   (d) Improved ground by light PC wall-piles

Figure 9.   Counter maps of ground vibration in the field model test.

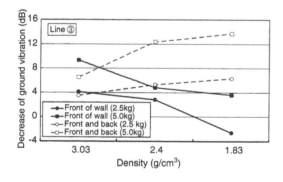

Figure 10. Comparison of isolation effect of every PC wall-piles with different weight.

Table 2.   Parameters in the numerical simulation.

Ground	
Wet density (ton/m^3)	2.5
S-wave velocity (m/sec)	153
P-wave velocity (m/sec)	700
Poisson ratio	0.475
Damping ratio (%)	2.0
PC wall-pile	
Density (ton/m^3)	2.3
Elasticity coefficient (kg/cm^2)	3.4×10^5

without PC wall-piles. This implies that PC wall-piles barrier has the isolation effect for higher acceleration level than 50 Hz, however, there were no changes in the low frequency area.

Therefore, the isolation effect of PC wall-pile barrier can be quantitatively evaluated by using the hybrid three-dimensional FEM analysis, and the PC wall-pile barrier is effective to isolate higher vibration acceleration level than 50 Hz.

5 CONCLUSIONS

In this study, the effectiveness of ground vibration isolation by a PC wall-pile barrier was investigated. Both field model tests and a hybrid three dimensional FEM analysis of PC wall-pile barrier were performed.

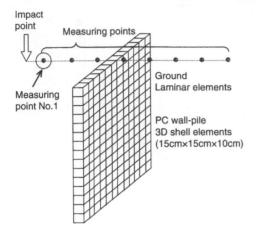

Figure 11. Model ground in the hybrid 3D FEM analysis.

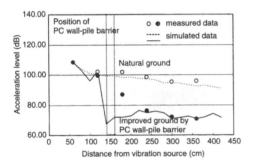

Figure 12. Comparison of measured and calculated acceleration level along the lone ③.

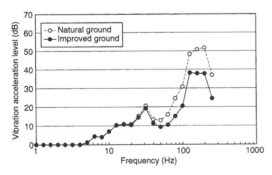

Figure 13. A result of frequency analysis in the numerical simulations with and without PC wall-piles.

From this study, the following conclusions can be made.

[1] The PC wall-pile barrier could reduce the vibration acceleration level regardless of the weight.

[2] The decrease in ground acceleration level of heavy PC wall-pile barrier was biggest among three different weight of PC wall-pile barriers.

[3] The vibration acceleration level in case of heavy PC wall-pile barrier decreases in the front of wall, and the acceleration level in case of light PC wall-pile decreases during transmission of PC pile walls.

[4] The reflection of ground vibration is dominant of the isolation mechanism in heavy PC-wall pile barrier, while the transmission is dominant in light PC-wall pile barrier.

[5] The isolation effect of the PC wall-pile barrier can be quantitatively evaluated by using the hybrid three-dimensional finite element method analysis.

[6] PC wall-pile barrier has the isolation effect for higher acceleration level than 50 Hz, however, there were no changes in the low frequency area.

REFERENCES

Ejima, A. 1980. Attenuation of surface waves by a trench and wall-in-earth, *Tsushi-to-Kiso*, 28(3): 49–55. (*in Japanese*)

Hayakawa, K. et al. 1998a. Ground vibration isolation by PC wall-piles, *4th Inter. Conf. on Case Histories in Geotech. Eng.*: 672–677, St. Louis.

Hayakawa, K. et al. 1998 b. The Effectiveness of Pre-cast Wall-Piles in Reducing Ground Vibration, *Building Acoustics*, Vol.5, No.3: 185–199.

Hayakawa, K. and Kani, Y. 1999a. An experimental study on ground vibration isolation using a model, *Inter-Noise 1999*: 871–874.

Hayakawa, K. 1999b. Background and trend for reduction method on propagation process of ground vibration, *Journal of the Acoustical Society of Japan*, 55(6): 449–454. (*in Japanese*)

Kani, Y., Hayakawa, K. and Maekawa, Y. 2000. An experimental study on ground vibration isolation using a model (second report), *Inter-Noise 2000*: 2020–2023.

Kani, Y. and Hayakawa, K. 2002. Reducing ground vibration around an elevated road with a PC wall-pile barrier, *Inter-noise 2002*: 1–6.

Kani, Y. et al. 2003. Vibration screening effect of PC wall-piles through full-size and medium-size field experiments, *Inter-noise 2003*: 1816–1823.

Matsubara, Y. et al. 1998. Isolation effects of ground vibration due to PC wall-piles on several sites, *Earthquake Geotech. Eng.*: 427–432.

Nabeshima, Y., Hayakawa, K. and Kani, Y. 2004a. *5th Inter. Conf. on Case Histories in Geotech. Eng.*: New York. (CD-ROM)

Nabeshima, Y., Hayakawa, K. and Kani, Y. 2004b. *15th Inter. Conf. on Geotech. Eng.*, Bangkok. (CD-ROM)

Ohbo, N. and Katayama, T. 1983. Numerical analysis on screening effect at an open trench by equivalent mass model, *Journal of Japan Society of Civil Engineering*, No. 335: 51–57.

Yoshioka, O. and Ishizaki, A. 1980. A study on reduction effects on ground vibration due to hollow trench and underground walls, *Railway Technical Research Report*, No.1147: 1–132.

Environmental Vibrations – Takemiya (ed.)
© *2005 Taylor & Francis Group, London, ISBN 0 415 39035 4*

Experimental study of railway-induced ground vibration reduction method

H. Suzuki & M. Kawakubo
Tokyu Construction Co., Ltd., Shibuya, Tokyo, Japan

S. Shibusawa & T. Nonaka
Tokyu Construction Co., Ltd., Shibuya, Tokyo, Japan

K. Sakai
Tokyu Construction Co., Ltd., Shibuya, Tokyo, Japan

ABSTRACT: The rail transit produces uncomfortable ground vibrations to the neighborhood along the railway. Especially in urban areas, while the railway improvement on convenience and transportation efficiency progresses, it is a critical issue to reduce environmental vibrations. The vibration reduction method was developed by using the plastic vibration isolators and the concrete vibration reflector. The experimental studies have been carried out and confirmed that the vibration acceleration level decreased by approximately 12 dB near the isolator after the isolator and reflector construction. Furthermore, it decreased by 9 dB at a point 5 m away from the isolator, and decreased by 4 dB at a point 10 m away from the isolator. The method thus proved to be much more effective for reducing vibration than conventional remedial methods. The method should therefore be applied to actual projects and the ease of application and the vibration reduction effect should be verified.

1 INTRODUCTION

Dense railway networks have recently been laid out in urban areas and more buildings are being constructed in the vicinity of railway tracks. An increasing number of theaters and halls are also built along railway tracks because of convenient locations. Train speeds have been increasing to enhance user convenience. Occupants of buildings along railway tracks, however, rarely allow railway-induced vibration although railways offer convenience and carry a large number of passengers. Creating quiet space along the route of the railway or further increasing train speed requires efforts for reducing vibration caused by the passage of trains.

Numerous methods have been developed to reduce vibration from the passage of trains. Research and development has been conducted to reduce vibration by such measures as trenches and barriers. These measures are beneficial because they involve no work at the source of vibration or in surrounding buildings. Their effectiveness has, however, not been verified explicitly. They have not been used widely because of their limited cost-effectiveness.

The experimental studies for developing a vibration reduction method using the plastic vibration isolators and the concrete vibration reflector have been carried out. This paper reports the results of experimental studies on the vibration reduction method.

2 OUTLINE OF METHOD

The method prevents the propagation of vibration by enclosing the source of vibration by reflector of higher rigidity (soil stabilization structure or concrete slab) installed in the ground under the rails and plastic isolators built on both sides of the track in the vibration propagation path. The downward vibration from the source is reflected by the reflector with a higher wave impedance than that of ground. The reflected vibration is attenuated by placing the reflector some distance below the subgrade. Hollow low-stiffness plastic isolators with wave impedance lower than that of ground are installed on both sides of the track to reflect lateral vibration. Lateral propagation of vibration is also controlled by the isolators. Figure 1 is a conceptual view of the vibration reduction method.

3 OUTLINE OF TEST

3.1 *Geological outline*

Ground is actually under complex geological conditions. It is rarely the case that a single soil type is distributed uniformly at any given test site. Vibration tests in actual ground therefore cannot accurately verify vibration reduction effects. In this study, field tests were conducted in a soil layer (an approximately

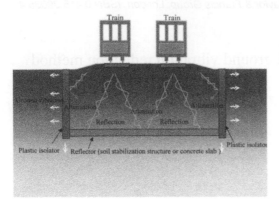

Figure 1. Conceptual diagram of vibration reduction method.

Table 1. Geological conditions of test ground.

Layer	Thickness	N-value
Fill up ground	2 m	2 to 4
Natural ground (mudstone)	–	20 to 25

Figure 2. General view of test site.

2 m thick layer of fill with an N-value of 2 to 4) laid over the natural ground to verify the vibration reduction effect of the proposed method. The geological conditions of the test ground are listed in Table 1. A general view of the test site is given in Figure 2.

3.2 Outlines of vibration attenuation isolator and reflector

The vibration attenuation isolator was made by assembling nearly hollow polypropylene units (with a breadth of 500 mm, a length of 500 mm and a thickness of 300 mm) with a void ratio of 95% into a wall. For the reflector, a 200 mm thick unreinforced concrete slab was constructed right below the source of vibration at a depth of approximately 1 m below the ground level. Figure 3 shows a plastic isolator under construction

Figure 3. Plastic vibration attenuation isolator (cross section and front view).

Figure 4. polypropylene unit (a breadth of 500 mm, a length of 500 mm and a thickness of 300 mm).

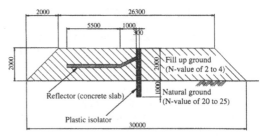

Figure 5. Cross sections of isolator and reflector.

(front view and cross section). Figure 4 shows a polypropylene unit. Figure 5 shows the cross sections of the isolator and reflector.

3.3 Outline of measurement

The ground was vibrated by dropping a heavy weight on the ground before and after the construction of the isolator and reflector. Then, the response to vibration was obtained at several measurement points in the ground to identify the vibration properties. Thus, the vibration attenuation effects of the isolator and reflector were evaluated.

Figure 6. Vibration using a heavy weight.

Figure 7. A 30 kg steel ball.

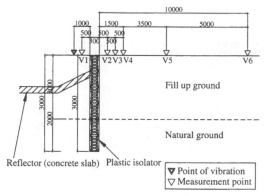

Figure 8. Measurement points.

3.3.1 *Vibration method*
The ground was vibrated by applying impact forces using a heavy weight (of constant energy). A 30 kg steel ball was subjected to free fall from a height of 30 cm. Figure 6 shows vibration using a heavy weight. Figure 7 shows a steel ball.

3.3.2 *Measurement points and items*
Measurements were taken at six points: (i) near the source of vibration (V1), (ii) 0.5, 1.0 and 1.5 m from the isolator (V2, V3 and V4) and (iii) 5.0 and 10.0 m from the isolator (V5 and V6). Measurements were made in three directions at each point: along the x-axis (transverse to the track), along the y-axis (parallel to the track) and along the z-axis (in the vertical direction) while the point of vibration was assumed to be at the origin. Accelerometers were used to obtain the level of acceleration of vibration.

The point of vibration and measurement points are shown in Figure 8. Table 2 shows outlines of measurement points. In addition, Table 3 lists measurement items.

Table 2. Outlines of measurement points.

Point	Outline of measurement points
V1	Near the source of vibration
V2	0.5 m from the isolator
V3	1.0 m from the isolator
V4	1.5 m from the isolator
V5	5.0 m from the isolator
V6	10.0 m from the isolator

Table 3. Measurement items.

Case	Measurement items
1	Before the implementation of control measures (in the natural ground)
2	After the construction of a 3-m-high isolator*
3	After the construction of a 3-m-high isolator and a reflector*

*Tests were also conducted for a 1.0-m-high isolator.

Vibration was identified on natural ground, after the construction of isolator, and after the construction of the isolator and reflector to verify the vibration reduction effect of the isolator and the combined effects of the isolator and reflector.

3.3.3 *Measurement method*
Acceleration pickups were used for measurement. They are small and light-weight and has a high mechanical strength, and provide for a wide frequency range and are widely applicable. Humans feel vibration as a mixture of various frequency components. Vibration frequency components and human sense of vibration vary greatly according to the conditions in the ground or at the source of vibration. To grasp vibration characteristics, it was important to know vibration frequency components. A frequency

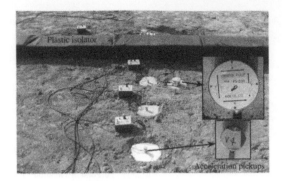

Figure 9.   Acceleration pickups.

Table 4.   Results of vibration measurement.

Point	Results of vibration measurement (dB)*		
	Case-1	Case-2	Case-3
V2	96.2	90.4	84.2
V3	94.6	90.0	82.2
V4	93.1	88.7	80.4
V5	85.3	81.2	75.9
V6	77.2	79.1	73.0

*Vibration acceleration level (dB).

analyzer (one-third octave band analyzer) was there-fore connected to the acceleration pickup for fre-quency analysis. Figure 9 shows acceleration pickups.

4   RESULTS OF MEASUREMENT

4.1   Confirmation of vibration attenuation effect

Acceleration was measured in the vertical direction (along the z-axis) at each measurement point before the implementation of control measures (in the natu-ral ground), after the construction of a 3-m-high iso-lator and after the construction of the isolator and a reflector. The results of measurement of acceleration (vibration acceleration levels) are shown in Table 4 and Figure 10. Measurements confirmed the following:

(1) Vertical vibration was reduced near the isolator (at V2, V3 and V4) by approximately 5 dB after the construction of the 3-m-high isolator, and by 12 dB after the construction of the isolator and the reflector.
(2) Vertical vibration was reduced at a point 5 m from the isolator (V5) by approximately 4 dB after the construction of the 3-m-high isolator, and by 9 dB after the construction of the isolator and the reflector.
(3) At a point 10 m from the isolator (V6), the vibration acceleration level was lower than near the isolator

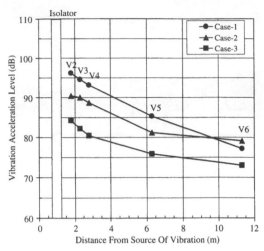

Figure 10.   Results of vibration measurement (in vertical direction along z-axis).

by 10 to 20 dB. Vibration, however, increased slightly after the construction of the isolator. Vibration was reduced by approximately 4 dB after the construction of the isolator and the reflector.

It was thus verified that installing isolators on both sides of the source of vibration was highly effective for attenuating vibration near the isolators. No signif-icant attenuation of vibration was, however, con-firmed at points away from the isolator probably because vibration form the source was transmitted below the bottom of the isolator. Enclosing the source of vibration by isolators and a reflector was found to be sufficiently effective for attenuating vibration not only near but also away from the isolators.

4.2   Results of frequency analysis

One-third octave band frequency analyses were made before the implementation of control measures and after the construction of a isolator and a reflector to identify the frequency components of vibration and the vibration attenuation effect of the proposed method. The results of one-third octave band frequency analy-ses in relation to vertical vibration acceleration levels obtained near the isolator (at V2), at point 5 m from the isolator (V5) and at a point 10 m from the isolator (V6) are shown in Figures 11 through 13 respectively. As a result of comparison of frequency analysis results before the implementation of control measures and after the construction of a isolator and a reflector, the following findings were obtained.

(1) More high-frequency components were reduced at points farther from the source of vibration either before the implementation of control measures

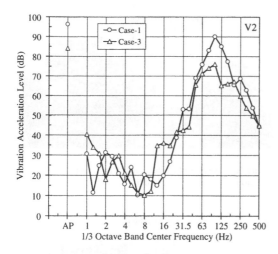

Figure 11. Results of frequency analysis (at V2 along z-axis).

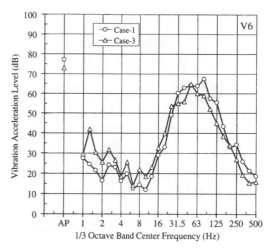

Figure 13. Results of frequency analysis (at V6 along z-axis).

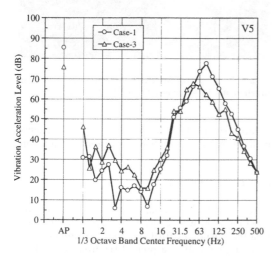

Figure 12. Results of frequency analysis (at V5 along z-axis).

(4) The predominant frequency near the isolator (at V2) was approximately 80 Hz either before the implementation of control measures or after the construction of the isolator and the reflector.

(5) At a point 5 m from the isolator (V5) and a point 10 m from the isolator (V6), the predominant frequency was nearly 80 Hz when no control measures were taken. Where the isolator and the reflector were constructed, the predominant frequency was 40 to 50 Hz, lower than when no control measures were taken.

5 TWO-DIMENSIONAL FINITE ELEMENT ANALYSIS

5.1 Outline of analysis

The results of measurement in the field tests were verified by response analysis using the two-dimensional finite element method (analysis software: SFLUSH). The physical property values of the test ground were determined based on the results of sounding and radioisotope measurement in the natural ground. Figure 14 shows view of sounding and radioisotope measurement. Table 5 lists the physical property values of the test ground used for analysis. The input ground motions used for analysis were obtained using load cells when a steel ball was dropped. The ground motions were input in the direction shown in the analysis model for making response analysis.

5.2 Results of analysis

The results of analysis made before the implementation of control measures, after the construction of the

or after the construction of the isolator and the reflector.

(2) Vibration near the isolator (at V2) was reduced more in a frequency zone of 31.5 Hz or higher after the construction of the isolator and the reflector than before such control measures were taken.

(3) A similar phenomenon to that described in 2) above was observed at a point 5 m from the isolator (V5) and at a point 10 m from the isolator (V6). The difference in vibration acceleration before and after the implementation of control measures became smaller at points farther away from the source of vibration (V5 and V6).

481

Figure 14.   View of sounding and radioisotope measurement.

Table 5.   Physical property values of test ground.

Layer	Unit weight γ_t (kN/m^3)	Shear wave velocity V_s (m/sec)	Poisson ratio ν
Fill up ground-1	14.0	92	0.45
Fill up ground-2	14.0	127	0.45
Natural ground-1	18.0	200	0.40
Natural ground-2	18.0	300	0.40

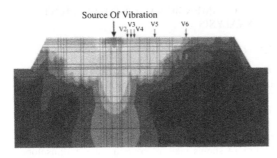

Figure 15.   Results of finite element analysis (before the implementation of control measures).

isolator and the reflector are shown in Figures 15, 16 and 17 respectively. The figures show contours of peak vertical accelerations obtained by response analysis. Where no control measures were taken, vibration induced by vertical vibration propagated horizontally and vertically. Horizontal propagation was controlled by the attenuation isolator. Installing a

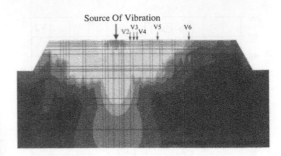

Figure 16.   Results of finite element analysis (after the construction of a 3-m-high isolator).

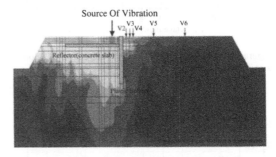

Figure 17.   Results of finite element analysis (after the construction of a 3-m-high isolator and a reflector).

reflector right below the source of vibration reduced both vertical and horizontal propagation of vibration.

6   CONCLUSIONS

As a result of field tests, it was confirmed that the vibration level was reduced by approximately 12 dB near the isolator, by 9 dB at a point 5 m from the isolator and by 4 dB at a point 10 m from the isolator owing to the construction of a isolator and a reflector. The method thus proved to be much more effective for reducing vibration than conventional remedial methods. The method should therefore be applied to actual projects and the ease of application and the vibration reduction effect should be verified.

REFERENCES

(1) Hiroshi Yamahara: Vibration isolation design for environmental conservation, Shokokusha Publishing Co., Ltd. (in Japanese).
(2) Masashige Kawakubo and Shin-ya Miwa: Facts about ground vibration due to the passage of trains, Summaries of Technical Papers of Annual Meeting, Architectural Institute of Japan, September 2003 (in Japanese).

(3) Masashige Kawakubo, Toshihiro Hayashida and Shin-ya Miwa: Characteristics of environmental vibration from the passage of trains and vibration attenuation methods, the Foundation Engineering & Equipment, Vol. 30, No. 1, pp. 74–76, January 2002 (in Japanese).

(4) Yoshiaki Nagataki and Norio Taguchi: Prediction and control of environmental ground vibration, Proceedings of the Symposium on Environmental Vibration, The Japanese Geotechnical Society, February 2001(in Japanese).

(5) Acoustic Materials Association of Japan: Noise and vibration control handbook, pp. 455–457, Gihodo Shuppan Co., Ltd., 1989 (in Japanese).

Environmental Vibrations – Takemiya (ed.)
© 2005 Taylor & Francis Group, London, ISBN 0 415 39035 4

Development of isolation barrier for ground vibration by employing group of embedded circular drain in the soil

K. Onoda & H. Nagasawa
Tokyo-Branch, Penta-Ocean Construction Co. Ltd., Tokyo, Japan

T. Ohshima & T. Tamura
Engineering Department, Penta-Ocean Construction Co. Ltd., Tokyo, Japan

ABSTRACT: Recently the ground vibration problems are prominent in the urban densely populated area because of the closeness of habited area with the environmental vibration sources. This paper shows the development of ground vibration isolation barrier consisting by group of circular drains embedded into the soil. Several field model test and experimental construction of isolation barrier near the commercial railway were done to verify the effectiveness of proposed technology. Effectiveness of this method was proved through these field tests. Numerical analysis by 2-D FEM with proper employment of the properties of barrier material explain the results of these field tests that shows the possibility of design of this vibration reduction barrier the site where ground vibration problems occurs.

1 INTRODUCTION

In Japan there are many cases that the houses are located very close to the major load and railway line because of the heavy human concentration in the urban area. Because the environmental situation in these area are apt to be so serious that the necessity of some countermeasures against noise and vibration are common problems. The countermeasures of noise and vibration reduction are one of the urgent and important problems for the urban area to maintain the comfortable environmental conditions. For the vibration reduction against railway source, many countermeasures have been done so far, for instance, introducing lightweight train and introducing high rigidity railway, low elasticity carriage are employed as the countermeasures to reduce the level of vibration source and realized the high efficiency to reduce the vibration. But the needs of speeding up are still important for the railway company to compete another company or transportation and that becomes the serious cause vibration problems. Due to these reason, especially in the densely populated area, further necessities for vibration reduction are not negligible.

Among the countermeasures for vibration reduction, vibration isolation barrier in the soil is considered to be one of effective methods. Most effective method of vibration isolation barrier is to construct the void trench at propagation path between source and receiver. But the construction and maintenance of void trench are

extremely difficult from the view point of cost and there is no example of employing this method to the railway vibration source. Several methods to construct the underground vibration isolation barrier were proposed. One of promising methods is to construct rigid wall type isolation barrier, for instance the concrete diaphragm wall, and efficiency of this methods are examined through the construction of demonstration experiment. There is some example of this method to employ the void within the concert wall. These methods usually cost high compared with other methods.

On the other hand, another type of underground isolation barrier with low rigidity and light weight such as EPS block wall are proposed. The reduction of vibration level is achieved by employing the light weight and low rigidity. In the case of EPS block wall, the construction of EPS block wall below under water line is extremely difficult and they proposed some special construction method. The efficiency of these methods is examined by demonstration experiments and numerical simulation.

In order to employ these methods to the site along the railway line, innovative construction method in little space and in very short construction time due to because railway business in daytime are usually continued are proposed. Most of the construction method of underground barrier are based on the use of usual large construction machine and cycle time of construction of underground wall are rather long and effect on surrounding ground deformation are inevitable. Due to this difficulty in the case for railway line, there is no actual construction of this type of underground wall as a countermeasure of vibration reduction.

This paper proposed the new type of underground vibration isolation wall which can be applied to the railway line with little space and short construction time available. Proposed method employs the group of embedded circular drain in the soil which composed of a kind of underground wall. Circular drain has light weight and low rigidity. Small scale demonstration experiment and field measurement by artificial vibration source have done to prove the efficiency of proposed method.

In this paper, demonstration experiment which was done at the site of real railway was presented. The train passing by the demonstration site was used as the moving vibration source and field measurement of vibration with and without underground vibration isolation barrier was done to examine the efficiency of this methods. In order to examine the effect of underground isolation wall, 2-D FEM was also employed to calculate the propagation of ground vibration and design method are considered.

2 PROPOSED ISOLATION BARRIER FOR GROUND VIBRATION

When we construct underground isolation barrier adjacent to the railway line, only night time when no scheduled trains are operating is available for construction. Especially the densely populated urban railway line, many trains are operating from early morning till late night, only several hours are available with additional two hours of preparing and clearance. Then it is necessary to complete the construction within two to three hours and cause no trouble for the successive train operation.

Proposed method employs the circular drain which have often used as the drain material for countermeasure of liquefaction. This circular drain material was placed vertically into the soil and forms a group of columns and act as the vibration barrier wall. In order to prevent the intrusion of underground soil into the drain, geotextile material is used to overlap the drain. Drain material was made of polypropylene and has lots of void. In case that the diameter of the drain material is 165 mm, the unit weight is 1.3 kg/m and has enough strength to resist the earth pressure. This material is chemically stable in the soil.

For the inserting of this drain material into the soil, digging of the ground was done at first and then the drain material were inserted into the digging hole and filling the void with appropriate material. Generally speaking this construction cycle is short and several drain can be constructed within limited construction time. It is possible to prevent unfavorable ground deformation by selecting appropriate material for filling and appropriate construction method during the train operation. Proposed methods employ the cement bentonite as the filling material.

Drain was placed in staggered mesh with appropriate spacing. In order to improve the vibration reduction capability increase the number of rows of drain should be concerned. Cement bentonite was used as the filling material. In order to fill completely the void, cement bentonite was filled from the bottom of the hole through special equipment prepared at the tip of the drain. Photo. 1 shows the drain and Photo. 2 shows the drain lapped by geotextile and special equipment for bentonite filling from the tip of the drain.

Conventional underground isolation wall employs the wall with homogeneous material such as concrete or construct the wall by mixing the in-situ soil with additional materials such as cement as cementing material and EPS chips or beads. Proposed method employs

Photo 1.   Drain used for proposed method.

Photo 2.   Surface appearance of drain with geotextile.

composition of soil and drain. Because the construction of vertical drain needs rather small construction machine and this methods can be applied to the site that has little space such as urban railway line. In addition to that, construction cycle for one vertical drain is short such as 2 hours, and can be applied to the tough conditions site of construction time.

3 IDENTIFICATION OF THE PROPOSED METHOD EFFICIENCY BY FIELD DEMONSTRATION EXPERIMENT

Demonstration experiment was done along the railway line and moving train was used as the vibration source. Before construction of isolation barrier wall measurement of vibration level were done to compare the results and to evaluate the vibration reduction level. The site is in Kanagawa prefecture and was covered by soft clay layer at the surface.

Figure 1 shows the soil conditions and the length of the drain. At the surface there is reclaimed layer of 2 m and 10 m soft clay layer locates under this layer. The length of isolation barrier is 30 m and the depth of the drain is 5 m which was previously determined by numerical simulation method to achieve the expected level of reduction of vibration. Figure 2 shows the layout of the drain. 300 mm spacing by staggered mesh was employed and 3 rows of drain were constructed. Thickness of the isolation barrier is about 650 mm. Measurement of ground vibration was done when the construction of the 3 row of drain was completed. In order to examine the effect of the thickness of the

isolation barrier, measurement of ground vibration of 2 rows of drain conditions was done.

Figure 3 shows the plan view of the site and relation of railway and isolation barrier and ground vibration measurement points are shown. Ground vibration points are the same for the before and after construction of isolation barrier in order to be able to compare the results for the evaluation of vibration reduction level. Moving train is considered as the vibration source. In order to obtain the consistent results, same rapid train with same speed and same number of train was employed for the measurement. Due to the limitation of the number of the acceleration meter, number of measurement points is limited and several times of measurement by different train are necessary to complete the entire measurement of the area. Measurement was done several times for each point to obtain the average value of the ground vibration. These measurement was done for three situation with initial condition without isolation barrier and with isolation barrier for both 2 rows and 3 rows of drains. As was mentioned before, careful selection of the train as the vibration source was done but the vibration source level fluctuate with some extents due to the difference of the number of peoples on the train and so on. Vertical acceleration meter were used to measure the ground vibration. Appropriate filtering was done to obtain the vibration level which was determined by Vibration Control Law. Horizontal acceleration was also obtained at some points. Following discussion were based on the results obtained from vertical acceleration meter and maximum vibration level was the main index for the following discussion and consideration.

Figure 4 shows the averaged maximum vibration level(over-all value) for each observation point. This figure shows cross-sectional distribution of each line.

G.L.	Soil classification	N value
-2.6 m	Filling soil	3~6 (Ave. 5)
	Silt	2~4 (Ave. 3)
-12.0 m		
	Sandy silt	3~17 (Ave. 8)
-22.0 m		
	Engineenng Bedrock	50 over

Vibration Transmission Barrier Wall

Figure 1.   Soil conditions at the site.

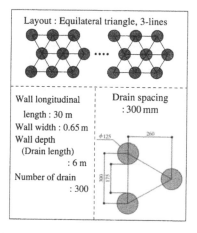

Layout : Equilateral triangle, 3-lines

Wall longitudinal length : 30 m
Wall width : 0.65 m
Wall depth (Drain length) : 6 m
Number of drain : 300

Drain spacing : 300 mm

Figure 2.   Spacing of the drain.

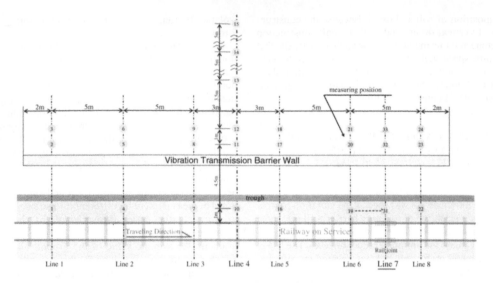

Figure 3. Plan view of railway and isolation barrier and observation points.

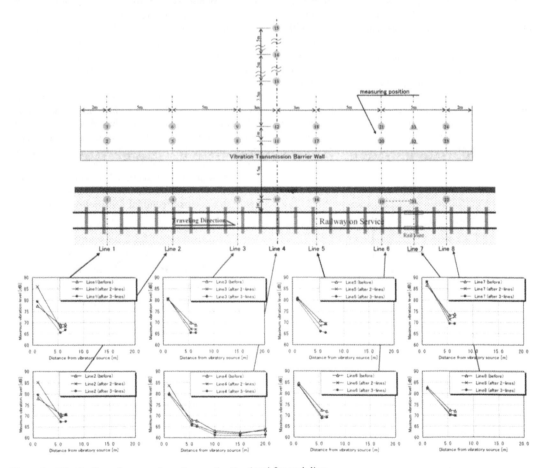

Figure 4. Distribution of averaged maximum vibration level for each line.

488

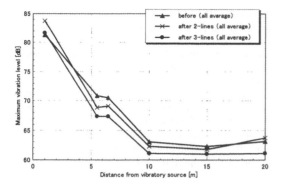

Figure 5. Distribution of maximum vibration level(over-all value) just at line of the center of the isolation wall.

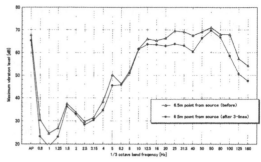

Figure 6. Distribution of 1/3 octave-band of vibration level at 5.5 m from the source at the center line.

Figure 5 pick-up the cross-sectional distribution of averaged maximum vibration level(over-all value) just at line of the center of the isolation wall. From these results, maximum reduction of ground vibration was appeared at the closest points from the isolation barrier. For the case of 2 row of drains, vibration reduction level of 2–3 dB was achieved and for the case of 3 row of drains, vibration reduction level of 2–3 dB was achieved. On the other hand, vibration reduction level decreased gradually as the observation points becomes far from the isolation barrier and 1–2 dB reduction level was observed for 3 row of drains. Figure 6 shows the distribution of 1/3 octave-band of vibration level at 5.5 m from the source at the center line. In this figure 2 row of drains and 3 row of drains are shown for the comparison. For both case, significant reduction of vibration level at around 10–20 Hz and around 40–60 Hz.

Because the length of the isolation barrier is 30 m and rather short, effect of the vibration from running train without isolation barrier is not negligible for the distant point from the isolation barrier. This is considered one of the reason of decrease of the vibration reduction capability for distant points.

4 NUMERICAL SIMULATION OF PROPAGATION OF GROUND VIBRATION BY 2D-FEM

2D-FEM program code of [SuperFLUSH/2D] was used to reproduce the observed field vibration data. This program can consider the ground-structure dynamic interaction and point excitation source can be considered. For the numerical simulation, one typical cross section was selected. This cross section was selected at the center of Isolation barrier at the vertical direction of the isolation wall. Because the excitation force are loaded at the rail and point excitation source should be put at this point. But the vibration measurement was

not done at this point and exact excitation level was unknown. Then we employ the nearest measurement point data of vibration as a point excitation source data.

Soil conditions were already shown in Figure 1 and soil parameter for the simulation were determined by SPT N-value with employing appropriate conversion formula. Because the proposed isolation barrier wall is a kind of composite wall with soil and drain, there is no appropriate conventional method to obtain the dynamic properties for the simulation. In this paper, the unit weight and stiffness of the wall are calculated by average of two values of soil and drain by weighing the cross-sectional area of soil and drain. Damping ratio of the wall are determined by trial and error approach to reproduce the observed data as precise as possible.

Figure 7 shows the analytical model based on the ground survey data. Ground was modeled as deep as the base layer where SPT N-value is more than 50. For horizontal direction, modeling was done 50 m from the point excitation source which was correspond to the length of twice as long as the vibration observation length. Boundary condition at the bottom is viscous boundary and horizontal boundary condition is energy transfer boundary conditions. Spacing of mesh is determined to satisfy the analytical upper bound frequency of 50 Hz.

Figure 8 shows the comparison of observed data and calculated data with and without the underground isolation barrier of 3 row of drain. Most appropriate damping value is obtained through this calculation procedure by trial and error approach. As was shown in this figure, accuracy of reproduction of observed data is not sufficient and priority for accordance was on the reproduction of observed data just behind the wall. Because of this procedure the accuracy of reproduction at the far side was not sufficient. Damping ratio of wall becomes 40%. In case of no isolation wall, calculated value reproduce the observed value well. On the other hand, for the case of existence of isolation wall calculated value at the far end shows smaller value compared with the observed ones.

As was shown in section 3, observation point far from the wall suffer the effect of train vibration source with no isolation barrier because of the limitation of the length of the wall. These effect increase the level of vibration at the site and 2D-FEM analysis that we employ could not taken into consideration of this effect. These explanation shows the analytical limitation of 2Dimensional analysis, but the most important is the reduction level at just behind the wall and 2D-analysis can be applied at least the reproduction near the wall where little wrap-around effect is expected.

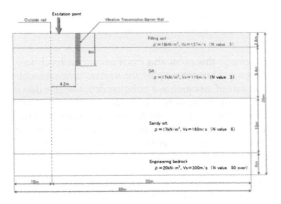

Figure 7.   Ground model and isolation barrier for the simulation of 2D-FEM based on the ground survey data.

That means 2D-FEM analysis gives sufficient accuracy for design purpose as long as the vibration reduction necessities are the nearest building.

In order to obtain the accurate numerical simulation results, importance of soil modeling and determination of dynamic properties are well known. But in this paper only damping ratio of isolation wall is the only adjusting parameters to improve the reproduction accuracy of the observed data and most appropriate damping value is obtained. We should be careful about the application of this value to other site, because the physical damping mechanism are not considered. But as long as the similar train sources are the main vibration source and the similar construction method are employed, the simulation process and the dynamic properties of isolation wall are applicable to another site. On the other hand, if the vibration source mechanism is different from the train source such as construction site vibration source and factory vibration source, then the analytical method and the dynamic properties of isolation wall should be reconsidered. For some case it might necessary to consider the another demonstration experiment that have done for this study and to determine the appropriate value of important parameters.

There are several important things that we should consider for analytical modeling. For instance, the effect of moving vibration sources with very high speed is not considered in the proposed analysis. Source

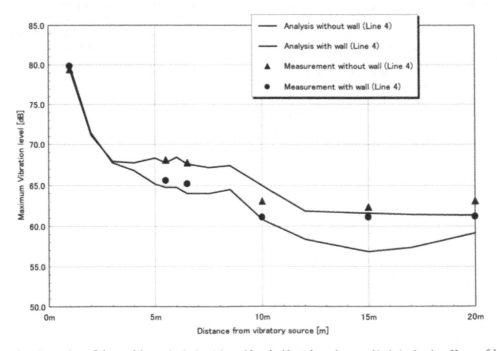

Figure 8.   Comparison of observed data and calculated data with and without the underground isolation barrier of 3 row of drain.

vibration level is inevitable to evaluate absolute value of vibration level.

5 CONCLUSIONS

In this report, new underground isolation barrier for ground vibration by employing group of embedded circular drain into the soil was proposed. The effectiveness of this mitigation measures were examined through the demonstration construction to the urban railway line by measuring the ground vibration before and after the construction of underground isolation barrier. 2D-FEM analysis was done to simulate the effect of underground isolation barrier on the propagation of ground vibration produced by the passing train. Following concluding remarks were obtained through the study.

(1) Although the construction length of demonstration construction of underground isolation barrier was limited, proposed underground isolation barrier shows

(2) 2D-FEM was done to simulate the observed ground vibration reduction by employment of underground isolation barrier. The simulated results shows good agreement with observed one especially near the underground isolation barrier.

(3) Because the horizontal length of underground isolation barrier was limited, train vibration outside the barrier wall was not negligible at far site from the wall. Those effect are not taken into account in to the 2D-FEM analysis, simulated results shows rather smaller value than observed ones.

Proposed underground vibration barrier was categorized by the methods that employs the wall material with light and small rigidity. In order to realize this condition, typical conventional method employed EPS as a wall material (Hayakawa 2004). According to the demonstration experiment shown by Hayakawa, ground vibration reduction of 5–7 dB was obtained by employment of EPS wall as the underground vibration reduction wall.

Above-mentioned EPS walls replaced the existing soil completely with the wall material. On the other hand, proposed method employ the group of circular drain materials which are partially embedded into the soil and was expected to show the similar capability of ground vibration reduction with the EPS wall. According to the field observation shown in the previous section, 3.6 dB was obtained as the average value of ground vibration reduction due to the employment of proposed underground vibration barrier. Increase of the number of row of embedded drains contribute the increase of the capability of vibration reduction.

This study examine the vibration reduction capability of proposed underground vibration barrier through the demonstrating experiment done at the soft ground condition site. Further study is needed to ensure the effectiveness of proposed underground vibration barrier on different ground condition with different vibration source conditions. Improvement of accuracy of numerical simulation methods are also important to design the proposed underground vibration barrier for various ground and vibration source conditions. Thank you very much for your kind information.

ACKNOWLEDGMENT

Demonstration construction of the proposed underground vibration barrier was done by Keihin Electric Express Railway Co. Ltd. Authors would like to express their sincere gratitude to Keihin Electric Express Railway Co. Ltd. for their cooperation and kind attention on the measurement of ground vibration and resultant analysis of obtained data.

REFERENCES

Asitani, K. & Yokoyama, H. 2002. Present state of R&D on mitigation measures on environmental vibration, *RTRI Report* Vol.16, No.12, (in Japanese)

Hashizume, H., Nagataki, Y. & Wakamei, Y. 1991. Mitigation of environmental vibration by diaphragm wall: 28 conference on soil mechanics and foundation engineering, JGS, 1245–1248. (in Japanese)

Hayakawa, K. 2004. Design and construction examples of ground vibration barrier by EPS: *Prediction and Mitigation of Environmental Vibrations: Proc. symp. Tokyo*, 96–99. JGS (in Japanese)

Ishi, K. et al. 2003. Study on the effectiveness of sheet pile wall and concrete wall as the ground vibration barrier, *58th JSCE national conference*, JSCE, 189–190. (in Japanese)

JSCE (eds) 1985. Handbook of vibration, *JSCE*: 13–16.

Kanda, M. et al. 2004. Development of traffic vibration barrier by soil-cement wall with mixture of foamed polystyrene beads: *Prediction and Mitigation of Environmental Vibrations: Proc. symp. Tokyo*, 209–214. JGS (in Japanese)

Lysmer, J., Udaka, T., Tsai, C.F. & Seed, H. B. 1975. FLUSH A COMPUTER PROGRAM FOR APPROXIMATE 3-D ANALYSIS OF SOIL – STRUCTURE INTERACTION PROBLEMS, *College of Engineering University of California Berkeley California*

Ohshima, T., Tamura, T. & Onoda, K. 2003. Proposal of underground vibration barrier by employing circular drain materials, *38th conference on Geotechnical engineering*: Tokushima, 2389–2390.

Yoshioka, O. & Kumagai, K. 1982. Simple procedure to evaluate the effectiveness of underground vibration barrier, *RTRI Report* No.1205, RTRI (in Japanese)

Yoshioka, O. & Ishizaki, A. 1980. A study on the effectiveness of blank groove and diaphragm wall as the mitigation measures to environmental vibration, *RTRI Report* No.1147, RTRI (in Japanese)

Environmental Vibrations – Takemiya (ed.)
© *2005 Taylor & Francis Group, London, ISBN 0 415 39035 4*

Environmental vibration control by active piezo-actuator system

H. Takemiya
Okayama University, Okayama, Japan

S. Ikesue
Mitubishi Heavy Industry, Ltd

T. Ozaki
RyomeiGiken, Co, Ltd

T. Yamamoto
Ryosen Engineering, Co, Ltd

Y. Fujitsuka
Randesu Co, Ltd

A. Shiraga
Gansui Kaihatsu, Co, Ltd

T. Morimitsu
Eito Consultants, Co, Ltd

ABSTRACT: Environmental vibrations are increasingly drawing attention in modern society. Protecting built-in areas from exposures to detrimental effects is desired. Herein, an attempt has been made to use an actuator forcing system to subdue the manmade ground vibrations as induced traffic and machine operation. Theoretically, it is demonstrated that another loading for the source loading has a good possibility to cancel out the transmitting vibration by the wave interferences. The validation of such phenomenon is proved by the field test. The key factor of generating tranquilized zone behind the actuator system is to adjust the phase between the two source force systems.

1 INTRODUCTION

Environmental vibrations as represented by high-speed train, massive traffic on highway, factory operations and massive events in halls, are increasingly drawing attention in modern society. These vibrations include mostly the low frequency nature from 3 to 8 Hz frequencies that give adverse affects to inhabitants and operation of vibration sensitive equipments in the neighborhood. According to he vibration legislation in Japan the limit of vibration acceleration level (VAL) is in the range of 60–65 dB during daytime while 55-d0 dB during nighttime. Protecting built-in areas from exposures to detrimental environment or vibration prolusion is strongly desired.

Some measures have been attempted for avoiding exposures to vibration sources in the built-in area. The are a row(s) of concrete piles or steel sheet-piles, open trenches of in-fill trenches with concrete or bentonite. These barriers aim to protect structures and facilities by creating a shadow zone. Recently, Takemiya (2004) Takemiya & Shimabuku (2005) developed a special cell-type WIB that can modulate the propagating waves and impeding them across its zone. They are classified all as passive type mechanism for vibration reduction.

Herein, an innovative idea is put forth to apply the active mechanism to control the wave field (Takemiya et al, 2005): An pair of shear force action is additionally introduced in the existing wave field by sources. By adjusting the frequency and phase angle, the total wave field can be reduced. Through the field test the authors confirmed a possibility to realize such a mechanism to application.

2 COMPUTER SIMULATION

2.1 Basic theory

The source loading from the traffic on highway emits vibrations that can be received along side residential houses. This wave mostly propagates as surface waves in the soft surface layers. In order subdue the field wave, we utilize wave inferences by imposing another wave to the original wave. Most effective way for this aim is to generate waves by a shear force action on the field. This mechanism is illustrated in Fig.1.

2.2 Model for computation

Computer simulation is taken first to get an insight into the present active WIB. The computation is based on the soil properties at the test site. Table 1 gives the predicted properties from the bore hall test. The values based on the N-values are used but later revised in comparison of the computation results with field test.

First, in order to get site characteristics for the wave field, the dispersion curves are computed by taking the thin layer method and are depicted in Fig.2 for the group velocity vs. frequency for the in-plane motions of the P and SV waves and for the out-of-plane motions of SH wave. The Airy phases that fall on the minimum group velocity are assessed for the respective wave fields. It is found that for the in-plane motions the first mode indicates it at 4 Hz and the second mode at 9 Hz and so on as marked by circles in Fig.2. This means that the first cut-off frequency is located at 4 Hz. The wave propagation occurs beyond this frequency. For the out-of-plane motions, the first mode cut-off frequency at 2 Hz and the second mode at 6 Hz and so on.

The simulation is conducted by using a finite element model for a pair of shear force action in addition to the source loading at the center. Fig.3 is the resultant response intensity map at the ground surface for the driving frequency at 4 Hz where only a

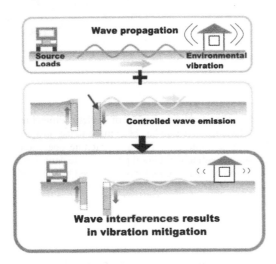

Figure 1. Subduction of wave field by additional shear force action on propagation field.

Table 1. Soil properties at site.

Layer depth [m]	Density [t/m³]	Shear velocity [m/s]	Poisson ratio	Damping ratio
3.0	1.8	120	0.45	0.03
7.0	1.8	102	0.45	0.03
3.0	1.8	120	0.45	0.03
2.0	1.8	144	0.45	0.03
2.0	1.8	180	0.45	0.03
8.0	2.0	400	0.45	0.03

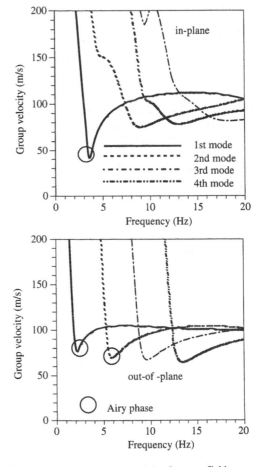

Figure 2. Dispersion characteristics for wave field.

quarter portion is depicted. It is clearly noted a tran-quilized zone is generated behind the active WIB in a widely spread out range. This confirms a good possibility of wave subduction theory by the wave superposition.

Secondly, another simulation is carried out to give some preliminary prediction for the succeeding field test. In this case, only one buried plate is excited on the top by an actuator harmonically. Therefore, the plate is modeled by a shell structure. The plate under-neath the active WIB is taken the same size with the field test. In Fig.4 the left-hand side shows a plan view of the consequent wave field. The wavelength is read off as around 15 m. The right hand side the wave field when a buried wall of 20 m long and 10 deep is used but in vain for screening waves because of the thickness 0.2 cm. Then, the wave impeding results only near area behind it.

3 FIELD TEST

3.1 *Test setup*

The Fig.5 illustrates a setup of the field vibration test. The source loading is given by an oil pressure type shaker of the 5 tf for the maximum capacity. The ver-tical loads are imposed directly on the ground surface through a concrete plate. The additional actuator is arranged on a buried PC plate on the measurement line. The measurement line connect these two apparatuses.

The apparatuses in Fig.6 show the loading systems: the front actuator is a 2.5 tf weight actuator which is put on a buried PC plate of 2 m width by 1 m height and 20 cm thickness is buried in turn: one has a smooth face and the other one has specially fabricated rough face as seen at the front in Fig.7. The convex height on the surface of the latter plate is 2 cm. The actuator gives a harmonics loading on either PC plate. The construction process, as seen also in Fig.7, is first to excavate soil and then refilled around the plate and compacted at every 30 cm to get a tight contact with the soil. The wave is therefore generated to emit from the face of this plate.

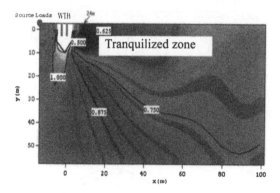

Figure 3. Plan view of ground response due to a simulta-neous loading at source and a pair of shear forces. A quarter portions is shown or 4 Hz loading.

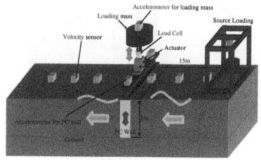

Figure 5. Field test setup.

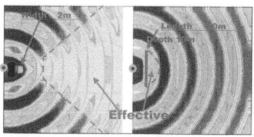

(a) due to a source (b) due to a simultaneous source and active WIB

Figure 4. Plan view of ground response, one-quarter portions.

Figure 6. Loading systems.

3.2 *Measurement*

The vibration measurement has been conducted on the lines that connect the source shaker and the active WIB. Fig.8 shows two arrays passing through the two buried PC plates. The inlet a photo shows three perpendicular velocity sensors (Max, velocity 10 kine) and one vertical accelerometer (Max. acceleration 1 G) at the distance of 30 m from the source shaker. The velocity sensors for micro-vibrations are used at the rest measuring points.

The experiment has been conducted by changing driving frequencies from 2 Hz to 10 Hz by the frequency step of 1 Hz. Fig.9 shows the velocity response at 30 m distance from the source shaker in a state of driving frequency of 5 Hz. Under this loading state, there inevitably included an additional high frequency due to the hydraulic pumping system. However, the high frequency around 9 Hz in this case might be caused by the second mode of soil vibration at the site.

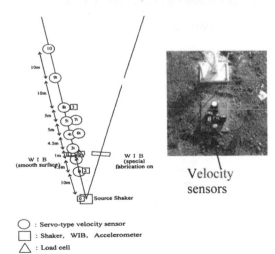

Figure 7.   Buried PC plate.

○ : Servo-type velocity sensor
□ : Shaker, WIB, Accelerometer
△ : Load cell

Figure 8.   Measurement array.

In the figure are also indicated the ground surface response when the active WIB are driven with the same frequency but with certainly adjusted phase difference that gives the minimum response. The phase angles for different response amplitudes are indicated in Fig.10 in which the phase angle that gives the minimum amplitudes are indicated by circles. The response reduction by use of the active WIB with such phase angles is noted significant. The response amplitudes are plotted in Fig.11 to evaluate the reduction rate by the active WIB in the frequency range 2 Hz through 10 Hz. The response amplitude grows drastically at 4 Hz and stays

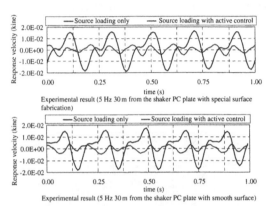

Experimental result (5 Hz 30 m from the shaker PC plate with special surface fabrication)

Experimental result (5 Hz 30 m from the shaker PC plate with smooth surface)

Figure 9.   Ground surface response (velocity) at 4 Hz.

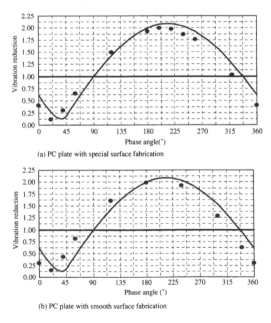

(a) PC plate with special surface fabrication

(b) PC plate with smooth surface fabrication

Figure 10.   Phase angles of driving the active WIB at 4 Hz.

496

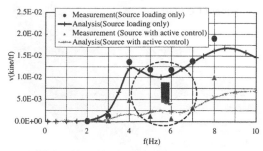

(a) PC plate with special surface fabrication

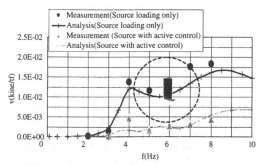

(b) PC plate with smooth surface fabrication

Figure 11. Response amplitudes vs. frequency at 30 m distance from the source shaker.

in the more ore less narrow banded zone above it until the amplitude increases again and attain another peak at 9 Hz and then decreases. These frequencies match with the cut-off frequencies of the in-plane motions from Fig.2. The computer simulation results are also depicted for comparison in Fig.11 by the lines. The fitting is excellent with the measurements both without and with operation of the active control apparatus.

The response reduction is evident in the frequency range beyond the 1st cut-off frequency through seemingly up to the 2nd one, being evaluated most effective at 5 Hz driving frequency. In case of the specially fabricated PC plate surface, the reduction is 90% at this frequency.

The Fig.12 is a representation of the response in terms of the vibration acceleration level (VAL) in the Japanese specification in order to give an easy understanding of the reduction effect by the active WIB in the environmental vibration assessment. In the frequency range 4 Hz through 8 Hz, the average reduction is around 10 dB.

The Fig.13 depicts the ground velocity response attenuation with the distance. The response reduction is attained beyond 20 m, as expected from Fig.3. The computer simulation results show a nice fitting with the experiment results along the substantial distance of vibration transmission. The similar trend appears for the frequency range above the cut-off frequency of 4 Hz.

The durability of the driving plate in the linear contact with the contact soil is of most concern. The

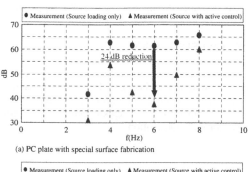

(a) PC plate with special surface fabrication

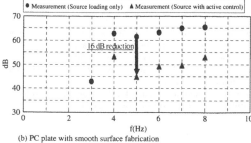

(b) PC plate with smooth surface fabrication

Figure 12. VAL vs. frequency at 30 m distance from the source.

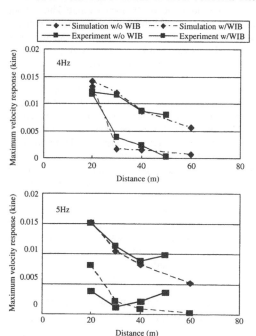

Figure 13. Ground velocity response vs. distance.

497

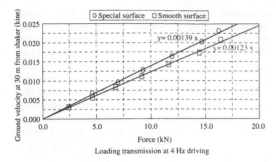

Figure 14. Ground response at 30 m for repeated loading for varying force level.

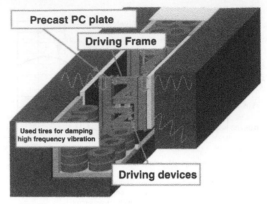

Figure 15. Vibration control apparatus.

friction state at the WIB plate is investigated by driving the active WIB for more than 5 minutes by varying the force to the maximum force capacity 17 kN. The cycles for repeating was over 1200. Fig.14 is the results for the ground response at 30 m for varying force levels. It is noted that the linear relationship is kept perfectly for both the smooth face plate and specially fabricated surface plate.

4 CONCLUSIONS

In this paper, an innovative idea of vibration control is developed that makes use of the superposition of another wave to the original wave in order to reduce the total wave field of the low frequency range. Throughout this investigation, only the vertical loading is considered. First, by using the finite element model, the computational approach was taken to get some preliminary information and demonstrate it. Then, by using two actuators, a field test was conducted to validate the theoretical wave interference in the actual ground.

The phase adjustment leads a significant response reduction nearly to one tenth or around 20 dB reduction. The apparent shift of the wave cut-off frequency toward the high frequency is interpreted both in computer simulation and field experiment. The effective zone for vibration amplitudes reduction extends in the long distance across the ground surface. The contact condition of the PC plate in the active driving system with the adjacent soil was proved almost linear.

The above mechanism can be put into practice for reducing the low frequency waves as encountered in the environmental vibrations in an effective way that cannot be achieved otherwise so far.

5 DESIGN CONSIDERATION AND FUTURE DEVELOPMENT

An actual system for the active WIB is devised as illustrated in Fig.15. A pair of shear force action can

be produced by using piezo-actuator that works independently for the horizontal as well as vertically. The high frequencies generated from the frame system used are absorbed by the used tire installation.

The points for the future research and development are placed on the following items.

(1) Robust control of frequency and phase control of the piezo-actuator system;
(2) Durability of driving frame structure, the PC plate and the contact condition with the adjacent soils;
(3) Handling for the transient random nature at the source vibrations;
(4) Maintenance during the long use of the apparatus.

ACKNOWLEDGEMENT

This research has been carried out in the frame work of cooperation among Industry, authority and academic for research seed raising program, 2005. An appreciation is extended to The Chugoku Technical Program Center for the coordination.

REFERENCES

Takemiya, H. 2004. Field vibration mitigation by honeycomb WIB for pile foundation of a high-speed train viaduct, Soil dyn. & Earthq. Eng. 24, 69–87.
Takemiya, et al. 2005. Patent Display -016020, Ground vibration mitigation system.
Takemiya, H. and Shimabuku, J. 2005. Honeycomb-WIB for mitigation of traffic-induced ground vibrations, ISEV2005.

Environmental Vibrations – Takemiya (ed.)
© 2005 Taylor & Francis Group, London, ISBN 0 415 39035 4

Ground vibration control using PANDROL VANGUARD in a tunnel on Guangzhou Metro Line 1

A. Wang & S.J. Cox
Pandrol Ltd, Addlestone, England

L. Liu, H. Huang & S. Chan
Guangzhou Metro, Guangzhou, P. R. of China

ABSTRACT: The paper describes the installation and testing of the PANDROL VANGUARD rail fastening system on Guangzhou Metro Line 1 in China. The system is designed to reduce the level of vibration transmitted from the rail into the ground. The fasteners were installed in a tunnel under a sensitive residential area in place of the existing rail fasteners. Measurements of deflection of the track and vibration in the tunnel and at the surface were made for normal service trains before and after the new fasteners were installed. The measurements show that a significant reduction in vibration level on the track and at the surface were achieved when the PANDROL VANGUARD fasteners were installed. Vibration acceleration levels decreased by 15.6 dB on the tunnel floor, and by 11.0 dB on the surface above the tunnel as the result of the installation of the PANDROL VANGUARD system.

1 INTRODUCTION

This paper describes the installation of a 192 m trial length of PANDROL VANGUARD vibration control rail fastening baseplates (Pandrol 2004a) on Line 1 of the Guangzhou Metro system in China in January 2005. The baseplates were installed in place of the existing fasteners in a tunnel on the southbound track between Changshoulu and Huangsha stations.

Measurements of deflection and vibration under traffic (Pandrol 2005) were made near the mid-point of the trial section before and after the installation. The existing Guangzhou Metro slab track has 60 kg rail fixed to cast-in concrete blocks with cast iron baseplates. The measurements were repeated after the installation of the PANDROL VANGUARD baseplates. All the measurements were made within one week. The PANDROL VANGUARD baseplates were fitted at mid-span between the existing fastenings on the concrete slab. The same rail remained in place throughout. All the measurements were made on one tunnel cross-section, at ¼ span relative to the existing fasteners, and therefore in the same position relative to the closest fastener location both before and after the changeover.

The deflections and accelerations of both rails were measured. The acceleration was also measured on the slab, on the tunnel wall, and at the surface above the tunnel. The measurements show that significant reductions in the vibration level on the track slab and at the surface immediately above the tunnel were achieved when the PANDROL VANGUARD fasteners were installed.

The motivation for the trials was that the PANDROL VANGUARD system was being considered as an alternative to floating slab track. As well as ground borne vibration, airborne noise in trains running in the tunnel needs to be considered. For comparison, track vibration levels with a floating slab track were also measured on the northbound track between Jiniantang and Yuexiugongyuan stations on the Guangzhou Metro Line 2. The results indicated that the rail vibration levels on the floating slab track and PANDROL VANGUARD tracks are very similar. Noise levels radiated from rail vibration on track fitted with PANDROL VANGUARD are likely to be similar to those on track with floating slab track. It was also noted that the slab vibration on the Line 2 floating slab track was quite high. Slab vibration may create a rumbling noise that is likely to be heard inside the vehicles.

2 RAIL FASTENING SYSTEMS AND TRACK VIBRATION

2.1 Existing Guangzhou Metro baseplate

The existing track consisted of Chinese 60 kg rail, spring clips with 10 mm thick rail pads, and cast iron

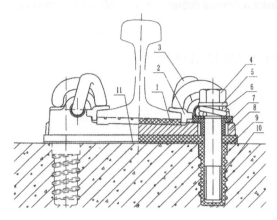

Figure 1.   Guangzhou Metro baseplate.

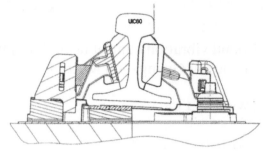

Figure 2.   Standard PANDROL VANGUARD.

baseplates fixed with M24 bolts through resilient 10 mm thickness baseplate pads on to concrete blocks which were cast directly into the concrete track slab as shown in Figure 1. The track has nominal 1435 mm gauge. The dynamic stiffness of the existing fastener has been measured and reported to be 58 kN/mm (CARS 2001).

2.2   PANDROL VANGUARD baseplate

PANDROL VANGUARD is a rail fastening system in which the rail is supported by elastic wedges under its head. The wedges are in turn held in place by cast iron brackets, which are fastened to a baseplate. The baseplate is rigidly fixed down to the track foundation. A typical configuration is shown in Figure 2. The principal advantage of the system over more conventional rail fastenings is that is allows significantly greater vertical deflections under traffic without an unacceptable accompanying degree of rail roll and without increasing the overall rail height.

The dynamic stiffness between 5 kN and 35 kN of the PANDROL VANGUARD baseplate designed for Guangzhou Metro is 6.0 kN/mm (Pandrol 2004b). This very low stiffness system reduces vibration transmission to the supporting structure and hence into the ground.

2.3   Track vibration and ground vibration control

There are two main noise and vibration problems associated with running trains in railway tunnels. The first is transmission of ground borne vibration into structures above the track. The second is airborne noise from the track, which is confined to the tunnel but may affect noise levels inside vehicles. Changes in vibration levels on the tunnel floor and walls are a good indicator of changes in ground-borne vibration levels, whereas vibration levels on the rail and track components are more relevant to airborne noise.

The following discussion outlines some of the basic principles involved in controlling railway vibration. It is provided in order to give a framework for understanding the results of the track tests given below.

Baseplates designed to control ground vibration from railways do so by offering a low dynamic track stiffness. Generally, the lower the stiffness which can be achieved, the lower the frequency at which effective vibration reduction results, and the better the overall performance. But there is a limit below which it has not previously been practical to reduce track stiffness through the use of resilient baseplates, because as the track is made softer the rail will also typically roll more under lateral loading, increasing the track gauge. Because the PANDROL VANGUARD system controls dynamic track gauge under lateral loading, it allows the track stiffness to be safely reduced much further than with conventional baseplates. This in turn leads to much improved control of ground vibration.

When a train runs along the track, it generates vibrations over a very wide range of frequencies. These are attenuated through the rail fastening system, and again as they are transmitted through the ground into nearby structures. The peak vibration level measured in buildings alongside the track is usually in the frequency bands between 20 Hz and 200 Hz. These are the frequencies that it is most important to control through the fastening system. The "insertion loss" is the change in vibration which is achieved when a new fastener is installed. The insertion loss for a fastener therefore depends not only on its own effectiveness but also on what it is replacing.

3   MEASUREMENTS

3.1   Test site and traffic

The test installations and measurements were made on the Line 1 of the Guangzhou Metro system in China. This line runs in a broadly north-east to south-west direction as shown in Figure 3. The PANDROL

Figure 3.   Guangzhou Metro system map.

Figure 4.   Surface measurement positions – Dadi Old Street.

VANGUARD baseplates were installed on the south-bound track in the tunnel between the Changshoulu and Huangsha stations. The northern end of the trial length is on a 400 m radius curve for 8 m, and runs through a transition of length 60 m and on to tangent track for 124 m. Maximum cant is 120 mm on the curve and averages 60 mm on the transition. The spacing of the existing baseplates is 595 mm on tangent track and 568 mm on transitions and curves. The same rail remained in place throughout the installation and testing. The tunnel has a round cross section and a single track.

The deflections and accelerations of both rails were measured. The vibration level was also measured on the slab, the tunnel wall and at the surface above the tunnel. Figure 4 shows the position of the surface measurements.

The traffic is 6 car EMUs with an axle load of approximately 16 tonnes and the maximum traffic frequency is 15 trains per hour during peak period operation. The track speed is about 70 km/h. All recordings were made under normal service passenger traffic at peak operating hours.

3.2   Measurement in the tunnel

Measurements in the tunnel were made at the centre of the installation and therefore on tangent track. Transducers were aligned at ¼-span from the existing fasteners for the "before" measurements, so that after the PANDROL VANGUARD baseplates had been installed at ½-span the same transducer mounting points also gave a ¼-span measurement during the "after" measurements. The measurement positions in the tunnel are shown in Figure 5.

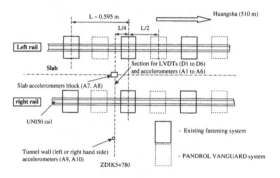

Figure 5.   Measurement positions.

3.2.1   Track deflection measurements

Deflections of the rail relative to the concrete slab were measured using strain gauge transducers. The transducers measure deflections of up to ±5 mm, with an accuracy of 0.1%. The transducers are mounted on studs, which are in turn fixed to metal plates glued down to the slab. Each transducer bears on to a target plate fixed to the measurement surface. Six recording channels were used, four channels to measure the vertical deflection of the rail relative to the slab on either side of each rail, and two channels to measure the lateral deflection of the rail relative to the slab for both rails. Positions of the sensors are shown in Figure 6. A National Instrument SCXI-1520 strain gauge amplifier was used to condition the signal from all six transducers.

3.2.2   Track vibration measurements

Measurements of vibration were made on the same fixed cross section at ¼-span between assemblies as for

Figure 6. Deflection transducer positions on rails relative to the slab at ¼-span between assemblies.

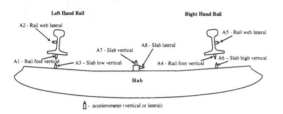

Figure 7. Accelerometer positions on the rail and slab at ¼-span between the assemblies.

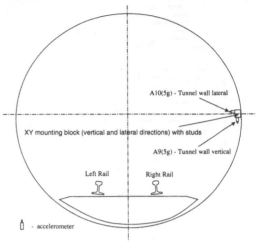

Figure 8. Accelerometer positions on the tunnel wall.

the deflection measurements. Measurements were made on both the rails in the vertical and lateral directions; on the concrete slab mid-way between the rails in the vertical and lateral directions; and on slab under each rail in the vertical direction. Tunnel wall vertical and lateral vibrations were also measured on the low rail side. The vibrations on the rails and concrete slab on the centre line of the track were measured using Kistler K-shear type 8702 accelerometers. The vibrations of slab beneath the rails were measured using PCB type 352C33 accelerometers. Tunnel wall vibration was measured using Kistler K-shear type 8712 accelerometers. A National Instrument SCXI-1531 coupler was used to condition the signal from all ten accelerometers. The positions of the transducers for vibration measurements are shown in Figures 7 and 8.

Measurements of track vibration were also made at kilometer post of K14 + 881 on the northbound track between Jiniantang and Yuexiugongyuan stations on the Guangzhou Metro Line 2. The test site has about 198 m of floating slab from K14 + 800 to K14 + 998. These measurements were made in the same way, and using the same equipment so that the vibration levels of the track with PANDROL VANGUARD system can be compared with a floating slab track.

3.3 Surface vibration measurements

Surface vibration measurements were also made near the mid-point of the trial section before and after the installation of PANDROL VANGUARD on Guangzhou Metro Line 1. The accelerometer positions on Dadi Old Street above the southbound tunnel in which the PANDROL VANGUARD installation

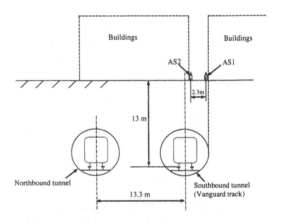

Figure 9. Approximately position of two tunnels.

was made are shown in Figure 4. Dadi Old street is a pedestrian street, and the nearest main road is the parallel Baohua Road, about 50 m distant.

Accelerometer AS2 was near the track centre line, with accelerometer AS1 a further 3 m away as shown. Both accelerometers were approximately 13 m vertically above the rail level. The approximate positions of the northbound and southbound tunnels are indicated in Figure 9.

The vibration measurements were made using two PCB type 393B12 accelerometers. The signal from the accelerometers was amplified and converted to voltage signal using PCB type 442B02 amplifiers. Acceleration levels from trains on both the northbound and southbound tracks were recorded, and were identified and separated using the timings found from the simultaneous recordings made in the southbound.

3.4 Data analysis

More than 20 trains were recorded and analysed in each measuring session. Given the nominally similar nature of the traffic, results from these were averaged all of across traveling at close to average speed.

The deflection records were first filtered with a low pass cut-off frequency of 30 Hz. This separates out the axle-passing component from the signals. The data is then reduced by identifying significant points such as peaks and troughs. The speed of trains is calculated from the time difference in the response to bogies on the train at a known distance apart.

Vibration records were analysed to obtain spectra in the frequency domain. One-third octave band spectra were found, and the linear and A-weighted levels calculated. Vibration levels of the track are presented here in terms of acceleration level. The vibration results presented here are of average vibration levels during train pass-bys at each site. Trains typically take eight or nine seconds to pass the measurement site.

4 RESULTS AND DISCUSSION

4.1 Deflection

Deflection measurements were made in the morning and evening traffic peaks, each for not less than two hours and a minimum of 20 trains. Train speeds were calculated from deflection records. The average speed of trains recorded over the period at the test site was about 70 km/h. All trains travelling at between 68 and 72 km/h at the test site were considered. Deflections have been analysed separately for the two rails and for leading and trailing axles. Since the tests were made on the tangent track, no significant difference was found between the leading axles and the trailing axles, so the deflections under all axles were averaged for each rail. The vertical deflections of the rail foot on the field side and the gauge side were first averaged to estimate the vertical deflections of the rail centre. Rail roll has been calculated by subtracting the gauge side deflections at the rail foot edge from field side deflections, dividing by two and multiplying by a geometry factor (rail foot divided by distance between field-side and gauge side vertical transducers). The lateral deflection of the rail head was estimated by multiplying the rail roll by an appropriate factor derived from the geometry of the rail section and adding the corresponding average lateral deflection of the rail foot. The geometry factor used in the calculation was the ratio between the height of the gauge corner of the rail and half the width of the base.

The mean deflections for the different tracks and both rails are given in Table 1. Negative values represent downward vertical deflection, or outward (gauge widening) lateral deflection and roll. The calculated lateral deflection of the rail head is shown in the same table.

Table 1   Average rail deflections under Line 1 traffic

System	Rail	Vertical (mm)	Rail foot Lateral (mm)	Roll (mm)	Rail head lateral (mm)
Existing	Left	−0.100	0.017	−0.026	−0.038
	Right	−0.161	0.019	0.060	0.146
	Average	−0.131			
VAN-GUARD	Left	−4.039	−0.224	0.208	0.220
	Right	−3.780	−0.377	0.263	0.184
	Average	−3.910			

With the existing fastening system, the net average rail vertical and maximum rail head lateral deflections are 0.131 mm and 0.146 mm respectively. With PANDROL VANGUARD baseplates, the net average rail vertical and maximum rail head lateral deflections are 3.91 mm and 0.220 mm respectively.

The data shows that the PANDROL VANGUARD fastening system is a lot more compliant in the vertical direction than the system it replaces, while rail roll and lateral deflection are only slightly increased and remain at quite acceptable levels.

This combination of low vertical stiffness and rail roll restraint with the PANDROL VANGUARD fastening system offers the potential for significant reductions in vibration transmission with a mechanically acceptable system.

The dynamic stiffness of the PANDROL VANGUARD assemblies measured in the track tests has been compared with the laboratory test results and the design objectives. A simple beam on the elastic foundation model (Grassie 1984) of the track has been used to estimate the track dynamic stiffness from the average vertical rail dynamic deflection and using other track parameters shown in Appendix A. The value obtained for the dynamic stiffness was 5.3 kN/mm per rail seat. This agrees quite well with the laboratory test result. The design objectives were met.

4.2 Track vibration

Vibration measurements may vary from one train pass to the next, due to the effects of speed and wheel condition. Therefore, trains were selected from the "before" and "after" recordings which were close to the average speed of 70 km/h and which appeared to give a clean signal on all channels (e.g. no impacts caused by wheel flats). 39 "before" trains and 33 "after" trains were identified and all vibration measurements presented here have been averaged across these selected train passes.

Total vibration levels for different measurement positions are shown in Figure 10. These were summed across the frequency ranges shown in the spectra shown below.

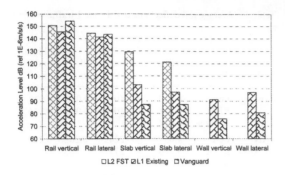

Figure 10.   Total acceleration levels.

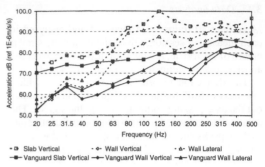

Figure 12.   Slab and tunnel wall acceleration spectra.

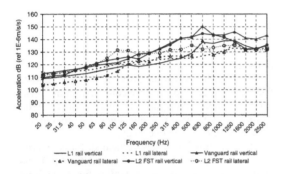

Figure 11.   Rail acceleration 1/3 octave band spectra.

Table 2   Track acceleration levels (dB ref $1 \times 10^{-6}\,\text{m/s}^2$)

Position	Existing system	VAN-GUARD	Insertion loss
Slab vertical	102.9	87.3	15.6
Tunnel wall vertical	91.2	75.7	15.5
Tunnel wall lateral	97.0	80.7	16.3

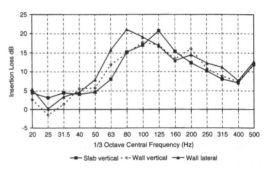

Figure 13.   Track insertion loss.

4.2.1   Rail vibration

The frequency range of greatest interest for rail vibration is up to about 2500 Hz. Figure 11 shows the acceleration spectra of the rails in the vertical and lateral directions for the frequency range 20 Hz to 2500 Hz. It was found that a peak at 630 Hz was caused by a pre-existing rail corrugation at a wavelength of 31 mm with the train speed of 70 km/h. The rail vertical acceleration increased after the installation of the standard PANDROL VANGUARD fastenings. The increase in vertical rail vibration is because of the lower stiffness of the track, and the lower wave decay rate along the rail that results.

To put these changes in rail vibration into perspective, they can be compared with the results of track vibration measurements made at kilometer post of K14 + 881 on the northbound track between Jiniantang and Yuexiugongyuan stations on the Guangzhou Metro Line 2. The test site has about 198 m of floating slab from K14 + 800 to K14 + 998. These measurements were made in the same way using the same equipment. The rail acceleration spectra recorded here are shown in Figure 11 for comparison. These indicate that the rail vibration levels on both the floating slab track and PANDROL VANGUARD tracks are very similar, except for a peak associated with the rail corrugation at 630 Hz on Line 1. Noise levels on track

fitted with PANDROL VANGUARD are likely to be no higher than for track with floating slab track.

4.2.2   Slab and tunnel wall vibration

The frequency range of greatest interest for the slab and tunnel wall vibration is from 20 Hz to 500 Hz. Acceleration spectra of the tunnel floor in the vertical direction and the tunnel wall in both the vertical and lateral directions on the track are shown in Figure 12. Total slab and tunnel wall acceleration levels are given in Table 2.

The insertion losses for each 1/3 octave frequency band are plotted in Figure 13. The total vibration acceleration level on the slab decreased by 15.6 dB as a result of the installation of the PANDROL VANGUARD baseplates. The tunnel wall acceleration decreased by 15.5 dB in the vertical direction and 16.3 dB in the lateral direction.

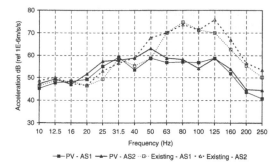

Figure 14. Surface acceleration spectra under Line 1 traffics.

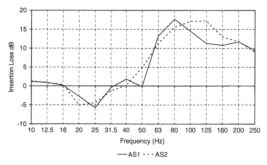

Figure 15. Surface insertion loss.

Table 3. Surface acceleration levels (dB ref 1×10^{-6} m/s^2)

Position	Existing system	VAN-GUARD	Insertion loss
Surface at AS1	78.4	67.8	10.6
Surface at AS2	79.9	68.6	11.3
Average	79.2	68.2	11.0

It should be noted that the slab vibration on the Line 2 floating slab track was quite high, more than 26 dB higher than that on the slab with the existing Guangzhou Metro rail fastening system on Line 1, and 42 dB higher than that on the slab with the standard PANDROL VANGUARD system. This high level of slab vibration will create a rumbling noise in the tunnel that is likely to be heard inside the vehicles.

4.3 Surface vibration

Recordings contaminated with road traffic vibration have been removed from the data analysis for each measuring session.

This process resulted in 17 "before" trains, and 17 "after" trains to consider for analysis. All vibration measurements presented here have been averaged across results from these selected train passes. Similar results were obtained for all of the selected trains with each set of data.

The frequency range of greatest interest for the surface vibration is from 20 Hz to 250 Hz. Vertical acceleration spectra at the surface at the two locations with the existing fastening and the PANDROL VANGUARD system installed are shown in Figure 14. Total surface acceleration levels are given in Table 3.

Vibration acceleration levels on the surface decreased by about 11.0 dB to 68.2 dB as the result of the installation of the standard PANDROL VANGUARD system. The reduced acceleration level is under the GZM environmental limit of 70 dB in the residential area. The original level exceeds the limit by a significant margin. Figure 15 shows the insertion loss between the PANDROL VANGUARD and the existing systems.

5 CONCLUSIONS

An installation of PANDROL VANGUARD baseplates has been successfully completed on the southbound track in a tunnel between Changshoulu and Huangsha stations on the Guangzhou Metro Line 1.

Vertical deflection of the rail relative to the slab was 3.910 mm with the standard PANDROL VANGUARD system under the Line 1 traffic. Rail head lateral movement was 0.220 mm. The dynamic stiffness of PANDROL VANGUARD baseplate estimated from the track tests agreed well with laboratory test results.

Vibration acceleration levels on the tunnel floor decreased by 15.6 dB as the result of the installation of the PANDROL VANGUARD assemblies. The tunnel wall acceleration decreased by 15.5 dB in the vertical direction and 16.3 dB in the lateral direction. Vibration acceleration levels on the surface decreased by about 11.0 dB to 68.2. The reduced acceleration level is under the GZM environmental limit of 70 dB in the residential area. The tests have demonstrated the level of vibration reduction that can be achieved in practice under operating conditions.

REFERENCES

CARS 2001, Tests on Guangzhou Metro low vibration fastening systems – final report, China Academy of Railway Sciences, Report No. 2001-0594-1.

Grassie S.L. & Cox S.J. 1984, The dynamic response of railway track with flexible sleepers to high frequency vertical excitation, Proceedings Institute of Mechanical Engineers, 198D.

Pandrol 2004a, Guangzhou Metro Line 3 and Line 4 – PANDROL VANGUARD Assembly, Pandrol Report No. 85171-20.

Pandrol 2004b, Testing of PANDROL VANGUARD assembly for Guangzhou Metro, China, Pandrol Report No. 85171-18.

Pandrol 2005, Installation and testing of the PANDROL VANGUARD system on Guangzhou Metro line 1, Pandrol Report No. 85171-24.

Moment of inertia in vertical $= 3217\,\mathrm{cm}^4$
Wheel load force $= 80\,\mathrm{kN}$
Support spacing $= 0.595\,\mathrm{m}$
Rail material Young's modulus $= 210\,\mathrm{GN/m}^2$

APPENDIX A – PARAMETERS FOR SIMPLE TRACK MODEL

Rail section: 60 kg
Cross section area $= 77.45\,\mathrm{cm}^2$
Theoretical weight per linear meter $= 60.64\,\mathrm{kg/m}$

Environmental Vibrations – Takemiya (ed.)
© 2005 Taylor & Francis Group, London, ISBN 0 415 39035 4

Characteristic change of traffic vibration due to road surface continuation on viaducts

Koichi Sugioka & Kiyoshi Yamamura
Osaka Business and Maintenance Department, Hanshin Expressway Corporation, Osaka, Japan

Yutaka Yamamoto & Yasuhiro Sanuki
Fuji Engineering Co., Ltd. Osaka, Japan

ABSTRACT: The Hanshin Expressway network in the Kansai Metropolitan Area linking the major cities of Osaka, Kobe and Kyoto currently extends 233.8 km in total length and is traveled by about 890,000 vehicles on average per day. This rapid increase of the traffic volume in urban areas, however, has caused amplification tendency of ground vibration when resonance happens between a large-sized vehicle and viaducts. An impact that is made when large-sized vehicles pass uneven expansion joint is one of main factors of vibration from viaducts. "Joint-less" is expected from a maintenance management side as well as environmental measures side very much. This report gives an outline by measurement results about characteristic change of traffic vibration with enforcement of three "Joint-less" methods. According to the measurement result, the vibration of the neighboring ground is reduced to 91–98% in all-pass (AP) by the three "Joint-less" methods.

1 INTRODUCTION

Since its foundation in May, 1962, the Hanshin Expressway Public Corporation (HEPC) has devoted itself to the development of an urban expressway network in the Kansai Metropolitan Area linking the major cities of Osaka, Kobe and Kyoto. The expressway network currently extends 233.8 km in total length and is traveled by about 890,000 vehicles or about 1.3 million people on average per day.

This figure which matches the population of Kobe or Kyoto City reveals that the Hanshin Expressway supports all social, cultural and economic activities in the area and plays an important role in providing vital bases for information and international exchanges in various fields.

This rapid increase of the traffic volume in urban areas, however, has caused amplification tendency of ground vibration when resonance happens between a large-sized vehicle and viaducts.

This may be because the urban expressway has been constructed as viaducts structure by the side of houses in order to use the ground effectively.

An impact that is made when large-sized vehicles pass uneven expansion joint is one of main factors of vibration from viaducts. In addition, expansion joints need to be inspected and repaired relatively often for the sake of safety and comfort for drivers. "Joint-less" is expected from a maintenance management side as well as environmental measures side very much.

This report gives an outline by measurement results about characteristic change of traffic vibration with enforcement of three "Joint-less" methods.

One of three methods is the buried expansion joint method applied to PC girders, and the second is the slab connection method applied to PC girders, and the third is the main girder connection method applied to I-beam steel girders.

According to the measurement result, vibration of the neighboring ground is reduced to 91–98% in all-pass (AP) by the three "Joint-less" methods.

2 VIBRATION CHARACTERISTIC OF VIADUCT

Figure 2 shows an example of average frequency characteristic for 10 minutes of ground vibration around the viaduct of 30 m I-beam structure. The dominant frequencies depend on types of superstructures. However, vibrations of three bands (2.5–5.0 Hz, 10–31.5 Hz, 40–80 Hz) tend to excel in general as shown in Fig. 2 expressed with 1/3 octave band analysis result.

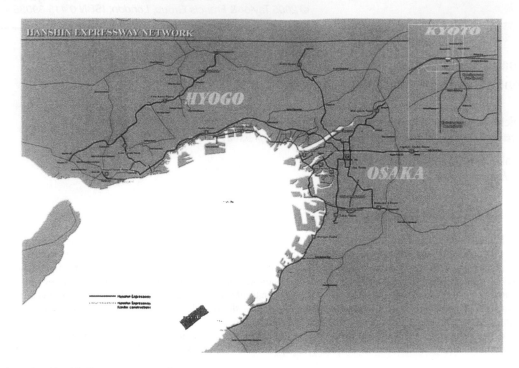

Figure 1.   Hanshin Expressway network.

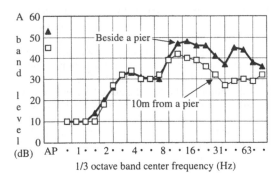

Figure 2.   One example of ground vibration spectrum of a viaduct.

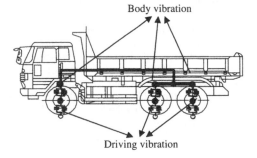

Figure 3.   The vibration of large-sized vehicles.

Vibration of a low frequency band of 2.5–5.0 Hz is caused by natural frequency of a low dimension mode of superstructure, and natural frequencies are different by the span lengths and structure models.

These vibrations are deeply related to body system's vibration (body vibration), and driving system's vibration (driving vibration) of large-sized vehicles as shown in Fig. 3. (Syoji et al. 1980)

The body vibration, that occurs when a large-sized vehicle passes uneven road surface, becomes power to generate vibration of this low mode. And it is estimated that the ground vibration occurs by a resonance phenomenon between the primary natural vibration and body vibration.

Vibration of 10–31.5 Hz almost agrees with the second or the third mode vibration of the superstructure in frequency. The driving vibration, that occurs when a large-sized vehicle passes uneven road surface, becomes power to generate vibration of these modes. Also it is estimated that the ground vibration propagates in outskirts area by a resonance phenomenon

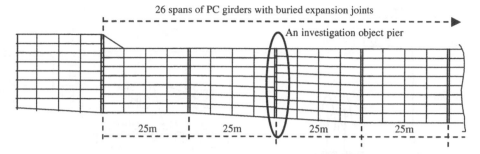

Figure 4. The bridge that buried expansion joints were introduced.

between the second or third natural vibration and driving vibration.

Moreover, the driving vibration generates impacting vibration of a high frequency band of 40–80 Hz at the time of the joint passage. Level and spectrum of the impacting vibration mainly depend on type of expansion joints, difference in level, material and type of bridges. This vibration decays greatly at the point that was 10 m away from a pier as shown in Fig. 2.

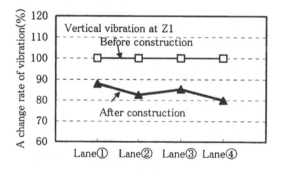

Figure 5. Buried expansion joint.

3 BURIED EXPANSION JOINT METHOD

3.1 *Investigation abstract*

Figure 4 shows the section where a post-tension PC girder continues. It was carried out pavement continuation 26 spans at the maximum in this section by the buried expansion joint method.

This buried expansion joint installs expanded metal fixed by a holing in anchor in the expansion and the contraction part of a base and uses guss asphalt as pavement materials as shown in Fig. 5. This is generally called the guss asphalt method (2) (Road Management Technology Center 1995a).

Investigation was conducted around a pier of Fig. 7, and the ground vibration measurement using a heavy vehicle (196 kN, 60 km/h) was carried out at night (21:00–6:00).

3.2 *Findings*

Figure 6 shows change rates of the ground vibration level at point Z1(pier side), point Z2(8.3 m from the pier side) as shown in Fig. 7, before and after the introduction of buried expansion joint. In addition, a vibration level of each point is a peak value when the heavy vehicle passed each lane.

The change rates of vibration with the buried expansion joint method are about 80–90% in any lanes.

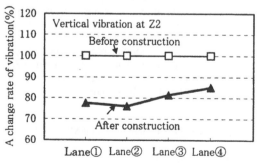

Figure 6. Characteristic change of vibration by the buried expansion joint method.

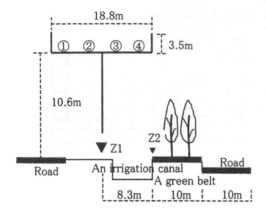

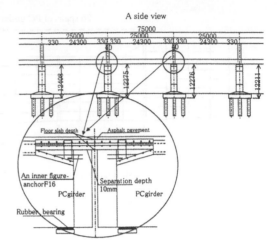

Figure 7.   Measurement points.

Figure 8.   PC slab connection method.

4   SLAB CONNECTION METHOD

4.1   *Investigation abstract*

The slab connection method means connecting two adjacent slabs by building a reinforced concrete slab. There are the steel slab connection method and the PC slab connection method (Road Management Technology Center 1995b). These methods are designed as simple girder bridges for live load, and designed as continuous girder bridges for earthquakes.

Figure 8 shows a general drawing of 24.3 m PC simple girder bridge which is carried out two parts of slab connections. Investigation was conducted around a pier of Figure 8, and the ground vibration measurement using a heavy vehicle (196 kN, 60 km/h) was carried out at night (21:00–6:00).

A peak vibration level in the heavy vehicle passage was measured by two places of the pier side and 8.8 m spot from the pier side.

4.2   *Findings*

Figure 10 shows a change of ground vibration characteristic by the slab connection method. A change rate of a frequency band of 8~10 Hz regarded as torsion vibration of superstructure is around 60%, and a change rate of a frequency band of 31.5–80 Hz is around 80%. In AP the vibration reduction percentage is, however, 91% at point Z1 and it is 93% at point Z2 as shown in Fig. 9, because contribution to AP of these frequency band levels is small.

5   MAIN GIRDER CONNECTION METHOD

5.1   *Investigation abstract*

The main girder connection method connects only web of the main girder with two moment plates and

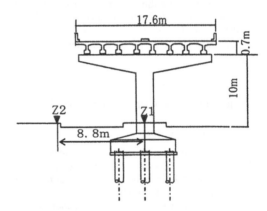

Figure 9.   Measurement points.

with a shear plate as shown in Fig. 11, and changes steel bearing to elastic bearing.

Figure 12 shows a general drawing of 30 m steel simple girder bridge which is carried out two parts of main girder connections. Investigation was conducted around a pier of Fig. 13, and the ground vibration measurement using a heavy vehicle (196 kN, 60 km/h) was carried out at night.

5.2   *Findings*

Figure 14 shows characteristic changes of ground vibration with main girder connections. This figure illustrates difference of vertical vibrations due to main girder connection for every frequency. The vibrations were measured before and after the connection at nearly the same traffic volume, and analyzed as average spectrums in ten minutes.

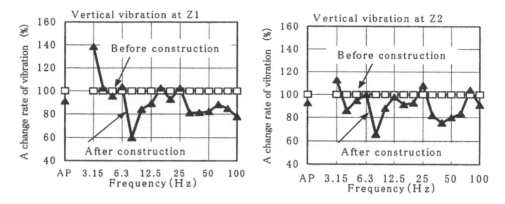

Figure 10.   Characteristic change of ground vibration with a PC slab connection.

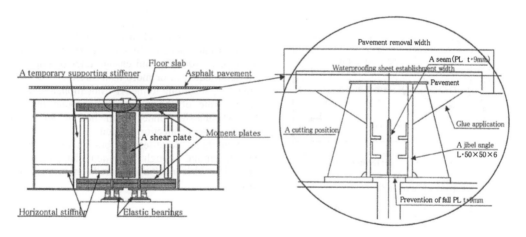

Figure 11.   Main girder connection method of steel I girder.

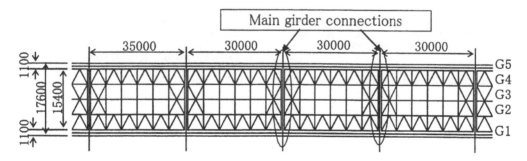

Figure 12.   Layout of main girder connections.

According to the analysis as shown in Fig.14, the vibrations of 2–4 Hz that mainly depend on natural frequency of the superstructure, are reduced to 80–90% owing to increase of stiffness as a result of the connection.

The change rate of frequency bands over 50 Hz shows 55–70% with the joint-less. But the change rate in AP remains in around 96–98% because these frequency bands are not effective for AP. In addition, the change rate of vibration at Z2 is smaller

511

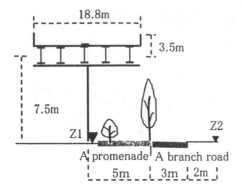

Figure 13. Measurement points.

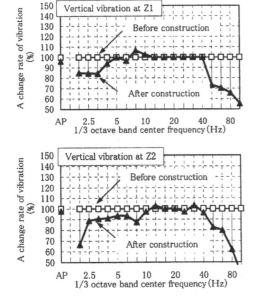

Figure 14. Change of a spectrum of ground vibration with a main girder connection method of a steel I girder.

than at Z1, as shown in Fig. 13 because of decay by distance.

6 CONCLUSION

The vibration characteristic change in each method is compiled as follows.

The change rate of peak ground vibration in the buried joint method was 80–90%.

On the peripheral ground near the pier which carried out a slab connection, a change rate of 8 Hz regarded as torsion vibration of the superstructure is around 60%, and change rates of 31.5 Hz–80 Hz that are impacting vibration in the joint passage of a large-sized vehicle is around 60% identically.

In addition, on the peripheral ground near the pier which is carried out a main girder connection method of steel I-beam bridge, change rates of 2–4 Hz regarded as the primary natural vibration of the superstructure are around 80–90%, and change rates of impacting vibration equal to or more than 50 Hz are 55–70%. However, according to the AP evaluation, a vibration change rate of the neighboring ground becomes 91–93% in a slab connection method and 96–98% in a main girder connection method.

With three "Joint-less", vibration of the neighboring ground brought vibration change rates of around 91–98% in AP and showed constant reduction effect.

REFERENCES

Road Management Technology Center 1995a, *Design and execution guide of joint-less method for existing bridges*, pp.1–77, Tokyo Japan

Road Management Technology Center 1995b, *Design and execution guide of joint-less method for existing bridges*, pp.2–80~2–111, Tokyo Japan

Syoji, H. & Yamamoto, T. & Hatakeyama, N. 1980, *Sanitary engineering handbook: Noise and vibration edition*, Asakura Syoten, Tokyo Japan

Environmental Vibrations – Takemiya (ed.)
© *2005 Taylor & Francis Group, London, ISBN 0 415 39035 4*

Effects of train-induced ground vibrations by stiffening structures

N. Iwata, H. Yokoyama & K. Ashiya
Railway Technical Research Institute, Tokyo, Japan

ABSTRACT: For the purpose of reducing environmental vibrations along the Shinkansen super-express track, we constructed trial concrete piers of a rigid-frame bridge to increase its cross-section. We made comparisons of data as measured and derived by the running of Shinkansen trains with dynamic simulation results by two or three-dimensional numerical analyses before and after the trial construction of piers. We studied the change of vibration property and adaptability of the numerical model. In these models we altered the parameters of composition members or inserting braces, and discussed about important factors to comprehend the characteristics of vibrations by stiffening structures of rigid-frame bridges.

1 INTRODUCTION

At present, current countermeasures for train-induced ground vibrations are mainly related to rolling stock and track facilities. For instance, reducing of car weights, softening track support suspensions and other related matters. Provided that additional countermeasures are needed, a measure such as provision of underground walls is to be carried out. Measures for reducing the level of vibrations of rolling stock, track and ground have been in progress based on the previous studies and numerical simulations. It is possible to estimate quantitative effects of vibration reduction.

On the other hand, vibration reducing measures for structures such as rigid-frame bridge etc, have been rarely studied up to now, including an area of road traffic vibrations because there are many problem areas such as economical aspect and seismic resistance to which consideration be given.

Where additional countermeasures are needed at the site that has already been executed, measures for increasing structural integrity are important. In this paper, to study a basic knowledge of a model method and measures for reducing vibrations by reinforcing the structure, we prepared models for the field-test working that is stiffened piers increasing its cross sectional areas and then executed two or three-dimensional dynamic analyses.

2 FIELD TEST TO EVALUATE VIBRATION PROPERTY BY STIFFENING PIERS

2.1 *Outline of field test*

In order to reduce the degree of ground vibration, we experimentally made piers of a rigid-frame bridge by concrete lining.

The outline of construction is that the elevated bridge piers were increased with the cross-section from 900* 900 (mm^2) to 1200*1200 (mm^2). The track condition is of ballasted track with ballast mats without rubber-coated sleepers. The structural condition is a rigid-frame bridge with a span of 8 m (Fig. 1). The ground condition is of an alluvial layer that has a contrast relatively between the surface layer and the foundation

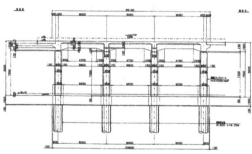

(a) Side view

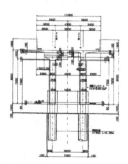

(b) Section view

Figure 1. Outline of field-test structure.

layer (Fig. 2). Measuring was made at three points on the ground with vibration level meters and at one point on pier with acceleration meter (Fig. 3). Ground surface condition is that 1ch and 2ch are paved with asphalt and 3ch is of a natural ground. Measured data are divided into three cases that are (1) before operation, (2) only upward operation and (3) after operation.

2.2 Results of field test

In order to evaluate the effect of reduced degree of vibrations by constructing massive piers, Figure 4(a) illustrates the train-speed dependency of before and after operation at 2ch, and Figure 4(b) and 4(c) illustrates the vibration acceleration level spectra that are averaged each car before and after operation. In addition to these, Figure 4(d) shows the changing spectrum in each car. In Figure 4(d), a minus value presents the reduced level of vibration while a plus value presents the increased degree of vibration, because the changing spectrum was calculated by subtracting averaged spectrum before from after.

From these figures, it is confirmed that changing vibration level is not so much between before and after the construction. Although it has some dispersion in

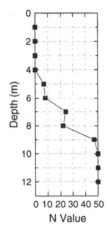

Figure 2.   N Value profile.

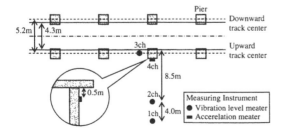

Figure 3.   Array of measured points.

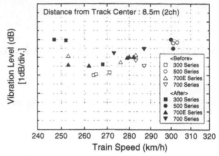

(a) Train-speed dependency

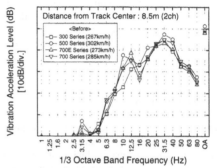

(b) 1/3 octave band spectrum (Before)

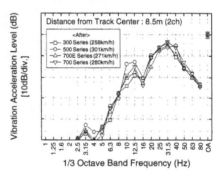

(c) 1/3 octave band spectrum (After)

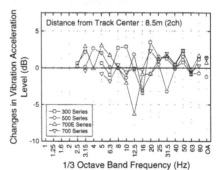

(d) Change of 1/3 octave band spectrum

Figure 4.   Results of field test at 2ch.

each car, it shows some reduced degree of vibration close to 16 Hz band.

The structure of this field test was much rigid before the execution. Because each pile was combined with underground beams, and each pile foundation adopted the large-scale diameter for soft ground condition. It was thought that the stiffening was not so much effective to change the system of natural vibration. As one of the causes does not change the vibration level so much, the frequency band of reducing vibration by this execution is not proposed the dominant frequency band of train-induced vibration.

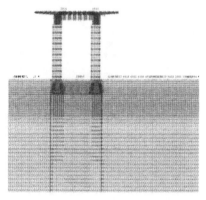

(a) An example of two-dimensional model

(b) An example of three-dimensional model

Figure 5. Diagrams of analysis models.

3 BASIC STUDY ON STRUCTURAL MEASURES BY TWO OR THREE-DIMENSIONAL DYNAMIC ANALYSES

3.1 Simulation models

A measure to reduce the degree of ground vibrations is necessary by improving structures considering the design thoroughly in advance, because there is much influence on economical and earthquake-resistant aspects. In this chapter, we executed the two or three-dimensional dynamic analyses that were modeled from the field-test working as mentioned in the second chapter, and studied basically the modeling adequacy of dynamic analyses.

Figure 5 illustrates example diagrams of two or three-dimensional models in this paper. In the two-dimensional models, ground and structure are modeled by a solid element. In the three-dimensional models, structures are modeled with beams and shell elements, and ground is modeled by a thin layer element.

The exciting force in two-dimensional simulation, it is modeled that the intensive force load vertically at the point of upward track center. Also in the three-dimensional one, there are three cases of exciting force models, (1) Line exciting force model that load vertically at the line of upward track center, (2) Point exciting force model that load vertically at the one of piers of upward (4ch in Fig. 3) and (3) Moving exciting force model that load vertically at the same position of Line with phase corresponding to train running at 260 km/h speed. With respect to the ground dynamic property, at first, the transfer characteristics from 8.5 m to 12.5 m were simulated by 2D model. In this pre-simulation, it is used that temporary parameters assumed from N value information at field-test site (Fig. 2) and empirical equations. Next, we modified the parameter Vs, a shear wave velocity, to correspond to actual measured data. Table 1 presents the simulation parameters.

3.2 Property of dynamic analysis models

In Figure 6(a), it is illustrated that the transfer properties of actual measured data in field test and that of analyzed data. In these figures, it is shown that two or three-dimensional dynamic analyses are able to

Table 1. Parameters of dynamic analysis models.

Unit weight (t/m³)	Vp (m/s)	Vs (m/s)	Damping constant (2D)	Damping constant (3D)	Shearing modulus (tf/m²)	Poisson's ratio	Note
1.5	1500	88	0.06	0.03	1184.5	0.49827	GL (0~−5 m)
1.6	1500	180	0.05	0.03	5286.2	0.49269	GL (−5~−7 m)
1.8	1500	260	0.05	0.03	12407.9	0.48451	GL (−7~−9 m)
1.8	1800	340	0.05	0.03	21218.2	0.48150	GL (−9~−10 m)
1.8	2000	400	0.04	0.03	29367.8	0.47917	GL (−10 m~)

reproduce the actual measured data approximately over 8 Hz band. The difference between actual measured data and each analyzed data shows a tendency that the precision is ranked, Line, Point, 2D and Moving exciting model.

In Figure 6(b) of transfer in ground, it is observed that the accuracy of Line exciting model and 2D models are relatively better than Point and Moving exciting model. In the case of Point exciting model, it is unable to re-create the actual data adequately. Regarding Moving exciting model, it has poor reproduction performance.

In order to compare the models of each exciting force in three-dimensional dynamic analyses, Figure 7 illustrates the relationship between actual measured data and analyzed data in each measured point. In this examination, the analyzed data are calculated by a vibration level adapted actual data at 3ch.

From these figures, it is presented that Line exciting model's dispersion from actual data is less, while Point and Moving are more. Train length is much longer than the one set of elevated bridges. Actual exciting force is almost steady during the passing of trains. Therefore it is thought that Line exciting model is able to simulate actual vibration state effectively.

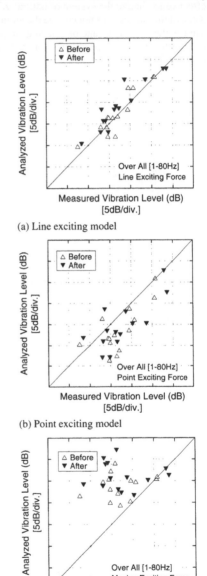

(a) Line exciting model

(b) Point exciting model

(c) Moving exciting model

Figure 7. Comparing with each exciting force model in three-dimensional analyses.

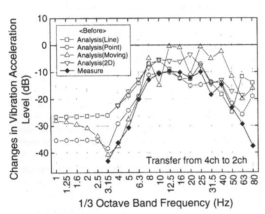

(a) Transfer from 4ch to 2ch

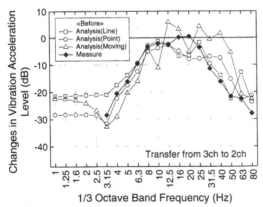

(b) Transfer from 3ch to 2ch

Figure 6. Comparing with the measure and the analysis transfer property.

Figure 8 illustrates the change in vibration characteristic before and after the operation. Approximately the analyses data are able to re-created the change of vibration property basically, that is a little reduction around 16 Hz band and a little increase around 40 Hz band. From such consequences, it is possible that the actual data is simulated roughly by either two or three-dimensional dynamic analyses. These methods are effective tools to simulate the structural measures in advance.

4 THE STUDY OF MEASURES IN STRUCTURE BY THREE-DIMENSIONAL DYNAMIC ANALYSES

4.1 The relationship between pre-structure and the change of vibration property

From the studies made until the third chapter, it was possible to simulate actual vibration property before operation by three-dimensional dynamic analyses. In this section, we study the change in vibration property by stiffing of rigid-frame bridge piers with or without underground beams. Figure 9 illustrates a modeling diagram of three-dimensional dynamic analyses. Besides the exciting force is Line one.

Figure 10(a) and 10(b) present the outcome of simulations. In these figures, a minus value presents the

Figure 8. Comparing with measures and analysis on changing vibration property at 2ch.

(a) With underground beams

(b) Without underground beams

Figure 9. Modeling diagram.

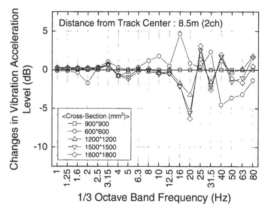

(a) With underground beams

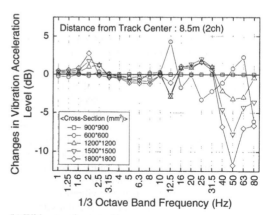

(b) Without underground beams

Figure 10. Change of 1/3 octave band spectra at 2ch.

(a) Parallel direction of track

(b) Perpendicular direction of track

Figure 11. Modeling diagram.

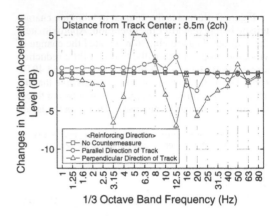

Figure 12. Change of 1/3 octave band spectrum at 2ch.

reduced level of vibration and a plus value presents the increased degree of vibration, because the varied spectra were calculated by subtracting from a models as measured model based on cross-section is 900*900 (mm^2). It is shown that the more rigid structures, the more remarkable change of vibration property by stiffening. In the case with underground beams, there is a tendency that the bigger the pier's cross-section, the more the reducing vibration.

4.2 Relationship between the reinforcing direction and the change of vibration property

When the structure is stiffened in the parallel direction or the perpendicular direction of track, there is a method to place a brace in each direction, except for increasing the pier's or beam's cross-section. In this section, it is reported that the outcome of simulation about steel H-brace placed in the parallel direction and the perpendicular direction of track. Figure 11 illustrates a modeling diagram. The intersections of the braces and the brace-structure are rigidly combined.

Figure 12 presents the change of vibration property in each reinforcing direction. In the case of inserted the braces perpendicularly, it is seemed that the change of vibration property is large, because the frequency band of rising vibration and that of reducing vibration

appears alternately. In the case of inserted the braces parallel that is not so large comparatively. In Line exciting force model, because of loading uniformly along railway, it is thought that the changing influence stiffened perpendicularly is more remarkable. When structures will be inserted braces additionally, changing the system of natural vibration is complex. Therefore it is necessary to study beforehand, including the structural kinds and reinforcing methods and so on.

5 CONCLUSIONS

In order to make good use of the structural countermeasures for reducing vibration that are still under development, we execute the two or three-dimensional dynamic analyses that are modeled in field-test site at first. Next it is studied basically the modeling methods of dynamic analyses and the improving method of structural countermeasures. The authors report the knowledge of these studies.

(1) Taking the field-test site as an object of study, we execute the simulation of two or three-dimensional dynamic analyses. Also, we studied the adaptability of these methods for evaluating the structural countermeasures. The effect of the studies, the change of vibration property before and after operation, for example constructing massive piers of elevated bridge is able to be re-created the actual measured data by these simulations.

(2) We studied a model method of three-dimensional dynamic analyses. Concerning the exciting force model, Line exciting force model is the most appropriate among others that are Point and Moving.

(3) In the three-dimensional dynamic analyses, we studied the relationship between the structural condition beforehand and the change of vibration property, and between reinforcing direction and the

change of vibration property. It is presented the fundamental knowledge of countermeasure to improve the structure. It is confirmed the adaptability of the three-dimensional dynamic analyses before the operation.

REFERENCES

Ashiya, K. 2003, Present State of Research and Development on Measures to Reduce Train-Induced Ground Vibrations, *Environmental Vibration*: 291–304.

Yoshioka, O. 1997, A Review of Studies on Ground Vibrations Induced by a Running Train (in Japanese), *RTRI REPORT*, 11(1): 17–26.

Hara, T. et al, 2004, Development of a New Method to Reduce Shinkansen-Induced Wayside Vibrations Applicable to Rigid Frame Bridges: Bridge-End Reinforcing Method (in Japanese), *Journal of Structural Mechanics and Earthquake Engineering (I)*, No.766/I-68: 325–338.

Yoshioka, O. 1996, Prediction Analysis of Train-Induced Ground Vibrations Using the Equivalent Excitation Force (in Japanese), *BUTSURI-TANSA*, 49(2): 136–146.

Environmental Vibrations – Takemiya (ed.)
© *2005 Taylor & Francis Group, London, ISBN 0 415 39035 4*

Vibration characteristics of buildings aiming at reduction of environmental vibrations and earthquake motions

T. Nakamura & M. Nakamura
Nakamura Bussan Ltd.

ABSTRACT: This paper discuss the "foundation substitution construction method" which uses foam resin materials in constructing building foundations which take into account of the effect of environmental vibration as well as earthquake energy input. The foam resin materials are installed at the foundation level, or next to the foundation, creating a subsurface level below the foundation. The foam resin ground materials are arranged according to targeted performance of buildings. The purpose and effect of the proposed construction method is to achieve the reduction of vibration and earthquake energy, as well as an equilibrium of weight distribution (reduction of the subterranean stress generated by the construction of the building), foundation heat insulation of the lower foundation, as well as to protect against frost damage and other factors. This method is applicable to wooden framed houses, steel framed buildings, as well as reinforced concrete buildings.

1 INTRODUCTION

This paper describes a "foundation substitution construction method" by using foaming resin construction for a foundation for houses. This reduces environmental vibration and earthquake motion. The substitution material used in this construction has to be light weight and durable. Besides the mitigation, some functions could be simply given to foaming resin by studying the design of construction.

Targeted performance can be improved by using artificially constructed ground which replaces forming resin material in conjunction with a subsurface layer. As result the artificial ground needs to be built in the lower part and the side part of each foundation in view of (1) environmental vibration, and earthquake motion, (2) control of different settlement, (3) foundation heat insulation and frost damage prevention. This construction method can solve these problems.

This construction method has been classified as the "foundation substitution construction method" under the construction standards of Japan. Form resin substitution materials were accepted as an artificially constructed ground in this construction method.

The outline of the design method of foaming resin artificially constructed ground is presented. Moreover, the dynamic characteristic of the houses and buildings built by this procedure was measured for earthquakes motions by the seismograph. There were case studies with this technology in Miyagi Prefecture and Niigata Prefecture hit by the big earthquake in last year

and the year before last. The measured velocity responses of the inside and the neighboring ground of houses has shown that the proposed method is effective in reducing environmental vibrations and earthquakes response.

When constructing a building on artificially constructed ground which contains foaming resin, the interaction of the foundation and the building must be taken into account in the design method. That is, when the measure which reduces environmental vibration and earthquake motion was designed, the interaction of inertia was evaluated and design modeling was carried out. The oscillating reduction effect is not quantitative. Therefore dissipation of the energy is decided by the type of foaming resin, composition, and thickness. This research recognized this interaction with the building when developing a building this interaction is important to the safety and the basic construction of the structure for living comfortably and functionally over a long period of time. It has been proposed that the artificial ground adopted as the reduction purpose of basic has validity by (which used foaming resin) floating and making it the foundation.

A light weight building is supported by the subsurface layer when constructing a building. If the allowable bearing capacity subsurface layer where long term is insufficient, the piles need to be driven until firm layer. In that case, the piles rise building bearing power to the foundation through a subsurface layer. This resembles the piloti structure in the building structure of the first floor, if seen from the foundation where the pile tip is firm. Moreover, the comparatively lightweight buildings are satisfied with the bearing power of a subsurface layer of the building resembles the float structure where it is floating in the subsurface layer. Since a ship floating on water is hardly influenced by an earthquake, more attention was paid to float foundation structure.

When the structure and the material of the artificial ground were considered, the amount of transfer of an environmental oscillatory wave and seismic waves imagined air and water as few materials. We envision designing a structure similar to a ship floating on water.

In the case of the direct foundation, the enforcement scope of the direct foundation using foaming resin which floats, studies the foundation and float material and is built on soft ground was extended from this image. The Building Center of Japan did examination proof of these fruits of work as construction technology and preservation technology, such as a building, as a foundation substitution construction method. As a result, it become that it is easy to be adopted as many buildings.

In the development of artificially constructed foundation design technology, more research into the factors of environmental vibration and earthquakes, is required especially in the following four areas.

(1) Designing a quantitative model for the amount of dissipation of energy.
(2) Extend the range of the application foundation of artificial ground, and buildings.
(3) Development of the foaming resin foundation materials which are used artificially constructed ground.
(4) Seismographs will be installed in many buildings build with this technology, will take measure various factors.

This research presentation which introduces the outline of the foaming resin artificially constructed ground design method accordance with the construction standards of Japan.

We have measured the velocity of response spectrum of various buildings constructed with this technology. In addition we conducted interviews with people living in houses with this type of technology.

2 LIGHT WEIGHT ARTIFICIALLY CONSTRUCTED GROUND

The material and physical properties which are used as lightweight materials are shown in Table 1, and chemical resistance is shown Table 2. Among the materials shown Table 1, foaming polystyrene resin is more economical than other resin and superior in the creep characteristic over a long period of time. Foaming polypropylene and

Name	Per unit weight (Kg/m³)	Density (Kg/m³)	Compress (KN/m²)
Polyethylene	30	30	176.4
	25	25	137.2
	20	20	98.1
	16	16	68.6
Polyproplyene	60	60	225.4
	50	50	176.4
	30	30	98.1

Table 2. Chemical resistance of lightweight materials.

Name	Polyethylene	Polypropylene
Machine oil	Δ	◎
Petrol	×	o
Kerosene	×	o
Heavy rude oil	Δ	o
Toluene	×	Δ
Acetone	×	◎
Ethyl alcohol	◎	◎
CCl$_4$	×	Δ
Acetic ethyl	×	o
Methyl ethyl ketene	×	o
Sulfuric acid	◎	◎
Acetic acid	◎	◎
Hydrochloric acid	◎	◎
Ammonium	◎	◎

* ◎No change o Slight changeΔExpand×Melt.

copolymer resin are effective in oscillating reduction. Range of applicability for lightweight materials for artificially constructed ground, and important factor inside usability. According to an inspection of the potential building site, as well as considerations of the proposed building, materials are chosen.

2.1 Scope of the proposed construction method

In adopting this construction method, it is important to do ground survey of the construction area, as well as a situation investigation of surface soil and a building investigation in advance. From that information, safety and applicability of a building is checked, and lightweight materials are chosen from Table 1.

In order to protect the lightweight foundation material, a concrete layer, above the materials, at least 150 mm of a thickness is required. Portions other than concrete protection are protected with sand and polyethylene sheet.

If chemicals which effect the lightweight foundation material at the construction ground are found, a method to reduce the harm of these chemicals and to check safety must be applied.

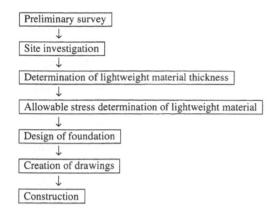

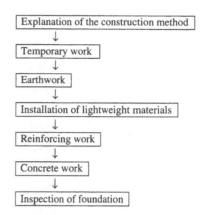

Figure 1. The design procedure of the artificially constructed ground.

Figure 2. Construction flows includings associated work.

2.2 *Design procedure*

Figure 1 shows the design procedure of the artificially constructed ground.

(1) Clarify the design details relating to the project, and their possible limiting factor.
(2) Investigation of the site.
(3) Before the construction can begin, a study of vibration of diffusion of the site must be performed, based on the desired value of oscillating reduction performance is decided.
(4) Compute the amount of substitution of the surface soil a ground level over long term due to allowable bearing capacity.
(5) Decide which type of foundation to use as well as the volume of artificially constructed ground.
(6) Submit the proper application paper to government authorities.
(7) Create the artificially constructed ground drawing, foundation drawing and construction working drawing. Make design drawing which explain how to construct the artificially constructed ground for the construction laborers.
(8) The component a cut into the proper size at the factory before being brought to the site.
(9) Construction.
(10) After construction is completed the effectiveness of the initial study is checked.

2.3 *Construction*

The relation of earth work and foundation work is important for construction of artificially constructed ground. Construction flows including associated work are shown in Figure 2. Artificially constructed ground is directly related to building support and the performance effect. Therefore, the following subjects are examined in enforcement and are explained by the designer to the execution site manager; (1) construction purpose (2) construction organization (3) scope (4) execution scheme (5) important point of construction (6) important construction of construction (7) execution management and quality control (8) control of maintenance (9) proper response when soil pollution occurs near building (10) proper response when differential settlement occur.

3 VIBRATION MEASUREMENTS

3.1 *Oscillating measurement method*

The velocity responses were taken at various points within the building, as well as the surrounding areas at time with different levels of traffic (heavy, light and no traffic). The velocity responses were measured three directions.

3.2 *Measurement*

The machine used for measurement can measure simultaneously two or more measurement points and in three directions. So that it will be known as follows; (1) Oscillating of propagation characteristic in the ground and response of the building foundation. (2) The feature of attenuation by the difference in building structure (e.g. wooden framework, steel framework and reinforced concrete). (3) The shape of the foundation inference the response of horizontal vibration. (4) The building's own natural vibration period. Through this measurement its possible to know the influence of sight movement and vibration from traffic.

Table 3 shows the building information that we measured oscillating reduction effect.

Table 3. Building information.

No.	Ground information	Structure information	Lightweight thickness and volume
1	SASW	RC2	700/400 mm
	BOR	660.5 m^2	865 m^3
2	SASW	S2	300 mm
	BOR	571.5 m^2	610 m^3
3	SASW	W1	200 mm
	SWS	219.4 m^2	122 m^3
4	SASW	S2	700 mm
	BOR	102.3 m^2	136 m^3
5	SASW	RC1	600 mm
	BOR	226.0 m^2	222 m^3
6	SASW	S1	600 mm
	SWS	72.7 m^2	77 m^3
7	MASW	S3	100/400/600 mm
	BOR	1193.2 m^2	1800 m^3
8	MASW	S2	150 mm
	BOR	448.0 m^2	449 m^3

Table 4. Oscillating reduction effect of environmental vibration(10^{-3} cm/sec). These values are maximum value of each measurement.

No.		Road	Site border	Ground floor	Top floor
1	X:	13.69	17.60	35.58	18.36
	Y:	16.57	21.69	18.11	21.37
	Z:	33.73	23.74	19.01	19.20
2	X:	7.04	16.57	24.96	17.73
	Y:	15.93	21.76	16.64	18.24
	Z:	22.40	18.88	18.94	18.05
3	X:	6.01	16.76	24.19	16.00*1
	Y:	14.08	18.81	15.16	17.85*1
	Z:	22.27	16.96	16.45	16.96*1
4	X:	7.29	30.72	16.89	15.48
	Y:	14.27	17.92	20.03	19.07
	Z:	23.23	19.33	16.89	18.68
5	X:	16.32	30.14	36.67	42.49
	Y:	24.70	33.21	27.84	58.04
	Z:	38.72	29.31	47.10	39.29
6	X:	12.54	29.88	616.32	29.95*2
	Y:	26.94	37.88	31.16	31.74*2
	Z:	41.47	32.19	36.48	26.81*2
7	X:	4.60	9.92	15.04	9.98
	Y:	9.02	12.54	10.94	10.62
	Z:	13.44	11.07	12.16	10.43
8	X:	30.53	16.64	25.28	
	Y:	51.58	21.37	17.53	
	Z:	37.69	38.72	18.36	

*1 ground floor; *2 a numerical value of basement.

4 FINDING AND FUTURE INVESTIGATION

4.1 Reconfirmation of problem

The following questions will be raised.

(1) Quantitative effect comparison with our construction method and other construction methods has not been performed.
(2) Comparison of the performance effect of building structure, as well as rigidity of form has not been performed.
(3) Standards for evaluating design value and a quality assessment value are not clear.

From this research we have realize the important. From Table 4 there is an oscillating reduction effect to environmental vibration. Through interviews with people living with this technology, we qualitatively measured their perception of the strength of the earthquake. This is an effective method of earthquake resistance construction.

4.1 Future subject for inquiry research

We would like many people to understand the safety, durability and effectiveness of this artificially constructive ground.

A seismograph should be install in many areas and building with this technology, then long term observation is needed. From this data, it will be possible to make model of data about oscillating reduction effect of this technology.

In addition to developing artificially constructed ground we need to develop a technology to reduce and absolve energy within building structure itself.

4.2 The scope and limit of this method

The technical range must be according to Japanese Building Standard Law when constructing a building. Further physical properties, such as the durability of foaming resin and chemical resistance are examined, and it becomes the construction in the building of those conditions within the limits and the site condition.

5 CONCLUSIONS

As a result this construction method resists environmental vibration and earthquake, as well as being effective for differential settlement, foundation heat insulation, of foundation.

Furthermore, in order to offer more effective technology, we think that further research will be necessary.

ACKNOWLEDGMENT

Gratitude is expressed to Dr. H.Yamanouchi, Dr. M.Futaki, Dr. M.Tamura and Dr. M.Fujii who evaluated this research technology for applying to architectural skills of Japan. Moreover, gratitude is expressed to Mr. T. Kobayasi of PLG Co., Ltd and

Mr. K. Chiyoda of JSP Co., Ltd who gave us precious advice and cooperation in enforcement of this construction method. And the authors are also thankful to Mr. T. Kajiwara of Ouyo Jishinn Keisoku Co., Ltd who cooperate and support in measurement.

REFERENCES

Nakamura, T. 2001. "Vibration Characteristic for foundation structure which used lightweight expanded resin ground" Proceedings of JGS Symposium, 120, 111–118

Nakamura, T. 2002. "The Characteristics of Floating Function using Expanded Resin Material Which Have Different Function" Proceedings to The Japan Geotechnical Society, 50. 4. 531, 7–9

The Building Centre of Japan. 2002, Ground Substitution Method "Columbus Method" JCB Examination proof 17, May 2002.

Building vibrations by traffic and other sources

Environmental Vibrations – Takemiya (ed.)
© 2005 Taylor & Francis Group, London, ISBN 0 415 39035 4

Traffic induced vibrations of ground environments and buildings

H. Xia, P.B. Wei, Y.M. Cao
School of Civil Engineering & Architecture, Beijing Jiaotong University, Beijing, China

G. De Roeck
Department of Civil Engineering, Catholic University of Leuven, Heverlee, Belgium

ABSTRACT: In this paper, the general summaries are made of the raise of the problem on traffic induced vibrations of ground environments and the nearby buildings; the research background of this subject in theoretical analyses, numerical simulations and experimental measurements; the characteristics of traffic-induced vibrations and their propagations from the aspects of duration, intensity distribution and the influences of vehicle load and speed; their influences on the daily life of people, the normal operation of instruments and the safety of buildings; the properties of traffic induced vibrations in several railway systems and buildings.

1 INTRODUCTION

As one of the main environmental pollutions, the vibration of environments that seemed to have been tolerated in the past is today increasingly being considered as a nuisance. With the rapid development of modern industry and the continuous expansion of big cities, the influences of vibrations on the living and working environments of the people have brought to the close attentions of many metropolitan governments as well as the engineers and researchers in China and abroad.

According to the statistical investigations in several countries, the traffic-induced vibration occupies about 14% of the total public complaints against environmental vibrations (see Fig. 1), which is nearly the most intense second only to those from industries and construction sites.

In the recent years, the consideration on the environmental influences of the traffic-borne vibrations is getting more and more important in designing and planning traffic systems.

The first reason is that in the past the residence blocks and buildings in cities were relatively sparse, while at present, with the rapid growth of modern cities, the multi-level roads, metro lines and urban railways are forming a multi-dimensional traffic system, extending and intruding from underground, ground and space into the crowded residence areas, commercial centers and industrial zones. In Tokyo and some other big cities, there have appeared some viaduct roads of 5 to 7 levels, with the minimum distances to the nearby buildings only several meters, or even inside the buildings (see Figs. 2 to 4).

The second reason is that the traffic flows are getting more and more intense, traffic loads are becoming heavier and heavier, and vehicles are running faster and faster.

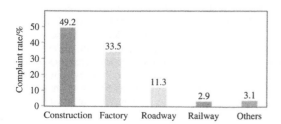

Figure 1.   Public complaints on environmental vibrations.

Figure 2.   Multi-level viaducts in Tokyo.

Figure 3. Viaduct road against a building (Guangzhou).

Figure 4. Monotrack passing through a building (Chongqing).

On the other hand, with the raise of people's living level, the requirements on the environments are becoming higher and higher. All of these are making the influence of the vibration problems more and more intense, and have brought new requirements on the study of the traffic-induced vibrations.

The generation mechanism of the vibrations, their propagation properties in the ground, the pollutions to the environments, the injuries to man's health, and their countermeasures against this problem, including the reduction of vibration sources, the attenuation and cutting of vibration transmission in ground soils and the isolation of buildings are all under study (Bata 1971, Crispino 2001, Degrande 2000, Hung 2001, Hunt 1996, Ju 2003, Metrikine 2000, Pan 1990, Takemiya 2003, Xia 2002, Xie 2001, Yang 2003, Yasutoshi 2003, Yoshioka 2000).

2 TRAFFIC INDUCED VIBRATIONS AND THEIR EFFECTS ON ENVIRONMENT

2.1 Generation of traffic induced vibrations

The vibration pollutions refer to the vibrations of ground, buildings, traffic vehicles and so on induced

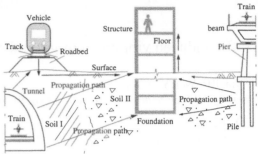

Figure 5. Generation and propagation of traffic induced vibrations.

by people's activities, which influences the routine life and the physical and psychological health of the residents living near the traffic roads, the working environments of the people, the safety of buildings and the normal operation of precise instruments.

It is an inevitable problem for a vehicle to induce vibration when it runs on a traffic structure, which may not only affect the riding comfort of passengers and the safety of the structure, but also cause the vibration of the environmental ground near the traffic line. The buildings on the ground or underground will be forced to vibrate (see Fig. 5).

2.2 Influence of traffic induced vibrations

According to the investigation of Japan, the highest complaint rate on traffic induced vibrations is their influence on people's sleeping, which is about 45% of the total complaints, secondly are the mental injuries to man and the damages to buildings, which are both about 20%, as is seen in Figure 6.

Generally, the environmental vibrations may not directly hurt a man's body, however they may disturb his normal life, cause him to feel uncomfortable or upsetting, or even influence his sleeping, working and studying. The experimental study shows that the higher the vibration level, the greater the influence on the sleep: A 60 dB's vibration can be just felt by a normal man, and will not affect his sleep, but it may affect a man sensitive to vibration or in unhealthy; 65 dB has slight effect on a sleeper; 69 dB may awake a light sleeper; 74 dB may awake all the men but the sound sleeper; 79 dB will wake up all sound sleepers. These values can be compared by the investigations on the environmental vibrations in several representative cities by the Institute of Labor Hygiene of the Railway Ministry of China. The results show that the vibrations measured at about 30 m from the railway lines in most of the places are as high as 80 dB.

Drivers are persons always working under vibration environments. An investigation on 370 tractor drivers

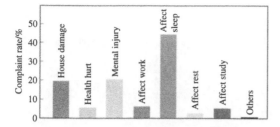

Figure 6. Public complaints on traffic-induced vibrations.

Figure 7. Vibration induced crack of an old church near road.

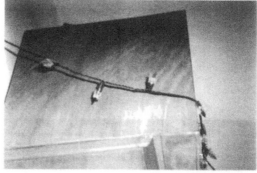

Figure 8. Vibration induced furniture shift near railway.

Figure 9. Metro lines through the city of Stockholm (Sweden).

reported the ratios of 71%, 52% and 8% of pathological changes of their bone joints, chest and lumber vertebras, and the ratio of pathological changes in both chest and lumber vertebras is as high as 40%.

The investigation also finds that the longer they have worked, the higher the ratios of pathological changes, and a very high pathological change ratio of 80% is reported for the drivers who have worked more than 10 years. Another comparison investigation by X-ray examination on 117 truck drivers and 95 band staffs reveals that the ratios of pathological change in their vertebra are 43% and 20%, respectively, and the ages of truck drivers when the pathological change occurred are much younger.

The vibrations induced by the traffic vehicles running underneath or close to the buildings may become considerable. They propagate underground or along the ground surface, and furthermore induce the secondary vibrations of the buildings and the furniture inside them. These vibrations seriously affect the structural safety of the ancient/old buildings and the working and daily life of the people inside the buildings near the traffic lines. In Czech, some ancient buildings of masonry structures beside the busy roadways and railways were cracked owing to the vibrations induced by the passing vehicles (see Fig. 7). With the enlargement of the cracks, some old churches and houses were damaged and collapsed in Plague, Hustopece and hrusov areas (Bata 1971).

In Beijing, near Xizhimen, the residents living in a 5-story building at 150 m to a railway line reflected that they could feel the strong vibrations in their rooms when the train passed. Moreover, after a period of time, the furniture inside the room shifted a bit because of the vibration (see Fig. 8).

In many big cities such as Paris, Brussels, Stockholm and Beijing, people can often feel the vibrations of the room floors when they are in their homes or stay in hotels, under which the metro trains or near which the urban trains are running (see Fig. 9). There have been several reports on the public complaints against this problem in Beijing and some other cities in China.

The environmental vibrations have big influences on the operations of precise instruments such as laser devices, electronic microscopes, electronic scales, the surgical operations, the manufacture of semiconductor integrated cores, photo-plate making, etc. The vibrations may result in inaccurate readings, decrease the precisions of the instruments, shorten their service lives or even interrupt their work.

Figure 10.   Vibration effect on electronic microscope images.

Figure 12.   High speed railway through a town (Belgium).

Figure 11.   The buildings affected by nearby metro trains.

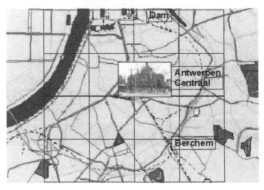

Figure 13.   High-speed railway through Antwerp Station.

Figure 10 shows the comparison between the image of a microscope of multiple 150000 under an ambient vibration and that without vibration, which shows that the environmental vibrations have great influence on the operation reliability of precise instruments, and can never be neglected.

It is a case in point that during the construction of the No.4 metro line in Beijing, the complaints have been raised about the possible effects of metro train induced vibrations on the precise instruments in the nearby buildings of the universities and the research institutes, see Figure 11. In this area, the bottom of the metro tunnel is 17 m below the ground surface, and the horizontal distances between the metro track and the affected buildings are from 39 m to 70 m.

In this case, the requirement of the precise instruments on the vibration environment is 0.5–100 μg for the frequency 0–10 Hz, and 2 μg for the frequency lower than 1 Hz. The countermeasures for solving this problem are now under the cooperative study of the Railway Construction Company and the Beijing Jiaotong University.

Japan is one of the countries where the environmental pollution of vibrations is serious. In its "Basic Law of Pollution Harms and Policies", it is stipulated that vibration is one of the seven main pollution harms, and measures should be taken to prevent the vibration

or to limit the vibration levels. In the "Law of Limiting Vibration", the traffic-induced vibration is specially mentioned to take measures to protect the environments and the people's health. Yasutoshi (2003) and Takemiya (2003) studied the generation mechanism, propagation laws both on and under ground of the traffic induced vibrations and their influences on the living of nearby residents, and proposed the predicting method of vibration levels of the surrounding environments.

In many European countries, the construction of high-speed railway network has brought great concern and study on the problem of traffic induced vibrations. In Belgium, some high speed railroads are constructed based on the reform of the existed railways, which often traverse the small towns, with some new built tracks closely against the foundations of existed buildings (see Fig. 12).

At the Antwerp station, the high-speed trains run under the ground (see Fig. 13), where the vibration effects similar to metro trains will occur. To study the transmission laws of the ground vibration induced by high-speed trains, Degrande (2000) carried out a dynamic experiment on the high-speed railway between Paris and Brussels, supported by the Belgium Railway Company. They also performed an impact-loading test

Figure 14.   Impact test of a truck (Belgium).

Figure 15.   Field test on train induced vibration (Beijing).

of a truck and the corresponding numerical simulations to study the properties of the vibration sources and the transmission relations between the vibration source and the ground (Fig. 14).

In Spain, Germany, Switzerland, United States and some other countries, efforts have also been bestowed on this field. The propagating and attenuating properties of the traffic-induced vibrations are systematically studied by the statistical analysis on the measured results of different ground soils. It has furthermore proposed the vibration control measures by decreasing vehicle speeds, reducing vehicle loads and improving the smoothness of road surfaces.

The field experiments on high-speed train induced environmental vibrations have also been carried out in Amsterdam-Utrent in the Netherlands ($40 \sim 160$ km/h), ICE in Germany ($100 \sim 300$ km/h) and X2000 in the western coast of Sweden, from which many useful results have been obtained.

In china, modern urban constructions started relatively late, but with the progress of industrialization, the development of large-scale traffic systems is extremely rapid.

Now in china, there are more and more cities where the metro systems are in operation or under construction. In the construction of traffic systems in urban areas, it has been predicted the influences of the traffic induced vibrations on the living and working environments of residents in the nearby buildings. These predictions include the vibration influences of the Metro Line No.4 from Xizhimen to the Summer Palace on the nearby universities and scientific institutes, the N-S axle metro line on the ancient buildings in Beijing, the planning Beijing-Shanghai high speed railway on the Huqiu Inclined Tower in Suzhou, and so on. Therefore, many Chinese researchers have studied the traffic induced vibrations through theoretical analyses, numerical calculations and field tests (see Fig. 15), with the combination of the construction of the metro and the urban rail transit lines in Beijing, Shanghai, Guangzhou and other big cities, and have published some useful

results. The environmental assessments including vibration effects have become one of the necessary programs in planning and designing of new railway systems.

In Taiwan, the possible influences of the high-speed railway between Taipei and Kaohsiung on the Science Park and technology in southern Taiwan have brought the attention of the researchers. Yang (2002) and Ju (2002) systematically studied the high-speed train induced vibrations and the countermeasures to reduce these influences.

3   CHARACTERISTICS OF TRAFFIC INDUCED VIBRATIONS

3.1   *Duration*

Long duration is the first characteristics of traffic induced vibrations. In a traffic system, there are always continuous flows of trains or automobiles, year after year, every day in a week, or even day and night.

For ordinary double track railways, the interval between trains can be as short as $5 \sim 10$ min, and the lasting period, according to the speed and the total length of the train, can be as long as $2 \sim 4$ min. Therefore, the influence time of train induced environmental vibrations can be $1/5 \sim 1/3$ of the total time.

For high-speed railways, the vibrations induced by trains are intermittent, while the time of each train's passage is short. The standard vibration level of the Shinkansen induced vibrations is of trapezoid forms, with the lasting duration being 7 sec under the train speed of 200 km/h. However, owing to its heavy traffic flows, such vibrations take place in every $5 \sim 10$ min, thus the influence is considerable.

For metro systems, according to the investigation in Beijing and other cities, there can be 30 or more couples of trains in two directions within one hour in a metro line. When a train passes, the induced vibration will last about 10 seconds. Therefore, the affecting duration of the vibration will occupy 15–20% of the total operation time of the metro system.

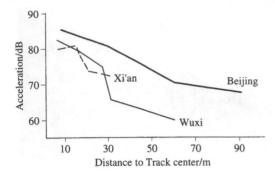

Figure 16. Vibration attenuation versus distance in several cities.

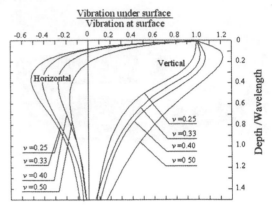

Figure 17. Vibration attenuation under ground surface.

For road transportation of living goods and production materials, especially the carriage of construction materials, borrow soils and spoils which are often carried by heavy lorries, the influences of environmental vibrations can be 24 hours a day since the urban roads are usually close to residences.

3.2 Intensity distribution

The Institute of Labor Hygiene of the Railway Ministry of China investigated the environmental vibrations in several cities. The results show that the traffic-born vibrations are of rather high levels. The vibrations measured at about 30 m away from the railway lines in most of the places are as high as 80 dB (see Fig. 16).

Owing to the energy dispersion in and absorption of the soil, the vibrations attenuate in the course of propagation. The characteristics of vibration attenuation depend on the types of the vibration sources and the properties of soil mediums. Generally the bigger the viscous coefficient of the soil is, the faster the vibration attenuates; while the higher the density of the soil, the slower the vibration attenuates.

The traffic-induced vibrations propagate in the ground in forms of shear wave, pressure wave and surface wave. The measured results show that the acceleration of the ground at 2 m under the surface is 20–50% of the surface acceleration, while at 4 m under is 10–30%. The surface wave occupies the main part of the vibrations induced by running vehicles.

Richart & Woods (1970) proposed the distribution curves of the vertical harmonic vibration attenuating in the soils versus the depth under the ground surface (see Fig. 17). It is shown that the vibration close to the ground surface is maximum; it attenuates rapidly with the increase of the depth, and almost disappears when the depth is bigger than vibration wavelength. Further research shows that at certain frequency, there occurred some resonance at certain depth in the soil.

As for horizontal direction, many investigations believe that the attenuation of train-induced ground

vibrations versus the distance to the railways is a general law. Some theoretical analyses show that for a point source, the vibration attenuates versus distance in the following law:

$$U_r = U_0 r^{-n} e^{-2\pi f h r / V} \tag{1}$$

where U_0 = the vibration amplitude near the source; U_r = amplitude at the place r (m) to the source; h = inner damping coefficient of the soil; f = frequency; V = propagation speed; n is a factor, $n = 0.5$ for surface wave and $n = 0.2$ for solid wave in infinite half space.

For line source of traffic-induced vibrations, the attenuation law can be expressed as:

$$A_{rz} = \frac{1}{2} K_0 r^{-k_z} e^{-\alpha_z r} V^{3/4} \tag{2}$$

where A_{rz} = the amplitude at the place r (m) to the source; V = running speed of train (km/h); K_0 = amplitude factor for different soils; K_z = comprehensive attenuation factor, $K_z = 0.75$ for vertical vibration and $K_z = 0.3$ for horizontal vibration; α = absorption factor of soil.

For road traffic induced vibrations, the vibration level also attenuates with the distance from the road. Generally, a distance of a lane width may show the attenuation of 3 dB, thus at the place of 50 m away from the road, the vibration may be 50 dB lower than that of the road center.

Many investigations support the above laws, while there are also some reports showing different patterns. In the distributions of vibrations versus distance from a railway site, there exists a zone where the vibrations are amplified (see Fig. 18). This law is in accordance with that from the theoretical analysis (Xia 2002).

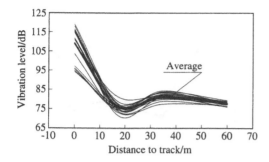

Figure 18. Ground vibration versus distance to track.

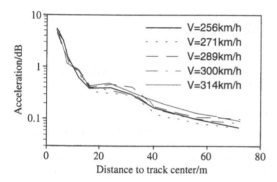

Figure 19. HSR induced vibrations (Antoing, Belgium).

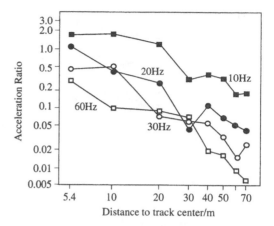

Figure 20. Distribution of vibration versus distance to railway.

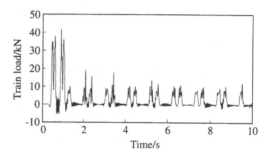

Figure 21. Train loads on the ground.

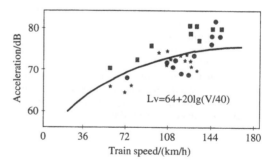

Figure 22. Ground vibration versus train speed (Spain).

In Belgium, Degrande (2000) carried out an *in situ* measurement of environmental vibrations induced by high-speed trains. The similar law was also found under the train speed of 256~314 km/h, in which the amplifying zone lies in the area of 20~40 m to the HSR line (see Fig. 19).

The measurement at a Shinkansen bridge in Japan shows that the ground vibrations attenuate in different laws with their frequency component: for those of $f > 30$ Hz, they monotonously decreases versus the distance to the track, while for $f = 10$~20 Hz, there occurred several zones where the ground accelerations were amplified (see Fig. 20).

In fact, careful observation on Figure 18 and Figure 19 may reveal the same law. This phenomenon thus seems to be universal, which can be explained by the multi-reflection of the S wave in the soft soil between the bedrock and the ground surface. The appearance of the amplifying zone is related to the depth of the bedrock, the properties of the soil and the velocity of waves, while Equations (1) and (2) do not show such a law.

3.3 Vehicle speed

As is shown in Figure 21, the forces of the train vehicles on the ground through the track are mainly the moving gravity loading of the train wheels. Therefore the ground vibration induced by the train should be influenced by the train speed.

In Spain, Volberg (1984) summarized the relation of the vibration level to the train speed (see Fig. 22):

$$L_V = 64 + 20 \lg(V / 40) \tag{3}$$

where L_V = the ground velocity level with the reference of 5×10^{-8} m/s, V = the train speed (km/h).

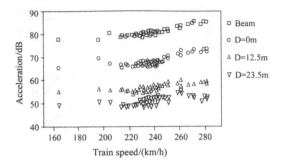

Figure 23. Distribution of vibration versus train speed (Japan).

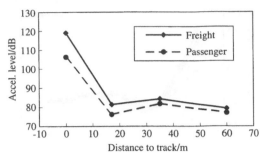

Figure 24. Ground vibrations of different train loads (China).

Yoshioka (2000) measured the ground vibration near a Shinkansen bridge. The results show that in the speed range of 160~280 km/h, the accelerations show a difference of 5~8 dB (see Fig. 23), and are expressed as:

$$L_V = 10n \lg(V/V_0) + L_{V0} \qquad (4)$$

where V_0 = the reference train speed (km/h); L_{V0} = average vibration level at V_0; n = coefficient having a choice range of 1.5~3.5.

For metro system, according to the investigations in Beijing and Tianjin, the vertical acceleration of the tunnel lining can be estimated by:

$$L_V = 67 + 20 \lg(V/65) + \Delta L \qquad (5)$$

where V = train speed (km/h); ΔL = correction value for wheel-rail conditions: $\Delta L = 0$ dB for CWR tracks with newly ground smooth wheels and rails; 5–11 dB for CWR tracks with worn wheels and rails; 10–22 dB for short rail tracks and worn wheels and rails.

The influence of roadway vehicle is greater than railway trains. Some investigations show that for the vehicle speed change of every 10 km/h, the ground acceleration can be influenced by 2~3 dB.

3.4 Vehicle load

It is believed that the loads of trains have big influences on the ground vibration. The heavier the loads are, the greater the impact of the train on the ground is. The measured results by Xia et al. (2005) shows that the freight trains with heavier axle loads produce higher ground vibrations than passenger trains by 2~10 dB, and the nearer the site to the track, the bigger the difference is (see Fig. 24).

During the recent experiment of the heavy-load trains on the Test Loop Railway of the China Academy of Railway Science in Beijing, the running of the train with 25 t heavy axle-loads excited much stronger ground vibrations than the ordinary trains, which incurred very serious complaints from the nearby residents.

4 VIBRATION CHARACTERISTICS OF SOME RAILWAY SYSTEMS

4.1 Viaduct railway system

For the viaduct railway system, the factors affecting the ground vibrations include train speeds, vehicle weights, types of bridge structures and foundations, bridge span lengths, stiffness and deflections. In addition, the dynamic interaction between the vehicles and the bridge will also enlarge the vibration.

From the experimental and theoretical analysis, the dynamic loads of the vehicles to the ground tracks and to the bridge piers are quite different (ref. Figs. 22 and 25), therefore the vibration responses are also different.

In addition to the general laws above mentioned, the vibration of viaduct railway systems under train loads have the following characteristics:

Figure 26 is the experiment site at the Daqinghe Bridge. The measured results show that the ground vibration induced by viaduct railway trains will in general tend to attenuate with the distance to the line. Compared to the common ground railways, the viaduct system can reduce the environmental vibration level, showing 5–10 dB lower at a place of 30 m to the line (Xia et al. 2002).

The types of the bridge foundations have great influences on the ground vibration: For pile foundations the vibrations are much smaller and attenuate more rapidly than those for common foundations (Xia et al. 2002).

Figure 27 shows the experimental results on the vibrations of the track-beam-pier-ground system on the Shinkansen high-speed railway (Yoshioka 2000), where D is the distance of the ground to the pier. From the figure one can clearly see the distribution laws of the high-speed train induced vibrations at different part of the system. The acceleration level is as high as 90 dB at the track, and attenuated by about 20 dB at the pier top. The property of the ground vibration attenuation

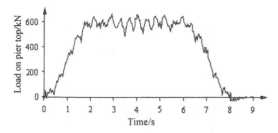

Figure 25.   Train loads acted on bridge pier.

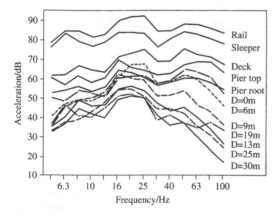

Figure 26.   Experiment site at Daqinghe Bridge (China).

Figure 27.   Measured accelerations of Shinkansen bridge.

versus the distance to the pier is obvious. For different points, the accelerations show peak values at the frequencies of 12~30 Hz.

The bridge structures of the viaduct light-rail system should be carefully designed to avoid the resonance of the vehicles and the structures so as to reduce the vibration.

4.2   Underground trains

For metro systems, the factors affecting the vibrations include train speeds, vehicle weights, types of tunnel foundations and linings, types of track, track isolation

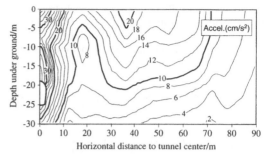

Figure 28.   Acceleration distribution in surrounding soils.

or damping measures, and so on. In addition, the dynamic interaction between the vehicles and tracks will also enlarge the vibration.

It is investigated that when a metro train is running in the tunnel, the average ground vibration levels are 81 dB at the place of 1.5 m to the track center, while 71.6 dB at 24 m away, which shows that the vibration attenuates with the distance to the track center. The train speed and the tunnel embedded depth also influence the ground vibration. Normally the higher the train speed, the higher the vibration level (when the train speed is doubled, the vibration level increases by 4–6 dB) and the larger the influencing area; while the deeper the embedded depth of the tunnel, the smaller the influencing area.

Based on the measured track accelerations, Pan (1990) proposed a mathematical expression to simulate the moving train load, and analyzed the vibration properties of a tunnel and the surrounding soils.

Xia et al. (2001) established a train-track-tunnel-ground system model to study the metro train induced ground vibrations and their propagation laws. Figure 28 illustrates the contour distribution of the maximum vertical ground accelerations. On the ground surface, there appear two local acceleration amplifying bands: one is right over the tunnel, the other is about 35 m away from the tunnel central.

In their study, Xia et al. (2001) also gives the acceleration distributions in different frequency components: for the vibration with low frequencies of 1~3 Hz, there occur three amplifying zones at 0 m, 36 m and 60 m away from the tunnel; for the vibration with middle frequencies of 5~6 Hz, there are only two amplifying zones at 0 m and 30 m, and the vibration attenuates rapidly at longer distances; for the frequencies higher than 8 Hz, the vibrations show monotone decrease with respect to the distance. The same laws were obtained in the *in situ* measurement by the Beijing Metro Company.

The structural type and the lining thickness have big influences on the metro train induced ground vibrations. There exist the relative vibration levels of

2–8 dB between different types of tunnel structures. The ground accelerations may reduce by 5~18 dB when the thickness of the lining is doubled.

The ground vibration induced by metro trains can be predicted by the following expression:

$$VL_1 = K_1 - 12\log L_1 + 25\log\frac{V}{50} - 24\log\frac{W_1}{30} + X \qquad (6)$$

$$VL_2 = K_2 - 16\log L_2 + 25\log\frac{V}{50} - 24\log\frac{W_2}{60} + X \qquad (7)$$

where VL_1 = vertical vibration level of the ground (dB); L = minimum distance from the lining side to the ground surface of the prediction point (m); V = train speed (km/h); W = unit weight of the tunnel structure (t/m); X = amendment to consider vehicle weight and other factors; K = basic vibration level for different track structures. The subscripts 1 and 2 in the equations stand for the tunnels with round and box cross sections, respectively.

4.3 Buildings close to traffic lines

The passage of traffic vehicles may induce the vibrations of buildings close to the traffic lines. A recent *in situ* measurement near a metro rolling stock depot in Beijing shows that the vibration level of the residence building over the metro lines was as high as 85 dB when the metro trains run through beneath the building at the speed of 15–20 km/h (see Fig. 29). It is estimated to have higher levels when the trains run at their normal operation speed of 70 km/h.

There have been several reports on the theoretical analyses and experimental studies on this subject, and many useful results obtained.

The types of the structure and the foundation of a building may affect the vibrations of the building. For a high-rise RC building with good foundation and huge mass and thus having low frequencies, the vibration level can be 10–20 dB lower than that of the base soil.

The *in situ* measurements on a high-rise building at 32 m horizontally to the center of the metro tunnel show that the vibration level at the basement of the building is not larger than 60 dB, and the vibrations at the first or higher floors can hardly be measured. For masonry buildings with ordinary foundations, the vibration can be attenuated by 5–10 dB. For buildings with poor foundations, or light material structures or shallow foundation structures, the vibration can hardly be attenuated. For such building, the measured floor vibrations are often almost the same as those of their base soil, even more, the case that the floor vibrations are greater than the soil ones can also be found.

Since the lateral vibration attenuates faster than the vertical one when propagating in the ground, the

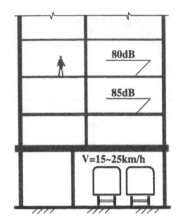

Figure 29. Vibration of building induced by subway trains.

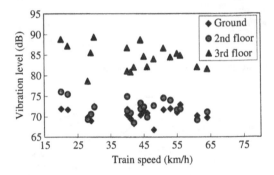

Figure 30. Measured vibrations at different buildings.

vertical vibration of the buildings near the railway is higher than the lateral component. Measurements show a difference of 10 dB between them.

For low and mediate height multi-story buildings, especially for those of four stories or lower, there is a trend that the vibration level will increase with the floor elevation. The measured and the analyzed results show that the vibration levels of the 3–5th floors are 3–5 dB higher than the 1st floor (see Fig. 30). According to the investigation on the buildings along the railway in Japan, the vibrations of floors are often bigger than the ground, with 50% of the differences being 5 dB and the maximum higher than 15 dB. As seen in Figure 31, the responses of wood buildings are much greater than the RC structures. The vibration of the 2nd floor of wood buildings may amplify the ground vibration by nearly 5 times (about 15 dB), while that of RC buildings are often smaller than the ground, with the maximum increase less than 2 times of the ground (about 5 dB).

For high-rise buildings, the distribution of the vibration versus the floors is different. Some measurements show that the vibrations at the 3rd to 7th

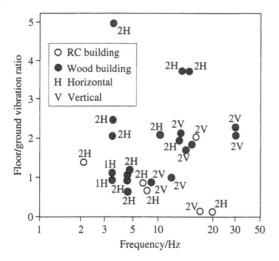

Figure 31. Measured vibrations at different building floors.

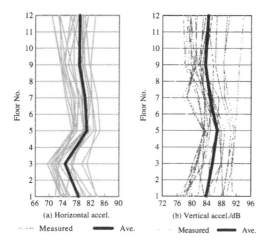

(a) Horizontal accel.　(b) Vertical accel./dB

⋯⋯ Measured　━━ Ave.　⋯ Measured　━ Ave.

Figure 32. Variety of acceleration levels of a 12-story building.

floors are relatively bigger than on the ground and the higher floors (see Fig. 32).

ACKNOWLEDGMENTS

This study is sponsored by the Flander-China Bilateral Project of Belgium (BIL 04/17) and the Key Research Foundation of Beijing Jiaotong University (No. 2004SZ005).

REFERENCES

Bata, M. 1971. Effect on buildings of vibrations caused by traffic. *Building Science* 6: 221–246.

Crispino, M. & D'apuzzo, M. 2001. Measurement and prediction of traffic-induced vibrations in a heritage building. *Sound & Vibration* 246(2): 319–335.

Degrande, G. & Lombaert, G. 2000. High-speed train induced free field vibrations: in situ measurements and numerical modeling. *Wave2000*, Okayama: 29–42.

Fujikake, T. 1986. A prediction method for the propagation of ground vibration from railway trains. *Sound & Vibration* 111(2): 357–360.

Hung, H.H., Kuo, J. & Yang, Y.B. 2001. Reduction of train-induced vibrations on adjacent buildings. *Structural Engineering and Mechanics* 11(5): 503–518.

Hunt, H.E. 1996. Modeling of rail vehicles and track for calculation of ground-vibration transmission into buildings. *Sound & Vibration* 193(1): 185–194.

Ju, S.H. 2003. Finite element analysis of building vibrations induced by high-speed train. *ISEV'2003*, Hangzhou: 3–23.

Metrikine, A.V. 2000. Ground vibration induced by a high-speed train in a tunnel: two-dimensional model. *Wave2000*, Okayama: 111–120.

Pan, C.S. 1990. Measurement and analysis of train induced vibrations in metro tunnel. Civil Engineering 23(2): 21–28.

Richart, F.E. 1970. *Vibration of Soils and Foundations*. New Jersey: Prentice-Hall Inc.

Takemiya, H. 2003. Vibration propagation and mitigation for a pile-supported foundation by honeycomb-shaped WIB. *ISEV'2003*, Hangzhou: 267–290.

Takemiya, H. & Kojima, M. 2003. 2.5-D FEM simulation for vibration prediction and mitigation of track and ground interaction under high-speed trains. *ISEV'2003*, Hangzhou: 130–138.

Volberg, G. 1984. Propagation of ground vibrations near railway tracks. *Sound & Vibration* 87(2): 371–376.

Xia, H. & Cao, Y.M. 2002. Vibration effects of light-rail train-viaduct system on surrounding environment. *Structural Stability & Dynamics* 2(2): 227–240.

Xia, H. & Zhang, N. 2005. Experimental study of train-induced vibrations of environments and buildings. *Sound & Vibration* 2005 (280): 1017–1029.

Xia, H. & Zhang, M. 2001. Characteristics of traffic induced vibrations and their effects on environments. *TIVC'2001*, Beijing: 83–90.

Xie, W.P. 2001. Train track-ground dynamics due to high speed moving source and ground vibration transmission. *Structural Mechanics & Earthquake Engineering (JSCE)* 682(56): 165–174.

Yang, Y.B. 2003. Ground vibration induced by high-speed trains over viaducts. *ISEV'2003*, Hangzhou: 147–157.

Yasutoshi, K. 2003. Simulation and observation of ground vibration caused by road traffic. *ISEV'2003*, Hangzhou: 202–216.

Yoshioka, O. 2000. Basic characteristics of Shinkansen-induced ground vibration and its reduction measures. *Wave2000*, Okayama: 219–239.

Environmental Vibrations – Takemiya (ed.)
© 2005 Taylor & Francis Group, London, ISBN 0 415 39035 4

Experimental study of moving train induced vibrations of a high-rise building

Y.M. Cao, H. Xia & J.W. Zhan

School of Civil Engineering & Architecture, Beijing Jiaotong University, Beijing, China

ABSTRACT: By means of a field experiment, the vibration characteristics of a high-rise building induced by moving trains are studied in this paper. The experimental results show that the structural vibrations of the building are of low frequency properties. The horizontal vibrations of the building show the same importance with the vertical ones, and cannot be neglected. The vertical vibrations of floors are higher than the horizontal ones, and the variety characteristics with the floor elevation of the horizontal and the vertical vibrations are different, showing very complex zigzag trends. Higher train speed induces greater building vibrations, and heavier freight trains induce stronger vibrations than lighter passenger trains. The nearer the railway track is to the building, the larger the structural vibrations will be.

1 INTRODUCTION

With the rapid development of urban railway traffic system, the influences of traffic-induced structural vibrations are becoming stronger and stronger. At the same time, the progress of the society and the raise of people's living levels requires better and better living and working environments. It is a case in point that during the constructing of the No.4 metro line in Beijing, the complaints have been raised about the possible influence of metro train induced vibrations on the precise instruments in the nearby buildings of the universities and the research institutes. The problem of traffic-induced vibrations of environments have roused the attention of the city authorities as well as the engineers and researchers. Some researches on this subject have been done in China and abroad (Bata 1971, Xia 2001, 2002, 2005, Ju 2003, Lincy 2002, Hunaidi 2000, Crispino 2001, Hunt 1996, Hung 2001).

For this problem, Ju (2003) and Lincy (2002) adopts numerical analysis, while Xia (2002, 2005) and Crispino (2001) emphasizes particularly on field experiments. However, these field experiments have been mainly concentrated on the vibrations of one-story houses or multi-story buildings, and few have been aimed at high-rise buildings.

In this paper, the vibration characteristics of a high-rise building induced by moving trains are studied through a field experiment. Some useful results have been obtained from the experiments and the measured data, which proved many of the previous theoretical conclusions about the propagating laws of the structural vibrations in buildings (Xia 2001, Cao 2004, Ju 2003), and can be referenced in the planning and designing of railway traffic systems in urban areas.

2 OUTLINE OF FIELD EXPERIMENT

2.1 *Experimental site*

The experiment was carried out at a high-rise building, which is very close to the Railway Station of Shenyang, the capital city of Liaoning Province in Northeast China (see Fig. 1).

The building is a 12-story RC frame structure. The total height of the building above the ground is 36.6 m, with the first floor height 3.6 m and the others 3.0 m, and a basement of 4.0 m. The total length of the building is 45.0 m with 10 rooms each having the width of 4.5 m, as is shown in Figure 2.

There are three railway tracks close to the building, with the shortest distance from the building to the railway track being only 12 m (see Fig. 3). The hotel staffs working and the guests accommodating there can always feel the vibration when a train is running by.

2.2 *Measuring method*

The evaluation code and the measuring method for environmental vibrations are respectively based on the national standard "Code for Environmental Vibration in Urban Areas [GB10070–88]" and "The Measurement Method for Environmental Vibration in Urban Areas [GB10071–88]" issued by the National Environmental

Protection Bureau of China (China Academe of Railway Sciences 1995).

Table 1 lists the evaluation code in which the allowances for environmental vibrations in different urban areas are regulated.

In the table, "residence districts A" means the residence districts where special silence is required, "residence districts B" means the districts of common residence, cultural sections and schools, and "mixed area C" is the mixed area of residence, industry, commerce and minor traffics.

The code is valid to continuous steady vibrations, shock vibrations and random vibrations. For shock vibrations that take place several times a day, the maximum accelerations shall be the tabled allowance plus 10 dB during the day and 3 dB at night.

2.3 *Layout of acceleration transducers*

To measure simultaneously the horizontal and the vertical vibrations of this high-rise building, both the

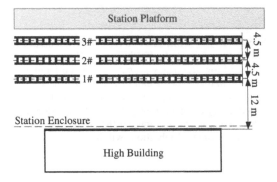

Figure 3. Schematic plan of experimental site.

Table 1. Allowance for environmental vibrations.

Area category	Daytime	Nighttime
Residence districts A	65	65
Residence districts B	70	67
Mixed area C	75	72
Commercial center	75	72
Concentrated industrial areas	75	72
Roadsides	75	72
Railway sides	80	80

Figure 1. The high-rise building near railway tracks.

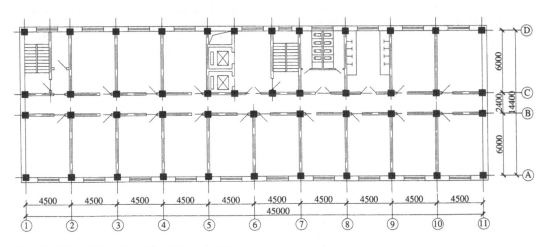

Figure 2. Typical floor plan of the high-rise building.

542

horizontal and the vertical accelerometers were laid on the floor of every other story, as shown in Figure 4. In order not to influence the normal business of the hotel, all measurement points were located at the floors of structural staircases.

2.4 Train loads of field experiment

The loads of this experiment were mostly the freight trains in normal operation (see Fig. 5), except for several passenger trains in the further track line at about 100 m away. The train speeds during the experiment were from 20 km/h~60 km/h. Totally there were 20 groups of data measured, which are listed in Table 2.

In the table, D is the distance between the building and the railway tracks, Loc means locomotives, F means cars with full load, and E without load.

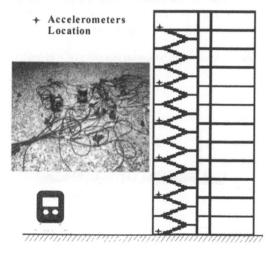

+ Accelerometers Location

Figure 4. Layout of accelerometers on building floors.

Figure 5. Freight train normally operating in field experiment.

It is stipulated that the maximum vibration of each measurement point should be recorded for each train's passage, and at least 20 trains be measured continuously. The arithmetical averages of the 20 maximum values are taken for the final vibration evaluations.

3 ANALYSIS OF EXPERIMENTAL RESULTS

3.1 Time-histories of floor vibrations

The 20th experimental data are taken as an example to illustrate the typical measured acceleration histories of the first and the top floors of the building, as plotted in Figure 6 and Figure 7, respectively.

Seen from the acceleration histories of floor vibrations of this group of data, the vertical vibration is higher than the horizontal vibration, and the attenuation of horizontal vibration is faster than that of the vertical during their transmission along the building structures. In addition, for the high-rise building, the vibration of top floor is no more always higher than that of first floor, which is very different from the vibration characteristics of multi-story building (Xia 2005). The detail variety trend with floors can be seen in Section 3.5.

3.2 Spectrum analysis of floor vibrations

Figure 8 and Figure 9 show the acceleration spectra of the 20th experimental data.

Table 2. Train loads of field experiment.

No.	D (m)	Speed (km/h)	Train load descriptions
1	16.5	50	Freight: 1/3 full, 2/3 empty
2	21	40	Passenger: 1 Loc + 17 car
3	12	20	Freight: 1 Loc + 39 car (full)
4	12	20	Freight: 1 Loc + 35 car (1/2 F, 1/2 E)
5	100	40	Freight: oil truck (1/2 E, 1/2 F)
6	16.5	36	Freight: 1 Loc + 62 car (1/4 E, 3/4 F)
7	16.5	60	Freight: 1 Loc + 52 car (1/2 F, 1/2 E)
8	100	20	Freight: 1 Loc + 63 car (full)
9	100	10	Freight: 1 Loc + 15 car (full)
10	16.5	40	Freight: 1 Loc + 54 car (empty)
11	100	10	Freight: 1 Loc + 19 car (full)
12	100	10	Freight: 1 Loc + 50 car (full)
13	100	40	Freight: 2 Loc + 62 car (full)
14	16.5	36	Locomotive: 1 Loc
15	100	10	Freight: 2 Loc + 60 car (full)
16	100	60	Freight: 2 Loc + 59 car (full)
17	16.5	70	Passenger: 1 Loc + 17 car
18	12	50	Freight: 1 Loc + 19 car (1/2 F, 1/2 E)
19	100	20	Passenger: 1 Loc + 20 car
20	12	60	Freight: 1 Loc + 36 car (full)

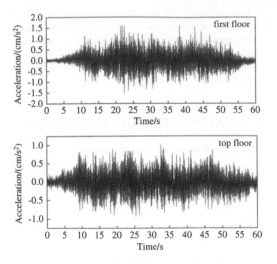

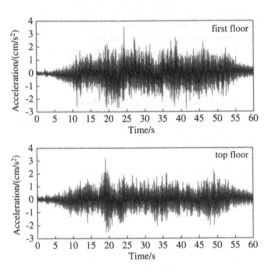

Figure 6. Measured horizontal acceleration histories of first floor and top floor ($V = 60$ km/h).

Figure 7. Measured vertical acceleration histories of first floor and top floor ($V = 60$ km/h).

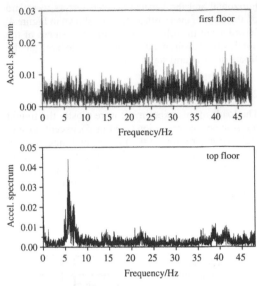

Figure 8. Spectrum of horizontal vibrations ($V = 60$ km/h).

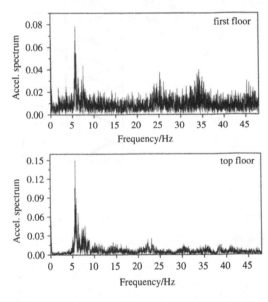

Figure 9. Spectrum of vertical vibrations ($V = 60$ km/h).

It is obvious from the figures that the frequency of the floor vibrations mainly concentrates on the range of 0~50 Hz, and the train-induced building vibrations are of low frequency properties.

Seen from the peak values of vibration spectra of the first floor, 25.27 Hz, 34.14 Hz and 43.46 Hz are the three dominant frequencies of the horizontal vibrations; while for the vertical vibrations, the dominant frequencies are 5.74 Hz, 25.27 Hz and 34.28 Hz, respectively, obviously lower than those of the horizontal ones.

While for the acceleration spectra of the top floor, the dominant frequencies of both the horizontal vibrations

and the vertical vibrations become 5.74 Hz, with few contributions of high frequencies. It means that the building structure itself played a role of filtering to the high frequencies as the vibration waves transmit from the first floor to the top floor, which can appropriately explain the laws responded from Figure 10 that with the floor height increasing, the variety extent of the high-rise building vibrations decreases instead.

3.3 Influence factors analysis of building vibrations

Through comparing and analyzing the 20 groups of experimental data, some influence factors of the building vibrations can be found.

Firstly, the distance between the building and the railway track has important effects on the vibrations of the building. With the increase of the distance, both the horizontal and the vertical vibrations of the building attenuate rapidly. So in the planning and designing of buildings, the minimum distance between the buildings and the railway tracks should be strictly controlled based on the performances of the buildings.

Secondly, within the range of the train speeds in this experiment, higher train speed induces greater floor vibrations of the building. As for whether there exists an critical train speed to the high-rise building vibrations, further investigation under wider train speed range should be done.

In addition, by comparing the 3rd group and the 4th group of the experimental data, it can be seen that the heavier (freight) trains induce greater vibrations than lighter (passenger) trains when other conditions are same.

3.4 Statistical analysis of floor vibrations

Because the moving trains on the normally operated railway lines were with different conformations, loads, and speeds, the statistical parameters of the vibration data were analyzed in study. The 20 groups of measured data were regarded as 20 samples, and the maximums, averages and standard deviations for the vibrations of the building were obtained, as are listed in Table 3.

It can be seen from Table 3 that the vertical vibrations of building floors are obviously higher than the horizontal ones. However, the horizontal vibrations cannot be neglected because they usually determine whether the structure is damaged, especially for the historical heritage buildings.

On the other hand, the standard deviations of the vertical vibrations are much greater than twice of the horizontal values, showing that the vertical vibrations of the floors are much sensitive than the horizontal ones to the loading conditions of trains. These characteristics can also be seen from Figure 10.

3.5 Variety characteristics of floor vibrations with floor elevation

In order to study the vibration characteristics of this building more conveniently, the acceleration levels are adopted in this section. They are calculated by the following formula:

$$VAL = 20\lg(a/a_0) \tag{1}$$

Table 3. Statistical analysis of measured floor accelerations.

Floor No.	Horizontal			Vertical		
	Max.	Ave.	Stand.	Max.	Ave.	Stand.
	cm/s²			cm/s²		
1	1.72	0.84	0.41	3.54	1.56	0.85
3	1.10	0.54	0.24	3.54	1.88	0.84
5	2.45	1.09	0.41	3.87	2.17	0.86
7	2.90	1.03	0.50	4.10	1.74	0.84
9	2.81	0.86	0.51	4.35	1.51	0.85
12	2.99	0.86	0.56	4.96	1.65	1.01

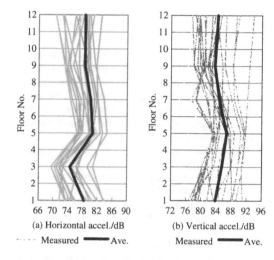

Figure 10. Variety of acceleration levels with floor elevation.

where VAL = acceleration level in dB; a_0 = reference acceleration with the value of 1×10^{-6} m/s²; and a = measured acceleration in m/s².

Figure 10 is the variety characteristics with floor elevation of measured maximum acceleration levels of this building, in which (a) refers to the horizontal vibrations while (b) refers to the vertical vibrations. The fine lines present the measured maximums of all tests, and the heavy lines correspond to the averages of all measured maximums.

It should be firstly noticed that for this high-rise building, the variety trend of the horizontal vibrations is different from that of the vertical ones, especially at lower floors.

Moreover, the variety characteristics of the floor vibrations with their elevations in both directions are very complex and show zigzag tendencies, which is different from those of the multi-story buildings (Xia 2005).

For the horizontal vibrations shown in Figure 10(a), there exist three alterations at the places of the third floor, the fifth floor and the ninth floor, presenting the trend of "decreasing-increasing-lightly decreasing-lightly increasing".

By comparing Figure 10(a) and Figure 10(b), it can also be found that the variety extent of the vertical vibrations is smaller than that of the horizontal ones. From the first floor to the fifth floor, the vertical vibrations are monotonously increasing; and from the fifth floors, the variety characteristics of the floor vibrations with their elevations in vertical direction are basically consistent with that in horizontal direction.

4 CONCLUSIONS

The following points have been concluded from the above experiment:

(1) For moving train-induced vibrations of the high-rise building, the vertical vibrations of building floors are obviously higher than the horizontal ones. But the horizontal vibrations have the same importance with the vertical vibrations, and cannot be neglected, especially for those historical heritage buildings.
(2) The vibrations of building floors are of low frequency properties.
(3) The variety characteristics with the floor elevation of the horizontal and the vertical vibrations are different, and show very complex zigzag trends, which are different from the multi-story buildings.
(4) When other conditions are the same, the heavier (freight) trains induce greater vibrations than the lighter (passenger) trains.
(5) Within the range of low speed, higher train speed induces stronger structural vibrations.
(6) With the increase of the distance between the building and the railway track, both the horizontal and the vertical vibrations of buildings attenuate rapidly. So in the planning and designing of buildings, the minimum distance between the buildings and the railway tracks should be strictly controlled based on the performances of the buildings.

ACKNOWLEDGMENTS

This study is sponsored by the Natural Scientific Foundation of Beijing (No. 8042017) and the Key Research Foundation of Beijing Jiaotong University (No. 2004SZ005).

REFERENCES

Bata, M. 1971. Effect on buildings of vibrations caused by traffic. *Building Science* 6: 221–246.
Xia, H. 2001. Characteristics of traffic induced vibrations and their effects on environments. *Proc. TIVC'2001*, Beijing: 83–90.
Xia, H. 2002. *Dynamic Interaction of Vehicles and Structures.* Beijing: Science Press.
Xia, H. & Cao, Y.M. 2002. Vibration effects of light-rail train-viaduct system on surrounding environment. *Structural Stability & Dynamics* 2(2): 227–240.
Xia, H. & Zhang, N. 2005. Experimental study of train-induced vibrations of environments and buildings. *Sound & Vibration* 2005 (280): 1017–1029.
Ju, S.H. 2003. Finite element analysis of building vibrations induced by high-speed train. *Proc. ISEV'2003*, Hangzhou: China: 3–23.
Lincy, P. & Degrande, G. 2002. Numerical modeling of traffic induced vibrations in buildings based on a dynamic soil-structure interaction formulation. *Proc. 15th ASCE Engineering Mechanics*, New York: 2–5.
Hunaidi, O. & Guan, W. 2000. Building vibrations and dynamic pavement loads induced by transit buses. *Soil Dynamics and Earthquake Engineering* 19(6): 435–453.
Crispino, M. & D'apuzzo, M. 2001. Measurement and prediction of traffic-induced vibrations in a heritage building. *Sound & Vibration* 246(2): 319–335.
Hunt, H. E. 1996. Modeling of rail vehicles and track for calculation of ground-vibration transmission into buildings. *Sound & Vibration* 193(1): 185–194.
Hung, H.H., Kuo, J. & Yang, Y.B. 2001. Reduction of train-induced vibrations on adjacent buildings. *Structural Engineering and Mechanics* 11(5): 503–518.
Cao, Y.M. & Xia, H. 2005. Experimental study and numerical analysis on moving train induced vibrations of high-rise buildings. *Engineering Mechanics*, in press.
China Academe of Railway Sciences. 1995. Analysis, measurement and precautions of environmental noises and vibrations induced by running trains on Guang-Shen Railway. Beijing: 38–71.

Environmental Vibrations – Takemiya (ed.)
© *2005 Taylor & Francis Group, London, ISBN 0 415 39035 4*

Centile spectra, measurement times, and statistics of ground vibration

H. Amick, M. Gendreau & N. Wongprasert
Colin Gordon & Associates, San Bruno, California, USA

ABSTRACT: Facilities for advanced technology – particularly nanotechnology – impose very stringent requirements on the quality of the site. Ambient ground vibrations are among the most critical factors to consider when selecting a site. Commonly-accepted vibration criteria are available and they, in turn, dictate many of the measurement parameters such as bandwidth, frequency range, etc. However, there are still many options available to the analyst for representing time-varying statistics. This paper reviews some of the statistically-based protocols for data representation for these facilities.

1 INTRODUCTION

A growing number of structures are being designed and constructed specifically to house and support advanced technology. Advanced technology (or "high tech") facilities can include those for semiconductor production, biotechnology R&D, physics research, and metrology (the technology associated with precise measurements). The newest entry into this group is nanotechnology.

Nanotechnology offers potential benefits that may be as revolutionary as the space program of the 1960s, the advent of computers, or even the Industrial Revolution. Almost everyone who has heard the word knows it deals with small things, but not everyone appreciates the extent to which it impacts advanced technology fields and, by default, the world of the design professional.

The sophisticated working environments required for nanotechnology facilities pose big challenges to their designers and constructors. The workplace environmental requirements may include temperature and humidity control, air cleanliness (i.e., particulate and chemical contamination), biohazard containment, limits on electromagnetic fields, special electrical power conditioning, and vibration and noise control. Most of these design aspects have evolved from the special needs of working at exceedingly small scales. Very few existing buildings can meet these demands – new construction is generally required.

Nanotechnology has been defined as research and technology development dealing with particles and systems with dimensions of approximately 1 – 100 nanometers. In many cases, the environment typical for semiconductor production is appropriate. In other cases, it is not stringent enough.

It has become common practice to use a limited set of published "generic" vibration criteria which may be selected for a facility or space within a facility based upon the most demanding equipment likely to be used in a given process. An assessment of a proposed site is then based upon the vibration criterion associated with the most demanding application.

The most popular of the generic criteria are defined numerically in Table 1 and shown graphically (as velocity spectra) in Figure 1 [IEST (2005), Amick, *et al.* (2002)]. Many of these criteria have been in use for 20 years in several advanced technology communities, providing an "experience base", and have been applied to the design several nanotechnology facilities.

The most highly-sensitive spaces – in which submicron and molecular-scale processes are carried out – use criterion VC-D or VC-E (both routinely used

Table 1. Numeric definition of common generic vibration criteria.

Category	Criterion	Definition
Human sensitivity	ISO Office	400–800 μm/s
Generic General Laboratory	VC-A	50 μm/s
	VC-B	25 μm/s
Highly sensitive	VC-D	6 μm/s
	VC-E	3 μm/s
	NIST-A	25 nm $1 \leqslant f \leqslant 20$ Hz
		3 μm/s $20 < f \leqslant 100$ Hz

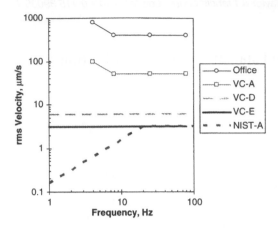

Figure 1. Common generic vibration criteria commonly used for nanotechnology facilities, in the form of velocity spectra.

worldwide for semiconductor facilities). There are some even more demanding spaces requiring the alternative NIST-A criterion, developed for metrology laboratory space at the Advanced Measurement Laboratory (AML) at the USA's National Institute of Standards and Technology (NIST) in Gaithersburg, Maryland. NIST-A is more stringent than VC-E at frequencies below 20 Hz [Amick, et al. (2002)].

The success of facilities to be designed to VC-D, VC-E or NIST-A will depend to a great extent upon the adequacy of their sites. Thus, it is customary to assess the site vibrations as part of the site selection or design process. The site survey may be simple or extensive, as dictated by the requirements of the project. Simply stated, it should capture the vibration statistics of the site that might impact the work to be done in the facility. It should include factors that will be present throughout the life of the facility, such as road and rail traffic, and exclude factors that will not be present, such as temporary construction [Ref. Gendreau & Amick (2005), Amick, et al. (2005), Gendreau, et al. (2004)].

The analyst performing a site survey may be presented with several decisions. Site vibrations will vary with time and will tend to be random. How is the frequency variation and temporal variation represented? The site may be large, so several measurement locations may be required to represent the site. How is the spatial variation represented? How are transient vibrations – such as those due to road or rail traffic – represented and compared with the "steady-state" vibrations? How might one compare several sites for a facility? How might the vibrations at one site be compared with a population of similar sites? All of these questions will be addressed in the sections that follow.

2 REPRESENTING FREQUENCY CONTENT

Frequency content is perhaps the easiest statistical variable to handle. It is usually defined by the criterion being used. The VC-D, VC-E and NIST-A criteria all use rms velocity spectra as measured in one-third octave bands of frequency, at frequencies between 1–100 Hz [Amick et al. (2002)]. In this representation, the bandwidth is 23 percent of the center frequency of each band. It approximates the half-power bandwidth associated with an oscillator with 10 percent of critical damping [Amick (1997)].

At a given site, there will tend to be predominant frequency content, often a mix of random and single-frequency components. In the absence of mechanical equipment – usually from nearby buildings – the site vibrations will be predominantly random. Quite often the spectrum will appear as a "hump", the frequency of which may depend on site conditions. It is not unusual for transient vibrations – such as those from trains – to have a different predominant frequency, especially at close proximity.

3 REPRESENTING TEMPORAL VARIATION

Random site vibrations are random over time as well as over frequency. Conventional signal processing assumes that a linearly averaged spectrum may be used to represent time-varying random vibrations if they are stationary, [Ref. Bendat & Piersol (1986)] with an integration time that is large enough to ensure that spectra taken over two different times are nearly identical. However, site vibrations may not conform to this definition of stationarity. In this instance, it may be useful to use centile spectra to represent both frequency content and statistical content. Figure 2 shows a representative set of centile spectra for the suburban site of a research facility, measured over a 30 min period during the middle of the day. There was a freeway about 0.5 km distant and on-site vehicle traffic. The six curves – starting at the top – represent the spectra that are exceeded 1%, 5%, 10%, 20%, 50% and 90% of the time. The predominant frequency at this site is 12.5 Hz.

Figure 3 shows the 20 Hz one-third octave band component of vibrations, as a function of time, at an urban site near three rail lines, processed with averaging times of 4, 32 and 60 s. The larger peaks centered on 50, 85 and 112 min represent three train passages. There are also some other unidentified transient events. The ambient spectrum is dominated by the 40 Hz component; the train passages produce more severe vibrations that are centered on 20 Hz (and to a lesser extent, 6.3 and 8 Hz). The segments between 0 and 30 min, and between 60 and 80 min are approximately stationary, and tend to be nearly constant with the longer integration times.

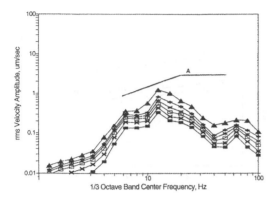

Figure 2. Typical centile spectra for a suburban site.

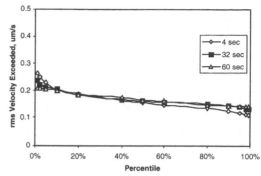

Figure 4. Statistical distribution of 20 Hz component in "steady-state" segment between 0 and 30 min.

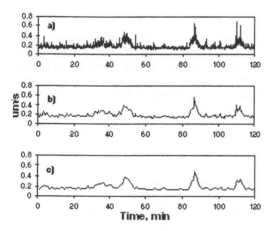

Figure 3. Vibrations measured over two hours, 20 Hz component, processed with integration periods of a) 4 s, b) 32 s, and c) 60 s. Broad peaks correspond to train passages.

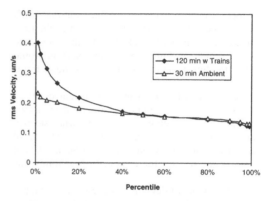

Figure 5. Comparison of statistical distribution of 2 hr period with the "ambient" or approximately "steady-state" segment between 0 and 30 min (20 Hz band 32 s integration time).

As noted earlier, the statistical distribution may be shown as a collection of centile spectra, as in Figure 1. However, it may also be presented in a distribution curve for a single frequency. This shows the percent of time a given amplitude has been exceeded. Figure 4 shows this representation for the three integration times in Figure 3, processing only the stationary ambient portion in the first 30 min of Figure 3.

The primary difference at the left end of the three distribution curves in Figure 4 is limited to the percentiles less than 10%. At percentiles greater than 50%, there is a notable difference between the 4 s integration curve and the other two, but the 32 s and 60 s integration times yield nearly identical curves. This indicates that the 32 s integration time is adequate to represent stationarity 90% of the time in this case.

It is also important to note that the maximum (leftmost) value of each distribution curve in Figure 4 is

different. The dependency of the maximum value on integration time will be discussed later.

The curve from Figure 4 for 32 s integration time is compared with the statistical distribution of the entire 2 hour record (also using 32 s integration time) in Figure 5. The impact of the rail passages is shown at percentiles less than 50%; the curves are nearly identical at higher percentiles. The periods in Figure 1 during which the vibrations rise above the background "ambient" represent about one-half of the total two-hour sample. The conclusion to be drawn in this case is that the trains have a measurable impact on the site about half of the time. If a particular vibration criterion applies to 20 Hz vibrations this analysis provides a means to show the percent of time that the criterion is exceeded (say, 5% of the time if the criterion is 0.3 μm/s).

The general conclusion to be drawn from the previous discussion is that the percentile at which a

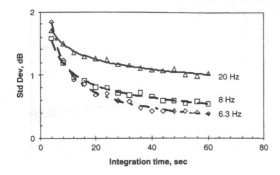

Figure 6. Effect of integration time on the log standard deviation of the 30 min "ambient" segment at the start of the 2 hr measurement period.

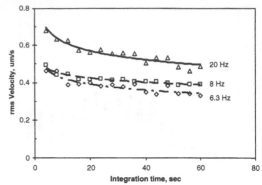

Figure 7. Effect of integration time on the maximum in the 2 hr measurement period.

distribution curve intersects with the ambient distribution curve is a function of the fraction of the total measurement time that is taken up by the transients. In the example in Figure 5, the intersection would shift to the left if there was only one of these trains present.

It should also be noted that a transient event by itself is a random environment, and will have its own distribution curve, tending to be somewhat steeper. Thus, Figure 5 really represents the combination of two distribution curves, one for the ambient (50% of the time) and one for the combination of trains (the other 50%).

The log standard deviation of the series of amplitudes at a single frequency, expressed as decibels (dB), may be used to define the statistical variation of the amplitude. Figure 6 shows the effect of integration time on this quantity. The 20 Hz component has a log standard deviation less than 2 dB at all integration times greater than 4 s, but it has decreased to only 1 dB at 60 s. The best-fit curve may be extrapolated to suggest that the curve comes close to 0 dB somewhere around 1000 s integration time. (We have actually found it necessary to use a 20 min average to achieve stationarity of a 6.3 Hz signal at one nanotechnology site, in order to compare daytime with nighttime vibration.) An integration time of about 500 s would produce the same result for the 8 Hz component in Figure 6.

The variation at the left end of the curves (1%) in Figure 4 indicates that the maximum is somewhat sensitive to the integration time. The integration time will also influence the observed maximum, as shown in Figure 7. An integration time of 4 s produced a maximum in the 20 Hz band of about 0.7 μm/s, but only about 0.055 μm/s for an integration time of 60 s.

In general, the maximum amplitude (or maximum spectrum), whether for ambient or transient conditions, will be dependent upon averaging time. When a

maximum is to be reported, a significant amount of care should be taken in the selection of an integration time, and the integration time should be included in documentation. However, it is common to use the energy average or some percentile equal to or greater than 10% when characterizing an ambient environment. The dependence on integration time becomes less significant, providing it is of adequate length.

4 REPRESENTING SPATIAL VARIATION

It is common for vibration amplitudes to vary with location over a large area. In some cases, the variation represents the proximity to sources. In others, the variation is a secondary effect due to the temporal variation in the vibrations of the surface and the difference in sampling times. Regardless, the objective is usually to compare to some criterion the statistical representation of the whole. A comparison using the average of the site would be unconservative. Using the maximum may be overly conservative.

First, however, it is important to define what is meant by "the site." It may represent a large piece of property, upon which the building to be placed. There may be a street on one or more sides, in which case the building may be set back to allow vibration attenuation with distance. Thus, an evaluation of the property as a whole should take into account the larger values that would be measured near the street. The spatial distribution of amplitudes over the site must be considered when deciding where to place the building.

On the other hand, if the evaluation is focusing on the suitability of the footprint of the proposed building (or area within the building), and the objective is to pass or fail the site, then the survey can be based on the statistics of measurements within that footprint. If there is an observable trend to the data within the footprint (*e.g.*, the vibrations are more severe on the side of the footprint nearer the street) then this should be reported as well, so it can be considered during design.

Our practice is to carry out measurements at a "statistically significant" number of locations. This is somewhat subjective, and depends in part upon the size of the area being evaluated, and the extent of variation to be expected. If we are evaluating a large property, between 5 and 20 locations might be appropriate. If the study is limited to the proposed footprint of a building of conventional size, then the quantity may be from the smaller end of this range. If there is a reasonable amount of uniformity to the data, the larger site, as a whole, may be represented by the spectrum calculated from the log mean plus the log standard deviation [Amick, *et al.* (2005)].

5   REPRESENTING MULTIPLE SITES

Occasionally during the site selection process there is a need to compare sites. This has arisen in two contexts. At times, it may be necessary to compare two sites in a manner more refined than a simple comparison of two spectra. In other instances it may be useful to know how a site compares to a group of other sites: Is it among the best? Among the worst? Average? The following two sections address approaches to these analyses

5.1   *Comparing two sites*

Figure 8 shows the statistical distribution of two candidate sites, A (dashed) and B (solid). The horizontal axis is a probability axis, which tends to stretch the extrema at each end, much as a logarithmic axis does at the left end. This comparison shows that the vibrations at Site B exceed those of Site A about 10% of the time, but are less than those of Site A about 90% of the time. This presentation format allowed the scientists who would occupy the facility to consider whether they were more concerned with short-term extrema or long-term stability. (They chose Site A. However, other applications might prefer the environment that was "quieter" the majority of the time, Site B.)

5.2   *Comparing one site to a family of sites*

Nanotechnology is currently one of the most heavily funded areas in research. A variety of research organizations are expending considerable funds to build the special facilities required for much of this work. Many of these are speculative – the institutions have only a nebulous idea of the potential areas for which they might be well positioned, and hope to use the new facility to attract desirable candidates. It is not unusual

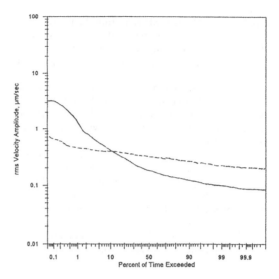

Figure 8.   Use of two centile distribution curves to compare two sites.

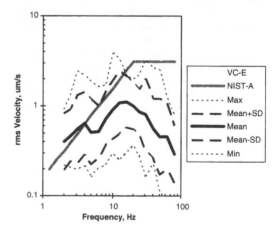

Figure 9.   Statistical representation of the vertical vibration performance of sites for 12 nanotechnology facilities.

to see the facility's environment (including vibrations) used as a magnet to draw potential researchers.

Two nanotechnology R&D facilities recently entering the design phase wanted to know how their sites compared to sites of other nano facilities. In order to preserve confidentiality of the twelve sites for which the authors had site data (each represented by a single log mean plus log standard deviation spectrum, which we denote as Mean + SD), a statistical approach was employed.

The statistical representation shown in Figure 9 was developed for the vertical data. (A similar representation was developed for horizontal data.) The population of Mean + SD data from the twelve sites is enclosed by the maximum and minimum curves. The mean, along with the mean plus/minus one standard deviation, are plotted between the upper and lower bounds.

It may be observed that all of the sites meet VC-E (with the exception of one site at 10 Hz), but that many sites do not meet NIST-A. This has become one of the points of pride with owners who can claim to meet the NIST requirement.

Upon completion of a site study, the Mean + SD spectrum from that study may be plotted on Figure 9, allowing a comparison of that site with the total population by means of the five statistical curves. Alternatively, the data from the 12 sites could be represented as percentiles of the total population (*e.g.*, 25%, 50%, and 75%, plus maximum and minimum, together defining quartiles of the population).

6 CONCLUSION

Ground vibrations measured at a site may vary in frequency, time and space, leading to a fairly complex representation of all the statistical variables. A variety of methodologies have been presented to illustrate how these three variables may be represented graphically, as well as means to compare the statistics of two sites and to compare a given site to a larger population.

When applying these methodologies, integration time (sometimes called "averaging time") is perhaps the most important variable under the analyst's control. It must be adequately long to represent stationarity, but its effect on a maximum value becomes significant. Documentation becomes important, particularly with regard to the integration time.

REFERENCES

Amick, H. 1997. On generic vibration criteria for advanced technology facilities: with a tutorial on vibration data representation. *J. Inst. Env. Sci.,* 40(5): 35–44.

Amick, H., Gendreau, M. & Gordon, C.G. 2002. Facility vibration issues for nanotechnology research. *Proc. symp. on nano device tech. 2002, May 2–3, 2002,* Hsinchu, Taiwan.

Amick, H., Gendreau, M. & Xu, T 2005. "On the appropriate timing for facility vibration surveys," *Semiconductor Fabtech,* 25.

Bendat, J. S. & Piersol, A. G. 1986. *Engineering Application of Correlation and Spectral Analysis.* 2E. New York: Wiley-Interscience.

Gendreau, M., Amick, H. & Xu, T. 2004. The effects of ground vibrations on nanotechnology research facilities. *Proc. 11th Intl. Conf. on Soil Dyn. & Earthquake Engng. (11th ICSDEE) & the 3rd Intl. Conf. on Earthquake Geotech. Engng. (3rd ICEGE), 7–9 January, 2004,* Berkeley, CA.

Gendreau, M. & Amick, H. 2005. Micro-vibration and noise. In Hwaihu Geng (ed.). *Semiconductor Manufacturing Handbook,* New York: McGraw-Hill.

Institute of Environmental Sciences (IEST) 2005. Considerations in clean room design. *IES-RP-CC012.2.*

Environmental Vibrations – Takemiya (ed.)
© 2005 Taylor & Francis Group, London, ISBN 0 415 39035 4

Measurements of vibrations in a wooden apartment house due to running trains

S. Yokoshima
Kanagawa Environmental Research Center, Hiratsuka, Kanagawa, Japan

K. Hiramatsu
R&D Headquarters, NTT FACILITIES INC., Toshima, Tokyo, Japan

Y. Sano
ACT Sound & Vibration Research Office Inc., Nagoya, Aichi, Japan

Y. Hirao
Kobayashi Institute of Physical Research, Kokubunji, Tokyo, Japan

Y. Nagaseki
Mitsui Home Co., Ltd., Kunitachi, Tokyo, Japan

T. Goto
Department of Architecture, College of Engineering, Hosei University, Koganei, Tokyo, Japan

ABSTRACT: In order to accumulate measurements of house vibration, WG on Measurement Technique of Environmental Vibration in AIJ and GOTO Laboratory in Hosei University jointly measured vibrations of running trains in a wooden house. Each maximum value of running root-mean-square values of vibration accelerations in 1/3 octave bands during a train passing was obtained. Results were the following. 1) Vertical ground vibration was larger than horizontal ground vibrations. 2) Vertical vibration at the center of the surveyed floor had a maximum of 63 Hz and increased with train velocity. 3) Vibration propagating from the foundation to the floor amplified. According to floor vibration modes determined using modal analysis, the gain can be attributed to the modes restrained by the joists which floor posts support. 4) We applied "Guidelines for the evaluation of habitability to building vibration" (AIJ, 2004) to the measurements and considered usage notes.

1 INTRODUCTION

"Guidelines for the evaluation of habitability to building vibration" in AIJ (Architectural Institute of Japan) was revised substantially in 2004. The new guideline involves traffic vibrations as well as existing floor and wind vibrations. Habitability to building vibration induced by motor vehicles or trains is evaluated by maximum amplitudes of 1/3 octave band vibration acceleration on the floor.

Vibration Regulation Law (1976) stipulates that vibration level of the operation of motor vehicle shall be measured at the boundary line of the road. A recommendation for Shinkansen railway vibration (1976) prescribes a guideline value, maximum value of vibration level on the ground. On the other hand, conventional railway vibration has not regulated yet.

Therefore, many measurements of vibration levels on the ground have been published; few cases referred to the results based on 1/3 octave band analysis in buildings. Hereafter, in order to utilize this guideline effectively, it is necessary to accumulate the measurements of 1/3 octave band value based on the method for the guideline.

WG on Measurement Technique of Environmental Vibration in AIJ and GOTO Laboratory in Hosei University jointly implemented an experimental measurement of house vibration caused by railway. This paper aimed at the following subjects. First, we released 1/3 octave band measurements of ground and floor vibrations induced by running trains. Furthermore, we clarified the characteristics of floor vibration by using vibration modal analysis. Finally, we applied "Guidelines for the evaluation of habitability to building

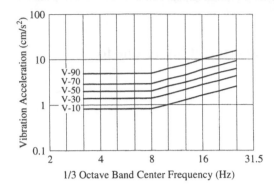

Figure 1. Evaluation curves of vertical traffic vibrations.

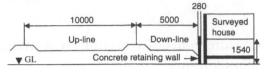

Figure 2. Positional relationship between railways and the surveyed house.

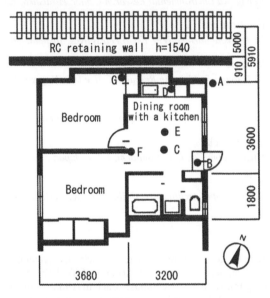

Figure 3. Ground plan of the surveyed house and measurement points.

Table 1. Outlines of measurement points.

Point	Surrounding of the points
A	Ground (concrete slab)
B	Center of the space inside the front door (concrete slab)
C	Center of a wooden floor
D	Center of a wall of the dining room
E	Neighborhood in the center of a wooden floor
F	Edge of a wooden floor (point adjacent to a pillar)
G	Edge of a wooden floor (point adjacent to a pillar)

vibration" to these measurements of house vibration and considered usage notes.

2 GUIDELINES FOR THE EVALUATION OF HABITABILITY TO BUILDING VIBRATION

We'd like to introduce outlines of the guidelines. The guidelines are applied to building vibrations with maintaining the habitability. Figure 1 illustrates evaluation curves for vertical traffic vibration. These curves are established by the probability of perceiving vibration. Thus "V-50" curve represents the vibration which people of 50% sense. The frequency range for vertical vibration is 3 to 30 Hz; horizontal vibration, 1 to 30 Hz. It is preferable that the habitability is evaluated by each maximum of vibration acceleration amplitudes in 1/3 octave band during the period of measurement. If unable to measure the amplitudes, it is possible to use running root-mean-square values (integration time of 0.01 s).

3 OUTLINES OF MEASUREMENT

3.1 Surveyed house and measurement points

A two-story wooden apartment house, containing 6 houses, along a conventional railway in Tokyo was targeted. Figure 2 shows the positional relationship between railways and the surveyed apartment house. Down and up lines were located at respective distances of 5 and 15 meters from the surveyed house. Figure 3 shows a ground plan of the surveyed house and measurement points.

The surveyed house was on the first floor and close to the down-line. 7 measurement points (1 on the ground and 6 in the house) were set up (see Table 1). At point A, a distance of about 5.9 m from the center of the down-line, horizontal vibrations (x: parallel direction to the railway, y: perpendicular direction to the railway) and vertical vibration were measured. At point D, a pickup was fixes on the wall and vibration in the y-direction was observed. At other points, vertical vibrations were measured on a wooden floor.

3.2 Measuring instruments

Figure 4 shows a block diagram of the measurement set-up. VM-53 and VM-52, which is in accordance with the requirements in JIS C 1510 (Vibration Level

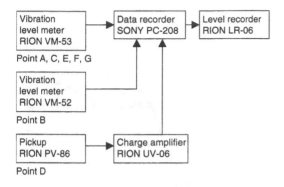

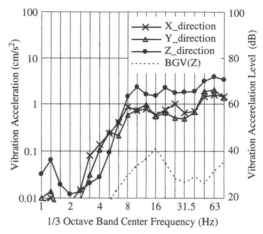

Figure 4. Block diagram of the measurement set-up.

Figure 5. Vibration acceleration from down-trains at point A.

Meter), include relative pickups of PV-83C and PV-83B. After vibration acceleration was recorded on-site, 1/3 octave band analysis was conducted in due course. An instrument used for the analysis is a 1/3 octave band real-time analyzer (ONO-SOKKI DS-9100).

3.3 Method of analysis

We obtained the highest running root-mean-square value of vibration acceleration during a train pass-by in every 1/3 octave band (1–80 Hz). The integration time of 0.01 s was used. The numbers of trains measured were 59 data in commuter trains (down-trains: 27; up-trains: 32), 7 in limited express trains (down-trains: 5; up-trains: 2) and 2 in trial running trains. Train velocities for down trains range 55 to 93 km/h; up trains, 38 to 91 km/h.

4 MEASUREMENTS

4.1 Ground vibrations

Figure 5 shows the mean values of vibration acceleration of 7 down commuter trains at point A. Figure 6 shows the similar results of 8 up commuter trains at the point. The range of train velocities were from 80 to 84 km/h. The left and right vertical axes mean vibration acceleration in cm/s^2 and vibration acceleration level in dB (re 10^{-5} m/s^2) respectively.

For the down-train, each horizontal vibration indicated the maximum value at 63 Hz; the vertical vibration, two peaks at 10 and 63 Hz. No difference in the 1/3 octave band spectra was observed between horizontal vibrations. The vertical vibration, however, indicated a larger value than horizontal ones. The values at the peak frequencies of the vertical vibration were about three times as large as that of the horizontal one.

We made comparisons of the vertical vibration between the up and down trains. Focused on the peak frequencies of the down-train, the up-train generated a peak at 10 Hz was confirmed; no peak at 63 Hz. This

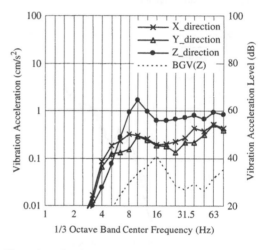

Figure 6. Vibration acceleration from up-trains at point A.

implies distance attenuation in vibration at higher frequencies is larger.

4.2 Floor vibrations

Figure 7 compares vertical vibration at point C according to the two directions of passing trains. Results were obtained from the averaged spectra of the commuter trains whose velocities range from 80 to 84 km/h. Not only the down-train but also the up-train showed maximum values at 63 Hz. In addition, at the frequencies from 8 to 80 Hz, each 1/3 octave band value of the down-train stood at up to 4 times as large as that of the up-train. However, for the up-train, it is observed that floor vibration was not dominant at 10 Hz, the peak of ground vibration. Thus, according to characteristics

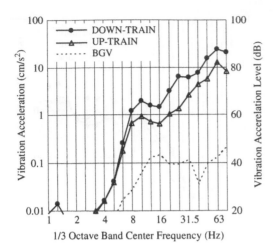

Figure 7. Comparison of vibration acceleration between down and up trains at point C.

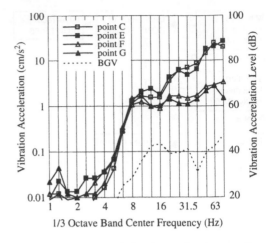

Figure 9. Comparison of vibration acceleration according to measurement points.

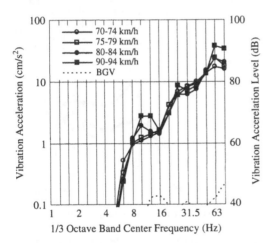

Figure 8. Comparison of vibration acceleration according to train velocities at point C.

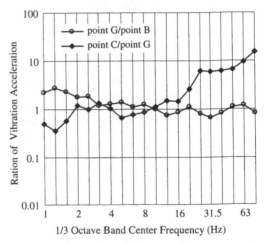

Figure 10. Amplification of floor vibration.

of a house, there are differences in 1/3 octave band spectra between the ground and floors.

4.3 Effect of train velocity

It is interesting to examine the relationship between train velocity and 1/3 octave band value for floor vibration. We compared vibration values of down commuter trains which are classified into four groups according to train velocities. Figure 8 indicates the averaged spectra at point C. At 10, 12.5, 63 and 80 Hz, observed values increased with train velocity. In particular, the velocities from 90 to 94 km/h generated large vibration.

4.4 Difference among measurement points

Figure 9 shows the difference in vertical vibration among measurement points: C, E, F and G. Results were obtained from the averaged spectra of commuter trains whose velocities range from 75 to 84 km/h. It is found that there was great difference in 1/3 octave band spectra between points in and near the center of the floor (point C and E), and points adjacent to pillars (point F and G). The difference at higher frequencies in particular was prominent.

In order to comprehend the change of acceleration in the floor vibration, Figure 10 demonstrates the ratios of acceleration values between points G and B as well as between points C and G. When vibration was transmitted from the foundation to the floor through floor

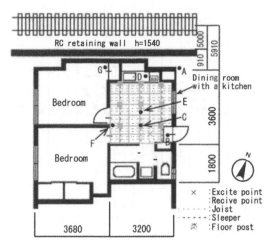

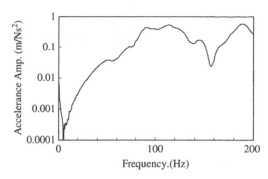

Figure 13. Result of frequency response function (point C).

Figure 11. Location of measurement points and house components.

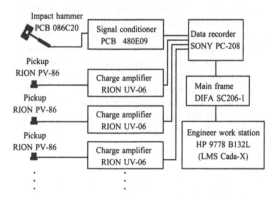

Figure 12. Block diagram of an experimental excitation test.

Table 2. Natural frequencies and damping ratios.

Mode	Frequency (Hz)	Damping ratio (%)
1	41	23.0
2	79	4.2
3	85	4.1
4	99	4.4
5	108	2.5

posts, little changes in 1/3 octave band values was observed. In contrast, the propagation of vibration from the edge to the center of the floor indicated huge gains at the frequencies from 31.5 to 80 Hz. The ratios at 63 and 80 Hz became about 10 and 15 times respectively. It occurs to us that floor resonances at higher frequencies provoke great amplification of the floor vibration.

4.5 Vibration modal analysis

With the primary objective of clarifying the cause of the gains in vibration at higher frequencies, an excitation test by using an impact hummer was carried out. Figure 11 locates excite and receive points and Figure 12 shows a block diagram of in this test.

The excitation test provided the measurements of exciting force and vibration values for acceleration at six receive points simultaneously. Since this test was intended for the wooden apartment house, the difference in exciting force adversely provoked nonlinear effects. Therefore, we tried to hammer a floor from a constant height about 30 times. In addition, we analyzed up to 10 out of 30 data, of which exciting forces were almost equal. A sequence of the test was repeated 7 times at different receive points. Consequently, frequency response functions were obtained by measurements at 42 points in total.

By determining vibration modal parameters, least square complex exponential method, one of time domain multiple degrees of freedom method, was used. The analysis was done with LMS CADA-X.

Figure 13 shows the result of frequency response function at point C. The spectrum of the broad peak around 85 Hz was observed. We, however, recognized that there was no sharply-peaked resonance in the case of concrete buildings.

Table 2 shows each of the natural vibration frequencies and the damping ratios of identified vibration modes. In addition, Table 3 illustrates modal overcomplexity value (MOV), which represents the accuracy of a modal model. The MOV is defined as the percentage of the response points for which a mass addition indeed decreases the natural frequency for a specific mode. This index should be high (near 100%) for high quality modes. At 41 Hz, where peaked resonance was not clearly recognized, the mode was identified. However, the damping ratio of 23% was very high and the

Table 3. Mode overcomplexity values.

Mode	Mode overcomplexity value (%)
1	39
2	85
3	79
4	84
5	72

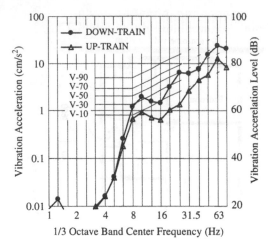

Figure 16. Application of house vibrations to the guidelines.

Each above left graph means the east-west cross-section of the mode shape. The relation between floor constructions and the mode shape was not clear. In contrast, above right graph of the north-south cross-section confirms that the positions around floor posts were reliable to become nodes. Hence the cause for gain in the floor vibration can be attributed to the modes restrained by the joists supported by floor posts. This suggests that the first-mode vibration was difficult to stand out because the joists are constrained due to floor posts.

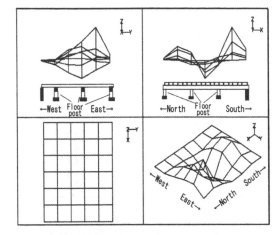

Figure 14. Vibration mode shape at 79 Hz.

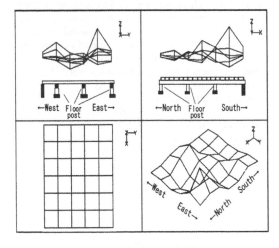

Figure 15. Vibration mode shape at 85 Hz.

MOV of 39% was low. For other modes, the damping ratios of less than 5% and the MOV of more than 70% confirmed the validity of these vibration modes.

We focused on the mode 2 and 3 where near frequencies indicated great gains in the floor vibration. Figures 14 and 15 shows vibration mode shapes at 79 and 85 Hz respectively.

4.6 Application to "Guidelines for the Evaluation of Habitability to Building Vibration"

Figure 16 compares the measurements shown in Figure 7 with evaluation curves of the guidelines ("V-10" to "V-90"). The up-train was estimated to be "V-10". The result corresponds to the fact that few participants in the measurement sensed vibrations caused by up-trains. On the other hand, the down-train was determined to be "V-50". The result, however, disagrees with the participants' responses during the measurement: all of them perceived vibrations induced by down trains.

Although the frequency range of the vertical vibration is 3–30 Hz, we enlarged the range up-to 80 Hz and assessed the down-train once again. The result was concluded to be "V-70". Since the surveyed house is made of wood, it is likely that effects of house vibration on human are strong compared to structure-borne sound. Therefore, it is likely that we felt vibrations beyond the upper-limit frequency.

As just described, in the case of assessing house vibration which is in the ascendant at a frequency of more than 30 Hz, evaluation is required for considering the other effects from the railway, e.g. structure-borne sound.

5 CONCLUSIONS

This paper described the results from experimental measurement of house vibrations caused by railway. The results are summarized as follows.

(1) Vertical ground vibration is larger than each of horizontal ground vibrations.
(2) Vertical vibration at the center of a floor shows a maximum of 63 Hz and increases with train velocity.
(3) Vibration propagating from an edge to the center of a floor amplifies. According to floor vibration modes determined using modal analysis, the gain can be attributed to the modes restrained by the joists which floor posts support.
(4) When "Guidelines for the evaluation of habitability to building vibration" is applied to the measurements of the house vibration which is in the ascendant at more than 30 Hz, evaluation is required for considering the other effects from the railway.

ACKNOWLEDGEMENTS

These measurements were supported by the members of WG on Measurement Technique of Environmental Vibration (see below). In addition, we are grateful that Dr. H. Naruse, professor of Aichi Institute of Technology, provides a clear guidance on vibration modal analysis.

Members of WG on Measurement Technique of Environmental Vibration (FY 2003–2004).

Shigenori YOKOSHIMA (Kanagawa Environmental Research Center), Kazutsugu HIRAMATSU (R&D Headquarters, NTT FACILITIES INC.), Riei ISHIDA, (Faculty of Engineering, Chiba University), Yasuhiko IZUMI (Railway Technical Research Institute), Toshihisa ISHIBASHI (Kajima Technical Research Institute), Michinari OKAZAKI (Rion Co., Ltd.), Masashige KAWAKUBO (Tokyu Construction Co., Ltd.), Yukio KOJIMA (SEKISUI HOUSE, LTD., Comprehensive Housing R&D Institute), Yasuyuki SANO (ACT Sound & Vibration Research Office Inc.), Toshio SUZUKI (Kokankyo Engineering Corporation), Ryuta TOMITA (College of Science and Technology, Nihon University), Yoshiki NAGASEKI (Mitsui Home Co., Ltd.), Toshikazu HANAZATO (Taisei Research Institute), Yoshihiro HIRAO (Kobayashi Institute of Physical Research), Hiroatsu FUKUHARA (ADVANCED ACOUSTIC RESEARCH).

Environmental Vibrations – Takemiya (ed.)
© 2005 Taylor & Francis Group, London, ISBN 0 415 39035 4

A study on the prediction and measures of floor vibrations from aerobics

Y. Tanaka
Technical Research Institute, Hazama Corporation, Tsukuba, Ibaraki, Japan

ABSTRACT: This paper presents a prediction method for floor vibrations from aerobics illustrating a case of two similar buildings one of which had a vibration problem at the lounge adjoining to an aerobics studio. The cause of the problem was quasi-resonance of the first mode of the floor and the second harmonics of the aerobics excitation. A countermeasure based on simulation results settled the problem. A study of how force is employed to the floor demonstrates that the floor vibration from aerobics with dozens of participants can be predicted accurately by the response to the force of around eight people in phase and multiplying it by the root of the number of participants divided by the number of people in phase.

1 INTRODUCTION

Excitation force from aerobics might cause a claim concerning habitability to building vibration though it does not affect the strength of the structure. In this paper, floor vibrations from aerobics are investigated in two structurally similar buildings only one of which had a vibration problem. The cause of the problem was quasi-resonance of the first mode of the floor and the second harmonics of the aerobics excitation. Though the rhythm of excitation changes somewhat according to music or movements, the resonance of the first to third harmonics of the excitation and the first mode of the floor is a major factor of vibration problem from human activities.

Illustrating investigation results of the problem, their evaluation and simulated or verified effects of countermeasures, a design technique to predict floor vibrations from aerobics with many participants precisely is proposed in conclusion.

2 INVESTIGATION OF FLOOR VIBRATION

Figure 1 shows the column and beam plan of the second floor of Building A and B. Their structures are similar. The problem was that very unpleasant vibrations occurred in the lounge of Building B from aerobics. There was not any claim in Building A.

Vibrations from aerobics are measured to grasp the cause of the problem. The measurements are about X, Y and vertical displacements at P1 and vertical displacements at P2 in both buildings. These data are recorded for 5 minutes each time the movements of aerobics became hard. Horizontal vibrations are less than a tenth

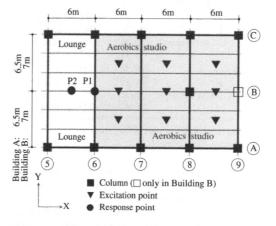

Figure 1. Column and beam plan of the second floor of Building A and B.

of vertical ones. Vertical vibrations at P1 (on girder) are larger than those at P2 (in the middle of floor).

Natural frequencies of the floors in Building A and B are 5.7 and 5.55 Hz respectively. They are obtained from Fourier spectra of ambient vibrations. As amplitudes of the ambient vibrations at P1 and P2 are almost the same in both buildings, the floors can be considered as unidirectional. Damping factors of floor estimated with the random decrement technique and ambient vibration data are about 0.03 in both buildings. The data are filtered by a fourth Butterworth filter whose bandwidth is 0.5 Hz.

Table 1 shows the maximum displacements at P1 from aerobics programs in both buildings. Each program lasts an hour. Maximum displacements of

Table 1. Maximum displacements of floor vibration during each program.

Building	Program	Time* minutes	Disp.** μm	Main movement
A	D 53 participants	15–20 20–25 35–40 43–48	236 253 289 210	 Steps
A	Intermission		76	
A	E 86 participants	15–20 21–26 30–35	419 639 949	Steps Knee bending Jump
B	A 22 participants	9–14 26–31	86 105	
B	B 22 participants	7–12	110	
B	C 53 participants	6–11 12–17 17–22	179 168 241	
B	E 77 participants	15–20 21–26 30–35 36–41	514 742 796 1687	Steps Knee bending Jump
After the measure B	E 76 participants	15–20 21–26 30–35 36–41	142 188 153 324	Steps Knee bending Jump

* Elapsed time from the start of the program.
** Maximum displacement.

program A to D with 22 to 53 participants and moderate movements are less than 300 μm. Those of program E with about 80 participants and hard movements are close to 1 mm in Building A and more than 1 mm in Building B. The largest vibration occurs about 35 minutes after the start of program E when participants are jumping with arms and legs shaking back and forth alternately.

Figures 2–5 show waveforms and Fourier spectra of vertical vibrations at P1 in both buildings during program E. Figure 2 tells the vibration of Building A exceeds 400 μm at times and 600 μm 25 or 34 minutes after the start. Figure 3 tells the vibration of Building B is about 500 μm from 22 minutes after the start and exceeds 1 mm in 36 minutes.

Figure 4 shows the second harmonics (5.0 Hz) of excitation from aerobics is about a half of the fundamental component (2.5 Hz) in Building A. In Building B (Fig. 5), the second harmonics (5.4 Hz) is 1.5 times as large as the fundamental (2.7 Hz). This suggests frequencies of the first mode of floor and the second harmonics of excitation in Building B are close (5.55, 5.4 Hz respectively) and more resonant than those in

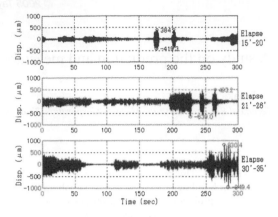

Figure 2. Waveform of vertical vibrations at P1 in Building A during program E.

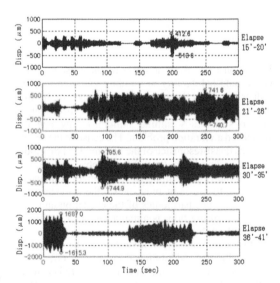

Figure 3. Waveform of vertical vibrations at P1 in Building B during program E.

building A (5.7, 5.0 Hz). Rhythms of same program E in both buildings are a little different.

One-third octave band peak holding spectra of vertical vibrations of both buildings during program E are shown in Figure 6 with foot-to-head vibration curves of ISO 2631–2 (1989). Peak holding values are calculated from one-third octave band spectra of vertical vibration data at P1 and P2 of both buildings divided every 20 seconds. The 5 Hz band component of Building B is more than four times of that of Building A. This seems to be the reason why the complaints occurred only in Building B.

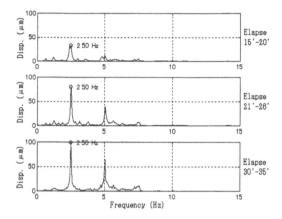

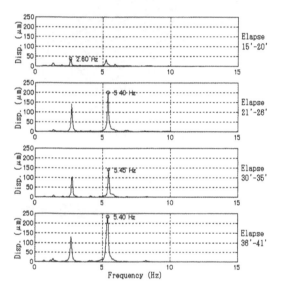

Figure 4. Fourier spectra of vertical vibrations at P1 in Building A during program E.

Figure 5. Fourier spectra of vertical vibrations at P1 in Building B during program E.

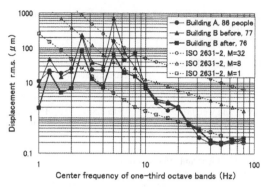

Figure 6. Peak holding spectra of vertical vibrations during program E.

Table 2. Multiplying factors in ISO 2631-2.

Place	Time	Continuous or intermittent vibration	Transient vibration excitation with several occurrences per day
Critical*	All	1	1
Residential	Day	2 to 4	30 to 90
	Night	1.4	1.4 to 20
Office	All	4	60 to 128
Workshop	All	8**	90 to 128**

* Critical working areas (for example some hospital operating-theaters, some precision laboratories, etc.).
** Doubling the suggested vibration magnitudes may result in adverse comment and this may increase significantly if the levels are quadrupled.

3 EVALUATION OF HABITABILITY

First of all, multiplying factor in ISO 2631-2 is considered. Factors M1 calculated by root-sum-square of main components (2.5 to 25 Hz band) of Figure 6 are 51 and 148 for Building A and B. Factors M2 calculated by root-sum-square of 2.5 to 25 Hz band components of one-third octave band ensemble average spectra whose original data are the same as Figure 6 are 16 and 41 for building A and B.

Table 2 indicates M1 and M2 of both buildings do not satisfy the habitability for workshop if vibrations from aerobics are considered continuous or intermittent. But the footnote ** of Figure 2 and M2 of both buildings suggest that the habitability of Building A is on the verge of claims and that of Building B is likely to suffer claims. If vibrations from aerobics are considered transient, M1 of Building A satisfies the habitability for office and workshop. But M1 of Building B does not satisfy the habitability for workshop.

Then allowable exposure time of reduced comfort boundary in ISO 2631-1 (1985) is considered. Allowable time T1 calculated from root-sum-square of 2.5 to 25 Hz band components of Figure 6 are 2 hours and 10 minutes for Building A and B. Allowable time T2 calculated from root-sum-square of 2.5 to 25 Hz band components of one-third octave band ensemble average spectra whose original data are the same as Figure 6 are 10 hours and 3 hours.

In addition, Dieckmann's K value and DIN 4025 (AIJ 1982) are considered. K values calculated from root-sum-square of 2.5 to 25 Hz band components of Figure 6 are 9 (sustainable up to an hour) and 25 (sustainable up to 10 minutes) for Building A and B.

Table 3. Acceleration limits for vibrations due to rhythmic activities.

Occupancies affected by the vibration	Acceleration limit	
	percent gravity	cm/s²
Office and residential	0.4 to 0.7	4 to 7
Dining and weightlifting	1.5 to 2.5	15 to 25
Rhythmic activity only	4 to 7	40 to 70

Table 4. Ideas and effects of measures.

Idea	Measure on 2nd floor	Ratio of vibration amplitude*			
		2.7 Hz	5.4 Hz	8.1 Hz	Maximum
A	2 braces in Section ⑥	0.38	0.10	0.37	0.20
B	3 posts in Section ⑥	0.49	0.14	0.65	0.28
C	6 posts in Section ⑤&⑥	0.52	0.22	0.66	0.29

* Compared to the case before measures.

Table 3 shows acceleration limits for vibrations due to rhythmic activities recommended in National Building Code of Canada (Allen 1990). Maximum accelerations during program E calculated from root-sum-square of 2.5 to 25 Hz band components of Figure 6 are 37 and 105 cm/s² for Building A and B. Therefore the habitability as lounge hardly seems to be satisfied in Building B. This implies the reason why vibration claims occurred only in Building B.

4 MEASURES TO FLOOR VIBRATION

As there was no problem of habitability in Building A, measures to suppress vibrations in Building B to those in Building A or less are needed. The floor of lounge in Building B is considered unidirectional and the vibration on girder is larger than that in the middle of floor as mentioned above. Then measures to raise stiffness of the girder between the lounge and aerobics studio are considered.

Table 4 shows ideas of the measure and their effects of vibration reduction predicted by finite element analyses. First in the prediction, mode analyses are performed. Boundary condition of finite element models is that the bottoms of columns and posts are clamped and ball jointed respectively. Vibration modes up to around 9 Hz (20 to 30 modes) are considered in the analyses. Table 5 shows common parameters of models. Figures 7–8 show the finite element model and critical vibration mode before measures. Double flooring in the aerobics studio is

Table 5. Parameters of finite element models.

Parameter	Steel	Reinforced concrete
Modulus of longitudinal elasticity(N/m²)	2.1×10^{11}	2.1×10^{10}
Poisson's ratio	0.29	0.17
Mass density (kg/m³)	7800	2400

Effective thickness of deck slab as concrete: 0.1235 m; Damping factor of all modes: 0.03.

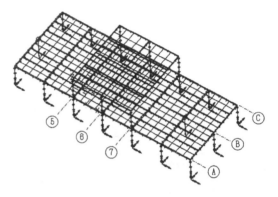

Figure 7. Finite element model before measures.

Figure 8. Critical vibration mode before measures (5.36 Hz).

omitted from the modeling. Natural frequency of the double flooring with 80 people assumed by static stiffness is more than 20 Hz. The effect of the flooring to aerobics excitation that is largely less than 10 Hz seems to be small.

Second in the prediction, excitation force from aerobics is made. Tables 6–7 show dynamic forces from human activities in ISO 10137 (1992) and dynamic load factors from aerobics (Allen 1990). Figure 9 shows the waveform of aerobics excitation force per person used in the analyses. Parameters of the force are described in the footnote of Table 6.

Third, displacement and acceleration responses at P1 from aerobics are calculated. Conditions that participants are evenly scattered except for near the walls

Table 6. Dynamic forces from human activities in ISO 10137.

$$F(t) = Q \left\{ 1 + \sum_{n=1}^{k} \alpha_n \cdot \sin(2\pi \cdot n \cdot f \cdot t + \phi_n) \right\}$$

where $F(t)$ = dynamic force at time t;
Q = static load of participating person;
n = integer designating harmonics of the fundamental;
k = number of harmonics that characterize the forcing function in the frequency range of interest;
α_n = numerical coefficient corresponding to the nth harmonic;
f = fundamental frequency component of repetitive loading;
ϕ_n = phase angle of nth harmonic.

Q = 588 (N) = 60 (kgf); k = 3; α_n: Table 7; f = 2.7(Hz); ϕ_1 = 0; ϕ_2 = $\pi/2$; ϕ_3 = 0; Δt = 1/200 (sec); t = 0 to 10 (sec) and adjusted so that $F(0) \fallingdotseq 0$.

Table 7. Dynamic load factors for aerobics.

Activity	Harmonic number, n	Forcing frequency, Hz	Dynamic load factor, α_n
Aerobics	1	2 to 2.75	1.5
	2	4 to 5.5	0.6
	3	6 to 8.25	0.1

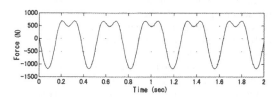

Figure 9. Waveform of aerobics excitation force per person.

and floors are considered unidirectional are taken into account. Excitation force and responses are all vertical. Force of Figure 9 is input at 9 excitation points in Figure 1 in phase. The number of participants of program E when large vibrations are observed is about 80. So based on the idea of energy sum, vibration response at P1 from aerobics with 80 participants is predicted by the response to the forces at the 9 points and multiplying it by the root of the number of participants divided by the number of forces in phase.

Figure 10 shows the predicted and measured responses at P1 in the cases of before and after measures in Table 4. Dynamic load factors in Table 7 used for the prediction correspond to when participants are jumping. Therefore the data during 36'10" to 36'30" in Figure 3 are selected as the measured response. The predicted and measured responses before measures are corresponding well.

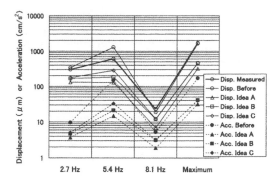

Figure 10. Vibration responses at P1 from aerobics with 80 participants in the cases of before and after measures.

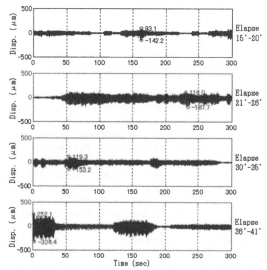

Figure 11. Waveform of vertical vibrations at P1 in Building B after the measure during program E.

According to the result of Figure 10 and Table 4, Measure A is implemented. Vibration measurements to verify the effects of the measure are also conducted. Figures 11–12 show the waveform and Fourier spectra of vertical vibrations at P1 in Building B after the measure during program E. Peaks of Figure 11 are indicated in Table 1 above.

Continuous vibrations from 22 minutes after the start of program E are less than 200 μm (Fig. 11). They were about 500 μm (Fig. 3) before the measure. Large vibrations from 36 minutes are less than 400 μm while they were over 1 mm before. Figure 12 describes that the fundamental and second components at 2.7 and 5.4 Hz decreased to a third to a quarter and a tenth

565

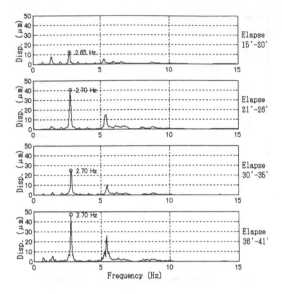

Figure 12. Fourier spectra of vertical vibrations at P1 in Building B after the measure during program E.

of those in Figure 5. This not only means that the resonance of the second harmonics and the floor is abated but also the stiffness of the floor is raised.

Figure 6 above also shows peak holding spectrum of vertical vibrations in Building B after the measure during program E. It describes that the 5 Hz band component after the measure is less than a tenth of that before the measure and less than a half of that of Building A. Other components of Building B after the measure are the same as those of Building A or less.

Displacement components at 2.7, 5.4 and 8.1 Hz and maximum displacement at P1 in Building B after the measure during 36'0" to 36'20" of program E are reduced to 0.38, 0.14, 0.29 and 0.19 times of those before the measure. This corresponds well with the predicted effect of Measure A in Table 4. The vibration problem in Building B is thus settled.

5 EXAMINATION OF DESIGN TECHNIQUE

Dynamic forces from rhythmic human activities have the most components at frequencies of 1, 2 and 3 times the fundamental rhythms as shown in Tables 6–7. In order to avert vibration problems from human activities, resonance of the floor and these harmonics should be avoided. It seems that these problems can be prevented if the natural frequency of the floor is set more than 10 Hz. But this may not be achieved from the restriction of the layout or the cost of the building. For such cases especially when the number of acting people is large, a technique to predict floor vibrations precisely is examined here.

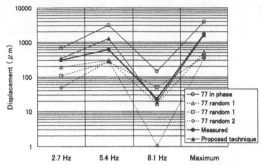

Figure 13. Vibration responses at P1 in Building B before the measure from aerobics jumping with 77 participants.

In order to set excitation forces to floors in structural design, it should be considered that the force per person decreases according to the increase of the number of people acting as the movement of all the people do not coincide perfectly. Using the fact that excitation forces from jumping can be considered in phase up to 8 people (Allen et al. 1985, Pernica 1990) and the concept of energy sum, the aerobics force of 77 participants in Building B before the countermeasure during program E is expressed by the force of about 8 people in phase and the root of the ratio of the number of all the people acting to the number of people acting in phase. As the situation that participants are evenly scattered in the rectangle aerobics studio is difficult to realize by 8 people, 9 is used as the number of people in phase.

Figure 13 shows the responses at P1 from aerobics on condition that the force of Figure 9 is input at 77 points evenly scattered in the aerobics studio (except for near the walls) in phase or with random phases of 3 cases. Figure 13 also describes the measured response in Figure 10 and the result predicted by the response to the force of Figure 9 input at 9 excitation points in Figure 1 in phase and multiplying it by the root of the number of participants divided by the number of people in phase ($\sqrt{(77/9)} \fallingdotseq 2.9$).

The response to 77 forces in phase is 2 to 3 times larger than the response measured. The responses to 77 forces with random phases are a half to a third of the response measured. The response to 9 forces in phase with consideration of 77 participants corresponds well with the response measured.

The technique proposed here is verified only for the case of Building B before the countermeasure with 77 participants. Therefore considering the number of aerobics participants 10 to 300, excitation condition of all the people in phase, proposed technique and all the people with random phases are compared in Figure 14 with the equivalent number of people exciting in phase. Tendency of each excitation condition

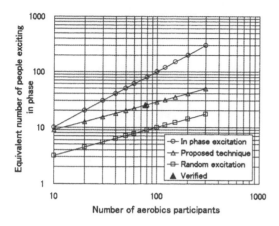

Figure 14.   Equivalent number of people exciting in phase.

and the verified situation, it is concluded that the proposed technique leads to more precise predictions of floor vibration than the excitation input of all the people in phase or all with random phases.

6   CONCLUSIONS

Floor vibrations from aerobics with about 10 to 300 participants can be predicted accurately by the response to the force of around 8 people in phase and multiplying it by the root of the number of participants divided by the number of people in phase. It is true on condition that the floors of the aerobics studio are affected by global vibration modes. If the floors are structurally independent, numbers of participants should be counted in the area of a span. This consideration of force input can be used for simulations with finite element method or dynamic equation of simple beam, plate and so forth.

REFERENCES

Allen, D. E. 1990. Building vibrations from human activities. Concrete International 12(6): 66–73.
Allen, D. E. et al. 1985. Vibration criteria for assembly occupancies. Canadian Journal of Civil Engineering 12(3): 617–623.
Architectural Institute of Japan. 1982. AIJ standard for structural calculation of reinforced concrete structures: 549.
ISO 2631-1. 1985. Mechanical vibration and shock. Evaluation of human exposure to whole-body vibration. Part1: General requirements.
ISO 2631-2. 1989. Evaluation of human exposure to whole-body vibration. Part2: Continuous and shock-induced vibration in buildings (1 to 80 Hz). Annex A.
ISO 10137. 1992. Bases for design of structures. Serviceability of buildings against vibration. Annex A.
Pernica, G. 1990. Dynamic load factors for pedestrian movements and rhythmic exercises. Canadian Acoustics 18(2): 3–18

Environmental Vibrations – Takemiya (ed.)
© 2005 Taylor & Francis Group, London, ISBN 0 415 39035 4

Traffic-induced vibrations and noises in elevated railway structures

R. Gao & H. Xia
School of Civil Engineering and Architecture, Beijing Jiaotong University, Beijing, China

ABSTRACT: Traffic induced structural vibrations and noises of railway station structures are analyzed and estimated in this paper. Two finite element models, train-bridge model and frame model, are used to simulate the dynamic behavior of the station structure. The analytical model for the train-structure system is presented. The method of statistical energy analysis is used to estimate the structural noises. The relationship between structural vibration and noise is determined through computer simulation on the basis of radiation ratio and statistical energy analysis. Several factors, such as structural forms, materials, support conditions, train speed, etc., affecting the noise are discussed. By comparison of the noise level expressed with sound power radiation, some conclusions are obtained. The results show that the method can be used as reference for the plan, design and material selection of elevated station structures.

1 INTRODUCTION

With the development of economy, more and more high-speed railway lines will be constructed in China. The railway for passenger transportation, connecting Qinhuangdao and Shenyang city, China, with the designing speed of 250 km/h, is going into operation. Other nine railway lines for passenger transportation, such as the lines from Zheng Zhou to Wu Han, from Beijing to Shi Jiazhuang, etc., are planned, designed and constructed. This means that the demand for the construction of transportation structures is absolutely to increase during the coming years. The elevated station structure, one of the transportation structures, often appears in big cities where the population density is very high. In the design of an elevated station structure, not only strength and deformation, but also vibration and noise must be considered, because traffic induced vibrations and noises affect the environment and human health and have aroused a great deal of public attention. Therefore, the analysis and the estimation of such traffic induced structural vibrations and noises have become one of the important considerations on the planning and designing of a high-speed railway station.

It is obvious that the vibrations and noises induced by rail transit system in city affect the environment. Schultz (1979) presented rating criteria of noise for elevated rapid transit structures in his report. Hanson (1983) investigated the vibration and noise control of transit vehicles running on the elevated structures, and proposed some control measures. Van Ruiten (1988) studied the mechanism of squeal noise generated by trams. Based on the finite element model of the rail and its supporting system, Chua et al. (1997) made the theoretical analysis on the vibrations produced by rail transit system. They got the good results compared with the test. Kraemer (1984), in his dissertation titled with rail roughness is expressed with velocity spectrum density function of vehicles, studied systematically structural vibrations produced by underground rail lines. Gao et al. (2001) analyzed vibrations of elevated railway station structures by using train-bridge model. Stimpson et al. (1986) use SEA method to predict sound power radiation from built-up structures.

The traffic induced vibrations and noises are affected by many factors, such as types of vehicles and structures, train speed, etc. Generally speaking, the noise level of railway bridges is greater than that of the rail lines on the ground surface, according to field test results, and the increment of the level is about from 0 to 20 dB. This is because of the second radiation – structural sound. Compared with the sound directly radiated by wheels and rails, the second sound radiation is induced by the vibrations of elevated structures. Therefore, the structural sound relates to the types and coefficients of structures.

Although there are many researches on vibrations and noises of rail transit system, most of them focus on elevated bridges and a few of them on elevated station structures that are absolutely necessary structural forms in the development of elevated rail transit system in city. Bridges and elevated railway stations are all belong to elevated structures, but their forms are different, the former is open, the latter is closed. It is necessary to analyze the vibrations and noises of elevated station structures.

For the existing structures, structural vibrations and noises can be evaluated on the test data. However, for the planning and designing, theoretical methods are the best way to estimate them.

In this paper, traffic induced structural vibrations and noises of railway station structures are analyzed and estimated. Two finite element models, train-bridge model and frame model, are used to simulate the dynamic behavior of the station structure. The analytical model for the train-structure system is presented. The method of statistical energy analysis (SEA) is used to estimate the structural noises. The relationship between structural vibration and noise is determined through computer simulation on the basis of radiation ratio and statistical energy analysis. Several factors, such as structural forms, materials, support conditions, train speed, etc., affecting the noise are discussed.

2 TYPES OF ELEVATED RAILWAY STATION STRUCTURES

Generally speaking, the elevated railway station structures can be divided into three types: spatial frame structure system; bridge structure system and composite structure system of frame and bridge. All of the three types are of three to four floors and mainly made of concrete or prestressed concrete.

(1) Frame structure system: the system is made up of spatial frame structure and continuous girder supported by the frame, shown in Figure 1a. In this system, the spatial frame structure works as both a building structure and bridge piers, by which the girder of the bridge is supported. For this system, its integrity is good, its rigidity is high, and its mass distributes regularly. Therefore, the earthquake-resistant ability of the system is relatively better. However, this system is also of some disadvantages too, for example, its foundation subsides unevenly because of irregular loads and the traffic induced vibrations of elevated railway station structure; besides, the design is more complicated because two types of specifications are required.

(2) Bridge structure system: the bridge structure consists of girders, piers and foundations, and then the platform is built on the bridge, shown in Figure 1b. For this system, the structural performance is similar to that of common bridges. Its mass center is on the upside of the structure, and so its earthquake-resistant ability is weak. Higher rigidity and better stability are also required for this system.

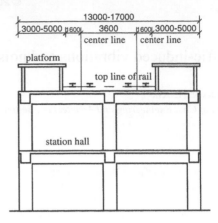

Figure 1a. Spatial frame structure.

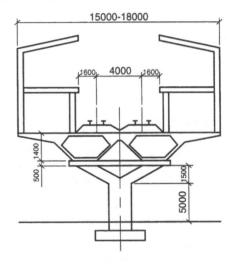

Figure 1b. Bridge structure system.

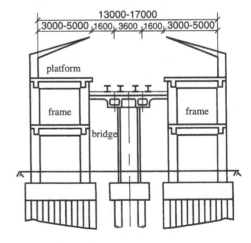

Figure 1c. Composite structure system.

(3) Composite structure system of frame and bridge: this system consists of two independent parts, the frame and the bridge. The frame is for buildings, and the bridge for railway, shown in Figure 1c. For this system, the vibrations induced by traffic are weaker than that of the other two types of elevated station structures since the traffic loads are only supported by the bridge structures.

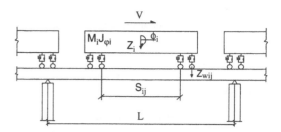

Figure 2. Train-bridge model.

3 DYNAMIC RESPONSES OF ELEVATED RAILWAY STATION STRUCTURES

According to the property of elevated station structures, the analysis method of 'two steps' is presented. That is to say, the train-frame system is separately considered. Firstly, a train-bridge model (Xia, H. & Chen, Y.J. 1992) is used to simulate the behavior of the structures, from which the loads, caused by moving train and acting on the structures, can be obtained. Secondly, a frame model consisting of three-dimensional beam elements is used to simulate the elevated station structure, where the above loads in turn act on it. Figure 2 shows the train-bridge model. From the model, three equations for dynamic analysis are given as follows:

The equation of vehicles:

$$
\begin{bmatrix} M_i & 0 \\ 0 & J_{\Phi_i} \end{bmatrix} \begin{Bmatrix} \ddot{Z}_i \\ \ddot{\Phi}_i \end{Bmatrix} + \begin{bmatrix} C_{zi} & 0 \\ 0 & C_{\Phi_i} \end{bmatrix} \begin{Bmatrix} \dot{Z}_i \\ \dot{\Phi}_i \end{Bmatrix} +
$$
$$
\begin{bmatrix} K_{zi} & 0 \\ 0 & K_{\Phi_i} \end{bmatrix} \begin{Bmatrix} Z_i \\ \Phi_i \end{Bmatrix} =
$$
(1)
$$
\sum_{j=1}^{2N_{wi}} \frac{1}{2N_{wi}} \begin{Bmatrix} K_{zi} Z_{wij} + C_{zi} \dot{Z}_{wij} \\ \frac{2\eta_j}{S_{ij}} \left(K_{\Phi_i} Z_{wij} + C_{\Phi_i} \dot{Z}_{wij} \right) \end{Bmatrix}
$$

where M_i = mass of the ith vehicle body; $J_{\phi i}$ = inertia moment of the ith vehicle body; N_{wi} = wheel number under a bogie of ith vehicle body; $K_{zi}, K_{\phi i}$ =combined stiffness of the ith vehicle body; $C_{zi}, C_{\phi i}$ = combined damping of the ith vehicle body, and Z_{wij} means the vertical displacement of the jth wheel of the ith vehicle body, given as follows:

$$
Z_{wij} = \sum_{n=1}^{N_q} q_n \phi_n \left(x_{ij} \right) + w\left(x_{ij} \right)
$$

in which $w(x_{ij})$ is the displacement of the jth wheel of the ith body due to wheel-rail roughness. x_{ij} is the location of jth wheel.

The equation of elevated station structures:

$$
[M]\{\ddot{v}\} + [C]\{\dot{v}\} + [K]\{v\} = \{F\}
$$
(2)

where $[M]$, $[C]$, $[K]$ are mass matrix, damping matrix and stiffness matrix of a structure respectively; $\{v\}$, $\{\dot{v}\}$, $\{\ddot{v}\}$ are displacement, velocity and acceleration of a structure; $\{F\}$ is the force produced by vehicles, acting on the structure.

Dynamic responses, such as acceleration and velocity, of a structure can be obtained by solving equation (1) and equation (2). The equations are differential equations varying with time, and can be solved by using Newmark-β method.

4 NOISE EVALUATION OF ELEVATED RAILWAY STATION STRUCTURES

Elevated railway station structures are generally composed of several individual substructures such as bars, beams, plates, etc. To predict the sound power radiation and the distribution of vibration energy of these complex structures, the method of Statistical Energy Analysis is used. SEA has been used for the estimation of the distribution of vibration energy throughout the structure, the total loss factors and for the study of sound transmission in room acoustics.

The structure is broken down as usual for SEA analysis into a number of separate coupled substructures, each substructure being considered as a separate source of noise. Supposing structural components are made of the same material, the total radiated sound power Π_{rad} can be written as follows:

$$
\Pi_{rad} = \rho_0 c \sigma_i S_i < \bar{v}_i^{-2} > \left[1 + \sum_{\substack{j=1 \\ j \neq i}}^{N} \frac{E_j}{E_i} \frac{\sigma_j}{\sigma_i} \frac{d_i}{d_j} \right]
$$
(3)

where $<v_i^{-2}>$ is the mean-square velocity of the ith substructure with respect to time and space. d_i is the average thickness of the ith substructure, S_i is the area of the ith substructure. σ is the radiation efficiency of the structure into the fluid media, c is the wave speed in the media, ρ is the density of the media material. E_j/E_i is the ratio of the stored energy in the jth substructure to that in the excited substructure. This ratio can be determined with the modal densities of the ith and jth substructures, the internal loss factor of the ith substructure and the coupling loss factor from the ith to the jth substructure.

5 SIMULATION OF AN ELEVATED RAILWAY STATION

5.1 Description of the elevated structure

This example is an elevated rail transit station in Shanghai, China. The elevated structure, with the size of $6 \times 1200\,\text{mm} = 72000\,\text{mm}$ in length, 24000 mm in width and 5600 mm in height, consists of reinforced concrete beams and plates supporting the rail system. The thickness of the plate is 700 mm, the area of the column cross section is $1.2 \times 10^6\,\text{mm}^2$, the elastic modulus is $3.1 \times 10^4\,\text{MPa}$.

In the dynamic analysis of an elevated railway station structure, the whole structure is regarded as a continuous girder bridge, and the frame structure as piers. By solving the equations, the bearing reactions (the loads caused by moving train and acting on the frame structures) can be obtained. Then a frame model consisting of three-dimensional beam elements is used to simulate the elevated station structure, on which the bearing reactions are applied. During the analysis, deferent speed conditions of the trains are considered.

5.2 Results and discussions

A typical time-history of bearing reactions when a train (with the speed of 60 km/h) passes the station is shown in Figure 3. By comparison, it is found that the peak values of bearing reactions are not sensitive to train speed if it is lower than 80 km/h. All bearing reactions are acted on the frame structure for further vibration and noise analysis of the elevated structure.

Figures 4–7 show the results calculated for the elevated railway station structure under several conditions, in which the noise level is expressed with sound power radiation (Eq.3). The first case is for the comparison of the noise level varying with train speeds. From Figure 4 it can be found that with the increase of the train speed, the noise level increases with some oscillations. The lowest level takes place at the train speed of 60 km/h.

The second case, Figure 5, is for the comparison of the noise level between two structures where their laterals are two spans and three spans respectively. The results show that the noise level radiated from the structure with two spans is a little higher than that from the structure with three spans. It can be deduced that if the vertical stiffness of the structure is higher, the noise level will be lower. The peak of the noise level for both structures appears at the frequencies of 34 Hz and 42 Hz, and the maximum of noise is about 72 dB.

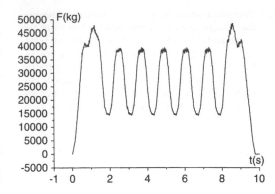

Figure 3.   Time history of bearing reaction.

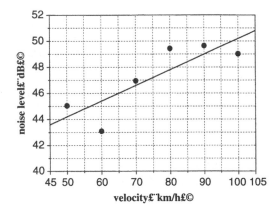

Figure 4.   Noise level varies with train speeds.

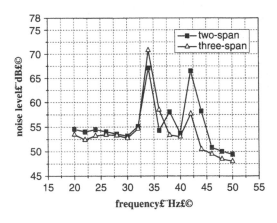

Figure 5.   Noise comparison for different spans.

The third case is for the comparison of the noise level between two kinds of supports, with and without rubber bearings on the top of the columns. Figure 6 shows that the noise level is reduced if rubber bearings are used

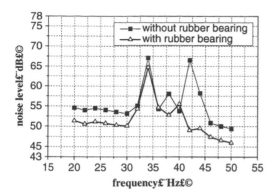

Figure 6. Noise comparison for different bearings.

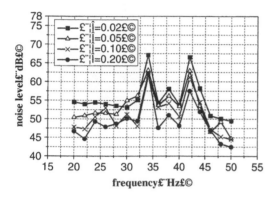

Figure 7. Noise comparison for different damping ratios.

and the decrease value of noise is from 2 dB to 10 dB. From the results it can be concluded that rubber bearings can be effectively used for controlling the noise radiated by elevated structures.

The final case is for the comparison of the noise level for different damping materials. Figure 7 shows that with the increase of damping ratio ξ of the structure the noise level decreases. If ξ increases to 0.05 from 0.02 (the damping ratio of the original structure), the reduction of noise level is from 2 to 5 dB; if ξ to 0.1, the reduction of noise level from 5 to 10 dB. It is clear that the structure with higher damping ratio will radiate noise in lower level.

6 CONCLUSIONS

Several conclusions can be drawn from the study described in this paper. (1) It is convenient to use the train-bridge model and the frame model to analyze structural vibrations of elevated railway stations. By using the conception of radiation ratio and SEA method, the relationship between vibration and noise can be established and the noise level expressed with sound power radiation can be estimated. The method provided in this paper can be used as reference for the plan, design and material selection of elevated station structures. (2) The peak values of bearing reactions are not sensitive to train speed if it is lower than 80 km/h. The noise level is reduced if rubber bearings are used and the decrease value of noise is from 2 dB to 10 dB. The rubber bearings can be effectively used for controlling the noise radiated by elevated structures. (3) The noise level increases with some oscillations, as the train speed increases. If the vertical stiffness of the structure is higher, the noise level will be lower. The structure with higher damping ratio will radiate noise in lower level.

REFERENCES

Schultz, T.J. 1979. Noise rating criteria for elevated rapid transit structures. Rep. No. UMTA-MA-06-0099-79-3, Washington D.C.
ISO2632/2. 1989. *Mechanical Vibration and Shock-Evaluation of Human Exposure to Whole Body Vibration-Part 2: Continuous and Shock Induced Vibration in Buildings (1–80 Hz).*
Hanson, C.E. 1983. "Noise control for rapid transit cars on elevated structures". *Journal of Sound and Vibration*, Vol.87, No.2, 285–294.
Van Ruiten, C.J.M. 1988. "Mechanism of squeal noise generated by trams". *Journal of Sound and Vibration*, Vol.120, No.2, 245–253.
Chua, K.H., Lo, K.W., & Koh, C.G. 1997. Performance of Urban Rail Transit System: Vibration and Noise Study. *Journal of Performance of Constructed Facilities*, No.5, 67–75.
Kraemer, S. 1984. Noise and vibration in buildings from underground railway lines. Ph.D. thesis, University of London.
Gao, R., Jiang, L.C. & Xia, H. 2001. "Types and Dynamic Analysis of Elevated Railway Station Structures", *International Symposium on Traffic Induced Vibration & Controls*. November 6–7, 2001. Beijing, China.
Stimpson, G.J., Sun, J.C. & Richards, E.J. 1986. "Predicting sound power radiation from built-up structures using SEA", *Journal of Sound and Vibration*, Vol.107, No.1, 107–120.
Xia, H. & Chen, Y.J. 1992. Dynamic interaction analysis of train-girder-pier system. *Journal of Civil Engineering*, Vol.25, No.2, 3–12. (in Chinese).

Environmental Vibrations – Takemiya (ed.)
© 2005 Taylor & Francis Group, London, ISBN 0 415 39035 4

Establishment of evaluation method of floor vibration caused by human motion and presentation of criterion on actual house floor

Y. Yokoyama
Tokyo Institute of Technology, Tokyo, Japan

H. Ono
Tohoku Institute of Technology, Sendai, Japan

ABSTRACT: The floor vibration caused by human daily motion such as walking is one of the most influential factor to make the residential environment discomfortable and uneasy. So the property of the vibration is usually very complicated, it is difficult to evaluate the vibration according to the previous studies or guidelines based on human sensation to the simple sinusoidal vibration. In this study, the original evaluation method was established.

At first, the load applied to the floor by human motion was measured. The load could be analyzed into two factors, the dynamic load by a foot landing, and the impacting load by a heel strike. Secondly, using sample floors of various vibration characteristics, the sensory test was carried out. From the relationship between the psychological scale constructed by the sensory test and the sample floor vibration, it was quantitatively confirmed that the human evaluation was influenced by two factors, the magnitude of the dynamic displacement, and the duration of the damping vibration generated by the impacting load. Based on this knowledge, the physical value *VI(2)* which consisted of the magnitude and the duration was proposed as the evaluation index.

Next, three kinds of apparatus were developed to measure *VI(2)*. These apparatus simulated the load by a heel strike or a foot landing, and vibration characteristics of human body. From the comparison with the vibration by human motion, it was made clear that the apparatus could reproduce the same vibration. Finally, the evaluation method was established which was to measure *VI(2)* using these apparatus and to evaluate the floor according to *VI(2)*.

Furthermore, in this study, the criterion to prevent the complaint from occupants was presented. The criterion, described by *VI(2)*, was proposed from the result of the evaluation according to the original method on actual house floors which had (had not) complained.

1 INTRODUCTION

Recently, under the influence of the use of light materials, the lowering of floor stiffness with long spans, it has become easy to occur the floor vibration caused by human motions with slight excitation force such as walking, and the problem by the effect adverse to the residential comfort has been increased. In order to present the evaluation method of the floor vibration, it is necessary to establish the reproducible measurement method of the floor vibration, and to investigate the evaluation index to verify the result of the measurement. This paper presents the evaluation index from a viewpoint of the residential comfort, and the measurement method using some apparatus simulating walking people characteristics. Furthermore, in this paper,

the criterion to prevent the complaint from occupants was presented.

2 FLOOR VIBRATION CAUSED BY WALKING

Figure 1 shows a typical example of the relationship between the time change of the vertical load applied to a floor by a foot while walking, and the condition of the foot when it contacts with the floor. Furthermore, Fig. 1 shows an example of the vertical floor vibration while walking as the displacement × time curve and the acceleration × time curve.

The load could be analyzed into two factors. One is a dynamic load which has two peaks, p_2 and p_3. The dynamic load is applied by a foot landing and stepping

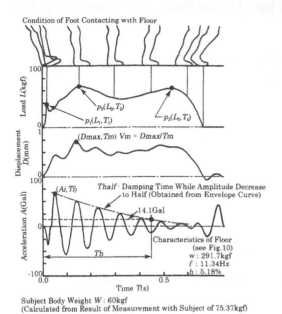

Subject Body Weight W: 60kgf
(Calculated from Result of Measurement with Subject of 75.37kgf)

Figure 1. Example of load applied to floor and floor vibration while walking.

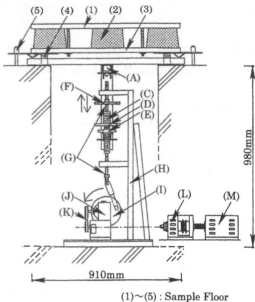

$(1)\sim(5)$: Sample Floor
$(A)\sim(M)$: Excitation Device

Figure 2. Outline of testing device for sensory test.

with the moving of a body weight. The other is an impacting load which has one peak, p_1. The impacting load is applied by a heel strike.

Then, the floor vibration while walking consists of two factors too. One is the dynamic behavior according to the dynamic load (see the displacement × time curve in Fig. 1). The other is the damping vibration at the natural frequency of the floor generated by the impacting load (see the acceleration × time curve in Fig. 1).

3 EVALUATION INDEX OF FLOOR VIBRATION

The author investigated two kinds of evaluation index, the index when the same person causes and perceives the vibration, and the index when a different person causes and perceives the vibration. In this paper, we explain the latter.

At first, we produced a testing device which included a sample floor and an excitation device. Fig. 2 shows the outline of the testing device. The sample floor consists of a variable weight panel (1)–(3), coil springs (4) and so on. The vibration system of single degree of freedom is constituted by the panel and springs, and vibration characteristics of the system (equivalent mass, natural frequency and damping ratio) can be controlled by changing weight of the panel and stiffness, damping characteristics of springs.

The excitation device consists of a motor (M), a brake and clutch (L), a reduction gear (J), a rotating

round plate (I), a vertical moving shaft (G), a coil spring to transmit the excitation force (C), a load cell (A) and so on. The device can apply the excitation force simulating the load while walking to the bottom of the panel. The excitation force when people stands or sits on the sample floor, is the same when anything is not placed on the sample floor. But damping characteristics of the vibration generated by the device are different by the effect of human body.

Secondly, using the testing device, a sensory test was conducted. Twenty-nine kinds of the vibration of the sample floor were selected as testing objectives. These objectives were actualized by the combination of twelve kinds of vibration characteristics of the sample floor and some kinds of excitation force. Twelve kinds of vibration characteristics were selected as the extent of characteristics of the sample floor enveloped the extent of actual floors.

The sensory test was conducted using the method of successive categories. The categories were shown in Fig. 4. Scores of testing objectives were obtained from the psychological responses of twenty panels standing or sitting on the sample floors. A scale constitution theory, based upon the assumption of normal distribution, was applied for the scaling of the psychological responses. According to the above process, two kinds of psychological scale, which indicated the degree of perception when the vibration was felt intermittently or continuously, were constructed. We called these evaluation scales.

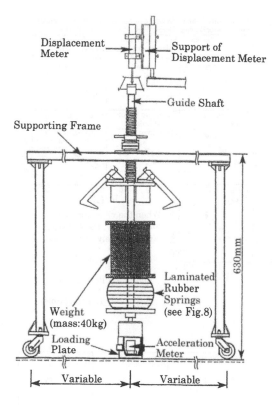

Displacement Meter

Support of Displacement Meter

Guide Shaft

Supporting Frame

630mm

Laminated Rubber Springs (see Fig.8)

Weight (mass:40kg)

Loading Plate

Acceleration Meter

Variable   Variable

Figure 3.   Outline of Vibration Measurement Apparatus.

Next, we developed "Vibration Measurement Apparatus" simulating vibration characteristics of human body. Fig. 3 shows the outline of the apparatus. The apparatus has the vibration system of single degree of freedom constituted by a weight and laminated rubber springs. Characteristics of the vibration system are as follows.

Equivalent Mass: 40 kg
Natural Frequency: 4 Hz
Damping Ratio: 20%

The apparatus can measure the floor vibration at the point on which the apparatus is placed. The vibration under the condition that people stands or sits on the point could be measured by a displacement meter set on the top of a guide shaft and an acceleration meter mounted in a loading plate (see Fig. 3).

Then, using the apparatus, twenty-nine kinds of the vibration employed the sensory test were measured, and the relationship between evaluation scales and the measured vibration was considered. As the result, it was confirmed that the evaluation of the vibration was influenced by two factors, the magnitude of the dynamic displacement as the greatness of the vibration at the beginning of the perception, and the duration of the damping vibration to the prescribed acceleration

amplitude as the length of the vibration. So, the physical value $VI(2)$ (*Vibration Index 2*) which consisted of the magnitude of the dynamic displacement and the duration of the damping vibration was proposed. $VI(2)$ was described as follows.

$$VI(2) = 0.2 \cdot \log(Dmax/Dref) + 0.5 \cdot \log(Vm/Vref) + \log(Th/Tref)$$

then $Dref = 1$ cm, $Vref = 1$ cm/s, $Tref = 1$ s.

$Dmax$ is the maximum displacement while a foot landing, and Vm is the deflection velocity in which $Dmax$ is divided by the deflection time Tm (see the displacement × time curve in Fig. 1). Th is the duration of the damping vibration generated by a heel strike to 14.1 Gal (see the acceleration × time curve in Fig. 1). Th could be obtained from the envelope curve of the acceleration × time curve illustrated by chain line in Fig. 1. According to "Meister Curve", 14.1 Gal is the value in which the vibration is clearly perceptible within the area of the natural frequency 3–30 Hz. If we treat when the same person causes and perceives the vibration, the value will be larger than 14.1 Gal.

Finally, from the relationship between evaluation scales and $VI(2)$, the evaluation index was investigated. Fig. 4 shows the evaluation index. From Fig. 4, it is confirmed that $VI(2)$ corresponds to evaluation scales thoroughly, and if we could measure $VI(2)$ reproducibly, we can evaluate the vibration quantitatively.

4   MEASUREMENT METHOD OF FLOOR VIBRATION

In order to establish the reproducible measurement method of $VI(2)$, we developed three kinds of apparatus, "Impacting Excitation Apparatus", "Dynamic Excitation Apparatus" and "Perception Apparatus".

Impacting Excitation Apparatus is the apparatus which can apply the impacting load simulating a heel strike including peak p_1. This apparatus is to measure Ai and Ti, the initial amplitude and the built-up time of the damping vibration (see Fig. 1).

Dynamic Excitation Apparatus is the apparatus which can apply the dynamic load simulating a foot landing including peak p_2. This apparatus is to measure $Dmax$ and Tm, and to calculate Vm.

Perception Apparatus is the apparatus which has the vibration system simulating vibration characteristics of walking human body after landing. This apparatus is to measure $Thalf$, the damping time while the amplitude of the damping vibration decreases to half (see Fig. 1). $Thalf$ can be measured by attaching this apparatus to a floor and striking the floor by a hammer.

Th can be calculated from Ai, Ti and $Thalf$.

Before the development of these apparatus, in order to obtain the extent and distribution of peak p_1 and p_2, the load applied to a floor while walking was measured

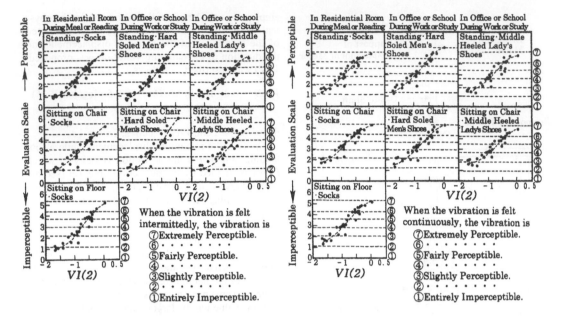

Figure 4. Evaluation index of floor vibration.

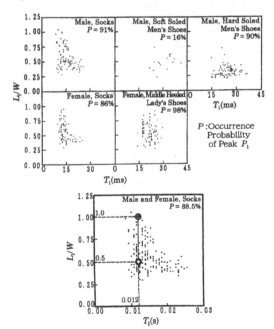

O : Impacting Excitation Apparatus (Mean Specification)
◉ : Impacting Excitation Apparatus (Maximum Specification)

Figure 5. Extent and distribution of peak p_1.

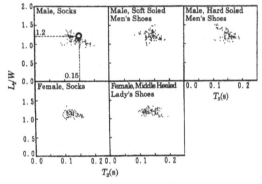

O : Dynamic Excitation Apparatus

Figure 6. Extent and distribution of peak p_2.

using four kinds of footwear and ten subjects of male and/or ten subjects of female.

Figure 5 shows the extent and distribution of peak p_1. Peak p_1 is distributed in comparative wide area. And when subjects wear some shoes, by the effect of the cushion of the sole, L_1/W becomes smaller, T_1 becomes longer and P, the occurrence probability of peak p_1 becomes smaller than socks according to the softness of the sole. From these results, peak p_1 is described as the impact of small mass of a foot and the floor. So peak p_1 is easy to be influenced by the speed and form of walking motions, and the softness of the sole of footwear.

On the other hand, Fig. 6 shows the extent and distribution of peak p_2. Peak p_2 is distributed in comparative small area. And L_2/W, T_2 are hardly changed by footwear. From these results, peak p_2 is described as the slow dropping of large mass of a whole body on

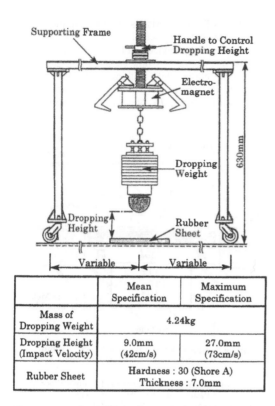

	Mean Specification	Maximum Specification
Mass of Dropping Weight	4.24kg	
Dropping Height (Impact Velocity)	9.0mm (42cm/s)	27.0mm (73cm/s)
Rubber Sheet	Hardness : 30 (Shore A) Thickness : 7.0mm	

Figure 7. Outline of Impacting Excitation Apparatus.

the soft absorbing system by joints of a leg and waist. So peak p_2 is difficult to be influenced by the difference of walking motions and footwear.

Based upon the results of the measurement, three kinds of apparatus were developed.

Figure 7 shows the outline of Impacting Excitation Apparatus. The apparatus consists of a dropping weight (mass: 4.24 kg), a rubber sheet and so on. The specification of the apparatus, mass of the dropping weight, dropping height (impact velocity), hardness and thickness of the rubber sheet, is shown in Fig. 7. Two kinds of specification were established by simulating mass of a foot, impact velocity at a heel strike and hardness of the skin of a heel. "Mean Specification" is to measure the mean value of Ai and Ti, and "Maximum Specification" is to measure the maximum value of Ai and the mean value of Ti. L_1/W ($W = 60$ kgf) and T_1 applied by the apparatus of these specifications are shown in Fig. 5. From Fig. 5, it is confirmed that L_1/W and T_1 by Mean Specification correspond to the mean area of L_1/W and T_1 by walking, and L_1/W and T_1 by Maximum Specification correspond to the maximum area of L_1/W and the mean area of T_1 by walking.

Figure 8 shows the outline of Dynamic Excitation Apparatus. Vibration Measurement Apparatus shown in Fig. 3 was applied for Dynamic Excitation Apparatus.

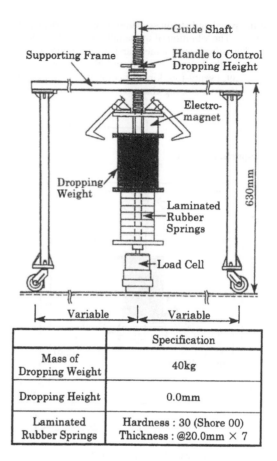

	Specification
Mass of Dropping Weight	40kg
Dropping Height	0.0mm
Laminated Rubber Springs	Hardness : 30 (Shore 00) Thickness : @20.0mm × 7

Figure 8. Outline of Dynamic Excitation Apparatus.

The apparatus consists of a dropping weight (mass: 40 kg), laminated rubber springs which are excellent both in absorbing and damping, and so on. The specification of the apparatus, mass of a dropping weight, the number and characteristics of rubber springs, were established by simulating mass of a whole body and the absorbing system by joints of a leg and waist. L_2/W ($W = 60$ kgf) and T_2 applied by the apparatus are shown in Fig. 6. From Fig. 6, it is confirmed that L_2/W and T_2 by the apparatus correspond to the mean area of L_2/W and T_2 by walking.

Figure 9 shows the outline of Perception Apparatus. After the consideration of all kinds, Perception Apparatus became the same with Vibration Measurement Apparatus shown in Fig. 3.

After the development, in order to verify the appropriateness of the apparatus, we produced a testing device shown in Fig. 10, and compared with the vibration caused by walking and the apparatus. A sample floor of the device consists of a variable weight panel, coil springs and so on, and vibration characteristics of the sample floor can be controlled by changing weight of the panel and stiffness, damping characteristics of springs. The vibration of the sample floor can be measured by a displacement meter and an acceleration meter set under the bottom of the sample floor. In this measurement, twenty-four kinds of combination of weight and stiffness, damping characteristics were selected as the extent of characteristics of the sample floor enveloped the extent of actual floors.

Using the testing device and five subjects, we measured the displacement × time curve and the acceleration × time curve while walking, and obtained the mean value of *Dmax*, *Vm*, *Thalf*, or the extent and distribution of *Ai*, *Ti*, *Th*. Next, using the apparatus,

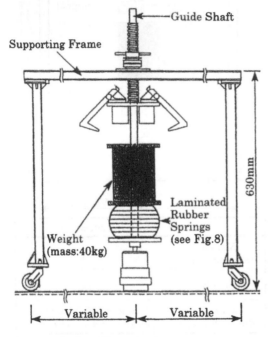

Figure 9.   Outline of Perception Apparatus.

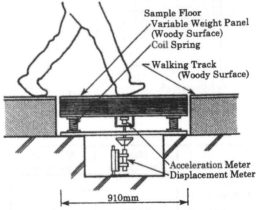

Width of Sample Floor and Walking Track : 910mm
Length of Walking Track : 1,800mm(One Side) × 2

Extent of Vibration Characteristics of Sample Floor
Weight of Panel   *w* : 30.7 ~ 1,073.5 kgf
Natural Frequency *f* : 5.18 ~ 32.6 Hz
Damping Ratio   *h* : 1.19 ~ 10.2 %

Figure 10.   Outline of testing device to verify appropriateness of apparatus.

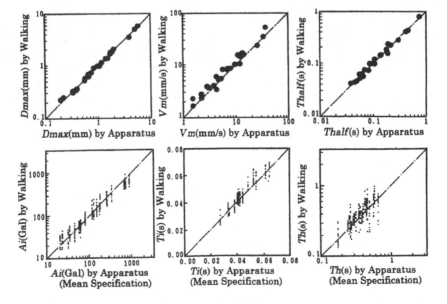

Figure 11.   Relationship between vibration caused by walking and apparatus.

580

Dmax, Vm, Thalf, Ai, Ti, Th were measured and calculated.

Figure 11 shows the relationship between the vibration caused by walking and the apparatus. From Fig. 11, it is confirmed that *Dmax, Vm, Thalf* measured by the apparatus correspond to the mean value while walking thoroughly, and *Ai, Ti, Th* measured by the apparatus of Mean Specification correspond to the mean area while walking thoroughly. The similar results were obtained about the apparatus of Maximum Specification.

From the above consideration, the appropriateness of the apparatus is verified. Using the apparatus, we can measure *VI(2)* reproducibly, and comparing *VI(2)* with the evaluation index shown in Fig. 4, we can evaluate the vibration caused by walking quantitatively.

5   EVALUATION METHOD OF FLOOR VIBRATION

In this paper, the evaluation method which consisted of the measurement method of *VI(2)* using some apparatus, and the evaluation index from a viewpoint of the residential comfort, was presented. The evaluation method of the floor vibration, when a different person walks and perceives the vibration, is as follows.

1. Vibration Measurement Apparatus shown in Fig. 3 is placed on the perception point of the floor.
2. Impacting Excitation Apparatus shown in Fig. 7 is placed on the excitation point of the floor, and *Ai, Ti* at the perception point are measured.
3. Dynamic Excitation Apparatus shown in Fig. 8 is placed on the excitation point of the floor, and *Dmax, Vm* at the perception point are measured and calculated.

4. Perception Apparatus shown in Fig. 9 is placed on the excitation point of the floor, and *Thalf* at the perception point is measured.
5. *Th* is calculated from *Ai, Ti* and *Thalf*.
6. *VI(2)* is calculated from *Dmax, Vm* and *Th*.
7. *VI(2)* is compared with the evaluation index shown in Fig. 4.

6   CRITERION ON ACTUAL HOUSE FLOOR

Applying the evaluation method established in this study, the criterion to prevent the complaint from occupants was investigated.

At first, in order to verify the applicability of the apparatus to actual house floors, the floor vibration caused by human walking and the apparatus were compared. Twenty actual wooden and light weight steel-frame floors were used in this experiment. The displacement × time curve and the acceleration × time curve while walking were measured using subjects from three to five. Fig. 12 shows a typical example of the displacement × time curve and the acceleration × time curve. In this figure, the curve when the subject steps only on the measurement point is shown too. From Fig. 12, it is comprehended that *Dmax, Vm, Th* by only one step is the same with those when the subject steps on the measurement point while continuous walking.

From the displacement × time curve and the acceleration × time curve while continuous walking, we picked up *Dmax, Vm, Th* by the way described in Fig. 12, and obtained the mean value of *Dmax, Vm,* or the extent and distribution of *Th, VI(2)*. Then, setting the apparatus on the measurement point, *Dmax, Vm,* Th, *VI(2)* were measured and calculated. Fig. 13 shows the relationship between the vibration caused by walking and the apparatus. From Fig. 13, it is confirmed

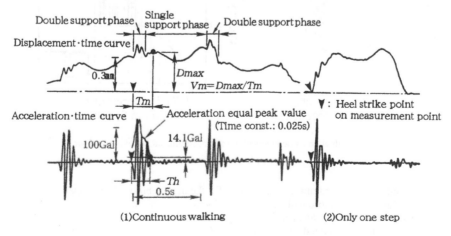

Figure 12.   Example of actual house floor vibration while walking.

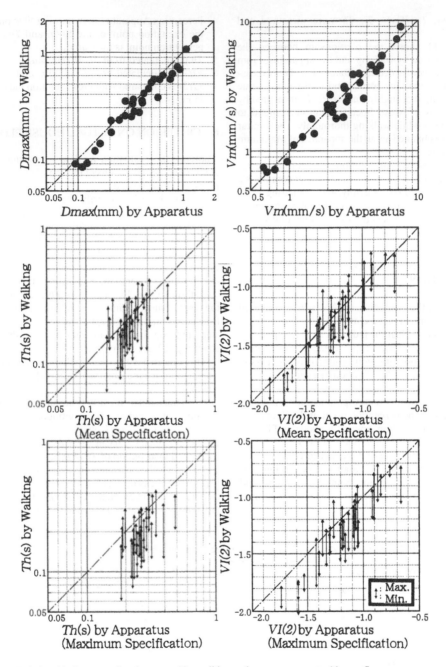

Figure 13.   Relationship between vibration caused by walking and apparatus on actual house floor.

that *Dmax*, *Vm* measured by the apparatus correspond to the mean value while walking thoroughly, and *Th*, *VI(2)* measured by the apparatus of Mean Specification or Maximum Specification correspond to the mean area or the maximum area while walking thoroughly.

So the applicability of the apparatus to actual house floors was verified.

Next, in order to propose the criterion described by *VI(2)*, the relationship between the vibration measured by the apparatus and the data on the complaint

582

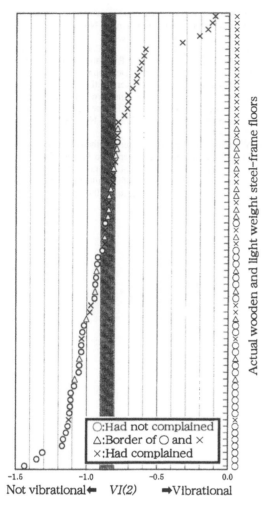

Figure 14. Relationship between *VI(2)* and complaint.

with no complaint. The symbol △ means the floor which had not complained, but the occupant has reported that the vibration was occurred. From Fig. 14, we can obtain the criterion as follows.

$-0.8 \leqq VI(2)$: the area of high probability to occur the complaint

$-0.9 \leqq VI(2) < -0.8$: the border area

$VI(2) < -0.9$: the area of low probability to occur the complaint

However, in Fig. 14, there are × or △ points in the area of $VI(2) < -0.9$. About these floors, we had more detailed investigation, and were suggested that the clang sound of furniture and tableware, or the shaking motion of plant and visual information monitor influenced to the occupant's sensitivity. The author presently studies on the evaluation method of the floor vibration which was influenced by acoustic or visual factors.

7 CONCLUSION

This paper presented the evaluation method of floor vibration caused by human walking from a viewpoint of the residential comfort. At first, from the relationship between the floor vibration and human reaction, the quantitative value *VI(2)* was proposed, which could be calculated from the displacement × time curve and the acceleration × time curve while walking, and corresponded to human evaluation. Next, by simulating the load applied to a floor by a foot, and simulating vibration characteristics of human body, three kinds of apparatus were developed. By using these apparatus, *VI(2)* which was equal to that while walking, could be measured reproducibly. Then, based on the evaluation index described by *VI(2)*, and the measurement method of *VI(2)*, the evaluation method was established. Furthermore, in order to verify the applicability of the result of the experimental study at the laboratory to the actual house, the relationship between *VI(2)* and the data on the complaint was investigated. From this investigation, criterion to prevent the complaint from occupants was presented.

The evaluation method and the criterion were referred by "Guidelines for the evaluation of habitability to building vibration" (Architectural Institute of Japan, 2004). We hope that the evaluation method and the criterion will be applied to the design, the selection and the development of reasonable floors.

REFERENCES

This paper is a summary of the result of our research published in the following papers.
1. Y. Yokoyama, H. Ono: Indicating Methods of Floor Vibrations Caused by Human Activities Based on Human

was investigated. Sixty-eight actual wooden and light weight steel-frame floors were selected as the object of this investigation. From one to three excitation points, and from four to twelve perception points were settled on each object floor. So *VI(2)* was measured and calculated by using the apparatus. From the relationship between *VI(2)* and the data on the complaint, it is comprehended that the most correlative *VI(2)* is *VI(2)* which is measured under the most vibrational condition on each object floor. In concrete terms, the excitation point is settled on the center of the span of the beam, the perception point is settled near the excitation point, and the specification of Impacting Excitation Apparatus is Maximum Specification.

Figure 14 shows the relationship between *VI(2)* and the data on the complaint. The symbol × means the floor which had complained, and ○ means the floor

Sensations, In the case of difference the vibration cause and the perceiver, Journal of Structural and Construction Engineering (Transactions of Architectural Institute of Japan), No. 390, pp. 1–9, Aug., 1988

2. Y. Yokoyama, H. Ono: Presentation of the Evaluation Method for Floor Vibration When a Different Person Causes and Perceives the Vibration, Study on method for evaluating vibrations of building's floors caused by human activities from a viewpoint of comfort (Part 2), Journal of Structural and Construction Engineering (Transactions of Architectural Institute of Japan), No. 418, pp. 1–8, Dec., 1990

3. Y. Yokoyama: Study on Excitation Apparatus and Perception Apparatus for Evaluating Floor Vibration Caused by Human Walking, Establishment of "Dynamic Excitation Apparatus" and "Perception Apparatus", Journal of Structural and Construction Engineering (Transactions of Architectural Institute of Japan), No. 466, pp. 21–29, Dec., 1994

4. Y. Yokoyama, M. Sato: Study on Excitation Apparatus and Perception Apparatus for Evaluating Floor Vibration Caused by Human Walking, Development of "Impactive Excitation Apparatus" and verification of appropriateness of method to compute duration of vibration, Journal of Structural and Construction Engineering (Transactions of Architectural Institute of Japan), No. 476, pp. 21–30, Oct., 1995

5. Y. Yokoyama: The Measurement Method of Actual House Floor Vibration and the Criterion to Prevent Complaint from Occupants, Journal of Structural and Construction Engineering (Transactions of Architectural Institute of Japan), No. 546, pp. 17–24, Aug., 2001

Environmental Vibrations – Takemiya (ed.)
© 2005 Taylor & Francis Group, London, ISBN 0 415 39035 4

Committee activities on environmental vibration in Architectural Institute of Japan

T. Hamamoto
Professor, Dept. of Architecture, Musashi Institute of Technology, Tokyo, Japan

T. Ishikawa
Professor, Dept. of Housing and Architecture, Japan Women's University, Tokyo, Japan

T. Goto
Professor, Dept. of Architecture, Hosei University, Tokyo, Japan

ABSTRACT: Committee activities on environmental vibration in Architectural Institute of Japan (AIJ) are reviewed on the basis of all proceedings of annual symposia on environmental vibration in buildings and related publications or reports in AIJ. Committee activities over two decades are summarized on the following points: the objective and scope of the committee, the sources, transmission path and receivers in environmental vibration, measurement procedures, vibration prediction methods, performance evaluation, countermeasures such as vibration control and mitigation, and a role of environmental vibration in performance-based design. Emphasis is placed on AIJ guidelines on environmental vibration in buildings. A future perspective of committee activities is also presented.

1 INTRODUCTION

In the basic law for environmental pollution that was enacted in 1967, vibration nuisance was recognized as one of seven major types of pollution. To regulate industrial vibration sources that cause vibration nuisance, vibration regulation law was enacted in 1976. The regulation standards for industrial vibration were specified at the site boundary of vibration sources. The disagreement between ground vibration to be legislated and building vibration to be not legislated has left an important issue to be solved in Architectural Institute of Japan (AIJ).

Committee activities on environmental vibration have continued since 1982 in AIJ. In the process of establishing the committee, the term "environmental vibration" was used first as far as we know. The environmental vibration was defined as ordinary vibration in both natural and artificial environments surrounding us. Unusual vibrations such as major earthquakes are not included within the scope. Over two decades, our committee has contributed to assure comfort and healthy vibration environments in buildings through a wide spectrum of committee activities.

So far, we have continuously opened to the public a number of fruitful outcomes of committee activities.

Among them, the first edition of "Guidelines for the evaluation of habitability to building vibration" was published in 1991 (AIJ, 1991). The second edition was recently published in 2004 (AIJ, 2004). Guidelines have been widely used in the design of various types of buildings. Another important role of the committee is submission of expert opinions to ISO 108 committee. The committee has significantly contributed to the first edition of ISO 2631-2 (building vibration).

Annual symposia on environmental vibration have been regularly held since the establishment of the committee. Each symposium has been organized under a number of attractive topics related to environmental vibration. The history of committee activities is strongly related to that of annual symposia. The central themes and contents of each symposium are shown in Table A-1 in Appendix. The table includes two pre-symposia (PS) before establishing the committee, a special symposium (SS) in 1987 and a panel discussion (PD) in the AIJ annual convention in 2004 in addition to regular symposia (RS). Roughly speaking, committee activities may be divided into three phases. A source-oriented approach was taken in the first phase, a subject-oriented approach in the second phase and a design-oriented approach in the current phase. In this paper, committee activities on environmental vibration in AIJ are overviewed through reviewing proceedings and related documents of all symposia that were held under committee activities.

2 OBJECTIVE

The objective of committee activities on environmental vibration in buildings in AIJ is to eliminate a number of unsatisfactory vibrations and, consequently, to assure comfort and healthy vibration environments in all kinds of building structure, such as residential houses, hotels, offices, department stores, schools, hospitals, public facilities, high-rise buildings, large-span arenas, observation towers and floating marine facilities. To achieve the objective, it is required to predict possible vibration at the design stage of new buildings and to assess the acceptability of vibration in existing buildings.

3 SCOPE

The scope of committee activities may be summarized using an expertise domain matrix as shown in Table 1. The column of the matrix contains vibration sources: industrial machinery, road and railway traffic, airplane and ship traffic, equipment machinery, human activity and miscellaneous sources. The row of the matrix consists of subject areas such as measurement procedure, source characteristics, transmission path, vibration prediction methods and performance evaluation.

To cover all the elements of the matrix, the committee has been organized by a wide variety of experts from universities, public agencies, design offices, construction and engineering companies. The environmental vibration can be recognized as an interdisciplinary field. In relation to vibration sources and subject areas, the required expertise contains environmental, structural, material, construction, mechanical and electrical engineering and, moreover, psychology, physiology and sociology.

The schematic diagram of vibration effects and countermeasures related to environmental vibration in buildings is shown in Figure 1. A major concern of environmental vibration is whole-body vibration. Sound effects can be induced by vibration and acoustic excitation and may emphasize the presence of vibration. Typical examples are the rattling of windows and ornaments. Visual effects can be observed in case of low-frequency vibration (<5 Hz) and may emphasize the presence of vibration. Typical examples are the swinging of suspended lights and the sloshing of water tank. Structure-borne noise is related to environmental vibration and may become audible as reradiated noise. Low-frequency airborne noise is also related to environmental vibration. Typical sources are elevated highway and railway bridges and building air-handling systems.

In addition to human response, satisfactory functioning of vibration-sensitive machinery that is located within the building is another important topic in the field of environmental vibration.

4 SOURCES

Committee activities during the first decade had mainly directed to a source-oriented approach. For various types of vibration source, measurement procedure, source characteristics, transmission path, vibration prediction methods and performance evaluation were investigated individually. The approach corresponds to the horizontal string (row) of the expertise domain matrix as shown in Table 1. The review of committee activities from a viewpoint of this approach have been published as "Current status and codes of evaluation of environmental vibration on habitability performance" in 2000 (AIJ, 2000).

Vibration sources produce the dynamic forces or actions. The sources can be inside or outside the building as shown in Figure 2. Vibration sources inside

Table 1. Expertise domain matrix for committee activities.

Vibration sources	Subject areas				
	0 Measurement procedure	1 Source characteristics	2 Transmission path	3 Vibration prediction	4 Performance evaluation
A Industrial machinery	A-0	A-1	A-2	A-3	A-4
B Road/Rail traffic	B-0	B-1	B-2	B-3	B-4
C Airplane/Ship traffic	C-0	C-1	C-2	C-3	C-4
D Equipment machinery	D-0	D-1	D-2	D-3	D-4
E Human activity	E-0	E-1	E-2	E-3	E-4
F Natural force	F-0	F-1	F-2	F-3	F-4
G Miscellaneous sources	G-0	G-1	G-2	G-3	G-4

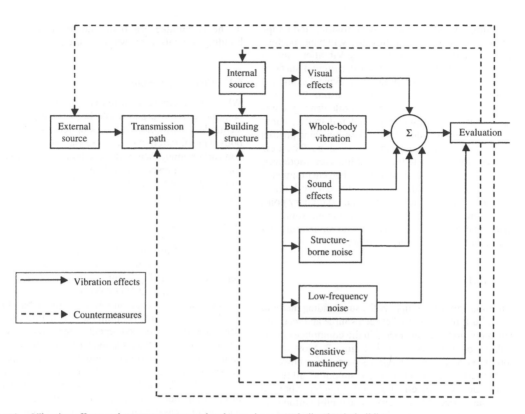

Figure 1. Vibration effects and countermeasures related to environmental vibration in buildings.

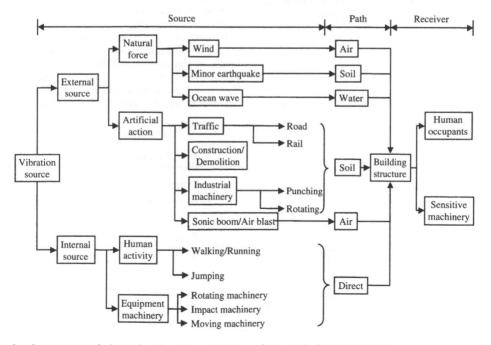

Figure 2. Sources, transmission path and receivers related to environmental vibration in buildings.

a building include human activity, rotating and reciprocating machinery, impact machinery, moving machinery, and repair or renovation activity. Vibration sources outside a building can occur on the ground surface, underground, in the air, or in water. Typical examples are construction, mining or quarry blasting, construction or demolition activity such as pile driving, compaction and excavation, road and rail traffic, sonic boom or air blast, wind or water flow, punching presses or other machinery, and impact of ships. The new sources of vibration acting on buildings are, moreover, emerging with the development of urban environments.

The forces or actions at the source vary both in time and in space. The sources can be stationary point source, line source, area source or moving source in space and may be continuous source, intermittent source or impulsive source in time.

5 PATH

Transmission path comprises the surrounding medium or the building structure between source and receiver. The medium to be considered in the committee can be soil, air or water as shown in Figure 2. Most of the committee activities on transmission path are, however, concerned with soil medium that supports buildings.

The path has the effect of modifying the vibration from the source to the receiver. The modification may be due to discontinuities in the medium, attenuation due to geometric spreading and material damping, and possible amplification or attenuation in certain frequency ranges.

6 RECEIVERS

Recent modern buildings can be regarded as man-machine systems. Therefore, receivers to be considered in the committee are both human occupants and building contents as shown in Figure 2.

6.1 *Human occupants*

Whole-body vibration is the central concern in the committee. Exposure to whole-body vibration causes a complex distribution of forces or actions within the body. The vibration may cause sensations such as discomfort or annoyance, influence human performance capability, or present a health and safety risk such as pathological damage or physiological change. Vibration effects on human occupants may be grouped into health, comfort, perception and motion thickness as stated in ISO 2631-1 (ISO, 1997).

The magnitudes and directions of excitation and the posture of human body are important factors of whole-body vibration. The human body axes are defined for seated position, standing position and recumbent position in ISO 2631-1.

The basic human response to vibration in buildings is adverse comment. As pointed out in ISO 2631-2 (ISO, 2003), adverse comments regarding building vibration in residential situations may arise from occupants when the vibration magnitudes are only slightly in excess of perception level. The degree of annoyance and complaint is not always explained by the magnitude of vibration alone. Complaints often arise due to secondary effects as shown in Figure 3. In this case, claims may arise even if whole-body vibration is lower than the perception level. This problem is an active research field in the committee.

Usually, satisfactory magnitudes in residential situations are not determined by factors such as short-term health hazard except in the case of sleepless night. Exceptions may be motion sickness in high-rise buildings against strong winds and in floating marine facilities against ocean waves. These problems are also

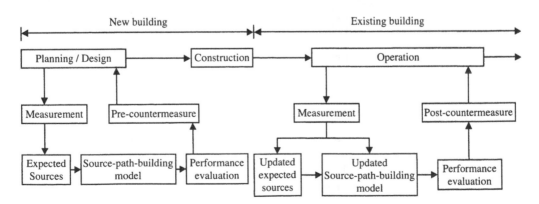

Figure 3.   Roles of vibration measurements and counter measures related to environmental vibration in buildings.

active research fields in the committee. Working efficiency is another important factor in working areas such as offices or factories. Even higher vibration magnitudes can be tolerated for temporary disturbances and infrequent events of short duration as in the case of construction projects.

6.2 *Building contents*

Buildings contain a variety of items such as furniture, machinery, computers and highly sophisticated instruments. The vibration susceptibility of the building contents varies greatly. Most common articles in buildings, such as furniture and operating machinery, have a relatively high tolerance to vibration. Most computers are only moderately sensitive to vibrations. However, highly sensitive to vibrations are, for example, optical instruments, crystal growing and computer chip manufacturing facilities. These sensitive machinery demands vibration-free environments. A number of research activities on this topic have been reported in the committee.

7 MEASUREMENT

The second phase of committee activities can be regarded as a subject-oriented approach. The approach corresponds to the vertical string (column) of the expertise domain matrix.

The roles of vibration measurement are shown in Figure 3 at the design stage of new buildings and at the renovation stage of existing buildings. Only ground vibrations are measured at design stage, whereas both building and soil vibrations are measured at the renovation stage.

At this moment, a measurement procedure has not been standardized yet in the committee. The review of measurement procedures related to environmental vibration has been published as "Measurement technical manual on environmental vibration and structure-borne sound" in 1999 (AIJ, 1999). From a point of view of vibration measurement, environmental vibration was grouped into three categories: whole-body vibration, vibration nuisance and structure-borne noise in the book. Measurement methods to be used conventionally are summarized in what follows.

A vibration measuring system consists of sensors, signal-conditioner and recording instrument. The primary quantity of vibration magnitudes is acceleration. Velocity measurement may be made and translated into acceleration in case of very low frequency and low vibration magnitudes. The measuring apparatus should be selected in consideration of the vibration parameter to be measured and the expected range of parameter.

In measuring vibrations in buildings, measurement locations and directions are chosen to be compatible with the geometry of the structure and excitation source. The directions of vibration are related to the structure rather than to the human body. When horizontal vibrations are measured on the existing building, measuring points are located on or near the walls or columns. When vertical vibrations are measured on floors, measuring points are usually located on the center of the floor slabs. Multiple measuring points are often used vertically in multistory buildings and horizontally in large-plan buildings. When forces acting on a building to be constructed are predicted, measuring points are located on the ground at the planning location of the building.

For human occupants, measured vibrations are frequency-weighted by vibration sensitive characteristics in ISO 2631-1. The frequency-weighted acceleration is used to express the vibration magnitude. It is recommended that the combined frequency weighting be used irrespective of the measurement direction. However, it is also possible to apply weightings for the vertical and horizontal directions individually. The vibration level meter can be used to measure the frequency-weighted acceleration.

8 PREDICTION

A variety of prediction methods are used in committee activities. The analysis of response requires a calculation model that includes the characteristics of the source, the transmission path and building structure. The type and complexity of the calculation model depends on the dynamical behavior to be reproduced and on the accuracy required in the prediction of the vibration response.

The simplifying assumptions are introduced to establish and solve the calculation model. It is usually assumed that building structures and the surrounding media respond linearly to the applied loads. This means that the structure and the surrounding media are not subject to significant non-linear effects. The analysis may concern whole structures or may be a part of the structures.

The exact characteristics of the vibration source, transmission path and receiver may not always be well defined. Empirical methods are employed when the complete analytical solution of a problem may not be practical. Empirical methods may be used when they have been derived from a large number of experimental or analytical results (Shioda, 1986).

When the vibration source does not move in space, many analysis methods can be employed to solve vibration problems. Common solution techniques involve the derivation of an equivalent single-degree-of-freedom system or modal analysis for both continuous and discrete systems.

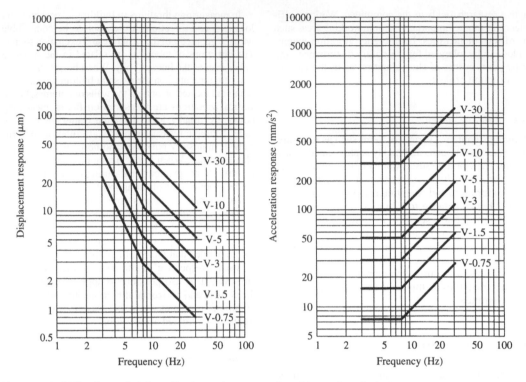

Figure 4. Guideline for vertical floor vibration against man- and machine-made excitations.

When the action varies both in time and space, vibration problems become very difficult to solve analytically. Examples are a vehicle moving along a street and a person walking across a floor. Because of the complexity of these problems many of them have been treated by empirical methods.

9   EVALUATION

To evaluate vibration environments in buildings, criteria for human occupants and sensitive machinery should be specified on the floor for both horizontal and vertical vibrations. The criteria may be described in terms of vibration amplitudes or vibration levels. The criteria are used to design new buildings and to assess existing buildings. Moreover, the criteria may be referenced in dealing with legal issues on unsatisfactory vibration in buildings.

9.1   Criteria for human occupants

Criteria for human occupants may be grouped into sensitive occupancies, regular occupancies and active occupancies. An example of sensitive occupancies is hospital operating rooms. Regular occupancies include

offices and residential areas. Active occupancies contain assembly areas or places of heavy industrial work. Floors in arenas, gymnasia and stadiums subjected to activities such as dancing, running, jumping and coordinated spectator movements belong to active occupancies. Guidelines on environmental vibration in AIJ pay attention to regular occupancies mainly.

9.1.1   The first edition of AIJ guidelines
The first edition of "Guidelines for the evaluation of habitability to building vibration" was published in 1991. The guidelines include vertical floor vibrations against man- and machine-made excitations and horizontal floor vibrations against wind excitations. Residential houses and offices are considered as occupancies. The evaluation criteria are described by maximum acceleration or displacement in combination to frequency. This is different from ISO standards that use root mean squared quantities.

Figure 4 shows the guidelines for vertical floor vibrations against man- and machine-made excitations. Although the guidelines of both response acceleration and velocity are given, either of them can be used. Table 2 explains the notation to be used in the figure. The number after V (vertical) in the figure corresponds

Table 2. Classification of performance evaluation for occupancies in Figure 4.

Occupancy	Vibration type I			Type II	Type III
	Rank I	Rank II	Rank III	Rank III	Rank III
House Living room & Bed room	V-0.75	V-1.5	V-3	V-5	V-10
Office Meeting room	V-1.5	V-3	V-5	V-10	V-30
Working room	V-3	V-5	≒V-5	≒V-10	≒V-30

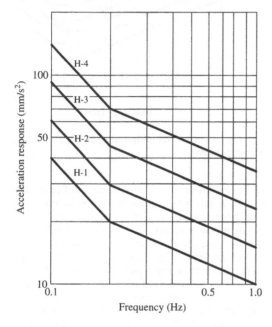

Figure 5. Guideline for horizontal floor vibration against wind excitations.

Table 3. Classification of performance evaluation for occupancies in Figure 5.

Occupancy	Rank I	Rank II	Rank III
House	H-1	H-2	H-3
Office	H-2	H-3	H-4

the notation to be used in the figure. The degree of habitability performance is divided into three ranks. Rank II is recommended in an average sense.

Horizontal accelerations of building with one-year return period are applied to the evaluation of habitability. The basic evaluation curve is specified with peak accelerations at the first natural frequencies in the principal structural directions of the building (normally, along-wind and cross-wind) and in torsion, and it gives for office and residence in relation to the average threshold of perception of horizontal motion by humans.

Response prediction should follow any one of the three methods. Method A: the response is calculated against possible wind force. Method B: the response is measured using wind tunnel test. Method C: the response is measured under real wind force.

to the root mean squared acceleration less than 8 Hz for each curve.

Floor vibration is classified into three types. Type 1 is a continuous or intermittent vibration. Type 2 is an impulsive vibration when the damping ratio is less than 3%. Type 3 is an impulsive vibration when the damping ratio is 3 to 6%.

The degree of habitability performance is divided into three ranks. Rank I is the level that response should be less than the corresponding curve. Rank II is the level that can be recommended in an average sense. Rank III is the level that response should not exceed the corresponding curve.

Response prediction should follow any one of the three methods. Method A: the response is calculated against possible excitations. Method B: the response is measured using vibration experiment. Method C: the response is measured under real excitations.

Figure 5 shows the guidelines for horizontal floor vibrations against wind excitations. Table 3 explains

9.1.2 *The second edition of AIJ guidelines*

The second edition of "Guidelines for the evaluation of habitability to building vibration" was published in 2004. In addition to floor vibrations and wind-induced motions, vertical and horizontal floor vibrations induced by road and rail traffic are included. It is noted that all guidelines are described by perception probabilities in a unified fashion. The number after V or H (horizontal) corresponds to the perception probability for each curve.

Figure 6 shows the guidelines for vertical floor vibrations against man- and machine-made excitations. Only the guidelines of response acceleration are given, which is different from the first edition. Response prediction methods are the same as the first edition.

Figure 7 shows the guidelines for horizontal floor vibrations induced by road and rail traffic. The guidelines for vertical floor vibrations are omitted because they are the same as those against man- and machine-made excitations.

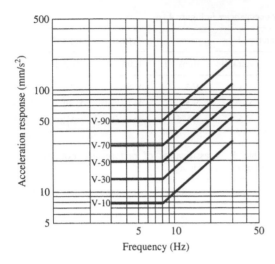

Figure 6. Guideline for vertical floor vibration against man- and machine-made excitations.

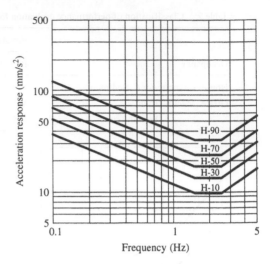

Figure 8. Guideline for horizontal floor vibration against wind excitations.

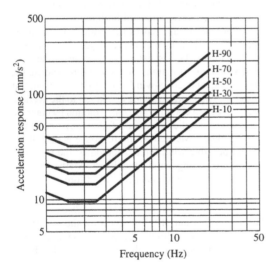

Figure 7. Guideline for horizontal floor vibration against road and rail traffic.

Figure 8 shows the guidelines for horizontal floor vibrations against wind excitations.

9.2 *Criteria for building contents*

Criteria for building contents are specified to assure satisfactory functioning of sensitive instruments or certain manufacturing processes that are located within the building.

In general, the vibration criteria for satisfactory performance need to be provided by the manufacturer of the equipment. Many sensitive pieces of equipment require a relatively vibration-free environment for proper functioning and performance. To achieve such an environment requires careful planning in building site location, choice of building type and materials, structural design, building layout, location of the instruments within the building, and operation of the building in terms of traffic flow of occupants and the location and operation of other machinery and vehicles within and outside the building. Vibration monitoring program is an essential component of such a process.

10 COUNTERMEASURES

Building vibration problems have been treated in both structural and environmental engineering committees in AIJ. Structural engineering committee is mainly concerned with a safety problem, whereas environmental engineering committee with a serviceability problem. To keep comfort and healthy vibration environments as a serviceability problem, however, a number of countermeasures have been proposed and developed in structural engineering field (Kushida, 1997).

These countermeasures may be divided into pre-countermeasures for newly constructed buildings and post-countermeasures for existing buildings as shown in Figure 3. In either case, countermeasures involve the change in vibration source characteristics, modification of transmission path, modification of building structures, changes or restriction of use and occupancy.

Especially, vibration control becomes an important topic in the committee. Vibration control can be achieved using passive or active control device. Vibration control concerns the monitoring of vibration effects at the source, in the transmission path, or at the receiver.

11 DESIGN

The current committee activities deal with all the elements of the expertise domain matrix totally. This may be called a design-oriented approach. Different from major earthquakes that are very rare events, environmental vibration is a daily phenomenon especially in the urban regions. The committee recognizes that occupants have their strong concern on the vibration performance because they are always exposed to environmental vibration.

In the context of a performance-based design, the criteria on environmental vibration in buildings should be characterized by accountability, disclosure, consent and liability. Accountability is achieved by explaining the vibration criteria in such a way that the client can easily understand it. Disclosure is achieved by opening all the information on the criteria to the client. Consent is achieved by establishing the agreement between designer and client on the criteria to be used at the design stage. Liability is achieved by taking responsibility of assuring the performance realization after completion.

The vibration effects represent a serviceability limit state in the structural design of buildings (ISO/CD 10137, 2004). The limit state is described by constraints, generally consisting of vibration amplitude in combination with frequency. AIJ guidelines can be used effectively within the framework of performance-based design.

12 PERSPECTIVE

With rapid development of urbanization, problems on environmental vibrations have been considerably complicated. Measurement, evaluation, prediction and countermeasure of environmental vibrations become essential for a group of buildings rather than for individual buildings. The reason may be due to diversification of vibration sources, intermingling of transmission path, and increase in the number of vibration-sensitive buildings, such as high-rise and large-span structures. Economic use of high-strength and lightweight materials has also resulted in a trend towards more dynamically responsive structure.

To deal with these problems, the committee has recently initiated to introduce innovative technologies, such as wireless sensor network, remote sensing, global positioning system (GPS) and geographic information system (GIS) in the field of environmental vibration.

REFERENCES

AIJ. 1991. Guidelines for the evaluation of habitability to building vibration, 1st ed., AIJ (*in Japanese*)

AIJ. 1999. Measurement technical manual on environmental vibration and structure-borne sound, Ohmsha (*in Japanese*)

AIJ. 2000. Current status and codes of evaluation of environmental vibration on habitability performance, AIJ (*in Japanese*)

AIJ. 2004. Guidelines for the evaluation of habitability to building vibration, 2nd ed., AIJ (*in Japanese*)

ISO 2631-1. 1997. Mechanical vibration and shock – Evaluation of human exposure to whole-body vibration – Part 1: General requirements, ISO

ISO 2631-2. 2003. Mechanical vibration and shock – Evaluation of human exposure to whole-body vibration – Part 2: Vibrations in buildings (1 Hz to 80 Hz), 2nd ed., ISO

ISO/CD 10137. 2004. Bases for design of structures – Serviceability of buildings and pedestrian walkways against vibration, ISO

Kushida, H. 1997. Engineering of environmental vibration, Rikotosho (*in Japanese*)

Shioda, M. 1986. Prediction methods of vibration nuisance, Inoueshoin (*in Japanese*)

APPENDIX

The main themes and contents of all symposia on environmental vibration in buildings that were held under committee activities in AIJ are summarized in Table A-1. In the first column, PS, RS, SS and PD represent the pre-symposium before establishing the committee, regularly organized annual symposium, specially organized symposium and the panel discussion that was held during the annual AIJ convention, respectively. The fourth column corresponds to the location in the expertise domain matrix as shown in Table 1.

Table A-1. Symposia held under committee activities on environmental vibration in buildings in AIJ.

Symposium	Year	Central theme	Contents	Elements of the matrix
1st PS	1981	As interdisciplinary field	Definition of environmental vibration, Effect on sleep, Motion sickness, Equipment vibration, Earthquake engineering, Habitability performance, Architectural planning	Columns 0 and 4
2nd PS	1982	Industrial machinery	Measurement procedure, External source, Source to soil, Soil to building, ISO standards	Row A
1st RS	1983	Road and rail traffic	Road traffic, Soil vibration, Prediction method, the Shinkansen, Vibration mitigation, Subway traffic, Building vibration	Row B
2nd RS	1984	Construction/Demolition	Vibration nuisance, Foundation construction, Demolition activity, Vibration risk, Vibration mitigation, Noise and vibration	Row A
3rd RS	1985	Performance evaluation	Vibration regulation law, Vibration nuisance, Floor vibration, Measurement procedure, ISO standards, Foreign codes	Column 4
4th RS	1986	Floor vibration	Gymnasium, Offshore structure, Sensitive machinery, Manufacturing facilities	Columns 0 and 4
5th RS	1987	Equipment machinery	Source characteristics, Real measurement, Structure-borne noise, Floating floor, Floor vibration, Vibration mitigation	Row D
SS	1987	Performance evaluation	ISO standards, Foreign codes, Vibration regulation law, Vibration nuisance, Measurement procedure, Floor vibration	Column 4
6th RS	1988	Role of structural engineering	Environmental engineering, Structural engineering, Structural design, ISO standards, Seismic isolation, Earthquake intensity	Columns 3 and 4
7th RS	1989	Measurement procedure	JIS standards, Whole-body vibration, Structure-borne noise, Natural force, Vibration sensor, Software tool, FFT analyser	Column 0
8th RS	1990	Explosion/Soil vibration	Tunnnel excavation, Explosive vibration, Vibration risk, Vibration mitigation, Subway traffic, Seismic wave propagation	G-1, G-2, B-1, B-2
9th RS	1991	Vibration diagnosis	Building diagnosis, Equipment diagnosis, Construction machinery, Vibration sensation, Residential house, Road traffic	A-4, B-4, D-4
10th RS	1992	10th anniversary	Medical science, Mechanical engineering, Civil engineering, Aerospace engineering, ISO standards, Vibration regulation law, Committee activities	All

(Continued)

Table A-1. (*Continued*)

Symposium	Year	Central theme	Contents	Elements of the matrix
11th RS	1993	Measurement procedure	Questionnaire survey, Vibration nuisance, Floor vibration, FFT analyser, Vibration level, Soil vibration, Vibration mitigation	Column 0
12th RS	1994	Vibration control	Active control, Vibration monitoring, Elevated highway, Rail traffic, Sensitive machinery, Vibration utilization	Columns 0 and 4
13th RS	1995	Future environment	Future housing, Super high-rise building, Marine floating structure, Space structure, Long-span bridge	Rows F and G
14th RS	1996	Research & Development	Railway traffic, Equipment machinery, Wind force, Soil vibration, Soil-structure interaction, Floor vibration, Whole-body vibration, ISO standards	All
15th RS	1997	Vibration mitigation	Railway traffic, Floating slab, Vibration insulation, Seismic isolation, Wind response control, Tunned mass damper, Magnetic levitation, Whole-body vibration, Structure-borne noise	Columns 0 and 4
16th RS	1998	Performance-based design	Residential house, Vibration control, Structural design, Vibration prediction, Vibration regulation law, Performance evaluation	Column 4
17th RS	1999	Vibration prediction	Vibration source, Transmission path, Soil vibration, Soil-structure interaction, Floor vibration, Frame vibration, Human activity, Road traffic	Column 3
18th RS	2000	Performance-based design	Performance evaluation, AIJ guidelines, Required performance, ISO standards, Performance evaluation	Column 4
19th RS	2001	Urban environment	Urban vibration environments, Future life-style, Urban transportation system, Change in modern buildings, Diversification of required performance, Performance indication	All
20th RS	2002	20th anniversary	ISO standards, Annoyance indicator, Perception probability, Railway traffic, Marine floating structure, Wind response control	All
21st RS	2003	Performance-based design	Structural design, Residential house, Hybrid wooden floor, Vibration comfort on train, Structural health monitoring, AIJ guidelines	Column 4
22nd RS	2004	Collaboration with partners	Complaint, Mechanical engineering, Geotechnical engineering, Civil engineering, the Shinkansen, Measurement procedure	All
PD	2004	Performance evaluation	AIJ guidelines, Troubles and claims, Structural design, Vibration control, Performance indication, New horizon of environmental vibration	Column 4
23rd RS	2005	Legal issues	Troubles and claims, Legal action, Structural design, Vibration control, Vibration mitigation, Performance evaluation, Tolerable limit	Column 4

Environmental Vibrations – Takemiya (ed.)
© *2005 Taylor & Francis Group, London, ISBN 0 415 39035 4*

A review of standards for evaluation of vibration in living environment and studies of human perception of whole-body vibration

Y. Matsumoto
Saitama University, Saitama, Japan

S. Kunimatsu
National Institute of Advanced Industrial Science and Technology, Tsukuba, Japan

ABSTRACT: In the assessment of vibration in living environment, it may be a primary concern to mitigate effects of vibration on people. Understanding of the human perception thresholds of whole-body vibration may be one of important information in the assessment of vibration in living environment. The evaluation methods of human exposure to vibration in relation to living environment currently defined in international standards and Japanese standards were reviewed first in the present paper. Major differences in the methods defined in those standards were identified. The experimental data of the perception threshold of whole-body vibration were then collected from the previous studies refereed to in the international standards. Finally, tentative evaluations of the perception threshold obtained by the methods defined in the standards were compared with the experimental data collected so as to illustrate the validity of the evaluation methods given in the standards.

1   INTRODUCTION

It has been one of the key issues in many individual countries, including Japan, and also in international bodies to develop a "reasonable" method to evaluate and assess human vibration exposure so as to mitigate problems caused by vibration in living environment. In Japan, the Vibration Control Law (Shindo-Kisei-Ho in Japanese) went into effect in December 1976 so as to control vibrations caused by factories and construction works and mitigate problems due to vibration caused by road traffic. The number of complaints caused by vibrations to which the law is applied has decreased since the implementation of the law, which indicates the effectiveness of the law. However, it has been recognized recently that there are problems caused by vibrations that are not covered by the law (Ministry of Environment 2004). This may be caused partly by changes in community life, technology, and other factors since the implementation of the law almost 30 years ago. The difficulty in defining a reasonable method of evaluation of effect of vibration might also be a cause of such a situation.

The Institute of Noise Control Engineering/Japan (INCE/J) has been involved in the evaluation and assessment of vibration in living environment in various ways. Its recent activities include the organization of the Subcommittee of Evaluation of Environmental Vibrations in 2000, which consists of about 30 members from central and local governments, industries, research institutes and universities. The subcommittee has exchanged information and discussed various aspects of the evaluation and assessment of vibrations in living environment. In the subcommittee, an issue concerned with vibrations in living environments has been dealt with by classifying it into one, or a few, of the following seven categories: (i) status of complaints, (ii) sources, (iii) propagation systems, (iv) receiving systems, (v) human perception, (vi) measuring systems, and (vii) trend in standards. The content of the present paper is an outcome of the activities of the subcommittee, particularly from the categories (v) and (vii).

The objective of the present paper was to improve understanding of the current evaluation methods of vibration in living environment. In the first half of the paper, international standards developed for the evaluation of vibration in living environment with respect to effects on human are summarized briefly. Comparison between those standards and the evaluation method used in Japan was also made. In the last half of the paper, the experimental results of the human perception thresholds of whole-body vibration are reviewed and compared with the evaluations obtained tentatively based on the current standards, which may provide with fundamental information about the validity of the standard methods in the context of the assessment of vibration in living environment.

2 STANDARDS FOR EVALUATION OF HUMAN EXPOSURE TO VIBRATION IN RELATION TO LIVING ENVIRONMENT

2.1 Review of international standards

International Standard 2631-2 (2003) specifies a method for measurement and evaluation of "human exposure to whole-body vibration and shock in buildings with respect to the comfort and annoyance of the occupants". This standard can be directly applicable to the evaluation of vibration in living environment. The part 1 of ISO 2631 series "defines methods of quantifying whole-body vibration in relation to human health and comfort, the probability of vibration perception, and the incidence of motion sickness". This standard may also be used to evaluate vibration in living environment as indicated in ISO 2631-2.

The first edition of ISO 2631-2 was published in 1989, and the recent revision was completed in 2003, based on the revision of ISO 2631-1 in 1997 and recent experiences and scientific findings obtained in member countries, apparently.

In the following part of this section, the methods of evaluation of human exposure to vibration defined in ISO 2631-1 and -2 are summarized briefly. For ISO 2631-1, the method in relation to the perception only is described because the perception may be the most appropriate factor to assess vibration in buildings. Those standards define the method of vibration measurement and the method of evaluation based on measured data. Any limit values or values of this kind are not provided in the standards.

2.1.1 Measurement method

The directions of vibration measurement defined in ISO 2631-2 (2003) are "three orthogonal directions" "related to the structure rather than the human body" (i.e., x- and y-axes for the horizontal directions and z-axis for the vertical direction). The location of measurement defined is a place in relevant room in a building "where the highest magnitude of the frequency-weighted vibration occurs on a suitable surface of the building structure".

In ISO 2631-1 (1997), the measurement of vibration is prescribed to be made at the principal interface between the human body and the surface supporting the body in three orthogonal axes. The directions of the two horizontal axes are dependent on the position of the body: the "chest-back" direction and the "right-left" direction for seated and standing positions, and the "head-feet" direction and the "right-left" direction for recumbent position.

In both standards, the basic quantity to express vibration magnitude defined is acceleration.

2.1.2 Evaluation

In both ISO 2631-2 (2003) and ISO 2631-1 (1997), the vibration measured is required to be frequency weighted for evaluation. The frequency weighting defined for the application to acceleration can be considered as a model of the frequency dependence of human responses to vibration.

International Standard 2631-2 defines the W_m frequency weighting that can be applied to the vibration measured irrespective of direction of measurement (Figure 1). It is noted that, if the position of the occupant in the building is defined, the frequency weightings given in ISO 2631-1 (1997), described below, can be used.

For the evaluation of the perception defined in ISO 2631-1 (1997), the W_k frequency weighting is given for the application to vertical vibration and the W_d frequency weighting is for horizontal vibration irrespective of the position of the body (Figure 1). When the vertical acceleration under the head of recumbent body is assessed, the W_j frequency weighting will be applied.

In ISO 2631-2, values representing the effect of vibration are defined to be obtained in the method given in ISO 2631-1. In ISO 2631-1, the root-mean-square (r.m.s.) value of the weighted acceleration is defined as the basic evaluation method:

$$a_w = \left[\frac{1}{T} \int_0^T a_w^2(t)\, dt \right]^{1/2} \tag{1}$$

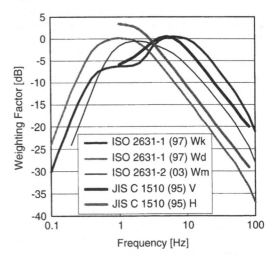

Figure 1. Frequency weightings used in the evaluations of vibration in living environment defined in ISO 2631-1, -2, and JIS C 1510. V: vertical characteristic, H: horizontal characteristic.

where $a_w(t)$ [m/s^2] is the weighted acceleration as a function of time and T [s] is the duration of the measurement. When the basic evaluation may underestimate the effect of vibration, which may be the case, for example, for transient vibration, additional evaluation methods are recommended to apply. The additional evaluation methods defined are the use of running r.m.s. acceleration (i.e., the maximum transient vibration value, MTVV):

$$MTVV = \max\left[\left\{\frac{1}{\tau}\int_{t_0-\tau}^{t_0} a_w^2(t)\,dt\right\}^{1/2}\right] \qquad (2)$$

where τ [s] is the integration time for running average, $\tau = 1$ s is recommended, and t_0 [s] is the time of observation, and the use of fourth power acceleration dose (i.e., the vibration dose value, VDV):

$$VDV = \left[\int_0^T a_w^4(t)\,dt\right]^{1/4} \qquad (3)$$

where the unit of the VDV is meters per second to the power 1.75 [m/s$^{1.75}$].

The evaluation of vibration given in ISO 2631-2 is defined to be based on the value obtained in the axis with the highest frequency-weighted vibration magnitude. The assessment of the perceptibility of the vibration given in ISO 2631-1 is made with respect to the highest weighted r.m.s. acceleration among all determined values.

2.2 Review of Japanese standards

For the evaluation of vibration in living environment, JIS C 1510 revised in 1995 and JIS Z 8735 (1981) has been used in Japan as the specification of the method of measurement and evaluation in accordance with the specification of the Vibration Control Law. This section summarizes briefly the evaluation method defined in those standards.

In 2004, the Japanese translation of ISO 2631-1 (1997) went into force as a Japanese standard: JIS B 7760-2 (2004). There may be a possibility in future for JIS B 7760-2 to be applied to the evaluation of vibration in living environment, particularly with respect to human exposure to vibration.

2.2.1 Measurement method

Although the Vibration Control Law requires the measurement of vertical vibration only for the evaluation of vibration in living environment, JIS C 1510 (1995) and JIS Z 8735 (1981) provides information about the measurement of both vertical and horizontal vibrations (i.e., three orthogonal directions). The directions of the measurement of horizontal vibrations (i.e., x- and y-axes) are specified with respect to the location of the source of vibration in JIS Z 8735. The location of the measurement defined is a location on the border of the premises in which the source of vibration is located (i.e., outside of a building).

2.2.2 Evaluation

As in the international standards, frequency-weighted acceleration is used for evaluation in the Japanese industrial standards. The "vertical characteristic" and the "horizontal characteristic" specified in JIS C 1510 (1995) are shown in Figure 1. These frequency weightings are in accordance with those specified in the previous version of ISO 2631-1.

The "Vibration Level" is then determined by:

$$L_v = 20\log_{10}(a_w/a_0) \qquad (4)$$

where a_w [m/s^2] is the root-mean-square weighted acceleration obtained by an integration circuit with a time constant of 0.63 s, and a_0 is the reference acceleration of 10^{-5} m/s^2.

2.3 Comparison between standard measurement and evaluation methods

As clearly seen in the above part of this section, there are differences between the methods of measurement and evaluation of vibration in living environment defined in the standards introduced here, even between the methods defined in different parts of the international standard.

The most significant difference between the international standards and the Japanese standards currently used for the evaluation of vibration in living environment is the measurement location: a location where vibration transmitted to the body is the measurement location specified for the international standards, while a location outside a building is the specified location in the Japanese methods. The selection of measurement location in the Japanese method may be based on the idea that the emission of vibration should be measured and controlled.

A major difference in the evaluation methods defined in the standards introduced in this section is the definition of frequency weighting. Upon the assumption that there is a common "limit" value for the assessment of the perception, there would be a maximum of about 6 dB difference for vertical vibration and of about 10 dB difference for horizontal vibration between the evaluations obtained by different standards only due to the differences in the frequency weighting. These significant differences between the evaluations by different standards are caused by the W_m frequency weighting. The W_m frequency weighting was originally required when individual frequency weightings were defined with respect to the position of the human body: the frequency weighting applied to, for example, vertical vibration was different whether the occupant was in a seated position or in a recumbent position. The definition of the W_m frequency weighting is equivalent to the combination of the vertical and horizontal characteristics used currently in JIS C 1510. After the last revision of ISO 2631-1 in 1997, it is not clear if the W_m frequency weighting is required in the standard in accordance with ISO 2631-1.

All standards introduced in this section defines a running average value for the evaluation of vibration, which may be useful in the evaluation of transient vibrations and shock. The time constant of the integration to obtain a running average is 1s in the international standard while 0.63s in the Japanese standard. This difference in the time constant may be significant if the phenomena to be evaluated has a very short duration, such as an impact caused by construction work.

3 HUMAN PERCEPTION THRESHOLDS AND STANDARDS

For the assessment of vibration in living environment, the human perception thresholds of vibration can be an important reference, which is also stated in ISO 2631-2 (2003). As described in the preceding section, the current standards for the evaluation of vibration involve the frequency weightings so as to take into account the frequency dependency of human responses to vibration. It is recognized that those frequency weightings were defined based on the experimental data of the equal sensation contours, or the equivalent comfort contours, which were determined with the vibration magnitudes higher than those that human can just perceive (i.e., perception thresholds). However, as observed in Miwa (1967), the frequency dependency of the subjective response tends to change with altering vibration levels. Therefore, it may not be reasonable to use the frequency weightings in the current standards to represent the perception thresholds which may be the most important reference for the assessment of vibration in buildings. In this section, the previous experimental data of the perception thresholds of whole-body vibration were compared with the evaluations obtained from the standards described in the preceding section so as to identify the differences between them.

3.1 Previous experimental data of perception thresholds

The previous experimental data of the perception thresholds of whole-body vibration used in this section were obtained from six publications. Four publications were extracted from the bibliography of ISO 2631-1

(1997) and ISO 2631-2 (2003) (Miwa 1967, 1969, Miwa et al. 1984, Persons & Griffin 1988). Reiher & Meister (1931) was also selected because it was used as the reference in the architectural engineering field in Japan for a long time. In addition, a recent set of results obtained in Japan was also reviewed (Yonekawa et al. 1999). The median or mean data obtained with more than several subjects were extracted from the figures presented in each publication and reproduced in the present paper for the comparison with the evaluations described later in this section. The input stimuli used to determine those experimental data were continuous sinusoidal vibration. An extensive review of the experimental results of the perception thresholds can be found in Griffin (1990).

3.2 Tentative evaluation of perception in standards

ISO 2631-1 (1997) states in Annex C that "Fifty percent alert, fit persons can just detect a W_k weighted vibration with a peak magnitude of 0.015 m/s^2". For sinusoidal vibrations, this value corresponds to approximately 0.011 m/s^2 r.m.s. Although there is no indication of the reference value for the perception for the W_d weighted acceleration (i.e., for horizontal vibration), the same value as the W_k weighted acceleration was used tentatively in the comparison with the experimental data in the later section. In ISO 2631-2 (2003), no reference is given to the perception thresholds for the W_m weighted acceleration. However, because the standard implies that either the W_m weighting or the W_k weighting can be applied to some types of vibration alternatively, it may be reasonable at this stage to use the reference value given in ISO 2631-1 as the reference for the W_m weighted acceleration also. The value equivalent to 0.0056 m/s^2 r.m.s. (i.e., the Vibration Level of 55 dB) has often been used as the indication of the perception threshold for the evaluation based on JIS C 1510 (1995), so that this value was used in the comparison with the experimental data.

The frequency weightings used to evaluate vibration with respect to the perception are presented in Figure 1 for ISO 2631-1 (1997), ISO 2631-2 (2003) and JIS C 1510 (1995). The reciprocals of those frequency weightings represented in linear scale were multiplied by the values representing the perception threshold described above for ISO and JIS, respectively, so as to obtain tentative evaluations of the perception thresholds at different frequencies in different vibration axes for each standard.

3.3 Perception thresholds and tentative evaluations

The perception thresholds and tentative evaluations obtained based on the standards as described in the

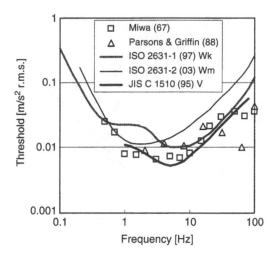

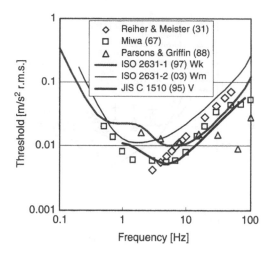

Figure 2. Perception thresholds and tentative evaluations for seated subjects exposed to vertical vibration. The evaluations indicated by lines were obtained by multiplying the reciprocals of each weighting factor by the value indicated in 3.2. (The numbers in brackets shown in the legend are the last two digits of publication year.)

Figure 3. Perception thresholds and tentative evaluations for standing subjects exposed to vertical vibration. The evaluations indicated by lines were obtained by multiplying the reciprocals of each weighting factor by the value indicated in 3.2. (The numbers in brackets shown in the legend are the last two digits of publication year.)

preceding two sections were compared each other. The comparisons were made for two vibration axes (i.e., vertical and horizontal) and three body positions (i.e., seated, standing and recumbent) separately because these factors appeared to have significant effects on the perception thresholds in the previous studies.

3.3.1 Vertical vibration, seated subjects

Figure 2 compares the median or mean perception thresholds determined by Miwa (1967) and Parsons & Griffin (1988) for seated subjects exposed to vertical whole-body vibration with the tentative evaluations obtained by the standards. The figure shows the variation in the experimental data obtained in different studies and also the differences between the evaluations obtained by different standards. The evaluation obtained by ISO 2631-1 (1997) was greater than Miwa (1967) in all frequency range except between 20 and 40 Hz and also greater than Parsons and Griffin (1988) in all frequency range except between 4 and 16 Hz. The evaluation obtained by ISO 2631-2 (2003) was greater than both experimental data in the frequency range above 1 Hz which is the principal frequency range of vibration problems in living environment. The evaluation obtained by JIS C 1510 (1995) showed an agreement with the experimental data, particularly Miwa (1967).

3.3.2 Vertical vibration, standing subjects

The median or mean perception thresholds reported by Reiher & Meister (1931), Miwa (1967) and Parsons &

Griffin (1988) for standing subjects when exposed to vertical vibration are compared with the tentative evaluations obtained by the standards in Figure 3. In general the observation that can be seen in Figure 3 is similar to that obtained from Figure 2 described above because the perception thresholds of vertical vibration of standing subjects were similar to those of seated subjects. However, for standing subjects, there was a set of data by Miwa (1967) that were less than any evaluations at low frequencies below 2 Hz.

3.3.3 Vertical vibration, recumbent subjects

The median or mean perception thresholds of vertical vibration determined for recumbent subjects in five studies (i.e., Reiher & Meister 1931, Miwa 1969, Miwa et al. 1984, Parsons & Griffin 1988, Yonekawa et al. 1999) are compared with the tentative evaluations obtained by the standards in Figure 4. The evaluation by ISO 2631-1 (1997) appeared to be close to the lowest values among the experimental data in the frequency range below 20 Hz, while, at frequencies above 31.5 Hz, the evaluation was greater than almost all experimental data. The evaluation by ISO 2631-2 (2003) was greater than almost all experimental data at frequencies above 8 Hz. The evaluations by JIS C 1510 (1995) showed significantly lower values than the experimental data at frequencies below 10 Hz. The observation for JIS C 1510 at higher frequencies was similar to that for ISO 2631-1.

3.3.4 Horizontal vibration, seated subjects

Figure 5 shows the median or mean perception thresholds of horizontal vibration determined for seated

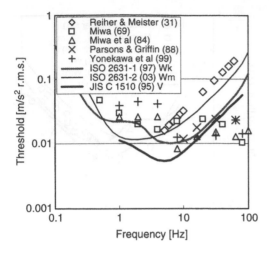

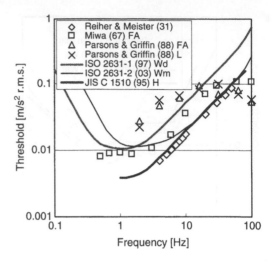

Figure 4. Perception thresholds and tentative evaluations for recumbent subjects exposed to vertical vibration. The evaluations indicated by lines were obtained by multiplying the reciprocals of each weighting factor by the value indicated in 3.2. (The numbers in brackets shown in the legend are the last two digits of publication year.)

Figure 6. Perception thresholds and tentative evaluations for standing subjects exposed to horizontal vibration. The evaluations indicated by lines were obtained by multiplying the reciprocals of each weighting factor by the value indicated in 3.2. (The numbers in brackets shown in the legend are the last two digits of publication year. FA: fore-and-aft; L: lateral.)

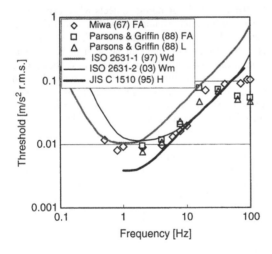

Figure 5. Perception thresholds and tentative evaluations for seated subjects exposed to horizontal vibration. The evaluations indicated by lines were obtained by multiplying the reciprocals of each weighting factor by the value indicated in 3.2. (The numbers in brackets shown in the legend are the last two digits of publication year. FA: fore-and-aft; L: lateral.)

subjects by Miwa (1967) and Parsons & Griffin (1988), together with the tentative evaluations obtained by the standards. It was found that the evaluations obtained by ISO 2631-1 (1997) exceeded almost all experimental data over the frequency range of interest. The evaluations obtained by ISO 2631-2 (2003)

showed an agreement with the experimental data in the frequency range between 8 and 31.5 Hz. At other frequencies, the evaluation was greater than the experimental data. The evaluation obtained by JIS C 1510 (1995) showed an agreement with the experimental data in the frequency range between 4 and 31.5 Hz. The evaluation was less than the experimental data at low frequencies while an opposite trend was observed at higher frequencies.

3.3.5 Horizontal vibration, standing subjects

The median or mean perception thresholds of horizontal vibration measured with standing subjects by Reiher & Meister (1931), Miwa (1967) and Parsons & Griffin (1988) are compared with the evaluations obtained by the standards in Figure 6. The evaluation by ISO 2631-1 (1997) was greater than the experimental data by Reiher & Meister and Miwa at all frequencies. At frequencies below 10 Hz, the evaluation was less than the data by Parsons & Griffin (1988). The evaluation by ISO 2631-2 (2003) agrees with the experimental data by Reiher & Meister and Miwa in all frequency range below 50 Hz, while the experimental data were less than the evaluation at frequencies above 63 Hz. The evaluation by JIS C 1510 (1995) showed a good agreement with the experimental data by Reiher & Meister but it is less than the newer data in general. At frequencies above 63 Hz, the experimental data shows lower values than the evaluation.

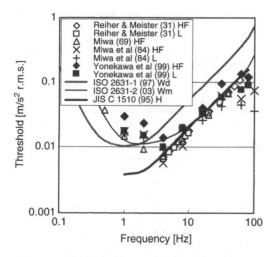

Figure 7. Perception thresholds and tentative evaluations for recumbent subjects exposed to horizontal vibration. The evaluations indicated by lines were obtained by multiplying the reciprocals of each weighting factor by the value indicated in 3.2. (The numbers in brackets shown in the legend are the last two digits of publication year. HF: head-to-feet; L: lateral.)

3.3.6 Horizontal vibration, recumbent subjects

The median or mean perception thresholds of horizontal vibration determined for recumbent subjects in four studies (i.e., Reiher & Meister 1931, Miwa 1969, Miwa et al. 1984, Yonekawa et al. 1999) are shown with the evaluations obtained by the standards in Figure 7. The evaluation by ISO 2631-1 (1997) was not in agreement with any experimental data and greater than all experimental data at frequencies above 4 Hz. The evaluation by ISO 2631-2 (2003) was close to the experimental data at frequencies below 2 Hz and between 10 and 50 Hz. At other frequencies, the evaluation was greater than most experimental data. The evaluations by JIS C 1510 (1995) also showed a good agreement with the experimental data in the frequency range between 4 and 50 Hz. At lower frequencies, the evaluation was significantly less than the experimental data, while an opposite trend was observed at higher frequencies.

3.3.7 Summary of comparison between perception thresholds and evaluations

It might not be reasonable to expect that the evaluation obtained by the standards represent well all perception thresholds obtained in different studies with different experimental conditions because of the effects of various factors including postures, direction of vibration and individual differences in the perception. The comparison made in this section may indicate, however, that there might be some cases in which the evaluation

by the standards may underestimate the effect of vibration on the perception (i.e., human may be more sensitive to vibration than the estimation based on the standards). This is observed for the perception of vertical vibration of standing subjects at very low frequencies, the perception of vertical vibration of recumbent subjects at frequencies above 31.5 Hz, and the perception of horizontal vibration of subjects in any positions at frequencies above 63 Hz. The discrepancy between the perception thresholds and the frequency dependence defined in the standards at higher frequencies was pointed out also by Griffin (1990). Additionally, for horizontal vibration with subjects in any positions, any experimental data are comparable with the evaluation obtained by JIS C 1510 (1995), which implies that the evaluation may tend to overestimate the human sensitivity to vibration (i.e., human may be less sensitive).

It was found in the experimental data of the perception thresholds collected in the present paper that the frequency dependence of the perception thresholds was similar in standing and seated positions, although there might be some differences observed in Parsons & Griffin (1988) for horizontal vibration. However, it was clear that there was a difference in the frequency dependence of the perception threshold between those two positions and recumbent position: in particular, the perception thresholds of vertical vibration measured with recumbent subjects tended to decrease with increasing vibration frequency. The shapes of the frequency weightings defined in the standards are closer to the perception thresholds of seated and standing subjects than those of recumbent subjects in general. Therefore, care must be taken if the evaluation method is applied to the assessment of vibration in an environment where people are expected to lie down.

In addition, it has often been the case that mean or median experimental data are used to determine a "limit value" for the assessment of vibration in living environment. However, it is reasonable to take into account significant variability in responses observed between individuals (i.e., inter-subject variability) in the definition of such a limit value so as to mitigate problems in practice. In this context, the range between about 0.01 and 0.02 m/s^2 for the W_k weighted peak acceleration is given as the inter-quartile range for the perception threshold in ISO 2631-1 (1997).

Vibrations to be evaluated in practice may often be transient and include more than one frequency components. The method to evaluate the perception thresholds of such types of vibration has not yet been established well, although there have been some scientific considerations. Further investigation may be required to understand the nature of the perception of transient vibration and random vibration so as to establish more appropriate evaluation method for the assessment of vibration in living environment.

4 SUMMARY

The present paper focuses on the human perception of vibration as an important factor in the assessment of vibration in living environment. The evaluation methods of such vibration with respect to effects of vibration on people, particularly the effect of the perception, currently used internationally and nationally were reviewed and differences between the standards were identified. Additionally, the experimental results of the perception thresholds of vibration reported in literature were collected and compared with the evaluations of the thresholds obtained tentatively with the standards reviewed.

A major difference between ISO 2631-1 and -2 and the Japanese standards was the location of vibration measurement: a location where vibration is transmitted to the body in ISO 2631s whereas a location on the border of the premises in which the source of vibration is located in the Japanese method. This might be due to the difference in their basic concepts: the Japanese method may focus on the emission and control of vibration from its source. Another significant difference between the standards was the definition of frequency weighting, which can result in up to 6–10 dB difference in the evaluated value. There is a discrepancy in the definition of the "combined" frequency weighting in the ISO 2631 series, which may be required to be considered in future revision of the standard.

The comparison between the perception thresholds measured in the previous studies and the tentative evaluations obtained by the standards showed that there was a possibility for the evaluation to underestimate the "median or mean" human sensitivity to vibration. The perception thresholds of recumbent subjects showed different frequency dependences from the thresholds of seated and standing subjects that were more closer to the frequency weightings defined in the standards.

REFERENCES

Griffin 1990. *Handbook of Human Vibration*. London: Academic Press.

International Organization for Standardization 1997. Mechanical vibration and shock – Evaluation of human exposure to whole-body vibration – Part 1: General requirements. ISO 2631-1. Geneva: International Organization for Standardization.

International Organization for Standardization 2003. Mechanical vibration and shock – Evaluation of human exposure to whole-body vibration – Part 2: Vibration in buildings (1 Hz to 80 Hz). ISO 2631-2. Geneva: International Organization for Standardization.

Japanese Industrial Standards Committee 1981. Methods of measurement for vibration level. JIS Z 8735. Tokyo: Japanese Standards Association. (in Japanese)

Japanese Industrial Standards Committee 1995. Vibration level meters. JIS C 1510. Tokyo: Japanese Standards Association. (in Japanese)

Japanese Industrial Standards Committee 2004. Whole-body vibration-Part 2: General requirements for measurement and evaluation method. JIS B 7760-2. Tokyo: Japanese Standards Association. (in Japanese)

Ministry of Environment 2004. Review of the execution status of Vibration Control Law in the year Heisei 15. Tokyo: Ministry of Environment. (in Japanese)

Miwa, T. 1967. Evaluation methods for vibration effect. Part 1. Measurements of threshold and equal sensation contours of whole body for vertical and horizontal vibrations. *Industrial Health* 5: 183–205.

Miwa, T. 1969. Evaluation methods for vibration effect. Part 9. Response to sinusoidal vibration at lying posture. *Industrial Health* 7: 116–126.

Miwa, T., Yonekawa, Y. & Kanada, K. 1984. Thresholds of perception of vibration in recumbent men. *Journal of Acoustical Society of America* 75(3): 849–854.

Parsons, K.C. & Griffin, M.J. 1988. Whole-body vibration perception thresholds. *Journal of Sound and Vibration* 121(2): 237–258.

Reiher, H. & Meister, F.J. 1931. Die Empfindlichkeit des Menschen gegen Erschütterungen. *Forschung* 2(11): 381–386.

Yonekawa, Y., Maeda, S., Kanada, K. & Takahashi, Y. 1999. Whole-body vibration perception thresholds of recumbent subjects – part 1: supine posture. *Industrial Health* 37: 398–403.

Environmental Vibrations – Takemiya (ed.)
© 2005 Taylor & Francis Group, London, ISBN 0 415 39035 4

The transition concerning the estimation of the ground environmental vibrations in Japan

M. Shioda
Department of Architecture, Kougakuin University, Tokyo, Japan

ABSTRACT: Studies on the environmental ground vibrations in Japan has stemmed from "the environmental vibration pollution" that was legitimated in 1976. The word "environmental vibrations" nowadays includes the structure vibrations due to earthquake and manmade sources as well as due to the wind loadings. The environmental vibration pollution is basically assessed by vibration regulation values based on the vibration perception by human body. This scientific evidence is clearly stated in the issue of ISO 2631.Concerning to the estimation of the environmental vibrations in Japan, reviewing the historical transition, this paper proposes a new estimation way of environmental ground vibrations.

1 INTRODUCTION

The vibration sense of human body to be exposed vibration corresponds the physical values of vibration usually. Therefore, the study to be made sure this relation is executed and the equivalent sense curve of vibration sense is got experimentally. However, the estimation methods concerning physiological effect to low level vibration does not yet make clear. The study of the environmental ground vibration in Japan begins from "environmental pollution vibration" on the base of the vibration regulation law and shows from the structure vibration to the swing of wind in these days. This environmental pollution vibration estimates by the vibration regulation values as base on the whole body vibration sense of human body. This scientific evidence depends to ISO 2631 series. This report introduces the transition concerning the estimation of the environmental ground vibration in Japan.

2 TRANSITION OF VIBRATION ESTIMATION

The study for the whole vibration of human body is executed from the1930. It is from the late 1960's for the study to be discussed internationally. The big scale study result did not come to the surface in Japan. However, the estimation for the whole vibration of human body comes to begin the change for the adoption of this amplitude values recently. The transition of vibration regulation in Japan is as follows:

1. the circumstance of vibration regulation before:
   Tokyo metropolitan public nuisance research institute and the Ministry of International Trade and Industry (MITI) corresponded to the vibration public nuisance. The public nuisance problems of vibration were the study theme of many researchers.
2. the circumstance of vibration regulation after:
   The vibration regulation law was enforced and the Environment Agency, MITI and a local government was corresponded by the whole country.
3. the present condition of vibration regulation:
   The observance of regulatory standard values, the mismatch between regulatory standard values and the increase tendency of complaint case (especially, subdivision construction works)

The researchers thought it is possible to estimate the similar thinking with noise and vibration too. That is to say, the estimation way is both dB unite. The difference between noise and vibration in the life environment shows Table 1.

3 THE EXTENT FROM THE PUBLIC NUISANCE VIBRATION TO THE ENVIRONMENTAL VIBRATION

The vibration regulation law has been enforced at eight years late since the enforcement of the noise

Table 1. Difference between noise and vibration in the life environment.

Noise regulatory law	Vibration regulatory law
Noise * Judgment from loudness within area audible	Whole body vibration * It is usual for human being to not feel vibration, it judge that it is abnormal for human being to feel vibration
* Indoor noise is lower than outdoor noise by insulation capability of structure buildings	* Indoor vibration is higher than outdoor vibration by amplification of structure buildings sometimes

regulation law. The vibration public nuisance in this age had same generation sources with noise public nuisance. This regulation ways of thinking was the same noise regulation law too. In the transition of age, what is called public nuisance vibration went to expand the defense area to the environmental ground vibration and it become to begin the target many vibration phenomena. This reasons is as follows:

1. diversification of vibration generation sources
2. alteration of life cycle and style
3. complexity of vibration propagation route
4. poverty and high-class of vibration receiver side
5. local environment to urban environment

4 AREA OF ENVIRONMENTAL VIBRATIONS

The definition of environmental vibrations in Japan is as follows.

4.1 Public nuisance vibrations

By the activity of human being and unnatural cause, within vibrations to be perceived as whole body vibration, human being has subjective characteristics and sensational characteristics in the vibration not proper to exist.

4.2 Environmental vibrations

It is said that environmental vibrations show the daily vibration condition of boundary that we surround with a certain spread, ground surface and buildings. That is to say, its vibration defines construction ground vibration, machine vibration, bridge vibration, blast vibration, public nuisance vibration, obscure vibration, floor vibration, swing to wind, structure borne sound. How about study area of environmental vibration?

1. public nuisance vibration concerning vibration regulation law
2. whole vibration of human body concerning labor environment
3. ground vibration concerning the old railroad line/the subway

4. whole vibration of human body concerning the car rides
5. structure vibration concerning swing to wind
6. structure borne sound from vibration propagation concerning building structure indoor

above, it may be "public nuisance vibrations concerning vibration regulation law" to be able to estimate as environmental vibrations. This reason is as follows.

1. Regulatory standard values for local division and time division exists
2. Japanese industrial standard for vibration measurement exists
3. How to reading way of measurement values exist

Vibration of this subject is ground vibration to be caused by industrial factory vibration and construction working vibration and road traffic vibration. Concerning the other vibration, vibration estimation is not standardized.

5 THE ESTIMATION OF THE ENVIRONMENTAL VIBRATIONS IN JAPAN

The vibration regulation law was enforced December 1976 and the revision of this law was enforced somewhat 1994 and 1995 and this law brings today. The subjects in this law are the environmental ground vibration of the industrial factories/the working offices, the construction working and the road traffic. The old railroad line and the structure borne sound indoor house does not take the objects of this law. Table 2 shows the estimation on the vibration regulation law.

The estimation on the vibration regulation law objects the vertical direction based on the scientific learning only. This scientific basis depends on the following reasons in Japan.

1. The vibration values of the vertical direction is bigger than the vibration values of the horizontal direction in the general ground surface.
2. Whole vibration of the vertical direction feels stronger than the horizontal direction in the frequencies band too.

Table 2. The estimation on the vibration regulation law.

Environmental ground vibration of objects	Condition	Estimation (dB)
Industrial factories/ The working offices	Differs from the characteristics of time changing	L_{10}, L_{Max}, 1.1.1 *The average of* L_{Max}
Construction working	Differs from the characteristics of time changing	L_{10}, L_{Max}, The average of L_M
Road traffic		L_{10}
Shinkansen		L_p

Table 3. The difference between JIS C 1510–1995 and ISO 2631-1/1997.

Terms	JIS C 1510	ISO 2631-1
Unit	dB	m/s^2
Frequency weighting curve(vertical direction) frat characteristics	4~8(Hz)	4~12.5(Hz)
Frequency weighting curve(horizontal direction) relative value	1~2(Hz) standard 3 dB	1~2(Hz) standard 0
Frequency weighting curve	curves of two	curves of six
Composition base frequency	not regulation	8 Hz
Frequency allowance error	1~80(Hz), +2., −5 dB	1~80(Hz), ±1 dB
Prescription for outside area of the subject frequencies	not regulation	filter characteristics and allowance error (±2 dB) prescription

3. The difference between the vertical direction vibration and the horizontal direction vibration is gained about 9 dB above 8 Hz by study results.

Therefore, the difference between JIS C 1510–1995 in Japan and ISO 2631 − 1/1997 makes clear as Table 3. This difference is the frequencies weighting curves.

1. JIS C 1510: vertical direction (foot-head), horizontal direction (back-chest, right side-left side)
2. ISO2631-1:vertical direction(seat, standing, vertical recumbent, head), horizontal reaction(seat, standing, vertical recumbent, head, motion sickness), rotation vibration(seat).

6 THE ISSUES FROM NOW ON

How thinking to research the problems of environmental ground vibration from multiple vibration sources ? The following vibration characteristics makes clear.

1. Road traffic vibration: influence from along the main road to life road
2. Railroad vibration: influence from along the old railroad line to along Shinkansen railroad
3. Manufacturing industrial machine/installation/equipment vibration: influence from industrial machines or many kind of motor in factory/business office, and medium and small plants 4) construction working vibration: influence from subdivision construction/new building construction due to big scale construction from medium and small construction

4. Indoor building vibration: influence from air conditioning equipments to electricity equipments

In addition, how thinking place and position of vibration influence estimation? For example, as follows.

1. The said person to perceive vibration indoor housing
2. The place and the position to perceive vibration indoor housing
3. The site boundary line and public/private boundary line due to vibration regulation law
4. The nearby of vibration generation sources

And it is the issue to think the vibration estimation? This thinking of way is as following.

1. Improvement of vibration regulation values
2. Innovation of three direction of vibration axis
3. Innovation of frequencies characteristics
4. Hybrid between old system and indoor building construction

REFERENCES

ISO2631-1/1997: "Mechanical vibration and shock-Evaluation of human exposure to whole body-Part 1: general requirement"

Law No.64 of 1976: Vibration regulation law in Japan

Masazumi SHIODA: "International tendency of noise and vibration evaluation", Vol. 40, No. 5, 2004, Japan Environmental Management Association for Industry (in Japanese)

Masazumi SHIODA: "The prediction and the reduction of the environmental ground vibration", pp. 549–554, Vol. 60, No. 9, (in Japanese)

Masazumi SHIODA: 2004.Symposium-"Thinking of the environmental ground vibration" INCE/Japan, Yamanashi University.

Environmental Vibrations – Takemiya (ed.)
© *2005 Taylor & Francis Group, London, ISBN 0 415 39035 4*

Method of evaluating of the influence of occurrence vibrations generated by heavy traffic noise on acoustic conditions in residential buildings

M. Niemas

Department of Acoustic, Building Research Institute, Warsaw, Ksawerów St. Poland

ABSTRACT: The paper presents measurement-calculation procedure, which may be use for estimation of increase in sound pressure level in dwellings radiated from vibrating partitions. The procedure is based on simultaneously twin channel measurements of vibration velocity on partitions and sound pressure level in dwellings. This method using properties of coherence function between has measuring signals. The calculations of the sound radiated from vibrating walls induced by vibro-acoustic sources (e.g. heavy traffic), which has been made by new measure-computational procedure directly from two channel measurements are presented as well.

1 THE ESSENCE OF THE PROBLEM

The phenomenon of the occurrence of noise in residential premises caused by vibrations on the surfaces of building partitions confining the premises (walls, ceiling, floor) is very complex (Fig. 1).

The final stage of the phenomenon of transmitting the vibrations energy to the building, that is: emission of acoustic energy by induced to vibrations building partitions confining the residential premises $-E_p$ is most important, as it includes the transformation of vibration energy into acoustic energy emitted to the premises. Understanding this phenomenon enables us to develop methods for evaluating the simultaneous occurrence of noise and vibrations in relation to the specific character of transportation sources.

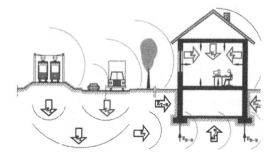

Figure 1. Diagram of the propagation of traffic-induced vibrations to a residential building.

The final effect of the occurrence of vibrations on the partitions in premises given above depends on four acoustic phenomena, namely:

1. the dispersion of vibrations on the building's structure, $-E_{B-B}$
2. inducing the building's structure to vibrations through the interaction of the base and the building's foundation, $-E_{T-B}$
3. propagation of waves in the base on the source $\Rightarrow$ building path, $-E_T$
4. inducing the base to vibrations by a passing heavy vehicle (tram, train, bus, truck tractor with semi-trailer, truck), that is generation of a vibration wave, $-E_{\dot{Z}}$

The analysis of the above phenomena will not be discussed in any more detail, as it is not the subject of the work.

2 THE PRESENT STATE OF KNOWLEDGE

The passing of vehicles, particularly heavy ones, is a source of both airborne as well as material sounds. The penetration of both types of sounds to premises in buildings located near transportation routes takes place:

- by air, i.e. by the weakest, from the acoustic point of view, element of the external wall, that is the window,
- by the material path, i.e. through the base to the building's structure and through the structure on to building partitions.

These issues are relatively well known (perhaps with the exception of low frequencies, which may appear during the passing of heavy vehicles).

After analyzing scientific literature, as well as on the basis of own research, it was found that:

- the joint appearance of vibrations and noise affects both the subjective perception of noise and vibrations by people, as well as the value of the sound pressure level found in the studied premises,
- the subjective final evaluation of the joint occurrence of the above mentioned phenomena should not be carried out for each phenomenon separately (usually from the physical point of view vibrations and noise are an inseparable whole), as this leads to incorrect conclusions,
- the correct carrying out of research tying the objective measurements of the parameters describing noise and vibrations with subjective studies connecting the influence of these phenomena may lead, when applying statistical methods on appropriately large samples, to determining the relations between vibrations and noise and the degree to which they are perceived by people,
- the existence of a dependence between the sound pressure levels in bands where the level of accelerations (velocity) of vibrations measured on partitions in the studied premises attained the highest values point to the existence of an energy relationship between the measured quantities,
- up to now no research has been done aimed at the quantitative evaluation of acoustic energy found in residential premises, occurring as a result of vibrations on partitions confining these premises with respect to traffic (particularly heavy).

3 THE ANALYSIS OF THE PHENOMENON, ASSUMPTIONS FOR METHOD DEVELOPMENT

In order to accurately define and determine the range of the work leading to solving the problem set out earlier on, the following assumptions were adopted:

- the source of simultaneous (occurring at the same time) noise and vibrations penetrating to premises was taken to be heavy road traffic in a city, including passing busses, truck and rail vehicles (trams),
- the studied premises are found in residential buildings, which means that:
- the average volume of the premises is $V = 25 \div 50 \, \text{m}^3$,
- the ratio between the surface of the partitions confining the premises (excluding floor surface) and the volume of these premises is $1.3 \div 1.8 \, \text{m}$ on average,
- the reverberation time in premises with furniture is $T = 0.5 \div 0.6 \, \text{s}$ on average,

- noise (airborne sounds) penetrating into the premises mainly through the weakest, from the acoustic point of view, element of the external wall, that is the window,
- the term "acoustic conditions in the premises" means the average level over time and space of acoustic pressure subsequently evaluated with correctly selected single-figure evaluation indices (in the case of residential premises with furniture, due to their geometrical parameters, it is possible to omit the spatial distribution of the acoustic field),
- residential buildings subjected to the study have a massive structure, which is important when analyzing the emitted sound by the building partitions in the studied residential premises.

3.1 Analysis of measurement methods

Methods which may be used to determine the value of the emitted acoustic power by the vibrating building partitions confining the analyzed premises include:

- classical methods (based on measuring acoustic pressure),
- intensity methods (based on measuring the normal component of sound intensity),
- statistical methods: correlation, coherence function, cepstrum (based on two-channel measurements of acoustic quantities or the vibration and acoustic signal).

When examining the phenomenon of vibrations appearing on the surface of the panel, such as a building partition (e.g. wall, ceiling), due to the emission of acoustic energy, the vibrating partition may be treated as a surface sound source. This way of describing the phenomenon makes it possible to apply the usual methods for determining the acoustic power emitted to the environment.

One of these methods is the method based on measurements of the sound pressure level in the range of frequencies of interest to us on a measurement surface surrounding the source (in every direction), called the classical method. But this method is not very accurate, particularly in real conditions, as it is based on the measurements of acoustic pressure being a resultant quantity of the examined phenomenon, which in light of the principles for carrying out measurements of acoustic power is an undesired quantity (interfering factor), affecting the measurement results.

Therefore this method cannot be applied for the phenomenon being the subject of deliberations here. A more precise method for determining the acoustic power of sound sources is the method based on measuring the normal component (perpendicular to the surface emitting the sound) of the sound intensity vector, called the intensity method. Sound intensity, contrary to acoustic pressure, is a parameter which includes more information about the quality of the acoustic field, as it is a vector quantity determining the direction and sense of the dispersion (flow) of acoustic energy. This is a very accurate method as it directly measures the sound intensity vector, but it has many disadvantages of a technical type. Considering the character of the phenomenon of simultaneous occurrence of noise and vibrations from passing heavy vehicles, the analysis should include the range of low frequencies, which in this method poses an additional hindrance due to the construction of the probe for measuring intensity (the type of microphones used, the distance between the microphones) and the resulting limitations in the measurement range.

Statistical methods are used to describe the vibroacoustic phenomena in which the generated acoustic signals may be considered as stochastic, stationary and ergodic processes. These methods use:

- correlation functions,
- the coherence function.

Assuming that the acoustic energy emitted by the given surface of the source depends mainly on the velocity of vibrations on this surface, it is possible to determine the mutual correlation function between the velocity of vibrations in surface points and the acoustic pressure in the near acoustic field of the sources. A measure of the share of the given vibrating element of the surface in the emitted acoustic pressure is the square of the normalized mutual correlation function between the vibrations velocity signal and acoustic pressure. This type of analysis makes it possible to determine the share of the vibrating surface in the emitted acoustic pressure in the domain of time. Subjecting time signals to Fourier's transformation we obtain an analysis of signals in the domain of frequency, where the square of the normalized mutual correlation functions corresponds to the value of the normalized coherence function.

The coherence function methods, owing to the type of the vibroacoustic signal used, may be divided into:

- coherence of the material vibrations – material vibrations type,
- coherence of the material vibrations – acoustic vibrations type,
- coherence of the acoustic vibrations – acoustic vibrations type.

The method of the coherence function of material vibrations – acoustic vibrations type and acoustic vibrations – acoustic vibrations type is used to study the influence of the various vibration or acoustic sources on acoustic pressure in the given point of the acoustic field. Assuming that the vibrating building partition radiates acoustic energy on its entire surface, it may be presented as a set "n" of substitute sources uniformly dispersed on the surface, and carry out the evaluation of the influence of the accelerations or vibrations

velocities at the locations of substitute sources on the value of acoustic pressure on the basis of the partial coherence function determined between accelerations or vibration velocities in given points, and the acoustic pressure in the premises.

The coherence function method in relation to real residential buildings may be the basis for evaluating the acoustic conditions in closed premises in the case of vibrations on building partitions confining the studied premises.

3.2 Assumptions to the developed method of measurement

In order to conduct an evaluation of the phenomenon of simultaneous occurrence of noise and vibrations on a measurement path using the properties of the coherence function, it is necessary to know the following quantities:

- averaging in time and space sound pressure level in the studied premises for the period of duration of vibrations for the passing of a single vehicle (i.e. for the time in which the value of the amplitudes of vibration velocities were greater than 0.2 of the maximum value – taken as 5 s),
- the average values of amplitudes or levels of vibration velocities on each of the building partitions confining the studied premises (for the same fragment of time as the registration of the sound pressure level),
- the values of the coherence function tying the sound pressure levels and vibration velocities in order to determine the existence of a mutual similarity of the examined signals.

Analyzing the needs and possibilities of measuring vibrations in the existing and inhabited premises from the point of view of determining the share of vibrations appearing on the partitions of the studied premises in the values of the sound pressure levels appearing in the studied premises, as well as from the point of view of non-invasion measurements, it was adopted that the measurement of the vibrations velocity in building partitions confining the residential premises will be done with a laser vibration meter. This complies with the principles of carrying out measurements in the building industry, as the parameter for determining the vibrations of building partitions is the vibrations velocity, which is also important considering the adopted coherence function method as the most appropriate one for evaluating the examined phenomenon.

The only problem with measuring the sound pressure level lies in the number of registrations needed in order to determine a reliable average value considering the type of noise measured in the residential premises (noise in the range of low frequencies).

Low-frequency noise is characterized by a low degree of sound dispersion in premises of standard dimensions (due to the length of the generated acoustic wave from several up to several tens of meters). The short duration of the measurement is connected with the need to register signals from single passage of the examined vehicles, is an additional hindrance. A certain solution comes in the large number of measurements connected with the need to register several times the passing of the same sources of traffic noise and vibrations, multiplied by the number of required registrations of the levels of vibration velocities on the partitions of the studied premises. Taking into consideration the variability of the same type of traffic sources from the technical point of view, it was assumed that at any one point measuring vibration velocity it will be necessary to register the signal at least three times.

The vibration velocity measurements should be conducted, as much as this is possible, on all of the uncovered surfaces of building partitions confining the studied premises (the tripod used to attach the laser head enables measurements to be carried out on the entire height of residential premises. Usually tests cannot be carried out on the floor (ceiling), as in most inhabited premises floors are covered with carpet coverings which prevent carrying out measurements, the so-called floating floors also give additional dampening of surface vibrations of the ceiling, deforming the results of measurements.

Analyzing the results of the measurements of soil vibrations and the measurements of the vibrations of facade on the side of the street in typical residential buildings during the passing of: a tram, bus, truck tractor with semi-trailer, it was observed that the dominating frequencies are found in the range up to 20 Hz (soil vibrations) and in the range up to 80 Hz (vibrations of external walls in apartments). This makes it possible to restrict the analysis of the sound pressure levels in residential premises to the range encompassing maximally the 2-harmonic of the vibrations on the partitions in the analyzed premises.

Due to the physical character of the measured phenomena, one common measurement range was determined for both quantities, namely the range encompassing the central frequencies of 1/3 octave bands 1 ÷ 160 Hz. This is the range of evaluating noise in the low-frequency band. This range includes frequencies applied in evaluating the influence of vibrations on building structures and inhabitants. Adopting the coherence function method for evaluating the acoustic conditions in residential buildings situated near traffic routes from simultaneously occurring noise and vibrations from traffic calls for the application of the narrow-band analysis using method FFT for the range of frequencies 1 ÷ 200 Hz with resolution Δf = 1 Hz.

4 DEVELOPMENT AND TESTING OF THE MEASUREMENT PROCEDURE

4.1 Model premises – laboratory conditions

A measurement experiment was carried out under laboratory conditions in order to develop a measurement procedure enabling to confirm the fact of occurrence of vibrations from heavy road traffic on building partitions confining the residential premises and estimate the influence of vibrations on the sound pressure level appearing in these premises. To this end, a series of measurements was programmed in a "model" of closed premises in the laboratory. In this case the word "model" means that a small room was erected, which represented closed premises, but during the tests the "model" was treated as the real small research room, i.e. the vibro-acoustic phenomenon was analyzed on a *1:1 frequency scale*. The room for tests consisted of a frame made of steel sections (angle bars) welded with each other to form a carrying structure with non-parallel walls. The walls were made of glass panels 4 mm thick, attached to the carrying structure on pads made of polyurethane foam in order to eliminate their contact with the chamber structure (energy bridges).

The excitation of the "model" structure to vibrations was carried out by supplying vibrations to the exciter flexibly attached to one of the partitions, a wide band signal (white noise) from the 20 ÷ 20000 Hz range. The excitation point (after testing several set-ups) was chosen non-symmetrically, in order to avoid the influence of the points of applying force on the response of the set-up (i.e. on the distribution of velocities of surface vibrations of the partitions).

The diagram of the measurement system used for analyzing the influence of the existence of vibrations on the sound pressure level inside the studied premises, as well as the way in which the measurements were carried out is presented on Figure 2.

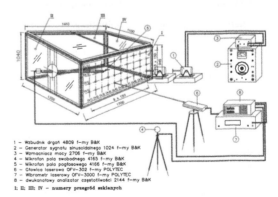

1 – Wzbudnik drgań 4809 f—my B&K
2 – Generator sygnału sinusoidalnego 1024 f—my B&K
3 – Wzmacniacz mocy 2706 f—my B&K
4 – Mikrofon pola swobodnego 4165 f—my B&K
5 – Mikrofon pola pogłosowego 4166 f—my B&K
6 – Głowica laserowa OFV–302 f—my POLYTEC
7 – Wibrometr laserowy OFV–3000 f—my POLYTEC
8 – dwukanałowy analizator częstotliwości 2144 f—my B&K
I; II; III; IV – numery przegród szklanych

Figure 2. Diagram of the measurement system for identifying the influence of vibrations on the sound pressure level (tests).

The obtained results of measurements and calculations during tests are presented on Figure 3.

4.2 Premises in the building – real conditions

Analyzing the location of buildings in Warsaw with respect to tram lines, transit routes, and bus lines, the building of the Museum of Independence was chosen as one of the extreme cases for conducting verification measurements. This building is located at Solidarności Avenue in Warsaw. It is found between traffic lanes, from which each one has a tram line. The tram lines are found at a distance of approx. 1 m from the external walls of the building.

The building of the Museum of Independence is a historical building (the Radziwiłł Palace) with a brick structure with 3 stories. The external walls have a thickness of approx. 1 m (this is a massive building). In order to verify the developed measurement procedure in real conditions, a measurement experiment had been programmed, which included:

- measuring the levels of vibrations velocities on the building's structural wall (in the cellar – below ground level), aimed at identifying the levels of vibration velocities mainly induced by tram traffic, received by the building structure,
- the proper measurement, i.e. simultaneous registration of the sound pressure levels and the levels of vibration velocities (on four building partitions) in a room specially chosen for this purpose (on the ground floor).

The room chosen for carrying out the measurements was found on the ground floor of the building, it had one window (wooden two-part double box) going out onto the street, the diagram of the room and the way in which the measurements were conducted is presented on Figure 4. Four partitions were chosen for registering the vibration velocities due to the technical conditions existing in the given room (the floor was covered with a short-pile carpet covering, and there were two door openings in one of the walls; the remaining free space was covered by a safe). The number of simultaneous registrations of the sound pressure levels and vibration velocities was determined by the number of measurement points of vibration velocities on the partitions confining this room. Measurement points of the vibration velocities were located on the remaining 4 partitions in a uniform oblong network at a distance of 60 cm from joint edges and a distance of 50 cm from each other, which for the dimensions of the partitions gave 16 measurement points on each partition. The exception here was the wall with the window and the wall with the door (12 points). The detailed plan of the placement of the points registering vibration velocities is given below.

Below two diagrams of the sound pressure level in the studied room are presented in order to demonstrate the influence of the registered acoustic background on the sound pressure level calculated using the value of the coherence function.

Carrying out the analysis of the influence of vibrations generated by heavy traffic on the sound pressure

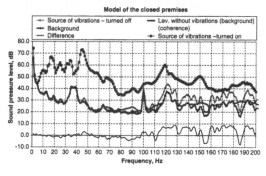

Figure 3. The values of the sound pressure levels in the "model" of the closed premises (a comparison of the real values with values obtained from calculations taking into consideration the coherence function).

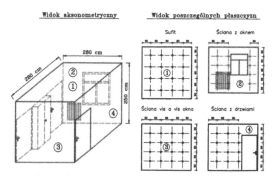

Figure 4. Diagram of the room being the subject of study in the building of the Museum of Independence, and the placement of points registering vibration velocities on chosen building partitions confining this room.

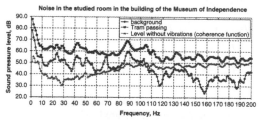

Figure 5. Sound pressure levels in a room in the building of the Museum of Independence, registered for the passing of a tram and calculated for a situation when the tram does not emit vibrations (not taking into consideration the acoustic background).

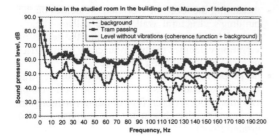

Figure 6. Sound pressure levels in a room in the building of the Museum of Independence, registered for the passing of a tram and calculated in a situation when the tram does not emit vibrations (taking into consideration the acoustic background).

level in closed premises one should take into consideration the level of the acoustic background as a reference point, as only this method of analysis is correct, i.e. enables the net evaluation of the passing of a heavy vehicle on the sound pressure level in the studied room, taking into consideration the sound transmitted to this room via the material path.

5 THE MEASUREMENT-CALCULATION PROCEDURE

Calculation procedures are following:

- The value of the increase in acoustic pressure $\Delta L_{p\,1\div200}$ (increase in the sound pressure level for sound transmission by material path by the examined partitions)

$$\Delta L_{p_i} = \gamma^2_i \cdot L_{p_i} \,, dB \;;$$

$$\begin{cases} \Delta L_{p_i} = \Delta L_{p_i} & L_{p_i} - \Delta L_{p_i} > L_{p,bgd_i} \\ \Delta L_{p_i} = L_{p_i} - L_{p,bgd_i} & L_{p_i} - \Delta L_{p_i} < L_{p,bgd_i} \end{cases} \qquad (1)$$

- The sound pressure level for airborne sound transmission $L_{air\,1\div200}$ (using the value of the coherence function)

$$L_{air_i} = (1 - \gamma^2_i) \cdot L_{p_i} \,, dB \;;$$

$$\begin{cases} L_{air_i} = L_{air_i} & ; L_{air_i} > L_{p,bgd_i} \\ L_{air_i} = L_{p,bgd_i} & ; L_{air_i} < L_{p,bgd_i} \end{cases} \qquad (2)$$

- The values: the real sound pressure level $L_{p\,1/3\,okt,}$ increase in the sound pressure level

$$L_{p_{k\,1/3\,okt}} = 20 \log\left[\sum_{f_{bk}}^{f_{tk}} (p_i \cdot \Delta f)/n_k \cdot p_0\right] ;$$

$$p_i = 10^{0.05 L_{p_i}} \cdot p_0 \,, dB \qquad (3)$$

- $\Delta L_{p\,1/3\,okt}$ (increase in the sound pressure level for sound transmission by material path by the examined partitions)

$$\Delta L_{p_{k\,1/3\,okt}} = 20 \log\left[\sum_{f_{bk}}^{f_{tk}} (p_i \cdot \Delta f)/n_k \cdot p_0\right] ;$$

$$p_i = 10^{0.05 \Delta L_{p_i}} \cdot p_0 \,, dB \qquad (4)$$

- The sound pressure level for airborne sound transmission $L_{air\,1/3\,okt,}$

$$L_{air_{k\,1/3\,okt}} = 20 \log\left[\sum_{f_{bk}}^{f_{tk}} (p_i \cdot \Delta f)/n_k \cdot p_0\right] ;$$

$$p_i = 10^{0.05 L_{air_i}} \cdot p_0 \,, dB \qquad (5)$$

The total real sound pressure level $L_{p\,(1\div160)},$

$$L_{p\,(1\div160)} = 10 \log\left[\sum_{k=1}^{23} 10^{0.1 \cdot L_{p_{k\,1/3\,okt}}}\right] \,, dB \qquad (6)$$

- The total increase in the sound pressure level $\Delta L_{p\,(1\div160)}$ (total increase in the sound pressure level for sound transmission by material path by the examined partitions)

$$\Delta L_{p\,(1\div160)} = 10 \log\left[\sum_{k=1}^{23} 10^{0.1 \cdot L_{p_{k\,1/3\,okt}}}\right]$$

$$- 10 \log\left[\sum_{k=1}^{23} 10^{0.1 \cdot L_{air_{k\,1/3\,okt}}}\right] \,, dB \qquad (7)$$

- The total sound pressure level for airborne sound transmission $L_{air\,(1\div160)}$ from the frequency range $1 \div 160$ Hz

$$L_{air\,(1\div160)} = 10 \log\left[\sum_{k=1}^{23} 10^{0.1 \cdot L_{air_{k\,1/3\,okt}}}\right] \,, dB \qquad (8)$$

where: γ^2_i, the value of the coherence function for i frequency band; L_{pi}, value of the real sound pressure level for i frequency band; L_{p,bgd_i}, the value of the level of acoustic background in the studied room for i frequency band, f_{bk}, bottom border frequency of k 1/3 octave band, Hz; f_{tk}, top border frequency of k 1/3 octave band; p_i, absolute value of acoustic pressure in the next constituent band of k 1/3 octave band; n_k, the number of measurement bands in the range of the analyzed 1/3 octave band; p_0, preference value for acoustic pressure; Δf, resolution of narrow band analysis (in our case $\Delta f = 1\,Hz$); i, next frequency band in k 1/3 octave band; $k,$-next 1/3 octave band.

616

6 SUMMARY AND CONCLUSIONS

The aim of the work was to demonstrate the possibility of developing and evaluation a method for estimation the influence of vibrations generated by heavy traffic on the acoustic conditions in residential buildings located near communication routes.

A measurement procedure was developed based on the statistical method of the coherence function, which made it possible to directly on the spot of the measurement determine the influence of occurring vibrations on the existing sound pressure level in the analyzed premises.

REFERENCE

Niemas M. "Ocena wplywu jednoczesnego wystepowania halasu i drgan od ruchu komunikacyjnego na warunki akustyczne w budynkach mieszkalnych". Ph.D. Thesis. Promotor prof. Jerzy Sadowski. Warszawa 1999, (in Polish).

Environmental Vibrations – Takemiya (ed.)
© 2005 Taylor & Francis Group, London, ISBN 0 415 39035 4

The evaluation of habitability to long period wind-induced horizontal torsional motion

S. Shindo & T. Goto

Department of Architecture, College of Engineering, HOSEI univ.

ABSTRACT: This paper consists of two chapters. At chapter 1, consideration of the experiment about human perception is developed. A human body detects motion with various feelings; the objects of this study are human visual and physical perception. The team of chapter 2 is motion analysis of existing buildings. The objects of observation are two buildings in TOKYO. The main observation item was the acceleration of translational motion along with wind direction. Furthermore, it is possible to calculate the velocity of the torsional motion by using the data of translational motion.

1 INTRODUCTION

High-rise buildings over 400 m in height have finally become reality thanks to high-technology architecture. Furthermore, it is theoretically possible to build 4,000 m tower-type buildings. Now, high-rise buildings are used not only for offices and shops but also for residences and hotels. But, high-rise buildings have some shortcomings, such as wind-induced long-period horizontal motion. In 1979, ten years after the first high-rise building emerges in Japan, Tokyo was visited by the large typhoon 7920. Many residents complained of headaches and giddiness, these symptoms were similar to seasickness.

Researchers are inquiring on the comfort in high-rise buildings. These searches can be divided into two fields: motion analysis of a building and motion feeling of a human body. Based on these researches, evaluation standards for high-rise buildings have been introduced in many countries. However, numerical values differ. In addition, there are no evaluation standards that take into account important factors such as the perception of torsional motion. Moreover, ISO6897, national building code of Canada, and the architectural institute of Japan do not mention torsional motion, and the numerical values for torsional motion in the European convention for construction steel work have not been confirmed by human perception. Therefore, in order to propose a new evaluation standard for torsional motion at AIJ, the author began studies.

This study consists of two sections; human perception of long-period torsional motion, motion analysis of real high-rise buildings.

2 CONSTRUCTION OF VISUAL SIMULATOR

For the experiment on human perception, it is important to take into account the relationship between three factors: experimental room, motion, and window view. The author developed the new technique in order to give a subject vision information. In this new method, the view from a window was recorded by a video camera moved at very slow speed and then from an elevated room at the same speed. After the recording, the original movie was edited, with the speed being changed by computer. In this way, a visual simulator can create any movie with perfect velocity and frequency, which permits examining human visual perception thresholds and obtaining some concrete results. This chapter describes the visual simulator used for the experiment.

2.1 *Movie-editing system*

The movie to be edited was given to participating subjects. The equipment used for editing included a computer (Apple: Macintosh 8100/100) and a periphery device (Radius: Video Vision System). Video Vision System is a video hardware interface that captures and

Table 1. Evaluation standards.

	Translational motion	Torsional motion	Visual perception
ISO	O	×	×
AIJ	O	×	×
ECCS	O	O	×

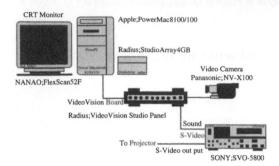

CRT Monitor
Apple;PowerMac8100/100
Radius;StudioArray4GB
NANAO;FlexScan52F
Video Camera
Panasonic;NV-X100
VideoVision Board
Radius;VideoVision Studio Panel
Sound
S-Video
To Projector
S-Video out put
SONY;SVO-5800

Figure 1.   Movie editing system.

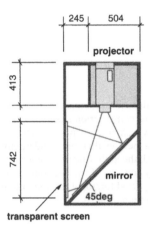

245 | 504
projector
413
742
mirror
45deg
transparent screen

Figure 2.   Projecting system.

plays back full-frame, full-motion composite, and S-video in either NTSC or PAL format. Figure 1 shows the equipment. The editing was conducted according to the following procedure:

2.2   *Projecting system*

Movies made using the filming equipment were projected from a video projector (SANYO: LP-9200N) via a video deck (SONY: SVO-5800) and reflected on a mirror to be displayed on a transparent screen. Figure 2 shows the projecting system.

2.3   *Experimental room*

The dimensions of the room were 2.33×2.33 m²; the window size was 80×60 cm². The angle of sight calculated in terms of the distance from subject to window was approximately 40 deg. Figure 3 shows a plan of the experimental room.

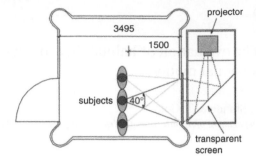

projector
3495
1500
subjects   40°
transparent
screen

Figure 3.   Plan of experimental room.

Table 2.   Questionnaire of during the experiments.

Imperceptible	Imperceptible
	Faintly felt motion
Perceptible	Vaguely felt motion
	Clearly felt motion

3   HUMAN PERCEPTION OF TORSIONAL MOTION

There are two methods of investigating the human perception of motion; questionnaire survey is conducted in an actual building, or in equipment of laboratory space. For this study latter has been chosen. A human body detects motion with various feelings; the objects of this study are human visual and physical perception.

3.1   *Summary of experiment*

3.1.1   *Contents of the questionnaire given to subjects*

Table 2 shows questionnaires to be filled out during the experiments. The selected subjects were healthy people who had answered "normal" to questions about their ordinary sensitivity to motion and their physical condition on the experiment day in a questionnaire given to them before the experiment. In reply to the questionnaire provided after the experiment, some mentioned the majority of subjects considered the quality of the movie excellent, although the need to improve the "reality" provided by the simulator.

3.1.2   *Experimental condition*

Taking into account the natural frequency of a real building, the experimental condition for visual experiments shown in Table 3, and physical experiments shown in Table 4 were chosen. Photos 1 and 2 shows the experimental view.

3.2   *Experiment results*

Of the predetermined response ratings, "clearly felt motion" level A and "vaguely felt motion" level B were

Table 3. Experimental conditions for visual.

Frequency [Hz]	Angular Velocity [mrad/s]	Frequency [Hz]	Angular Velocity [mrad/s]
0.076	0.40	0.202	0.40
	0.60		0.60
	0.80		0.80
	1.20		1.20
	1.60		1.60
	–		3.20
0.101	0.40	0.303	0.40
	0.60		0.60
	0.80		0.80
	1.06		1.20
	1.60		1.60
	3.20		3.20
0.121	0.40	0.606	0.40
	0.60		0.60
	0.80		0.80
	1.20		1.20
	1.60		1.60
	3.20		3.20
0.152	0.40		
	0.60		
	0.80		
	1.20		
	1.60		
	3.20		

Table 4. Experimental conditions for physical.

Frequency [Hz]	Angular Velocity [mrad/s]	Frequency [Hz]	Angular Velocity [mrad/s]
0.076	0.83	0.200	2.19
	1.66		4.38
0.101	1.11	0.303	3.32
	2.22		6.65
0.121	1.32	0.588	6.45
	2.64		12.90
0.152	1.66		
	3.32		

Photo 1. Distant window view.

Photo 2. Closed window view.

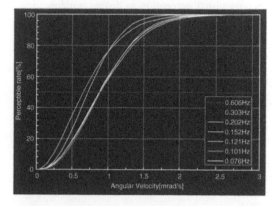

Figure 4. Results of torsional visual perception experiments at Weibull curves (distant window view).

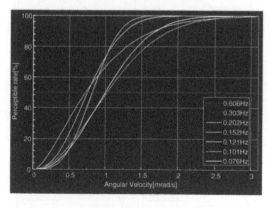

Figure 5. Results of torsional visual perception experiments at Weibull curves (closed window view).

considered to be evidence that the subjects had perceived motion. Weibull distribution curves were used to approximate levels of visual perception against peak angular velocity for the experiments of visual perception, as shown in Figs 4 and 5.

Weibull distribution curve: $Y = 1\text{-exp}\,\{-(x-r)^{m/b}\}$
Y: Perception Ratio [%], x: Period [sec]

Figure 4 and 5 show the collected distribution curves, with the same tendency toward every frequency. Consequently, measuring the degree of motion with a scale in angular velocity, the human perception threshold appears to be independent of the frequency, and the average perception threshold attains the same velocity. And Table 5 and 6 show the visual perception threshold at the experiments.

3.2.1 *Comparison of visual and physical perception*

Figure 6 compares the results of the experiments of torsional visual perception as distant window view

Table 5. Perception thresholds (distant window view).

Frequency [Hz]	0.076	0.101	0.121	0.152	0.202	0.303	0.606
50% [mrad/s]	0.83	0.84	0.67	0.84	0.61	0.90	0.83

Table 6. Perception thresholds (closed window view).

Frequency [Hz]	0.076	0.101	0.121	0.152	0.202	0.303	0.606
50% [mrad/s]	0.95	0.97	0.71	0.77	0.68	0.71	0.77

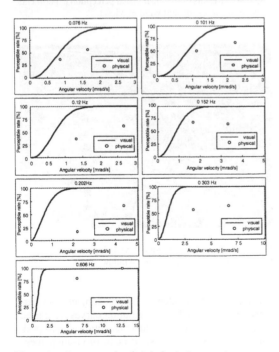

Figure 6. Compares visual and physical perception.

and torsional physical perception. The figure indicates that visual perception is more sensitive than physical perception except in the case of translational motion. Therefore, the visual factor should be taken into account when proposing a new guideline for torsional motion.

3.2.2 *Comparison with the evaluation standards*

This experiment was conducted to evaluate effects of torsional motion on habitability in existing high-rise buildings. Table 7 lists the motion-related evaluation standards for habitability applicable in ECCS and Isyumov's value. The Isyumov's value sets a standard for torsional motion on the highest floor of a building, with emphasis on the effect of visual perception on habitability. Specifically, the value is set at 1.5 mrad/s for wind load over 1year and 3 mrad/s for that of 10 years at the Isyumov's value, and 1.0 mrad/s for that of 1 year at ECCS. However, there are no reports on experiments where subjects were put in an artificial environment of the stated torsional motion levels in order to evaluate habitability. Consequently, the results of experiments are compared with the evaluation standards. Figure 7 shows that the results of the experiments are lower than the numerical values of standards. Moreover, the numerical value of line A corresponds to the 100% perception threshold of the experiments. Similarly, line B corresponds to 95%, and line C corresponds to 70%.

Table 7. Evaluation standards of motion related habitability.

	Recurrence period	Angular velocity	mark
ECCS	1 year	1.0 mrad/s	C
N. Isyumov	1 year	1.5 mrad/s	B
	10 years	3.0 mrad/s	A

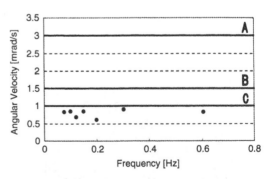

Figure 7. Compares experiment result and evaluation standards.

4 MOTION ANALYSIS OF EXISTING BUILDINGS

The objects of observation are 2 buildings existing in Shinjuku, TOKYO. The structural outlines are shown in Table 8. Observation periods were 2 years and 4 months at building B, and for 1 year at building A. The main observation items were acceleration of the translational motion for every 10 minutes, in addition to wind direction and power at that time. Furthermore, it is also possible to calculate the velocity of the torsional motion by using the data of translational motion. The relation between angular velocity of the torsional motion and the average wind power are shown in Fig. 8 (building A), 9 (building B).

Large-scale motion is observed at the wind direction SSE, S, and SSW, at building A. Therefore; it is considered that the above-mentioned three wind directions in A building. Figure 10 shows the relation

Table 8. Structural outlines of 2 buildings.

	Building A	Building B
Structural type	Steel frame (RC for floors 1to3)	Steel frame (RC for floors 1to5)
Maximum height	203 m	226.5 m
Use purpose	Office	Office (partly retail)
Primary natural frequency	0.328 Hz (Torsional motion)	0.233 Hz (Torsional motion)

between angular velocity of the torsional motion and the average wind power.

Especially in a wind direction SSW, distribution of data is large. In order to explore the cause which distribution becomes large, the gust-factor of the wind is shown in the following Fig. 11.

At the wind from south, the gust-factor is large. Also, a harder motion has occurred in the wind direction with a large gust-factor. Therefore, a possibility is suggested that a harder motion will arise when the wind is confused.

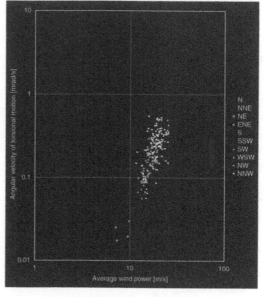

Figure 9. Relation between the torsional motion and the wind power (Building B).

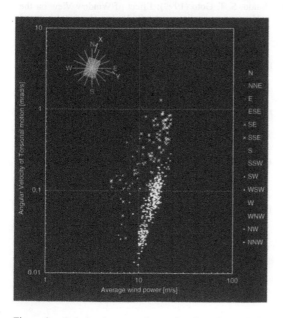

Figure 8. Relation between the torsional motion and the wind power (Building A).

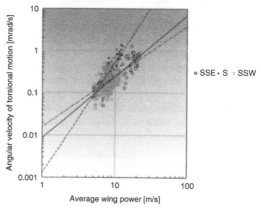

Figure 10. Relation between the torsional motion and the wind power (Building A: SSE, S, SSW).

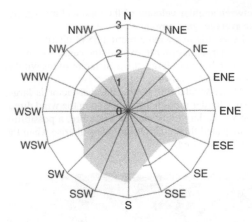

Figure 11. Gust-factor.

5 CONCLUDING REMARKS

The previous researches on human perception of the motion were analyzed. The following five opinions were acquired as a result of breakdown.

1. Most researches are carrying out for translational motion.
2. There are few researches for torsional motion and visual perception.
3. There is little research for the motion analysis of existing buildings.
4. A difference is in a numerical value for the evaluation standards.
5. There is only one guideline in consideration of torsional motion.

Consideration of the experiment about human perception is developed. The authors come to a following conclusion at this experiment:

6. Measuring the degree of motion with a scale in angular velocity, the human perception threshold appears to be independent of the frequency.
7. Average visual perception threshold for the torsional motion is 0.79[mrad/s].
8. Visual perception is more sensitive than physical perception, except in the case of translational motion.
9. Visual perception threshold of the experiments are lower than the numerical values of standard and existing proposal value.

The team of chapter 4 was motion analysis of existing buildings. The authors come to a following conclusion at this analysis:

10. Motion exceeding average visual perception threshold is observed only at A building.
11. Large-scale motion is observed in the wind direction SSE, S and SSW, at building A.
12. The cause of a large motion is that the scale of gust-factor is large with the wind coming from south.

ACKNOWLEDGEMENT

We would like to express our special thanks to many subjects and member of laboratory.

REFERENCES

John W. Reed (1971): Wind-induced Motion and Human Discomfort in Tall Buildings, Massachusetts Institute of Technology

Irwin A.W. and T. Goto (1984): Human Perception, Task Performance and Simulator Sickness in Single and Multi-axis Low Frequency Horizontal Linear and Rotational Vibration

Ishikawa, T. C. Noda, F. Kumazawa, T. Okada (1994): A Study on Serviceability Limit for Wind-Induced Motions in Consideration of Visual Stimulus, Journal of Structural Engineering, 40B, pp. 1–6, March

Shindo, S. T. Goto (1999): Effect of Window View on the Human Perception to Tall Building's Wind-Induced Torsional Motion, The 1999 International Conference on Tall Buildings & Urban Habitat, Kuala Lumpur, Malaysia

ISO6897 (1984): Guidelines for the evaluation of the response of occupants of fixed structures, especially buildings and off-shore structures, to low frequency horizontal motion (0.063 to 1Hz)

National building code of Canada (1990): Part 4 Structural Design, Supplement 4

Architectural Institute of Japan (1991): The guidelines for the Evaluation of Habitability to Building Vibration

European Convention for Constructional Steelwork (1987): Recommendations for the calculation of wind effects on buildings and constructions, Technical Committee T12

Isyumov N. (1993): Criteria for Acceptable Wind-induced Motion of Tall Buildings, International Conference on Tall Buildings and Urban Habitat, Rio de Janeiro, 17–19 May.

Environmental Vibrations – Takemiya (ed.)
© *2005 Taylor & Francis Group, London, ISBN 0 415 39035 4*

Experimental research relating to responsive appraisal of compound vibrations and noise stimuli

T. Goto
Professor, Dept. of Architecture, Faculty of Eng. Hosei Univ., Tokyo, Japan

Y. Nagaseki
Mitsui Home Co., Ltd., Tokyo, Japan

M. Shimura
Director, Civil Engineering and Eco-Technology Consultants Co., Ltd., Tokyo, Japan

R. Endo
Associate Professor, Dept. of Architectural system eng., The Polytechnic University, Kanagawa, Japan

Y. Kawakami
Lecturer, Dept. of Architectural system eng., The Polytechnic University, Kanagawa, Japan

ABSTRACT: This research deals with the problem of evaluating building floor vibrations caused by railway operations from a livability point of view. It is distinct in that it deals with vibrations and noise as compound stimuli, as opposed to independently as has always been the case. The aim of this research is to further improve evaluating guidelines (AIJ)[1] that are already public.

1 INTRODUCTION

With increasing urbanization in recent years, the incidence of residences (such as condominiums, apartments and houses) built adjacent to railway lines is increasing, and this has been raised as a cause of a worsening residential environment. Residents living near railway are not just exposed to vibrations, but also to noise. However, vibrations and noise are presently evaluated according to independent criteria. Thus, when living spaces subject to vibrations and noise are evaluated, it is preferable that vibrations and noise are evaluated in terms of the relationship between themselves, as with the relationship between temperature and humidity when evaluating a thermal environment. Thus, this research focuses on railway vibrations and conducts response experiments with the following two types of subjects.

Experiment I is an experiment whereby subjects were simultaneously exposed, in an experimental chamber, to residential floor vibrations and noise of railway operations really measured, to evaluate the degree of response. The evaluation was comprised of three categories: an evaluation limited to just vibrations, an evaluation limited to just noise, and a general response combining vibrations and noise. Experiment II is an experiment for evaluating the degree to which the subject responds to sinusoidal oscillations extracted from the floor response acceleration spectrum described above.

2 EXPERIMENT I: EXPOSURE TO BOTH VIBRATIONS AND NOISE SIMULTANEOUSLY

2.1 Experiment objective

The objective of performing Experiment I was to confirm (1) the effect of noise on the vibration response evaluation, (2) the effect of vibrations on the noise response evaluation and (3) the effect of vibrations and noise on the evaluation of the compound response to vibrations and noise.

2.2 Experiment outline

2.2.1 Devices used in the experiment
The experiment used a 3D vibration exciter (IMV: TE-2000S-15L, Electro Dynamic Drive, 1,500 × 1,500 mm) at The Polytechnic University. An

experimental chamber was constructed on the vibration exciter. The experimental chamber was made of wood. Figure 1 shows the exterior appearance of the experimental chamber. Figure 2 shows the interior appearance of the experimental chamber. The devices shown in the photograph are described below.

To shut out noise from the vibration exciter that can influence the experiment, that could be considered to influence, the walls of the experimental chamber were constituted from the outside, by 12 mm plasterboard, 12 mm laminated wood, a 2 mm soundproof rubber sheet and 12 mm plasterboard. The inside of the experimental chamber was provided with a speaker, a fan, an internal phone and lamps. The speaker generates the railway noise and the internal phone and lamps are used for communications between the subject in the experimental chamber and the outside experimenter.

2.2.2 *Experimental conditions*

For the object vibration, a vertical vibration[2),3)] indicated by Measurement data line(due to a real measurement) in Figure 3, was outputted to the vibration exciter. The vibrations exposed to the subject were vibrations with their frequency characteristics unaltered, but the level of acceleration altered in six steps. The other curves in the same figure show the results of reproducing the conditions at the six experimental levels, and results measured in the center of the floor of the experiment chamber in a one third octave band analysis. The values used were the maximum values based on root mean square. When these were compared to the Measurement data line, there was general matching of the frequency characteristics, although the experimentally conditioned values were slightly lower in the low frequency range of 16 Hz or less. However, the conditions were judged to be sufficient for a comparative experiment since they showed substantially comparable frequency characteristics.

The following concerns noise to which a subject was exposed. Railway noise inside the real wooden house was recorded with a digital recorder at the same time as when the Measurements data as a floor vibrations were being taken. These were exposed to the subject through the speakers in the experimental chamber. Figure 4 shows the exposure conditions at a vibration acceleration level of 91 dB. Comparing the observed values, the generation of solid born noise was confirmed near 63 Hz.

As seen above, the experimental conditions consisted of the following. Seven conditions where six conditions were vibrations varied between a vibration acceleration level of 61 to 91 dB (VAL) in 5 dB (VAL) pitch units and one condition of "no amplified vibration (background vibration 54 dB (VAL))"; and six conditions where five conditions were noises varied between a noise level of 62 to 94 dB (A) in 8 dB (A) pitch units and one "no noise (background noise 60 dB (A))"

Figure 1.　Exterior appearance of the experimental chamber & vibration exciter.

Figure 2.　Interior appearance of the experimental chamber.

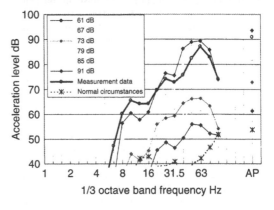

Figure 3.　Measurement vibration and objective vibration.

626

condition. The total experimental conditions in the end comprised of 41 conditions: 42 combinations of vibrations and noise, minus one condition, the combination of "no vibration, no noise". It should be noted that the dotted line in Figure 4 shows, for reference, the equal loudness contour. Furthermore, in the notation to follow, symbols within the brackets of vibration level dB (VAL) and noise level dB (A) are omitted.

2.2.3 Questionnaire content

The subjects were a mix of 20 men and women in their 20s. The questionnaire relating to personal information was obtained after the completion of both Experiment I and Experiment II.

The vibration response evaluation performed in Experiment I was comprised of seven levels: "1. Felt nothing", "2. Uncertain", "3. Felt slightly", "4. Felt clearly", "5. Felt strongly", "6. Felt severely" and "7. Unbearable". The noise response evaluation was comprised of seven levels: "1. Heard nothing", "2. Uncertain", "3. Heard slightly", "4. Heard clearly", "5. Loud", "6. Very loud" and "7. Unbearable". The evaluations noted above are principally perceptive evaluations. The general response evaluation was comprised of seven levels: "1. Absolutely no problem", "2. Basically no problem", "3. Slightly annoying", "4. Annoying", "5. Very annoying", "6. Incredibly annoying" and "7. Unbearable". The evaluation here may be considered a ranking of interference reaction.

The values shown on the response evaluation axis of the graphs below mean the content of the evaluations given above.

2.2.4 Experiment method

After explaining the overview of the content and proceedings of the experiment to the subject, the experimenters directed the subject to sit directly on the floor of the experimental chamber. Firstly, the subject was exposed once each to the maximum and the minimum level of vibrations and noise of the experimental conditions. The experiment was then started after one practice run, in which one of the experimental conditions was randomly selected. The order of the experimental conditions was arranged so that 'magnitude' was random and the same for all subjects. The response was evaluated after the subject was exposed twice to a single condition. After the test of a single condition, the experimenters contacted the subject via the internal phone mentioning that the experiment was progressing to the next condition. Furthermore, with consideration to fatigue of the subject, the experiment was performed with a 15 minute break after every fifteen conditions.

2.3 Experiment results and considerations

2.3.1 Regarding personal information
Due to lack of page space, answers to personal information, have been hereby omitted from this paper.

2.3.2 Vibration response evaluation of vibration stimulus without noise
Here, the discussion relates only to those vibration response evaluations of the 41 conditions in which vibrations were exposed without noise.

The average values and standard deviations of the vibration levels that correspond to the vibration response evaluations supplied by the subjects are shown in Figure 5. When the average values that are based on the results in the column on the left of the figure (normal circumstances) are considered, the vibration level values that correspond to the range of "1. Felt nothing" to "7. Unbearable" of the response evaluation are distributed in the range from just over 60 dB to about 90 dB. As opposed to vibration level changes within the range of "1. Felt nothing" to "5. Felt strongly"; "6. Felt severely" and "7. Unbearable" are distinctly close. That is to say, the tolerance level was quite wide in the low range of the response evaluation, but narrowed as the response evaluation increased. At the same time, the spread of the standard deviation was wide with a small vibration response evaluation, and narrowed as the vibration response evaluation

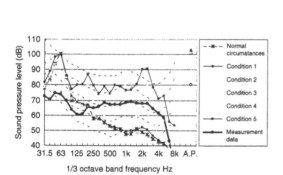

Figure 4. Exposure conditions at a vibration acceleration level of 91 dB.

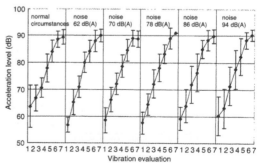

Figure 5. The average values and standard deviations of vibration response evaluation.

increased. Therefore, the evaluation was ambiguous while the exposure value was small, but the results showed a tendency toward constricted responses as the exposure level increased.

2.3.3 Vibration response evaluation of exposure to both vibrations and noise simultaneously

When a change due to added noise is viewed on the same figure: the acceleration level value decreases for response evaluations that are low, "1. Felt nothing" or "2. Uncertain". Put another way, a tendency to response sensitivity may be seen with added noise, but in the reactions relating to middle to high response evaluations of "3. Felt slightly" and above, no substantial change was observed. Accordingly, it is difficult to conclude that there is an overall effect. In relation to the standard deviation, there is the same tendency toward less variability with a greater vibration response evaluation, as there is in the vibration exposure without noise.

The vibration exposure without noise and the exposure to both vibrations and noise simultaneously in Figure 6 are compared to the average values of the vibration response evaluation. No tendency toward a higher vibration response evaluation in accordance with increased noise exposure value under any vibration condition can be seen. Accordingly, it cannot be concluded that vibration response evaluations increase with added noise.

2.3.4 Noise response evaluation of exposure to noise stimulus without vibrations

Figure 7 shows the average values and standard deviations of noise levels according to noise response evaluations during noise exposure without vibrations.

According to the results in the column on the left of this figure (normal circumstances), with regards to noise, the first condition (62 dB) already exceeded the perceptive level, thus there are no results corresponding to the evaluation "1. Heard nothing", and accordingly, the evaluation of 62 to 94 dB corresponds to the range of "2.Uncertain" to "7. Unbearable". For noise, the noise level values corresponding to the low response evaluation of "2. Uncertain" and "3. Heard slightly" are very close, and this is the opposite result to the case of vibration exposure without noise. It would seem that this is because sensitivity toward sound is greater than sensitivity toward vibrations.

2.3.5 Noise response evaluation of exposure to both vibrations and noise simultaneously

The response "1.No sound", that was not seen during the evaluation of the noise exposure without vibrations, appeared with the addition of vibrations when the response evaluation to noise exposure without vibrations was compared to the response evaluations to noise relating to simultaneous exposure to vibrations and noise. Therefore, it can be seen that the noise is masked by the vibrations when vibrations are added, and shut out.

When considering the reaction of the low response evaluation "2. Uncertain", as opposed to the level of noise exposure without vibrations being 64 dB, the level corresponding to "2." of the vibrations and noise simultaneous exposure, is lower than the noise level of 60 dB, apart from the case when it was combined with the 91 dB vibrations. It can be seen also from this fact that a certain effect may be caused by the addition of vibrations. At the same time, in the response evaluations of "6. Very loud" and "7. Unbearable", overall, the values where vibrations were added were larger, and a tendency toward damping of the values can be seen.

However, when the stimulating level of vibrations increase (85 dB or 91 dB), the noise level value decreases, indicating that vibrations may lead to greater annoyance. Regarding the noise with noise exposure without vibrations and exposure with both vibrations and noise simultaneously, even if the average values of the response evaluation are plotted in a similar manner to Figure 6, the content described above is substantially corroborated. However, all in all,

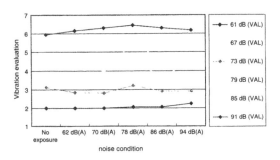

Figure 6. Compared to the average values of the vibration response evaluation.

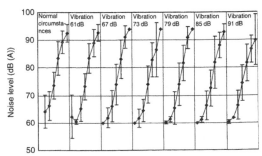

Figure 7. The average values and standard deviations of noise levels.

628

a correlating correspondence between the increased ratios of the vibration stimulus value to the response evaluation could not be confirmed.

It should be noted that when the standard deviations are considered, a tendency for the variability to decrease can be seen when the noise response evaluation is large, with the exception of the vibration conditions 73 dB and 91 dB. However, overall, the variability of the standard deviations is larger than in the vibration response evaluation of Figure 5.

2.3.6 The effect of vibrations and noise on the compound response evaluation

Here, the general response evaluation that includes both vibrations and noise is described in relation to the same conditions given thus far.

The subjects were asked for a combined response evaluation with both vibrations and noise, in relation to the 41 conditions, and the results of average values and standard deviations of the vibration acceleration level are shown in Figure 8 in the same way as before. According to this, the relative correspondence between the response evaluation and the level value, relating to the compound conditions of exposure to both vibrations and noise simultaneously as opposed to vibration exposure without noise is very clear, differing from those of Figure 6 and Figure 7. Thus, as the exposed stimuli of the compound conditions increase, the general tendency is for the level of the value of the compound response evaluation to decrease is clearly expressed (downward slope). If looked at more closely, although the average value of the vibration acceleration level was 92 dB when the compound response evaluation was "7" in the conditions that included no noise, the acceleration level was about 83 dB when the exposed noise was 94 dB, 9 dB smaller.

Although the width of the standard deviations decreases with increasing magnitude of the compound evaluation with exposure to vibrations without noise, this tendency is observed only with noise exposure up to 78 dB. For noise levels of 86 dB 94 dB and the like,

there is a converse increase in the width of the standard deviations, and it can be assumed that the perceptive difference between individuals is larger.

2.3.7 Consideration

When speculating the results noted above, it can be inferred that compositely evaluating the compound stimuli, rather than focusing on and evaluating stimuli individually, a more accurate degree of degradation in the environment is possible.

In order to further substantiate the content above, below is a multiple regression analysis performed on the results of the compound evaluation, wherein the compound response evaluation is the object variable, and vibrations and noise are explanatory variables. These results are shown in Table 1. The results show a high multiple correlation coefficient of 0.899. Since the multiple correlation coefficients are the correlation between the compound evaluation and the vibrations and noise, it can be stated that the numerical value of the compound evaluation in this experiment is not affected by other conditions, and that the result is a pure assessment of the vibrations and noise of the experimental conditions.

Accordingly, regression analysis was attempted to verify the extent to which the vibrations and sound, the exposure to which was the object of the compound evaluation, individually affect the results. Simple regression analysis was performed wherein the compound evaluation is the explanatory variable, and the vibrations and noise are the object variables.

Table 1. Results of multiple regression analysis.

	Correlation R	0.899
	Decision R2	0.809
	Correction R2	0.799
	Standard error	0.681
	Number of observations	41.000
Cut	Coefficient	−6.674
	Standard error	0.924
	t	−7.221
	P-Value	0.000
	Lower bound 95%	−8.545
	Upper bound 95%	−4.803
Vibration	Coefficient	0.071
	Standard error	0.009
	t	7.834
	P-Value	0.000
	Lower bound 95%	0.052
	Upper bound 95%	0.089
Noise	Coefficient	0.084
	Standard error	0.008
	t	10.391
	P-Value	0.000
	Lower bound 95%	0.067
	Upper bound 95%	0.100

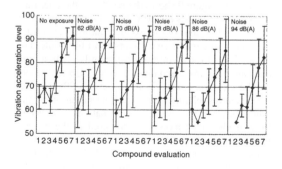

Figure 8. Average and values and standard deviations of the vibration acceleration level.

Table 2. Vibration.

	Correlation R	0.516
	Decision R2	0.266
	Correction R2	0.247
	Standard error	1.317
	Number of observations	41.000
Cut	Coefficient	−0.057
	Standard error	1.296
	t	−0.044
	P-Value	0.965
	Lower bound 95%	−2.679
	Upper bound 95%	2.564
Vibration	Coefficient	0.065
	Standard error	0.017
	t	3.759
	P-Value	0.001
	Lower bound 95%	0.030
	Upper bound 95%	0.101

Table 3. Noise.

	Correlation R	0.707
	Decision R2	0.500
	Correction R2	0.487
	Standard error	1.087
	Number of observations	41.000
Cut	Coefficient	−1.225
	Standard error	0.972
	t	−1.261
	P-Value	0.215
	Lower bound 95%	−3.191
	Upper bound 95%	0.740
Noise	Coefficient	0.080
	Standard error	0.013
	t	6.249
	P-Value	0.000
	Lower bound 95%	0.054
	Upper bound 95%	0.106

These results are shown in Table 2 and Table 3. Thus, since the determinant coefficient of the vibrations shows a low value of 0.266, the application of the line of regression is not particularly good. Since the P-value is 0.001, it can be seen from an examination of the results that the simple regression coefficient of vibrations is significantly not zero at a 5% significance level.

At the same time, when the determinant coefficient of noise is considered, it is approximately twice the value of that of the vibrations, at 0.50, and it can be seen that the application of a straight line is improved. Furthermore, the numerical value of the regression coefficient is 0.08. Since the P-value shows a value of 0.001 or less, it can be stated that the regression coefficient is significantly not zero at a 5% significance level.

The correlation coefficient between the exposed vibrations and noise is −0.02. This indicates that there is no relationship between the exposed vibrations and noise, and can be stated that since the vibrations and the noise do not affect each other, it is possible to measure the strength of their influence simply by comparing the size of the simple regression coefficients. Since the simple regression coefficient of noise is 0.08, and is 0.015 larger than the simple regression coefficient of the vibrations at 0.065, it can be seen that there is a tendency to be affected by noise rather than vibrations when measurements are taken in an environment having exposure to both vibrations and noise simultaneously. This is consistent with the developments described in Sections 2.3.2 to 2.3.5.

3 EXPERIMENT II: SINUSOIDAL WAVE VIBRATIONS RESPONSE EXPERIMENT

3.1 Objective of the experiment

This was an experiment to identify which frequency components of those frequency components of floor vibrations caused by railways are felt. Thus, the suitability of the evaluated range of vertical vibrations (3 to 30 Hz) specified in the guidelines for the evaluation of habitability[*1] was investigated.

3.2 Outline of the experiment

In this experiment, a vibration response test was carried out using sinusoidal vibrations. This experiment was performed, based on the results of floor vibration measurements[2),3)] in a wooden home adjacent to railway tracks. These measurements were carried out jointly between the Goto Research Laboratory of Hosei University, and the Vibration Measurement Technical Sub-Committee, under the auspices of the Environmental Vibration Committee of the AIJ.

An outbound train line is located 5 m from the object home on the other side of a concrete retaining wall, and an inbound train line is positioned another 10 m away. Figure 9 shows the results of the one third octave band analysis of vertical vibrations measured at point C in the floor of the wooden home when a train runs along the tracks five meters from the home. For the analysis, measurements were recorded with a time constant of 10 msec and the maximum effective value.

The solid lines in the figure (V-10 to V-90) are evaluation curves relating to vertical floor vibrations in the guidelines for the evaluation of habitability. When these evaluation curves are compared to the experimental measurement results, the vibrations generated at the floor surface correspond to V-50. However, there doubts emerged about the suitability of V-50 to handle the evaluation because the floor vibrations were felt by all 10 members of the measurement group.

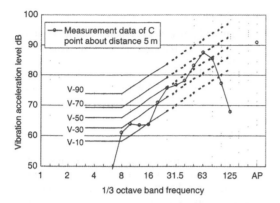

Figure 9. Results of the one-third octave band analysis of vertical vibrations measured on a real wooden floor and the guideline curves of AIJ.

Therefore, as can be seen by the dotted line in the figure, when the evaluation curve is extended to beyond 63 Hz, which is the prevailing vibration, the evaluation corresponds to V-70, and the application of the evaluation nears. Because of such reasons, the real measurement vibration values at the C point were set to be the foundation of this experiment. Paying particular attention to the frequency, the authors of this experiment have examined which vibrational frequencies corresponded to the response evaluation, and have investigated the need to widen the evaluated range to frequencies higher than 30 Hz, which is the upper limit specified in the evaluation guidelines.

3.3 Experimental conditions

The frequencies for the object of this experiment were selected, and the experimental level values were determined based on the experimental measurement data seen in Figure 9. These were 7 points selected from the frequencies according to the evaluation guideline values V-30, and from the frequency region of the range of vibration levels above these. Sinusoidal waves, in which those vibration acceleration levels were set as the peak, were generated in the vibration exciter used in Experiment I.

3.4 Experiment method

The subjects sat directly on the floor of the experimental chamber, and were exposed to the seven sinusoidal wave conditions. A response evaluation was performed with respect to those vibrations. Seven response levels were provided for the vibration response evaluation in a similar manner to Experiment 1, from "1. No sensation" to "7. Unbearable". The subjects were 11 men and 9 women, giving a total of 20.

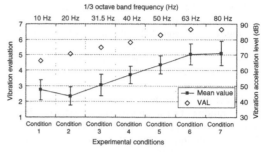

Figure 10. Average values and standard deviations of the vibration response evaluations.

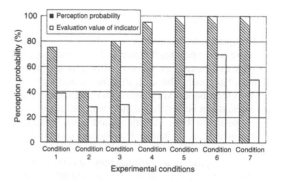

Figure 11. Comparison of the vibration perceptive probability.

3.5 Experiment results and considerations

Figure 10 shows the average values and standard deviations of the vibration response evaluations for the vibration conditions. By observing this, since the vibration response evaluations for 63 Hz and 80 Hz show "5. Felt strongly", it can be understood that of those floor vibrations generated at the point C while measuring vibrations in the experimental chambers, the 63 Hz and 80 Hz frequency region, are those regions in which the vibrations are most strongly felt. Figure 11 shows a comparison of the vibration perceptive probability obtained in this experiment, and the perceptive probability values provided by the evaluation guidelines. The perceptive probabilities of this experiment were calculated by dividing the responses into: vibrations felt when the response evaluation given by the subject was a reply of "3. Felt slightly" and above, and vibrations not felt when the reply was "2. Uncertain" and below. Furthermore, the probability values for 30 Hz and above for the evaluation guidelines were calculated according to the dotted lines in the previous Figure 9. It can be seen from these results (Figure 11) that the perceptive probability during this experiment is greater than with the guidelines under all conditions. Moreover, in relation to the conditions of 40 Hz or more, which is outside the range of the designated evaluation guidelines, all the results show a large perceptive probability of 95% or greater. Thus, it can be ascertained that it is necessary to also include the evaluations of frequencies of 30 Hz or greater within the range of frequencies designated by the guidelines. In particular, in relation to the responsive vibrations of structures caused by external vibrations such as railways, the frequencies to be evaluated should also include a range of frequencies exceeding the most prominent frequency.

4 CONCLUSIONS

From experiment I

1. Compared to exposures to vibrations without noise and exposures to noise without vibrations, in regards to exposure to both vibrations and noise simultaneously, the individual noise and individual vibration has very little effect when evaluated.
2. Thus, it can be concluded that even with regards to exposure to both vibrations and noise simultaneously, if an evaluation that focuses on only one or the other of these is sought, an evaluation by extracting the desired stimuli may be carried out.
3. However, with regards to combined exposure, both vibrations and noise simultaneously, a definitive influence can be seen enough to infer that both vibrations and noise will be considered and used for the evaluation.
4. The definitive difference appears as a difference of approximately 10 dB between the exposure to vibrations without noise and exposure to both vibrations and noise simultaneously (dB (A)), relating to the response evaluation "7. Unbearable".
5. In regards to the effect of the combination of vibrations and noise in the compound response evaluation, exposure to both vibrations and noise simultaneously, the combined conditions, it can be seen that the effect of noise is greater than that of vibrations(due to regression analysis).
6. For the standard deviations, most cases corresponded to the general suggestion that variability decreases in accordance with increasing response evaluation.

From experiment II

7. The human body is very responsive to floor vibrations caused by railways, even when frequencies exceed 30 Hz.
8. The probability of perceiving sinusoidal vibrations that real measurement floor vibrations of a wooden house may reach a response of 95% or greater.

FINAL REMARKS

From the results of Experiment I it can be stated that it is not only the compound effect of sensory reactions that are large, but the effects of stimuli on annoyance evaluations is also significant. And from this it is also suggested that it is necessary to carry out more detailed research in the future. Environmental stimuli in many living spaces are becoming more complex. Accordingly, compound evaluations are close to evaluations that are actually made during daily life, and the need for the development toward compound evaluations has been keenly felt. Furthermore, this experiment proposes a re-investigation into not just the range of objects for evaluations given by the guidelines, but also including their levels.

Note: This research is a part of a research carried out based on Hosei University's Domestic Researcher System of the 2004 fiscal year.

REFERENCES

1. AIJ: Vertical and Horizontal vibration induced by traffic. Guide- lines for the evaluation of habitability to building vibration, 2nd ed., AIJ, 2004
2. Nagaseki, Y. et al: Part 1: Outlines of the measurement and characteristics of the railway vibration, Measurements of vibration in a wooden house due to running trains, Summaries of Technical Papers of Annual Meeting AIJ, pp. 297–298, 2004
3. Shimura, M. et al: Part 2: Application of guidelines for the evaluation of habitability to building vibration, Measurements of vibration in a wooden house due to running trains, Summaries of Technical Papers of Annual Meeting AIJ, pp. 299–300, 2004

Environmental Vibrations – Takemiya (ed.)
© 2005 Taylor & Francis Group, London, ISBN 0 415 39035 4

Influence of visual sensation on sense of horizontal vibration

C. Noda & T. Ishikawa

Dept. of Housing and Architecture, Japan Women's University, Tokyo, Japan

ABSTRACT: Experiments have been carried out to determine the sense of horizontal vibration where subjects can see the view form a window. The visual situation has the greatest influence on vibration perception, and this is closely connected with vibration amplitude. Vibrations cannot be perceived physically in the low frequency and small acceleration range, but they can be recognized visually as movement of the view. Therefore, many subjects can perceive vibrations when they can see the view. It is therefore necessary to assess the perception threshold in consideration of the influence of visual sensation. The visual sensation does not influence perception in the high frequency range, because the vibration amplitude is imperceptible in that range. It is therefore possible to assess vibrations on the basis of perception threshold scatter by physical feeling in the body.

1 INTRODUCTION

Existing vibration evaluation standards, including the AIJ Guidelines (Architectural Institute of Japan 2004), use perception thresholds based on bodily sensation. In real life, however, people sometimes realize there is a vibration by seeing the movement of furniture or the landscape outside a window or hearing the sound of wind or squeaking noise generated by the building, as can be seen from questionnaire surveys relating to existing high-rise apartment buildings and base-isolated buildings. (Ishikawa, Uekusa, Ichiriki & Noda 1993, Nakamura, Kanda, Shioya & Nagaya 1995)

Although the relevant international standard (ISO2631–2 2003) refers to visual effects such as one of the parameters to be considered in its annex, it does not quantify their effects, and therefore does not provide concrete evaluation criteria. To bring habitability assessment more in line with reality by taking into account the effects of circumstantial factors, it is necessary to elucidate the influence of visual sensation, audible sensation and other circumstantial factors on the sense of horizontal vibration.

Of these circumstantial factors, we focused on visual sensation in this study. In actual living environments, there are various objects that could give you a visual clue about vibration, including interior lighting fixtures or furniture, a water surface and the landscape outside a window. (Ishikawa, Uekusa, Ichiriki & Noda 1993, Nakamura, Kanda, Shioya & Nagaya 1995) This paper discusses experiments conducted to elucidate subjects' sense of horizontal vibration when the view outside the window is visible.

Although the authors had conducted various experiments involving vertical stripes (Ishikawa, Noda, Kumazawa & Okada 1994a) and a model high-rise apartment building (Ishikawa, Noda, Kumazawa & Okada 1994b) before, there was still room for improvement in terms of the investigation of the impact of visual sensation based on more realistic visual objects. In our latest experiments, therefore, we provided the subjects with such visual objects by making the actual view outside a window visible to them. Our aim here is to elucidate the influence of visual sensation on subjects' perception thresholds for horizontal vibration through a comparison of cases in which the outside view was visible (Noda & Ishikawa 1999b) and those in which it was not visible (Noda & Ishikawa 1999a).

2 OUTLINE OF EXPERIMENTS

Prior to experiments, a $3\,m^2$ and 3 m high simulated habitable room was built on the shake table. To reproduce an environment that is close to an actual domestic indoor space, the habitable room was furnished with sash windows, wall materials, fittings and other common housing features. To minimize the impact of circumstantial factors other than visual sensation, special care was given to the acoustic environment. The mechanical noise from the vibration source was kept to a minimum through the use of an electrodynamic shaker. The walls were filled with a 10 cm-thick sound absorbing material to reduce the transmission of outside noise by about 15 dB. In addition, music was played at a set volume during all experiments.

Figure 1 shows the overall layout of the experimental site and setup of the habitable room interior for experiment (1), which was conducted by keeping the outside view visible to the subjects. Three subjects were instructed to sit on chairs by facing a 140 cm by 100 cm window mounted on one of the walls at a height of about 100 cm. Outside the habitable room, there was a single-story structure at a distance of about 9.3 m, and the subjects were able to view the structure and background landscape through the window. Figure 2 shows the view outside the window as seen from the middle chair, though there may have been a slight difference depending on the actual location of the chair. To the subjects experiencing a left-to-right horizontal vibration, this view would have appeared to be swaying from left to right. This paper examines the influence of visual sensation on the sense of horizontal vibration by analyzing experimental results obtained under those conditions.

Input vibrations were basically horizontal sinusoidal vibrations applied in a left-to-right direction relative to the subjects. In setting the input vibration range, wind vibration in high-rise and super high-rise buildings, wind vibration in low and mid-rise buildings, traffic vibration and equipment vibrations in which high frequency components dominated were mainly considered. 14 frequency options are set ranging from 0.1 to 40 Hz.

During experiments, input vibrations were measured using strain-gauge accelerometers and an oscillographic recorder, with input reproducibility verified each time. A vibration with a gradually increasing acceleration was applied at a frequency selected from 14 options between 0.1–40 Hz. The subjects were instructed to signal the experimenter when they felt a vibration while looking at the outside view, and the experimenter gathered vibration data recorded before and after the signal while constantly monitoring the interior of the habitable room from outside via a CCD camera. The experimenter also gave instructions on the commencement of a vibration run, etc. from outside via a microphone.

By changing frequency in a random sequence, experiments were conducted over the whole spectrum of frequencies. A total of 45 subjects, all females aged between 19 and 24, took part in the experiments.

In experiment (2) (Noda & Ishikawa 1999a), a control experiment in which the view outside the window was blocked, a steel cover was placed over the window of the habitable room, forcing the subjects to detect vibration relying on bodily sensation alone. Unlike experiment (1), in which the view outside the window was visible, experiment (2) involved a single subject squatting on the floor for each session, although other conditions, such as the vibration input method, were the same. From previous experiments, we had learned that subjects' posture and number of

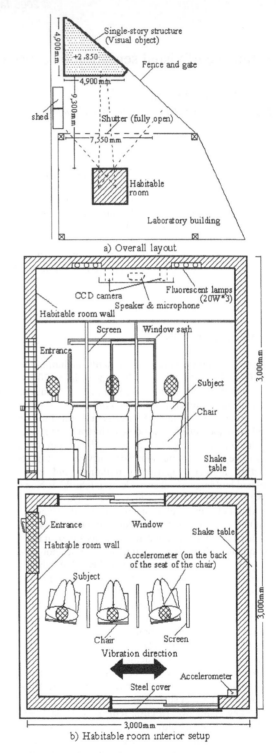

a) Overall layout

b) Habitable room interior setup

Figure1.   Experimental setup.

Figure 2.   View from habitable room window.

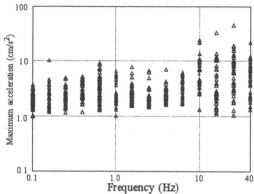

Figure 3.   Distribution of perception thresholds when view outside window is visible.

subjects present in each session had a small impact on the results compared to the assessment method, so that experiment (2) was chosen as a control experiment. This experiment involved a total of 35 subjects, all females aged between 19 and 24.

3   INFLUENCE OF VISUAL SENSATION ON PERCEPTION THRESHOLD

From the above observations, it is easy to see that visual sensation exerts a great influence on perception thresholds. In this chapter, we evaluate perception thresholds in terms of signals raised by subjects upon sensing a vibration, and discuss the impact that visual sensation has on perception thresholds.

In experiment (1), in which the view outside the window was visible, acceleration measured at the surface of the seat of the chair was used, while in experiment (2), in which the view outside the window was not visible, acceleration measured at the floor surface was used.

Figure 3 shows the distribution of personal perception thresholds recorded over the whole spectrum of frequencies when the view outside the window was visible. When the view outside the window was not visible (Fig. 4), many subjects did not feel vibration in the low frequency range below 0.16 Hz even at the largest acceleration used in the experiment.

When the view outside the window was visible (Fig. 3), on the other hand, all the subjects felt vibration at all frequencies. This shows that the availability of visual sensation has a great impact in the low frequency range.

When the view outside the window was not visible, the scatter of perception thresholds varied across frequencies, but when the view outside the window was visible, it remained more or less the same up to around 6.3 Hz. When the view outside the window was visible, the scatter of perception thresholds was generally

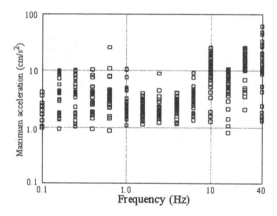

Figure 4.   Distribution of perception thresholds when view outside window is not visible.

small, and this tendency was somewhat stronger in the low frequency range, where displacement was large. This can be interpreted that when the view outside the window was visible, displacement as straightforward vibration information provided by vision kept person-to-person variations in perception thresholds small. In the high frequency region, where displacement was small, the scatter of perception thresholds increased as was the case when the view outside the window was not visible because of the difficulty in detecting vibration from a small displacement.

To elucidate the influence of visual sensation on perception thresholds, perception probability curves are calculated and compared for each visual condition. Figures 5 and 6 show regression expression of degree 3 of every 20th percentile perception thresholds for the two cases.

Variations in perception thresholds attributable to differences in visual conditions were observed in the

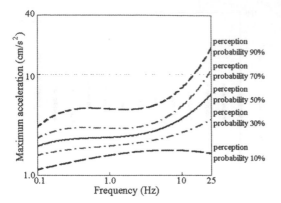

Figure 5. Perception probability curve when view outside window is visible.

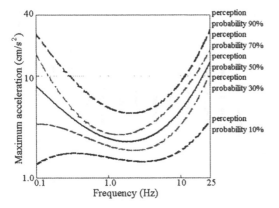

Figure 6. Perception probability curve when view outside window is not visible.

low-frequency region below 0.4 Hz. When the view outside the window was visible, the perception threshold fell as frequency decreased, and the gap between the two curves widened in the low frequency region. This can be interpreted that as the frequency decreases, people have greater difficulty in detecting vibration relying only on bodily sensation, while displacement, the vibration information provided by vision, becomes larger, thus increasing the impact of visual sensation on perception thresholds in the low-frequency region. Namely, subjects detected vibration from the movement of the view outside the window even when their bodies did not feel it.

In the medium to high frequency region above 0.63 Hz, the two sets of perception thresholds were roughly equal. As frequency increased, displacement decreased, making it difficult to detect vibration through visual sensation even if the view outside the window was visible. Namely, visual sensation had a

minimal impact in this region, and perception thresholds were derived mostly from bodily sensation.

Although a discrepancy between perception thresholds appeared again in the high frequency region, this appears to have been due to the interference of the resonance of the chair, which started at around 10 Hz. In experiment (1), in which the view outside the window was visible, the acceleration measured at the surface of the seat of the chair was used in the analysis, but other factors, such as the vibration of the floor and the response of the back of the chair, may also have been involved.

Compared to past experiments conducted by the authors, et al. using vertical stripes as the visual object (Ishikawa, Noda, Kumazawa & Okada 1994a), the impact of visual sensation was small. As has been discussed in a study on perception based only on visual sensation (Tsurumaki, Goto & Inoue 1995), the impacts of near and distant views on perception differ in translational vibration, and people are more sensitive to the near view. Compared to looking at a building standing a fair distance away as was the case in our experiments, looking at a nearby object, such as vertical stripes placed just outside the window, appears to result in a stronger impact of visual sensation because of the ease of gauging the displacement. The impact of visual sensation tended to be stronger in the large displacement region above 1.0 cm in our experiments, but this boundary displacement value is likely to change according to visual conditions.

4 COMPARISON WITH PERCEPTION THRESHOLDS BASED ON BODILY SENSATION IN PAST STUDIES

Figure 7 shows the impact of visual sensation on perception thresholds identified in this study in comparison with perception thresholds measured without showing the outside view to subjects in our past experiments. (Ishikawa & Noda 1998, Ishikawa & Noda et al. 1999, Noda & Ishikawa 1999a) Perception thresholds measured without showing the outside view to subjects are represented by approximation cubic curves plotted from 50th percentile perception thresholds obtained in the experiments after sorting it according to experimental conditions applicable to them so as to show the scatter of perception thresholds based on bodily sensation alone. (Ishikawa & Noda et al. 1999)

As has been discussed in our past study (Ishikawa & Noda et al. 1999), perception thresholds based on bodily sensation exhibit a scatter of the magnitude shown in the figure due to differences in the subjects' state and level of vibration consciousness. When subjects have no prior knowledge of the occurrence of a vibration and are relaxed, perception thresholds tend to be

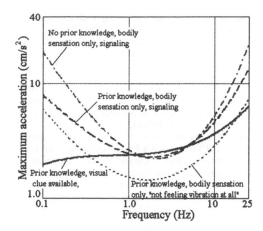

Figure 7. Comparison with perception thresholds based on bodily sensation in past studies.

high. In contrast, when subjects anticipate the occurrence of a vibration and are predisposed to believing "it must be shaking" due to exposure to reply choices in a questionnaire, perception thresholds tend to be low. Various situations in which building occupants are placed in daily lives are likely to encompass all the conditions considered in our experiments, and this gives rise to a need for the thorough incorporation of the resulting scatter of perception thresholds into habitability assessment.

Perception thresholds based on both bodily sensation and visual sensation were lower than those based on bodily sensation alone in the low-frequency region below around 0.25 Hz. Namely, despite the significance of the impact of the subjects' states and levels of vibration consciousness on perception thresholds based on bodily sensation, there is a region in which the impact of visual sensation even exceeds it. In the low-frequency region, perception thresholds tend to be low as subjects are able to pick up weak vibrations that their bodies do not feel through visual sensation, as displacement is large in that region despite small acceleration. In the low-frequency region, therefore, habitability assessment needs to properly take into account the impact of visual sensation on perception thresholds in consideration of its significance.

In the medium to high frequency range above 0.4 Hz, perception thresholds are mainly determined by bodily sensation, as the impact of visual sensation is small due to small displacement. In this region, the scatter of perception thresholds measured while showing the view outside the window to subjects is in line with that of perception thresholds based on bodily sensation alone. Namely, habitability assessment needs to mainly focus on the scatter of perception thresholds based on bodily sensation despite the fact that visual sensation may exert some influence.

As can be seen from the above, in the lower acceleration region that is within the scope of a habitability assessment, the impact of visual sensation on perception thresholds differs according to the frequency, so that it is necessary to give different weights to bodily sensation and visual sensation depending on the target building. For example, bodily sensation exhibits high sensitivity to vibrations in the frequency range of 2.0–4.0 Hz, common in low-rise buildings such as three-story prefabricated houses (Otsuki, Tamura, Nakata, Naito & Kiriyama 1994), and the impact of visual sensation is small compared to other frequency regions. In such a region, therefore, it is appropriate to focus on bodily sensation when conducting a habitability assessment. On the other hand, bodily sensation is relatively insensitive to vibrations in the low-frequency range, common in super high-rise apartments and other buildings, and it is highly likely that visual sensation impacts on the sense of horizontal vibration. In such a region, it is necessary to increase the weight given to visual sensation when conducting a habitability assessment.

5 CONCLUSIONS

In this study, we conducted experiments by making the view outside a window visible to the subjects and analyzed the results in comparison with those of earlier experiments based only on bodily sensation so as to elucidate the impact of visual sensation on the sense of horizontal vibration.

The findings of this paper are listed below.

1. Visual sensation has a significant impact on perception thresholds, and this tendency becomes stronger as the frequency decreases. In the low-frequency region, displacement is large despite small acceleration, and this lowers perception thresholds as the subjects are able to detect vibration their body cannot feel by obtaining sufficient information from vision in the form of a large displacement. In contrast, since displacement decreases as frequency increases, visual sensation has little impact in the high frequency region, so that subjects assess vibration by relying on bodily sensation.

2. The impact of visual sensation on the sense of horizontal vibration is closely related to displacement as the form of vibration information afforded by vision. This impact may therefore be analyzed by dividing the displacement spectrum into three regions. At a certain threshold level, displacement just becomes visually recognizable, though the actual value varies depending on various conditions. If displacement is below this level, people assess vibration by relying only on bodily sensation even if the view outside the window is visible.

If it is above this level, people assess vibration by relying on both bodily sensation and visual sensation in an integrated manner. In a certain displacement region further up the displacement scale, visual sensation intensifies the body's assessment of vibration.

3. Although the perception threshold based on bodily sensation varies depending on the subject's state, level of vibration consciousness, and the like, the impact of these factors is overtaken by that of visual sensation in the low frequency region, and this results in the lowering of the perception threshold. In contrast, the impact of visual sensation is small in the high frequency region, so that habitability can be evaluated by focusing mainly on the scatter of the perception threshold based on bodily sensation despite the fact that visual sensation may still have some impact.

The frequency range in which visual sensation has a significant impact as established in this study is subject to a set of visual conditions, and more universal research will be needed in the future research. Despite such restrictions, this study has shown the necessity of assessing habitability for horizontal vibration after identifying regions in which the impact of visual sensation must be taken into consideration and regions in which it is sufficient to consider bodily sensation.

In an actual living environment, various other factors, such as whether there is a window near an occupant, whether objects prone to shaking are present and if so whether the occupant looks at them, will have to be covered in an evaluation. The authors wish to present a statistical evaluation method for the scatter of perception thresholds that fully takes into account personal and circumstantial variabilities, including the impact of visual sensation as shown above.

In conducting experiments and analyzing data, students were working on their graduation theses at our laboratory at the time, provided valuable assistance. The authors acknowledge their contribution and express their gratitude. The authors would also like to thank all those who took part in the experiments. This study has been partially funded under the Ministry of Education Science Research Grant Program, Basic Research (C) (1998–2000).

REFERENCES

Architectural Institute of Japan 2004. *Guidelines for the evaluation of habitability to building vibration.* Tokyo: AIJ

Ishikawa, T. & Noda, C. 1998. Sensory Assessment of Horizontal Vibration over a Wide Frequency Range. *Journal of Architecture Planning and Environmental Engineering(Transactions of AIJ)* (506): 9–16

Ishikawa, T., Noda, C., Kumazawa, F. & Okada, T. 1994a. A Study on Serviceability Limit for Wind-Induced Motions in Consideration of Visual Stimulus. *Journal of Structural Engineering* 40B: 1–6

Ishikawa, T., Noda, C., Kumazawa, F. & Okada, T. 1994b. Effects of Difference in Visual Objects on Sense of Horizontal Vibration. *13th National Symposium on Wind engineering papers*: 107–112

Ishikawa, T. & Noda, C. et al. 1998. Evaluation of Sense of Horizontal Vibration in a Condition Approached Living Environment – Part 1 Outline of the Experiment and Characteristics of Psychological Evaluation–; – Part 2 Difference in Perception Thresholds Depending on Experimental Conditions –. *Summaries of Technical Papers of Annual Meeting, the A.I.J.(Environmental Engineering I)*: 323–326

Ishikawa, T. & Noda, C. et al. 1999. Influence of Circumferential Factors on Perception Threshold of Horizontal Vibration- Part I Outlines of the Experiments–; – Part 2 Differences in the Subjects' Consciousness – & – Part 3 Presentation of Guidelines on Perception – *Summaries of Technical Papers of Annual Meeting, the A.I.J. (Environmental Engineering)*: 319–324

Ishikawa, T., Uekusa, T., Ichiriki, Y. & Noda, C. 1993. A study on vibration sensation considering habitability of high-rise dwelling – Part 1 Outline of the research and the result –. *Summaries of Technical Papers of Annual Meeting, the A.I.J. (Environmental Engineering)*: 95–96

ISO2631–2 2003. *Mechanical vibration and shock – Evaluation of human exposure to whole-body vibration Part 2: Vibration in buildings (1 Hz to 80 Hz)*

Nakamura, T., Kanda, J., Shioya, K. & Nagaya, M. 1995. Statistical Study of Perception of Vibration in Base Isolated Buildings during Earthquakes. *Journal of Architecture Planning and Environmental Engineering (Transactions of AIJ)* (472): 185–192

Noda, C. & Ishikawa, T. 1999a. Influence of the subjects' situation on perception threshold of horizontal vibration. *Journal of Architecture Planning and Environmental Engineering (Transactions of AIJ)* (524): 9–14

Noda, C. & Ishikawa, T. 1999b. Influence of visual sensation on sense of horizontal vibration. *Journal of Architecture Planning and Environmental Engineering (Transactions of AIJ)* (525): 15–20

Otsuki, T., Tamura, Y., Nakata, N., Naito, S. & Kiriyama, S. 1994. Characteristics of Road Traffic Vibration in Steel Frame Housing. *Summaries of Technical Papers of Annual Meeting, the A.I.J. (Environmental Engineering)*: 1885–1886

Tsurumaki, H., Goto, T. & Inoue, J. 1995. Research on Effects of Window View on Sense of Vibration for Low-Frequency Vibrations – Part 2 Outline of Vibration-induced Visual Sensation Simulator Experiment and Results –. *Summaries of Technical Papers of Annual Meeting, the A.I.J. (Environmental Engineering)*: 361–362

Environmental Vibrations – Takemiya (ed.)
© *2005 Taylor & Francis Group, London, ISBN 0 415 39035 4*

Indoor noise levels, annoyance and countermeasures in Assiut urban sites, Egypt

S.A. Ali
Faculty of Engineering Assiut University, Assiut, Egypt

ABSTRACT: A field study has been carried out in urban Assiut city, Egypt. The goals of this study are: (1) to carry out measurements to evaluate indoor noise levels, (2) to determine if these levels exceeds permissible levels, (3) to examine people's attitudes towards indoor noise, (4) to ascertain the relationship between indoor noise levels and degree of annoyance. 24 hour equivalent continuous noise levels (L_{Aeq}) were measured for indoor-domestic noise and comparison with current western standards show that the levels measured exceeded the limits of dissatisfaction given by those standards. Overall sound pressure levels measured inside typical university offices indicate that the presence of individual room-units of air-conditioning impairs the acoustic quality in those environments. The attitude survey showed that 29.4% of the respondents reported felling highly annoyed from indoor noise. By increasing the noise level, the percentage of respondents who reported felling highly annoyed is also increased. The results of this research demonstrate the need for attention to be paid to the indoor-acoustic quality of homes and offices.

1 INTRODUCTION

Environmental noise pollution has been recognized in recent years as a serious threat to the quality of life in Egyptian cities (Minster of Egyptian Environment, 2003). As a result the protection of people against problems caused by noise – for example, task and sleep interference. Temporary and permanent induced threshold shifts of hearing and damage to the hearing system – is now considered a national goal for central governments, as well as local health protection agencies. Residents have individual difference in reactions to noise (Weinstern, 1978). In Assiut city, a rapid rate of urbanization, accompanied by effective contributions to community noise, is well recognized. This includes the installation of industrial and constructional plants, the introduction of inter-city roads and highways, public transport and huge numbers of private cars and commercial vehicles and the use of modern domestic appliance.

2 FIELD MEASUREMENTS

The (A) weighted sound pressure level was measured by a precision sound level meter type ONO SOKKI LA-5120. The measurement time interval was adjusted for 10 minutes (an equivalent level measured each 10 min). 24 hour equivalent continuous noise levels (L_{Aeq}) were obtained automatically in the sound level meter. The sound level meter was positioned at a height of 1.2 m above the ground. The microphone was placed in the center of the room (Hay, 1972). The output of the sound level meter was fed to a digital printer type ONO SOKKI RO-110. The relative humidity and temperature of sites varied from 50% to 60% and 20° to 30°C, respectively, at the time of measurements.

2.1 *Indoor noise of university offices*

Noise levels were recorded in twenty-four offices in Assiut University for a period of 10 min in the middle of the working day, with the normal activities continuing. The recorded levels were used to determine an equivalent average sound pressure level for every office.

The offices were characterized by the nature of the work carried out in them and the numbers of occupants. It was noticed that all offices under consideration (like all other offices in the University) were air-conditioned. Some of these offices had individual air-conditioning units (group A), while others were centrally air-conditioned (group B). However, for the group A offices, the sound pressure level measurements were repeated to cover the following conditions:

- The air conditioner is switched off.
- The air conditioner is switched on – high speed.
- The air conditioner is switched on – low speed.

The measurements of sound pressure level, in-group B offices were all carried out under one appropriate constant temperature and normal working conditions.

2.2 Indoor home noise

Thirty flats representing three types of flats were considered for the evaluation of indoor-acoustic quality. These flats were classified into three groups as follows:

- Group A is low class economic flats, area of flat ranged from 60 to 80 m², numbers of rooms ranged from two to three rooms and hall, 10 flats were selected to represent this group. All rooms of flats have normal painted concrete walls with the same type of fitting, and ceiling fans. Radios and television sets are located in the living rooms and the rooms of flats have removable carpets. One flat (flat No. 7) has individual air-conditioning units in the living room.
- Group B is intermediate class economic flats, area of flat ranged from 80 to 120 m², numbers of rooms ranged from three to four rooms and large hall, 10 flats were selected to represent this group. All rooms of flats have normal painted concrete walls with the same type of fitting, and ceiling fans. Radios, television sets, TV games, and computers are located in the living rooms and the rooms of flats have removable carpets. Two flats (flats No. 3 and 6) have individual air-conditioning units in the living room.
- Group C is high class economic flats, area of flat over 120 m², numbers of rooms are more than four rooms and large halls, 10 flats were selected to represent this group. All rooms of flats have normal painted concrete walls with the same type of fitting, and ceiling fans and individual air-conditioning units. Radios, television sets, musical instruments, TV games, and computers are located in the living rooms and the rooms of flats have removable carpets. One flat (flat No. 9) didn't have individual air-conditioning units in the living room.

Also, measurements were taken in reception of five flats in Group C where there wasn't domestic appliance. For all flats, sound pressure levels were measured during a 24-h period at 10 min intervals, at the center of living rooms of the flats.

Day–night indoor noise levels L_{dn} were calculated for all sites as follows:

Day–night indoor noise levels L_{dn}

$$= 10 \log_{10} 1/24(15(10^{Ld/10}) + 9(10^{(Ln + 10)/10})).$$

where L_d and L_n stand for the daytime and night–time average sound levels, respectively. Day–night sound level is abbreviated as DNL, and symbolized in equation as L_{dn} (Schultz, 1982).

Measurements were carried out from midnight-to-midnight. The daytime period is 07:00 to 22:00 and the night–time period is 22:00–07:00.

3 RESULTS OF INDOOR NOISE MEASUREMENTS

Results of measurements discussed as follows:

- Indoor noise of university offices
- Indoor noise of homes.

3.1 Indoor noise of university offices

The instant values of sound pressure levels taken at the centre points of twenty four university offices, used for various activities, are shown in Tables 1 and 2. It can be noticed for group A offices, that levels of the order of 63.1 to 73.5 dB usually prevail when air-conditioning units are operated at high speed. However, these levels are very much reduced when the individual air-conditioning units are switched off. Unfortunately, because hot weather dominates almost all days of summer, and cold weather in winter, these air-conditioning units work on a high speed setting almost continuously and thus emit high noise levels. However, when air conditioners are set on low speed the sound levels are substantially reduced to the order of 58.1 dB to 67.2 dB with an exception in the office 4 where a higher level of 71.4 dB was recorded due to noise from computers. It is also noticeable that low levels of the order 49.3 to 56.3 dB were recorded inside this group of offices when air conditioners were switched off with an exception of discussion offices. Beranek et al (Beranek, 1971) have reported a design objective for indoor A-weighted sound levels in large offices of 42 to 52 dB to avoid speech interference. It can, however, be noticed that unless the air conditioners are switched off in the group A offices, the range of noise levels will never comply with such an acceptable objective.

In the group B offices, the sound levels are normally in the range of 41.7 to 58 dB except when working people are holding discussions when the level rises beyond this range (in room 4, the level was recorded as 65 to 71 dB due to a normal discussions and in room 11 a range of 57 to 70 dB was recorded for a professor teaching in his office).

The fact that the group B offices are all centrally air conditioned and well carpeted would, however, explain the compatibility of the levels recorded in these rooms with the acceptable levels of the design given in reference 4.

3.2 Indoor noise of homes

Values of L_{dn} of all flats were determined and are given in Table 3. It is noticeable that L_{dn} dB in flats of

Table 1. Sound pressure levels inside university offices, group A offices, with individual air-conditioning units.

| | SPL dB | | | | |
Room No.	Conditioner on/high	Conditioner on/low	Conditioner off	Number of occupants	Type of office
1	66.2	59.4	53.5	1	Professor's office
2	69.1	64.5	54.7	3	Secretary's office
3	64.7	58.1	52.6	2	Professor's office
4	73.5	71.4	68.2	2	Computer room office
5	64.2	61.7	56.7	3	Student affairs office
6	71.3	67.1	65.1	4	Sitting room
7	63.5	61.4	49.3	2	Office in small laboratory
8	63.7	59.1	53.4	2	Librarian office
9	63.1	58.3	54.2	3	Museum office
10	72.4	67.2	65.3	25	Class room
11	65.1	59.7	53.9	2	Professor's office
12	66.3	63.2	56.8	4	Laboratory office

Table 2. Sound pressure levels inside university offices, group B with central air-conditioning systems.

Room No.	SPL dB	Number of occupants	Type of office
1	49.1	1	Professor's office
2	54.9	4	Laboratory office
3	42.3	3	Librarian office
4	65–71	6	Sitting room
5	52.1	2	Professor's office
6	41.7	4	Museum office
7	50.6	2	Professor's office
8	53–58	5	Student affairs office
9	49.1	2	Secretary's office
10	56–64	25	Class room
11	57–70	4	Professor's office
12	43.5	3	Laboratory office

group A ranged from 61.7 to 67.3 dB, but in flat No. 7 the noise level is 77.4 dB where individual air-conditioning was found. In-group B, the noise level ranged from 68.3 to 71.6 dB, but in flat No. 8 the noise level is 73.6 dB where TV game and computer were found. In flat No. 3 the noise level is 76.6, and in flat No. 6 the level is 77.1, where individual air-conditionings were found. In-group C, the noise level ranged from 75.2 to 80.1 dB where individual air-conditionings were found, but in flat No. 9 the noise level is 70.4 where individual air-conditioning wasn't found.

The results in Table 3 indicate that the noise levels inside living rooms in Assiut's homes are, very much higher than those specified by international agencies to protect public health and welfare. A recommended figure of $L_{dn} \leqslant 45$ dB indoor for residential areas (Anon, 1974), (Egyptian Ministry of Environment, 1994), (Gottob, 94) is of course, very low compared with corresponding values that appear in Table 3, ranging between 61.7 and 80.1 dB. The main reason

for the present higher values was the noise emitted from individual air-conditioning units, radio, TV, TV games, computers, and musical instruments.

Domestic appliances and outdoor noise, however, contribute to the noise and could, in some instances, be more intruding, but not as much as air conditioner noise. The lower noise levels in Table 3, respectively, refer to rooms with no air conditioner in operation.

However, it was noticeable during measurements that the operation of the usual TV sets, radios and stereo units was at relatively high levels. It was also noticed that room interior design is not adequate for sound conditioning and noise pollution control. For instance, acoustic plaster coating of the walls and floor carpeting could help to reduce reverberation indoor and thus reduce standing wave patterns of noise. Also, properly designed double glazed windows would certainly help in reducing extraneous noise intrusion.

The results showed that L_{dn} in reception in the measured five flats ranged from 51.3 to 53.7 dB. So

Table 3. Values of L_{dn} for living rooms of 30 flats inside typical Assiut homes.

Flat No. of Group (A)	L_{dn} dB	Flat No. of Group (B)	L_{dn} dB	Flat No. of Group (C)	L_{dn} dB
1	64.5	1	70.1	1	76.9
2	62.4	2	71.4	2	78.5
3	66.1	3	76.6	3	77.3
4	63.2	4	70.2	4	80.1
5	61.7	5	69.7	5	79.4
6	63.9	6	77.1	6	76.8
7	77.4	7	68.3	7	77.1
8	63.0	8	73.6	8	75.2
9	64.1	9	69.4	9	70.4
10	67.3	10	71.5	10	79.2

the minimum background noise level inside the homes under investigation is 51.3 dB whilst, due to domestic activities, the overall sound pressure level varies between this minimum level and a maximum of 80.1 dB. Through the measurements, it is noticed that substantial increase in sound pressure level occurs during the day between 8 am and 12 noon and during the evening between 6 and 10 pm. This reflects the fact that, in eastern countries, people usually stay awake and active to fairly late hours.

4 SOCIAL SURVEY

The subjective response to indoor noise was measured by means of a social survey. The survey was carried out in the same three types of flats, where measurements were carried out (low, intermediate and high class economic), and it was performed to investigate the individual's attitude to the residence. The investigation was carried out by using the techniques generally applied in studies of this kind (Namba, 1996).

The questionnaire contained questions about demographic data, social status, educational level and annoyance.

The questionnaire was distributed by hand. The respondents completed the questionnaire themselves. A total of 1000 questionnaires were distributed in the three types of flats equally. A person returned later to collect the completed questionnaires. Some questionnaires were returned through the post office to our address. A total of 691 questionnaires were finally collected.

5 RESULTS OF SOCIAL SURVEY

The respondents were male (56%) and female (44%). The ages of interviewed people were as follows: 20–30 years (19%), 30–40 years (27%), 40–50 years (29),

Table 4. Noise, which disturbs people at home.

Description of noise	Percentage of respondents disturbed when at home
Domestic appliance	48
Neighbours' impact noise (Knocking, walking, etc.)	33
Children	42
Adults voices	17
Radio/TV	44

Table 5. Percentage of respondents who reported felling highly annoyed due to indoor noise.

Group No.	Noise levels L_{dn}	Average of noise levels L_{dn}	Percentage of respondents who reported felling highly annoyed
A	61.7–67.3	64.5	17.9
B	68.3–71.6	70	21.5
C	75.2–80.1	77.6	43.5

50–60 years (17) and older than 60 (8). The education levels of interviewed people were: 15% educated in prep schools, 43% educated in intermediate schools (technical and secondary schools), and 42% in high educations (universities). The social statuses of interviewed people were: 25% single, 51% married, 11% divorced, and 13% widow people.

To the question were they annoyed by indoor noise or not, 51.3% of respondents were annoyed from indoor noise.

Table 4 indicates the result of the question regarding which sources of indoor noise disturb them when they are at home. Noise emitted from domestic appliance was the highest percentage (48%), followed by noise emitted from radio/TV (44%).

Attitudes to indoor noise were elicited by means of a five step semantic scale. About 29.4% of the interviewed people declared to be "highly annoyed," 19.7% "rather annoyed," 28.5% "moderately annoyed," 9.3% "little annoyed," 13.1% "not annoyed at all". Table 5 indicates the percentage of respondents who were highly annoyed in Assiut areas.

6   RELATIONSHIP BETWEEN LEVELS AND ANNOYANCE

As shown in Fig. 1, there was a strong relationship between levels in L_{dn} and percentage of respondents who reported feeling highly annoyed. The percentage of respondents who reported feeling highly annoyed increased with the increasing level of indoor noise.

7   COMPARISON OF ANNOYANCE IN ASSIUT WITH ANNOYANCE IN OTHER SURVEYS

Figure 1 indicates a comparison of the annoyance in Assiut with the annoyance in other survey in relation to noise levels (L_{dn}) as in the Schultz study, the annoyance reported in Assiut city seems lower than annoyance in other cities. A contributory factor might be that some residents consider that they are accustomed to indoor noise whereas others consider that some sources of indoor noise are out of their control, as noise emitted from air – conditioning units, because they are obliged to operate them due to bad weather, so they must be patient with it.

In general annoyance in Assiut agrees with the annoyance in other surveys. There was a strong

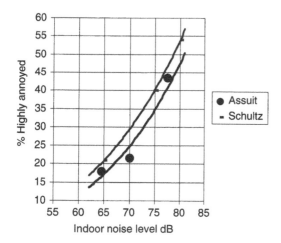

Figure 1.   Comparison of annoyance in Assiut with annoyance in Schultz study.

relationship between noise level and percentage of respondents who reported feeling highly annoyed. The percentage of respondents who reported feeling highly annoyed increased with increasing level of noise.

8   COUNTERMEASURES

- The results of this research demonstrate the need for attention to be paid to the indoor-acoustic quality of homes and offices.
- Interior design acoustics must be done and must be adequate for sound conditioning and noise pollution control. For instance, acoustic plaster coating of the walls and floor carpeting could help to reduce reverberation indoor and thus reduce standing wave patterns of noise. Also, properly designed double glazed windows would certainly help in reducing extraneous noise intrusion.
- It is recommended that central air-conditioning systems are used in University offices to provide an acceptable, quite pleasing environment.
- Control the time of using domestic appliances such as TV, radio sets, electro-mechanical kitchen facilities especially at night time.

9   CONCLUSIONS AND REMARKS

- Typical noise levels inside Assiut's flats under consideration are, in general, exceeding the limits of acceptability. The main sources of noise are air conditioners. Domestic appliances such as TV, radio sets, electro-mechanical kitchen facilities, etc., would also contribute to the noise levels.
- The acoustic environment of typical government offices in Assiut is dominated by excessively high noise levels due to the operation of individual air-conditioning system units. However, central air-conditioning systems provide an acceptable, quite pleasing environment.
- Through the measurements, it is noticed that a substantial increase in sound pressure level occurs during the day between 8 am and 12 noon and during the evening between 6 and 10 pm. This reflects the fact that, in eastern countries, people usually stay awake and active to fairly late hours.
- The attitude survey showed that 29.4% of the respondents reported felling highly annoyed from indoor noise.
- By increasing the noise level, the percentage of respondents who reported felling highly annoyed is also increased.
- The results of this research demonstrate the need for attention to be paid to the indoor-acoustic quality of homes and offices.

ACKNOWLEDGEMENTS

The authors are grateful to Assiut University and Assiut city hall for their help in carrying out this study. This study was financially supported by The Egyptian Ministry of Higher Education.

REFERENCES

[1] Report of Minster of Egyptian Environment about environmental noise pollution in Egyptian cities, Al-Ahram Newspaper (Egyptian Newspaper), 23, 11, 2003.

[2] N.D. Weinstern, Individual differences in reactions to noise, a longitudinal study in a college dormitory, J. Psychology, Vol. 63 (1978), pp 462–466.

[3] B. Hay, M.F. Kemp, Measurements of noise in air conditioned landscaped offices, Journal of Sound and Vibration, Vol. 23 (3), (1972) pp 363–373.

[4] J. Schultz, Community noise rating, Applied Science Publishers, London, 1982.

[5] L.L. Beranek, W.E. Blazier and J.J. Figwer, Preferred noise criteria (PNC) curves and their application to rooms, Journal Acoustical Society of America, Vol. 50 (1971), pp 1223–8.

[6] Anon, Information on levels of environmental noise requisite to protect health, EPA, March (1974).

[7] Egyptian environmental law No. 4 of year 1994, Egyptian Ministry of Environment, Cairo, Egypt, 1994.

[8] D. Gottob, Regulations for community noise, inter-noise 94, Yokohama, Japan, August 29–31, pp 43–55.

[9] Seiichiro Namba, Juichi Igarashi, Sonoko Kuwano, Kazuhiro, Minoru Sasaki, Hideki Tachibana, Akihiro Tamura, and Yoshiaki Mishina, "Report of the committee of the social survey on noise problems", The Journal of the Acoustical Society of Japan (English), 1996; 17 (2): 109–113.

Author index

Printed and bound by CPI Group (UK) Ltd, Croydon, CR0 4YY

01/11/2024

01782599-0010